Mastering Azure Virtual Desktop

A practical guide to designing, implementing, and managing Azure Virtual Desktop environments

Ryan Mangan | Neil McLoughlin | Marcel Meurer | Shabaz Darr

Mastering Azure Virtual Desktop

Copyright © 2024 Packt Publishing

All rights reserved. No part of this book may be reproduced, stored in a retrieval system, or transmitted in any form or by any means, without the prior written permission of the publisher, except in the case of brief quotations embedded in critical articles or reviews.

Every effort has been made in the preparation of this book to ensure the accuracy of the information presented. However, the information contained in this book is sold without warranty, either express or implied. Neither the authors, nor Packt Publishing or its dealers and distributors, will be held liable for any damages caused or alleged to have been caused directly or indirectly by this book.

Packt Publishing has endeavored to provide trademark information about all of the companies and products mentioned in this book by the appropriate use of capitals. However, Packt Publishing cannot guarantee the accuracy of this information.

Group Product Manager: Preet Ahuja

Publishing Product Manager: Vidhi Vashisth

Book Project Manager: Ashwini Gowda

Senior Editor: Roshan Ravi Kumar

Technical Editor: Arjun Varma

Copy Editor: Safis Editing

Proofreader: Roshan Ravi Kumar

Indexer: Tejal Soni

Production Designer: Ponraj Dhandapani

Senior Developer Relations Marketing Executive: Rohan Dobhal

First published: March 2022

Second edition: July 2024

Production reference: 1190724

Published by Packt Publishing Ltd.

Grosvenor House

11 St Paul's Square

Birmingham

B3 1RB, UK

ISBN 978-1-83588-414-0

www.packtpub.com

The past few years have been challenging, marked by both personal and professional hurdles, including a health scare. Throughout this whirlwind, I found strength and inspiration from my unwavering pillars of support. To my beloved wife, Alexandra: your constant encouragement and profound understanding made this journey possible. Your love and patience provided me with the foundation I needed to persevere. To my precious daughters, Sienna and Gabriella: your curiosity and love filled each day with purpose and joy. Your boundless energy and bright spirits have been a constant source of motivation for me. Thank you for being my rock, my inspiration, and my joy. In the end, it is family that truly matters, and I am profoundly grateful for mine.

– Ryan Mangan

To my beloved wife, Vivianne, who has been my unwavering support and the beacon of love in my life while co-writing this book. Behind every successful man stands a woman, and without you this would not have been possible.

– Neil McLoughlin

For all those who love technology as much as I do: never stop learning and never give up. With the biggest thanks to my family, friends, supporters, and the community.

– Marcel Meurer

Foreword

In 1985, Microsoft envisioned a future where every home and desk would be equipped with a computer. This ambitious dream seemed far-fetched at the time, yet it has become a reality we are familiar with today. Windows has continuously evolved, and since the 1990s, its operating systems have been remotely accessible. In 1994, Microsoft introduced the **Remote Desktop Protocol 4.0 (RDP 4.0)** in Windows NT4, a game-changer for IT administrators as it enabled remote server management. The following year, Windows NT debuted with the user interface featuring the Start menu and taskbar, elements that still serve billions of users globally.

The journey of Windows virtualization progressed with the advent of **Remote Desktop Services (RDS)**. However, it still necessitated a control plane comprising a web server, gateway, and broker, in addition to the session host for user sessions. This architecture persisted until Microsoft Azure's emergence in 2012, which presented fresh possibilities such as hosting RDS on Azure through **Infrastructure as a Service (IaaS)**.

The control plane's transformation into a cloud-based service materialized with Windows Virtual Desktop (now Azure Virtual Desktop), and the acquisition of FSLogix in 2018 marked another milestone (this was the same year I embarked on my journey with Microsoft). As the virtualization and cloud industry matured, virtualization-specific expertise became a staple in every enterprise.

With applications transitioning to software-as-a-service models, offering ease of purchase through subscriptions, maintenance simplicity, and scalability, cloud virtualization sought a similar simplicity. Virtualization required a solution as seamless as the shift from Office to Office 365, leading to the creation of Windows 365—a cloud service with Cloud PC as the endpoint managed by Microsoft, heralding a new revolution. Recognized by Gartner in the 2023 Magic Quadrant for desktop-as-a-service, Windows 365 and Azure Virtual Desktop positioned us as leaders in virtualization, in just 4.5 years with Azure Virtual Desktop and 2.5 years with Windows 365.

Windows 365 ignited a computing revolution, transitioning PCs to the cloud while preserving "like-local" experiences, manageable via Microsoft Intune without the need for specialized skills. These principles guide Windows 365's latest end user experience features, such as Boot and Switch, enabling effortless logins for anyone familiar with Windows, unlike traditional VDI, which often requires manuals and assistance.

In the future, you could expect more experiences coming that bring cloud and client computing with AI together within Windows 365. Microsoft is uniquely positioned to deliver a combined solution of Azure, Windows, Microsoft 365, and Copilot integrated into every aspect of the customer experience.

With this book, you will learn all you need to know to implement Azure Virtual Desktop as you embark on your journey to the Microsoft cloud. Windows 365 and Azure Virtual Desktop complement each other. Use Azure Virtual Desktop for greater flexibility coming from on-premises VDI and use Windows 365 Enterprise or Frontline to modernize your virtualization experience for your users with Microsoft Intune and features such as Windows 365 Boot, Switch, and much more to come soon!

Before I close, I want to personally thank Ryan, Neil, Marcel, and Shabaz for being such amazing community contributors. Writing a book takes a lot of time away from friends and family. It shows their true dedication to what community means to me personally and to Microsoft.

This book supports anyone's journey to Windows in the cloud. Have fun and enjoy the new IT superpowers you gain with this amazing book!

Christiaan Brinkhoff

Principal PM, Azure Virtual Desktop and Windows 365

Contributors

About the authors

Ryan Mangan, with over 17 years of experience in the tech industry, specializes in cloud and end user computing. His extensive understanding of end user computing, applications, networks, and virtualization has earned him recognition as a Microsoft MVP. Additionally, he was a VMware vExpert for over a decade. As a Chartered Fellow of the British Computer Society and holder of other esteemed titles, Ryan's professional dedication is evident. A humble author and speaker, he generously shares his knowledge, fostering growth in the tech community. In 2023, he established Efficient Ether, focusing on cloud optimization, sustainability, and generative AI (EtherAssist), showcasing his commitment to advancing sustainable technology solutions.

Check out Ryan's blog: `https://www.democratising-clouds.com/`

You can also follow him on LinkedIn: `https://www.linkedin.com/in/ryanmangan01/`

You can find Ryan on Twitter/X: `@RyMangan`

Neil McLoughlin is a seasoned IT professional based in Manchester, UK, with over 20 years of diverse experience in the industry. For a decade, he provided Citrix and VDI consultancy to large enterprise clients. About five years ago, Neil shifted his focus to cloud technologies and **Desktop as a Service (DaaS)**, becoming a specialist in cloud-based desktop solutions, particularly Azure Virtual Desktop, Windows 365, Intune, and Microsoft 365.

Neil is deeply committed to community engagement and co-organizes the EUC Forum, an in-person conference in the UK. He is also an active Microsoft MVP. Currently, he serves as the principal technical account manager at Nerdio. His previous roles included senior consultant and architect positions at New Signature, Computacenter, and Capgemini, where he specialized in end user computing.

Connect with Neil on Twitter/X: `@virtualmanc`.

Marcel Meurer is responsible for the professional IT services business unit at sepago GmbH in Cologne and is the founder of the development company ITProCloud GmbH. In this role, he leads a team of consultants who provide their expertise in Microsoft and Citrix technologies for customers and partners. His technical focus is Microsoft Azure platform services, and he has been a Microsoft Azure MVP since 2016. He loves working in the community. Besides his blog, he publishes tools that simplify working with the Azure cloud – especially in the context of Azure Virtual Desktop. His well-known tools include WVDAdmin and Hydra for Azure Virtual Desktop. Marcel Meurer graduated as an engineer in electrical engineering from the University of Applied Science, Aachen.

Guest author

Shabaz Darr is an infrastructure master from the UK who has close to 20 years of experience within the IT industry. His main areas of focus are Azure IaaS, PaaS, AVD, and Microsoft Security. He is a Microsoft MVP in three categories, including Azure, Enterprise Mobility, and Security, as well as a Microsoft Certified Trainer. In his spare time, he enjoys creating cloud-focused content on his YouTube channel, *I Am IT Geek*.

Firstly, I would like to thank Ryan Mangan for inviting me to be part of this amazing project. To my wife, Reema, and children, Zoya and Mikaeel, thank you for supporting me throughout my career and giving me the motivation to achieve what I have so far. Finally, to the cloud community, who I owe a huge debt for helping me and encouraging me along the way.

About the reviewers

Toby Skerritt has worked in the technology industry for over 20 years, with the last decade spent helping clients design and implement end-user-focused solutions to remote and flexible working challenges. He has architected and delivered services across a range of technologies, including VMware Horizon, Microsoft Configuration Manager, Azure Virtual Desktop, and Intune. As a senior product manager for Nerdio, Toby works with the product and development teams, designing and implementing features to enhance desktop and application management in the enterprise space.

Michel Roth leads the AVD GBB specialist team for EMEA at Microsoft, a role he has embraced since the launch of AVD. In this position, he collaborates with Microsoft's largest customers, partners, and colleagues to ensure they receive the best AVD experience possible. With 25 years of experience in the VDI industry, Michel has held various roles ranging from technical, including a seven-year tenure as a Microsoft MVP for RDS, to product management at RES Software (now Ivanti), and as director for FSLogix in EMEA, which Microsoft acquired in 2018. Passionate about the EUC community, he shares his knowledge extensively, especially on LinkedIn. He would love to connect with you at `https://aka.ms/mroth`.

Shabaz Darr is an infrastructure master from the UK who has close to 20 years of experience within the IT industry. His main areas of focus are Azure IaaS, PaaS, AVD, and Microsoft Security. He is a Microsoft MVP in three categories, including Azure, Enterprise Mobility, and Security, as well as a Microsoft Certified Trainer. In his spare time, he enjoys creating cloud-focused content on his YouTube channel, *I Am IT Geek*.

Table of Contents

Preface — xxi

Part 1: Introduction

Introduction to Azure Virtual Desktop — 3

Desktop virtualization — 4
Azure Virtual Desktop – What is it? — 4
Providing the best user experience — 4
Enhanced security — 5
Simplifying management — 5
Managing AVD performance — 5
What licenses do I need? — 6
Comparing Windows 365 and Azure Virtual Desktop — 6
How does Azure Virtual Desktop work? — 7
What's managed by Microsoft and what you manage — 7
What Microsoft manages — 8
What does the customer manage? — 8
Azure Virtual Desktop Stack HCI — 10
Summary — 10

Part 2: Planning an Azure Virtual Desktop Architecture

Designing the Azure Virtual Desktop Architecture — 13

Assessing existing physical and virtual desktop environments — 14
Assessing AVD deployments — 14
User personas — 16
Application groups — 18

Assessing the network capacity and speed requirements for AVD — 19
- Applications — 20
- Display resolutions — 21
- AVD experience estimator — 22
- RDP bandwidth requirements — 23
- Estimating bandwidth utilization — 24
- Estimating the bandwidth that's used by remote graphics — 24
- Dynamic bandwidth allocation — 27
- Limit network bandwidth use with throttle rate limiting — 27
- Reverse connect transport — 27
- Session host communication channel — 28
- Client connection sequence — 28
- Connection security — 29

Identifying an OS for an AVD implementation — 29
- Supported Azure OS images — 29
- What is Windows 11 multi-session? — 30
- Customizing the Windows 11 multi-session image for your organization — 31

Planning and configuring name resolution for AD and Microsoft Entra Domain Services — 32

Planning a host pool architecture — 32
- App groups — 33
- End users — 34
- Registering the DesktopVirtualization resource provider — 34
- Registering the provider using Azure PowerShell (optional) — 35

Resource groups, subscriptions, and management group limits — 36

Configuring the location for the AVD metadata — 38

Calculating and recommending a configuration for capacity and performance requirements — 39
- Multi-session recommendations — 39
- Recommendations on sizing VMs — 40
- General recommendations for VMs — 41
- Testing workloads — 42

Summary — 42
Questions — 42
Further reading — 43

3

Designing for User Identities and Profiles — 45

Selecting a licensing model for your Azure Virtual Desktop deployment — 45
- Applying Azure Virtual Desktop licensing to virtual machines — 47
- Azure Files tiers — 52

Planning for user profiles — 53
- User profiles — 53
- Challenges with previous user profile technologies — 53
- An introduction to FSLogix profile containers — 54
- Azure Files integration with Entra ID DS — 59
- Microsoft Entra Kerberos for hybrid identities — 60
- Planning for user identities — 61

Planning for Windows Remote Desktop client deployment — 65
- Installing the Windows Remote Desktop client — 66
- Windows App for the Windows Store (currently in public preview) — 68

Subscribing to a workspace	69	Summarizing the prerequisites for Azure Virtual Desktop	74
Accessing client logs	70	Summary	75
Connecting to Azure Virtual Desktop using the web client	71	Questions	75
Setting up email discovery to subscribe to the Azure Virtual Desktop feed	72		

4

Implementing and Managing Networking for Azure Virtual Desktop 77

Implementing Azure VNet connectivity	**77**	**Implementing and managing network security**	**88**
Azure VNet	78	Azure network security overview	88
What is an Azure VNet?	78	Understanding AVD network connectivity	88
Communication between Azure resources	78	**Managing AVD session hosts using Azure Bastion**	**100**
Communication with on-premises networks	79	What is Azure Bastion?	100
Filtering and routing Azure network traffic	80	Setting up Azure Bastion	101
Understanding what VNet integration is for Azure services	81	Connecting to a VM using Azure Bastion	104
Managing connectivity to the internet and on-premises networks	**81**	**Monitoring and troubleshooting network connectivity**	**105**
Types of VPNs available to you	81	Summary	108
Internet access and outbound connections	87	Questions	108

5

Implementing and Managing Storage for Azure Virtual Desktop 109

Configuring storage for FSLogix components	**110**	Step 2 – Configuring the basics	116
		Step 3 – Configuring advanced settings	118
FSLogix profile container storage options	110	Step 4 – Configuring networking	120
The different Azure Files tiers	112	Step 5 – Configuring data protection	122
Best practices for Azure Files with AVD	113	Step 6 – Configuring encryption	125
Configuring storage accounts	**114**	**Configuring file shares**	**128**
Step 1 – Creating a new storage account	114	**Configuring disks**	**132**

Ephemeral OS disks	134	Dynamic disks versus fixed disks	145
Creating a VHD image	135	**Summary**	**145**
Creating a VM	136	**Questions**	**145**
Creating a local image	144	**Further reading**	**146**

Part 3: Implementing an Azure Virtual Desktop Infrastructure

6

Creating Host Pools and Session Hosts — 149

Creating a host pool by using the Azure portal	**149**	**Automating the creation of AVD hosts and host pools**	**164**
Host pool creation	150	Setting up PowerShell for AVD	164
Workspace information	161	Creating an AVD host pool with PowerShell	167
		Summary	**174**
		Questions	**175**

7

Configuring Azure Virtual Desktop Host Pools — 177

Windows Server session host licensing	**178**	Configuring direct assignment using PowerShell	201
Configuring host pool settings	**182**	Applying OS and application updates on an Azure Virtual Desktop host	204
Customizing RDP properties	182	Configuring a validation pool	205
Using PowerShell to customize RDP properties	186	Applying security and compliance settings to session hosts	207
Using PowerShell to configure load-balancing methods	191	**Summary**	**212**
Assigning users to host pools via PowerShell	198	**Questions**	**213**
Configuring automatic assignment	198		
Re-assigning a personal desktop	200		

8

Entra ID Join for Azure Virtual Desktop — 215

Prerequisites	215	Enabling Microsoft Entra authentication for RDP	225
Deploying an Entra ID-joined host pool	216	Configuring local admin access	229
Enabling user access	221	Summary	232
Connecting to Entra ID-joined session hosts using the Remote Desktop client	223	Questions	232

9

Creating and Managing Session Host Images — 233

Creating a golden image	233	Creating an image definition from the ACG	269
Creating a VM	233	Creating an image version	271
Connecting to the VM	237	**Troubleshooting OS issues related to AVD**	**274**
Custom image templates	240	VMs are not joined to the domain	274
Creating your first custom image	240	AVD Agent and AVD bootloader are not installed	275
Modifying a session host image	246	AVD Agent is not registering with the AVD service	275
Disabling automatic updates	247	Basic performance troubleshooting in AVD	278
Installing language packs in AVD	247	Networking troubleshooting	285
Optimizing an image	250	Summary	287
Capturing an image template	256	Questions	287
Creating and using an ACG	259		
Creating your first ACG	260		
Capturing an image in an ACG	263		

Part 4: Managing Access and Security

10

Managing Access   291

Introduction to Azure RBAC	291	Managing local roles, groups, and rights assignments on AVD session hosts	305
Planning and implementing Azure roles and RBAC for AVD	293		
The delegated access model	296	Configuring user restrictions by using Entra ID Domain Services group policies	309
Assigning RBAC roles to IT admins	297		
The PowerShell way to assign role assignments	299	Summary	309
Creating a custom role using the Azure portal	300	Questions	309

11

Managing Security   311

Introduction to MFA	311	Configuring Microsoft Defender Antivirus for session hosts	339
How does Entra MFA Work?	312		
Security defaults	313	What's the difference between Microsoft Defender Antivirus and Microsoft Defender for Endpoint?	340
Conditional Access	314		
Planning and implementing MFA	317	Getting the latest updates	344
Creating a Conditional Access policy for MFA	320	Setting the scheduled task to run the PowerShell script	344
Managing security by using Microsoft Defender for Cloud	326	Manually downloading and unpacking	344
		Configuring quick scans	345
Securing AVD using Microsoft Defender for Cloud	330	Suppressing notifications	346
		Enabling headless UI mode	348
Using Microsoft Defender for Cloud and AVD	332	Summary	350
		Questions	350
Enabling enhanced security for AVD	335		

Part 5: Managing User Environments and Apps

12

Implementing and Managing FSLogix — 353

Installing and configuring FSLogix	354	Cloud Cache	368
License requirements for FSLogix profile containers	354	Configuring Cloud Cache	369
FSLogix key capabilities	355	Microsoft Teams integration	372
FSLogix installation and configuration	355	Teams exclusions	372
Configuring antivirus exclusions	358	FSLogix profile container best practices	372
Configuring exclusions using PowerShell	359	Summary	373
Configuring profile containers	360	Questions	374

13

Configuring User Experience Settings — 375

Configuring Universal Print	376	Enabling screen capture protection for AVD	406
Prerequisites for Universal Print	378	Enabling screen capture protection via Intune	410
Universal Print administrator roles	379		
Setting up Universal Print	380		
Registering printers using the Universal Print connector	385	Enabling watermarking	410
Assigning permissions and sharing printers	388	Prerequisites for watermarking	411
Adding a Universal Print printer to a Windows device	391	Enabling watermarking using Group Policy	411
		Enabling watermarking via Intune	413
Configuring user settings using Microsoft Intune	394	Troubleshooting FSLogix profile issues	414
Start VM on Connect	401	Troubleshooting AVD client issues	416
Configuring with the Azure portal	401	Testing connectivity	417
Supported VM sizes	405	Resetting the Remote Desktop client	419
Prerequisites for enabling hibernation mode	405	The Remote Desktop client is showing no resources	419
Integrating hibernation into AVD	405	Summary	420

| Further reading | 420 | Questions | 420 |

14

MSIX App Attach 421

Configuring dynamic application delivery by using MSIX app attach	422	Uploading MSIX images to Azure Files	458
What is MSIX?	422	Configuring MSIX app attach	461
What does it look like inside MSIX?	423	Publishing an MSIX app to a RemoteApp application group	465
What is MSIX app attach?	424	Troubleshooting MSIX app attach	472
MSIX app attach terminology	427	Published MSIX app attach applications not showing in the Start menu	474
An overview on how MSIX app attach works	427	App attach (public preview)	474
Prerequisites	428	How app attach works	475
Creating an MSIX package	430	Key differences between MSIX app attach and app attach	475
Packaging a simple application in an MSIX container	431	Creating an app attach package	477
Creating an MSIX image	442	Summary	480
Configuring Azure Files for MSIX app attach	446	Further reading	481
Importing the code-signed certificate	455	Questions	481

15

Configuring Apps on a Session Host 483

Application masking	484	Implementing and managing multimedia redirection	507
Rule types available	484	Managing internet access for Azure Virtual Desktop sessions	512
Deploying an application as a RemoteApp application	493	What are VM applications?	515
Implementing and managing OneDrive for Business for a multi-session environment	500	Summary	515
Implementing and managing Microsoft Teams AV redirection	503	Questions	516

Part 6: Monitoring and Maintaining an Azure Virtual Desktop Infrastructure

16

Planning and Implementing Business Continuity and Disaster Recovery 519

Designing a backup strategy for Azure Virtual Desktop	519	Configuring backup and restore for FSLogix user profiles, personal virtual desktop infrastructures (VDIs), and golden images	530
Planning and implementing a disaster recovery plan for Azure Virtual Desktop	522	Virtual machine backup and restore	530
Virtual network	522	Zone-redundant storage	544
Virtual machines	523	Azure file backup and restore	544
Managing user identities	526	Replicating virtual machine images between regions	549
Configuring user and app data	528		
Disaster recovery considerations for MSIX app attach	529	Summary	550
Application dependencies	529	Questions	550

17

Automating Azure Virtual Desktop Management Tasks 551

Creating an Automation account for Azure Virtual Desktop	551	Autoscale – scaling plans	568
Giving the Automation account permissions	555	Giving Microsoft access to start and stop VMs	568
Automating the management of host pools, session hosts, and user sessions using PowerShell	556	Creating a pooled scaling plan (multiuser)	570
		Creating a personal scaling plan (assigned user)	578
Configuring an Azure automation runbook	557	Summary	585
Testing a PowerShell runbook in Azure	563	Questions	585
Creating a schedule	565		

18

Monitoring and Managing Performance and Health — 587

Configuring Azure Monitor for AVD	587	Setting up alerts using alert rules	607
Creating a Log Analytics workspace	588	**Introduction to Kusto**	**615**
Configuring the monitoring of AVD	592	Connecting Log Analytics to Kusto.Explorer	616
Configuring performance counters and event logs	598	Creating queries for AVD using Kusto.Explorer	618
Using Insights	**601**	Some additional Kusto queries	622
Differences between AVD Insights and host pool insights	601	**Using Azure Advisor for AVD**	**625**
Using the host pool insights	601	**Summary**	**627**
		Questions	**628**

19

Azure Virtual Desktop's Quickstart Feature — 629

How the Quickstart feature works	**629**	Using the Quickstart feature without an identity provider	638
Prerequisites	630	Post-deployment cleanup	640
Using the Quickstart feature with Entra ID Domain Services (Entra ID DS)	**631**	Troubleshooting the Quickstart feature	642
		Summary	**643**
		Questions	**643**

Final Assessment — 645

Questions	645	Answers	656

Appendix — 665

Microsoft Resources and Microsoft Learn	665	Introducing EtherAssist – the premier AI technical assistant	669
Azure Virtual Desktop community shout-outs!	666	Specialized support for Azure Virtual Desktop	669
Cool vendors	667		

Level up at AVD TechFest	670	Summary	670
Why attend AVD TechFest?	670		

Index 671

Other Books You May Enjoy 688

Preface

Mastering Azure Virtual Desktop, Second Edition offers complete coverage of Azure Virtual Desktop as well as up-to-date coverage of the AZ-140 exam so that you can sit the exam with confidence. With this book, you will learn the steps for planning, implementing, and managing an Azure Virtual Desktop environment. You will also find hints, tips, and advice on common issues you may face with configuration and day-to-day management.

Who this book is for

This book is for IT professionals who wish to attain the Microsoft Certified: Azure Virtual Desktop Specialty certification and those who work in the end user computing field, whether as an administrator, consultant, or architect. Readers should already be familiar and comfortable with cloud computing and Microsoft Azure principles. You should also have experience administering core features and services within a Microsoft 365 tenant.

What this book covers

Chapter 1, Introduction to Azure Virtual Desktop, provides an introduction to Azure Virtual Desktop, giving a high-level overview of the service offering and the associated benefits.

Chapter 2, Designing the Azure Virtual Desktop Architecture, provides guidance on the requirements to plan and design an Azure Virtual Desktop environment. You will also learn about sizing, network guidelines, and Azure Virtual Desktop connectivity.

Chapter 3, Designing for User Identities and Profiles, covers everything you need to know for designing user identities and profiles.

Chapter 4, Implementing and Managing Networking for Azure Virtual Desktop, looks at the considerations and techniques for implementing and managing networking for Azure Virtual Desktop.

Chapter 5, Implementing and Managing Storage for Azure Virtual Desktop, details the requirements and storage options required for FSLogix components as well as teaching you how to create storage accounts and configure disks and Azure file shares.

Chapter 6, Creating Host Pools and Session Hosts, teaches you how to create and configure Azure Virtual Desktop host pools and session hosts.

Chapter 7, Configuring Azure Virtual Desktop Host Pools, guides you through the configuration of host pools, the use of a **Remote Desktop** (**RD**) license server for those using server-based session hosts, custom **Remote Desktop Protocol** (**RDP**) properties, and applying security and compliance settings to a session host.

Chapter 8, Entra ID Join for Azure Virtual Desktop, teaches you how to join session hosts to Entra ID and how to carry out basic troubleshooting.

Chapter 9, Creating and Managing Session Host Images, looks at the configuration of host pools including the configuration of a gold image, Azure Compute Gallery optimization, and basic performance troubleshooting.

Chapter 10, Managing Access, discusses how to plan and implement Azure roles and **role-based access control** (**RBAC**) and how to manage local roles, groups, and rights assignments for Azure Virtual Desktop session hosts.

Chapter 11, Managing Security, provides a clear understanding of Azure multi-factor authentication and its benefits, how to configure conditional access policies, the use of Azure Defender for Cloud, and the configuration of Microsoft Defender Antivirus for Azure Virtual Desktop.

Chapter 12, Implementing and Managing FSLogix, shows how to install, configure, and manage FSLogix profile containers and Cloud Cache.

Chapter 13, Configuring User Experience Settings, looks at some of the features and functions you can configure with Azure Virtual Desktop, including Universal Print, Start VM on Connect, Screen Capture Protection, FSLogix troubleshooting, and Remote Desktop client troubleshooting.

Chapter 14, MSIX App Attach, teaches you how to implement and manage MSIX app attach for Azure Virtual Desktop.

Chapter 15, Configuring Apps on a Session Host, teaches you how to configure app masking, deploy RemoteApp applications, configure Microsoft Teams AV Redirect, multimedia redirection, and manage internet access for Azure Virtual Desktop.

Chapter 16, Planning and Implementing Business Continuity and Disaster Recovery, discusses the options available to you when planning and designing business continuity and disaster recovery for Azure Virtual Desktop.

Chapter 17, Automating Azure Virtual Desktop Management Tasks, teaches you how to automate repeated maintenance tasks, implement custom autoscaling scripts, and configure and deploy scaling plans.

Chapter 18, Monitoring and Managing Performance and Health, teaches you how to configure Azure Virtual Desktop insights to monitor user experience and overall environment performance. This chapter also discusses setting up alerts and an introduction to Kusto.

Chapter 19, Azure Virtual Desktop's Quickstart Feature, teaches you how to use the getting started feature to deploy an Azure Virtual Desktop environment.

Appendix, Microsoft Resources and Microsoft Learn, contains useful information and other interesting content from Microsoft, communities, and MVPs on Azure Virtual Desktop.

To get the most out of this book

If you are an IT professional, an end user computing administrator, an architect, or a consultant looking to learn about implementing and managing Azure Virtual Desktop, this book is for you.

Download the example code files

You can download the example code files for this book from GitHub at https://github.com/PacktPublishing/Mastering-Azure-Virtual-Desktop-2nd-Edition. If there's an update to the code, it will be updated in the GitHub repository.

We also have other code bundles from our rich catalog of books and videos available at https://github.com/PacktPublishing/. Check them out!

Conventions used

There are a number of text conventions used throughout this book.

`Code in text`: Indicates code words in text, database table names, folder names, filenames, file extensions, pathnames, dummy URLs, user input, and Twitter handles. Here is an example: "The client stores the connection configuration for each available resource in a set of `.rdp` files."

A block of code is set as follows:

```
{ "joeclbldhdmoijbaagobkhlpfjglcihd": { "installation_mode": "force_installed", "runtime_allowed_hosts": [ "*://*.youtube. com" ], "runtime_blocked_hosts": [ "*://*" ], "update_url": "https://edge.microsoft.com/extensionwebstorebase/v1/crx" } }
```

Any command-line input or output is written as follows:

```
Install-Module -Name Az.DesktopVirtualization
```

Bold: Indicates a new term, an important word, or words that you see onscreen. For instance, words in menus or dialog boxes appear in **bold**. Here is an example: "You will then see the **Sign in to your account** popup."

> **Tips or important notes**
> Appear like this.

Get in touch

Feedback from our readers is always welcome.

General feedback: If you have questions about any aspect of this book, email us at `customercare@packtpub.com` and mention the book title in the subject of your message.

Errata: Although we have taken every care to ensure the accuracy of our content, mistakes do happen. If you have found a mistake in this book, we would be grateful if you would report this to us. Please visit `www.packtpub.com/support/errata` and fill in the form.

Piracy: If you come across any illegal copies of our works in any form on the internet, we would be grateful if you would provide us with the location address or website name. Please contact us at `copyright@packtpub.com` with a link to the material.

If you are interested in becoming an author: If there is a topic that you have expertise in and you are interested in either writing or contributing to a book, please visit `authors.packtpub.com`.

Share Your Thoughts

Once you've read *Mastering Azure Virtual Desktop*, we'd love to hear your thoughts! Scan the QR code below to go straight to the Amazon review page for this book and share your feedback.

https://packt.link/r/1-835-88415-6

Your review is important to us and the tech community and will help us make sure we're delivering excellent quality content.

Download a free PDF copy of this book

Thanks for purchasing this book!

Do you like to read on the go but are unable to carry your print books everywhere?

Is your eBook purchase not compatible with the device of your choice?

Don't worry, now with every Packt book you get a DRM-free PDF version of that book at no cost.

Read anywhere, any place, on any device. Search, copy, and paste code from your favorite technical books directly into your application.

The perks don't stop there, you can get exclusive access to discounts, newsletters, and great free content in your inbox daily

Follow these simple steps to get the benefits:

1. Scan the QR code or visit the link below

```
https://packt.link/free-ebook/978-1-83588-414-0
```

2. Submit your proof of purchase
3. That's it! We'll send your free PDF and other benefits to your email directly

Part 1: Introduction

This section offers an introduction to Azure Virtual Desktop with a view to providing a high-level overview of the subject before we start to delve into the details of the core functions and features of the product.

This part of the book comprises the following chapter:

- *Chapter 1, Introduction to Azure Virtual Desktop*

Introduction to Azure Virtual Desktop

In today's world, many businesses have been building out or are planning to adopt remote work strategies to ensure continuity. These strategies enhance security and reduce infrastructure costs. **Azure Virtual Desktop** has emerged as a key technology in this transformation. This second edition of the book, takes a deeper look into the expanded capabilities and advancements of Azure Virtual Desktop since the first edition.

We explore how Azure Virtual Desktop has adapted and provide guidance on how you can implement, manage, and support it within your organization. This book has been updated to serve as an all-encompassing manual on planning, designing, implementing, and supporting an Azure Virtual Desktop environment in today's context. We will first start with the fundamentals of understanding the key components of Azure Virtual Desktop and then advance through the chapters into broader topics.

The following topics will be covered in this chapter:

- Desktop virtualization
- Azure Virtual Desktop – what is it?
- How does Azure Virtual Desktop work?
- What's managed by Microsoft and what you manage
- Azure Virtual Desktop Stack HCI

Desktop virtualization

Desktop virtualization, also known as **Virtual Desktop Infrastructure (VDI)**, refers to virtualization and **Virtual Machines (VMs)** that provide and manage virtual desktops. Users access these VMs remotely from any supported device, and the compute processing is completed on the host server. Users connect to their virtual desktop sessions through a **connection broker**, also known as just a **broker**. This broker functions as a software layer that serves as an intermediary between the user and the server, facilitating the orchestration of user sessions to virtual desktops or published applications.

VDI is usually deployed in an organization's data center and managed by its IT department. Typical on-premises providers include Citrix, Omnissa (previously VMware), and **Remote Desktop Services (RDS)**. VDI can be hosted on-premises or in the cloud. Some organizations use the cloud to scale virtual desktop environments, enabling a hybrid capability that allows IT admins to meet changing organizational demands quickly.

Azure Virtual Desktop – What is it?

Azure Virtual Desktop is a desktop and app virtualization service that runs on Microsoft Azure. Azure Virtual Desktop works across devices, including Windows, Mac, iOS, Android, and Linux, with apps that you can use to access remote desktops and applications. Modern browsers can also be used to access Azure Virtual Desktop.

Providing the best user experience

Users have the freedom to connect to Azure Virtual Desktop from any capable device over the internet. You can use an Azure Virtual Desktop client to connect to published Windows desktops and applications. There are three flavors of client that you can use to connect: a native application on the device, a mobile app, or the Azure Virtual Desktop HTML5 web client.

You can improve application performance on session host VMs by running apps near cloud services, connecting to your data center or the cloud. Running apps near cloud services will reduce the risk of long loading times and keep your users productive.

User sign-in to Azure Virtual Desktop is much faster because user profiles are containerized using FSLogix profile containers. The user profile container is dynamically attached to the session host or VM in question at user sign-in. The user profile is made available and appears in the system exactly as a local user profile would.

You can provide individual ownership to session desktops using personal (persistent) desktops for those specific use cases. For example, you may want to offer personal remote desktops for members of a web development team. They would be able to add or remove programs without impacting other users on that virtual desktop.

Enhanced security

Azure Virtual Desktop provides centralized security for users' desktops with **Microsoft Entra ID**. You can further enhance security by enabling **multi-factor authentication (MFA)** to provide secure user access. You can also secure access to data by using Azure's granular **role-based access control (RBAC)** for users.

Azure Virtual Desktop separates the data and apps from the local hardware and runs both resource types on a remote server. The risk of confidential data being left on a personal device is significantly reduced when using Azure Virtual Desktop.

User sessions can be isolated in both single- and multi-session virtual desktop deployments.

Azure Virtual Desktop improves security by using reverse connect technology, a more secure connection type than the **Remote Desktop Protocol (RDP)**. However, the session hosts do open inbound ports to the session host VMs.

Simplifying management

Azure Virtual Desktop is a Microsoft Azure service that's familiar to Azure admins. You use Microsoft Entra ID and RBAC to manage access to resources. With Microsoft Azure, you are provided with the tools to automate VM deployments, manage VM updates, and provide disaster recovery.

As with other Microsoft Azure services, Azure Virtual Desktop uses Azure Monitor for monitoring and alerts. This allows IT admins to identify issues through a single interface.

Managing AVD performance

Azure Virtual Desktop provides you with options to load balance users on your VM **host pools**. Host pools are collections of VMs with the same configuration assigned to multiple users.

You can configure session load balancing to occur as users sign in to session hosts, also known as **breadth mode**. Breadth mode essentially means that users are sequentially allocated across the host pool for your workload. You also have the option to configure your VMs for depth mode load balancing to save costs, where users are fully allocated on one VM before moving to the next. In addition, Azure Virtual Desktop provides the tools and the capability to automatically provision additional VMs when incoming demand exceeds a specified threshold.

> **Multi-session Windows 11**
>
> Azure Virtual Desktop enables and headlines Windows 11 and 10 Enterprise multi-session since they are the only Windows OSs (client-based) that enable multiple concurrent users on a single Windows 11/10 VM.

Azure Virtual Desktop also provides a familiar experience with broader application support than the traditional Windows Server-based remote desktop solutions.

What licenses do I need?

Azure Virtual Desktop is available at no additional cost if you have an eligible Microsoft 365 license. However, it is important to note that you pay for the Microsoft Azure resources that are consumed by Azure Virtual Desktop:

- You must have a Windows or Microsoft 365 license to be able to use Windows 11 Enterprise and Windows 10 Enterprise desktops and apps (eligible)
- You must be a Microsoft RDS **Client Access License** (**CAL**) customer for Windows Server RDS desktops and apps (eligible)

We now move on to take a look at the comparison of Azure Virtual Desktop and Windows 365, looking at the similarities and differences between each.

Comparing Windows 365 and Azure Virtual Desktop

Windows 365 is a relatively new cloud-based virtual desktop service from Microsoft that allows organizations to provision and assign cloud PCs to individual users through the Microsoft 365 portal. Each organizational user licensed for Windows 365 gets a dedicated cloud PC. Operating as a **Software as a Service** (**SaaS**) model, Windows 365 is managed via the **Microsoft Intune admin center**. It is distinct from other solutions such as Azure Virtual Desktop regarding resource allocation.

Windows 365 establishes a 1:1 relationship between the user and their cloud PC. This contrasts with Azure Virtual Desktop, which provides consumption-based virtual desktops and multiple virtual desktop provisioning types, the most common being **pool-based virtual desktop offerings**.

Turning to Azure Virtual Desktop, this operates as a **Platform as a Service** (**PaaS**) model and offers more flexibility and options for organizations, especially those with dynamic resource requirements or specific needs. The pricing structure varies based on usage and configuration, making it suitable for businesses with fluctuating resource demands. Azure Virtual Desktop is managed through the Azure portal, Azure CLI, PowerShell, and REST API.

Both Windows 365 and Azure Virtual Desktop maintain similar security standards. However, it's important to note that Azure Virtual Desktop offers greater configuration flexibility, including security and platform configuration, which helps organizations with strict or complex security requirements.

For those wanting to learn more about Windows 365, check out the book *Mastering Windows 365* by Christiaan Brinkhoff, Morten Pedholt, and Sandeep Patnaik.

Let's now delve deeper into the world of Azure Virtual Desktop.

How does Azure Virtual Desktop work?

Azure Virtual Desktop simplifies deployment and management compared to traditional RDS or VDI environments. You don't have to provision and manage servers and server roles such as the gateway, connection broker, diagnostics, load balancing, and licensing.

What's managed by Microsoft and what you manage

The following diagram shows what services Microsoft manages and what you manage:

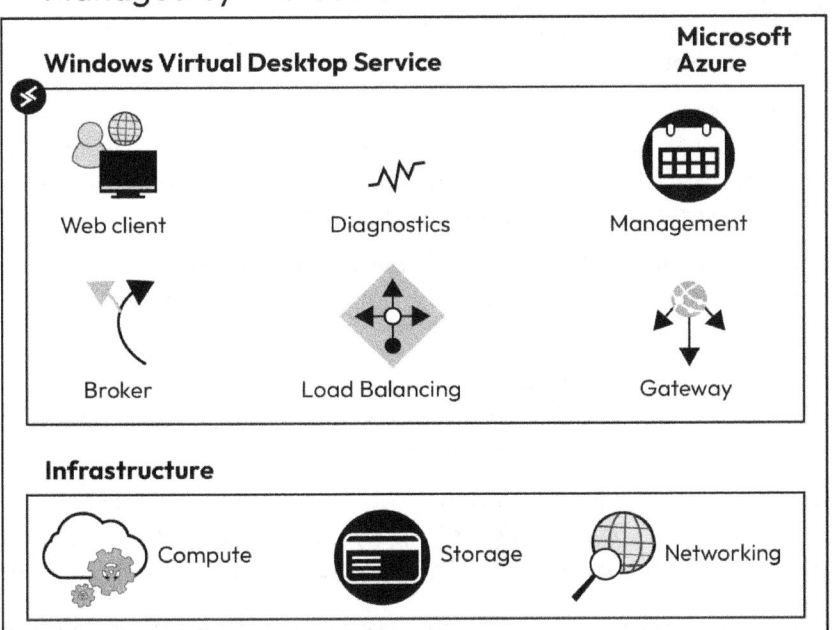

Figure 1.1 – Services managed by Microsoft and you

What Microsoft manages

Azure Virtual Desktop provides a virtualization infrastructure as a managed service. Azure Virtual Desktop's core components are as follows:

- **Web client**: The web access service feature in Azure Virtual Desktop management allows users to access virtual desktops and remote apps via any HTML5-compatible web browser, similar to a local PC experience. This can be accessed from any location and on any device. Enhanced security for web access is achievable through MFA in Microsoft Entra ID.
- **Diagnostics**: Azure Virtual Desktop includes an event-based aggregation service, Remote Desktop Diagnostics. This service logs each user or administrator's action within the deployment, categorizing them as successes or failures. Administrators can review these event logs to pinpoint any problematic components.
- **Management**: Azure Virtual Desktop configurations are manageable directly through the Azure portal. This includes the ability to manage and publish resources in host pools. The platform also supports various extensibility components, with management options via Windows PowerShell or REST APIs, facilitating integration with third-party tools.
- **Broker**: The Connection Broker service in Azure Virtual Desktop oversees user connections to both virtual desktops and remote apps. It is responsible for load balancing and facilitating reconnections to existing sessions.
- **Load balancing**: Azure Virtual Desktop provides session host load balancing, which can be configured either depth-first or breadth-first. The connection broker manages the distribution of new sessions across the VMs in a host pool.
- **Gateway**: The Remote Connection Gateway service enables remote user connectivity to Azure Virtual Desktop's remote apps and desktops from any internet-enabled device capable of running an Azure Virtual Desktop client. It establishes a connection from the VM to the gateway.

Windows Virtual Desktop uses Azure infrastructure services for compute, storage, and networking.

What does the customer manage?

Now, let's look at what you, as the customer, manage. First, we'll look at the desktop and remote apps part of Azure Virtual Desktop.

Desktop and remote apps

With these options, you can create application groups to group, publish, and assign access to remote apps or desktops:

- **Desktop**: Remote Desktop application groups give users access to a full desktop. You can provide a desktop where the session host's VM resources are shared or pooled. You can give dedicated personal desktops to those users who need to add or remove programs without impacting other users.

- **Apps**: RemoteApp application groups provide users with access to the applications you individually publish to the application group. You can create multiple RemoteApp app groups to accommodate different user scenarios. For example, you can use RemoteApp to virtualize an app that runs on a legacy OS or needs secured access to corporate resources.

- **Images**: When you configure session hosts for application groups, you have a choice of images. You should use a recommended image such as Windows 10 Enterprise multi-session and Office 365. Alternatively, you can choose an image in your gallery, or an image provided by Microsoft or other publishers.

Management and policies

Now, let's look at the customer responsibilities for management and policies:

- **Profile management**: Configure FSLogix profile containers with a storage solution such as Azure Files to containerize user profiles and provide users with a fast and stateful experience.

- **Sizing and scaling**: Here, you can specify session host VM sizes, including GPU-enabled VMs, as well as specify depth or breath load balancing when you create a host pool. Finally, you can configure automation policies for scaling.

- **Networking policies**: Define a network topology to access the virtual desktop and virtual apps from the intranet or internet based on the organizational policy.

- **Connectivity options**: Connect your Azure Virtual Network to your on-premises network by using a virtual private network. Alternatively, you can use Azure ExpressRoute to extend your on-premises networks into the Microsoft cloud platform over a private connection.

- **User management and identity**: Use **Microsoft Entra ID** and RBAC to manage user access to resources. Take advantage of Microsoft Entra ID security features such as Conditional Access, MFA, and Intelligent Security Graph. Azure Virtual Desktop requires **Microsoft Entra Domain Services (MEDS)**. Domain-joined sessions host VMs on this service. You can also sync MEDS with Microsoft Entra ID so that users are associated between the two. Once you've done this, you can use **Microsoft Entra join** to deliver virtual desktops to your users.

Azure Virtual Desktop Stack HCI

At the time of writing, **Azure Virtual Desktop for Azure Stack HCI (AVD HCI)** is in preview. It offers a versatile solution for deploying session hosts onpremises or as an extension of Azure Virtual Desktop deployments in the Azure cloud. This service is particularly beneficial for enhancing performance in regions with limited Azure public cloud connectivity and for meeting specific data locality requirements, as it allows for the keeping of app and user data on-premises.

Some of the key benefits of AVD HCI include improved performance and connectivity and cost-effective multi-session options with Windows 10 and Windows 11 Enterprise. It also offers simplified deployment and management via the Azure portal. It also supports RDP Shortpath for low-latency access and allows the use of Azure Marketplace images for deploying up-to-date, fully patched virtual machines.

However, it is important to note that this service is not Azure Arc-enabled and is not supported as a standalone service outside Azure, in a multi-cloud environment, or on Azure Arc-enabled servers other than when in use with Azure Stack HCI VMs.

Some of the limitations of AVD HCI include the non-support of some Azure Virtual Desktop features, such as autoscale and multimedia redirection, and restrictions on mixing session hosts from Azure and Azure Stack HCI in the same host pool. At the time of writing, there may be performance and user density variations due to the diverse hardware and networking capabilities supported by Azure Stack HCI.

With this, we have come to the end of the chapter.

Summary

This chapter provided an introduction to Azure Virtual Desktop, some of the key benefits of the service, and an overview of its components and capabilities. We also briefly covered the differences between Azure Virtual Desktop and Windows 365, and finally, we provided an insight into AVD HCI. In the next chapter, we will look at designing an Azure Virtual Desktop architecture.

Part 2: Planning an Azure Virtual Desktop Architecture

This section takes a look at the planning and design of Azure Virtual Desktop's architecture, which covers both the core architecture and the design of user identities and profiles.

This part of the book comprises the following chapters:

- *Chapter 2, Designing the Azure Virtual Desktop Architecture*
- *Chapter 3, Designing for User Identities and Profiles*
- *Chapter 4, Implementing and Managing Networking for Azure Virtual Desktop*
- *Chapter 5, Implementing and Managing Storage for Azure Virtual Desktop*

2
Designing the Azure Virtual Desktop Architecture

We will start this book by looking at the design of **Azure Virtual Desktop** (**AVD**). Design is an integral part of any suitable technology solution, and in this chapter, we will look at the areas you should consider when designing AVD.

In this chapter, we'll cover the following topics:

- Assessing existing physical and virtual desktop environments
- Assessing the network capacity and speed requirements for AVD
- Identifying an **operating system** (**OS**) for an AVD implementation
- Planning and configuring name resolution for **Active Directory** (**AD**) and Microsoft Entra Domain Services
- Planning a host pool architecture
- Resource groups, subscriptions, and management groups
- Configuring the location for the AVD metadata
- Calculating and recommending a configuration for capacity and performance requirements

Assessing existing physical and virtual desktop environments

Before we can look at the components of AVD, we need to understand the current environment, the requirements, and other information that would be useful.

Assessing AVD deployments

When designing AVD, we must first examine the current desktop estate within the organization. The desktop estate could be purely physical, virtual, or a mixture of both physical and virtual. Suppose your organization is deploying a virtual desktop environment for the first time (**greenfield deployment**). In this case, you should still assess your physical desktop estate to understand the applications, data, and profile usage within your organization.

The desktop assessment should evaluate areas including user persona, consistent host pool types of **virtual machines** (**VMs**), applications, and user profiles. The data gathered from this assessment can be used to plan the deployment of new infrastructure and to guide the AVD migration process.

When preparing your planning methodology, you can use Microsoft's best practices (given in the following list) on cloud adoption. They will help you document your technology strategy and current desktop state:

- Inventory and rationalize your desktop estate based on assumptions that align with motivations and business outcomes with the digital estate guidance from Microsoft: `https://docs.microsoft.com/azure/cloud-adoption-framework/digital-estate/rationalize`.

- Establish a plan for initial organizational alignment to support the proposed adoption plan using the organizational alignment: `https://docs.microsoft.com/azure/cloud-adoption-framework/plan/initial-org-alignment`.

- Create a readiness plan for addressing any skills gaps that may be present. You can find this in the skills readiness plan: `https://docs.microsoft.com/azure/cloud-adoption-framework/plan/adapt-roles-skills-processes`.

- Develop a cloud adoption plan to manage change across the digital estate, operational and technical skills, and overall organization using the cloud adoption plan: `https://docs.microsoft.com/azure/cloud-adoption-framework/plan/plan-intro`.

Another helpful tool that can be used to plan is a **digital estate assessment**, which enables you to measure your desktop estate changes based on the organization's desired outcomes:

- **Infrastructure**: For organizations that are inward facing and seek to optimize costs, operational processes, agility, or other aspects of their operations, the digital estate focuses on VMs, servers, and workloads.
- **Applications**: It is recommended that you focus on applications, APIs, and transactional data that support the customers.
- **Data**: It is somewhat challenging to launch new products/services without some data. You should also focus on the silos of data across the organization.
- **Operational**: Businesses require stable technologies to operate efficiently and effectively. Where possible, businesses need to aim to be as close as possible to zero downtime. Service reliability is critical in today's competitive markets. When stability is a priority, the digital estate should be measured on the positive or negative impact on stable operations. The reliability of workloads, disaster recovery, and business continuity are good measures you can use for operational stability per asset.

Using the data you have collected and analyzed will help you create the migration plan. You may need to carry out workload assessments to capture specific requirements. For an AVD migration deployment plan, you will require data about the desktops, users, and workloads used by each user.

Let's look at some tools available to you.

Azure Migrate

Azure Migrate is used to discover, assess, and migrate on-premises servers, apps, and data to Microsoft Azure, highlighting its principal utilities:

- Azure Migrate's appliance is used to discover installed applications (software inventory)
- Agentless VMware migration supports concurrent replication of 500 VMs per vCenter
- Azure Migrate installs the Azure VM agent automatically on the VMware VMs while migrating them to Azure using the agentless VMware migration method

> **Important note**
>
> **Movere** was acquired by Microsoft in 2019 and is now a part of Azure Migrate. You can read more here: `https://learn.microsoft.com/movere/integrate-azure-migrate#add-movere-in-azure-migrate`.

Lakeside is also integrated with Azure Migrate within the virtual desktop infrastructure migration goals section to assess your current state. This vendor can help you map out an AVD deployment plan, including personas, host pools, applications, and user profiles specific to a virtual desktop environment.

User personas

User personas are the specifications for a particular group of users within a physical or virtual desktop environment with common characteristics or working methods. You may have multiple user personas in an AVD environment.

Once you have completed the required data capture, you will see the resources and workloads being used within your environment. You can then use this data to group user personas based on the following criteria:

- **Personal pools**: Some users may require dedicated desktops (personal pools). For example, security, compliance, high-performance, or noisy-neighbor requirements might lead to some users running on dedicated desktops that aren't part of a pooling desktop strategy. You can use one-to-one here, though one of the biggest benefits of using personal pools is the ability to give local administrator rights and solve the challenge of those apps that do not support pooled multi-session deployments. You would enter this information by specifying a personal host pool type during the AVD host pool deployment (`https://docs.microsoft.com/azure/virtual-desktop/create-host-pools-azure-marketplace#begin-the-host-pool-setup-process`).

- **Density**: Power users may benefit from fewer users per session host for the more intensive workloads. For example, heavier density (applications/users/load) may require two users per **virtual central processing unit** (**vCPU**) instead of the typical six users' light-user assumption per vCPU. You must enter the required density information in the pool settings of the AVD host pool deployment (`https://docs.microsoft.com/azure/virtual-desktop/create-host-pools-azure-marketplace#begin-the-host-pool-setup-process`).

- **Performance**: High-performance desktop requirements for workloads or specific user scenarios. Some users may need more memory per vCPU than the assumed 4 GB of RAM per vCPU. You must enter the VM sizing in the AVD host pool deployment in the VM details section (`https://docs.microsoft.com/azure/virtual-desktop/create-host-pools-azure-marketplace#virtual-machine-details`).

- **Graphical processing**: Some users may require a **graphic processing unit** (**GPU**) for CAD or other graphical applications/workloads. Some users may require vGPU-based VMs in Azure, as demonstrated in this guide for configuring GPU VMs: `https://docs.microsoft.com/azure/virtual-desktop/configure-vm-gpu`.

- **Azure region**: Localized regional requirements to mitigate any **latency** and connectivity issues. Before configuring the host pool, it is recommended that a user from each region should test latency to Azure by using the Azure experience estimation tool: `https://azure.microsoft.com/services/virtual-desktop/assessment/#estimation-tool`. The test user should provide details for the lowest-latency Azure region and the latency in milliseconds for the top three Azure regions. Additionally, if a local backend is needed for the applications that are served via AVD, the latency between the application and the backend can be more important than the latency between the user and the session.
- **Business functions**: Department grouping for billing or specific operational requirements. This type of grouping will help you align corporate costs in the later stages of operations. You can use different subscriptions per department or use tagging to allocate costs to different business cost centers.
- **User count**: One question you should consider is, "How many users will be in each distinct persona?"
- **Max session counts**: Consider the geography and operational hours to predict the peak concurrent user count for each persona.

The following table shows responses to populating a completed assessment or design document:

Example User Groupings			
Criterion	**User Group 1**	**User Group 2**	**User Group 3**
Pool Type	Shared	Shared	Personal (security concerns)
Density	Light (6 users/vCPU)	Heavy (2 users/vCPU)	Dedicated (1 user/vCPU)
Performance	Low	High memory	Low
GPU	N/A	Required	N/A
Azure Region	UK West	Western Europe	UK South
User Count	1,000	50	20
Session Count	200	50	10

Table 2.1 – Table showing example user groupings

The table is from the following link: `https://learn.microsoft.com/azure/cloud-adoption-framework/scenarios/azure-virtual-desktop/migrate-assess#user-personas`.

Each persona/grouping, or each group of users with unique/specific/individual business functions and technical requirements, would require a specific host pool configuration.

The end-user assessment helps you realize the required data: pool type, density, size, CPU/GPU, landing zone region, and so on.

> **Important note**
> An Azure landing zone is a collection of best practices guided by eight key design principles to ensure optimal implementation and management.
>
> You can read more about Azure landing zones here: `https://docs.microsoft.com/azure/cloud-adoption-framework/ready/landing-zone/`.

The host pool configuration assessment aligns/correlates/translates that data to a deployment plan. Aligning the technical requirements, business requirements, and cost will improve the ability to determine the host pools' proper number and configuration.

Pricing examples are available for Microsoft Azure in the East US (`https://azure.com/e/448606254c9a44f88798892bb8e0ef3c`), West Europe (`https://azure.com/e/61a376d5f5a641e8ac31d1884ade9e55`), and Southeast Asia (`https://azure.com/e/7cf555068922461587d0aa99a476f926`) regions.

Application groups

Both Movere and Lakeside assess the current on-premises environment and provide data about the applications running on end-user desktops. Using the data you have collected, you can create a list of all the applications that are required for each persona. For each required application, the answers to the following questions will help shape deployment iterations:

- Which applications need to be installed for the persona to use this desktop? (group/departmental applications)

    Unless the persona uses 100% web-based software as a service application, you will most likely need to create and configure a custom master VHD image (`https://docs.microsoft.com/azure/virtual-desktop/set-up-customize-master-image`) for each persona. You will then need to work out which applications are typical applications and group/departmental applications. Common applications should be installed on the master image.

> **Tip**
> You can create custom images within the Azure portal or use Hyper-V, as suggested in the preceding paragraph.

- Is this application compatible with a Windows 11 Enterprise multi-session (W11EMU)?

    If an application isn't compatible, a personal pool (`https://docs.microsoft.com/azure/virtual-desktop/configure-host-pool-personal-desktop-assignment-type`) may be required to run the custom VHD image.

- Will mission-critical applications suffer from latency between the AVD instance and any backend systems?

    If this is the case – and it is likely to be the case – you may want to consider migrating the backend systems that support the application to Azure.

These answers may require the plan to include remediation to the desktop images or support application components before desktop migration or deployment.

In this section, we looked at application groups and some of the tools available, as well as some questions to help you gather the requirements for your future AVD solution.

Now, let's look at the network capacity and speed requirements for AVD.

Assessing the network capacity and speed requirements for AVD

In this section, we will look at assessing the network requirements for AVD and some of the considerations you should factor into your design.

A **Remote Desktop Protocol** (**RDP**) session relies on network bandwidth. Problems with bandwidth will impact your user experience within a Windows session. Depending on the applications and display resolutions, you may require different network configurations for specific groups within your organization. Incorrectly configuring your network to meet your remote desktop needs and requirements can lead to project failure and users not being able to carry out their required tasks within AVD.

Applications

Before we understand how applications affect bandwidth, let's look at what user-specific bandwidth recommendations are available:

Application Recommended Minimum Bandwidths	
Workload	Bandwidth Recommendation
Light User	1.5 Mbps
Medium User	3 Mbps
Heavy User	5 Mbps
Power User	15 Mbps

Table 2.2 – Application recommended minimum bandwidths

The preceding table provides guidance on the minimum recommended bandwidths for an acceptable user experience. The listed recommendations are based on the guidelines in Microsoft's guide to remote desktop workloads (`https://docs.microsoft.com/windows-server/remote/remote-desktop-services/remote-desktop-workloads`).

> **Important note**
> The recommendations in the table apply to networks with less than 0.1% loss. These recommendations apply regardless of how many user sessions you are hosting on your VMs.

Remember, application workload outputs, frame rate, or display resolutions will apply stress to your network. As framerates increase, your bandwidth requirement will change.

A good example is when you add Microsoft Teams video conferences without any audio/visual redirection to a typical light workload with a high-resolution display – your bandwidth requirement will increase. We will look at how to improve user experience and performance using Teams AV redirect in *Chapter 15, Configuring Apps on a Session Host*.

Some of the other typical use cases that have changing bandwidth requirements are as follows:

- Voice
- Real-time communication
- Streaming 4K video

Load testing the applicable use cases and scenarios in your deployment using tools such as Login VSI and PC Mark, which simulate load, is recommended. When load testing or benchmarking, it's important to vary the load sizes and run specific stress tests to simulate your future environment. It is also recommended that you test typical user scenarios in remote sessions to understand your network's requirements and capabilities.

Display resolutions

Your required display resolution will determine the required bandwidth. The following table provides an example of the required bandwidths Microsoft recommends you have for a display resolution with a frame rate of 30 **frames per second (FPS)**. Following these guidelines will help you provide a smooth user experience. The same recommendations apply to both single- and multiple-user scenarios. Remember, scenarios involving a frame rate under 30 FPS, such as reading static text, will require less available bandwidth than any graphically intensive applications:

Typical Display Resolutions at 30 FPS	Recommended Bandwidth
About 1,024 × 768 px	1.5 Mbps
About 1,280 × 720 px	3 Mbps
About 1,920 × 1,080 px	5 Mbps
About 3,840 × 2,160 px (4 K)	15 Mbps

Table 2.3 – The recommended bandwidth for different display resolutions at 30 FPS

We now move on to looking at the experience estimate for AVD.

AVD experience estimator

Latency can be defined as *the delay before data transfer begins after following an instruction for its transfer*. Latency does not have to be geographic – it can also be based on your company's network topology. The distance you are from a Microsoft Azure data center can have an impact on the user experience. You can check the **round trip time** (**RTT**) of each Azure region using the AVD Experience Estimator (`https://azure.microsoft.com/services/virtual-desktop/assessment/`):

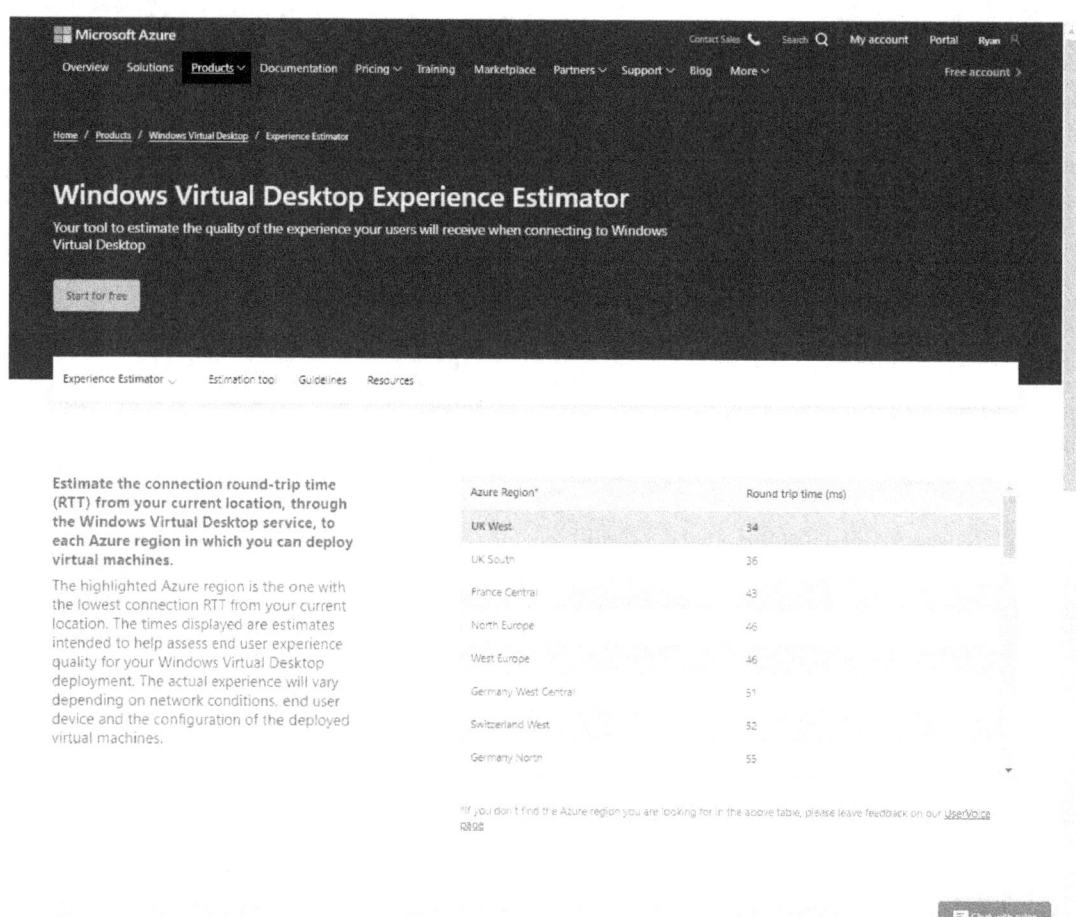

Figure 2.1 – AVD Experience Estimator

The AVD Experience Estimator is a great tool to help you pick the appropriate Azure regions in terms of the closest region to deploy your AVD environment in. You can also use this tool to understand the best location for a possible AVD disaster recovery solution.

> **Assistive technologies**
>
> The phrase "assistive technology" is used to describe products or systems that support and assist individuals with disabilities, restricted mobility, or other impairments.
>
> It is important to note that, when using assistive technology workloads, you will need to ensure that the round trip time is less than 20 **milliseconds (ms)** to achieve a good user experience.

RDP bandwidth requirements

AVD uses RDP as the connection method to provide remote display and input capabilities over network connections. RDP was initially released with **Windows NT 4.0 Terminal Server Edition** and has continuously evolved with every future Microsoft Windows and Windows Server release.

From the outset, RDP was designed to be independent of its underlying transport stack, and today, it supports multiple types of transport.

RDP is a sophisticated technology employing various techniques to transmit the server's remote graphics to the client device. Depending on the use case, scenario, availability of computing resources, and network bandwidth, RDP dynamically adjusts several parameters to provide the best remote user experience.

RDP multiplexes (multiple messages) multiply **dynamic virtual channels (DVCs)** into a single data channel sent over different network transports. Did you know that there are separate DVCs for remote graphics, input, device redirection, printing, and others? The total amount of data that's sent over RDP depends on the user's activity – for example, a typical user working with basic textual content for most of the user's session. The bandwidth is low until the user generates a printout of a 100-page document to a printer. The print job will use more network bandwidth compared to the typical textual content that's used within the session.

A network's available bandwidth impacts the remote session's quality of experience. It is important to note that each application and display resolution can require different network configurations. It is essential to ensure that your network configuration meets your needs. It is recommended that you profile your network requirements based on user activity.

Estimating bandwidth utilization

RDP uses various compression algorithms for different types of data. The following table details some estimates for the data transfers:

Data type	Direction	How to Estimate
Remote Graphics	Session host to client	See the detailed guidelines (https://docs.microsoft.com/azure/virtual-desktop/rdp-bandwidth?WT.mc_id=modinfra-17152-thmaure#estimating-bandwidth-used-by-remote-graphics).
Heartbeats	Both directions	~ 20 bytes every 5 seconds.
Input	Client to session host	The amount of data is based on the user activity, which is less than 100 bytes for most operations.
File transfers	Both directions	File transfers use bulk compression. Use .zip compression for approximation.
Printing	Session host to client	Print job transfers depend on the driver and use bulk compression. Use .zip compression for approximation.

Table 2.4 – Details of some estimates relating to data transfers

This table has been taken from Microsoft (https://docs.microsoft.com/azure/virtual-desktop/rdp-bandwidth#estimating-bandwidth-utilization).

As discussed in the previous section, *Assessing the network capacity and speed requirements for AVD*, you need to consider bandwidth changes when you're doing the following:

- Streaming 4 K video
- Voice or video conferencing
- Real-time communication

Now, let's look at estimating bandwidth for users who require remote graphics.

Estimating the bandwidth that's used by remote graphics

It is difficult to predict remote desktop bandwidth usage. The bandwidth usage depends on the user's activities and those activities that generate the most remote desktop traffic. Every individual user is different, and differences in their work patterns may change network usage.

One of the recommended ways to assess network bandwidth requirements is to monitor real user session connections. You can monitor connections using the built-in performance counters, network equipment, or third-party tooling.

In most cases, you would estimate network utilization by understanding how RDP works by analyzing your organization's user work patterns.

The RDP delivers the remote server's graphics to be displayed on a local client monitor. This process's technical definition is *the remote protocol provides the desktop bitmap entirely composed on the server*. Sending a desktop bitmap may seem like a simple task. However, it does require a significant amount of resources to achieve. For example, a 1.080 px desktop image in an uncompressed format is circa 8 MB in size. Displaying this image on a locally connected monitor with a screen refresh rate of 30 Hz requires approximately 240 MB/s of bandwidth.

RDP uses a combination of different techniques to reduce the amount of data that's transferred over a network, including, but not limited to, those mentioned in this table:

SN	Different RDP Techniques
1.0	Framerate optimization
2.0	Screen content classification
3.0	Content-specific codecs
4.0	Progressive image encoding
5.0	Client-side caching

Table 2.5 – Different RDP techniques

To understand remote graphics, you should consider the following:

- The more complex/richer the graphics, the more bandwidth it will take:
  - Text, window UI elements, and solid color consume less bandwidth than anything else.
  - Natural images are the more significant contributors to bandwidth use. Client-side caching can help with reducing the natural image bandwidth.
- It is important to note that RDP only transmits changed parts of the screen. When there are no visible updates on the session screen, no updates are sent.
- Image slideshows are also known as video playbacks and other high-framerate content. RDP dynamically uses the required video codecs to deliver content close to the original frame rate. However, graphics are the most significant contributors to bandwidth utilization.
- An idle remote desktop uses minimal bandwidth during idle times.
- When the Remote Desktop client window is minimized, no graphical updates are sent from the session host.

Please note that the stress you apply to your network depends on both the application workload's output, framerate, and display resolution. If the framerate or display resolution increases, the bandwidth requirement increases. One example is a light workload with a high-resolution display requiring more available bandwidth than a light workload with regular or low resolution. When using high display resolutions, expect to see the bandwidth requirements increase.

The following table provides examples of the data used by different graphic scenarios. These figures apply to a single monitor configuration with a 1,920 x 1,080 resolution and with both default graphics and H.264/AVC 444 graphics modes:

Scenario	Default Mode	H.264/AVC 444 Mode	Description of the Scenario
Idle	0.3 Kbps	0.3 Kbps	The user has paused their work, and there are no active screen updates.
Microsoft Word	100-150 Kbps	200-300 Kbps	The user is actively working with Microsoft Word, typing, pasting graphics, and switching between documents.
Microsoft Excel	150-200 Kbps	400-500 Kbps	The user is actively working with Microsoft Excel. Multiple cells with formulas and charts are updated simultaneously.
Microsoft PowerPoint	4-4.5 Mbps	1.6-1.8 Mbps	The user is actively working with Microsoft PowerPoint, typing, and pasting. The user is also modifying rich graphics and using slide transition effects.
Web Browsing	6-6.5 Mbps	0.9-1 Mbps	The user is actively working with a graphically rich website that contains multiple static and animated images. The user scrolls the pages both horizontally and vertically.
Image Gallery	3.3-3.6 Mbps	0.7-0.8 Mbps	The user is actively working with the image gallery application, browsing, zooming, resizing, and rotating images.
Video Playback	8.5-9.5 Mbps	2.5-2.8 Mbps	The user is watching a 30 FPS video that consumes half of the screen.
Fullscreen Video playback	7.5-8.5 Mbps	2.5-3.1 Mbps	The user is watching a fullscreen, 30 FPS video.

Table 2.6 – Examples of the data used by different graphic scenarios

This table has been taken from Microsoft (https://docs.microsoft.com/azure/virtual-desktop/rdp-bandwidth#estimating-bandwidth-used-by-remote-graphics).

Dynamic bandwidth allocation

RDP is designed to dynamically adapt to changing network conditions. Rather than relying on fixed bandwidth limits, RDP employs a feature called **continuous network detection**, which actively monitors available network bandwidth and packet round-trip time.

Based on what's been detected, RDP dynamically chooses the graphic encoding methods and assigns bandwidth for device redirection and other virtual channels.

This technology enables RDP to utilize the full network bandwidth when available and quickly reduce its usage when the network is needed for other services. RDP can detect such changes and adjust image quality, frame rate, or compression algorithms accordingly if other applications require the network bandwidth.

Limit network bandwidth use with throttle rate limiting

There's no need to limit bandwidth utilization as limiting may affect the overall user experience in most scenarios. However, in constrained networks, you may want to restrict or limit network utilization to prevent service degradation. Another good example is leased networks (mobile hotspots or pay-as-you-consume connectivity), which charge for the amount of traffic/bandwidth used.

In these cases, the advice is to limit RDP outbound network traffic by specifying a throttle rate in a **quality of service** (**QoS**) policy.

> **Important note**
>
> Throttle rate limiting is not supported for reverse connect transport with AVD. Microsoft details how to enable this here: https://docs.microsoft.com/azure/virtual-desktop/shortpath.

Reverse connect transport

AVD uses reverse connect transport to establish the remote session and carries RDP traffic from the session host to the broker. The client then connects to the broker. Unlike the traditional on-premises **Remote Desktop Services** (**RDS**) deployments, reverse connect transport doesn't use a TCP listener to receive incoming RDP connections. Instead, it uses outbound connectivity to the AVD infrastructure over the HTTPS connection, hence the term reverse connect.

Session host communication channel

Upon starting the AVD session host, the installed Remote Desktop Agent service establishes a connection to the AVD broker's persistent communication channel. This communication uses a secure **Transport Layer Security** (**TLS**) connection. This serves as a bus for service message exchange between the session host and AVD infrastructure. Essentially, this allows AVD to know when a session host is available, offline, or has any issues.

Client connection sequence

The following client connection sequence is used for AVD:

1. Using a supported Remote Desktop client, the user *subscribes* to the AVD workspace or workspaces. We will cover this in *Chapter 3, Designing for User Identities and Profiles*, and *Chapter 6, Creating Host Pools and Session Hosts*.
2. Microsoft Entra ID authenticates the user and returns the token that's used to enumerate resources available to authenticate the user.
3. The client passes the token to the AVD feed subscription service.
4. The AVD feed subscription service then validates the token.
5. The AVD feed subscription service then passes the list of available desktops and remote apps back to the client in a digitally signed connection configuration.
6. The client stores the connection configuration for each available resource in a set of .rdp files.
7. When a user selects the resource that they want to connect to, the client uses the associated .rdp file, which establishes a secure TLS 1.2 connection to the closest AVD gateway instance and passes the connection information.
8. The AVD gateway validates the request and asks the AVD broker to orchestrate the connection.
9. The AVD broker service identifies the session host and uses the previously established persistent communication channel to initialize the connection.
10. The Remote Desktop stack initiates the TLS 1.2 connection to the same AVD gateway instance that's used by the client.
11. Once both the client and session host are connected to the AVD gateway, the gateway can then start relaying the raw data between both endpoints; this establishes the base reverse connect transport for the RDP connection.
12. Once the base transport has been set, the client initiates the RDP handshake.

You can read more here: `https://docs.microsoft.com/azure/virtual-desktop/network-connectivity#client-connection-sequence`.

Connection security

The AVD infrastructure, session hosts, and clients all use TLS 1.2 to initiate connections.

AVD uses the same TLS 1.2 ciphers, just like Azure Front Door (https://docs.microsoft.com/azure/frontdoor/front-door-faq#what-are-the-current-cipher-suites-supported-by-azure-front-door).

> **Important note**
> Make sure both the client computers and session host have the required TLS 1.2 ciphers; otherwise, you will have connection issues.

Reverse transport works in the following way: the session host and the client connect to the AVD gateway. Once the TCP connection is established, the session host or the client proceeds to validate the AVD gateway's certificate. Once the base transport has been established, RDP establishes a nested TLS connection between the client and session host using the session host's certificates. By default, a self-generated certificate from the OS is used for RDP encryption when the system is deployed. Nonetheless, it is possible for customers to opt for the use of centrally managed certificates, which are issued by their enterprise certification authority.

For more information about configuring certificates, see Microsoft's Windows Server documentation: https://docs.microsoft.com/troubleshoot/windows-server/remote/remote-desktop-listener-certificate-configurations.

This section covered how to assess and plan for the required network bandwidth. This could be in main offices and home offices. Ensuring you have the correct bandwidth is important to ensure your users have a good user experience when using AVD.

Now, let's look at the OS images you can use for an AVD implementation.

Identifying an OS for an AVD implementation

In this section, we will cover the different OSs available to you.

Supported Azure OS images

AVD supports the following OS images:

- Windows 11 Enterprise multi-session
- Windows 11 Enterprise
- Windows 10 Enterprise multi-session
- Windows 10 Enterprise

- Windows Server 2022
- Windows Server 2019
- Windows Server 2016

Importantly, AVD does not support x86 (32-bit) OSs.

What is Windows 11 multi-session?

Windows 11 multi-session is a Remote Desktop session host offering that allows multiple concurrent remote user sessions. Previously with on-premises desktop virtualization solutions such as RDS, you had no option but to use multi-session using Windows Server OSs.

This new version gives users the same Windows 11 desktop experience rather than a server desktop UI. Organizations can also benefit from multi-session cost benefits and use existing per-user Windows licensing rather than the traditional RDS **Client Access Licenses** (**CALs**).

> **Note**
>
> For more information about licenses and pricing, see the AVD pricing page at `https://azure.microsoft.com/pricing/details/virtual-desktop/`.

Windows 11 multi-session is a virtual edition of Windows 11 Enterprise for use only on Microsoft Azure. The key difference when using Windows 11 multi-session compared to traditional Windows 11 is that this OS reports `ProductType` as having a value of 3. This is the same value that's used by the Windows Server OS. This property allows the OS to be compatible with existing host management tools, host multi-session-aware applications, and performance optimizations for session hosts.

It is important to note that some application installers can block the installation on Windows 11 multi-session through the installer. If your app doesn't install, contact your application vendor for an updated version or packaging company to repack it into a suitable format.

Windows 11 multi-session is not to be used with on-premises production environments. This is because it has been optimized for AVD using Microsoft Azure.

> **Important note**
>
> It's against the Microsoft licensing agreement to run Windows 11 Enterprise multi-session outside of Azure for production purposes. Furthermore, Windows 11 Enterprise multi-session will not activate against on-premises **Key Management Services** (**KMS**).
>
> It is currently impossible to upgrade an existing VM running Windows 11 Professional or Enterprise to Windows 10 Enterprise multi-session. When using Windows 11 Enterprise multi-session, you may decide to update the product key to another edition. You will not be able to switch back to Windows 11 Enterprise multi-session. You will need to redeploy the VM.

Customizing the Windows 11 multi-session image for your organization

You would customize the image just like any image outside of AVD. You would deploy the VM in Azure with Windows 11 multi-session and customize it with the required **line-of-business** (**LOB**) applications, settings, and customizations, including optimizations. You can then proceed with Sysprep/generalize, ready to create an image template.

You can find the Windows 11 multi-session images in the Azure Compute Gallery. To find the available image templates, navigate to the Azure portal and search for `Windows 10 Enterprise for Virtual Desktops`. There are two options for integration with Microsoft 365 Apps for Enterprise and a plain image, and Microsoft Windows 11 + Microsoft 365 Apps for Enterprise and Microsoft Windows 11.

Windows 11 multi-session supported versions

These Windows 11 releases follow the same support life cycle policy as the traditional Windows 11 Enterprise OS. This means that the March release is supported for 18 months and that the September release is supported for 30 months.

> **Important note**
> Windows 10 will reach the end of its support on October 14, 2025. The current version, 22H2, will be the final version of Windows 10, with all editions continuing to receive monthly security updates until that date. Existing LTSC releases will continue to receive updates beyond October 14, 2025, according to their specific life cycles (`https://docs.microsoft.com/en-us/lifecycle/products/windows-10-enterprise-and-education`).

Profile management solution for Windows 11 multi-session

You should consider using **FSLogix** profile containers when deploying AVD environments, especially on non-persistent deployments or other specific use cases requiring a centrally stored profile.

One of the key benefits of using FSLogix is that you can centralize your user profiles and provide a seamless experience for users accessing AVD resources.

AVD entitled users can use FSLogix at no additional cost. FSLogix comes pre-installed on all Windows 10 Enterprise multi-session images. However, you still need to configure the FSLogix profile container via the registry or group policy for a storage share location and any customizations required specifically for your environment.

We will cover FSLogix profile containers in more detail in *Chapter 12, Implementing and Managing FSLogix*.

Check out the AVD pricing page for a complete list of applicable licenses: `https://azure.microsoft.com/pricing/details/virtual-desktop/`.

Planning and configuring name resolution for AD and Microsoft Entra Domain Services

AVD can be used in a cloud-only organization or a hybrid environment. Hybrid is an approach that's used within IT where some of your IT resources are in-house, and others are cloud-based services. You need to ensure that the following is configured/set up before proceeding with an AVD project.

For hybrid environments that use Entra ID services, you'll need the following:

- **A Microsoft Entra ID domain**: A domain controller that is configured and synced with Microsoft Entra ID. You can configure this with one of the following:
  - **Microsoft Entra Connect**
  - **Microsoft Entra Domain Services: A Microsoft Entra ID domain**: A Microsoft subscription that contains a virtual network connected to Windows Server AD or **Microsoft Entra ID Domain Services (Entra ID DS)**.

For cloud-only organizations, you'll need the following:

- A Microsoft Entra ID organization
- Microsoft Entra Domain Services setup and configured
- An Azure subscription that contains a virtual network connected to the Entra ID DS

The Azure VMs you create for AVD must be as follows:

- Standard domain-joined or hybrid AD-joined. VMs can also be Entra ID-joined.
- Run a supported OS image.

We will cover AD and the various deployment configurations in more detail in *Chapter 3, Designing for User Identities and Profiles*, and *Chapter 8, Entra ID Join for Azure Virtual Desktop*.

Planning a host pool architecture

A **host pool** is one or multiple VMs within Azure using the same image. These are registered as AVD session hosts that have a configured AVD agent. It is recommended that all session host VMs be deployed from the same template image for optimal user experience in a host pool.

There are two types of host pools:

- The **personal host pool** type is used for individual desktop user assignments.
- A **pooled host pool** type is where session hosts can have multiple users connecting to multiple session hosts, sharing the compute resources of hosts within that host pool.

There are several additional properties available for configuration for host pools. You can also configure the host pool's load balancing behavior, which includes how many sessions each session host can handle and what the user can do while connected to session hosts in the host pool. You control the resources that have been published to users through app groups, as mentioned previously.

App groups

The best way to describe an app group is that it's a logical grouping of applications configured for use with a session host pool. There are two app group types:

RemoteApp	This is used to provide user access to RemoteApps you individually select and publish to the app group.
Desktop	This option provides users with access to a full desktop.

Table 2.7 – Table showing the two types of app groups

The host pool deployment automatically creates a desktop app group (named **Desktop Application Group**). App groups can be removed at any time. It is important to note that you cannot have multiple desktop app groups in a host pool.

When publishing a RemoteApp, first, you need to create a RemoteApp app group. Once created, you can create multiple RemoteApp app groups to cater to the possible different user scenarios required. You can have multiple RemoteApp app groups with overlapping RemoteApps.

When publishing remote resources to users, you should assign the users to specified app groups. When assigning users to app groups, you should consider the following things:

- You can assign users to both a desktop app group and a RemoteApp app group in the same host pool. It is important to note that users can only launch one type of app group per user session from the same host pool. It is not possible to launch both types of app groups within a single session simultaneously.
- Users can be assigned to one or many app groups within the same host pool. The user workspace feed will show a mixture of both app group types.

> **Important note**
> A workspace is described as *a logical grouping of application groups* in AVD. Each AVD app group must be associated with a workspace to see the remote apps and desktops published to users.

End users

Once you've assigned users to the required app groups, they can connect to an AVD deployment with any supported AVD clients.

Registering the DesktopVirtualization resource provider

Before you can provision AVD resources, you need to register the required subscriptions with the `Microsoft.DesktopVirtualization` resource provider once.

> **Important note**
> You must have permission to register a resource provider, which requires the `*/register/action` operation. This is included if your account is assigned the contributor or owner role on your subscription.

Follow these steps to complete this process:

1. Open the Azure **Subscriptions** services menu:

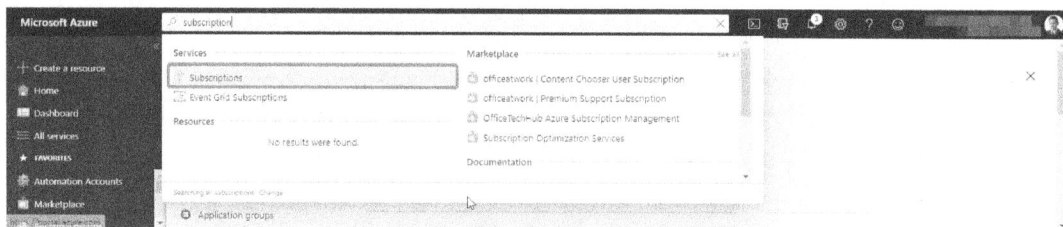

Figure 2.2 – Subscriptions

2. Select the subscription you would like to register:

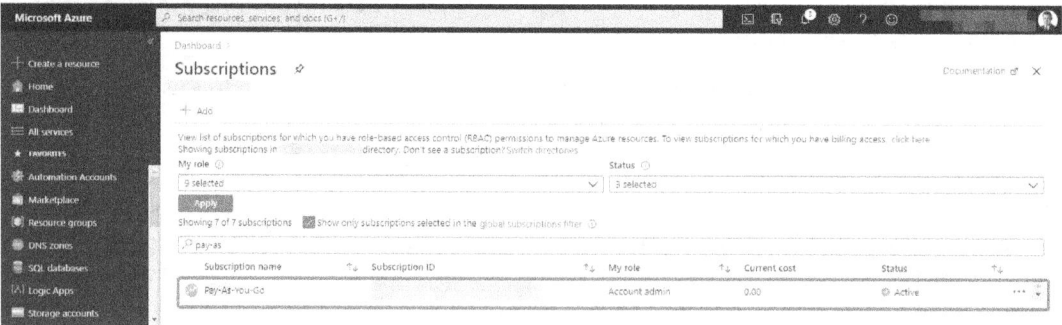

Figure 2.3 – Selecting the required subscription

3. Search for the `Microsoft.DesktopVirtualization` provider and click **Register**:

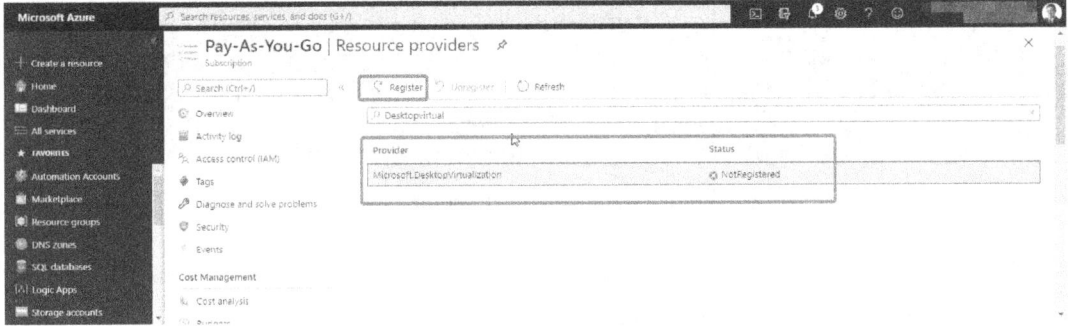

Figure 2.4 – Registering the chosen subscription with the Microsoft.DesktopVirtualization provider

4. Once registered, the status will change to **Registered**, as shown in the following screenshot:

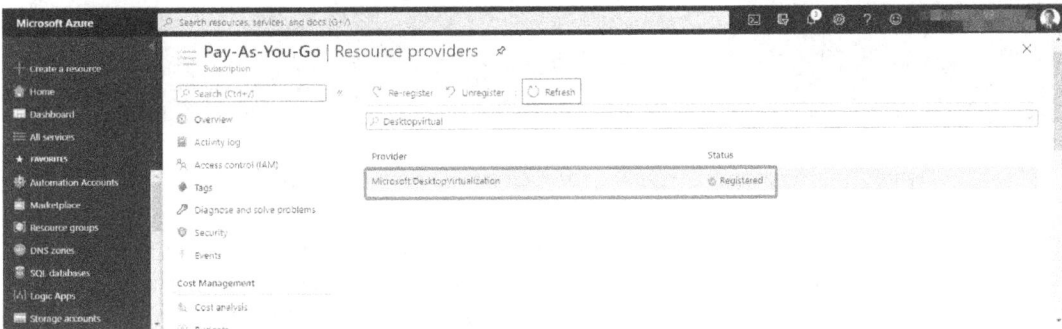

Figure 2.5 – The Microsoft.DesktopVirtalization provider showing as Registered

Once the registration has been completed and shows as **Registered**, you can deploy AVD components within your subscription.

Registering the provider using Azure PowerShell (optional)

The following steps detail the process of registering the provider using PowerShell:

1. Check the registration status of the provider:

    ```
    Get-AzResourceProvider -ProviderNamespace "Microsoft.
    DesktopVirtualization"
    ```

2. Register the `Microsoft.DesktopVirtualization` provider:

    ```
    Register-AzResourceProvider -ProviderNamespace "Microsoft.
    DesktopVirtualization"
    ```

You'll see the following output:

```
PS /home/ryan> Register-AzResourceProvider -ProviderNamespace "Microsoft.DesktopVirtualization"

ProviderNamespace  : Microsoft.DesktopVirtualization
RegistrationState  : Registering
ResourceTypes      : {workspaces, applicationgroups, applicationgroups/applications, applicationgroups/desktops...}
Locations          : {East US, East US 2, West US, West US 2...}
```

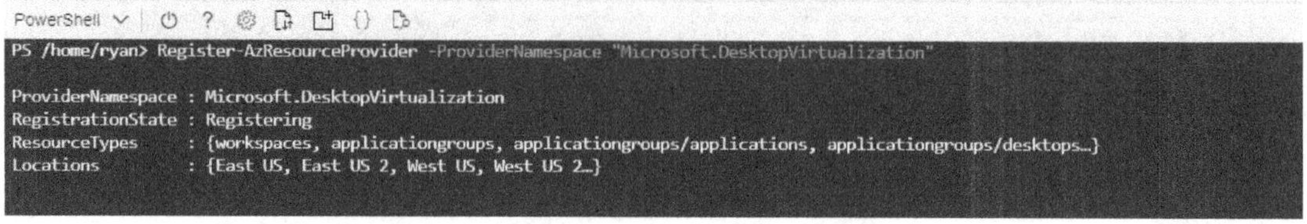

Figure 2.6 – Powershell cmdlet to register the Desktop Virtualization provider

3. Rerunning the same PowerShell cmdlet will confirm the status of the registration:

    ```
    Register-AzResourceProvider -ProviderNamespace "Microsoft.
    DesktopVirtualization"
    ```

    You'll see the following output:

```
PS /home/ryan> Register-AzResourceProvider -ProviderNamespace "Microsoft.DesktopVirtualization"

ProviderNamespace  : Microsoft.DesktopVirtualization
RegistrationState  : Registered
ResourceTypes      : {workspaces, applicationgroups, applicationgroups/applications, applicationgroups/desktops...}
Locations          : {East US, East US 2, West US, West US 2...}
```

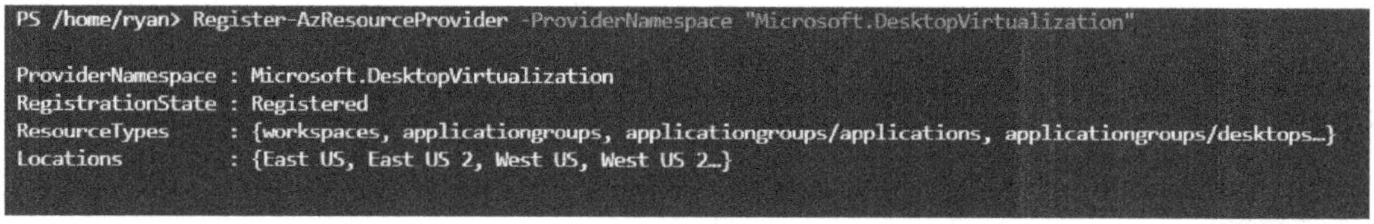

Figure 2.7 – The output of running the register-AZResourceProvider cmdlet

In this section, we showed you how to register the `Microsoft.DesktopVirtualization` provider using both the GUI within the Azure portal and using Powershell. In the next section, we will look at the resource groups, subscriptions, and management group limits.

Resource groups, subscriptions, and management group limits

You can scale an AVD environment to over 10,000 sessions hosted per workspace.

It is advised that you consider the AVD control plane limitations during the initial design to limit any changes that may be required later. The following details four considerations to be aware of:

- Microsoft recommends that you do not deploy 5,000+ VMs per Azure subscription per region. This recommendation is for both host pool types – personal and pooled – for a Windows 11 Enterprise single and multi-session deployment. It is suggested that most organizations will want to use Windows 11 multi-session, which allows multiple users to log on to a session host VM. You can increase the resources (compute and storage) of an individual session host VM to facilitate additional remote sessions.

- There are limits of circa 1,200 VMs per Azure subscription per region for automated session host scaling tools.
- You should create multiple Azure subscriptions in a hub and spoke architecture and connect them via **virtual network peering** to manage environments with more than 5,000 VMs per Azure subscription in the same region. You can also deploy VMs in different regions within the same subscription to increase your total number of VMs.
- The API throttling limit prevents you from exceeding 600 Azure VM reboots per hour via the Azure portal when using the **Azure Resource Manager (ARM)** subscription.

> **Tip**
> You can reboot all your VMs via the OS by using scripting or third-party products that do not consume any ARM subscription API calls.

- You can deploy 399 VMs per AVD (ARM) template deployment without using availability sets at the time of writing. Alternatively, you can deploy 200 VMs per availability set.

> **Tip**
> You can increase the number of VMs per deployment by turning off availability sets in the ARM template or the Azure portal host pool enrolment.

- VM session hostname prefixes within Azure cannot exceed 11 characters due to auto-assigning instance names. It is also important to note that the NetBIOS limit of 15 characters per computer account still applies.
- You can deploy 800 instances of resource types within a resource group.

> **Important note**
> For more information about Azure subscription limitations, see Azure subscription and service limits, quotas, and constraints at https://docs.microsoft.com/azure/azure-resource-manager/management/azure-subscription-service-limits.

Configuring the location for the AVD metadata

AVD is currently available for deployment in all Azure regions. IT admins can select the geographical location to store metadata when creating their host pool VMs and associated services. You would want to store the metadata locally in your organizations for data sovereignty purposes.

You can find out more about Azure geographies using the Azure data center map: `https://azure.microsoft.com/explore/global-infrastructure/geographies/#overview`.

There is no region restriction or limit to where you can access user- and app-specific data.

AVD metadata is collected, and the information that's captured includes host pool names, app group names, workspace names, and user principal names in a data center.

When an admin creates a service object, they will be asked to select a location for the service object. The location that's selected essentially determines where the metadata for the object will be stored. As part of its design, you will need to decide where you would like to store your metadata.

For a list of all Azure-associated regions and geographies, see Azure geographies: `https://azure.microsoft.com/global-infrastructure/geographies/`.

> **Important note**
> When you select a region to create AVD service objects in, you will see regions under both US and EU geographies. It is recommended that you understand which region would work best for your deployment. To find out more, take a look at Microsoft's Azure global infrastructure map: `https://azure.microsoft.com/global-infrastructure/geographies/#geographies`.

Metadata is encrypted and stored at rest, and georedundant mirrors are kept within the geography.

It is understood that Microsoft is working on more geographies that will become available as the service grows.

> **Important note**
> Microsoft replicates service metadata within the Azure geography for disaster recovery purposes.

Calculating and recommending a configuration for capacity and performance requirements

This section provides a high-level set of recommendations for sizing AVD. This will help you to plan and also calculate the costs of what the compute resource may look like. It is advised that you load test all the workloads, even when following guided recommendations.

Multi-session recommendations

The following examples should be used as guidance for initial performance estimates. As you build out your AVD environment, you will need to tweak and adjust these configurations to meet the requirements of your user base.

The following table shows the number of users per vCPU and the suggested VM hardware configuration for each workload. It also provides an example of session host sizing recommendations based on the number of users per vCPU and workload type.

	Multi-Session Recommendations			
Workload Type	Maximum users per vCPU	vCPU/RAM/OS storage minimum	Example Azure instances	Profile container storage minimum
Light User	6	8 vCPUs, 16 GB RAM, 32 GB storage	D8s_v5, D8s_v4, F8s_v2, D8as_v4, D16s_v5, D16s_v4, F16s_v2, D16as_v4	30 GB
Medium User	4	8 vCPUs, 16 GB RAM, 32 GB storage	D8s_v5, D8s_v4, F8s_v2, D8as_v4, D16s_v5, D16s_v4, F16s_v2, D16as_v4	30 GB
Heavy User	2	8 vCPUs, 16 GB RAM, 32 GB storage	D8s_v5, D8s_v4, F8s_v2, D8as_v4, D16s_v5, D16s_v4, F16s_v2, D16as_v4	30 GB
Power User	1	6 vCPUs, 56 GB RAM, 340 GB storage	D16ds_v5, D16s_v4, D16as_v4, NV6, NV16as_v4	30 GB

Table 2.8 – Multi-session recommendations

This table was taken from the *Remote Desktop workloads* Microsoft document: `https://docs.microsoft.com/windows-server/remote/remote-desktop-services/remote-desktop-workloads`.

The following article shows some real-life examples of different performance results from testing different Azure VM sizes for use with AVD: `https://www.democratising-clouds.com/blog/results-from-benchmark-tests-completed-on-avd-and-cloud-pc-virtual-machines`.

Microsoft recommends using VM sizes between 4 vCPUs and 24 vCPUs. It is not recommended to use less than two cores or more than 32 or more cores per session host.

> **Tip**
>
> All VMs require a minimum of two parallel threads (cores) for the heavier rendering operations. Having multiple users on a two-core or fewer VM could lead to a poor user experience. It's advised for stability that you use four or more cores when using a multi-user VM.

Recommendations on sizing VMs

When increasing the number of VM cores, it's important to note that the overhead of the VM will increase (system synchronization). For most workloads, it's advised that you stick to no more than 16 cores per session host. As the CPU number increases, the lower the return on investment. The additional capacity is negated by the synchronization overhead.

Microsoft, as per their best practices, states that you should stay in the range of 4 to 24 vCPU cores per session host.

> **Tip**
>
> It is suggested that a 1.75 ratio scenario offers improved burst capacity for applications that have short-term CPU demands.
>
> When you have 20 or more user session connections on a single VM, it is advised that multiple smaller VMs would show better results than one or two larger VMs. It is also recommended that you use smaller VMs instead of larger ones so that shutdown management is simpler. This is particularly useful for update management. From my testing and field experience, I have always found AMD Azure VM skews to offer better performance and user experience compared to the Intel equivalent skews.

A common resource that is sometimes forgotten when sizing resources is the session host virtual disk and associated **input/output operations per second** (**IOPS**, pronounced as *eye-ops*). Poor performing storage can result in a degraded user experience.

For persistent VMs (single user/single session), it is recommended that you use VMs with a minimum of two CPU cores per VM. However, for the best results, Microsoft recommends four vCPUs with hyperthreading enabled.

General recommendations for VMs

Microsoft recommends using premium storage (SSD) for AVD session hosts as they offer high-performance and low-latency disk support for the session host VM.

> Tip
> For a production workload requiring a **service-level agreement** (**SLA**), you will need to assign your VM's premium storage (SSD). For more details, see the SLA for VMs: `https://azure.microsoft.com/support/legal/sla/virtual-machines/v1_8/`.

You can enhance the user session experience even further using GPUs, which let you use graphic-intensive programs such as 3D design, video rendering, and simulation software. You can read more about graphics acceleration by going to the *Remote Desktop Services – GPU acceleration* guide: `https://docs.microsoft.com/windows-server/remote/remote-desktop-services/rds-graphics-virtualization`.

You can read more about the Microsoft Azure graphics acceleration deployment options that are available, as well as the GPU VM sizes offered, by going to the *GPU optimized virtual machine sizes* page: `https://docs.microsoft.com/azure/virtual-machines/windows/sizes-gpu`.

> Tip
> B-series burstable VMs are helpful for users who don't always need maximum CPU performance. However, these are not so good if you use all the burstable resources. These are also useful for older applications that are delivered via RemoteApp. Just remember to benchmark usage before using it as burstable storage has a very limited use case.

You can find out more about VM types and sizes by going to the *Sizes for Windows virtual machines in Azure* document: `https://docs.microsoft.com/azure/virtual-machines/windows/sizes`.

You can find the pricing information on Microsoft's *Virtual Machine Series* page: `https://azure.microsoft.com/pricing/details/virtual-machines/series/`.

Testing workloads

Before rolling out any new desktop virtualization project to production, it is recommended that you run several tests, including simulation tests, to stress test for real-life usage. This will help provide a baseline and ensure that the environment you plan to release offers the resilience and responsive performance required to meet current and future needs.

Load testing is a really good way to understand the limitations of the desktop environment, which helps with capacity planning as well as managing and maintaining a good user experience.

> **Tip**
> You can use tools such as Login VSI, which can automate load tests for your AVD environment, to help you understand performance, capacity management, instance selection, and potential user experience issues. You can read more about Login VSI here: `https://www.loginvsi.com/solutions/azure-virtual-desktop/`.

Summary

In this chapter, we discussed assessing your current desktop estate to understand current baselines for performance, data, and user experience. We then looked at the considerations for AVD, including network bandwidth planning, sizing, and Windows 11 multi-session. Throughout this chapter, tips and recommendations were provided for you to get the most out of your AVD environment.

The next chapter will look at the design for user identities and profiles, where we will look at selecting an appropriate licensing model for AVD based on our requirements. Here, we will recommend an appropriate storage solution, plan for AVD client deployments and user profiles, and recommend a solution for network connectivity and planning for Microsoft Entra Connect for user identities.

Questions

Here are a few questions to test your understanding of this chapter:

1. Which tools can you use for assessing your requirements for AVD?

    Lakeside SysTrack and Azure Migrate.

2. Before you can start to deploy AVD, what do you need to register for the subscription?

    Register the Microsoft.DesktopVirtualization provider.

3. What is the name of the tool that's used to gauge the round trip time (latency) of AVD?

    AVD Experience Estimator.

4. What is the benefit of Windows 11 multi-session over other OS types?

   Windows 11 multi-session offers a familiar Windows 11 experience that's exclusively offered to Microsoft Azure.

Further reading

You can refer to the following links for more information regarding the topics that were covered in this chapter:

- *Azure Virtual Desktop for the enterprise* – Azure Example Scenarios | Microsoft Docs (https://docs.microsoft.com/azure/architecture/example-scenario/wvd/windows-virtual-desktop).
- *Microsoft Cloud Adoption Framework for Azure* – Cloud Adoption Framework | Microsoft Docs (https://docs.microsoft.com/azure/cloud-adoption-framework/).
- For more information on the Azure subscription limitations, see the following links for management groups (https://docs.microsoft.com/azure/governance/management-groups/overview) and general limits (https://docs.microsoft.com/azure/azure-resource-manager/management/azure-subscription-service-limits).
- *Deliver remote desktops and apps from Azure with Azure Virtual Desktop* – Learn | Microsoft Docs (https://learn.microsoft.com/learn/paths/m365-wvd/).

3
Designing for User Identities and Profiles

In this chapter, we will explore how to properly manage user identities and profiles, which is critical to ensuring a smooth and secure user experience within Azure Virtual Desktop (AVD).

In this chapter, we'll cover the following topics:

- Selecting a licensing model for your Azure Virtual Desktop deployment
- Planning for user profiles
- Planning for Azure Virtual Desktop client deployment
- Summarizing the prerequisites for Azure Virtual Desktop

Selecting a licensing model for your Azure Virtual Desktop deployment

This section will introduce Azure Virtual Desktop licensing and its entitlements. You need to ensure you have the correct license to ensure you are appropriately licensed.

> **Important note**
> Before getting started, please note that the previously named Microsoft 365 for Business is now known as Microsoft 365 Business Premium. Users can use Azure Virtual Desktop from their non-Windows Pro devices if they have a Microsoft 365 E3/F3/E5/Business/A3/Student user or Windows 10 VDA per-user license.

The following table details the licenses and their eligibility for Azure Virtual Desktop:

Type	Description	Eligibility
Virtualize Windows 11 and 10	Windows 11 Enterprise and Windows 10 Enterprise desktops and apps are included if you have an eligible Windows or Microsoft 365 license. Support for Windows 7 ended on January 10, 2023. Windows 10 Enterprise will be retired on October 15, 2025, so for any new deployments, it is advised to use Windows 11 Enterprise.	You are eligible to access Windows 11 and Windows 10 with Azure Virtual Desktop if you have one of the following per-user licenses*: • Microsoft 365 E3/E5 • Microsoft 365 A3/A5/Student Use Benefits • Microsoft 365 F3 • Microsoft 365 Business Premium** • Windows 11 & 10 Enterprise E3/E5 • Windows 11& 10 Education A3/A5 • Windows 11& 10 VDA per user
Virtualize Windows Server	Access desktops powered by Windows Server **Remote Desktop Services** (**RDS**) desktops and apps at no additional cost if you are an eligible Microsoft RDS **Client Access License** (**CAL**) customer.	You are eligible to access Windows Server 2016 and newer desktops and apps if you have a per-user or per-device RDS CAL with active **Software Assurance** (**SA**).

Table 3.1 – Virtualizing Windows desktops and servers

Please note that the preceding table has been adapted from the one on Microsoft Licensing page: https://azure.microsoft.com/en-gb/pricing/details/virtual-desktop/.

The preceding table details the options and eligibility criteria for virtualizing Windows 11, Windows 10, and Windows Server desktops and apps using Azure Virtual Desktop. It covers the specific Windows and Microsoft 365 licenses required for virtualizing Windows 11 and 10, the end of support for Windows 7, the retirement plans for Windows 10 Enterprise, and the types of licenses that grant access to these virtual desktops. Additionally, it outlines the eligibility for virtualizing Windows Server through Microsoft RDS CALs with active SA.

> **Note**
> Windows 10 Enterprise will retire on October 14, 2025. For new deployments, Windows 11 Enterprise is advised.

You will also need an Azure subscription to deploy and manage your desktop virtualization environment. The following Azure components are key areas when you're looking at Azure Virtual Desktop deployment costs:

- Networking and any third-party **Network Virtual Appliances** (**NVAs**)
- **Virtual Machines** (**VMs**) and **Operating System** (**OS**) storage
- Data disk (optional)
- Storage for user profiles

> **Important note**
> Session host VMs on Azure deployments are billed at Linux compute rates for the following OS versions: Windows 10 Single, Windows 10 multi-session, and Windows Server. You can read more on this here: https://azure.microsoft.com/pricing/details/virtual-desktop/.

Applying Azure Virtual Desktop licensing to virtual machines

This section covers how to check whether your Azure Virtual Desktop session hosts have the correct licensing applied.

> **Important note**
> This process is typically taken care of for you when creating VMs from the wizard. However, it is worth checking that the license type has been set correctly. You should also check this when you deploy from a custom or imported image.

One way you can apply a license to a VM is by using the following PowerShell cmdlet:

```
$vm = Get-AzVM -ResourceGroup <resourceGroupName> -Name <vmName>
$vm.LicenseType = "Windows_Client"
Update-AzVM -ResourceGroupName <resourceGroupName> -VM $vm
```

Here's what the output looks like:

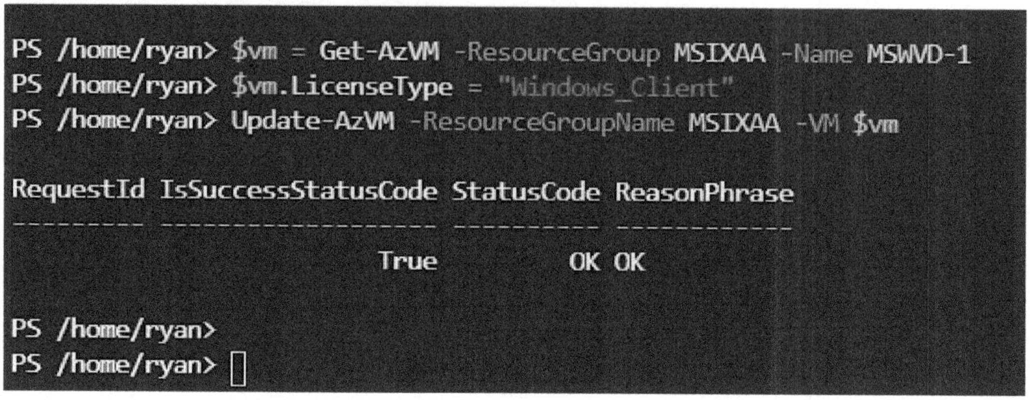

```
PS /home/ryan> $vm = Get-AzVM -ResourceGroup MSIXAA -Name MSWVD-1
PS /home/ryan> $vm.LicenseType = "Windows_Client"
PS /home/ryan> Update-AzVM -ResourceGroupName MSIXAA -VM $vm

RequestId IsSuccessStatusCode StatusCode ReasonPhrase
--------- ------------------- ---------- ------------
                    True                  OK OK

PS /home/ryan>
PS /home/ryan>
```

Figure 3.1 – PowerShell script showing how to apply AVD licensing to VMs

Once you have set the license type, you can check it using the following cmdlet:

```
Get-AzVM -ResourceGroupName <resourceGroupName> -Name <vmName>
```

Here's what it looks like:

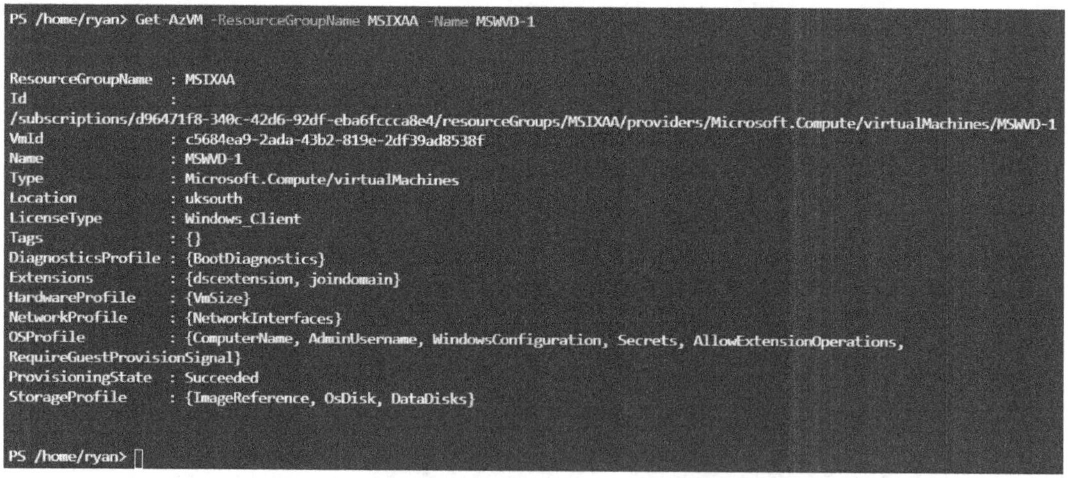

```
PS /home/ryan> Get-AzVM -ResourceGroupName MSIXAA -Name MSWVD-1

ResourceGroupName : MSIXAA
Id                :
/subscriptions/d96471f8-340c-42d6-92df-eba6fccca8e4/resourceGroups/MSIXAA/providers/Microsoft.Compute/virtualMachines/MSWVD-1
VmId              : c5684ea9-2ada-43b2-819e-2df39ad8538f
Name              : MSWVD-1
Type              : Microsoft.Compute/virtualMachines
Location          : uksouth
LicenseType       : Windows_Client
Tags              : {}
DiagnosticsProfile : {BootDiagnostics}
Extensions        : {dscextension, joindomain}
HardwareProfile   : {VmSize}
NetworkProfile    : {NetworkInterfaces}
OSProfile         : {ComputerName, AdminUsername, WindowsConfiguration, Secrets, AllowExtensionOperations, RequireGuestProvisionSignal}
ProvisioningState : Succeeded
StorageProfile    : {ImageReference, OsDisk, DataDisks}

PS /home/ryan>
```

Figure 3.2 – The output of running the get-AZVM cmdlet

If you run the following cmdlet, this will provide you with a list of all the VMs licensed under `Windows_Client`:

```
$vms = Get-AzVM
$vms | Where-Object {$_.LicenseType -like "Windows_Client"} | Select-Object ResourceGroupName, Name, LicenseType
```

Here's what the output looks like:

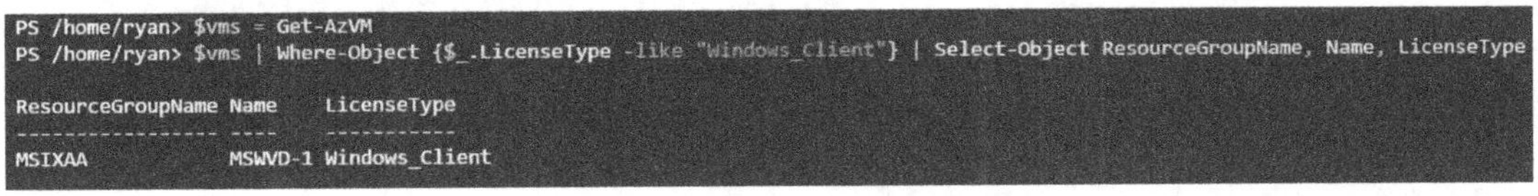

```
PS /home/ryan> $vms = Get-AzVM
PS /home/ryan> $vms | Where-Object {$_.LicenseType -like "Windows_Client"} | Select-Object ResourceGroupName, Name, LicenseType

ResourceGroupName  Name      LicenseType
-----------------  ----      -----------
MSIXAA             MSWVD-1   Windows_Client
```

Figure 3.3 – The licensed VMs

Let's now look at the recommended storage solutions for Azure Virtual Desktop.

Recommended storage solutions

FSLogix is a profile container technology included with Azure Virtual Desktop at no extra cost. FSLogix lets you roam profiles in desktop virtualization environments such as Azure Virtual Desktop.

Multiple storage options are available for storing your FSLogix profile containers. The most common option organizations use is Azure Files.

> **Important note**
> Microsoft recommends Azure Files for typical FSLogix profile container deployments, though it's recommended that you use Azure NetApp Files for larger enterprise environments.

When a user signs in to an Azure Virtual Desktop session, the container (`container FSLogix (profile container)`) is attached (dynamically) to the allocated session host using a natively supported virtual disk, known as **Virtual Hard Disk/Hyper-V Hard Disk (VHD/VHDX)**.

The user's profile is made available and provides the same user experience, just as a native user profile would.

The following table provides a comparison between the three common storage solutions available for FSLogix profile containers with Azure Virtual Desktop:

Features	Azure Files	Azure NetApp Files	Storage Spaces Direct (S2D)
Use case	General-purpose storage	Ultra performance (enterprise) or migration from NetApp on-premises storage solutions	Cross-platform
Platform service type	Azure solution	Azure solution	No, self-managed
Azure regional availability	All Azure regions	Selective regions (`https://azure.microsoft.com/global-infrastructure/services/?products=netapp&regions=all`)	All Azure regions
Redundancy	Locally redundant/ zone-redundant/ geo-redundant/ geo-zone-redundant	Locally redundant/ cross-region replication	Locally redundant/ zone-redundant/ geo-redundant
Storage tiers and performance	Standard (transaction optimized) or Premium up to a maximum of 100K IOPS per share with 10 GBps per share at about 3 ms latency	Standard, Premium, and Ultra Up to 320k (16K) IOPS with 4.5 GBps per volume at about 1 ms latency	Standard **Hard Disk Drive (HDD)**: Up to 500 IOPS per-disk limits Standard **Solid-State Drive (SSDs)**: Up to 4k IOPS per-disk limits Premium SSD: Up to 20k IOPS per-disk limits It is recommended that premium disks are used for S2D

Features	Azure Files	Azure NetApp Files	Storage Spaces Direct (S2D)
Capacity capability	100 TiB per share, up to 5 PiB per general-purpose account	100 TiB per volume, up to 12.5 PiB per subscription	Maximum 32 TiB per disk
Required infrastructure	Minimum share size 1 GiB	Minimum capacity pool 2 TiB, minimum volume size 100 GiB	Two VMs on Azure IaaS (plus Cloud Witness) or at least three VMs without any costs for disks
Protocols	SMB 3.0/2.1, NFSv4.1 (preview), REST	NFSv3, NFSv4.1, SMB 3.x/2.x	NFSv3, NFSv4.1, SMB 3.1

Table 3.2 – Azure Storage platform comparison

The preceding table has been adapted from the one on Microsoft (https://learn.microsoft.com/azure/virtual-desktop/store-fslogix-profile).

The preceding table provides a detailed comparison of Azure Files, Azure NetApp Files, and S2D. It highlights their features, use cases, platform types, regional availability, redundancy options, storage tiers, performance, capacity capabilities, required infrastructure, and supported protocols.

The following table provides a comparison between the features of these three storage solutions available for FSLogix profile containers with Azure Virtual Desktop:

Features	Azure Files	Azure NetApp Files	S2D
Type	Cloud, on-premises, and hybrid (Azure File Sync)	Cloud, on-premises (via ExpressRoute)	Cloud, on-premises
Backup capabilities	Azure backup snapshot integration	Azure NetApp Files snapshots	Azure backup snapshot integration
Security and compliance	All Azure-supported certificates	ISO completed	All Azure-supported certificates
Directory integration	Entra ID Domain Services (Entra ID DS), and Entra Domain Services	Entra ID Domain Services	Entra ID Domain Services support only

Table 3.3 – Azure Storage feature comparison

The preceding table has been adapted from the one on Microsoft (https://learn.microsoft.com/azure/virtual-desktop/store-fslogix-profile).

This table contrasts Azure Files, Azure NetApp Files, and S2D in terms of deployment types (cloud, on-premises, and hybrid options), backup capabilities (including snapshot integration), security and compliance certifications, and directory integration services (including support for various domain services).

You can review the pricing for each storage method using Azure Calculator: https://azure.microsoft.com/pricing/details/virtual-desktop/.

Azure Files tiers

In this section, we will look at the different Azure Files tiers and the benefits of each.

There are two types of storage for Azure Files: standard and premium. These two tiers offer different performance requirements and costs. The following will help you weigh up which is more suited to your needs.

The key difference between the two tiers is the type of disks used. Premium uses SSDs, while standard uses traditional HDDs:

- Premium file shares provide high performance and low latency for **Input and Output (I/O)**-intensive workloads. This is useful for larger organizations or user profiles that require higher I/O.
- Standard file shares are suitable for less intensive I/O workloads, such as general-purpose file shares and dev/test environments. These are more suitable for smaller Azure Virtual Desktop environments.

The following table provides guidance when it comes to choosing a recommended file tier:

Workload	Recommended File Tier
Light (less than 200 users)	Use standard file shares
Light (more than 200 users)	Use premium file shares or standard with multiple file shares split between user groups/departments
Medium workloads	Use premium
Heavy workloads	Use premium
Power workloads	Use premium

Table 3.4 – Choosing a recommended file tier

The preceding table has been adapted from the one on Microsoft (https://learn.microsoft.com/azure/virtual-desktop/store-fslogix-profile).

This table outlines the recommended file storage solutions tailored to different workload intensities, from light workloads involving fewer than 200 users to power workloads, highlighting the transition from standard to premium file shares based on user volume and operational demands.

This section provided an overview of the storage options available when using FSLogix Profile Containers with Azure Virtual Desktop.

Planning for user profiles

In this section, we will discuss user profiles and some of their associated considerations for Azure Virtual Desktop.

User profiles

A user profile comprises several data elements of a user. This can include desktop settings, persistent network connections, and application configuration settings. On the first login, Windows creates a user profile (local) from the default template. This user profile is tightly integrated within the Windows OS and is required for the user session to function correctly.

When comparing a remote user profile to a local profile, the key difference is that the remote profile allows the Windows OS to be replaced without impacting the user data. This means you can log in to another session host and still have the same settings you previously did. Remote profiles are typically used in a pooled (non-persistent) desktop virtualization environment.

Microsoft has several products available for delivering remote user profiles, including the following:

- **User Profile Disks (UPDs)**
- **Enterprise State Roaming (ESR)**
- **Roaming User Profiles (RUPs)**

UPDs were previously the most widely used profile technology for desktop virtualization. There has been a significant shift toward FSLogix Profile Containers since their acquisition in 2019.

Challenges with previous user profile technologies

Traditional user profile solutions present a few challenges. For example, many organizations that wanted to migrate to **Microsoft 365** struggled with UPDs. Some of the issues with traditional profile disks were related to caching mailbox problems and to other applications not functioning correctly, as the redirection was not seamless.

When using Azure Virtual Desktop, it is recommended that user profiles be configured using FSLogix Profile Containers. FSLogix has been designed to enable roaming profiles for desktop virtualization environments such as Azure Virtual Desktop. It stores the user profile in a single VHD/VHDX container. When a user initiates the connection to a virtual desktop, the user profile is made available and appears exactly like a local user profile would.

Now, let's take a deeper look into FSLogix and build on the information we've covered in this and the previous chapter.

An introduction to FSLogix profile containers

Microsoft acquired the vendor FSLogix on November 19, 2018. It came with a set of products and solutions to improve desktop virtualization products. One of the main driving forces for the acquisition was that Microsoft could replace UPDs with FSLogix profile containers. This was due to the limitations within the UPD, which had an impact on those who wanted to use Microsoft 365.

The FSLogix product portfolio includes the following:

- Profile container
- Office container
- Application masking

Some of the key benefits when using FSLogix include the following:

- The ability to minimize sign-in times for pooled virtual desktop environments, also known as non-persistent desktops
- Using FSLogix's filter driver enables a much better profile experience and removes many of the traditional incompatibilities with UPDs, relating to what some describe as visible redirection
- Application masking enables you to hide applications on a central desktop image from users who have not been assigned permission via a security group

FSLogix capabilities

This section provides an introduction to some of FSLogix's capabilities and the benefits associated with them:

- Redirecting user profiles to a network storage location. Profile containers store user profiles in a Virtual Hard Disk (VHD) or Virtual Hard Disk Extended (VHDX) file, which are attached to the user's session at login. As the profiles are accessed via a network share, this eliminates delays typically seen with profile solutions that copy files from the share to the target device.
- FSLogix can also maintain the Windows index, resulting in a smoother login on new virtual machines (there is no requirement to recreate the index).
- FSLogix offers Office containers. These containers allow you to solve Office issues within a non-persistent desktop virtualization environment. One common use case is how to handle the Outlook .ost files or the Microsoft Teams cache, which we cover in *Chapter 12, Implementing and Managing FSLogix*.

- FSLogix's filter driver architecture redirects the profile so that applications and the Windows OS do not recognize that the user profile is stored outside the host. Typically, most applications won't work on user profiles stored on remote storage.

- FSLogix Cloud Cache enables you to create a highly available desktop virtualization environment. The technology works by placing a portion of the user profile on the host's local disk. With Cloud Cache, you can also configure multiple remote profile locations, which protects users from network and storage issues/failures.

- Application masking enables you to control access to applications and other items, such as fonts. Application masking helps you segregate apps and other items via user groups or other IP addresses to provide better segregation on a central gold image.

FSLogix requirements

To be entitled to use profile containers and the other suite of tools by FSLogix, you will need to ensure you have one of the following licenses:

- Microsoft 365 E3/E5
- Microsoft 365 A3/A5/Student Use Benefits
- Microsoft 365 F1/F3
- Microsoft 365 Business
- Windows 11 Enterprise E3/E5
- Windows 11 Education A3/A5
- Windows 11 VDA per user
- RDS CAL
- RDS SAL

You can use the FSLogix products on any public or private data center with the correct license. The following link goes to the Microsoft official documentation on the eligibility requirements: `https://docs.microsoft.com/fslogix/overview`.

FSLogix filter driver architecture

Traditional profile technologies copy the user profile to and from the network during sign-in and sign-out. The problem is that this can cause delays, and if profiles are larger, which can often be the case, the sign-in and sign-out times could reach unacceptable time frames. One of the key differences with FSLogix is that the user profile is stored centrally on a network share rather than being copied. The profile is essentially redirected when the profile disks are mounted when the user logs in. As the profile is being mounted, this reduces the traditional delays associated with the copy.

The following diagram depicts the architecture of how FSLogix works within a Windows OS. Two filter drivers are injected into the OS, which are then installed when you run the FSLogix installer. We will learn how to install FSLogix in *Chapter 12*, *Implementing and Managing FSLogix*. The diagram depicts Azure Storage. Using the OS registry, group policies (ADMX), or Intune policies, you can specify the location of where to store your VHD or VHDX container:

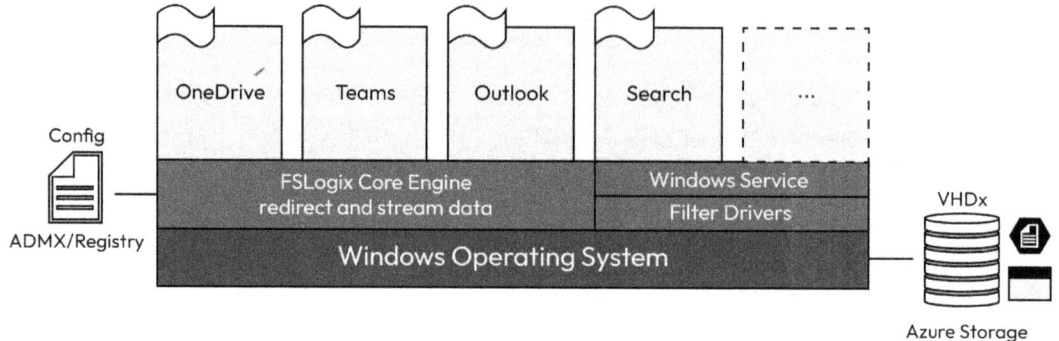

Figure 3.4 – FSLogix architecture

Exploring the differences between profile containers and Office containers

FSLogix offers two types of containers for profiles: profile containers and Office containers. You should note that the Office container is essentially a subset of the profile container. There are use cases when you should use these together, particularly in larger organizations where you want to split a profile from the Office data. It is important to fully understand the process of configuring them together to avoid impacting user profiles.

The key difference between the two containers is that the Office container is specifically used for improving the performance of Microsoft Office within a non-persistent desktop virtualization environment. The benefit of using both is that it allows you to segment office profile components into one office container disk and everything else into a profile container disk.

> **Important note**
> When using FSLogix in an Azure Virtual Desktop environment, it is recommended that you only use profile containers, unless there is a specific requirement for **Business Continuity and Disaster Recovery (BCDR)**.

Let's briefly take a look at the concept of multiple profile connections.

Multiple profile connections

This refers to a user connected to multiple desktop virtualization sessions. This could be multiple hosts or one concurrently using the same profile. It's important not to get confused with the term multi-session.

> **Important note**
> Concurrent or multiple connections are not recommended when using Azure Virtual Desktop. Microsoft suggests that you create a different profile location for each host pool.

Some of the limitations associated with multiple connection deployments are as follows:

- Microsoft OneDrive is not supported with multiple connection environments.
- You could experience data loss when attempting to use OneDrive with multiple profile connections.
- There is limited support with Microsoft Outlook for multiple connections.
- End user training would be required as the user experience changes. Using a read, the only container changes how they should use the virtual desktop session. Without training or proper context, users could experience data loss.

FSLogix performance requirements

When planning for an FSLogix profile container deployment, you need to consider a few things. The overall profile size and limitations will depend on the storage type you use, as well as the chosen disk format type – VHD or VHDX.

You should plan for 5-30 IOPS per user profile, as profile requirements can vary widely depending on their exact usage. The performance requirements are most likely different for each organization. Each user is different, and it is suggested that you review a subset of users, their applications, and their activity to gain an understanding of true profile performance needs.

It is also recommended that you conduct profile assessments to establish your current profile utilization, which will help you scope more accurate storage requirements. There are several community tools available to help you assess your current profile size state. These tools can also shrink the profiles, reducing the overall storage size. As of FSLogix v2210, the shrink functionality is built in, and it will automatically shrink the profile upon log off. More can be found here - https://learn.microsoft.com/fslogix/concepts-vhd-disk-compaction

> **Tip**
> Jim Moyle created a community tool that can be used to shrink FSLogix and O365 dynamically expanding disks. You can read more here: https://github.com/FSLogix/Invoke-FslShrinkDisk.

The following table provides an example of the typical resources a profile needs to have to support each user:

Performance requirements	
Resource	Requirement
Steady-state IOPS	10
Sign-in/sign-out IOPS	50

Table 3.5 – FSLogix storage IOPS requirements

For example, let's say that you have 100 users within an organization. Using the preceding table as a baseline, you may require around 1,000 IOPS. You would also need a suggested 5,000 IOPS at login or signup. You should also factor in logon storms. If multiple users log in simultaneously, it may create a logon storm. To avoid this, you should factor in any additional IOPS to reduce the likelihood of a login storm.

Now, let's recap the storage options from the previous chapter.

Storage options for FSLogix profile containers

There are multiple storage solutions available to you for use with FSLogix profile containers. Azure Files is recommended for most scenarios; however, there are multiple storage options available for FSLogix profile containers. It is recommended that enterprises use Azure NetApp Files to benefit from high-performance storage in larger deployments.

> **Tip**
> It is not recommended for production use, but you can use S2D with FSLogix and Azure Virtual Desktop.

FSLogix storage best practices

This section will specify the best practices you should remember when deploying and configuring FSLogix profile containers:

- Ensure that your chosen storage and host VMs are within the same data center location for optimal performance
- Ensure you set exclusions for antivirus scanning on VHD and VHDX disks to prevent antivirus-induced performance bottlenecks
- Exclude the VHD(X) files for profile containers from antivirus scanning to avoid performance bottlenecks
- Use separate profile containers per host pool when you want to use multiple active sessions

The following table details the recommended Azure Storage permission for use with an FSLogix profile container:

User Account	Folder	Permissions
Users	This folder only	Modify
Creator/owner	Subfolders and files only	Modify
Administrator (optional)	This folder, subfolders, and files	Full control

Table 3.6 – Azure Files NTFS permissions

The preceding table details the access permissions for user accounts, specifying the scope of access (such as *this folder only*, *subfolders and files only*, or *this folder, subfolders, and files*) and the level of permissions (*modify* or *full control*) granted to users, the creator/owner, and administrators (optional).

The following table shows the required permissions you should set for Azure Files:

Role	Group
Storage file Data SMB Share contributor	Domain users or AVD users
Storage file Data SMB Share Elevated contributor	Storage admin

Table 3.7 – Azure Files role permissions

This table delineates the access roles within SMB share management, specifying the groups associated with each role – from standard contributors (Domain or AVD users) to those with elevated privileges (Storage admins).

Azure Files integration with Entra ID DS

Microsoft announced the general availability of Azure Files authentication with Entra ID on August 7, 2019. This lets you use NTFS permissions with Azure Files.

The following diagram details the steps of on-premises Entra ID Authentication to Azure files using **Server Message Block** (**SMB**). The Entra ID environment must be synced with Entra ID using Microsoft Entra Connect Sync. Only configured hybrid users in Entra ID and Entra ID can access and be authorized to use Azure file shares. This is because the share-level permissions are configured against the identity represented in Entra ID, where the directory or file is enforced with Entra ID DS. Therefore, you need to ensure that you configure the permissions correctly for hybrid users.

The following diagram shows the process of Azure Files using AD credentials:

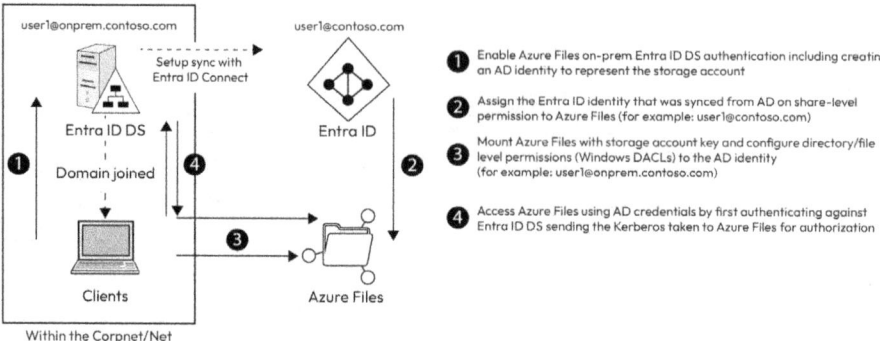

Figure 3.5 – Azure Files integration with Entra ID

This provides seamless integration for those who want to integrate Azure Files into Active Directory, including the use of FSLogix and MSIX app attach. It is also important to note that using Azure Files with ADDS integration offers improved security, as it allows you to configure **Access Control Lists** (**ACLs**), similar to a traditional Windows file server. Without Entra ID integration, you will only have basic security features, which may not be suitable for the organization you are deploying Azure Files to.

Microsoft Entra Kerberos for hybrid identities

Microsoft recently announced the capability to allow Microsoft Entra users to access Azure file shares using Kerberos authentication. Using this configuration, Microsoft Entra ID issues the necessary Kerberos tickets to access the file share.

This removes the requirement to deploy domain controllers inside your Azure subscriptions, as the authentication is against the Entra ID service.

Currently, this only supports hybrid identities, that is, those users who have been synced to Entra ID from Entra ID.

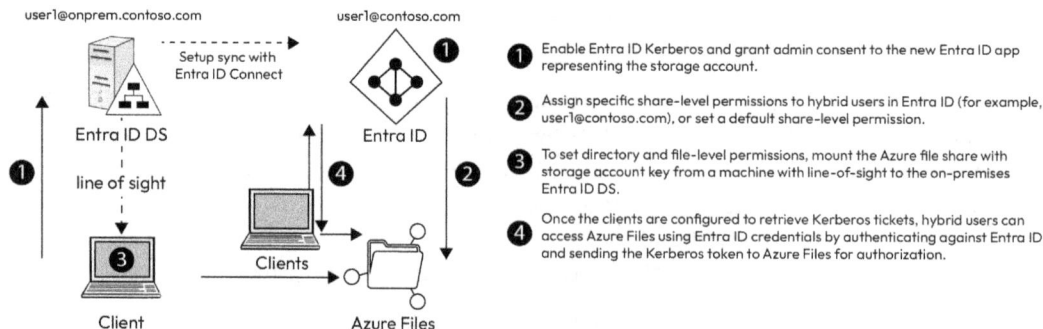

Figure 3.6 – Azure Files integration with Entra ID

> **Important note**
>
> You can use **Entra ID Domain Services (Entra ID DS)** for domain joins. The best practice would be to use Entra ID DS where possible, as Entra ID DS does have limitations.

This section discussed the topic of planning for user profiles; this included using Azure Files storage, FSLogix profile containers, recommendations, and best practices.

Planning for user identities

Before you can get started with Azure Virtual Desktop, you need to choose an identity strategy. The following identity strategies could apply to your Azure Virtual Desktop and Azure infrastructure:

Option	Pros	Cons
Deploy a Domain Controller in your Azure subscription	You can sync with on-premises domain controllers using a VPN or ExpressRoute. A familiar AD Group Policies experience.	An additional layer of management is added through the use of VMs and Active Directory in Azure is added.
Use Entra ID DS (cloud-only organizations)	This is great for testing or lab/isolated environments that do not require connectivity to on-premises resources.	Entra IDDS is a service that is always running, resulting in a **fixed charge per month**. Entra ID DS doesn't provide all the functions that native AD does. It's been known that a lot of organizations that have deployed Entra ID DS have been removed due to some of the limitations they have experienced.
Use VPN or ExpressRoute and make sure your on-premises domain controllers can communicate in Azure (hybrid organizations)	There is no requirement for Entra ID DS or a domain controller in Azure.	Increased latency could add delays to VMs during user authentication.

Option	Pros	Cons
Use Entra ID Join for Azure Virtual Desktop	There is no requirement for Entra ID DS or a domain controller in Azure.	Entra ID-joined VMs don't currently support external users and only support local user profiles. The Windows Store client is not supported at the time of writing, and you cannot access Azure Files file shares for FSLogix or MSIX app attach unless you're using third-party solutions.

Table 3.8 – Comparison of Entra ID deployment options

This table contrasts various Entra ID deployment options, outlining their advantages and disadvantages. It includes deploying a domain controller in Azure, utilizing Entra ID DS for cloud-only setups, leveraging VPN or ExpressRoute for hybrid organizations, and employing Entra ID Join for Azure Virtual Desktop. Each option is evaluated based on its integration capabilities, management complexity, cost implications, and specific limitations or benefits.

> **Important note**
> Most organizations choose the first or third option when choosing an identity strategy.

If you're using native Active Directory, identities must be synchronized with Entra ID. Microsoft Entra Connect is used to link on-premises Active Directory to Microsoft Azure, providing a seamless single sign-on experience between cloud and on-premises services.

Some of the key benefits of using Microsoft Entra Connect are as follows:

- You can use a single identity to access cloud services, such as Microsoft 365, and on-premises applications
- It is a simple tool that takes care of the deployment, sign-in, and synchronization
- It replaces older versions of identity synchronization, such as Entra ID Sync and DirSync
- Microsoft Entra Connect is included in your Azure subscription at no cost

The following diagram depicts how Microsoft Entra Connect works logically:

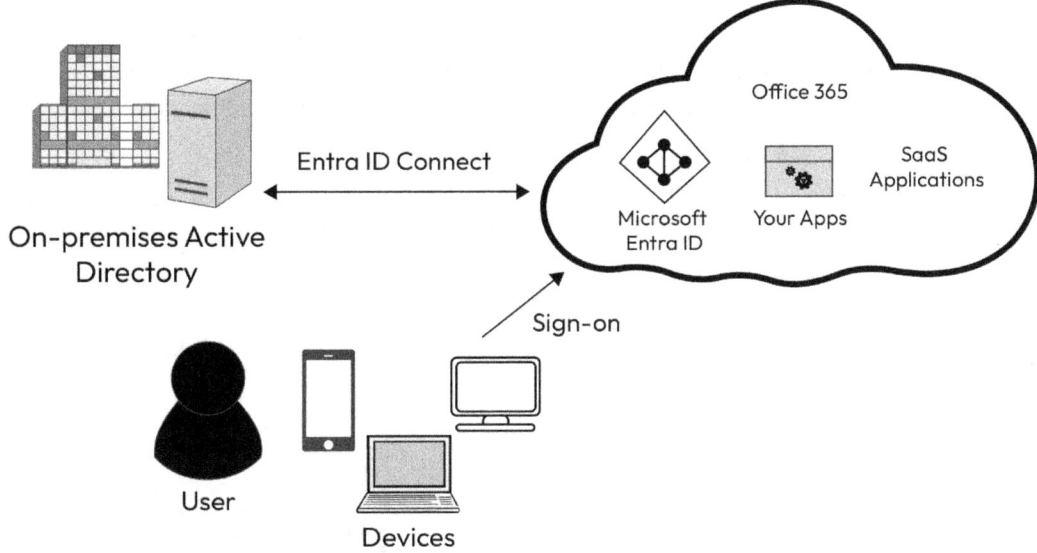

Figure 3.7 – A logical diagram showing how Microsoft Entra Connect works

Domain-joined VMs

Azure Virtual Desktop does not support the use of workgroups. Here, you will need an Active Directory domain.

To join the VMs to a domain, you must provide the complete domain name of the Active Directory instance to join, such as `RMITBLOG.com`. If you've set up a test environment with Entra ID DS, use the DNS domain name that's on the properties page for Entra ID DS, for example, `RMITBLOG.onmicrosoft.com`.

You must also provide an administrator account for the provisioning process to join the VMs to the domain. This account should be delegated with the correct permissions so that you can join session hosts to the domain.

Firewalls and other network requirements

Azure Virtual Desktop depends on specific firewall rule configurations for proper operation. Neglecting to enforce these rules on the virtual machine, Azure Firewall, or any third-party firewall might lead to networking communication problems with Azure Virtual Desktop. For instance, blocking the outbound `1688` TCP port for `kms.core.windows.net` could cause issues with Windows activation.

There are mandatory firewall rules for Azure Virtual Machines:

Address	Outbound TCP Port	Purpose	Service Tag
`*.wvd.microsoft.com`	443	Service traffic	Azure Virtual Desktop
`mrsglobalsteus2prod.blob.core.windows.net`	443	Agent and SXS stack updates	Azure cloud
`*.core.windows.net`	443	Agent traffic	Azure cloud
`*.servicebus.windows.net`	443	Agent traffic	Azure cloud
`prod.warmpath.msftcloudes.com`	443	Agent traffic	Azure cloud
`catalogartifact.azureedge.net`	443	Azure Marketplace	Azure cloud
`kms.core.windows.net`	1688	Windows activation	Internet
`avdportalstorageblob.blob.core.windows.net`	443	Azure portal support	Azure cloud

Table 3.9 – Firewall rules for Azure Virtual Desktop communication

The following are not mandatory; however, it advised that these URLs be added:

Address	Outbound TCP Port	Purpose	Service Tag
`*.microsoftonline.com`	443	Authentication to Microsoft Online Services	None
`*.events.data.microsoft.com`	443	Telemetry service	None
`www.msftconnecttest.com`	443	Detects whether the OS is connected to the internet	None
`*.prod.do.dsp.mp.microsoft.com`	443	Windows Update	None
`login.windows.net`	443	Sign in to Microsoft Online Services, Microsoft 365	None

Address	Outbound TCP Port	Purpose	Service Tag
`*.sfx.ms`	443	Updates for OneDrive client software	None
`*.digicert.com`	443	Certificate revocation check	None

Table 3.10 – Outbound TCP ports and their purposes for Microsoft Online Services

The mandatory Remote Desktop client URLs are as follows:

Address	Outbound TCP Port	Purpose	Client(s)
`*.wvd.microsoft.com`	443	Service traffic	All
`*.servicebus.windows.net`	443	Troubleshooting data	All
`go.microsoft.com`	443	Microsoft FWLinks	All
`aka.ms`	443	Microsoft URL shortener	All
`docs.microsoft.com`	443	Documentation	All
`privacy.microsoft.com`	443	Privacy statement	All
`query.prod.cms.rt.microsoft.com`	443	Client updates	Windows Desktop

Table 3.11 – Outbound TCP ports and their purposes for various clients

The preceding tables has been adapted from the one on Microsoft (https://learn.microsoft.com/azure/virtual-desktop/required-fqdn-endpoint?tabs=azure).

In this section, we looked at planning for user profiles comprising licensing, storage, profile options, and Entra ID. In the next section, we will look at the Windows Remote Desktop client.

Planning for Windows Remote Desktop client deployment

The **Windows Remote Desktop client** can access Azure Virtual Desktop on devices within **Windows 10/11** and **Windows 10/11 IoT Enterprise**.

In this example, we will deploy a Windows Remote Desktop client.

Installing the Windows Remote Desktop client

In this section, we will cover how to install and set up a Windows Remote Desktop client:

1. Open the **Remote Desktop Setup** client and click **Next**:

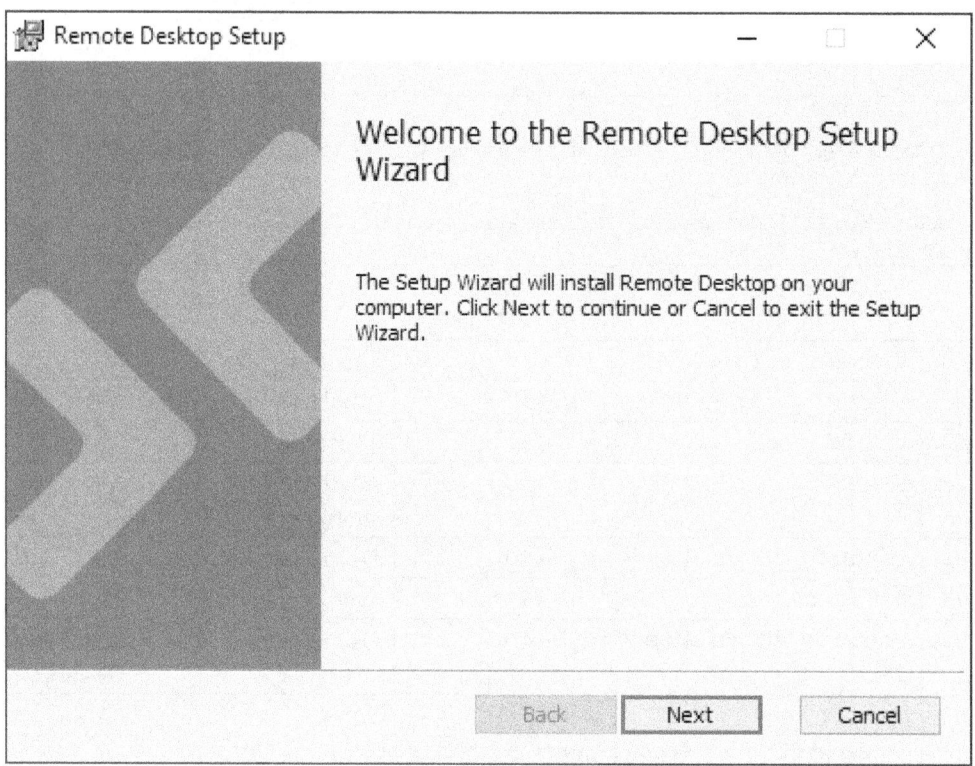

Figure 3.8 – Remote Desktop Setup wizard

2. Check the box next to **I accept the terms in the License Agreement** and click **Next**:

Planning for Windows Remote Desktop client deployment | 67

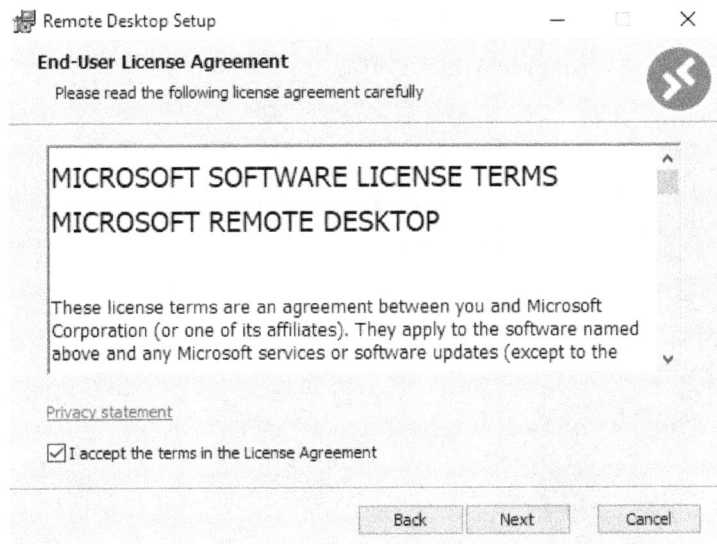

Figure 3.9 – License agreement

3. After downloading the client installer, when you run it, you will have two options:

   • **Install just for you**

   • **Install for all users of this machine** (requires admin rights):

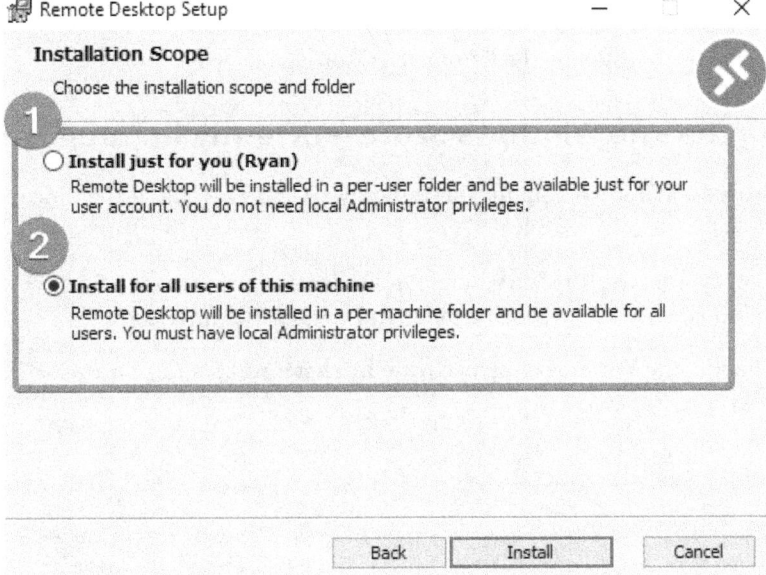

Figure 3.10 – Installation scope

4. If more than one user uses the client device, make sure option 2 is selected and click **Install**.
5. When the installation is completed, click **Finish**:

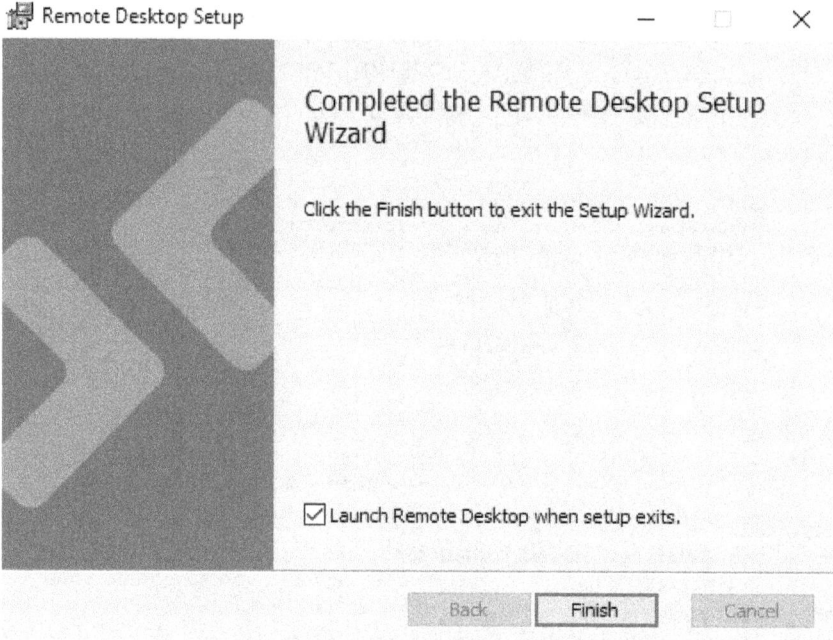

Figure 3.11 – Remote Desktop Setup wizard completed

With that, we are done with installing the Remote Desktop client.

Windows App for the Windows Store (currently in public preview)

Microsoft has released a Windows App for the Windows Store, which is, at the time of writing, in public preview.

This is available directly from the Windows Store. It also allows connections to Windows 365 desktops and AVD desktops and comes with a much cleaner and crisper user interface.

More information about the Windows App can be found here: https://learn.microsoft.com/windows-app/overview.

Figure 3.12 – Windows App GUI Overview

Subscribing to a workspace

When you subscribe to a workspace or to multiple workspaces, you will be provided with a list of managed resources that you can access by the admin. After subscribing, the resources become available on your supported device. The Windows Remote Desktop client supports Windows 365 and Azure Virtual Desktop published resources.

There are two options for subscribing to a workspace. One is that the client can discover the resources available to you using a work or school account. The second is by directly specifying the URL where the resources are. The second option is typically used when the client cannot find the resources.

Once you are subscribed to a workspace, you will be able to launch resources using one of the following two methods:

- Using the Start menu, navigate to the workspace's name or enter the resource name in the search bar
- Within Connection Center, double-click the required resource to launch it

The following URLs are used for subscribing to a workspace:

Available Resources	URL
Azure Virtual Desktop	`https://rdweb.wvd.microsoft.com`
Azure Virtual Desktop (US Gov)	`https://rdweb.wvd.azure.us/api/arm/feeddiscovery`

Table 3.12 – URLs for Azure Virtual Desktop web access

Accessing client logs

If you do experience issues with the Remote Desktop client, you can investigate problems using the client logs.

To collect the client logs, follow these steps:

1. Close all the sessions on the client's device.
2. Ensure that the client process isn't running in the background. You can complete this by right-clicking on the Remote Desktop icon in the system tray and selecting **Disconnect all sessions**.
3. Open **File Explorer**.
4. Navigate to the **%temp%\DiagOutputDir\RdClientAutoTrace** path:

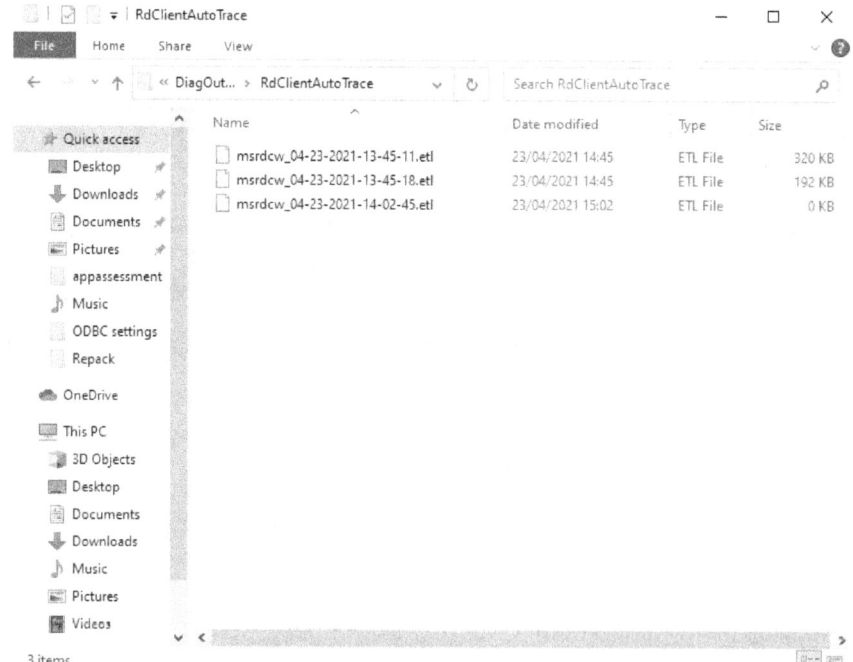

Figure 3.13 – Remote Desktop client log files

You will then see the **Event Trace Log** (ETL) files for the Remote Desktop client.

Connecting to Azure Virtual Desktop using the web client

Azure Virtual Desktop lets users connect to resources using a supported browser.

> **Important note**
> At the time of writing, the web client does not provide mobile OS support.

The following table details the supported web browsers for accessing Azure Virtual Desktop:

Browser	Supported OS	Notes
Microsoft Edge	Windows, macOS, Linux, Chrome OS	Version 79 or later
Apple Safari	macOS	Version 11 or later
Mozilla Firefox	Windows, macOS, Linux	Version 55 or later
Google Chrome	Windows, macOS, Linux, Chrome OS	Version 57 or later

Table 3.13 – Supported web browsers for Azure Virtual Desktop

This table provides a summary of the major supported web browsers, including Microsoft Edge, Apple Safari, Mozilla Firefox, and Google Chrome, along with their supported OSs and minimum version requirements for optimal functionality.

To view the latest supported browsers, please visit `https://learn.microsoft.com/azure/virtual-desktop/virtual-desktop-fall-2019/connect-web-2019`.

To access the web client, you must navigate to `https://client.wvd.microsoft.com/arm/webclient/` using one of the supported browsers, as shown here:

Figure 3.14 – The Remote Desktop web client's All Resources page

In this section, we covered the Azure Virtual Desktop client and web client. Microsoft does offer mobile client support and a Microsoft Store client version. You can find out more here: `https://learn.microsoft.com/azure/virtual-desktop/users/connect-ios-ipados`.

Setting up email discovery to subscribe to the Azure Virtual Desktop feed

This section will show you how to set up a DNS record to enable email discovery to subscribe to an Azure Virtual Desktop feed.

You can improve the user experience when connecting to a workspace by configuring a DNS record so that users can use their email address to discover the subscription. This makes it very easy for users to access their desktops and RemoteApp instances.

> **Important note**
>
> Azure Virtual Desktop requires the following URL: `https://rdweb.wvd.microsoft.com/api/arm/feeddiscovery`.

To set up email discovery, you need to configure a few things.

Within your domain registrar or DNS, you will need to enter a new DNS record with the following properties:

- **Host**: `_msradc`
- **Text**: `<RD Web Feed URL>`
- **TTL**: `300 seconds`
- **TXT**: `https://rdweb.wvd.microsoft.com/api/arm/feeddiscovery`

The names of the DNS records fields can vary depending on the domain name registrar. The generic process will result in a TXT record named `_msradc.<domain_name>` (such as `_msradc.rmitblog.com`) that has a value of the full Remote Desktop web feed.

The following screenshots depict the steps within Azure DNS:

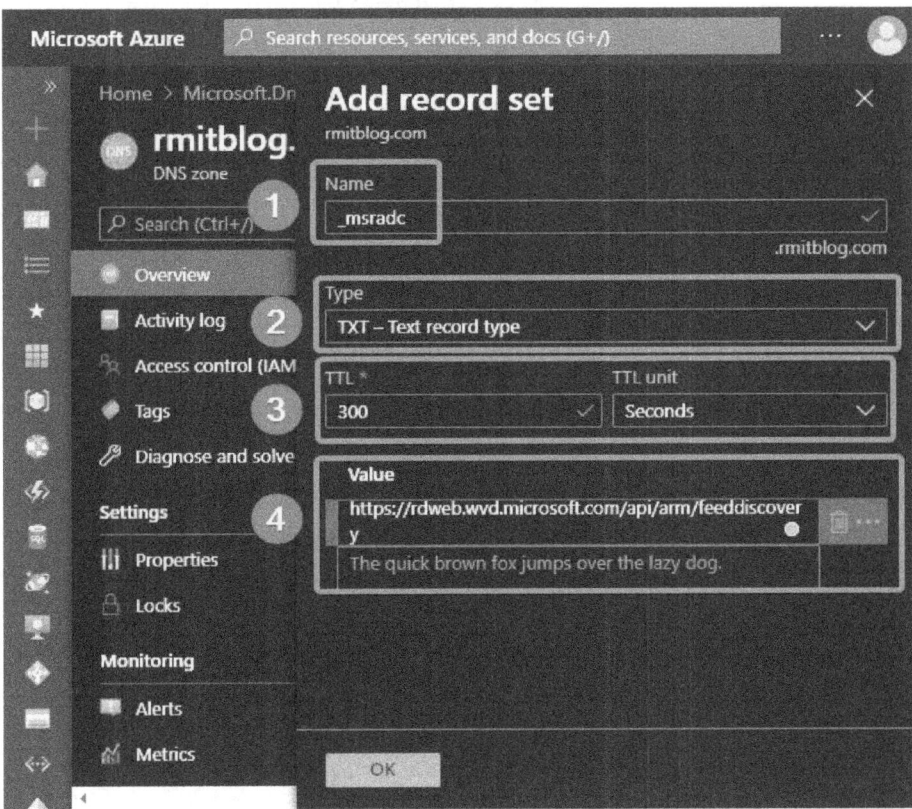

Figure 3.15 – Adding a DNS record for email discovery

In this section, we showed you how to configure DNS to enable email discovery within workspaces. Next, we'll summarize the prerequisites for Azure Virtual Desktop.

Summarizing the prerequisites for Azure Virtual Desktop

The following table summarizes the prerequisites for deploying an Azure Virtual Desktop environment:

Infrastructure prerequisites	Azure Virtual Desktop can only be used in a cloud-only organization or a hybrid environment.
Hybrid Active Directory environments/cloud organization component prerequisites	<ul><li>An Entra ID organization.</li><li>A domain controller that's synced with Entra ID. You can configure this with one of the following:<ul><li>Microsoft Entra Connect</li><li>Entra ID Domain Services</li><li>A virtual machine in Azure acting as a domain controller</li><li>An Azure subscription that contains a virtual network that either contains or is connected to Windows Server Active Directory or Entra ID Domain Services</li><li>Entra ID DS</li></ul></li></ul>
Azure VM requirements for Azure Virtual Desktop	<ul><li>Standard domain-joined or hybrid AD-joined</li><li>Running one of the supported OS images</li></ul>
Supported Remote Desktop clients	<ul><li>Windows Desktop</li><li>Web</li><li>macOS</li><li>iOS</li><li>Android</li><li>Linux thin clients that connect using the official SDK</li></ul>
Supported VM OS images	<ul><li>Windows 11 Enterprise multi-session</li><li>Windows 11 Enterprise</li><li>Windows Server 2022</li><li>Windows Server 2019</li><li>Windows Server 2012 R2</li></ul>

Table 3.14 – Prerequisites for deploying Azure Virtual Desktop

This table outlines the essential infrastructure components, supported Remote Desktop clients, and compatible OS images required for deploying Azure Virtual Desktop in both cloud-only and hybrid environments. It details the necessities, such as Entra ID organization, domain controller synchronization options, Azure subscription, virtual network requirements, VM configuration, and the range of supported client devices and OSs.

Summary

In this chapter, we covered the requirements for designing user identities and profiles. This included the licensing options available for Azure Virtual Desktop, the available storage solutions, and the recommended solutions, such as Azure Files and Azure Virtual Desktop client deployments. It further covered planning for user profiles, identity strategies, firewalls, and other network requirements.

In the next chapter, we will look at implementing and managing networking for Azure Virtual Desktop.

Questions

Here are a few questions to test your understanding of this chapter:

1. What is the recommended profile container technology for Azure Virtual Desktop?

    A. *FSLogix profile containers*

    B. UPD

    C. Local profiles

2. Is there a license you should apply to an Azure Virtual Desktop virtual machine?

    A. *Yes*

    B. No

3. What types of storage can you use for Azure Virtual Desktop?

    A. *Azure Files, Azure NetApp Files, and Storage Spaces Direct*

    B. Local storage

4. When configuring an Azure Storage account for FSLogix, which two privileges do you need to assign administrators (storage admins)?

    A. *Storage File Data SMB Share Contributor role, Full Control NTFS permissions*

    B. Blob Data Contributor role, Modify NTFS permissions

4
Implementing and Managing Networking for Azure Virtual Desktop

In this chapter, we will take a look at how to implement and manage networking for **Azure Virtual Desktop** (**AVD**). First, we will provide an overview of designing an AVD environment and the differences and requirements between a single and multi-region deployment. We will then look at what the AVD networking considerations are and understand **virtual machine** (**VM**) sizing and scale. Then, we will understand the Active Directory considerations with AVD, its maintenance, and how to support AVD.

The following topics will be covered in this chapter:

- Implementing Azure **virtual network** (**VNet**) connectivity
- Managing connectivity to the internet and on-premises networks
- Implementing and managing network security
- Managing AVD session hosts using Azure Bastion
- Monitoring and troubleshooting network connectivity

Implementing Azure VNet connectivity

Before building an AVD environment, we need to review the network requirements and the different components and configuration options available.

Azure VNet

When deploying an AVD environment, one of the first things you need to ensure that you have configured correctly is the VNet.

The VNet enables Azure resources, such as Azure VMs, to access the internet and on-premises networks.

The VNet is similar to a traditional network you would operate within your environment. However, it allows you to take advantage of Microsoft Azure's infrastructure, including availability, scaling, and isolation.

What is an Azure VNet?

Azure VNets allow resources to securely connect to services and access on-premises resources. VNets have several features and configurable options for filtering and routing traffic and integrating with Azure services. The following sections detail some of the features and what they do.

Communication between Azure resources

Azure resources communicate with each other through a VNet using network interfaces and service endpoints. Different VNets could be connected via VNet peering.

The following table lists the various Azure resources and their descriptions:

Type	Description
VNet	Deploy VMs and other types of Azure resources to a VNet, such as Azure App Service, **Azure Kubernetes Service** (**AKS**), and Azure **Virtual Machine Scale Sets** (**VMSS**).
VNet service endpoint	VNet service endpoints allow you to connect your VNet's private address space to Azure service resources. These resources include Azure Storage accounts, Azure Key Vault, and Azure SQL databases. Service endpoints enable you to securely connect to Azure service resources directly without passing the public internet. Using private endpoints can also offer potential performance benefits.
VNet Peering	You can connect different VNets with network peering. VNets can connect in the same or different Azure regions, thereby providing flexibility. When peering between two or more subscriptions, you can associate against the same or different Active Directory tenants.

Table 4.1 – Network resources

Communication with on-premises networks

The following table specifies the options you can use when you want to connect your on-premises computers and networks to an Azure VNet:

Type	Description
Point-to-site **virtual private network (VPN)**	Point-to-site connections are similar to a typical client VPN. They allow a client device to connect to an Azure network.
	The communication between the client and a VNet is sent through an encrypted tunnel over the internet.
Site-to-site (S2S) VPN	An S2S VPN connects the VPN device and an Azure VPN gateway that has been deployed on a VNet.
	This type of connection allows you to connect your Azure network to on-premises resources (for example, the local data center).
	The connection between your on-premises VPN device and an Azure VPN gateway communicates through an encrypted tunnel over the internet.
Azure ExpressRoute	ExpressRoute is a connection between your network and Azure that uses an ExpressRoute ISP partner. The ExpressRoute connection is private, and traffic does not communicate over the internet.
	This type of connectivity allows you to connect your Azure network to on-premises resources (for example, the local data center).
	You can share ExpressRoute connections across multiple subscriptions.

Table 4.2 – Connection options

Filtering and routing Azure network traffic

You can filter network traffic between subnets using one of the following features within Azure:

Filtering Network Traffic	
Network security groups (NSGs)	NSGs and application security groups can be configured with multiple inbound and outbound security rules. These allow you to filter traffic to and from resources by source and destination IP address, port, and protocol. NSGs can be applied to a network interface or a subnet.
Network virtual appliances (NVAs)	An NVA is a VM that performs a network function, such as a firewall, WAN optimization, or other network functions. You can deploy these from the Azure Marketplace. Vendors include Checkpoint, Watchguard, and SonicWALL, to name a few.
Azure Firewall	Azure Firewall allows you to centrally create allow and deny network filtering rules by source and destination IP address, port, and protocol.

Table 4.3 – Securing network

By default, you can route Azure traffic between subnets, connected VNets, on-premises networks, and the internet. You can configure both options shown in the following table:

Routing Network Traffic	
Route tables	Create and configure custom route tables with routes that control the routing of traffic for each subnet
Border gateway protocol (BGP) routes	VNets connected to an on-premises network using an Azure VPN gateway or ExpressRoute connection can propagate on-premises BGP routes to the Azure VNets

Table 4.4 – Routing options

Understanding what VNet integration is for Azure services

One of the features you are most likely to use within AVD is the integration of Azure services with an Azure VNet. This is because of the requirement for Azure Files and the ability to create a private endpoint. This allows traffic from the service to communicate on the VNet rather than publicly.

The following options are available to you when you're integrating Azure services into your VNet:

- You can deploy a dedicated service instance into a VNet. The services can then be accessed within the VNet and from on-premises networks that are connected to Microsoft Azure.
- A private link is used to access a specific instance of a service privately from your VNet and on-premises networks.
- You can also access services using public endpoints by extending a VNet to the service. This is done through service endpoints. Service endpoints allow service resources to be secured to a VNet.

> **Important note**
> There are limits regarding the number of Azure resources you can deploy. Check out https://docs.microsoft.com/azure/azure-resource-manager/management/azure-subscription-service-limits#networking-limits, which shows the Azure networking limits and maximum values.

This section provided an introduction to this chapter, specified why there is a requirement to use Azure VNets, covered communication between Azure and on-premises, and provided an overview of filtering and routing. The next section will explain how to manage connectivity to the internet and on-premises networks.

Managing connectivity to the internet and on-premises networks

In this section, we will look at the options available for VPN gateway design. This should enable you to make an informed choice that best suits your deployment needs. Please note that a VPN is only required for connecting resources between on-premises and cloud platforms. For client connectivity to the AVD platform, you would use the reverse connect feature built into the AVD management plane.

Types of VPNs available to you

The following sections cover several different configurations available for VPN connections. I have provided a summary of each type and a diagram to help you decide which topology meets your requirements.

S2S

An **S2S** VPN gateway connection is an IPsec/IKE (IKEv1 or IKEv2) VPN tunnel. This type of connection can be used for cross-premises and hybrid connectivity. An S2S connection requires an on-premises VPN device with an assigned public IP address.

Figure 4.1 shows an S2S VPN:

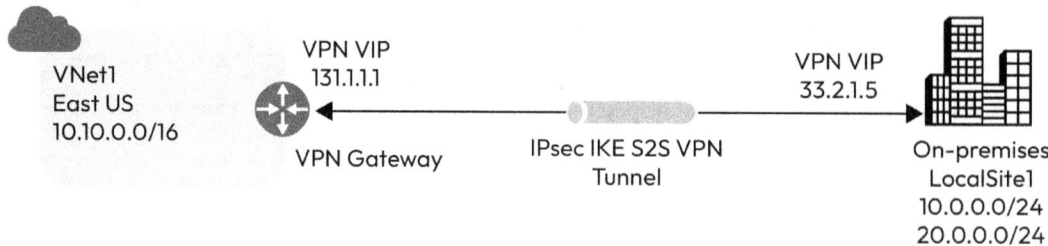

Figure 4.1 – S2S connection

You can configure a VPN gateway in active-standby mode using one public IP address or you can configure an active-active connection using two public IP addresses. Here are some further details regarding these:

- **Active-standby**: This allows you to switch traffic to the standby tunnel in the event of an issue.
- **Active-active**: This is the recommended option in which both tunnels are active, allowing higher throughputs and active-active mode to prevent noticeable failure. Throughput may be constrained in a partial failure scenario.

Multi-site

A **multi-site** VPN connection allows you to have more than one VPN connection from a VNet gateway. This allows you to connect to multiple sites, which are typically on-premises. Note that you must use a route-based VPN (https://docs.microsoft.com/azure/vpn-gateway/vpn-gateway-connect-multiple-policybased-rm-ps#about) type when using multiple connections, also known as a **dynamic gateway**. This is because each VNet can only have one VPN gateway and all connections through a gateway share the bandwidth:

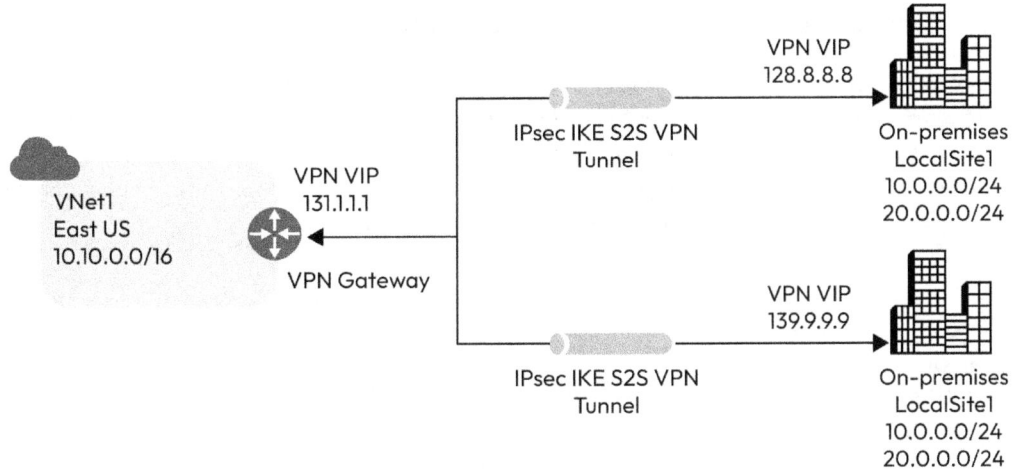

Figure 4.2 – Multi-site connection

Point-to-site VPN

Point-to-site (**P2S**) VPN connections are established from a client computer. This solution is helpful for telecommuters who need to connect to an Azure VNet from a remote location such as a home office or a conference center. P2S connections do not require a public-facing IP address or a VPN device:

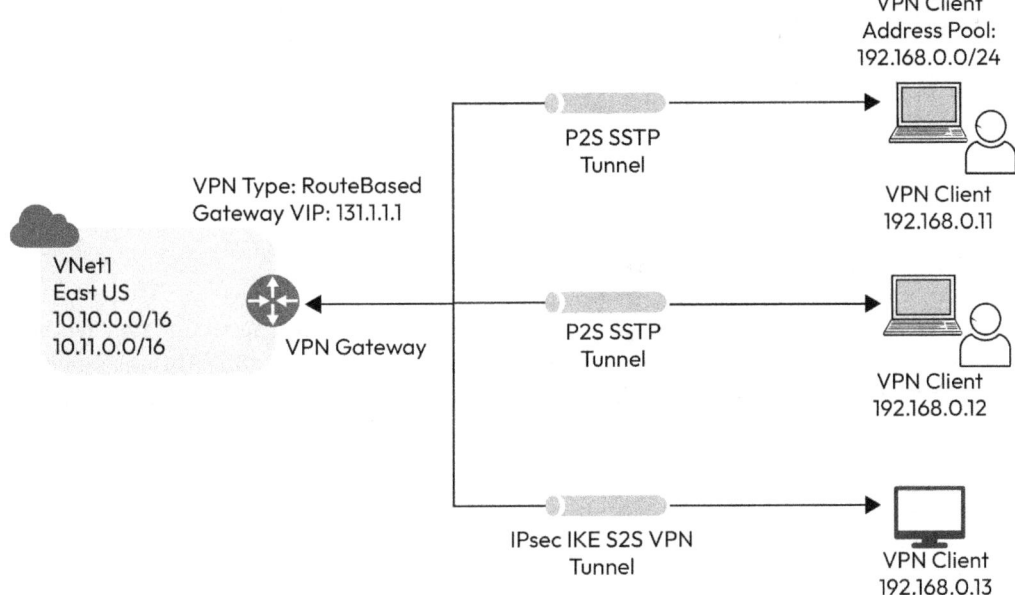

Figure 4.3 – P2S connection

VNet-to-VNet connections (IPsec/IKE VPN tunnel)

VNet-to-VNet connections provide connections from one VNet to another. This type uses **Internet Protocol Security/Internet Key Exchange (IPsec/IKE)** (https://docs.microsoft.com/azure/vpn-gateway/vpn-gateway-ipsecikepolicy-rm-powershell) to provide a secure tunnel between the same or different regions/subscriptions/deployment models. You can also combine VNet-to-VNet communication with multi-site connection configurations:

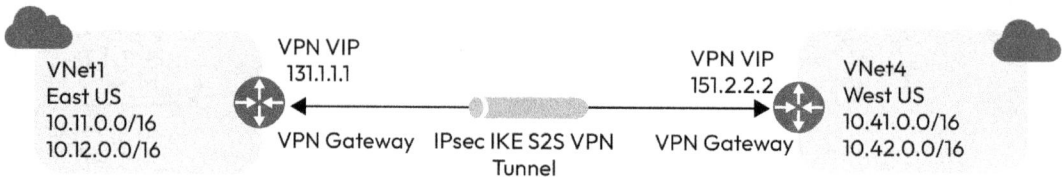

Figure 4.4 – VNet-to-VNet connection

VNet peering

VNet peering is a way of connecting two VNets without using a VNet gateway. You can connect VNets in the same and different subscriptions. Traffic is routed through Microsoft's backbone infrastructure, meaning no public network is involved:

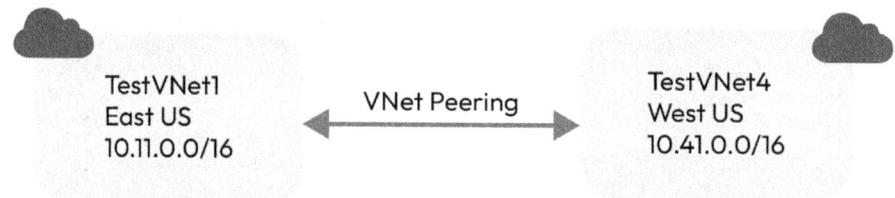

Figure 4.5 – VNet peering

ExpressRoute (private connection)

ExpressRoute is a typical deployment option within an enterprise. This connection type lets you extend your on-premises network to Microsoft Azure over a private connection. This is something that an **internet service provider (ISP)**, also known as a network service provider, can facilitate:

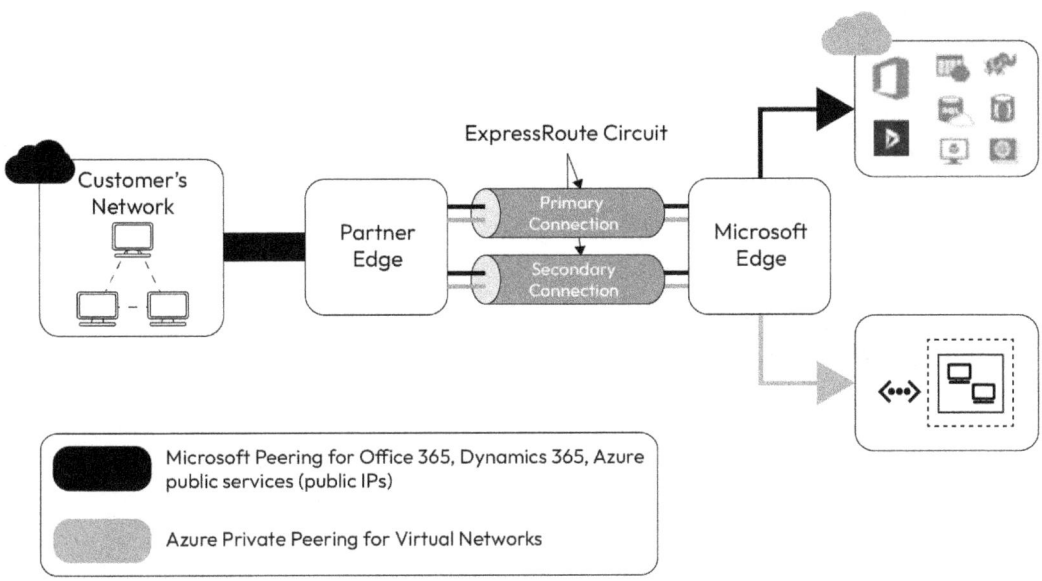

Figure 4.6 – ExpressRoute connection

ExpressRoute allows connections to Microsoft Cloud services such as Microsoft Azure, Microsoft 365, and CRM Online. There are several connectivity options available, including any-to-any, point-to-point, or virtual cross-connection options.

ExpressRoute is a private connection that is slightly different from other connection types that use the public internet. It provides better reliability, faster speeds, lower latency, and higher security than the traditional connection types that use the internet. It's also important to note that ExpressRoute offers a 99.95% uptime SLA for each ExpressRoute that's deployed. You can read more about the ExpressRoute SLA here: `https://azure.microsoft.com/en-gb/support/legal/sla/expressroute/`.

> **Important note**
> You can have multiple VNets in different subscriptions that connect using the same ExpressRoute. This offers a benefit compared to using VPN connections. You can read more about this here: `https://docs.microsoft.com/azure/expressroute/expressroute-howto-linkvnet-portal-resource-manager#connect-a-vnet-to-a-circuit---same-subscription`.

When configuring an ExpressRoute, you must set the gateway type to **ExpressRoute** instead of **VPN**.

> **Important note**
>
> ExpressRoute traffic is not encrypted by default; however, it is possible to send encrypted traffic over an ExpressRoute circuit.

S2S and ExpressRoute coexisting connections

As well as an ExpressRoute, you can configure an S2S VPN as a secure failover path for the ExpressRoute. The capability to configure both S2S VPN and ExpressRoute connections for a VNet has various benefits.

You can connect to sites that are not part of your ExpressRoute network as well as adding a failover path using the S2S VPN:

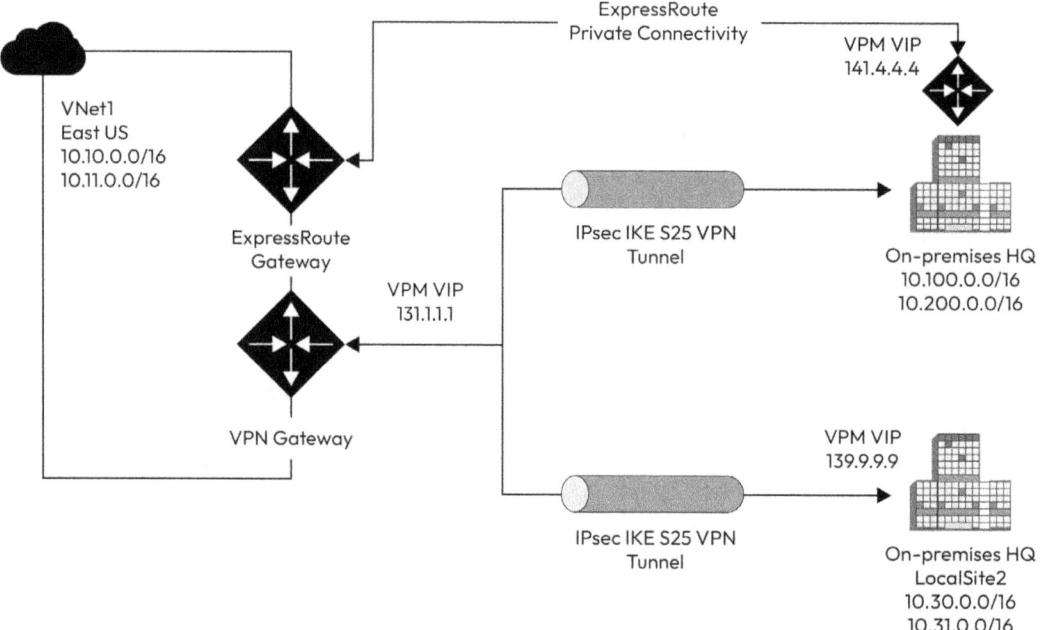

Figure 4.7 – S2S and ExpressRoute connections

Internet access and outbound connections

By default, each Azure VM can connect directly to the internet via a VNet. For this, the VM uses a random public IP to connect to the internet. That is comparable to your computer at home behind a router. The IP you are using on the internet is not the local IP address – it's an IP address from your provider (typically). It's the same in Azure: a VM with a private IP address in a VNet can connect to the internet with the "router" Microsoft. This will change in October 2025. So, it makes sense to directly use one of the following options to organize internet traffic.

Option 1 – using a Firewall

A **firewall** (that is, **Azure Firewall** from Microsoft or another vendor) can be used to filter the network traffic to the internet and other network resources, such as the on-premises data center.

Option 2 – using a NAT gateway

An **Azure NAT gateway** is a service in Azure that configures explicit public IP addresses to access the internet. While your public IP addresses are yours, you can also use these source addresses as an additional parameter to configure trusted locations in Entra ID conditional access.

Option 3 – public IP addresses for Azure VMs (not recommended for security purposes)

More information on this can be found at `https://learn.microsoft.com/azure/nat-gateway/nat-overview`.

> **Important note**
>
> On September 30, 2025, default outbound access connectivity for new VMs in Azure will be retired. New VMs cannot access the internet without a NAT gateway, public IP, load balancer, or firewall. You can learn more about this at `https://azure.microsoft.com/en-us/updates/default-outbound-access-for-vms-in-azure-will-be-retired-transition-to-a-new-method-of-internet-access/`.

This section provided an overview of the different connectivity options available within Azure to connect on-premises infrastructure. In the next section, we will look at AVD network security.

Implementing and managing network security

This section covers some of the network security features available for AVD.

Azure network security overview

Microsoft Azure offers a wide range of protecting resources to prevent unauthorized access or attacks by applying network controls to the Azure VNet. These are as follows:

- Azure networking
- Network access control
- Azure Firewall
- Secure remote access and cross-premises connectivity
- Availability
- Name resolution
- Perimeter network (DMZ) architecture
- Azure DDoS protection
- Monitoring and threat detection

This section provides a high-level overview of network security. We'll start by looking at specific network connectivity requirements for AVD.

Understanding AVD network connectivity

AVD uses **remote desktop protocol** (**RDP**) to connect to an AVD deployment. The RDP connection is responsible for providing display and input functions over a network connection.

When we look at the connection data flow for AVD, the client starts with a **domain name server** (**DNS**) lookup for the nearest Azure data center. The lookup uses **round-trip time** (**RTT**) to identify the nearest data center. The gateway is used as an intelligent reverse proxy that manages all session connectivity, with nothing but the RDP bitmap (pixels) reaching the client.

The following are the five connection flow process steps for AVD:

1. The user authenticates in Entra ID; a token is returned to the remote desktop client.
2. The AVD gateway checks the token with the AVD connection broker.
3. The broker queries the database for resources assigned to the user.
4. The AVD gateway and the AVD broker select the session host for the connected client.
5. The session host creates a reverse connection to the client by using the AVD gateway.

The following figure visualizes the connection flow:

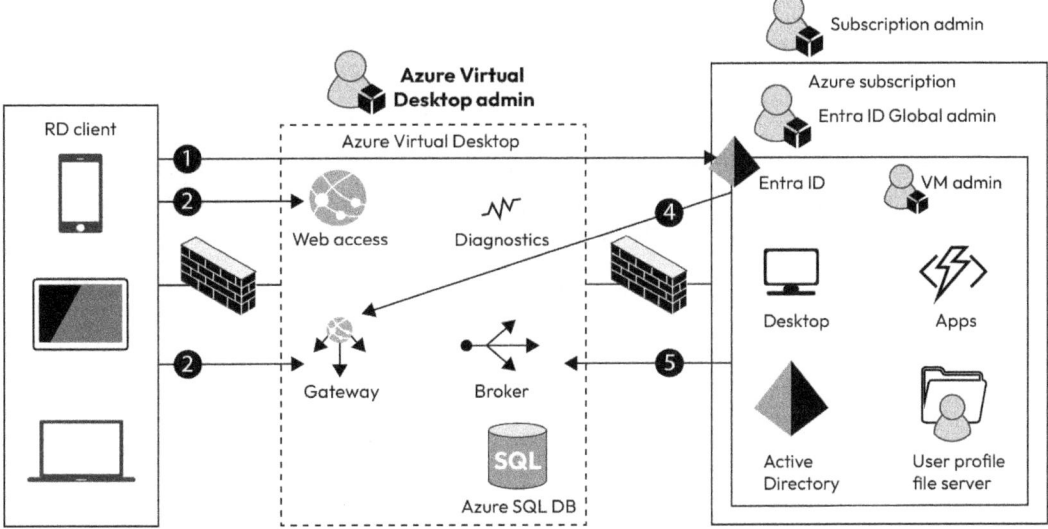

Figure 4.8 – Connection flow process

> **Important note**
> AVD uses the same TLS 1.2 ciphers as Azure Front Door. You'll need to make sure that the client computers and session hosts can use these ciphers.

Reverse connect

Reverse connect is a technology that removes the requirement for a session host within AVD to have any inbound ports opened. By default, the RDP port, `TCP/3389`, doesn't have to be open. Instead, the agent on the session host creates an outbound connection to the **Azure Virtual Desktop Management** plane using `TCP/443`. This is a reverse proxy for RDP traffic.

VMs in AVD are not publicly exposed to the internet directly. They use a private IP address within an Azure VNet and run isolated from other workloads or the internet if required.

RDP-Shortpath

RDP-Shortpath enables users to use low-latency UDP connection to session hosts. This is important if users are working with video applications or engineering applications where an engineer must rotate 3D objects in an application.

RDP-Shortpath for public networks is the default connection method for supported clients. TCP-based reverse connect is used as a fallback. So, each AVD deployment with native clients prefers RDP-Shortpath. With RDP-Shortpath, a client will directly connect to a session host using UDP.

While the client and the host are mostly not facing the internet (but should be), Microsoft uses the following technologies to allow a direct connection:

- **Direct connection via STUN**: STUN is a technology that gets the interfacing IP address (which is normally not known by a client behind a NAT/firewall). If all prerequisites are full-filled, the client and hosts are connected directly via the interfacing IPs with UDP.
- **Indirect connection via TURN**: TURN is an extension of STUN that uses an intermediate server between the client and host. It's used if STUN isn't working in the network environment. TURN also comes with a TCP fallback for UDP.

The state of the connection is shown in the remote session. Click on the network icon on top of a connected desktop and click **Show details**:

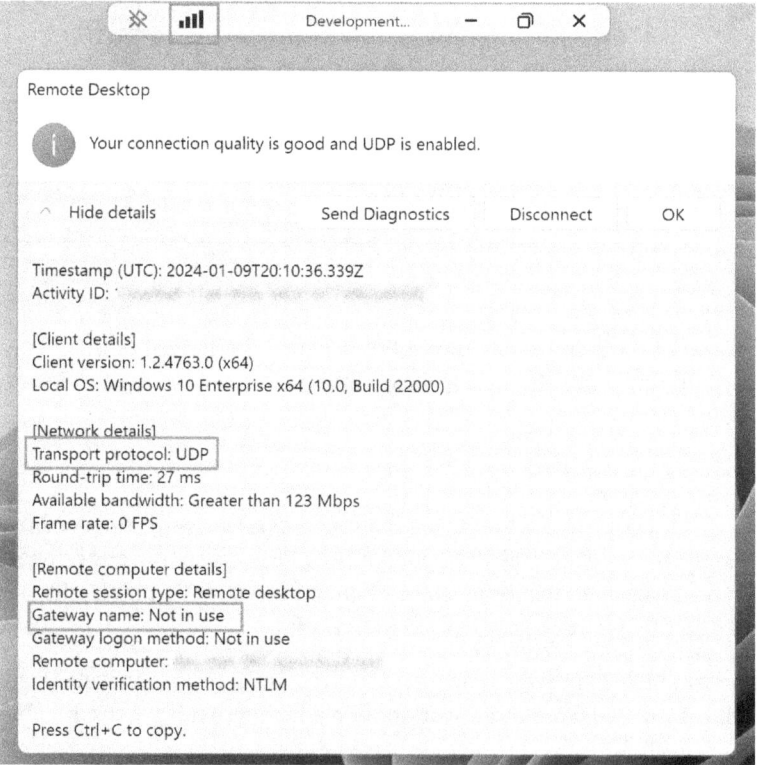

Figure 4.9 – Connected to the RDP-Shortpath public network

RDP-Shortpath for managed networks is another option that lets clients connect directly to the host using a routed network or VPN. It's typically used in environments where the clients are in an office network, which is routed to the Azure network of the hosts. If RDP-Shortpath is configured for the hosts and the client can access the hosts by UDP port 3390, a direct connection is established. TCP-based reverse connect is also used as a fallback.

Implementing and managing network security

RDP-Shortpath for managed networks must be configured on each host. Run the following PowerShell code with administrative privileges on each deployed host to enable RDP-Shortpath for managed networks (restart the host after running this script):

```
$regPath="HKLM:\SYSTEM\CurrentControlSet\Control\Terminal Server\
WinStations"
if (-not (Test-Path $regPath)) {

New-Item -Path $regPath -Force -ErrorAction SilentlyContinue
}
New-ItemProperty -Path $regPath -Name "fUseUdpPortRedirector" -Value 1
-force
New-ItemProperty -Path $regPath -Name "UdpPortNumber" -Value 3390
-force
# Configuring RDS settings: Configuring Windows Firewall
New-NetFirewallRule -DisplayName "Remote Desktop - Shortpath
(UDP)" -Action Allow -Description "Inbound rule for the Remote
Desktop service to allow RDP traffic on UDP 3390" -Group "@
FirewallAPI.dll,-28752" -Name 'RemoteDesktop-RDP-Shortpath-
UDP' -PolicyStore PersistentStore -Profile Any -Service TermService
-Protocol udp -LocalPort 3390 -Program "%SystemRoot%\system32\svchost.
exe" -Enabled:True -ErrorAction SilentlyContinue
```

You can easily run the script in an administrative ISE, like this:

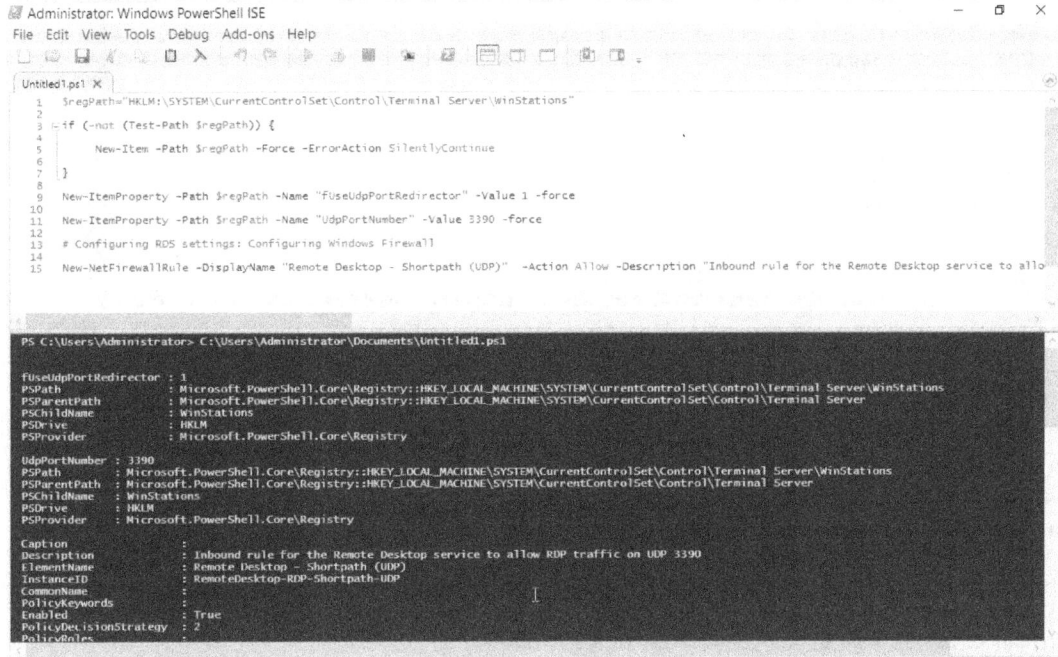

Figure 4.10 – Running a script in ISE

The state of the connection is shown in the remote session. Click on the network icon on top of a connected desktop and click **Show details**:

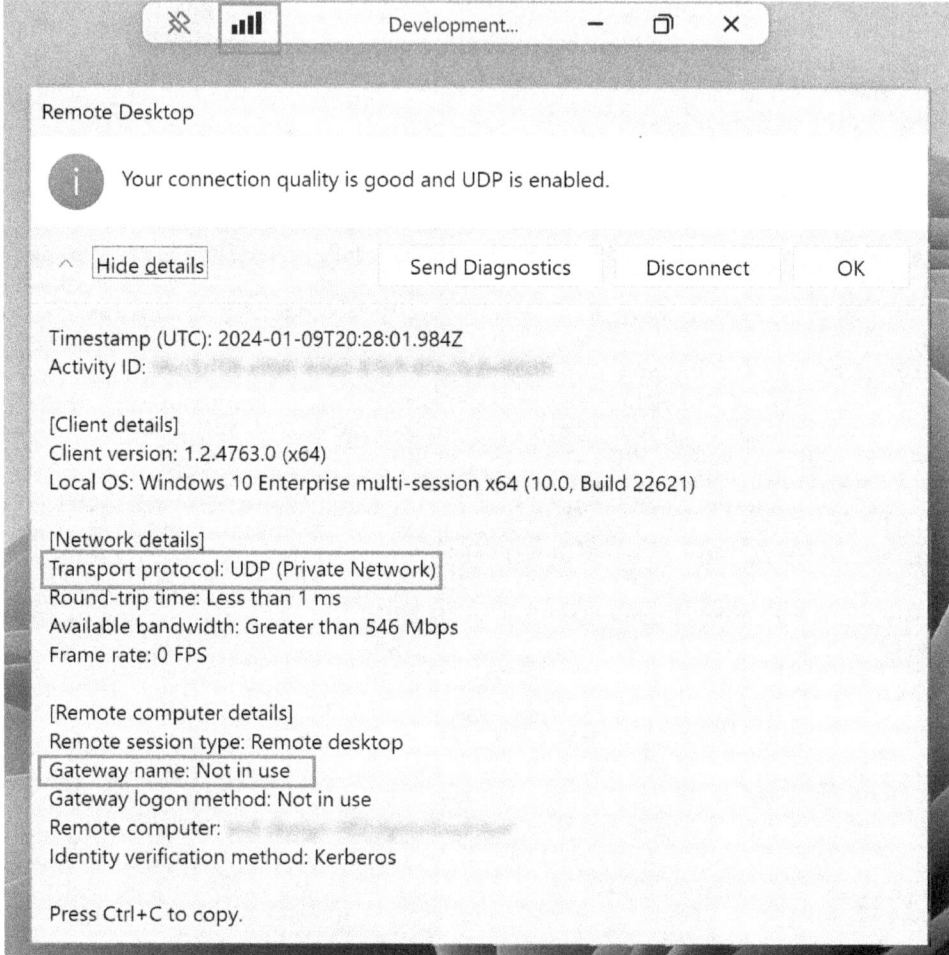

Figure 4.11 – Connected with the RDP-Shortpath managed network

You can read more about RDP-Shortpath at `https://learn.microsoft.com/azure/virtual-desktop/rdp-shortpath`.

> **Important note**
>
> In Azure, customers have to pay for outgoing data. Data sent to the clients using the RDP gateway service is included and doesn't count as part of the outgoing data. RDP-Shortpath bypasses the RDP gateway service and results in additional costs.

Private Link with AVD

Like other services, Microsoft also offers Azure Private Link for AVD. You can force that connections and feeds (information about available apps and desktops for a user) don't traverse the internet. In this case, only users connected to the Azure VNet have access. This is typically done to prevent external users from using AVD.

Administrators can configure what kind of data must be used by Private Link:

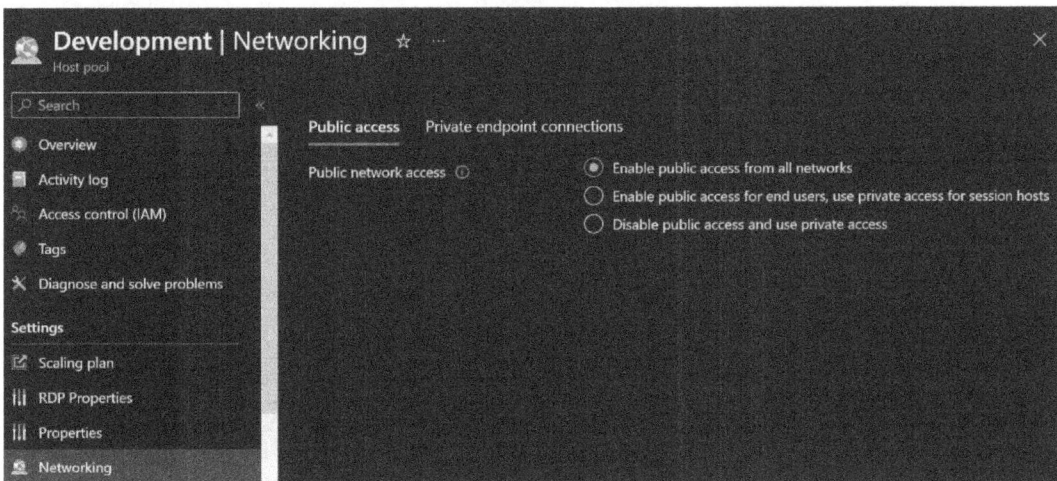

Figure 4.12 – Configuring Private Link for a host pool

The options shown in the preceding screenshot are described in the following table:

Enable public access from all networks	Users can get feeds and access AVD using public and private networks.
Enable public access for end users, use private access for session hosts	Users can get feeds over public and private networks. Connections to session hosts must use private networks.
Disable public access and use private access	Feeds and access to the host must use private networks.

Table 4.5 – Options for Private Link for a host pool

If you want to use private networks, a private endpoint must be configured for the host pool.

You can learn more at `https://learn.microsoft.com/azure/virtual-desktop/private-link-overview`.

> **Important note**
> At the time of writing, you cannot use Private Link and RDP-Shortpath at the same time.

NSGs and service tags

NSGs are an excellent way to control/limit AVD traffic. You can configure network security on both a VM and an Azure VNet, making it easy to restrict network access.

You can also use service tags within NSG rules to allow or deny traffic to a specific Azure service per Azure region or globally.

An NSG is essentially a collection of rules for inbound/outbound traffic, allowing you to control network traffic flow in or out of Azure resources.

Each rule can be configured with the following properties: name, priority, source or destination, protocol, direction, port range, and action.

> **Important note**
> NSGs are Layer 3 and Layer 4 network security services. An NSG consists of several security rules (allow or deny). This is achieved using a five-tuple hash. A five-tuple hash is based on the source and destination IPs, the protocol, and destination ports.

When you create an Azure VM for AVD, you must configure security network filtering using NSGs or other chosen configurations to grant access to the following URLs:

Address	Outbound TCP Port	Purpose	Service Tag
`login.microsoftonline.com`	443	Authentication to Microsoft Online Services	
`*.wvd.microsoft.com`	443	Service traffic	WindowsVirtualDesktop
`*.prod.warm.ingest.monitor.core.windows.net`	#	Agent traffic Diagnostic output	AzureMonitor
`catalogartifact.azureedge.net`	443	Azure Marketplace	AzureFrontDoor.Frontend
`gcs.prod.monitoring.core.windows.net`	443	Agent traffic	AzureCloud
`kms.core.windows.net`	1688	Windows activation	Internet
`azkms.core.windows.net`	1688	Windows activation	Internet
`mrsglobalsteus2prod.blob.core.windows.net`	443	Agent and **side-by-side** (**SXS**) stack updates	AzureCloud
`wvdportalstorageblob.blob.core.windows.net`	443	Azure portal support	AzureCloud
`169.254.169.254`	80	Azure Instance Metadata service endpoint	N/A
`168.63.129.16`	80	Session host health monitoring	N/A
`oneocsp.microsoft.com`	80	Certificates	N/A
`www.microsoft.com`	80	Certificates	N/A

Address	Outbound TCP Port	Purpose	Service Tag
`login.windows.net`	443	Sign in to Microsoft Online Services and Microsoft 365	N/A
`*.events.data.microsoft.com`	443	Telemetry Service	N/A
`www.msftconnecttest.com`	80	Detects if the session host is connected to the internet	N/A
`*.prod.do.dsp.mp.microsoft.com`	443	Windows Update	N/A
`*.sfx.ms`	443	Updates for OneDrive client software	N/A
`*.digicert.com`	80	Certificate revocation check	N/A
`*.azure-dns.com`	443	Azure DNS resolution	N/A
`*.azure-dns.net`	443	Azure DNS resolution	N/A

Table 4.6 – Necessary addresses and ports for AVD

If you're using AVD in the US government cloud, other endpoints are needed. These can be found at https://learn.microsoft.com/azure/virtual-desktop/required-fqdn-endpoint.

What is Azure Firewall?

Azure Firewall is a Microsoft-managed network security service that protects your Azure VNet resources. This is a fully stateful firewall with the added benefits of cloud high availability and scalability:

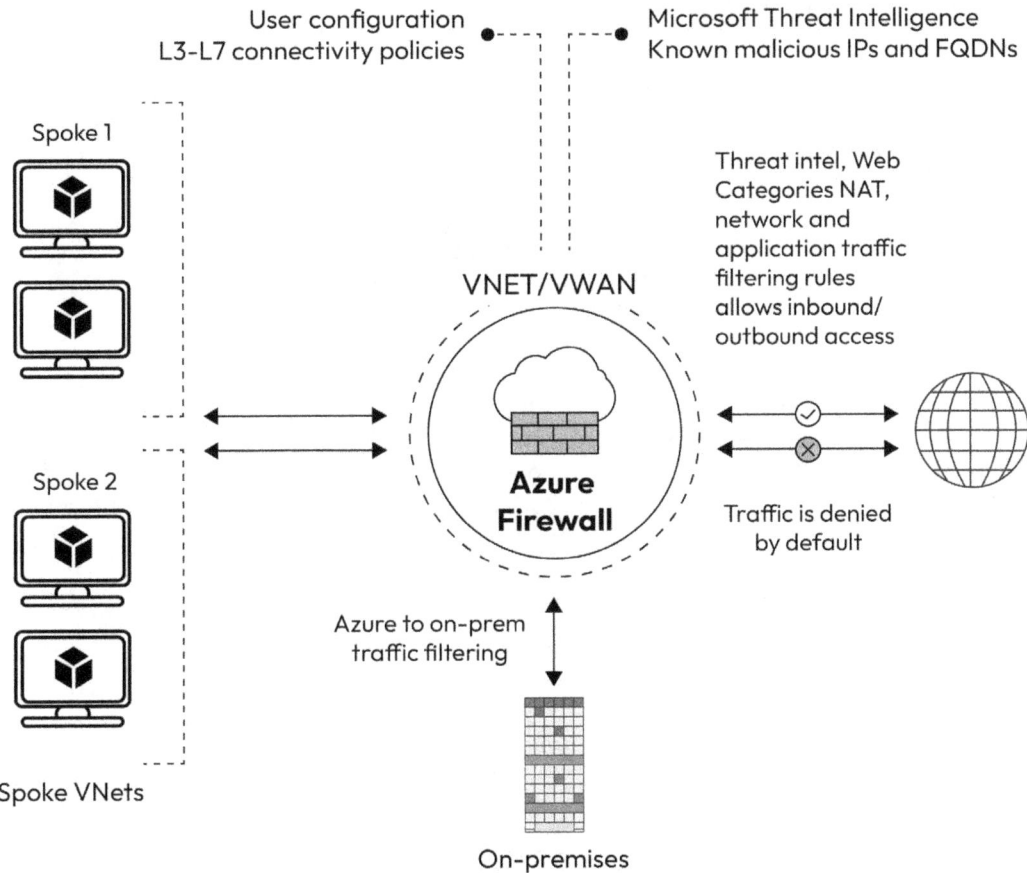

Figure 4.13 – Azure Firewall standard

It is also important to note that there is an Azure Firewall Premium version that offers advanced capabilities, including signature-based IDPs:

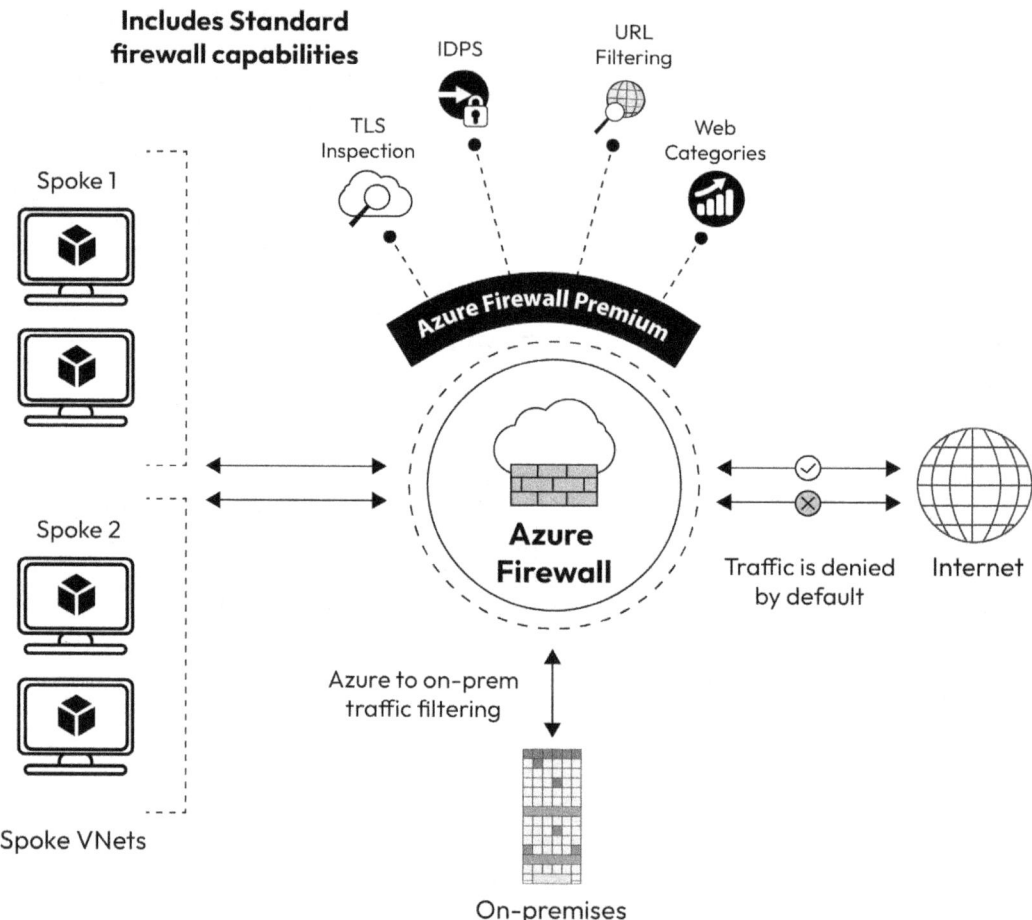

Figure 4.14 – Azure Firewall Premium

There's also Azure Firewall Basic, which comes with a reduced feature set and is recommended for SMB customers with less than 250 Mbit/s.

You can read more about the different SKUs here: https://learn.microsoft.come/azure/firewall/choose-firewall-sku.

Azure Firewall allows you to create, configure, and log network and application connectivity policies across Azure subscriptions and VNets. You can even integrate with Azure Monitoring Service for logs and analytics.

> **Important note**
> Azure Firewall uses a static public IP address for VNet resources for external network devices to identify traffic originating from your Azure VNet.

Azure Firewall for application-level protection

Azure Firewall provides an AVD **fully qualified domain name (FQDN)** tag to simplify the deployment and configuration of allowed URLs.

The following four steps will guide you through allowing outbound AVD traffic using Azure Firewall:

1. Deploy Azure Firewall and configure your AVD host pool subnet **user-defined route (UDR)** to route all traffic via Azure Firewall.
2. Create an application rule collection and add a rule to enable the AVD FQDN tag.
3. As the set of required storage and service bus accounts for your AVD host pool is unique to your deployment, this information will not be captured in the AVD FQDN tag by default. You can resolve this by allowing HTTP access from your host pool subnet to the specific service using wildcard FQDNs – for example, `*app.blob.core.windows.net`.
4. Create a network rule collection and allow traffic from your Entra ID DS private IP address to `*` for TCP and UDP port `53` and allow traffic from your AVD VMs to Windows Activation Service TCP port `1688`.

> **Tip**
> In *Step 3*, you can also use a log analytics query to list the exact required FQDNs and then allow them explicitly in your firewall application rule. This provides more granular control to access.
>
> These wildcard FQDNs enable the required access but are less restrictive.

Host pool outbound access to the internet

There may be a need to enable secure outbound internet access for users. In scenarios where the list of allowed destinations is defined, you can use the Azure Firewall application and network rules to control the required access granularly. You may also want to include whitelists for required destinations, such as web services that the organization's users may use.

> **Important note**
>
> You can configure AVD host pools using an explicit proxy configuration using an existing on-premises secure web gateway. However, this may impact performance. You may want to consider using Azure **Network Virtual Appliances** (**NVA**) to offer content filtering/proxy-based filtering services. You can also use Microsoft Defender for endpoints and Azure Firewall using the web categories feature.

NVAs

To benefit from standardization and comply with any organizational technology policies, you can deploy an NVA to Azure Virtual Network. These NVAs can be firewalls that are deployed as VMs on Azure. Routing can be configured so that the NVA controls inbound and outbound traffic flow, as well as publishing network services through the use of public IP addresses.

NVAs can be used with AVD, which can add benefits such as advanced network security to your VNet and content filtering and threat detection capabilities.

In this section, we looked at how to implement network security for AVD, NSGs, reverse proxy, Azure Firewall, and NVAs. We will now look at how to manage session hosts using Azure Bastion.

Managing AVD session hosts using Azure Bastion

In this section, we will look at Azure Bastion as an additional layer of security for accessing VMs securely through an HTML5 browser within the Azure portal.

What is Azure Bastion?

Azure Bastion is a platform-managed service that enables admins to connect to VMs in Azure using their web browser. The service makes it easy to connect securely to your VMs directly from the Azure portal over a **transport layer security** (**TLS**) connection. Azure Bastion also removes the need for public IPs or remote desktop services ports to open on your NSGs for the internet.

When using Azure Bastion, you use the Azure portal to connect the VM, which is essentially an HTML5 TLS connection. The Bastion deployment then connects to the resources securely inside the VNet using RDP or the **Secure Shell** (**SSH**) protocol, depending on the VM remote protocol requirement.

> **Important note**
>
> Azure Bastion uses port 443; ensure that you configure the NSG for this port when using Bastion deployments.

Azure Bastion is deployed per VNet, which means that if you use multiple VNets for AVD, you will need multiple Bastion deployments.

The benefit of Bastion is that standard RDP/SSH ports are not exposed over the internet, providing a secure connection to Azure resources. This also reduces the threat surface from an attacker. The following figure shows the connection flow:

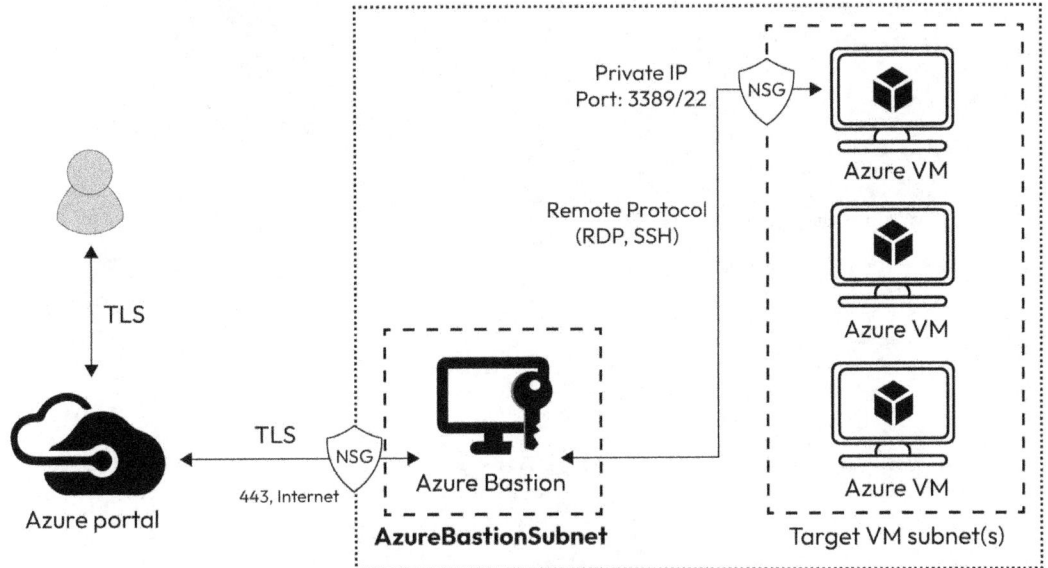

Figure 4.15 – The communication that's followed for Azure Bastion

Setting up Azure Bastion

This section provides a high-level deployment guide for setting up your first Bastion object within the Azure portal.

> **Important note**
> You will need to deploy a Azure Bastion for each VNet you wish to connect to.

The steps are as follows:

1. From the Azure portal's **Home** page, select **+ Create a resource**.
2. On the **New** page, in the **Search** box, type `Bastion`, and then click **Enter** to get to the search results. On the result for `Bastion`, verify that the publisher is Microsoft.
3. Click **Create**.

4. On the **Create a Bastion** page, configure a new bastion resource, as shown in the following screenshot:

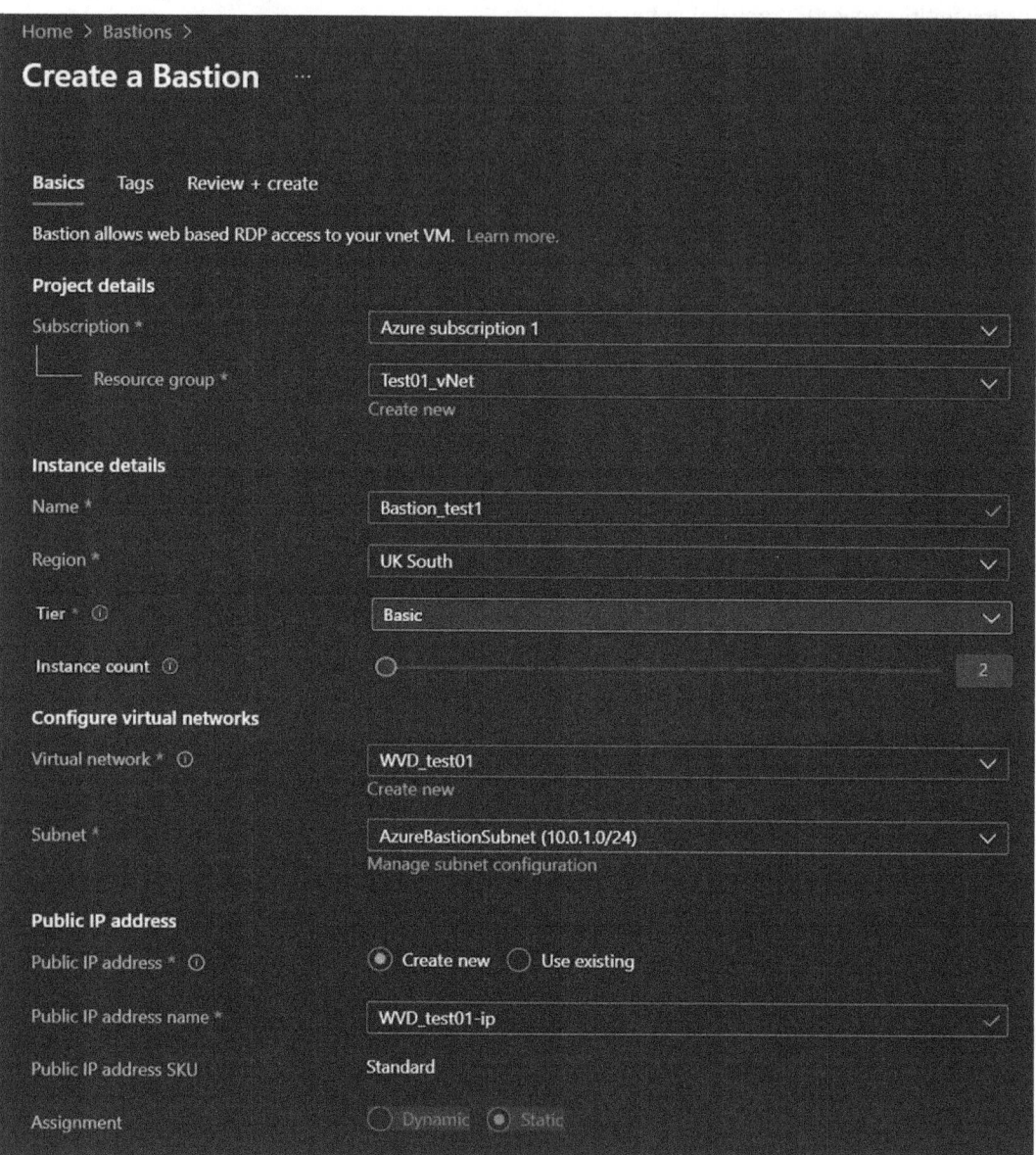

Figure 4.16 – The Create a Bastion page

You will be presented with the **Create a Bastion** page. Ensure that you fill out the following details:

- Select the required Azure subscription.
- Choose the resource group in which you will deploy the Bastion resource.
- Give the Bastion resource a name.
- Select the required Azure region.
- Select the VNet where the Bastion will be created. You can create a new VNet or use an existing one.
- Ensure the Bastion host is deployed to its own dedicated subnet.
- Use an existing public IP address or a new one.
- Note that you cannot change the public IP address SKU.
- **Assignment** should be the default setting of **Static**.

5. Once you have finished specifying the required settings, click **Review + create**:

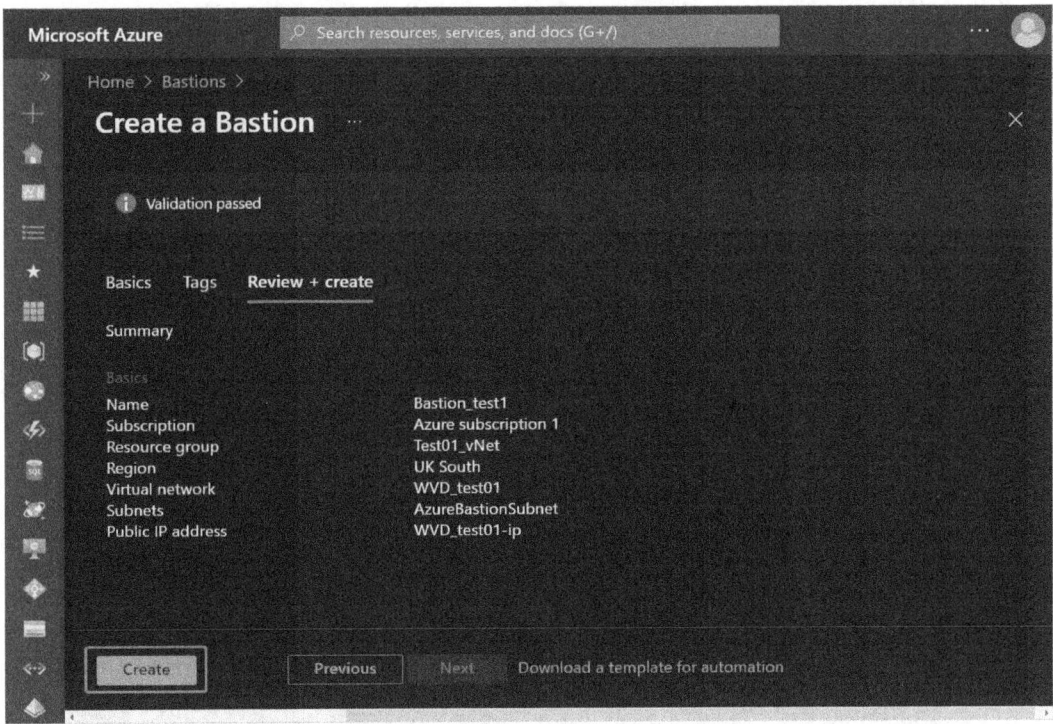

Figure 4.17 – The Review + create page

6. Review the settings and then click **Next** at the bottom, as shown in the preceding screenshot.

Connecting to a VM using Azure Bastion

The steps for this are as follows:

1. Navigate to the VM you wish to connect to using Bastion. Click **Connect**, then **Connect via Bastion**:

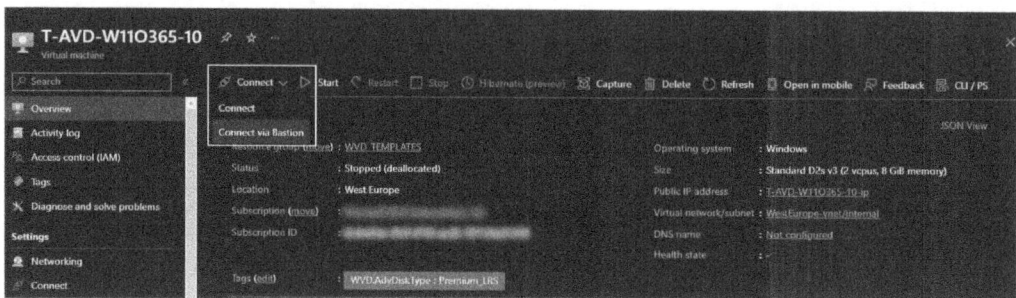

Figure 4.18 – Connecting to a VM using Bastion

2. You should see the **Connect using Azure Bastion** page. Enter the username and password for the VM and then click **Connect**:

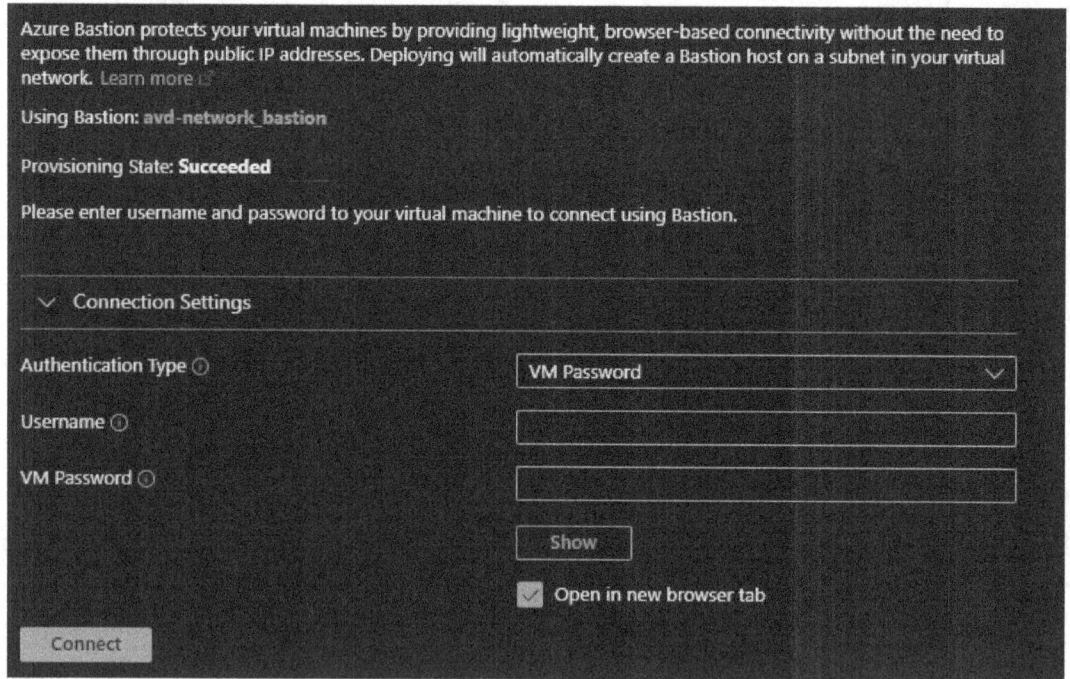

Figure 4.19 – Connecting using the Connect using Azure Bastion page

3. The administrative RDP connection to the VM will be completed using Bastion, and then the session will be streamed over HTML5 using port 443. The following screenshot shows a session using the Bastion service:

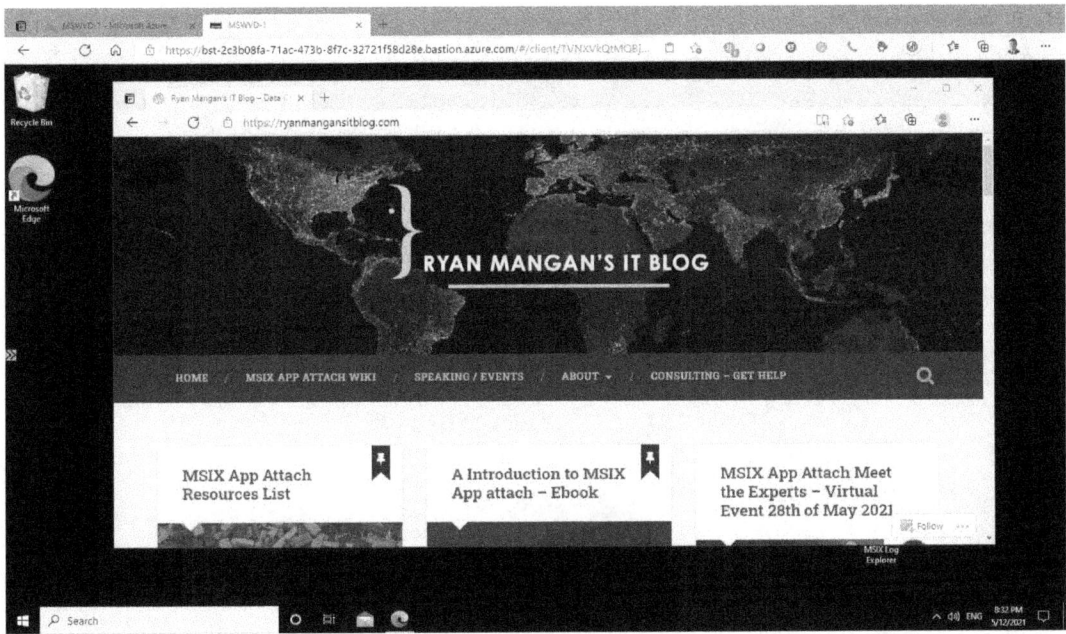

Figure 4.20 – A session using Bastion

In this section, we looked at what Azure Bastion is and how to configure/deploy Azure Bastion for AVD. In the next and final section, we will provide an overview of monitoring and troubleshooting network connectivity.

Monitoring and troubleshooting network connectivity

This section briefly covers some of the tools you can use to monitor and troubleshoot network connectivity issues for AVD.

Confirming all required URLs aren't blocked

To confirm that you have set up your Azure VNet correctly and allowed the required URLs, run the built-in tool called `WVDAgentURLTool.exe`. You can find this command-line tool in the RD agent folder on the session host.

> **Tool requirements**
>
> The RD Agent needs to be version 1.0.2944.400 or higher. You can find the latest version at https://docs.microsoft.com/azure/virtual-desktop/safe-url-list.

You can find your Agent version from the host pool within the Azure portal, as shown in the following screenshot:

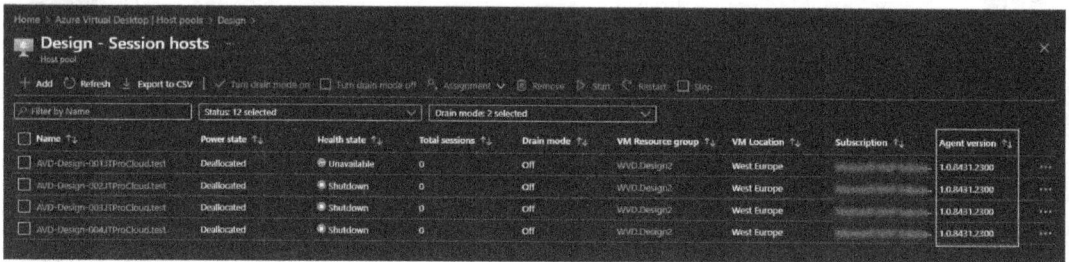

Figure 4.21 – AVD's Session hosts page

Using the URL check tool

Before starting, please check your Agent version as the folder may vary depending on the AVD Agent version.

To use the tool, perform the following steps:

1. On the session host, open a command prompt as an administrator.
2. Run the following command to change directories to the same folder as the AVD build agent:

   ```
   cd "C:\Program Files\Microsoft RDInfra\RDAgent_1.0.7755.1800"
   ```

3. Then, run the tool:

   ```
   WVDAgentUrlTool.exe
   ```

4. Once you run the command-line app, you'll see a list of accessible and inaccessible URLs for AVD, as shown in the following screenshot:

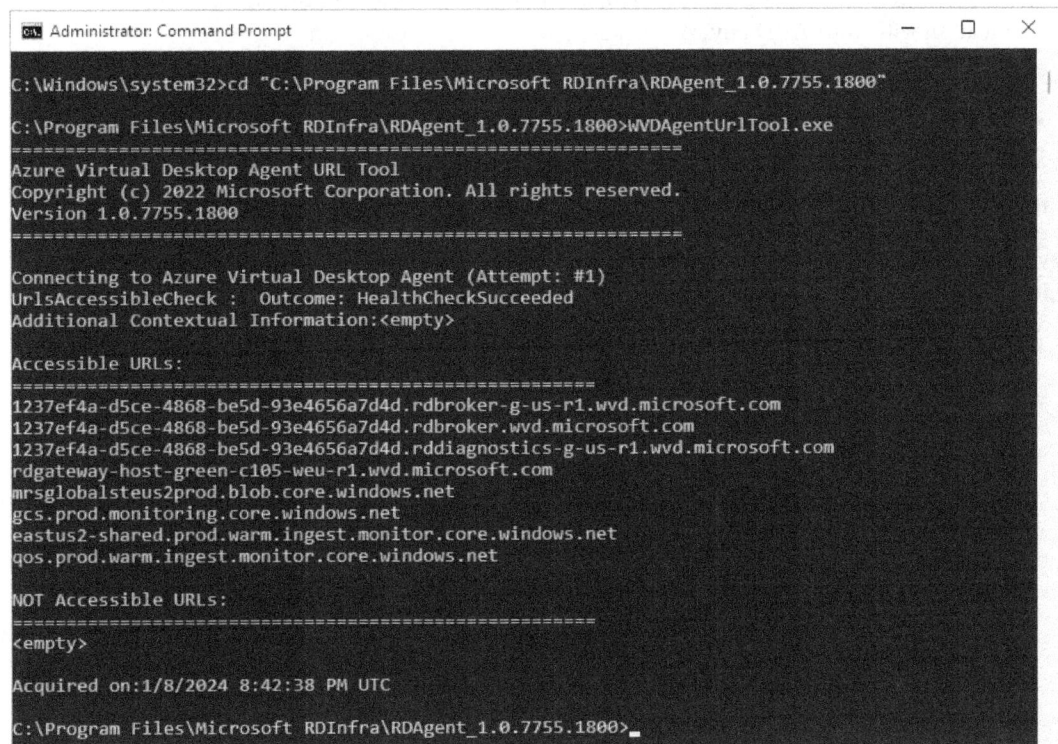

Figure 4.22 – List of accessible and inaccessible URLs for AVD

As shown in the preceding screenshot, the tool has listed all accessible URLs, meaning the network configuration is correct.

In this section, we looked at how to test to ensure that the URLs for AVD are accessible using the **WVDAgentURL** tool. Based on the results, this will confirm whether everything is working correctly or identify missing URLs that you need to add, check, and configure.

Summary

Networking is essential for AVD functionality. Proper design and configuration ensure effective communication with your AVD environment.

In this chapter, we looked at Azure VNet connectivity for AVD. We covered implementing Azure VNet connectivity, managing connectivity between on-premises and the internet, implementing network security, Azure Bastion, and troubleshooting network connectivity within Azure.

In the next chapter, we will look at implementing and managing storage for AVD.

Questions

Answer the following questions to test your knowledge of this chapter:

1. What tool can you use to check the AVD URLs?

    WVDAgentURLTool.exe

2. Is Azure Bastion configured per subnet or VNet?

    VNet

3. What can you use to secure the networking of AVD?

    Azure Firewall

4. What kind of configuration allows a low-latency UDP connection to access a session host?

    RDP Shortpath

5
Implementing and Managing Storage for Azure Virtual Desktop

In this chapter, we'll learn how to implement and manage storage for **Azure Virtual Desktop** (**AVD**). Storage is the base technology in **Infrastructure as a Service** (**IaaS**). It is used to store documents, profiles, and virtual computer disks. Size is not the only important factor regarding storage; different types of storage also significantly impact performance and costs.

We will create a storage account and configure Azure Files specifically for FSLogix profile containers.

The following topics will be covered in this chapter:

- Configuring storage for FSLogix components
- Configuring storage accounts
- Creating file shares
- Configuring disks

Configuring storage for FSLogix components

This chapter looks at storage options that are available for FSLogix profile containers when preparing and configuring AVD. We will focus on Azure Files as the storage option of choice as this is the most commonly used storage option for AVD.

FSLogix profile container storage options

There are three common storage options for AVD. This section provides a comparison of these options.

> **Important note**
> Microsoft recommends storing FSLogix profile containers in Azure Files unless there is a specific requirement not to. However, this may not meet all organizations' requirements.

FSLogix, acquired by Microsoft, provides AVD with roaming profiles by dynamically attaching a **virtual hard disk** (**VHD**) at sign-in. The user profile stored on the virtual disk becomes immediately available and appears in the system like a typical profile.

> **Important note**
> You can use the FSLogix profile solution outside of AVD.

The following table provides a comparison of the different storage options and features:

Features	Azure Files	Azure NetApp Files	Storage Spaces Direct
Use case	General-purpose	Ultra performance or migration from NetApp on-premises	Cross-platform
Platform service	Yes; Azure-native solution	Yes; Azure-native solution	No, self-managed
Regional availability	All regions	Select regions	All regions
Redundancy	Locally redundant/ zone-redundant/ geo-redundant/ geo-zone-redundant	Locally redundant/ cross-region replication	Locally redundant/ zone-redundant/ geo-redundant

Features	Azure Files	Azure NetApp Files	Storage Spaces Direct
Tiers and performance	Standard (transaction optimized) Premium Up to a maximum of 100K IOPS per share with 10 Gbps per share at about 3 ms latency	Standard Premium Ultra Up to 4.5 Gbps per volume at about 1 ms latency. For IOPS and performance details, see the Azure NetApp Files performance considerations and the FAQ.	Standard **hard disk drive** (**HDD**): Up to 500 IOPS per-disk limits Standard **solid-state drive** (**SSD**): Up to 4K IOPS per-disk limits Premium SSD: Up to 20K IOPS per-disk limits We recommend Premium disks for Storage Spaces Direct
Max capacity	100 TiB per share; up to 5 PiB per general-purpose account	100 TiB per volume; up to 12.5 PiB per subscription	Maximum 32 TiB per disk
Required infrastructure	Minimum share size 100 GiB for Premium	Minimum capacity pool 4 TiB; minimum volume size 100 GiB	Two **virtual machines** (**VMs**) on Azure IaaS (plus Cloud Witness) or at least three VMs without and costs for disks
Protocols	SMB 3.1.1/3.0/2.1, NFSv4.1 (only Premium), REST	NFSv3, NFSv4.1 (preview), SMB 3.x/2.x	NFSv3, NFSv4.1, SMB 3.1

Table 5.1 – Storage options and features

This table was taken from the following site: `https://docs.microsoft.com/azure/virtual-desktop/store-fslogix-profile?WT.mc_id=modinfra-17152-thmaure#azure-platform-details` and `https://learn.microsoft.com/azure/storage/files/storage-files-netapp-comparison`.

As shown in the preceding table, Azure Files is the likely candidate for AVD deployments, while Azure NetApp Files offers high performance. There is also an Azure VM option for using Storage Spaces Direct.

The following table details the features available for Azure Files, Azure NetApp Files, and Storage Spaces Direct:

Features	Azure Files	Azure NetApp Files	Storage Spaces Direct
Access	Cloud, on-premises, and hybrid (Azure File Sync)	Cloud, on-premises	Cloud, on-premises
Backup	Azure backup snapshot integration	Azure NetApp Files snapshots Azure NetApp Files backup	Azure backup snapshot integration
Security and compliance	All Azure-supported certificates	All Azure-supported certificates	All Azure-supported certificates
Entra ID integration	Native Active Directory and Entra Domain Services	Entra Domain Services and Native Active Directory	Native Active Directory or Entra Domain Services support only

Table 5.2 – Storage options and additional features

This table was taken from the following site: `https://docs.microsoft.com/azure/virtual-desktop/store-fslogix-profile?WT.mc_id=modinfra-17152-thmaure#azure-management-details`.

This section looked at the three storage options that are available when you're planning to configure FSLogix profile containers. In the next section, we will look at the two different Azure Files tiers.

The different Azure Files tiers

Azure Files has two different tier types of file storage: standard and premium. The key difference between the two is performance, as premium uses SSDs and these are deployed in the file storage account type. Premium file share types are helpful in larger organizations where higher performance and low latency are required due to the number of users accessing the file share storage.

Standard file shares use HDDs (or have a software-defined reduction of performance) and are deployed as **general-purpose version 2** (**GPv2**) storage account types. Therefore, you should expect to use standard file shares in small environments or organizations with low I/O needs.

> **Important note**
>
> Standard file shares are only available in pay-as-you-go billing models. This means that billing is based on the total storage used, whereas when you're using premium file share storage, you pay for the configured capacity.

The following table provides examples of when you should use standard file shares versus premium file shares:

Deployment Type	Recommended Storage Tier
Fewer than 200 users	Standard file shares
Greater than 200 users	Premium file shares or multiple standard storage accounts
Medium	Premium file shares
Heavy	Premium file shares
Power	Premium file shares

Table 5.3 – Recommended storage tiers

This section explored the two different Azure File storage tiers and when to use each type. The next section looks at Azure Files integration.

Best practices for Azure Files with AVD

The following are some best practices associated with Azure Files when you're configuring it for use with AVD:

- It is advised that you create your storage accounts in the same region as the session host VMs. This is to ensure that latency is kept to a minimum. This also applies to optimal performance when you're using FSLogix profile containers.
- It is recommended that you use Active Directory or Microsoft Entra Kerberos (with hybrid identities) integrated file shares for security and that the following permissions should be set:

User Account	Folder	Permissions
Users	This folder only	Modify
Creator/Owner	Subfolders and files only	Modify
Administrator (optional)	This folder, subfolders, and files	Full control

Table 5.4 – FSLogix New Technology File System (NTFS) permissions

- To allow users or user groups general access to the share (share permission), add users or groups with the **Storage File Data SMB Share Contributor** role in access control (**identity and access management**; **IAM**) on the share in the Azure portal.

- When you're storing images in Azure Files, it is advised that you store the master image in the same region as where the VMs are being provisioned.

This section looked at the different storage options available to you, including Azure Files, Azure NetApp Files, and Storage Spaces Direct. We also looked at the different storage tiers for Azure Files, **Microsoft Entra ID Domain Services (Entra ID DS)** integration, and storage best practices when configuring FSLogix profile containers.

Now, let's learn how to configure a storage account.

Configuring storage accounts

This section will look at creating a storage account and configuring data protection.

To create a storage account, you need to follow the stepwise procedure detailed in the following subsections.

Step 1 – Creating a new storage account

From the left menu within the Azure portal, select **Storage accounts** to display a list of your storage accounts. You can also search for `storage accounts` in the top search bar. This is shown in the following screenshot:

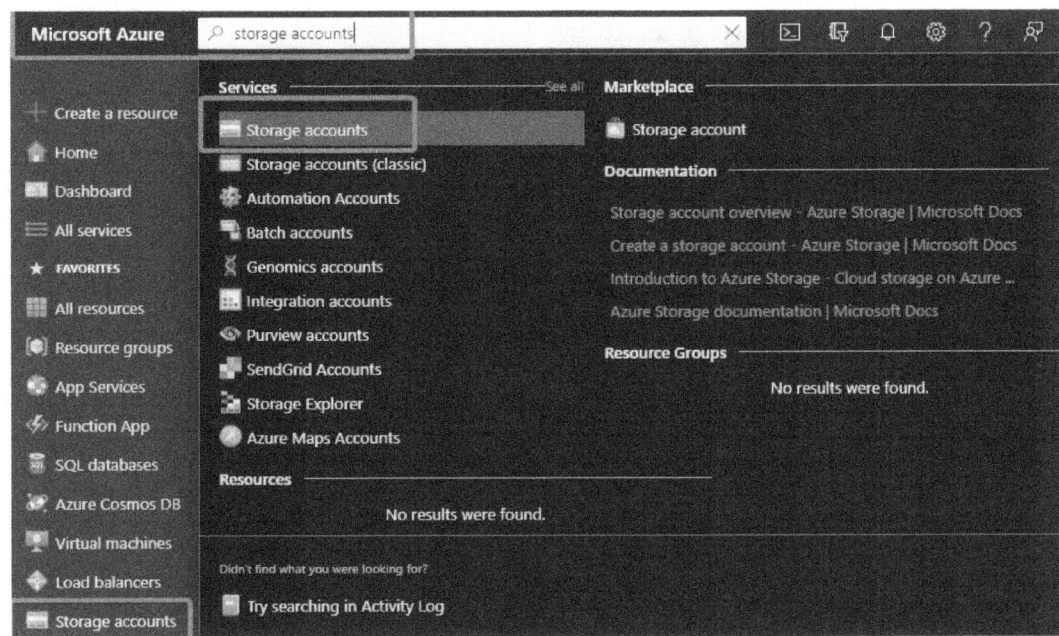

Figure 5.1 – Using the search bar to show the Storage accounts service in the Azure portal

Once on the **Storage accounts** page, you will see all storage accounts and an icon to create one in the page's navigation bar. To create a new storage account, click **Create**:

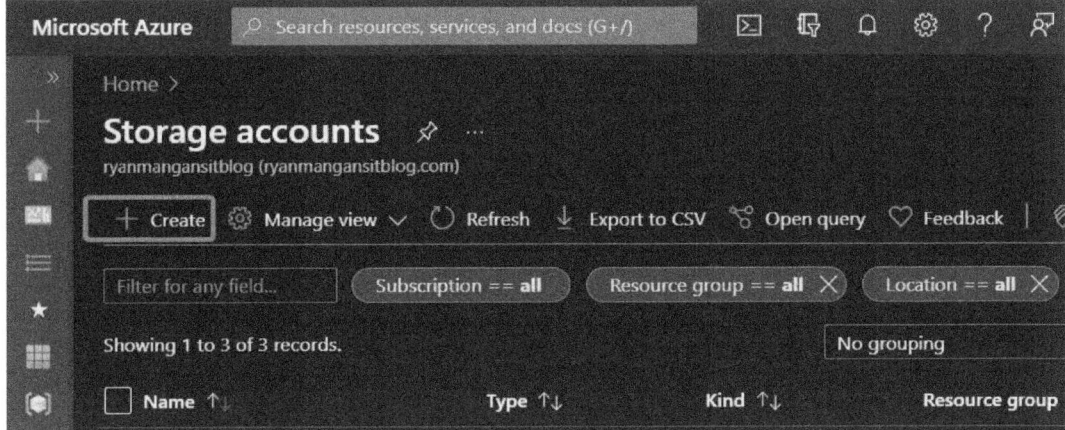

Figure 5.2 – Storage accounts

Once you have clicked **Create**, we can move on to the next section.

Step 2 – Configuring the basics

Once you have selected **Storage accounts** and clicked **Create**, you will see a basic **Create a storage account** page:

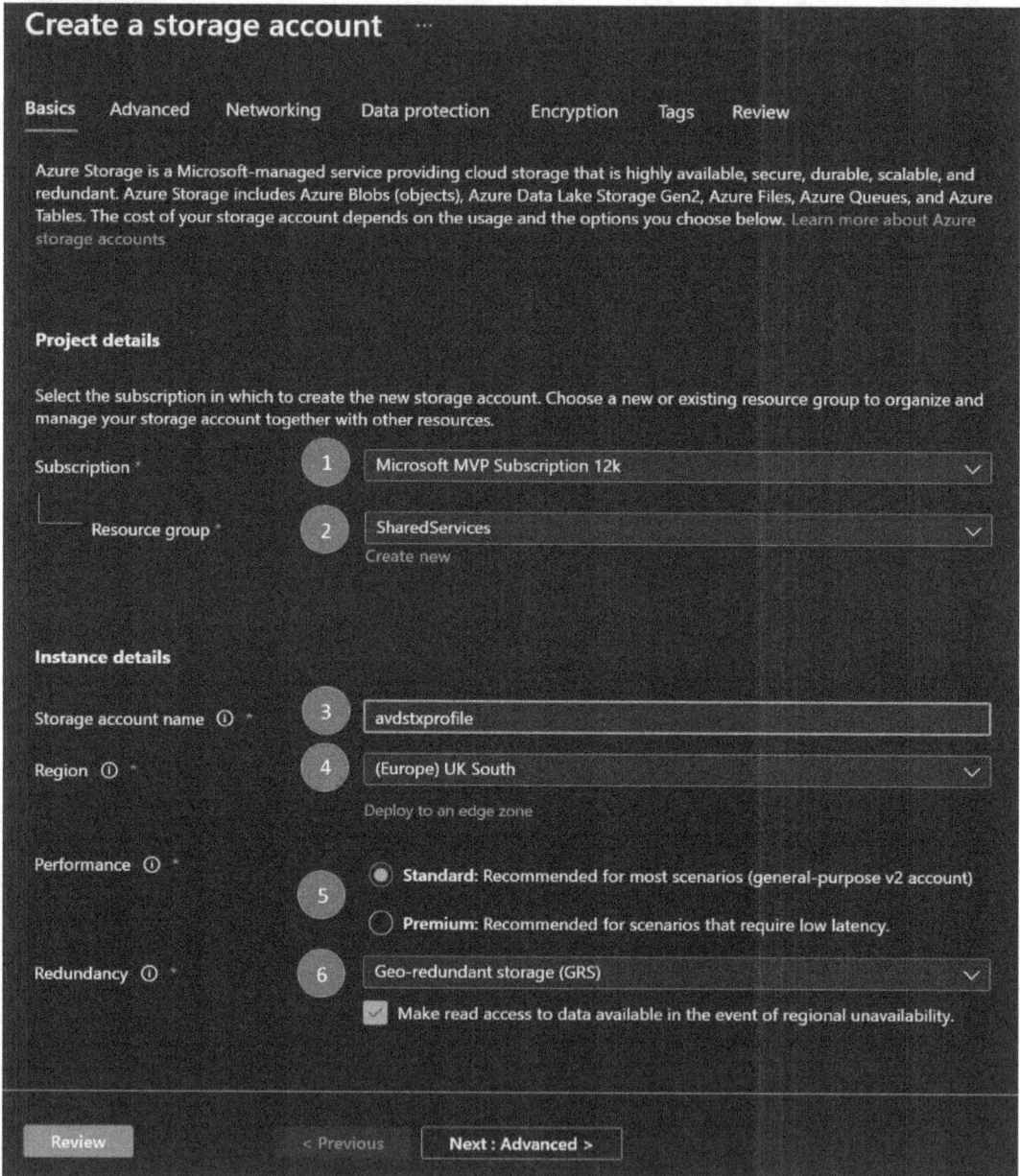

Figure 5.3 – Create a storage account

The following table details the steps shown in the preceding screenshot. You are required to complete these steps before progressing to the **Advanced** tab:

Step number	Name	Required or Optional	Description
1	**Subscription**	Required	Select the subscription that's required for the storage account.
2	**Resource group**	Required	Select an existing resource group for this storage account, or create a new one.
3	**Storage account name**	Required	Choose a name for your storage account; this must be unique. The **Storage account name** value must be between 3 and 24 characters in length and contain numbers and lowercase letters only.
4	**Region**	Required	Select a region for your storage account.
5	**Performance**	Required	Select **Standard** performance for GPv2 storage accounts; this is the default. Microsoft recommends this type of account for most scenarios. Use **Premium** storage for low latency.
6	**Redundancy**	Required	Select the required redundancy configuration. Remember – not all redundancy options are available in all regions. By selecting a geo-redundant configuration (GRS or GZRS), your data is replicated to a data center in a different region. For read access to data in the secondary region, ensure you select **Make read access to data available in the event of regional unavailability**.

Table 5.5 – Storage account basic properties

Once you have configured the **Basics** section of creating a new storage account, we can look at configuring advanced settings.

> **Important note**
> Not all regions are supported for all types of storage accounts or redundancy configurations. The choice of region can also have a billing impact.

Step 3 – Configuring advanced settings

Once you're in the **Advanced** tab, you will see several security and storage configuration options. You can leave these as is or customize them as required:

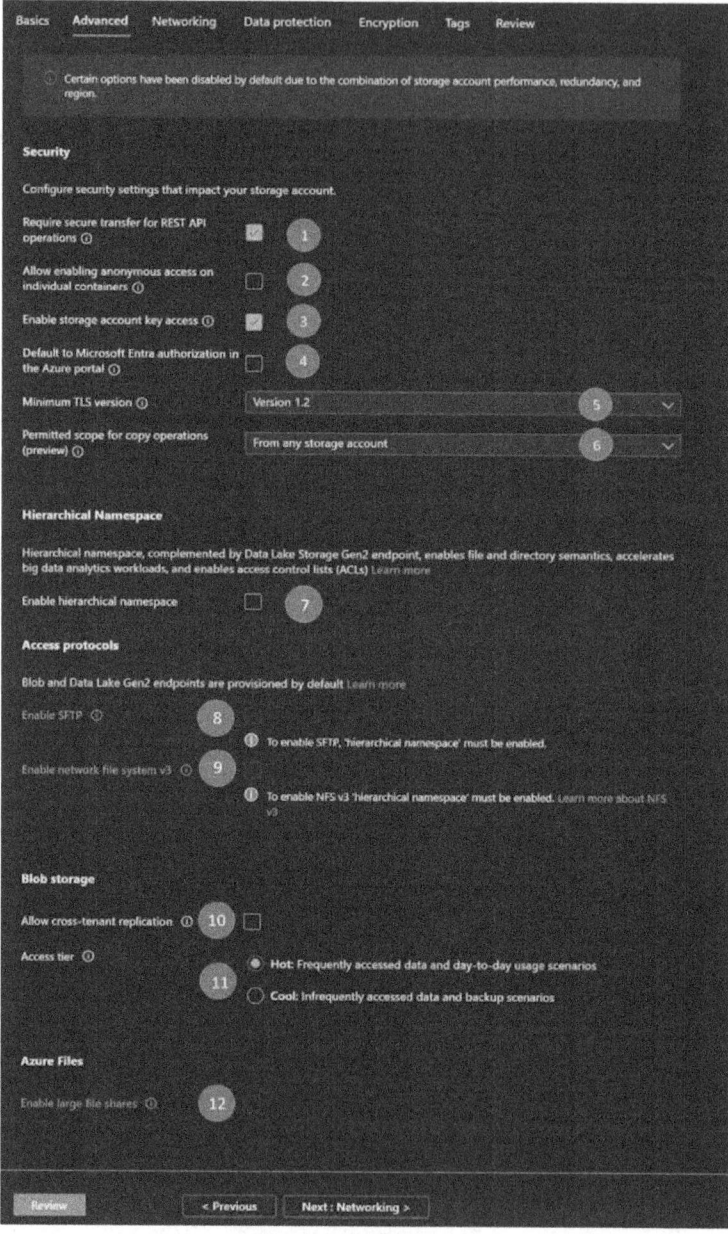

Figure 5.4 – Advanced tab: Create a storage account

The following table details the 12 configuration options. These configuration settings are cross-referenced in the preceding screenshot:

Step number	Name	Required or Optional	Description
1	Require secure transfer for REST API operations	Required	Enabling secure transfer requires that incoming requests to this storage account be made only via HTTPS (default). This is recommended for optimal security. It's also required for encrypted **Server Message Block (SMB)** file service operations.
2	Allow enabling anonymous access on individual containers	Optional	Users with the appropriate permissions can enable anonymous public access to a container in the storage account when enabling this setting. If you disable this setting, it prevents all anonymous public access to the storage account, making it private.
3	Enable storage account key access	Optional	When enabled, this setting allows clients to authorize requests to the storage account using either the account access keys or an Entra ID account. Disabling this setting prevents authorization with the account access keys.
4	Default to Microsoft Entra authorization in the Azure portal	Optional	The default authentication is set to Entra ID if you create a new container, share, or table. This can be overwritten by the administrator for each container, share, or table.
5	Minimum TLS version	Required	Select the minimum version of **Transport Layer Security** (**TLS**) for incoming requests to the storage account. The default value is TLS version 1.2. When set to the default value, incoming requests made using TLS 1.0 or TLS 1.1 are rejected.
6	Permitted scope for copy operations (preview)	Optional	Configures the scope of copy operations from/to other storage accounts.
7	Enable hierarchical namespace	Optional	You need to configure a hierarchical namespace to use this storage account for **Azure Data Lake Storage Gen2** (**ADLS Gen2**) workloads.

Step number	Name	Required or Optional	Description
8	**Enable SFTP**	Optional	Blob and Data Lake Gen2 endpoints only: Enables Secure FTP (FTP using SSH) to encrypt FTP communication over the network.
9	**Enable network file system v3**	Optional	Blob and Data Lake Gen2 endpoints only: **Network File System** (**NFS**) v3 provides Linux filesystem compatibility at object storage scale and enables Linux clients to mount a container in Blob storage from an Azure VM or a computer on-premises.
10	**Access tier**	Required	Blob access tiers enable you to store blob data cost-effectively, based on usage.
11	**Allow cross-tenant replication**	Optional	Allows blob replication to storage accounts in other tenants.
12	**Enable large file shares**	Optional	This is only available for premium storage accounts for file shares.

Table 5.6 – Storage account advanced properties

Once you have chosen the required advanced settings, you can start configuring the **Networking** section.

Step 4 – Configuring networking

This step is where you configure specific network connectivity requirements, including public and private endpoints. You can also specify the required routing option:

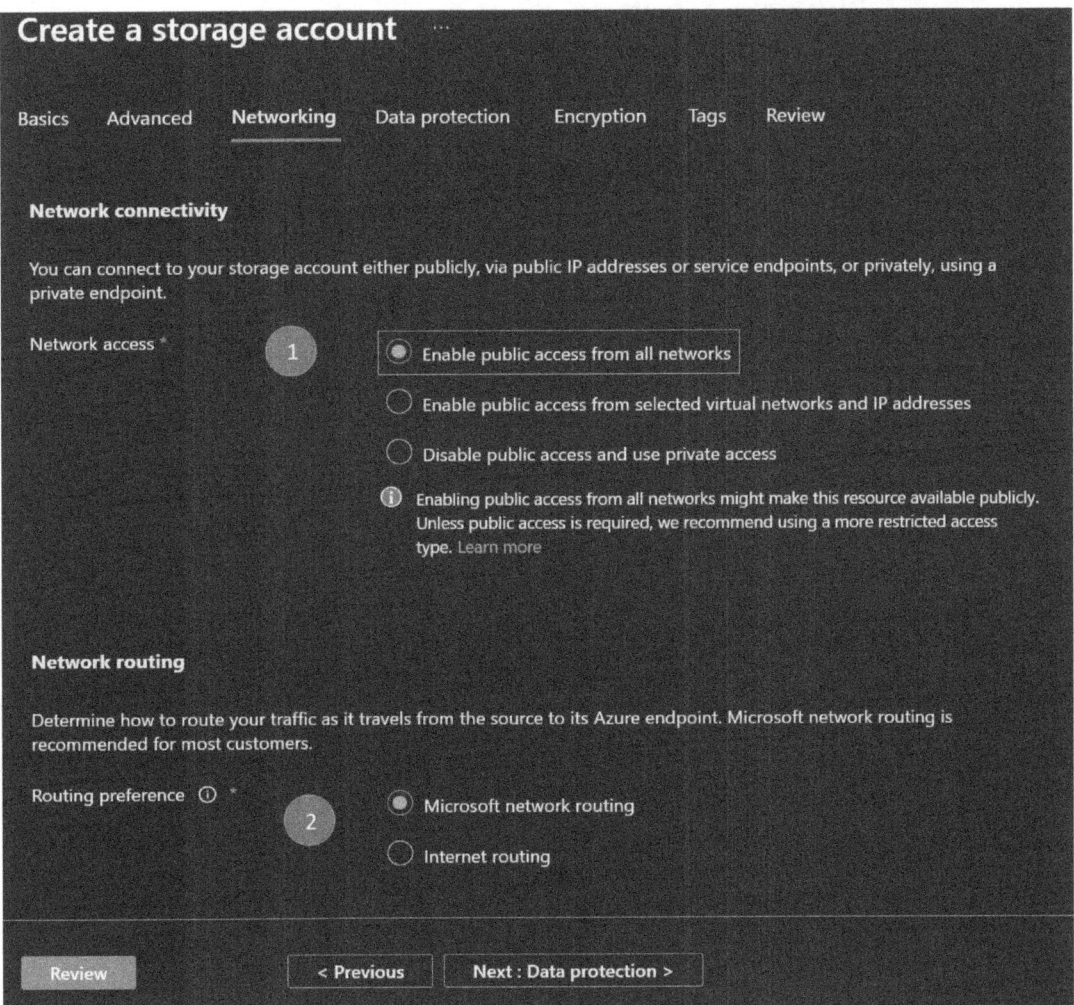

Figure 5.5 – The Networking tab

The preceding screenshot is numbered to reference the **Network access** and **Routing preference** areas shown in the following table:

Step number	Name	Required or Optional	Description
1	**Network access**	Required	Incoming network traffic is routed to the public endpoint for your storage account by default. You can specify that traffic must be routed to the public endpoint through an Azure virtual network. You can also configure private endpoints for private network communication to the storage account.
2	**Routing preference**	Required	This setting specifies how network traffic is routed to the public endpoint of your storage account from clients over the internet. A new storage account uses Microsoft network routing by default. However, you can also configure how network traffic is routed through the **point of presence** (**POP**) closest to the storage account, which may lower networking costs.

Table 5.7 – Storage account network properties

Now that you have configured the **Networking** section of **Create a storage account**, we can move on to *step 5*, where we will configure the data protection settings for the storage account.

Step 5 – Configuring data protection

Within this tab, you can configure various recovery and tracking options for your storage account. The following screenshot, whose numbers are referenced in the following table, shows several options that are available to you:

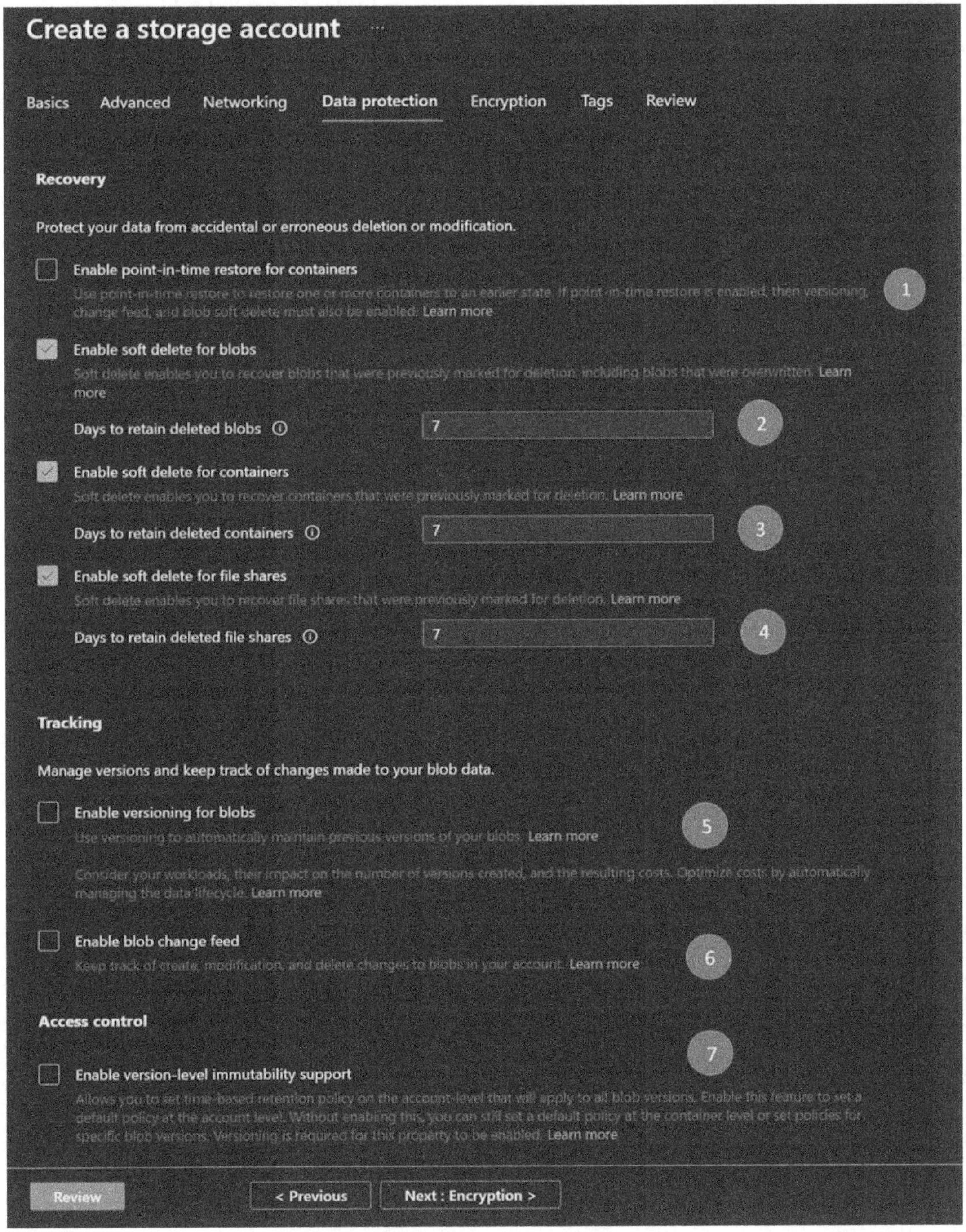

Configuring storage accounts

Create a storage account

Basics   Advanced   Networking   **Data protection**   Encryption   Tags   Review

Recovery

Protect your data from accidental or erroneous deletion or modification.

☐ **Enable point-in-time restore for containers** (1)
Use point-in-time restore to restore one or more containers to an earlier state. If point-in-time restore is enabled, then versioning, change feed, and blob soft delete must also be enabled. Learn more

☑ **Enable soft delete for blobs**
Soft delete enables you to recover blobs that were previously marked for deletion, including blobs that were overwritten. Learn more

Days to retain deleted blobs ⓘ   [7]   (2)

☑ **Enable soft delete for containers**
Soft delete enables you to recover containers that were previously marked for deletion. Learn more

Days to retain deleted containers ⓘ   [7]   (3)

☑ **Enable soft delete for file shares**
Soft delete enables you to recover file shares that were previously marked for deletion. Learn more

Days to retain deleted file shares ⓘ   [7]   (4)

Tracking

Manage versions and keep track of changes made to your blob data.

☐ **Enable versioning for blobs**   (5)
Use versioning to automatically maintain previous versions of your blobs. Learn more

Consider your workloads, their impact on the number of versions created, and the resulting costs. Optimize costs by automatically managing the data lifecycle. Learn more

☐ **Enable blob change feed**   (6)
Keep track of create, modification, and delete changes to blobs in your account. Learn more

Access control

☐ **Enable version-level immutability support**   (7)
Allows you to set time-based retention policy on the account-level that will apply to all blob versions. Enable this feature to set a default policy at the account level. Without enabling this, you can still set a default policy at the container level or set policies for specific blob versions. Versioning is required for this property to be enabled. Learn more

[Review]   [< Previous]   [Next : Encryption >]

Figure 5.6 – The Data protection tab

The preceding screenshot is annotated with numbers *1* to *7*; this correlates with the following table, which shows the options for configuring data protection for the new storage account:

Step number	Name	Required or Optional	Description
1	**Enable point-in-time restore for containers**	Optional	Point-in-time restore protects against accidental deletion or corruption by enabling you to restore block blob data to an earlier state. Enabling point-in-time restore enables blob versioning, blob soft delete, and blob change feed. These prerequisite features have a cost impact, so be sure to check their pricing first.
2	**Enable soft delete for blobs**	Optional	Blob soft delete protects an individual blob, snapshot, or version from accidental deletes or overwrites by maintaining the deleted data in the system for a specified retention period. *During the retention period, you can restore a soft-deleted object to its state when it was deleted.*
3	**Enable soft delete for containers**	Optional	Container soft delete protects a container and its contents from accidental deletes by maintaining the deleted data in the system for a specified retention period. *During the retention period, you can restore a soft-deleted container to its state at the time it was deleted.*

Step number	Name	Required or Optional	Description
4	**Enable soft delete for file shares**	Optional	Soft delete for file shares protects a file share and its contents from accidental deletes by maintaining the deleted data in the system for a specified retention period. *During the retention period, you can restore a soft-deleted file share to its state at the time it was deleted.*
5	**Enable versioning for blobs**	Optional	Blob versioning automatically saves the state of a blob in a previous version when the blob is overwritten. It is recommended that you enable blob versioning for optimal data protection for the storage account.
6	**Enable blob change feed**	Optional	The blob change feed provides transaction logs of all changes that have been made to all blobs in your storage account, as well as their metadata.
7	**Enable version-level immutability support**	Optional	Configures a time-based retention policy at the storage account level.

Table 5.8 – Storage account data protection properties

Step 6 – Configuring encryption

Within this tab, you can configure various encryption options for your storage account. The following screenshot, whose numbers are referenced in the following table, shows several options that are available to you:

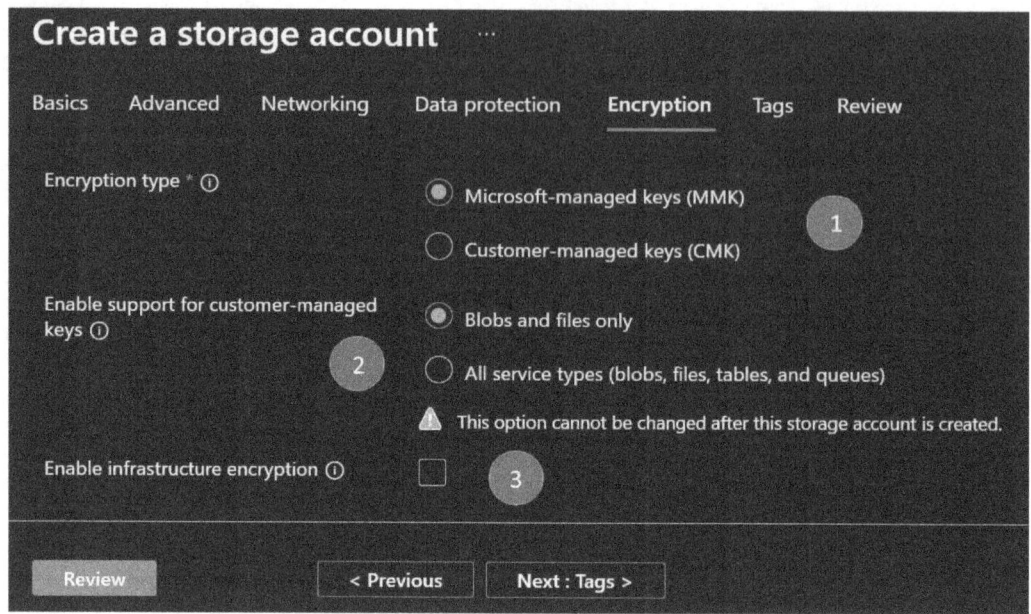

Figure 5.7 – The Encryption tab

The preceding screenshot is annotated with numbers *1* to *3*; this correlates with the following table, which shows the options for configuring data protection for the new storage account:

Step number	Name	Required or Optional	Description
1	**Encryption type**	Optional	Data is encrypted by default by Microsoft using **Microsoft-managed keys (MMK)**. Customer can choose to use their own keys (**customer-managed keys; CMK**) typically stored in a key vault.
2	**Enable support for customer-managed keys**	Optional	To enable support for CMKs for tables and queues, it is advised that you ensure this setting is enabled while creating the storage account.
3	**Enable infrastructure encryption**	Optional	Infrastructure encryption is not enabled by default. You can enable infrastructure encryption to encrypt your data at both the service level and the infrastructure level.

Table 5.9 – Storage account encryption properties

Once you have selected the required options for data protection, you can set tags or proceed to the **Review** tab.

Within the **Review** tab, check if all the settings are as you require, then proceed to create a storage account:

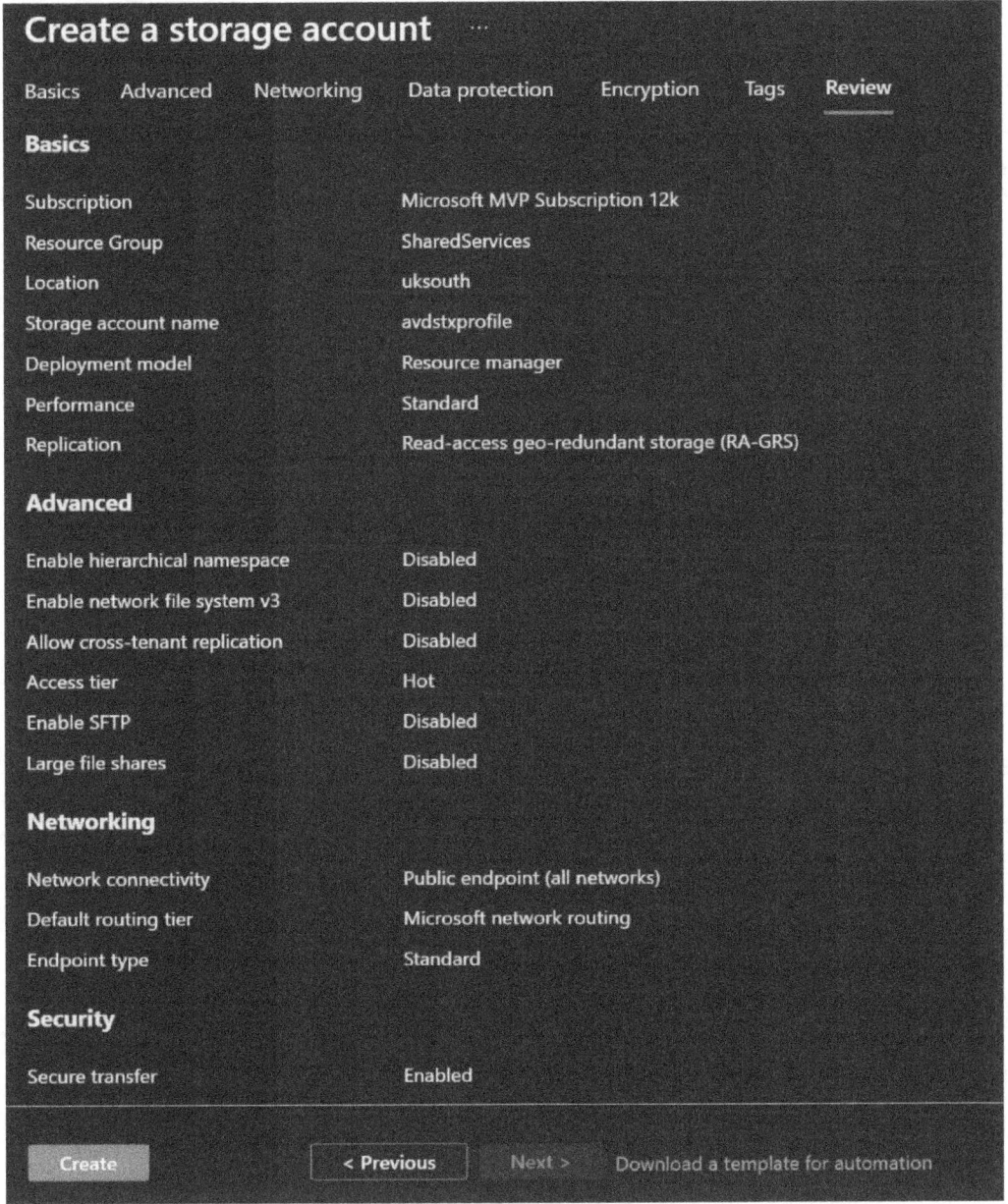

Figure 5.8 – The Review + create tab

Once the storage account has been created, you will see it appear on the **Storage accounts** page. In the next section, we will look at configuring an Azure file share.

Configuring file shares

Once you have created your storage account, you need to create a file share for FSLogix profile containers. This section will look at configuring a file share in a storage account ready for use with FSLogix.

Before we get started with Azure file shares, let's have a look at the different tiers that are available per share:

- **Premium** file shares use SSDs, which provide higher constant performance and lower latency than standard storage. This file share tier type is beneficial for larger shares or high I/O workload requirements.
- **Transaction**-optimized file shares, similar to standard storage, use HDDs. This is suitable for heavy workloads but does not provide the required latency that premium file shares offer.
- **Hot** file shares provide storage optimized for general-purpose file sharing for items such as department shares. Hot files use HDDs.
- **Cool** file shares provide cost-effective storage for archive storage requirements. This type of storage tier uses HDDs.

> **Important note**
> For larger organizations and high I/O workloads, it is recommended that you use the premium storage tier for Azure file shares.

Creating an Azure file share is quite simple. You need to make sure you have created a storage account before proceeding. Within the storage account, you need to navigate to the **File shares** icon within the table of contents for the storage account:

Configuring file shares

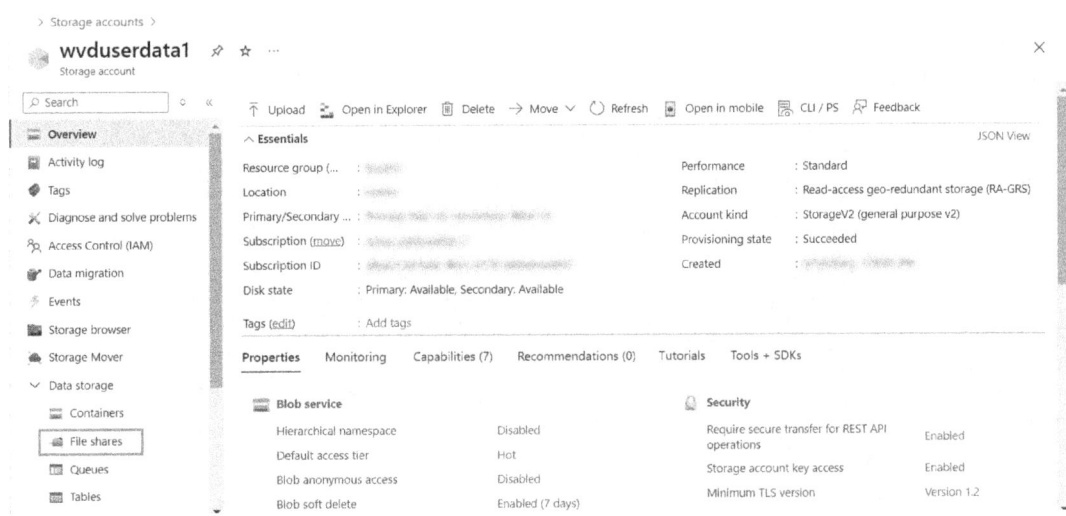

Figure 5.9 – The File shares link within the storage account

On the **File shares** page, click the **File share** button, as shown in the following screenshot:

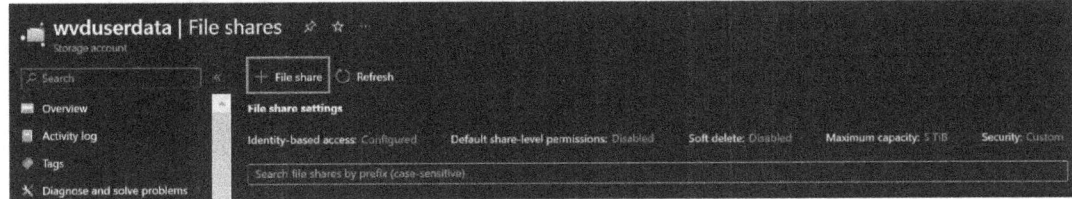

Figure 5.10 – The File share button

Once you have clicked the **File share** button, you will see a **New file share** blade appear. Fill in the following fields in this blade to create a new file share:

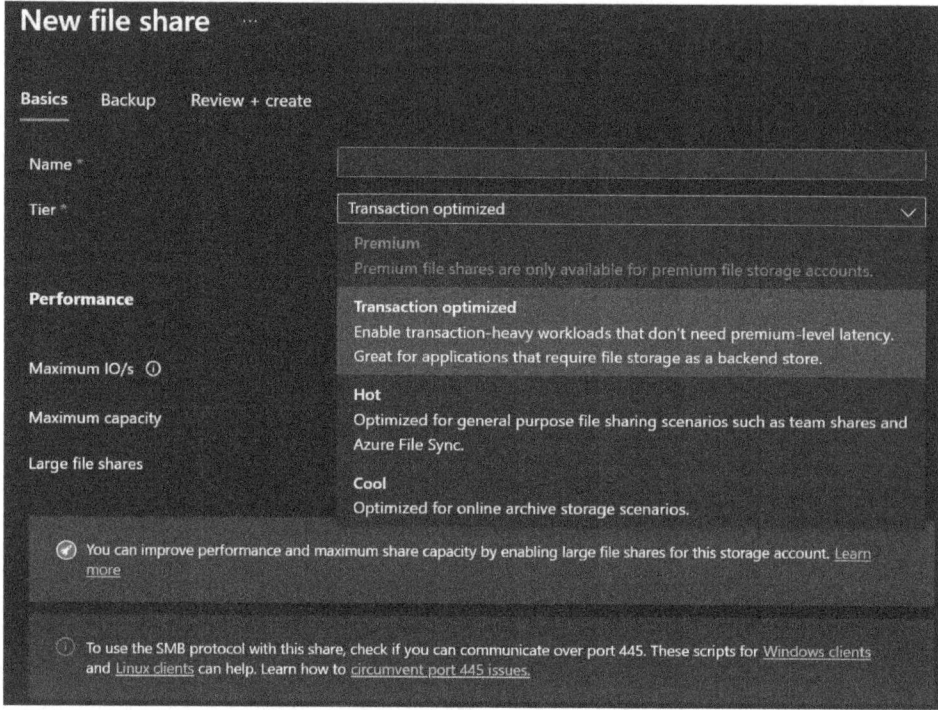

Figure 5.11 – The New file share configuration

You will need to enter a name for the share and a quota size and choose the tier you would like.

Once you have entered the required details, click **Review + create** to finish creating the new share:

Figure 5.12 – The newly created file share

Note that the experience within a storage account using *premium* storage for file storage has a slightly different UI experience, as shown in the following screenshot:

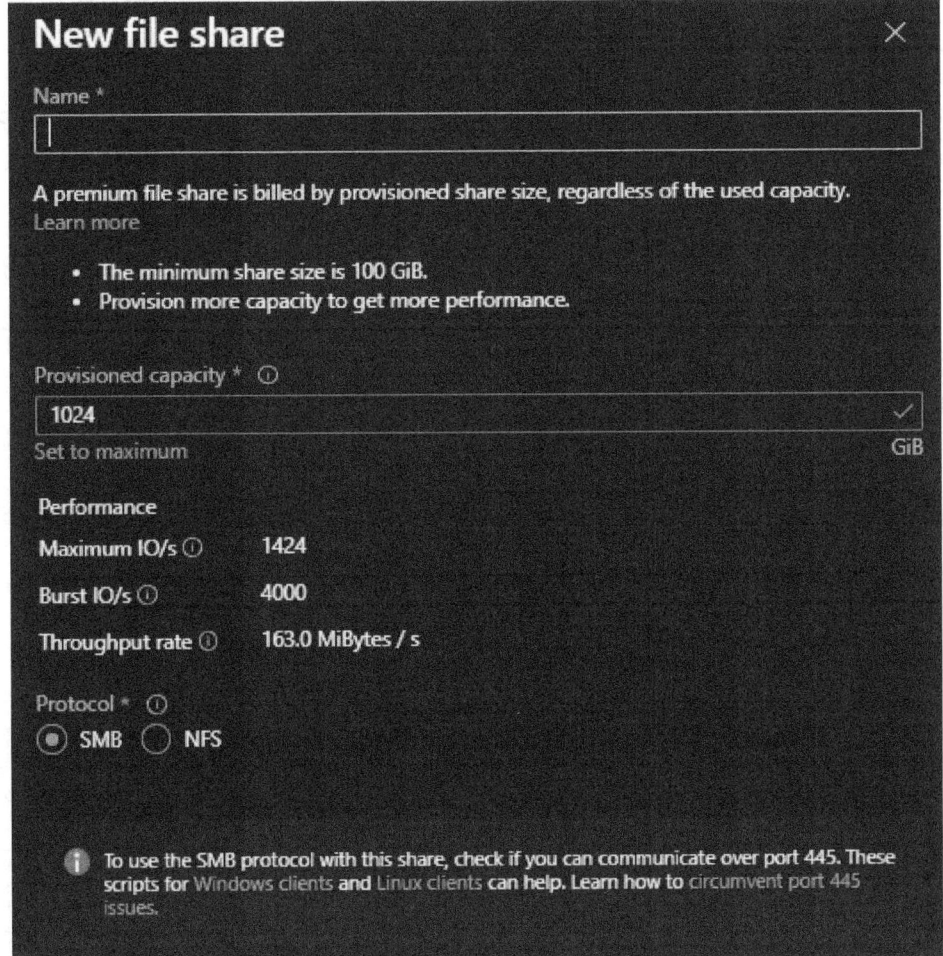

Figure 5.13 – Configuring premium file shares

This section summarized the different Azure file share storage tier options and how to create a new Azure file share. In the next section, we will look at Azure managed disks and ephemeral OS disks and learn how to prepare a custom image.

Configuring disks

This section will look at Azure managed disks, the different available options, and how to prepare a custom VHD image.

An Azure managed disk is a virtual disk (block-level storage volume) with Azure VMs. Managed disks are designed to provide an availability of 99.999%. This is achieved by providing three replica copies of your data, which provides high durability.

The following table details the different types of managed disks that are available:

Detail	Ultra Disk	Premium SSD v2	Premium SSD	Standard SSD	Standard HDD
Disk type	SSD	SSD	SSD	SSD	HDD
Scenario	I/O-intensive workloads such as SAP HANA, top-tier databases (for example, SQL, Oracle), and other transaction-heavy workloads	Production and performance-sensitive workloads that consistently require low latency and high IOPS and throughput	Production and performance-sensitive workloads	Web servers, lightly used enterprise applications, and dev/test	Backup, non-critical, and infrequent access
Max disk size	65,536 gibibytes (GiB)	65,536 GiB	32,767 GiB	32,767 GiB	32,767 GiB
Max throughput	4,000 MB/s	1,200 MB/s	900 MB/s	750 MB/s	500 MB/s
Max IOPS	160,000	80,000	20,000	6,000	2,000
Usable as OS disk?	No	No	Yes	Yes	Yes

Table 5.10 – Azure disk types and performance

This table was taken from the following site: `https://docs.microsoft.com/azure/virtual-machines/disks-types`.

As shown in the preceding table, each type of disk has a specific use case. For AVD multi-session deployments, it is recommended that you use premium SSDs to avoid any IOPS bottlenecks. Please note that not all disk types are usable for the OS disk. You can use standard SSDs for personal desktop deployments. It is not recommended to use standard HDD disks for AVD deployments as performance could be degraded:

> **Important note**
> It is recommended that premium SSDs be used for session hosts.

Premium SSD Sizes	P10	P15	P20	P30	P40	P50	P60	P70	P80
Disk size in GiB	128	256	512	1,024	2,048	4,096	8,192	16,384	32,767
Provisioned IOPS per disk	500	1,100	2,300	5,000	7,500	7,500	16,000	18,000	20,000
Provisioned throughput per disk	100 MB/s	125 MB/s	150 MB/s	200 MB/s	250 MB/s	250 MB/s	500 MB/s	750 MB/s	900 MB/s
Max burst IOPS per disk	3,500	3,500	3,500	30,000*	30,000*	30,000*	30,000*	30,000*	30,000*
Max burst throughput per disk	3,500	3,500	3,500	30,000*	30,000*	30,000*	30,000*	30,000*	30,000*
Max burst duration	30 min	30 min	30 min	Unlimited*	Unlimited*	Unlimited*	Unlimited*	Unlimited*	Unlimited*
Eligible for reservation	No	No	No	Yes, up to 1 year	Yes, up to 1 year	Yes, up to 1 year	Yes, up to 1 year	Yes, up to 1 year	Yes, up to 1 year

Table 5.11 – Azure disk sizes and performance

This table was taken from Microsoft's documentation site: `https://docs.microsoft.com/azure/virtual-machines/disks-types#premium-ssd-size`.

Typically, Azure managed disks are **locally redundant storage** (**LRS**). This means that the storage is replicated three times within a single data center in the region where you deployed the VM.

You can also configure **zone-redundant storage** (**ZRS**) for managed disks. ZRS replicates Azure managed disks synchronously across three Azure availability zones within a selected Azure region. Each zone is a separate physical location with independent networking, cooling, and power.

There is no difference in latency or performance; the only improvement when using ZRS is the improved data protection.

Ephemeral OS disks

Ephemeral operating system disks, also known as stateless disk storage, are created on the Azure hypervisor's local storage as part of the VM cache. One benefit of using ephemeral disks over Azure managed disks is that ephemeral disks are free. This allows the stateless disk storage to provide lower latency and faster reads and writes.

The following table details the differences between Azure managed disks and ephemeral disks:

	Persistent OS Disk	**Ephemeral OS Disk**
Size limit for OS disk	4* TiB	Cache size or temp size for the VM size or 2040 GiB, whichever is smaller. For the cache or temp size in GiB, see DS, ES, M, FS, and GS.
VM sizes supported	All	VM sizes that support premium storage such as DSv1, DSv2, DSv3, Esv3, Fs, FsV2, GS, M, Mdsv2, Bs, Dav4, Eav4
Disk-type support	Managed and unmanaged OS disk	Managed OS disk only
Region support	All regions	All regions
Data persistence	OS disk data written to OS disk is stored in Azure Storage	Data written to OS disk is stored on local VM storage and isn't persisted to Azure Storage
Stop-deallocated state	VMs and scale set instances can be stop-deallocated and restarted from the stop-deallocated state	Not supported
Specialized OS disk support	Yes	No
OS disk resize	Supported during VM creation and after VM is stop-deallocated	Supported during VM creation only
Resizing to a new VM size	OS disk data is preserved	Data on the OS disk is deleted; OS is reprovisioned
Redeploy	OS disk data is preserved	Data on the OS disk is deleted; OS is reprovisioned

Table 5.12 – Comparing persistent and ephemeral disks

This table was taken from the following Microsoft site: https://docs.microsoft.com/azure/virtual-machines/ephemeral-os-disks.

Note that you cannot start and stop/deallocate an Azure VM that's been configured with an ephemeral OS disk (OS cache). The only options that are available to you are to restart, reimage, and delete.

> **Important note**
> If you want to use ephemeral disks, you need to use a custom **Azure Resource Manager** (**ARM**) template, third-party tooling, or PowerShell.

In this section, we looked at what ephemeral disks are, their pros and cons, and the differences between Azure managed disks and ephemeral disks. In the next section, we will create a custom master VHD image.

Creating a VHD image

In this section, you will learn how to prepare a master VHD image for Azure. Note that Microsoft recommends that you use an image from the Azure image gallery. However, this section covers both options, giving you the ability to customize an image offline and upload it to Azure when you're finished. You can also use **Microsoft Deployment Toolkit** (**MDT**) and **enterprise content management** (**ECM**) to create images for AVD. To upload these images, you can use the following tools:

- **Azure portal**: Use the upload feature within the storage account.
- **Azure Storage Explorer**: https://azure.microsoft.com/features/storage-explorer/
- azcopy: https://docs.microsoft.com/azure/storage/common/storage-ref-azcopy

> **Important note**
> Ensure your image does not have the AVD agent installed on the VM. The agent can cause issues, including blocking registration and preventing user session connections.

Creating a VM

There are two options for creating a VM. First, you can provision the VM in Azure and then customize and install the required software. Alternatively, you can create an image locally using Hyper-V and customize it to your requirements.

First, let's look at deploying a VM in Azure. Go to the **Virtual machines** section and click **Create | Azure virtual machine**:

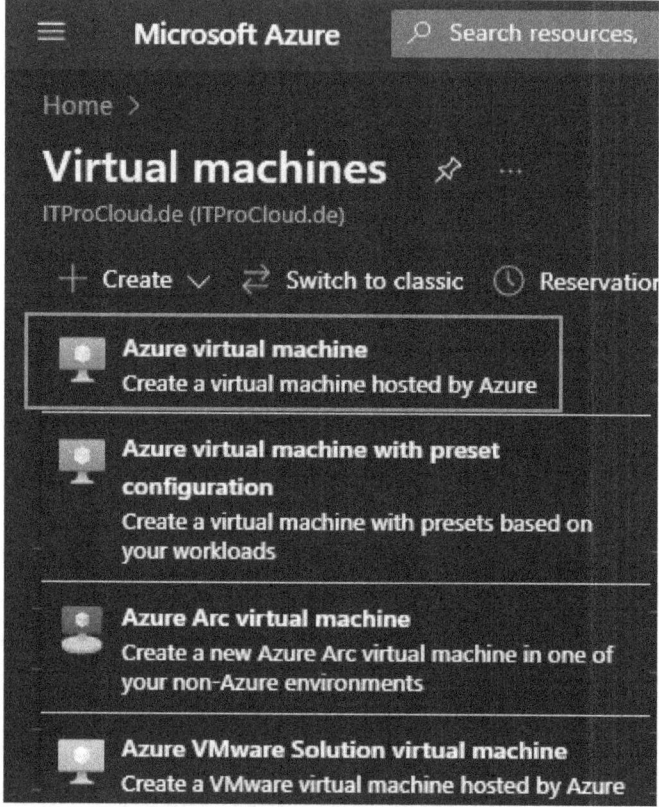

Figure 5.14 – Creating a new VM

In the basic settings, we're configuring the following topics:

Configuration	Value
Subscription	Our AVD subscription
Resource group	A resource group to store our new golden master VM to
Virtual machine name	The name of our new VM; for example, `T-GM-M365-01`
Region	Region of the new resource; in this case, **West Europe**
Availability options	In this case, **No infrastructure redundancy required**
Security type	**Trusted launch virtual machines** (recommended)
Image	Click on **See all images** and search for your image. In our case, we want to use Windows 11 with Microsoft 365. Search for `Windows multi-session + Microsoft 365 Apps` and select **Windows 11 Enterprise multi-session + Microsoft 365 Apps, version 23H2**
Size	Select a VM size for your golden master. A 4-core 16 GB RAM VM size would work well for non-GPU software. We will take `Standard_D4as_v5`.
Username	Enter a username for the local administrator (`vmAdmin`)
Password	Enter a password for the local administrator
Public inbound ports	Choose **Allow selected ports**, then select **RDP (3389)** to allow logon to the VM
Licensing	Confirm if you have an eligible Windows 10/11 license with multi-tenant hosting rights

Table 5.13 – Creating a VM: Basic settings

The following figure shows the basic settings configuration:

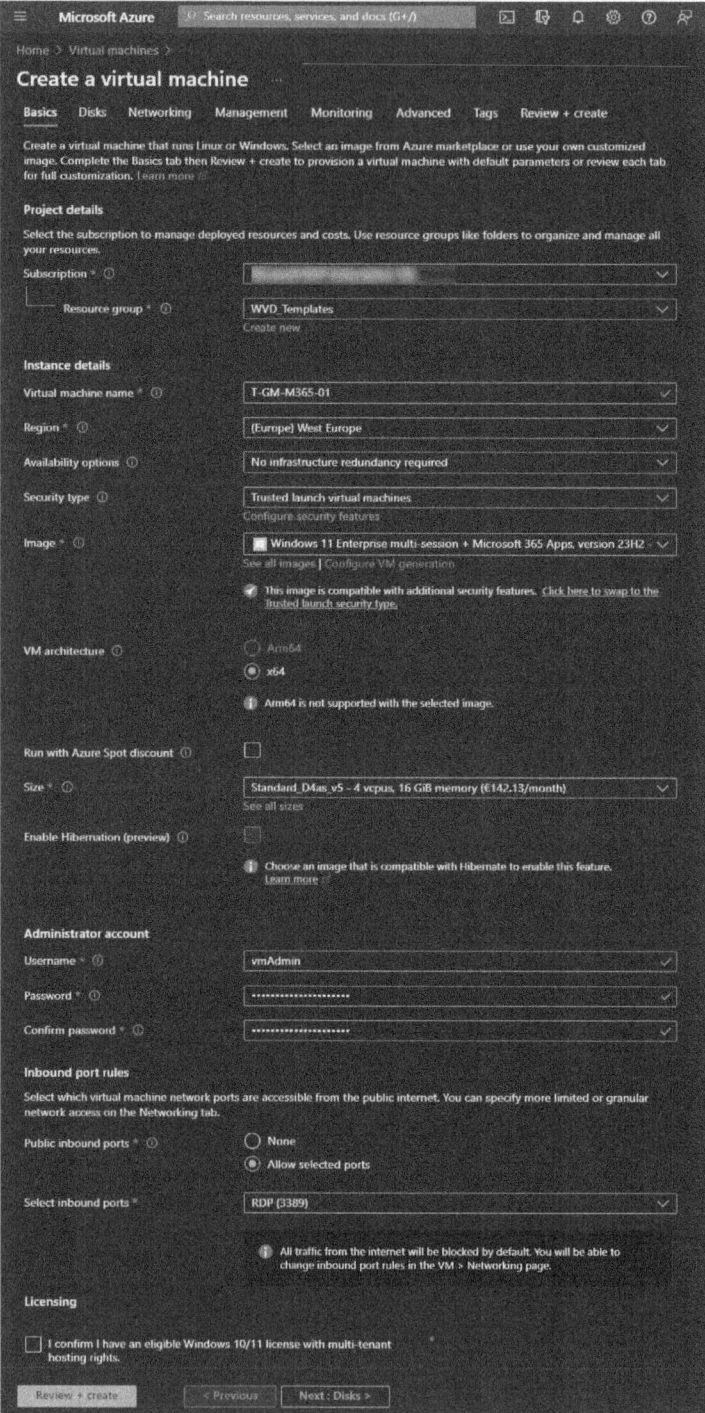

Figure 5.15 – Creating a new VM: Basic settings

Click on **Next: Disks** to configure the disk parameters. In the **Disks** section, we have to choose the right OS disk type. A standard SSD works well for this VM. If you plan to use the VM as a multi-session host, a premium disk is recommended:

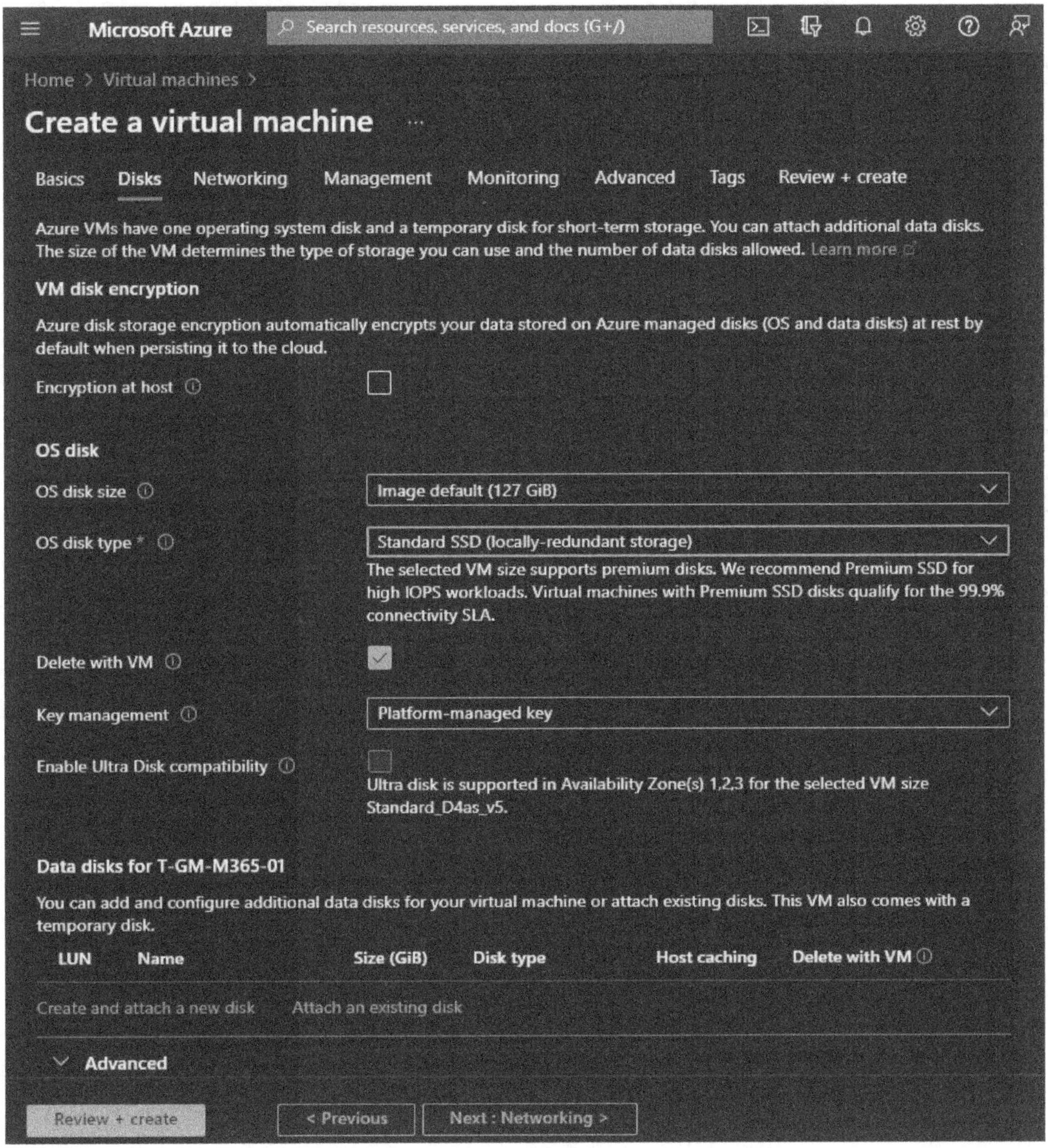

Figure 5.16 – Creating a new VM: Disks settings

We can configure network-related settings in the **Networking** tab:

Configuration	Value
Virtual network	Select your virtual network from the list
Subnet	Select the subnet of the virtual network
Public IP	None (for an AVD session host, a public IP is not needed; see also *Chapter 3, Designing for User Identities and Profiles*) Note: I recommend accessing the template VM via a connected (VPN) network or a bastion host. If a public IP is needed, ensure that a network security group (NSG) is configured.
NIC network security group	None
Delete public IP and NIC when VM is deleted	Check the box to delete the network interface card (NIC) with the VM
Enable accelerated networking	No – check the box if your running application needs low latencies
Load balancing options	None

Table 5.14 – Create a virtual machine: Network settings

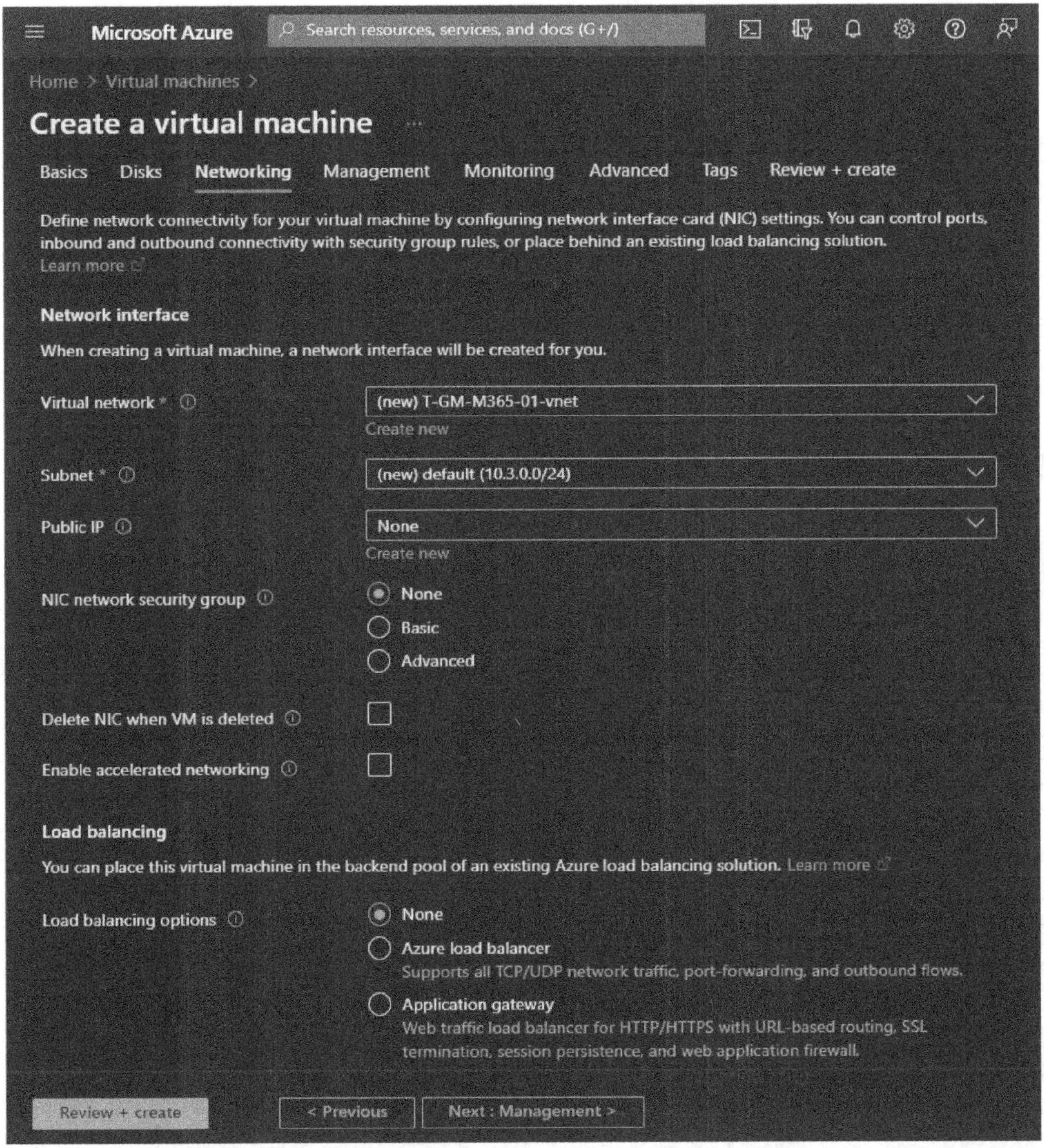

Figure 5.17 – Create a virtual machine: Networking settings

The other settings in **Management**, **Monitoring**, **Advanced**, and **Tags** can stay as the default. After clicking on **Review + create**, we can validate the configuration and start the rollout with a click on **Create**:

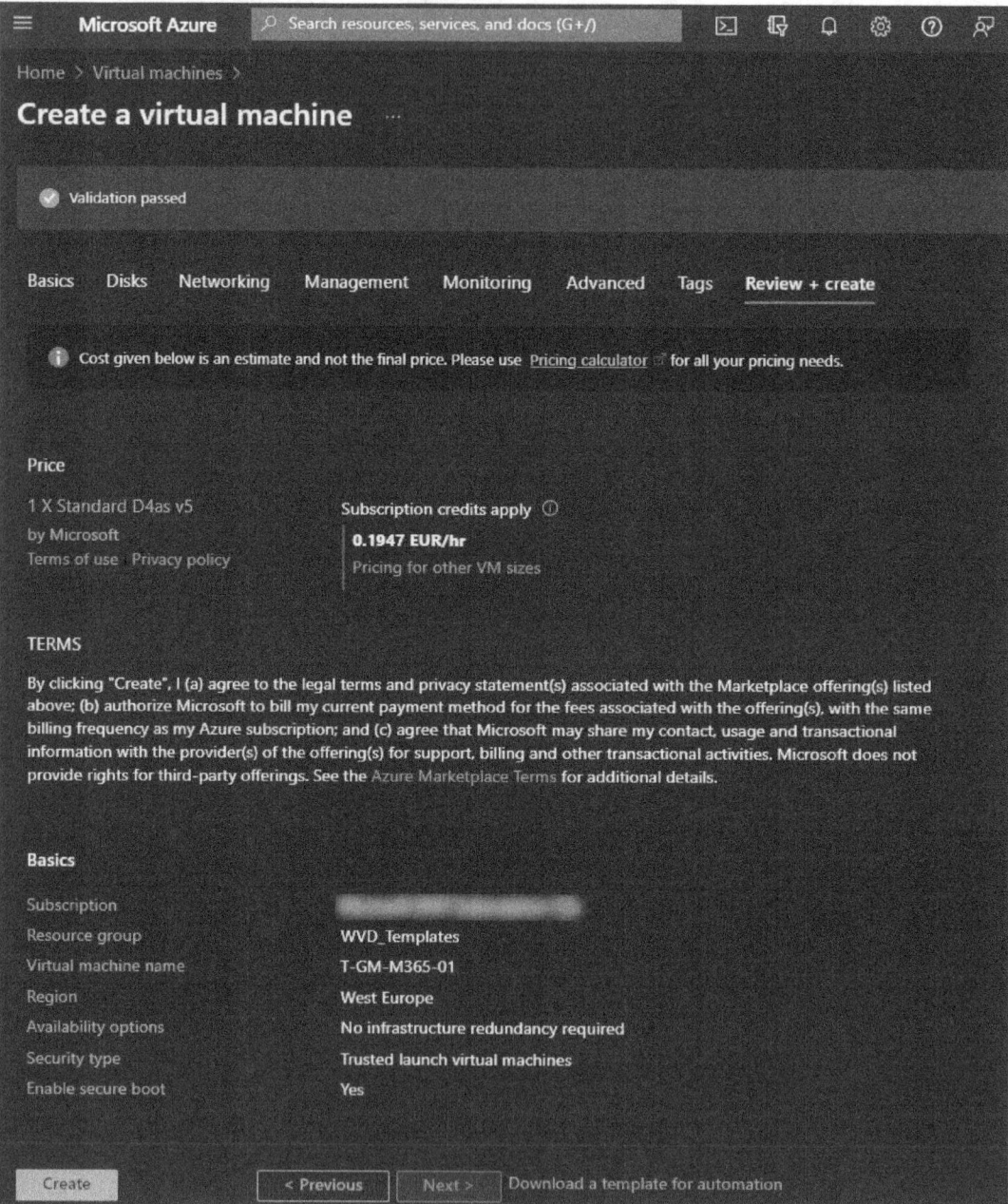

Figure 5.18 – Creating a new VM: Review + create

After the creation, you can connect to the VM with **Remote Desktop Protocol** (**RDP**) (`mstsc.exe`). Use the internal IP address to connect to the VM with the local administrator credentials. The internal IP is listed in the **Overview** tab of the VM:

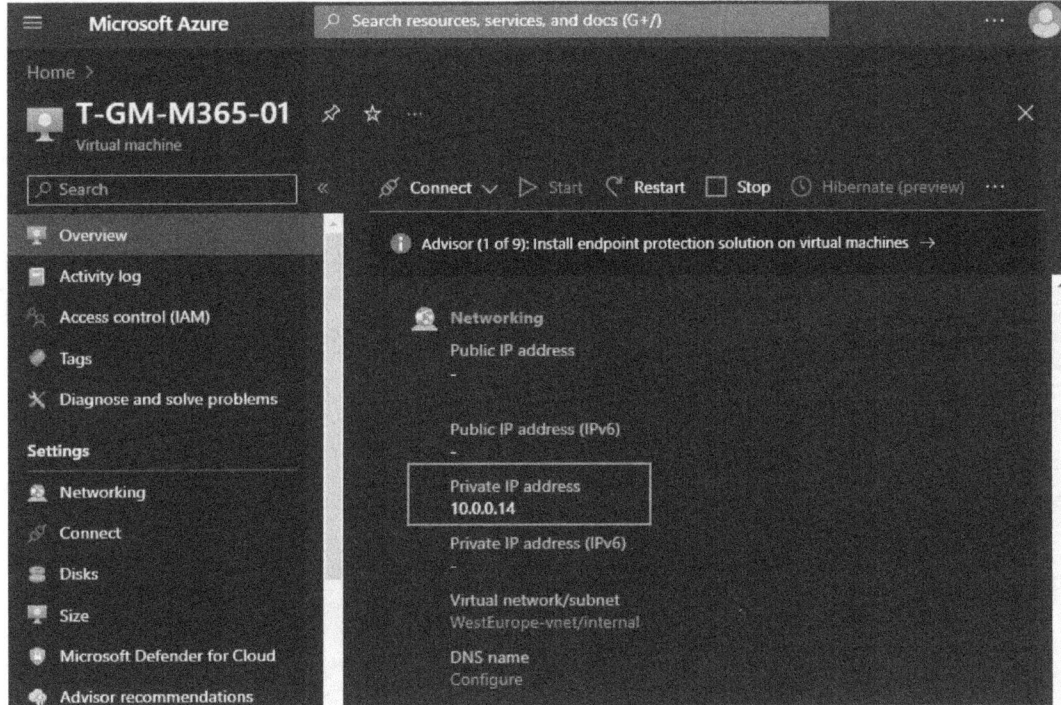

Figure 5.19 – Creating a new VM: Private IP address

> **Important note**
> If you need access via a public IP address (not recommended), you must have an NSG and enable incoming traffic to the RDP port.

Creating a local image

First, you will need to download the required operating system image. Then, using Hyper-V, you must create a VM using the downloaded VHD. You need to ensure that you complete the following steps:

1. Specify the generation as **Generation 1**:

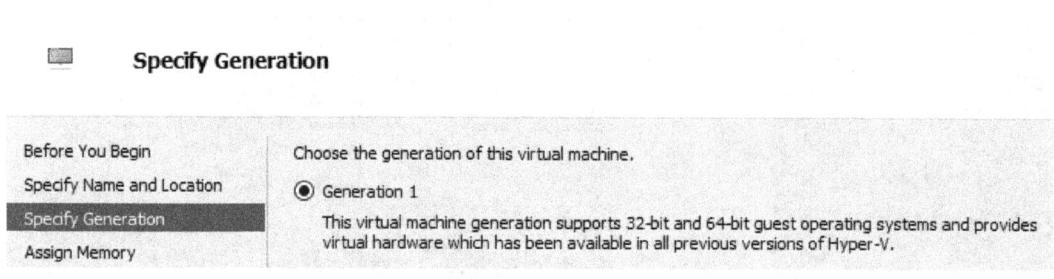

Figure 5.20 – Choosing Generation 1 in Hyper-V

2. Disable the checkpoints for the VM:

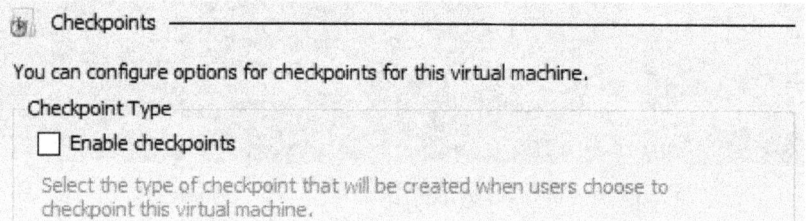

Figure 5.21 – Disabling the Enable checkpoints box

The following PowerShell cmdlet allows you to disable checkpoints:

```
Set-VM -Name <VMNAME> -CheckpointType Disabled
```

Now, let's look at the difference between dynamic and fixed disks since Azure only supports the fixed disk format.

Dynamic disks versus fixed disks

When creating a VM from an existing VHD, it creates a dynamic disk by default. However, you can change this by selecting the **Edit Disk…** option within Hyper-V.

You can also use PowerShell to change a dynamic disk to a fixed disk, as follows:

```
Convert-VHD -Path c:\test\MY-VM.vhdx -DestinationPath c:\test\
MY-NEW-VM.vhd -VHDType Fixed
```

This section detailed the options available to you when creating an image. We also covered some of the requirements if you decide to customize an image outside of AVD using Hyper-V.

Summary

In this chapter, we looked at implementing and managing storage for AVD. First, we explored the requirements for storing FSLogix profile containers, storage account tiers, Azure Files storage tiers, and Azure Files integration with Entra ID DS. Next, we looked at creating a new storage account and configuring Azure file shares. Then, we reviewed the differences between Azure managed disks and ephemeral OS disks and finished by looking at the options available for creating a VM with Azure. We are also walked through the process of creating a custom image in Azure Compute Gallery without losing our golden master.

In the next chapter, we will look at creating and configuring host pools and session hosts.

Questions

Here are a few questions to test your understanding of this chapter:

1. What is the recommended storage solution for FSLogix profile containers?

    Azure Files

2. Do all regions support all types of storage accounts and redundancy configurations?

    No

3. When it comes to storage accounts for larger organizations and high I/O workloads, what is the recommended storage tier?

    Premium tier

4. What is the recommended disk type for session hosts?

    Premium SSD

5. What disk format does a VHD need to be to upload and function correctly within Azure?

    Fixed disk

Further reading

Please refer to the following links for more information regarding the topics that were covered in this chapter:

- Using Azure NetApp Files for FSLogix: `https://docs.microsoft.com/azure/architecture/example-scenario/wvd/windows-virtual-desktop-fslogix#storage-options-for-fslogix-profile-containers`
- Setting up an Azure NetApp Files account for FSLogix: `https://docs.microsoft.com/azure/virtual-desktop/create-fslogix-profile-container`

Part 3: Implementing an Azure Virtual Desktop Infrastructure

In this section, we look at how to implement an Azure Virtual Desktop environment, including the creation of Azure host pools, Entra ID Join, customization, and managing session host images.

This part of the book comprises the following chapters:

- *Chapter 6, Creating Host Pools and Session Hosts*
- *Chapter 7, Configuring Azure Virtual Desktop Host Pools*
- *Chapter 8, Entra ID Join for Azure Virtual Desktop*
- *Chapter 9, Creating and Managing Session Host Images*

6
Creating Host Pools and Session Hosts

This chapter takes a look at creating host pools and deploying session hosts. We will cover the creation process first using the Azure portal in the web browser. Then, we will use PowerShell, which is great for automation and repeat processes

The following topics are covered in this chapter:

- Creating a host pool by using the Azure portal
- Automating the creation of **Azure Virtual Desktop** (**AVD**) hosts and host pools

Creating a host pool by using the Azure portal

In this section, we'll look at creating and configuring host pools and session hosts. **Host pools** are collections of one or more **virtual machines** (**VMs**) within AVD environments. These are typically identical and are created from a central image, an Azure Compute Gallery, or a custom image. We'll also take a look at Azure Compute Galleries for template deployment in *Chapter 9*, *Creating and Managing Session Host Images*. In addition, each host pool can contain app groups that are used for user assignments. This section looks at creating a host pool for AVD through the Azure portal.

Before we get started, you need to ensure that you have the following host pool prerequisites ready:

- The VM image name
- VM configuration
- Domain and network properties
- AVD host pool properties

> **Important note**
> You need to ensure that a virtual network exists in the Azure region of your choice and that it has a "line of sight" with the domain unless you are using Entra ID Join.

You also need to consider the following:

- Are you using an operating system image from the Azure Compute Gallery or a custom image?
- Domain join credentials to connect your session hosts to your domain

> **Tip**
> Make sure you have registered the `Microsoft.DesktopVirtualization` resource provider before attempting to deploy a host pool. When using an account with Global Administrator admin rights, the registration is done automatically during host pool creation.

Host pool creation

In this section, we will create our first host pool using the Azure portal:

1. Sign in to the Azure portal at `https://portal.azure.com`.

> **Important note**
> If you are signing in to the US government portal, go to `https://portal.azure.us/` instead.

2. Enter `Azure Virtual Desktop` into the search bar. Once shown, select **Azure Virtual Desktop** under **Services**.
3. On the **Azure Virtual Desktop** overview page, select **Create a host pool**.
4. You will then be presented with a **Basics** tab. Select the correct subscription under **Project details**.
5. Select **Create new** to create a new resource group or select an existing resource group you may have created previously in the drop-down menu.
6. Enter a unique name for your new host pool.
7. Select the Azure region to create the host pool using the drop-down menu in the **Location** field.

The Azure region you've chosen dictates the geographic location where the metadata for this host pool and its associated objects will reside. Ensure that you select a region within the desired geography for storing the service metadata:

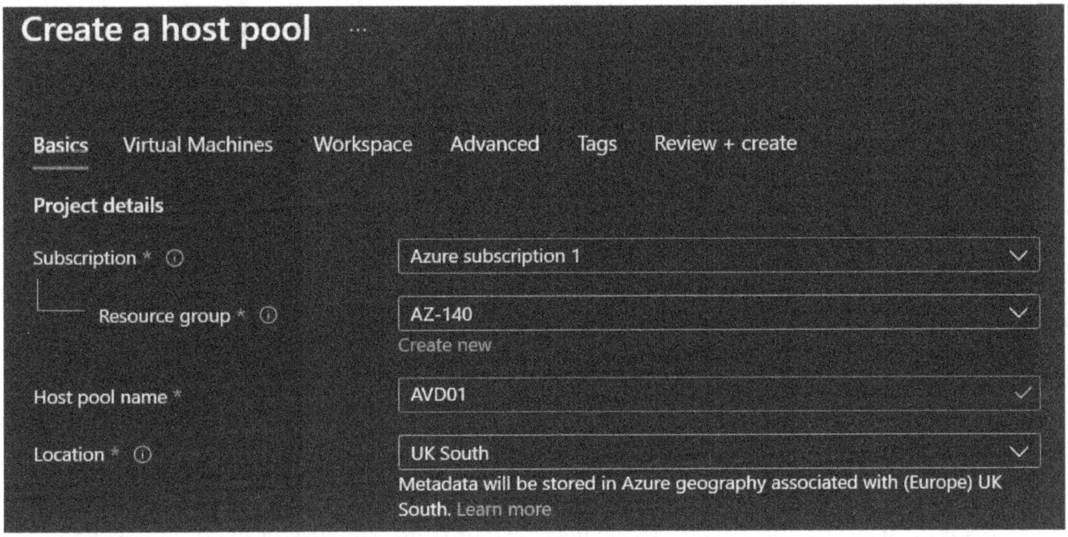

Creating a host pool by using the Azure portal

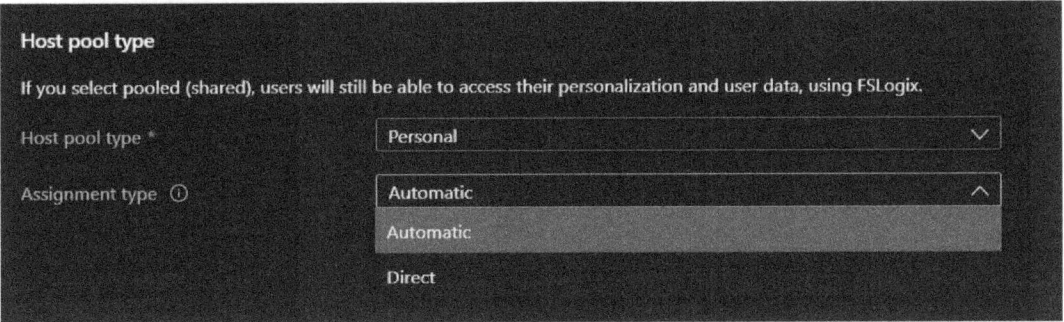

Figure 6.1 – Screenshot showing the Create a host pool page

> **Important note**
> The AVD service metadata is independent of the VM location. It is also important to understand that the metadata locations are not available for all Azure regions.

1. Under **Host pool type**, select whether you want to create a **Personal** or **Pooled** host pool. If you choose **Personal**, you will need to select **Automatic** or **Direct** in the **Assignment type** field:

Figure 6.2 – Screenshot showing Personal host pool type options

If you choose **Pooled**, you need to specify the following information:

- For **Max session limit**, enter the maximum number of users you wish to load balance on a single session host.

152 Creating Host Pools and Session Hosts

- For **Load balancing algorithm**, select either **Breadth-first** or **Depth-first**, based on your usage pattern:

Host pool type

If you select pooled (shared), users will still be able to access their personalization and user data, using FSLogix.

Host pool type *	Pooled
Load balancing algorithm	Breadth-first
	Breadth-first
Max session limit	Depth-first

Figure 6.3 – Screenshot showing pooled host pool options

Load balancing algorithm has the following two options:

- **Breadth-first**: Breadth-first load balancing is a strategy that employs an algorithm to identify the session host with the minimum active sessions for assigning new user sessions. For instance, when a user attempts to connect to an AVD host pool configured with **Breadth-first**, the system initiates a search across all available session hosts in the pool as part of the login procedure. This approach ensures the new session is allocated to the host currently hosting the fewest sessions. In scenarios where multiple session hosts are tied with an equal number of active sessions, the load-balancing algorithm defaults to selecting the first session host that appears in the search results. This ensures that the users get a good user experience as the sessions will be spread out evenly among all available session hosts:

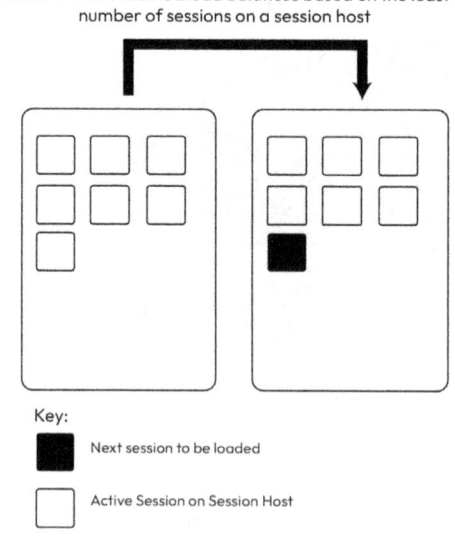

Figure 6.4 – Diagram showing an example of how breadth-first load balancing works

- **Depth-first**: The depth-first load-balancing approach focuses on fully utilizing each session host before distributing sessions to the next one in line. This strategy is particularly beneficial for organizations aiming for an active/passive AVD deployment model or looking to minimize operational costs. In employing the depth-first method, the system assesses all available session hosts to determine the placement of new sessions. Should a session host reach its predefined maximum session capacity, the algorithm then moves to allocate new sessions to the next available host. In cases where multiple session hosts are at the same session level, the algorithm selects the first one identified in the assessment. The following diagram outlines this procedure. By integrating depth-first load balancing with scaling plans, organizations can activate additional session hosts as they approach full capacity, thereby managing costs through optimized compute resource use. Essentially, this means session hosts are activated as necessary and deactivated during off-peak periods for efficient resource management:

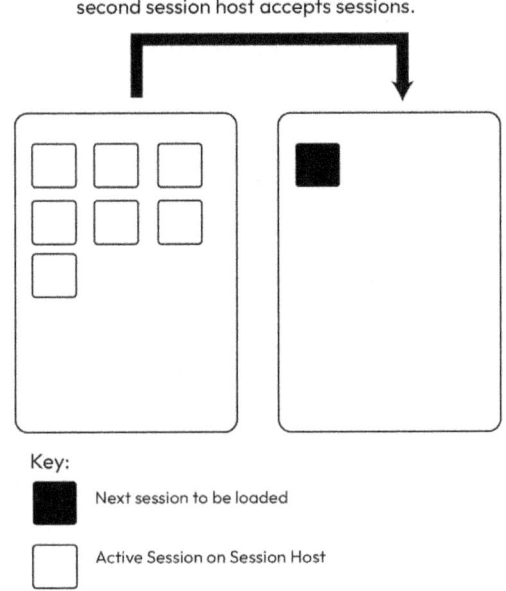

Figure 6.5 – Diagram showing an example of depth-first load balancing

2. Select **Next: Virtual Machines**. If you have already created VMs and want to use them with this new host pool, select **No** then **Next: Workspace** and move on to the **Workspace information** section. However, if you want to create new VMs and register them to the new host pool, click **Yes**, as highlighted in *Figure 6.6*:

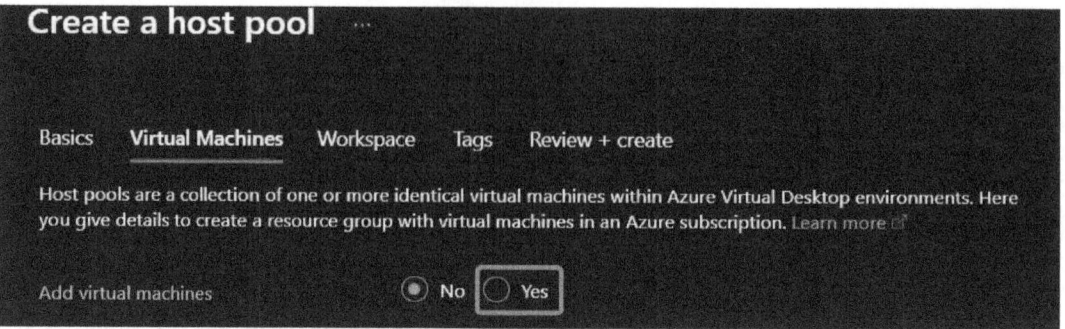

Figure 6.6 – Screenshot showing the tab to create VMs in the host pool

> **Important note**
> A workspace is a logical grouping of application groups within AVD. Each AVD application group must be associated with a workspace for users to see the remote apps and desktops published to the workgroup.

We'll now take a look at creating VMs within the **Create a host pool** tab.

Creating VMs within the Create a host pool tab

When setting up your VMs within the host pool setup process, you need to complete the following:

1. Under the **Resource group** section, choose the resource group where you want to create VMs (session hosts). This can be a different resource group than the one you used for the host pool or the same.
2. After that, you will need to provide a name prefix, which is used to assign names to the VMs during the creation process. The suffix will be – with numbers starting from 0.
3. Choose the VM location where you want to create VMs. The location can be the same as your host pool or different. It is advised that you deploy VMs in the same region unless there is a specific reason not to do so.
4. Next, choose the availability option that best suits your needs:

> **Important note**
> **Availability sets** offer a 99.95% Azure **Service Level Agreement** (**SLA**) and essentially provide a logical grouping of VMs. Microsoft recommends that you use two or more VMs within an availability set to provide **high availability** (**HA**).
>
> **Availability zones** allow you to control where in the Azure region your VMs are stored. There are three availability zones per supported Azure region. Each zone has what Microsoft describes as a distinct power source, network, and cooling. This enables you to split session host deployments between different zones.

Creating a host pool by using the Azure portal   155

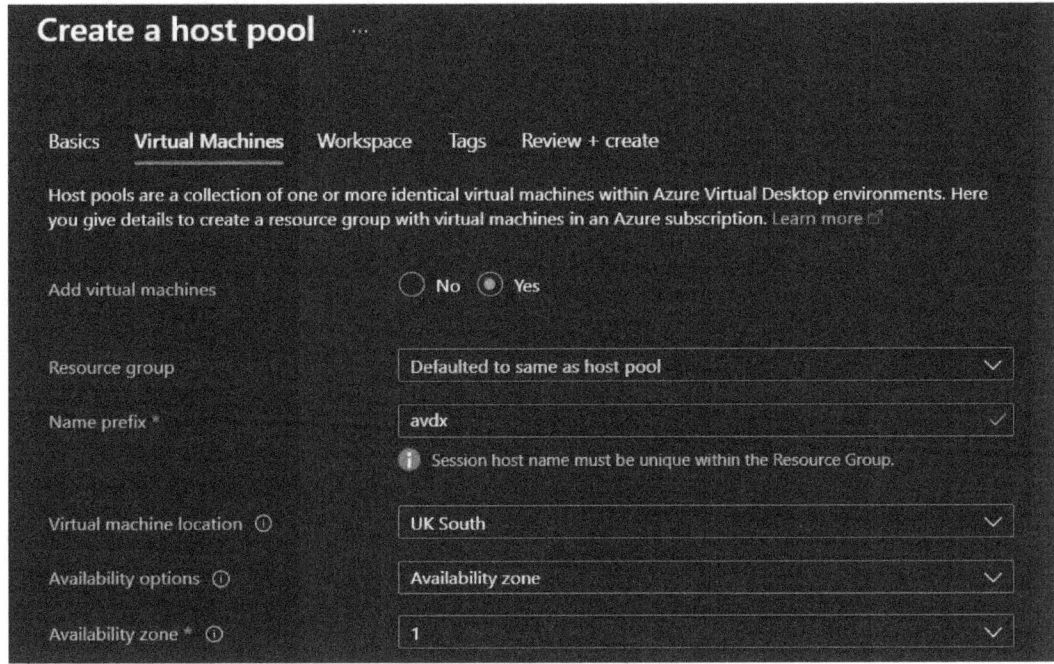

Figure 6.7 – Screenshot showing the Virtual Machines tab

5. Next, we need to choose the security type of the VM. We have three security types to choose from:

   - **Standard**: These are standard VMs.
   - **Trusted launch virtual machines**: These are Gen2 VMs that enable secure boot and **virtual Trusted Platform Modules (vTPMs)** to be used.
   - **Confidential virtual machines**: Confidential VMs offer higher confidentiality and integrity:

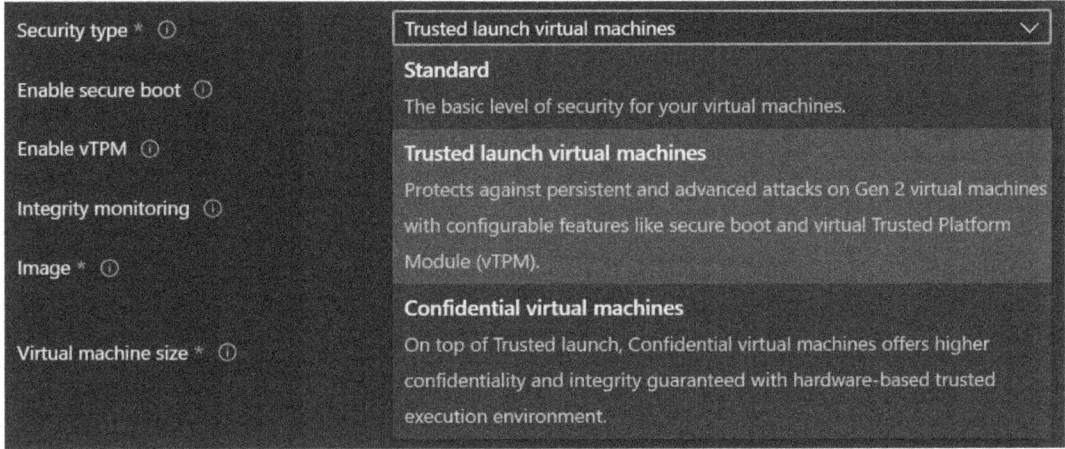

Figure 6.8 – Screenshot showing different security types

> **Important note**
> If you want to deploy trusted launch-enabled or confidential VMs as session hosts, then the image used on those VMs must be configured here accordingly.

6. Next, choose the VM image that needs to be used to create the VM. You can choose either **Gallery** or **Storage blob**.

   If you choose **Gallery**, select one of the images from the drop-down menu:

   - **Windows 11 Enterprise multi-session**
   - **Windows 11 Enterprise**
   - **Windows 10 Enterprise multi-session, Version 21H2**
   - **Windows 10 Enterprise multi-session, Version 21H2 + Microsoft 365 Apps**
   - **Windows Server 2019 Datacenter**
   - **Windows 10 Enterprise multi-session, Version 23H2**
   - **Windows 10 Enterprise multi-session, Version 23H2 + Microsoft 365 Apps**

> **Important note**
> Gallery images created by Microsoft include the FSLogix agent. Microsoft maintains these and keeps them updated with the latest patches and updates.

   If you do not see the image you require, select **See all images** and browse to another image in your Gallery or an image provided by Microsoft or other third-party publishers. Make sure that the image you choose is one of the supported images:

   - **Windows 11 Enterprise multi-session**
   - **Windows 11 Enterprise**
   - **Windows 10 Enterprise multi-session**
   - **Windows 10 Enterprise**
   - **Windows Server 2022**
   - **Windows Server 2019**
   - **Windows Server 2016**

7. You can also navigate to **My Items** and choose a custom image you may have already uploaded:

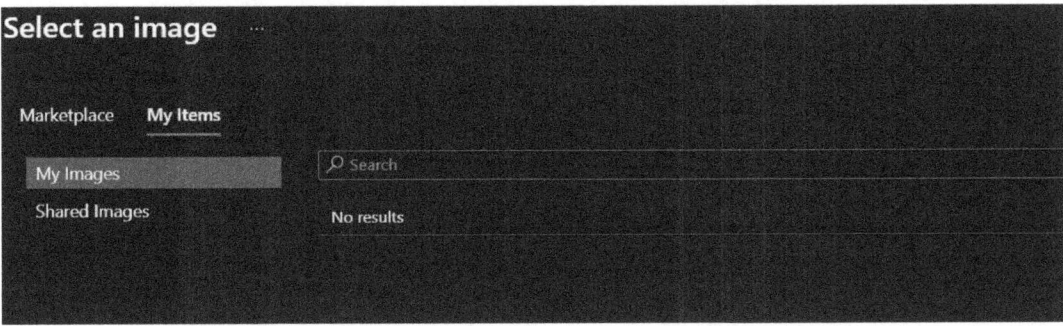

Figure 6.9 – Screenshot showing selecting an image from My Images

If you choose **Storage Blob**, you can use your image build through Hyper-V or an Azure VM. To do this, you need to enter the location of the image in the storage blob as a URI.

> **Tip**
> The location of the image is not influenced by the availability option. However, its zone resiliency dictates whether it can be utilized within the availability zone. Therefore, if you opt for an availability zone when creating your image, ensure that you're employing an image from the Gallery with zone resiliency enabled.

8. After that, choose the **Virtual machine size** value you want to use. You can keep the default size as is or select **Change size** to change the size. If you select **Change size**, choose the VM size suitable for your workload in the window that appears.

9. Under **Number of VMs**, provide the number of VMs you want to create for your host pool.

> **Important note**
> During the setup of your host pool, the deployment process can generate a maximum of 400 VMs, with each VM setup generating four objects in your resource group. The creation process does not verify your subscription quota, thus it's crucial to ensure that the number of VMs you specify adheres to the Azure VM and API limits for your resource group and subscription, staying within the maximum limits. Additional VMs can be added after completing the creation of your host pool.

10. Choose the type of operating system disks you want your VMs to use: **Standard SSD**, **Premium SSD**, or **Standard HDD**.

> **Tip**
> Microsoft recommends using **Premium SSD** for AVD session hosts. See *Chapter 5, Implementing and Managing Storage for Azure Virtual Desktop*.

11. Choose the OS disk size. The default is 128 GiB.

> **Tip**
> The size of the image will be the minimum size of the storage that you deploy to your session hosts, so if possible pick the minimum size to reduce the cost of your session hosts.

12. Microsoft recommends using **Premium SSD** for AVD session hosts. See *Chapter 5, Implementing and Managing Storage for Azure Virtual Desktop*.
13. Under the **Network and Security** section, select the virtual network and subnet where you want the VMs to reside.

> **Tip**
> Ensure the virtual network can connect to the domain controller as the VMs need to join the domain. You should configure the virtual network DNS to be configured with the IP address of your domain controller.

The following screenshot shows where you would configure the virtual network DNS to point to the domain controller:

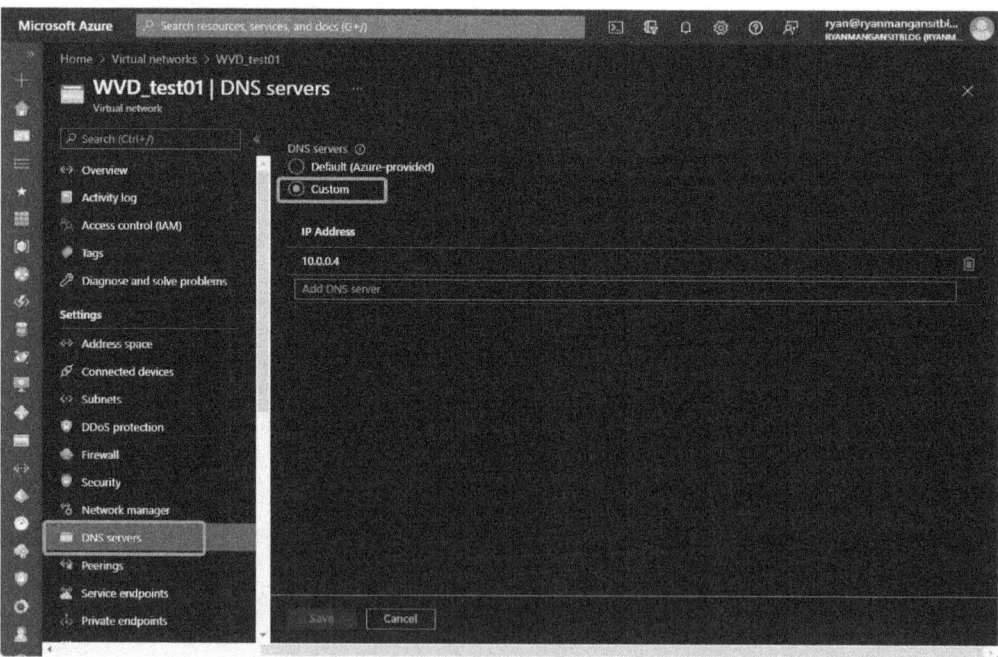

Figure 6.10 – Screenshot showing the domain controller added as a custom DNS entry on the Azure virtual network

14. Select the security group type you want: **Basic**, **Advanced**, or **None**. If you select **Basic**, you will need to select whether you want any inbound ports open. If you select **Yes**, select from the list of standard ports to allow inbound connections:

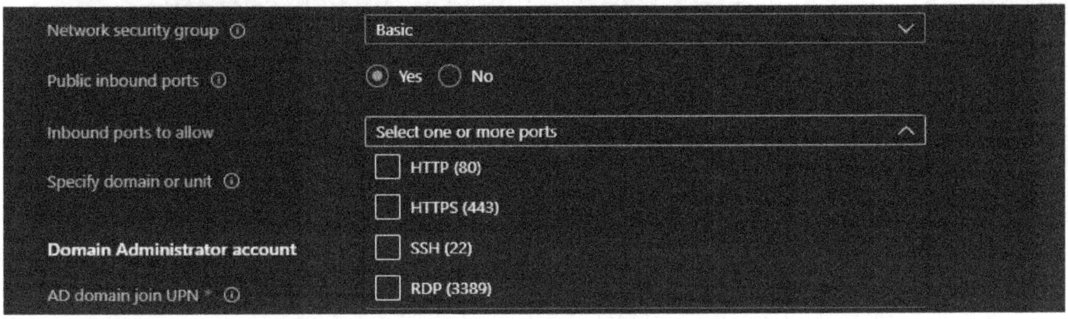

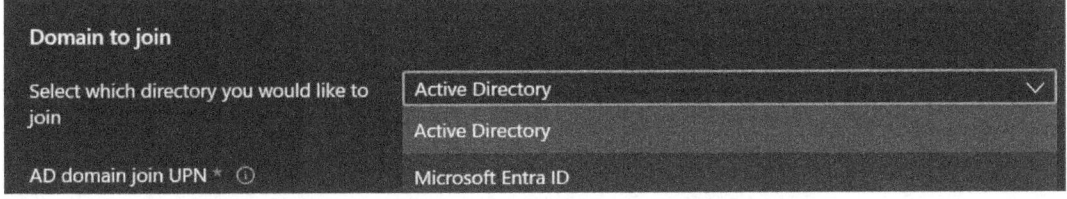

Figure 6.11 – Screenshot showing network security options

If you choose the **Advanced** option, select an existing network security group that you have already configured.

> **Important note**
> For enhanced security, Microsoft recommends that you don't open public inbound ports.

15. Next, we need to determine whether we will join our session hosts to Active Directory or Microsoft Entra ID:

Figure 6.12 – Screenshot showing domain join options

16. When joining Active Directory, you must decide whether you want the VMs to join a particular domain and **organizational unit** (**OU**). If you opt for **Yes**, ensure to specify the domain for joining. Optionally, you can specify a particular OU for the VMs. If you choose **No**, the VMs will automatically join the domain corresponding to the suffix of the AD domain join **User Principal Name** (**UPN**):

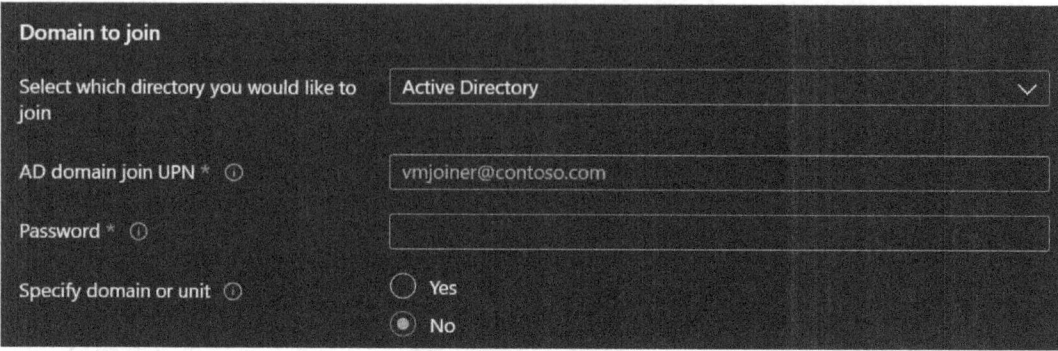

Figure 6.13 – Screenshot showing Active Directory domain join options

If you are joining Entra ID, you will need to specify whether you want the hosts to register with Intune:

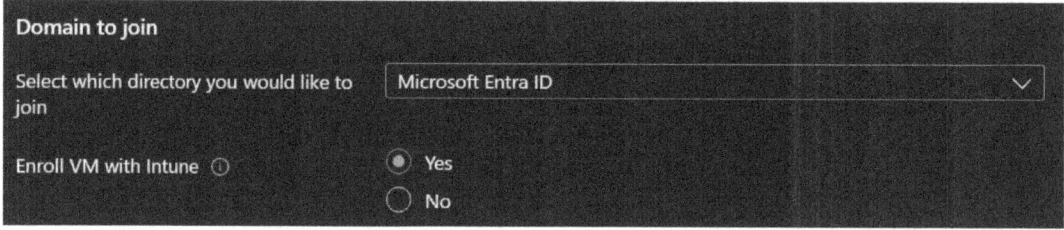

Figure 6.14 – Screenshot showing Entra ID join options

> **Tip**
> When you specify an OU, make sure you use the full path known as the **distinguished name (DN)**. You can find a DN by enabling advanced settings within Active Directory and navigating to the **Attributes** section of an OU.

1. Under the **Domain Administrator account** section, enter credentials for the Active Directory domain admin account of the virtual network you selected. This admin account can't have **multi-factor authentication (MFA)** enabled as this will cause the deployment to fail. It is recommended that you use a service account with specific permissions on the target OU.

   When joining an **Entra ID Domain Services (Entra ID DS)** domain, the account must be part of the Entra ID DC Administrators group. Additionally, the account password must also work in Entra ID DS.

2. Provide a VM administrator account. The password needs to be at least 12 characters long:

Figure 6.15 – Screenshot showing the Virtual Machine Administrator account section

3. As an optional setting, you can set post-update custom configurations using **Azure Resource Manager (ARM)** templates:

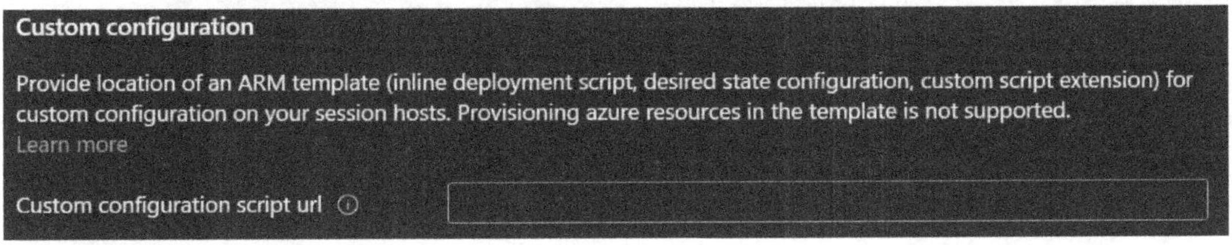

Figure 6.16 – Screenshot showing the option to add post-update custom configurations

4. Select the **Next: to the Workspace** tab.

We'll now move on to setting up your new host pool: registering an app group to a workspace.

Workspace information

The host pool deployment process creates a desktop application group by default. This is the default app group of a host pool. For the host pool to function correctly, you must publish this app group to users or user groups and register the app group with a workspace. To register the desktop app group to a workspace, you need to complete the following steps:

1. Select **Yes** under **Register desktop app group**.

    If you select **No**, you can register the app group later. However, Microsoft recommends you complete the workspace registration during the host pool deployment.

2. Next, choose between creating a new workspace or selecting from existing workspaces. Only workspaces created in the same location as the host pool will register the app group:

Creating Host Pools and Session Hosts

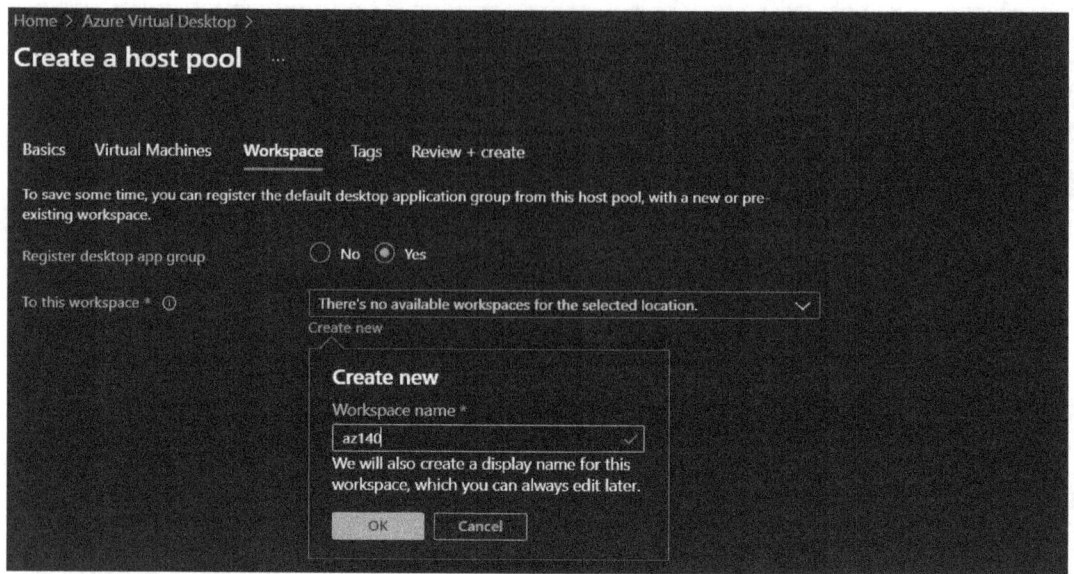

Figure 6.17 – Screenshot showing the creation of a new workspace

3. Optionally, you can select **Next: Tags**.

> **Tip**
> Adding tags to group objects with metadata is helpful for IT admins.

4. When you have finished, select **Review + create**.

> **Important note**
> The **Review + create** validation process doesn't check whether your admin password meets security standards or your architecture is correct. You must check your deployment before continuing.

5. Before proceeding with the deployment, you should review the information about your deployment to make sure everything looks correct. When you have finished, select **Create**.

   This starts the deployment process, which creates the following Azure objects:

   - The new host pool.
   - The desktop app group.
   - A workspace if you choose to create it.
   - Registration of the desktop app group; registration will be completed.

- If you choose to create VMs, they are joined to the domain and registered with the new host pool.
- A download link for an ARM template based on your configuration:

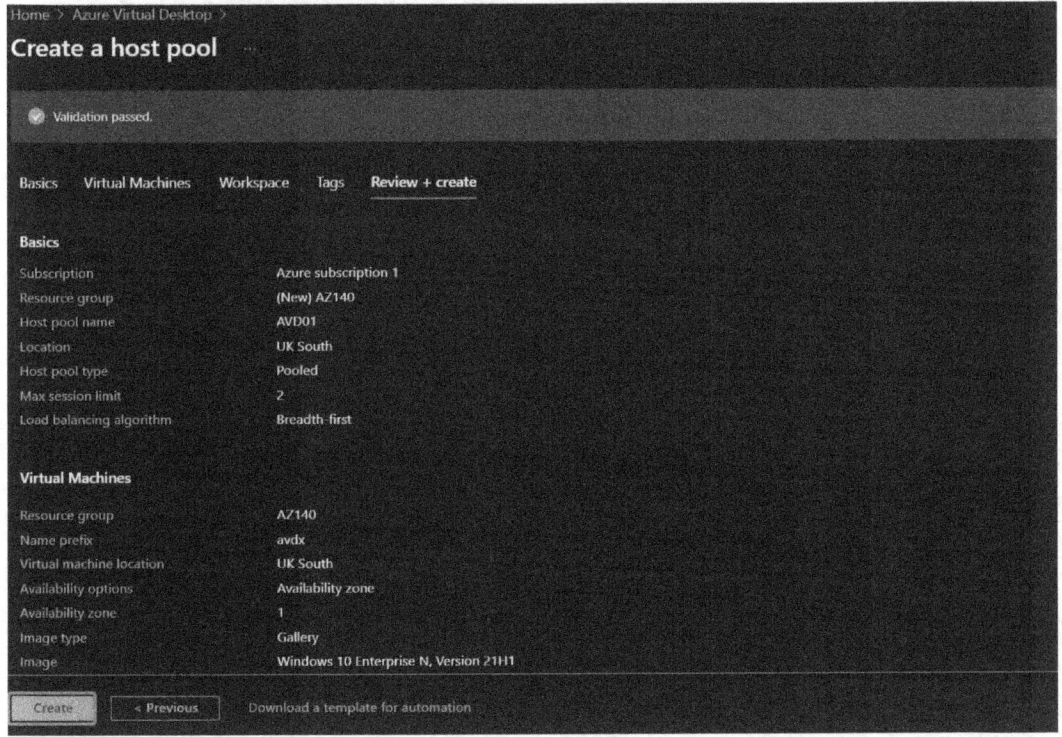

Figure 6.18 – Screenshot showing the Review + create page

After that, you just need to wait for your deployment to finish:

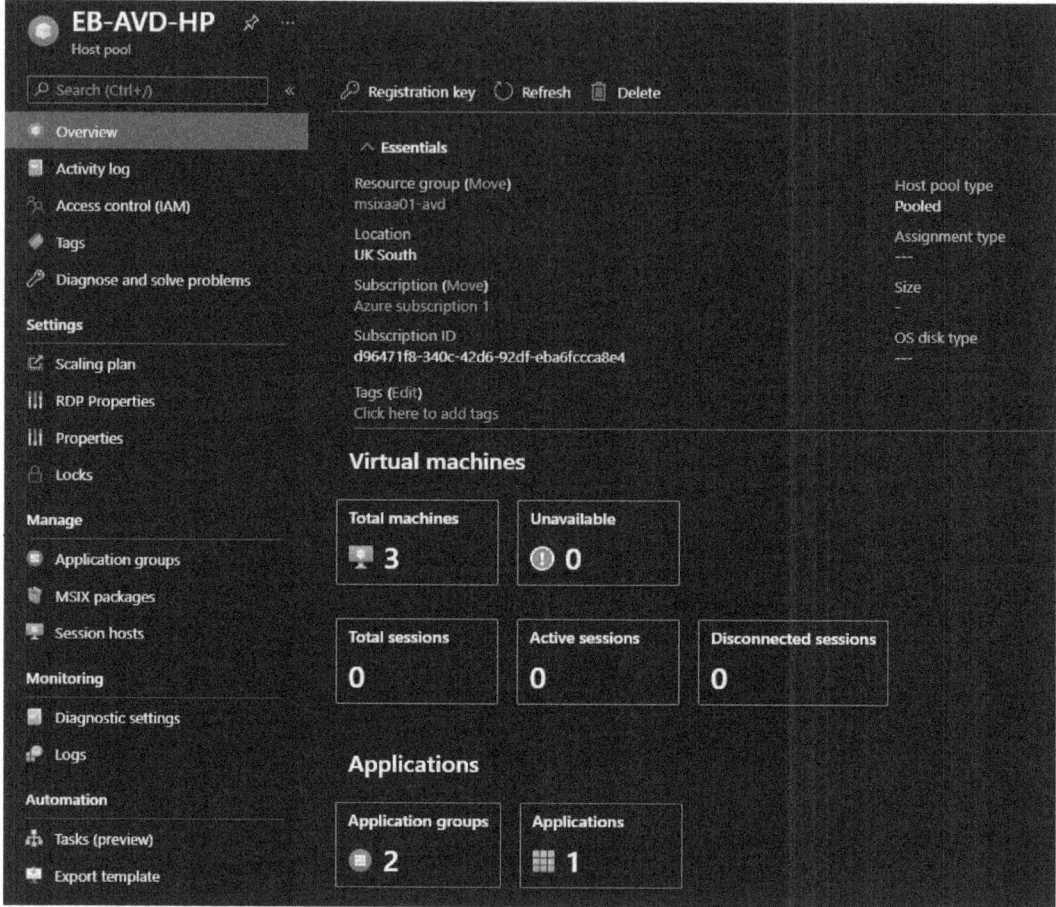

Figure 6.19 – Screenshot showing an example of a created host pool

This section looked at creating a host pool and deploying VMs into the new host pool. We will now take a look at an automated approach to creating AVD host pools using PowerShell.

Automating the creation of AVD hosts and host pools

This section looks at setting up PowerShell for AVD and deploying a new host pool using PowerShell.

Setting up PowerShell for AVD

Before we can get started, you first need to install the PowerShell module for AVD. You can do this by opening PowerShell in elevated mode.

> **Tip**
>
> Make sure you install the `Az` module. If you haven't already done so, run the `Install-Module -Name Az -Force` command.

Once you have opened PowerShell in elevated mode, run the following cmdlet:

```
Install-Module -Name Az.DesktopVirtualization
```

The following screenshot shows you how to install the `Az.DesktopVirtualization` PowerShell module:

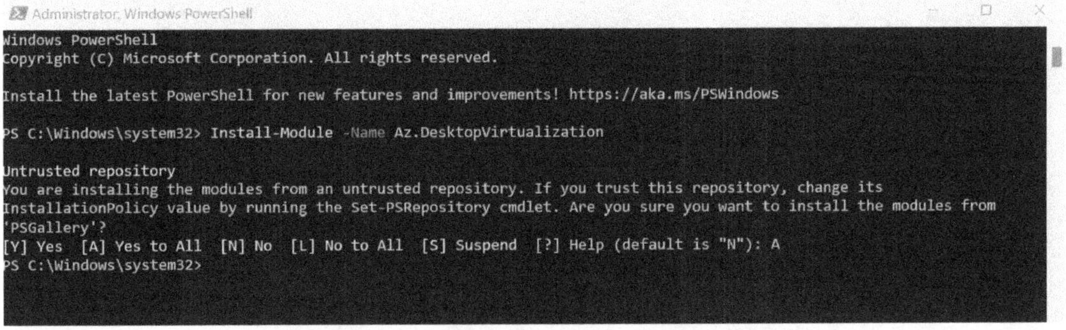

Figure 6.20 – Screenshot showing the installation of the PowerShell module for Az.DesktopVirtualization

As shown in *Figure 6.20*, you may be asked to confirm whether you trust the repository.

Next, you will need to connect to Microsoft Azure using the following cmdlet:

```
Connect-AzAccount
```

Once you have run the cmdlet, you will then see a **Sign in to your account** popup:

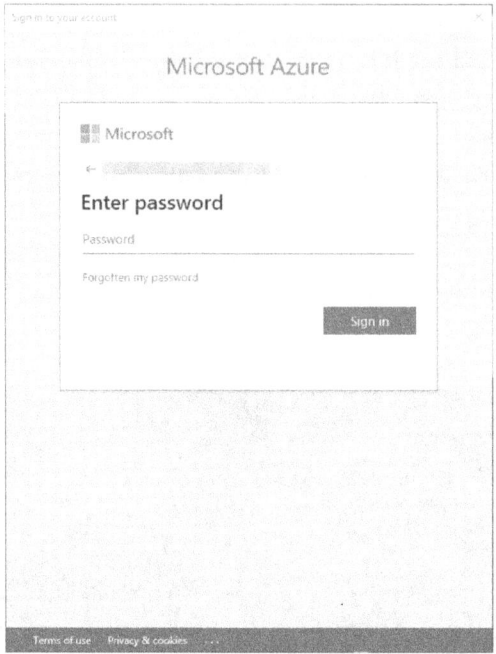

Figure 6.21 – Screenshot showing the Sign in to your account popup

Enter your username/password and any MFA details that may be requested.

The output from completing this process is shown in *Figure 6.22*:

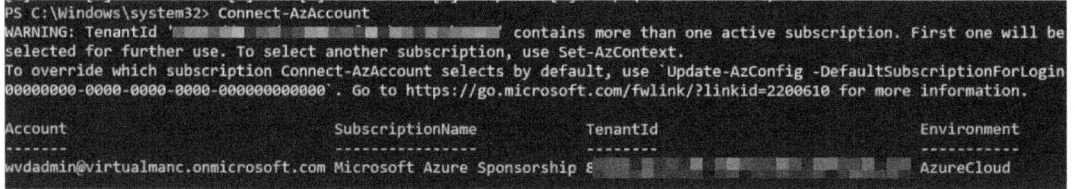

Figure 6.22 – Screenshot showing cmdlet to connect to Azure

As shown in *Figure 6.22*, once you have connected to Azure, you should see details of any subscriptions in the tenant.

The final step is to select the required subscription you plan to deploy AVD resources to.

Select the subscription you want to use; you can use the `out-gridview` cmdlet to select the one you want:

```
Get-AzSubscription | Out-GridView -PassThru | Select-AzSubscription
```

Once you have selected the subscription you require, click **OK**. In this example, there is only one subscription available:

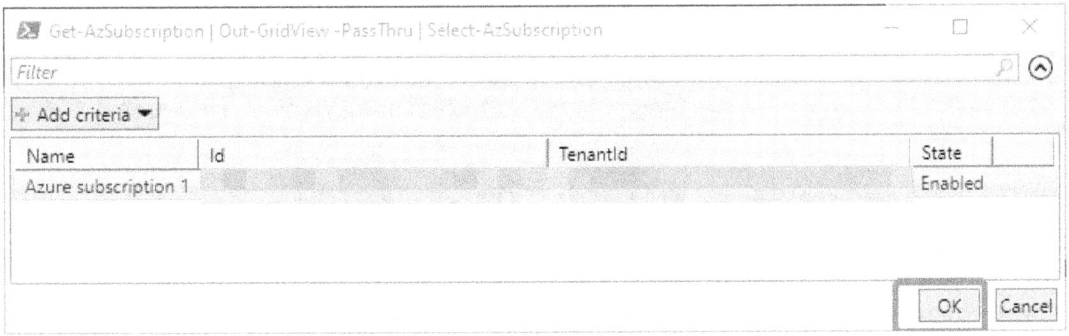

Figure 6.23 – Screenshot showing the grid view for selecting an available Azure subscription

In this section, we looked at setting up PowerShell for AVD. It is important to ensure you have the correct PowerShell modules and Azure connectivity via PowerShell working before continuing. We will now move on to deploying a host pool using PowerShell.

Creating an AVD host pool with PowerShell

In this section, we'll look at creating some resources for AVD using PowerShell.

We will use PowerShell to do the following:

- Create a resource group
- Create a host pool
- Create a registration token
- Assign users to a host pool application group
- Assign groups to a host pool application group

Once connected to Azure via PowerShell following the instructions set out in the *Setting up PowerShell for AVD* section, you can start to deploy your AVD host pool using PowerShell.

First, we need to create a resource group.

To do this, you can use the following cmdlets:

```
New-AzResourceGroup -Name <Resource Group Name> -Location <Region>
Example:
#Create Resource Group
New-AzResourceGroup -Name az140pw -Location "UK South"
```

Once we have created the resource group, we can then proceed with creating a host pool:

```
New-AzWvdHostPool -ResourceGroupName "<Resource Group Name>"
-Name <Host Pool Name> -WorkspaceName <Workspace Name> -HostPoolType
<Host Pool Type>
-LoadBalancerType <Load balancer method> -Location <Region>
-DesktopAppGroupName <App group Name> -PreferredAppGroupType <App
group type>
```

Examples of cmdlets used to deploy a host pool using PowerShell are as follows:

```
#Create Host Pool
New-AzWvdHostPool -ResourceGroupName "az140pw"
-Name "pwdeployment" -WorkspaceName "workspacename1" -HostPoolType
"Pooled"
-LoadBalancerType "BreadthFirst" -Location "UK South" -
DesktopAppGroupName "az140pw" -PreferredAppGroupType "Desktop"
```

Figure 6.24 shows the output from running the creation of both a resource group and host pool:

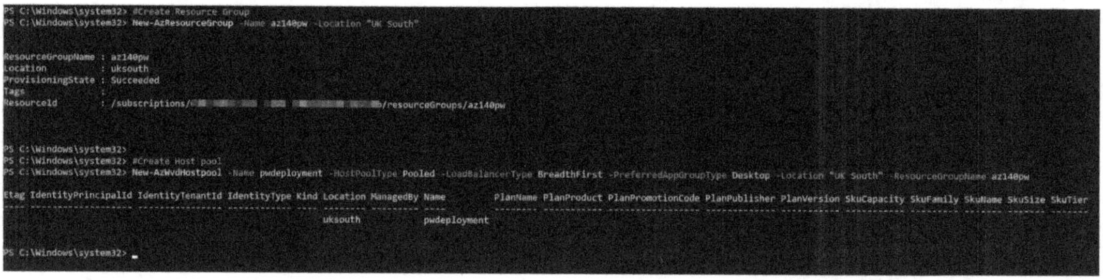

Figure 6.24 – Screenshot showing creating a resource group and creating a host pool

As shown in *Figure 6.24*, we have now deployed our host pool using PowerShell. You can find out more about the cmdlets for creating a host pool at `https://learn.microsoft.com/powershell/module/az.desktopvirtualization/new-azwvdhostpool?view=azps-11.2.0`.

You can check that the resource group and host pool have been created by navigating to the path in the Azure portal. As shown in the following screenshot, you can see the AVD resources have been deployed into the resource group:

Automating the creation of AVD hosts and host pools   169

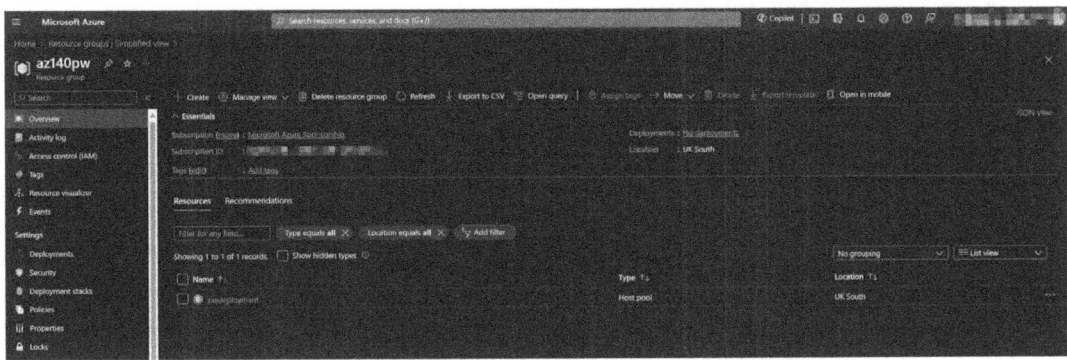

Figure 6.25 – Screenshot showing resources created using PowerShell for AVD

The next step would be to create a registration token for the deployment of session hosts into a host pool. This can be done by using the following cmdlets:

```
New-AzWvdRegistrationInfo -ResourceGroupName <Resource group
name> -HostPoolName <Host pool name> -ExpirationTime $((get-date).
ToUniversalTime().AddHours(<hours eg (2)>).ToString('yyyy-MM-
ddTHH:mm:ss.fffffffZ'))
```

The following provides an example of what the cmdlet should look like:

```
# create a registration token
New-AzWvdRegistrationInfo -ResourceGroupName az140pw -HostPoolName
pwdeployment -ExpirationTime $((get-date).ToUniversalTime().
AddHours(2).ToString('yyyy-MM-ddTHH:mm:ss.fffffffZ'))
```

The following screenshot shows the output of generating a new registration token:

Figure 6.26 – Screenshot showing the creation of a registration token

As shown in *Figure 6.26*, you can see that a new token has been generated ready for deploying VMs in a custom deployment. Token generation is taken care of natively.

Before we recap on deploying session hosts into a host pool, let's look at adding users and groups to a host pool via app groups.

To assign a user to an app group, you would use the following cmdlets:

```
New-AzRoleAssignment -SignInName <User UPN name@company.com>
-RoleDefinitionName "Desktop Virtualization User" -ResourceName <App
group Name> -ResourceGroupName <Resource Group Name> -ResourceType
'Microsoft.DesktopVirtualization/applicationGroups'
```

The following example shows the assignment of a user to an app group within a host pool:

```
New-AzRoleAssignment -SignInName wvdadmin@virtualmanc.onmicrosoft.
com -RoleDefinitionName "Desktop Virtualization User" -ResourceName
az140pw -ResourceGroupName az140pw -ResourceType 'Microsoft.
DesktopVirtualization/applicationGroups'
```

The following screenshot shows the output of running the `New-AzRoleAssignment` cmdlet:

Figure 6.27 – Screenshot showing user assignment to AVD resource group

Figure 6.27 shows the assignment of a user to the app group for the newly created host pool.

You can also assign a group rather than a user using the object ID of the group. You would use the following cmdlets to add a group to an app group:

```
New-AzRoleAssignment -ObjectId <Group Object ID> -RoleDefinitionName
"Desktop Virtualization User" -ResourceName <App Group Name>
-ResourceGroupName <Resource Group Name> -ResourceType 'Microsoft.
DesktopVirtualization/applicationGroups'
```

An example showing the cmdlets to add a group to an app group is as follows:

```
New-AzRoleAssignment -ObjectId c203d0fa-a05a-40be-acd5-d203b252435a
-RoleDefinitionName "Desktop Virtualization User" -ResourceName
az140pw -ResourceGroupName az140pw -ResourceType 'Microsoft.
DesktopVirtualization/applicationGroups'
```

The screenshot in *Figure 6.28* shows the use of the `New-AzRoleAssignment` cmdlet to assign a group via object ID:

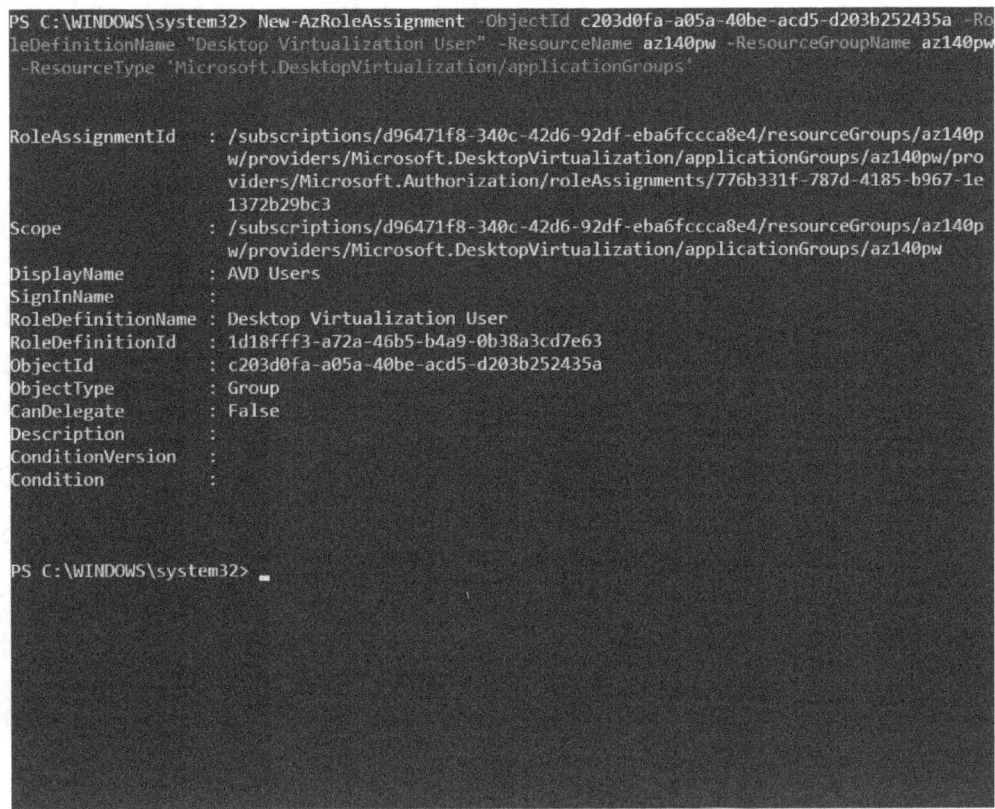

Figure 6.28 – Screenshot showing group assignment to AVD resource group

As shown in *Figure 6.28*, the group has been assigned to the app group.

Once we have finished setting up the host pool and assigning users and groups, we can add session hosts to the host pool. This is typically done using the UI. You can follow the steps detailed in the *Creating VMs within the Create a host pool tab* section.

To add VMs, you need to make sure that you have created a registration token and the previous one has not expired. If it has expired, you will need to run the **Create registration token cmdlets** script again to enable the ability to add VMs to the host pool.

The following steps detail the process of adding or expanding a host pool by adding VMs:

1. Sign in to the Azure portal.
2. Search for and select **Azure Virtual Desktop**.
3. Head to the menu on the left side of the screen and select **Host pools**. Following that, select the name of the host pool you want to add VMs to.
4. Select **Session hosts** from the menu on the left side of the screen.
5. Select + **Add** to start creating your host pool, as shown in *Figure 6.29*.

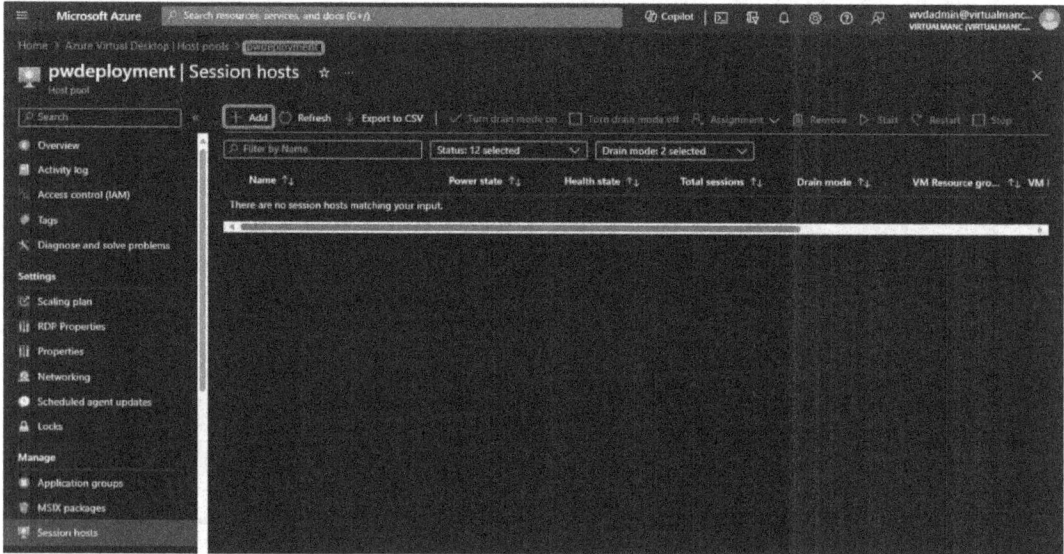

Figure 6.29 – Adding session hosts to the new deployment

Figure 6.30 shows the **Add virtual machines to a host pool** page within the Azure portal for deploying VMs into a host pool:

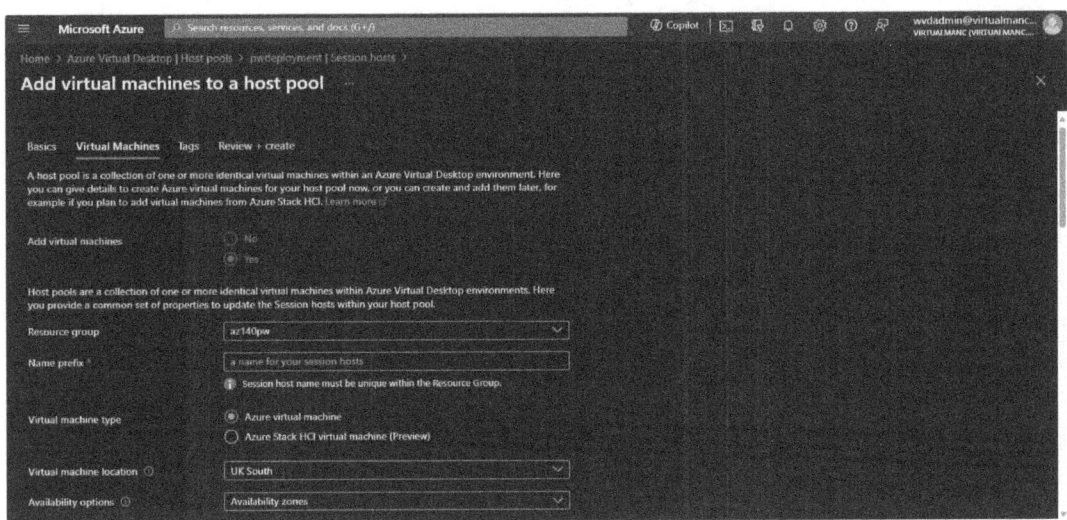

Figure 6.30 – Screenshot showing the Add virtual machines to a host pool page

6. Ignore the **Basics** tab as this is not required; instead, select the **Virtual Machines** tab. Under this tab, you can view and edit the details of the VM you want to add to the host pool.

7. Select the resource group you want to create the VMs under, then select the Azure region. You can also choose the current Azure region you're using or a new Azure region.

8. Enter the required number of session hosts you wish to add to your host pool. For example, if you're expanding your host pool by two hosts, enter 2.

> Important note
> Although it's possible to edit the image and prefix of the VMs, Microsoft does not recommend editing them if you have VMs with different images in the same host pool. Edit the image and prefix only if you plan to remove VMs with older images from the affected host pool.

9. For **Virtual network information**, select the virtual network and subnet to which you want to join the VMs. You can choose either the virtual network your existing machines currently use or opt for a different one that better suits the Azure region you selected in *step 7*.

10. For **Administrator account**, enter the username and password associated with **Entra ID Domain Services (Entra ID DS)** on the virtual network you selected. These credentials will be used to join the VMs to the virtual network. Ensure you check the password as the deployment will fail if incorrect.

> **Tip**
> Ensure your admin names comply with the information given on screen and MFA is not enabled on the account.

11. Select the **Tag** tab or skip if not required.
12. Select the **Review + create** tab. Review your configuration, and if everything looks fine, select **Create** to start the deployment.
13. Once the deployment has finished, check your VMs, and you should be ready to test.

> **Tip**
> You can also deploy the host pool and create session host VMs using an ARM template. You can download an example from `https://github.com/Azure/RDS-Templates/tree/master/ARM-wvd-templates`.
>
> It is also important to note that if you are using an automated process to build your AVD environment, you will need to use the latest configuration JSON file available. You can download this from `https://wvdportalstorageblob.blob.core.windows.net/galleryartifacts/armtemplates/Hostpool_10-13-2021/CreateHostpoolTemplate.json`.

You are now skilled in configuring PowerShell, then creating a resource group, deploying a host pool, generating a registration token, and configuring users/groups.

Summary

In this chapter, we looked a creating a host pool using the Azure portal and PowerShell. We looked at some of the requirements and gotchas associated with creating a host pool and how to overcome these. We also deployed a host pool using PowerShell and assigned users access to the host pool via app groups.

In the next chapter, we will continue our journey through AVD and take a look at configuring host pools and session hosts. This includes creating Windows Server session hosts, configuring host pool settings, assigning users to host pools, and finally, applying updates and security and compliance settings for session hosts.

Questions

Here are a few questions to test your understanding of this chapter:

1. Before creating a host pool, what is the first thing you should check?

    That the Microsoft.DesktopVirtualization resource provider is registered.

2. Which load-balancing method is used to consolidate sessions onto a session host before allocating sessions to a new session host?

    Breadth-first load-balancing method.

3. You plan to deploy your first AVD host pool. What do you need to configure on the virtual network before you deploy your first host pool?

    Modify the DNS settings of the virtual network ("line of sight" with the Active Directory domain).

7
Configuring Azure Virtual Desktop Host Pools

In this chapter, we will look at configuring **Azure Virtual Desktop** (**AVD**) host pools. This includes configuring licenses for server-based multi-session deployments, host pool settings, using PowerShell to customize RDP properties, load-balancing methods, personal desktop assignment, and securing host pools and session hosts.

We start this chapter by looking at deploying Windows Server licensing for those who want to move existing Windows Server session hosts from on-premises to AVD. This may not be a common use case for AVD deployments; however, it's important that we cover this concept for those who require the use of Windows Server session hosts within AVD.

We will cover the following topics in this chapter:

- Windows Server session host licensing
- Configuring host pool settings
- Configuring Azure Virtual Desktop load-balancing methods
- Assigning users to host pools
- Applying OS and application updates to a running Azure Virtual Desktop host
- Applying security and compliance settings to session hosts

> **Important note**
> It is important to note that Windows 10 and Windows 11 including multi-session do not require a licensing server.

Windows Server session host licensing

This section covers installing the **Remote Desktop** (**RD**) licensing role and activating the licensing server to use server OSs in AVD. To use a server OS in AVD, you will need to deploy a Windows Server with the RD licensing role installed and configured on the Azure virtual network. The following steps summarize the installation and configuration of the RD licensing role:

1. Sign in to the Windows Server with an administrator account.
2. In the **Server Manager** console, click **Roles Summary**, and then click **Add Roles**. Click **Next** on the first page of the roles wizard.
3. Select **Remote Desktop Services**, click **Next**, and then click **Next** on the **Remote Desktop Services** page.
4. Select **Remote Desktop Licensing**, and then click **Next**:

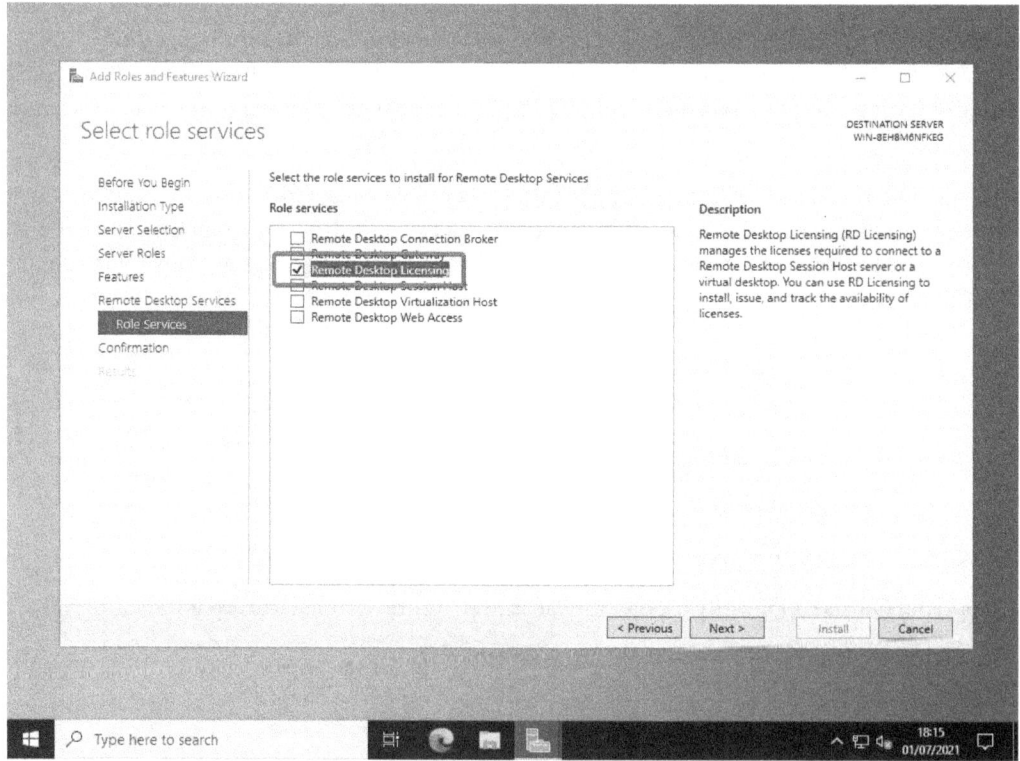

Figure 7.1 – The Add Roles and Features Wizard window for the server OS

5. To configure the domain, select **Configure a discovery scope for this license server**, click **This domain**, and then click **Next**.

6. Then, click **Install**:

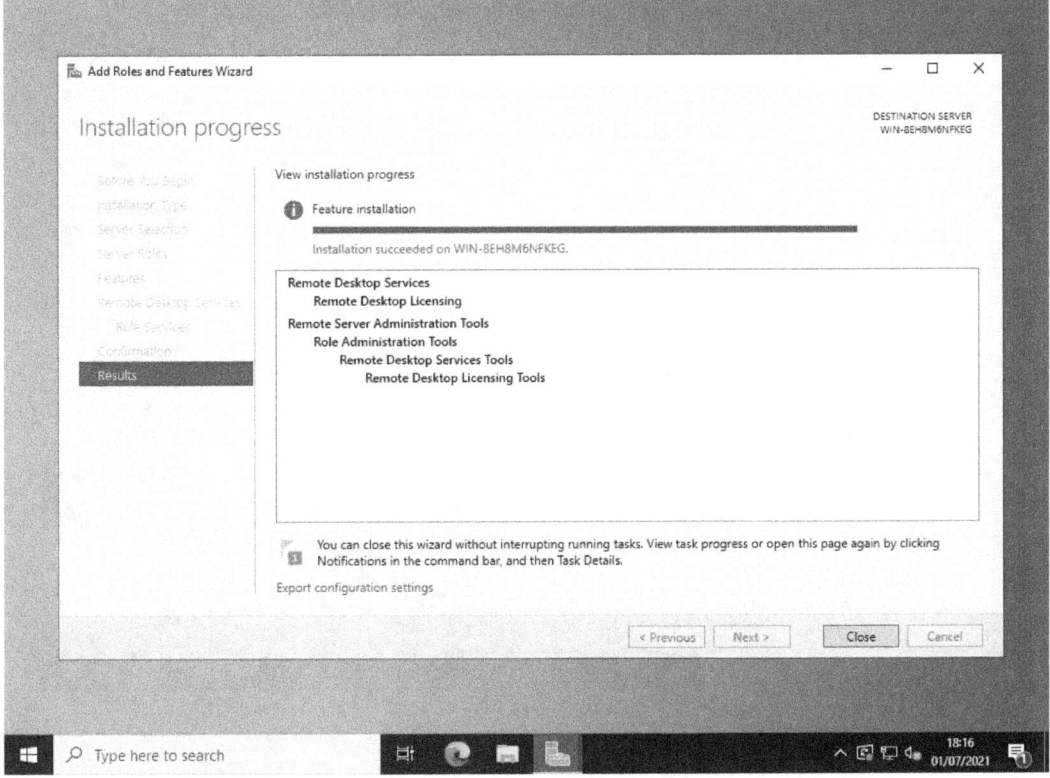

Figure 7.2 – Installation progress of the RD licensing server role

Once installed, the **Remote Desktop Licensing Manager** app will appear in the Start menu.

To activate the licensing server, please use the following steps:

1. Open the **Remote Desktop Licensing Manager** app, then go to **Start | Administrative Tools | Remote Desktop Services | Remote Desktop Licensing Manager**. You will get what is shown in the following screenshot:

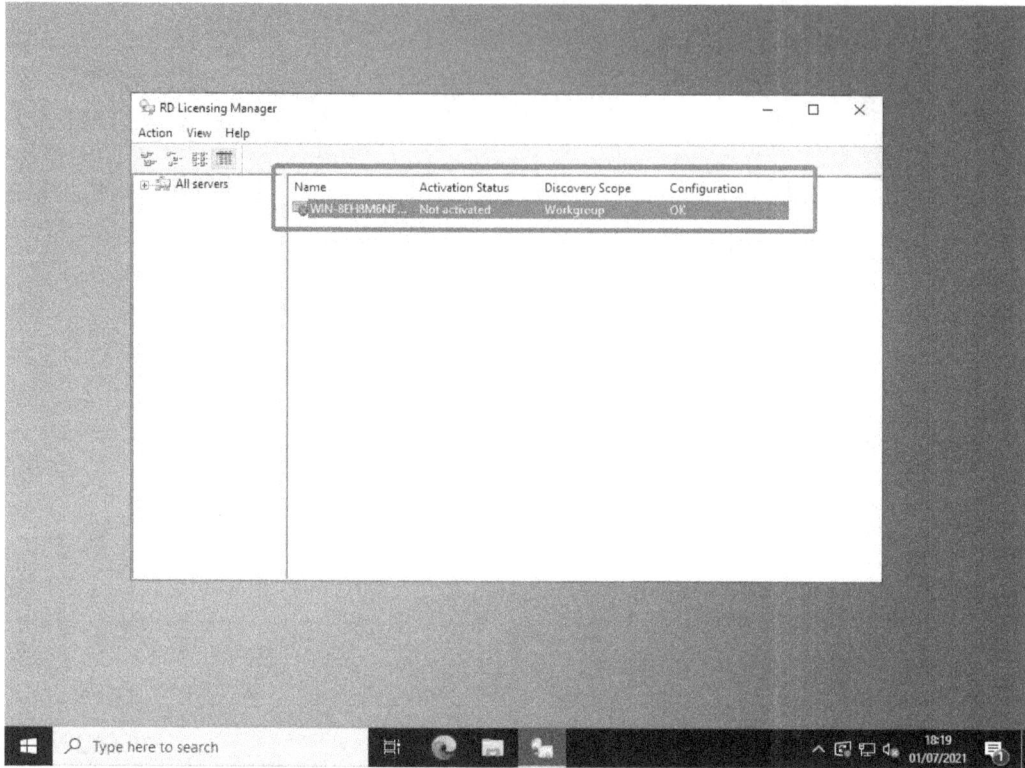

Figure 7.3 – License server role installed but not activated

2. Right-click the license server, and then click the **Activate Server** option.
3. Click **Next** on the welcome page.
4. For the connection method, select **Automatic connection** as recommended by Microsoft, and then click **Next**:

Windows Server session host licensing 181

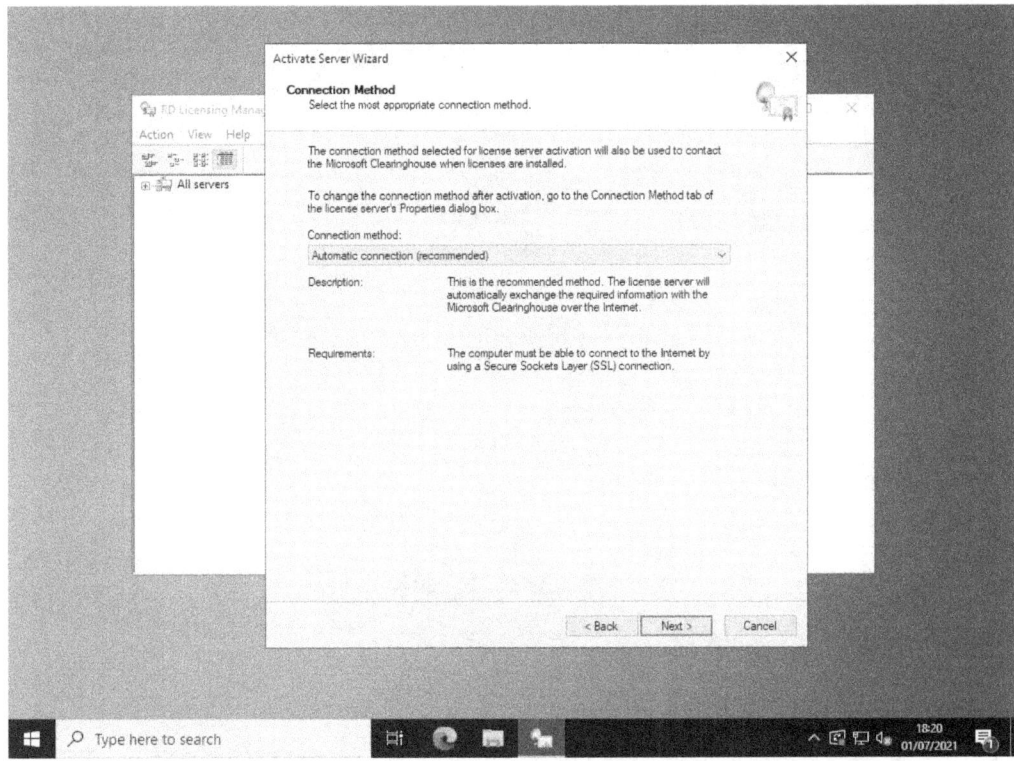

Figure 7.4 – Active RD license server wizard

5. Enter your company details (your name, the company name, and geographic region), and then click **Next**.

6. You can also enter optional information (for example, email and company addresses) and then click **Next**.

7. Ensure that **Start Install Licenses Wizard** is not selected as this is done later, and then click **Next**. Your license server is ready to start issuing and managing licenses:

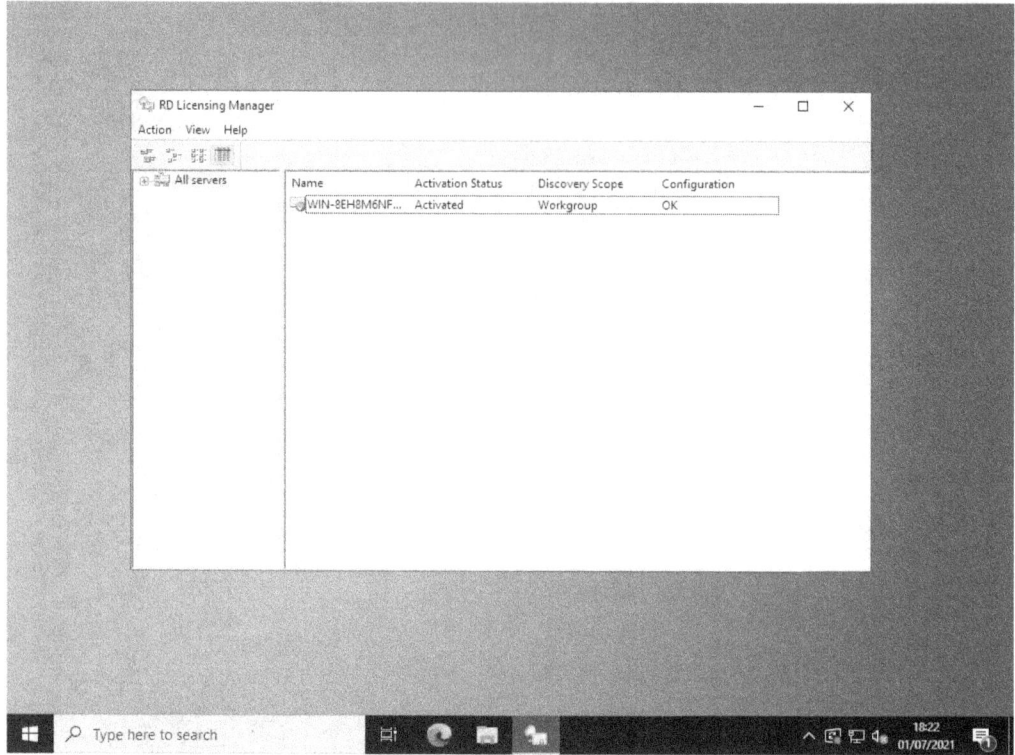

Figure 7.5 – RD license server activated

For information on installing the RDS CAL, please use the following link to the Microsoft article: https://docs.microsoft.com/windows-server/remote/remote-desktop-services/rds-install-cals

There you have it; we have now configured an RD licensing server for use when using server OSs as session hosts on our AVD environment. In the next section, we take a look at configuring host pool settings.

Configuring host pool settings

This section looks at customizing a host pool, including RD properties, load-balancing methods, and configuring personal host pool assignment types.

Customizing RDP properties

AVD allows you to configure and customize host pool settings using the RD protocol properties. This allows you to configure things such as audio redirection, video playback, and drive redirections.

The supported RDP settings are split into five different areas:

- Connection information
- Session behavior
- Device redirection
- Display settings
- RemoteApp

These different categories contain a wide range of settings you can apply to your host pool configuration.

You can find a complete list of settings that can be applied or changed here: `https://learn.microsoft.com/azure/virtual-desktop/rdp-properties`

You can change the RDP settings by navigating to the host pool within the Azure portal and selecting **RDP Properties**. You will then see the following five tabs:

- **Connection information**
- **Session behaviour**
- **Device redirection**
- **Display settings**
- **Advanced**

> **Important note**
>
> You can read more about redirection support for each different client type here: `https://docs.microsoft.com/windows-server/remote/remote-desktop-services/clients/remote-desktop-app-compare#redirection-support`

The tabs are shown in the following screenshot:

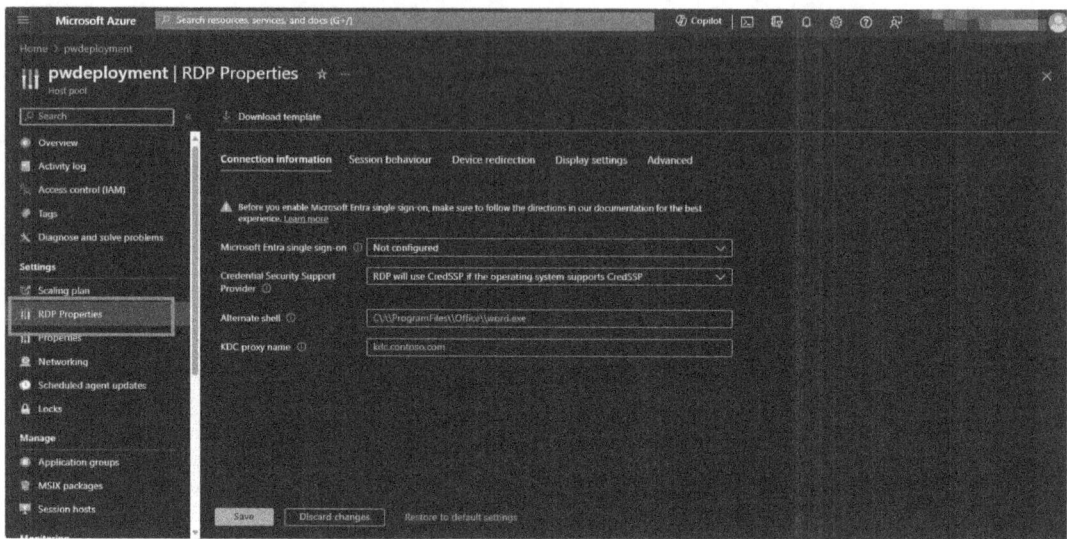

Figure 7.6 – RDP properties within a host pool

The Azure portal makes it easy for IT admins to make changes to the RDP properties as you can set parameters in the **Advanced** section or use the specific tabs that have drop-down boxes. You can see in the following screenshot an example of what to expect in the tabs:

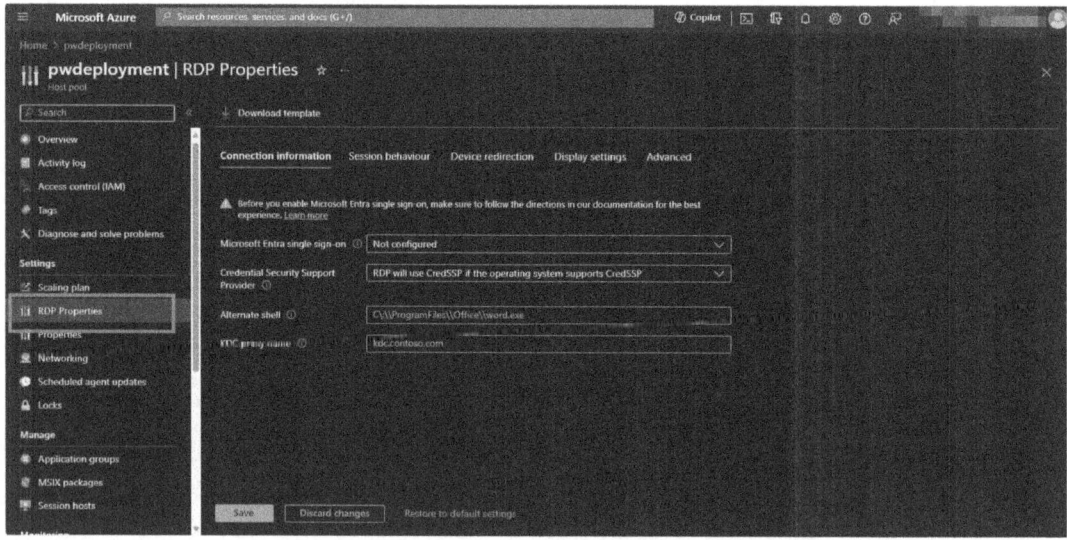

Figure 7.7 – The Connection information tab for RDP properties with a host pool

If you are looking to make more specific changes and customizations to the host pool's RDP properties, you can make changes within the **Advanced** tab, as shown in *Figure 7.8*. It is also important to note that there are three buttons at the bottom of this Azure blade – **Save**, **Discard changes**, and **Restore to default settings**:

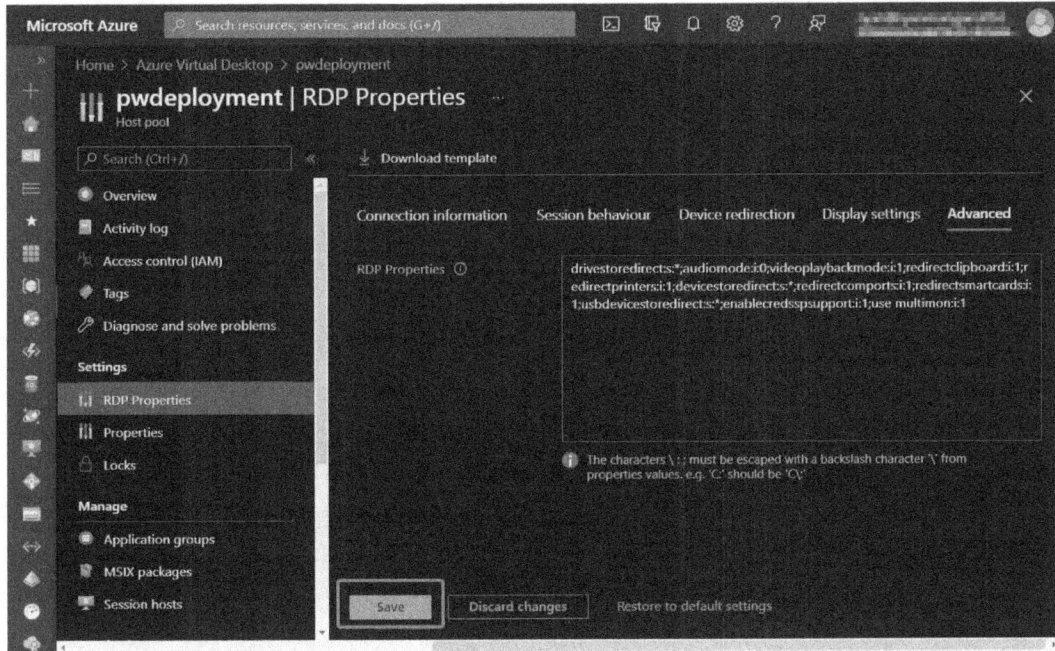

Figure 7.8 – Advanced tab of RDP properties of a host pool

Remember to click **Save** to apply customized settings and **Discard changes** if you want to revert before choosing customizations. It's also important to note that you can also use the **Restore to default settings** link to revert the custom RDP properties to the defaults.

The step-by-step process of configuring RDP properties for a host pool is as follows:

1. Sign in to the Azure portal using `https://portal.azure.com`.
2. Enter `Azure Virtual Desktop` into the search bar located at the top of the page.
3. Under **Services**, select **Azure Virtual Desktop**.
4. Within the **Azure Virtual Desktop** page, select **Host pools** on the left-hand side of the screen.
5. Choose the host pool you want to update.
6. Select **RDP Properties** in the menu on the left-hand side of the screen.
7. Set the property you want.

8. As mentioned, you can also use the **Advanced** tab to add/customize RDP properties in a semicolon-separated format.

9. Select **Save** to save changes when you have finished.

In the next section, we will examine using PowerShell to configure RDP properties for a host pool.

Using PowerShell to customize RDP properties

Before you can use PowerShell to customize your host pool RDP properties, you will need to configure the PowerShell module for AVD as a prerequisite. I showed you how to do this in *Chapter 6, Creating Host Pools and Session Hosts*.

Adding or editing a single RDP property

In this subsection, we will look at adding and editing single/multiple RDP properties.

The following shows you the cmdlet parameters for updating an RDP property. This is useful for when you want to change an existing configuration:

```
Update-AzWvdHostPool -ResourceGroupName <resourcegroupname> -Name <hostpoolname> -CustomRdpProperty <property>
```

In this example, I have applied the redirect clipboard custom RDP property by specifying `redirectclipboard:i:1`. This property enables the clipboard feature when copying and pasting files between the remote session and the local desktop:

```
Update-AzWvdHostPool -ResourceGroupName az140pw -Name pwdeployment  -CustomRdpProperty redirectclipboard:i:1
```

Here's the output:

Figure 7.9 – Cmdlets and output from running custom RDP property cmdlets

To check what RDP properties are applied to the host pool, you can use the following:

```
Get-AzWvdHostPool -ResourceGroupName <resourcegroupname> -Name
<hostpoolname> | format-list Name, CustomRdpProperty
```

The following provides an example of the usage:

```
get-AzWvdHostPool -ResourceGroupName az140pw -Name pwdeployment |
format-list name, customRdpProperty
```

The following screenshot shows the Get-AzWvdHostpool cmdlet output; as you can see, the redirect clipboard property we added is now set:

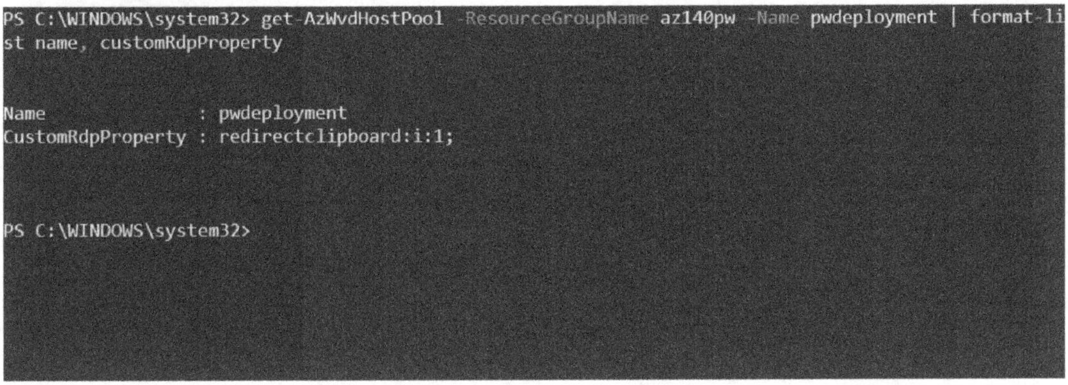

Figure 7.10 – get-AzWvdHostPool cmdlets for checking the custom RDP properties

Adding or editing multiple custom RDP properties

To add multiple custom RDP properties, you can use the following PowerShell variables to add multiple properties with a semicolon, as follows:

```
$properties="<property1>;<property2>;<property3>"
Update-AzWvdHostPool -ResourceGroupName <resourcegroupname> -Name 
<hostpoolname> -CustomRdpProperty $properties
```

The following provides an example of the usage:

```
$properties="redirectclipboard:i:1;use 
multimon:i:1;drivestoredirect:s:*"
Update-AzWvdHostPool -ResourceGroupName az140pw -Name pwdeployment 
-CustomRdpProperty $properties
```

The following screenshot shows the output from running the PowerShell cmdlet for adding multiple RDP properties:

```
PS C:\WINDOWS\system32> $properties="redirectclipboard:i:1;use multimon:i:1;drivestoredirect:s:*"
PS C:\WINDOWS\system32> Update-AzWvdHostPool -ResourceGroupName az140pw -Name pwdeployment -CustomR
dpProperty $properties

Location Name         Type
-------- ----         ----
uksouth  pwdeployment Microsoft.DesktopVirtualization/hostpools



PS C:\WINDOWS\system32>
```

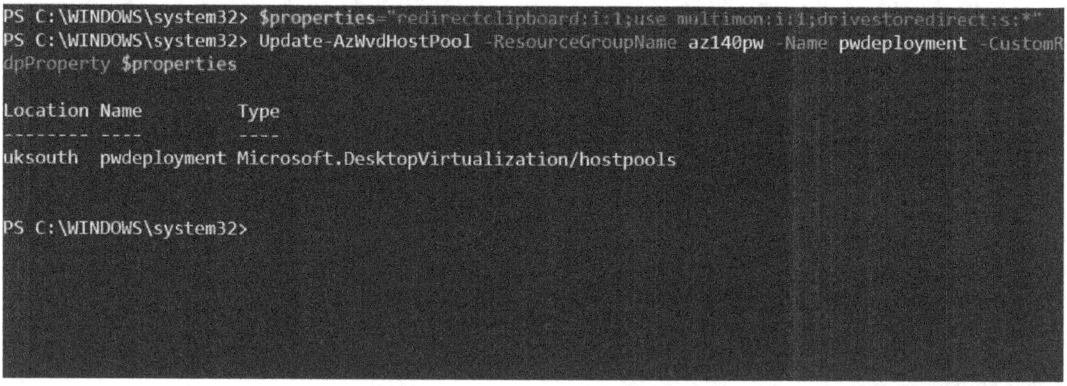

Figure 7.11 – Cmdlets for assigning multiple RDP properties using PowerShell

Using the same `get-AzWvdHostPool` PowerShell cmdlet, you can check to see whether these properties have been applied.

The following screenshot shows the `get-AzWvdHostPool` cmdlet being run to confirm that the RDP properties have been set:

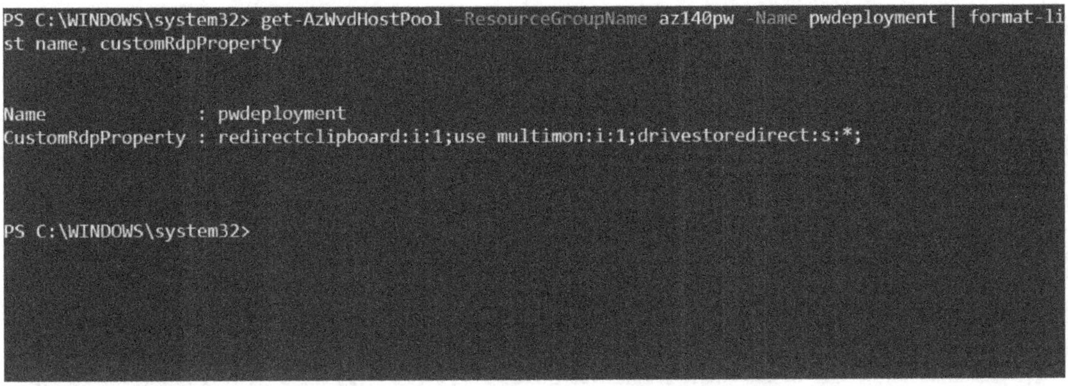

```
PS C:\WINDOWS\system32> get-AzWvdHostPool -ResourceGroupName az140pw -Name pwdeployment | format-li
st name, customRdpProperty


Name             : pwdeployment
CustomRdpProperty : redirectclipboard:i:1;use multimon:i:1;drivestoredirect:s:*;


PS C:\WINDOWS\system32>
```

Figure 7.12 – get-azwvdhostpool cmdlets for checking the custom RDP properties have been set

Resetting all custom RDP properties using PowerShell

To reset the RDP properties to the default, you would use the following cmdlets:

```
Update-AzWvdHostPool -ResourceGroupName <resourcegroupname> -Name
<hostpoolname> -CustomRdpProperty ""
```

The following provides an example of the usage:

```
Update-AzWvdHostPool -ResourceGroupName az140pw -Name pwdeployment
-CustomRdpProperty ""
```

The following screenshot shows the reset PowerShell cmdlets being run:

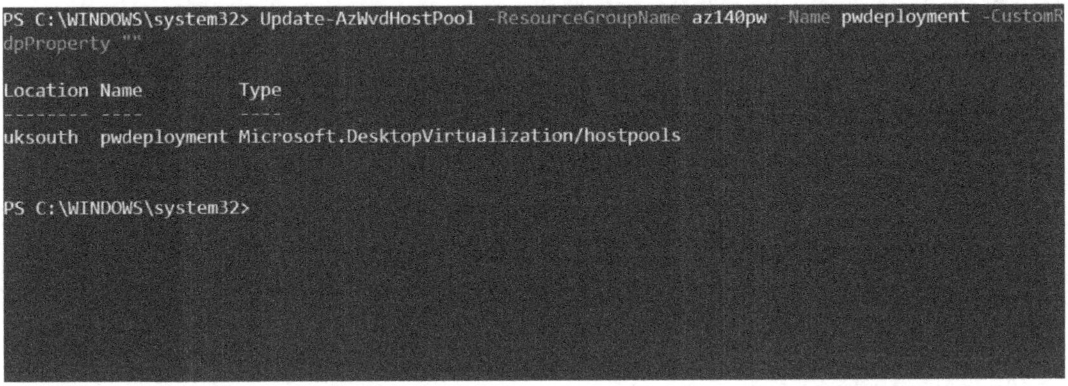

```
PS C:\WINDOWS\system32> Update-AzWvdHostPool -ResourceGroupName az140pw -Name pwdeployment -CustomR
dpProperty ""

Location Name         Type
-------- ----         ----
uksouth  pwdeployment Microsoft.DesktopVirtualization/hostpools


PS C:\WINDOWS\system32>
```

Figure 7.13 – Cmdlets for resetting the custom RDP properties with PowerShell

Again, you can use the same `get-AzWvdHostPool` cmdlet; you can check to see whether these properties have been applied. The following screenshot shows the RDP properties being reset:

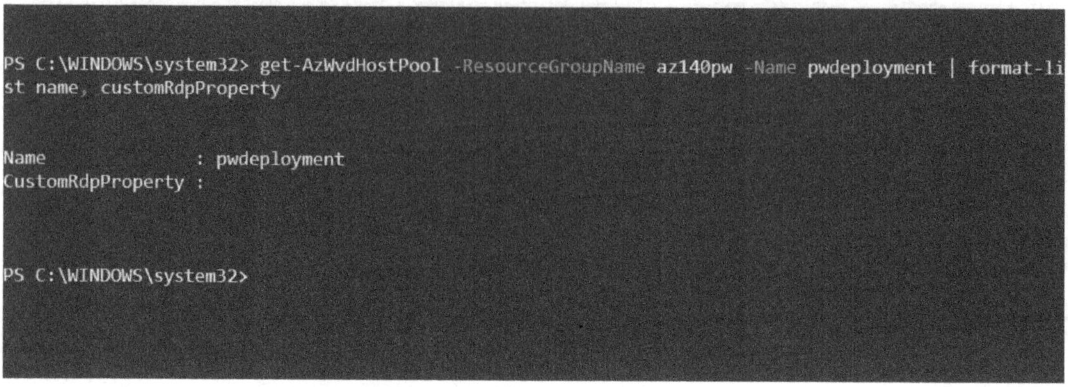

```
PS C:\WINDOWS\system32> get-AzWvdHostPool -ResourceGroupName az140pw -Name pwdeployment | format-li
st name, customRdpProperty


Name               : pwdeployment
CustomRdpProperty :



PS C:\WINDOWS\system32>
```

Figure 7.14 – get-azwvdhostpool cmdlets for checking the reset has been completed

In this section, we looked at configuring custom RDP properties for host pools. This covered both the configuration through the Azure portal and using PowerShell. In the next section, we take a look at configuring load-balancing methods on host pools.

Configuring Azure Virtual Desktop load-balancing methods

In the previous chapter, we looked at the different types of session load balancing. We will now take a look at configuring load balancing within the Azure portal and PowerShell.

This section examines configuring the load-balancing method for AVD using both the Azure portal UI and PowerShell.

Let's now take a look at how to configure load balancing.

Configuring load balancing

We'll now look at configuring host pool load balancing. The following steps walk you through configuring load balancing within the Azure portal:

1. Sign in to the Azure portal using the following URL: https://portal.azure.com.
2. Search for and select **Azure Virtual Desktop** under **Services**.
3. On the **Azure Virtual Desktop** page, select **Host pools**.
4. Select the name of the host pool you want to edit.
5. Select **Properties**.
6. Select the load-balancing algorithm you want for this host pool in the drop-down menu.
7. Enter the required **Max session limit** value into the field:

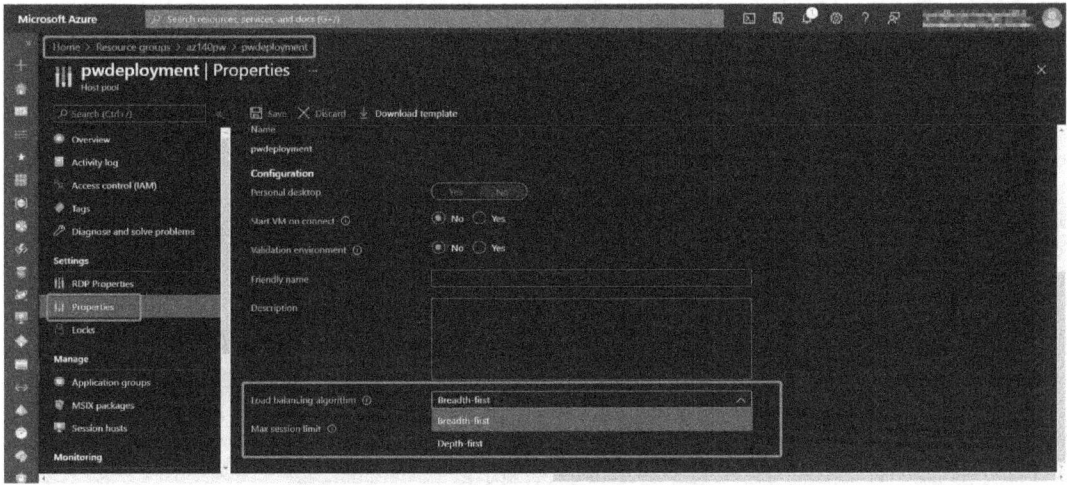

Figure 7.15 – Change the load-balancing method of a host pool in the Azure portal

8. Select **Save**. This applies to the new load-balancing settings.

You have now configured your load-balancing settings. Let's move on to the next section, where we will take a look at the option to configure load-balancing settings using PowerShell.

Using PowerShell to configure load-balancing methods

Before you can use PowerShell to customize your host pool load-balancing methods, you will need to set up and load the PowerShell module for AVD as a prerequisite. I showed you how to do this in *Chapter 6, Creating Host Pools and Session Hosts*.

Breath-first load balancing

In this subsection, we take a look at the **breadth-first load-balancing method** using PowerShell. Breadth-first is the default configuration when deploying pooled session hosts in a host pool. As a recap, this method distributes connections evenly to available session hosts within the host pool.

> **Important note**
> `MaxSessionLimit` is used to control the maximum number of sessions allowed per session host in the AVD host pool. When using depth-first load balancing, the value is used to determine when to stop allocating users to one host and start sending user sessions to the next host.

The following cmdlets are used to configure the load-balancing methods:

```
Update-AzWvdHostPool -ResourceGroupName <resourcegroupname> -Name
<hostpoolname> -LoadBalancerType 'BreadthFirst' -MaxSessionLimit ###
```

The following example demonstrates how to change/update the load-balancing method:

```
Update-AzWvdHostPool -ResourceGroupName az140pw -Name pwdeployment
-LoadBalancerType 'BreadthFirst' -MaxSessionLimit 999999
```

Here's the output:

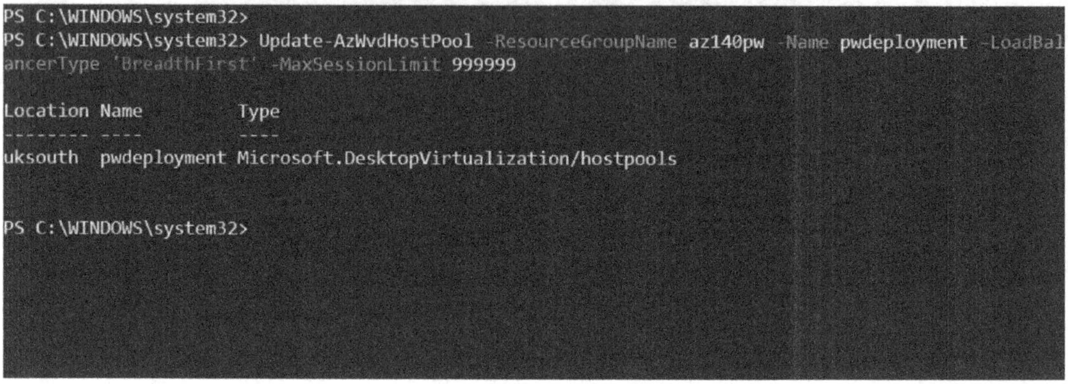

```
PS C:\WINDOWS\system32>
PS C:\WINDOWS\system32> Update-AzWvdHostPool -ResourceGroupName az140pw -Name pwdeployment -LoadBal
ancerType 'BreadthFirst' -MaxSessionLimit 999999

Location Name         Type
-------- ----         ----
uksouth  pwdeployment Microsoft.DesktopVirtualization/hostpools


PS C:\WINDOWS\system32>
```

Figure 7.16 – Cmdlets for setting the BreadthFirst load-balancing method

To check that the setting has been applied, you can run the following cmdlet:

```
Get-AzWvdHostPool -ResourceGroupName <resourcegroupname> -Name
<hostpoolname> | format-list Name, LoadBalancerType, MaxSessionLimit
```

The following example demonstrates how to check that breadth-first has been set:

```
Get-AzWvdHostPool -ResourceGroupName az140pw -Name pwdeployment |
format-list Name, LoadBalancerType, MaxSessionLimit
```

Here's the output:

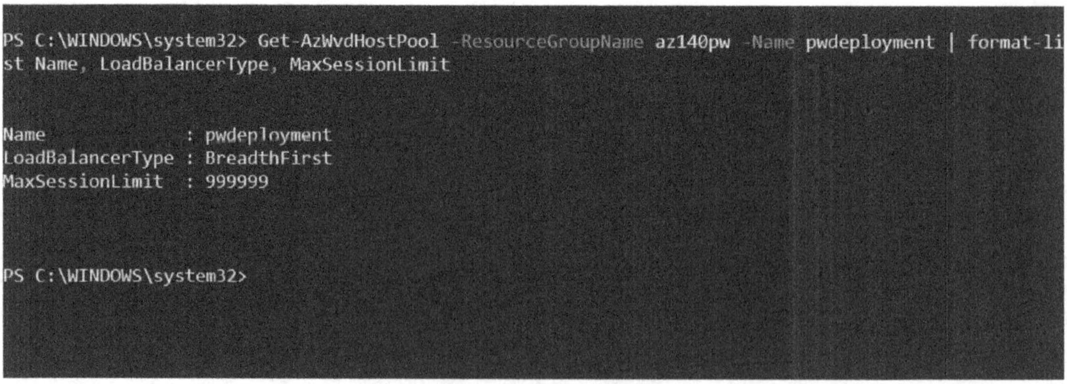

```
PS C:\WINDOWS\system32> Get-AzWvdHostPool -ResourceGroupName az140pw -Name pwdeployment | format-li
st Name, LoadBalancerType, MaxSessionLimit


Name             : pwdeployment
LoadBalancerType : BreadthFirst
MaxSessionLimit  : 999999



PS C:\WINDOWS\system32>
```

Figure 7.17 – Cmdlets for checking the load-balancing method set on a host pool

Depth-first load balancing

Depth-first load balancing is used to populate session hosts with the highest number of connections that have not reached the maximum session limit threshold.

> **Important note**
> You must enter a maximum session limit per session host within the host pool when configuring depth-first load balancing.

To set depth-first load balancing to your host pool, you would need to use the following cmdlets:

```
Update-AzWvdHostPool -ResourceGroupName <resourcegroupname> -Name
<hostpoolname> -LoadBalancerType 'DepthFirst' -MaxSessionLimit ###
```

The following provides an example of the usage:

```
Update-AzWvdHostPool -ResourceGroupName az140pw -Name pwdeployment
-LoadBalancerType 'DepthFirst' -MaxSessionLimit 999999
```

Here's the output:

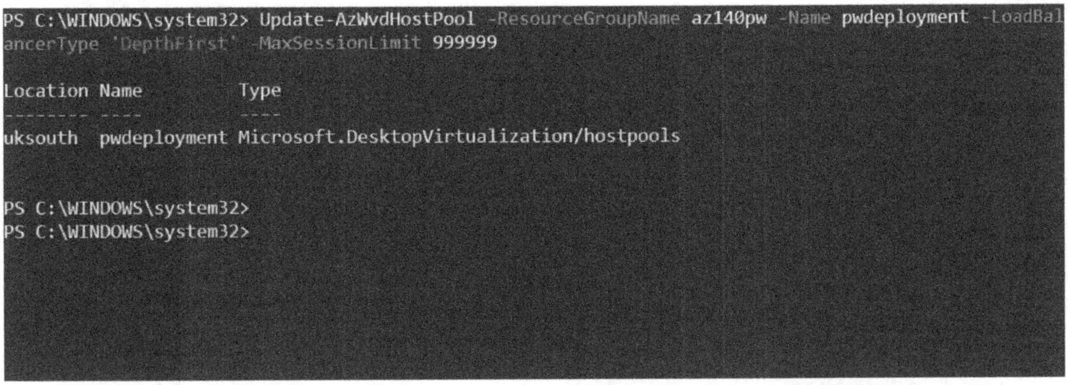

```
PS C:\WINDOWS\system32> Update-AzWvdHostPool -ResourceGroupName az140pw -Name pwdeployment -LoadBal
ancerType 'DepthFirst' -MaxSessionLimit 999999

Location Name          Type
-------- ----          ----
uksouth  pwdeployment  Microsoft.DesktopVirtualization/hostpools


PS C:\WINDOWS\system32>
PS C:\WINDOWS\system32>
```

Figure 7.18 – Update cmdlet for changing the load-balancing method of a host pool

To confirm settings have been applied, use the following cmdlets, as shown in the breadth-first example:

```
Get-AzWvdHostPool -ResourceGroupName az140pw -Name pwdeployment |
format-list Name, LoadBalancerType, MaxSessionLimit
```

Here's the output:

```
PS C:\WINDOWS\system32> Get-AzWvdHostPool -ResourceGroupName az140pw -Name pwdeployment | format-li
st Name, LoadBalancerType, MaxSessionLimit


Name             : pwdeployment
LoadBalancerType : DepthFirst
MaxSessionLimit  : 999999



PS C:\WINDOWS\system32>
```

Figure 7.19 – Cmdlets and output of checking the load-balancing method of a host pool

This section looked at the configuration and updating of the load-balancing method within a host pool. I covered both ways – the Azure portal UI and PowerShell. We'll now move on to assigning users to host pools.

Assigning users to host pools

Multi-session is great for when you want to provide a desktop experience to users where they all share the same compute resource of a session host. In this section, we will take a look at assigning personal desktops to users. Personal host pools allow you to commit compute resources per user as you are allocating a session host per user. When you deploy a personal host pool, users are not assigned to specific session hosts. Instead, personal session hosts are used for specific use cases, such as developers or users who need administrator access to the desktop. Other user cases include software license requirements that do not allow multiple users to connect to the device.

Assigning users to a session host in the Azure portal

In this subsection, we will look at assigning users to personal session hosts using the Azure portal.

> **Important note**
> Please note this example is for assigning users and groups to personal desktops; however, the process is the same for both personal and multi-session-based deployments.

The following steps show you how to assign a user to a personal desktop:

1. Sign in to the Azure portal using the following UI: https://portal.azure.com.
2. Enter Azure Virtual Desktop into the search bar located at the top of the page.

3. Under **Services**, select **Azure Virtual Desktop**.
4. Within the **Azure Virtual Desktop** page, on the window's left-hand side, select **Host pools**.
5. Select the host pool you want to update.
6. Next, go to the menu on the left-hand side of the page and select **Application groups**:

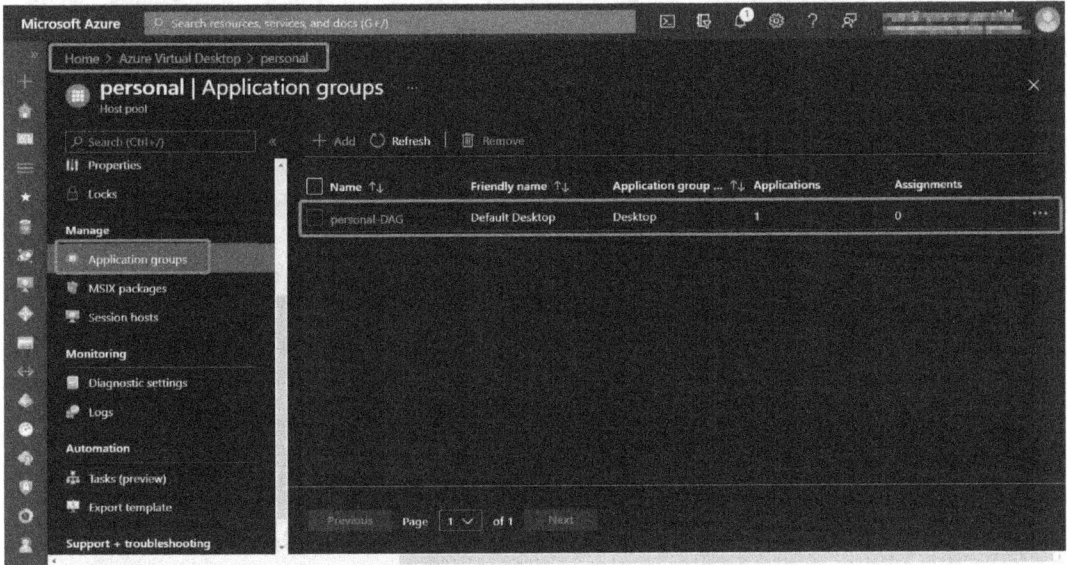

Figure 7.20 – The application groups within a host pool

7. Select the desktop app group name you want to edit, then select **Assignments** in the menu on the left-hand side of the window.
8. Select + **Add**, then select the users or user groups you want to publish in the desktop app group:

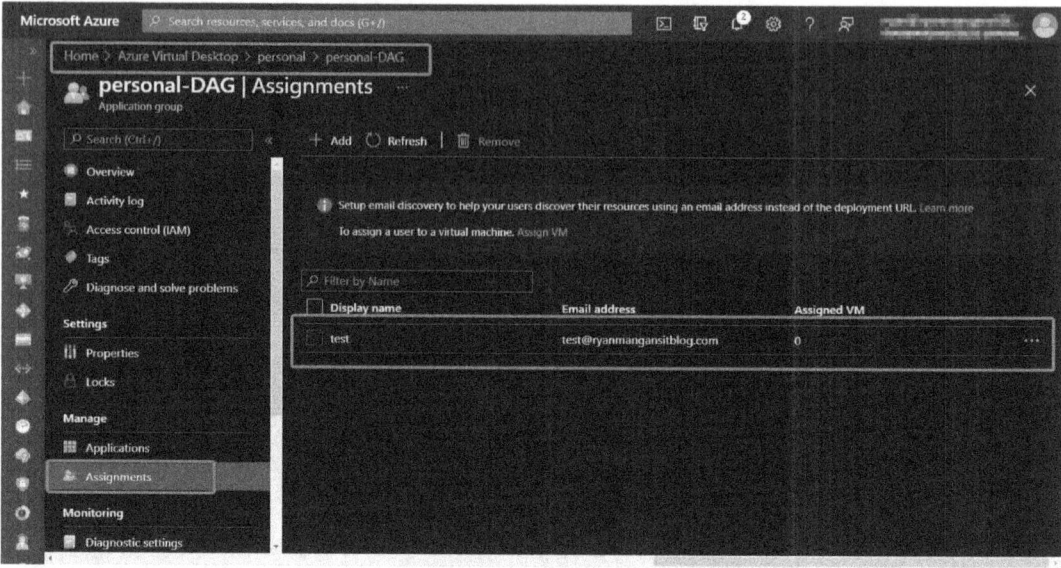

Figure 7.21 – User assignments within an app group

9. Select **Assign VM** in the information bar to assign a session host to a user.
10. Select the session host you want to assign to the user, then select **Assign**:

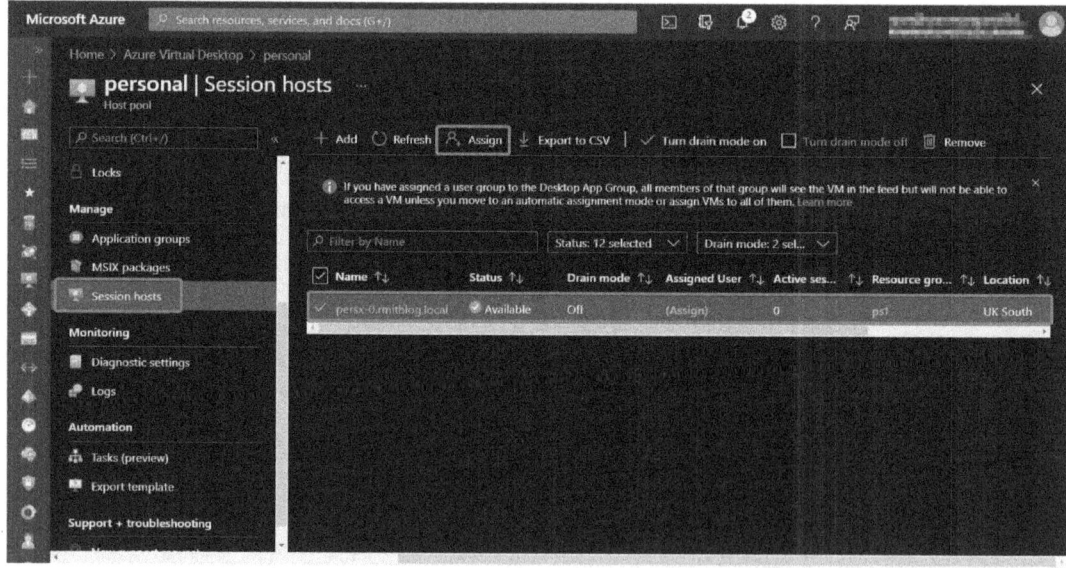

Figure 7.22 – The Assign button for assigning a user a personal session host within the host pool

11. Select the user you want to assign the session host to from the list of available users, as shown in the following screenshot:

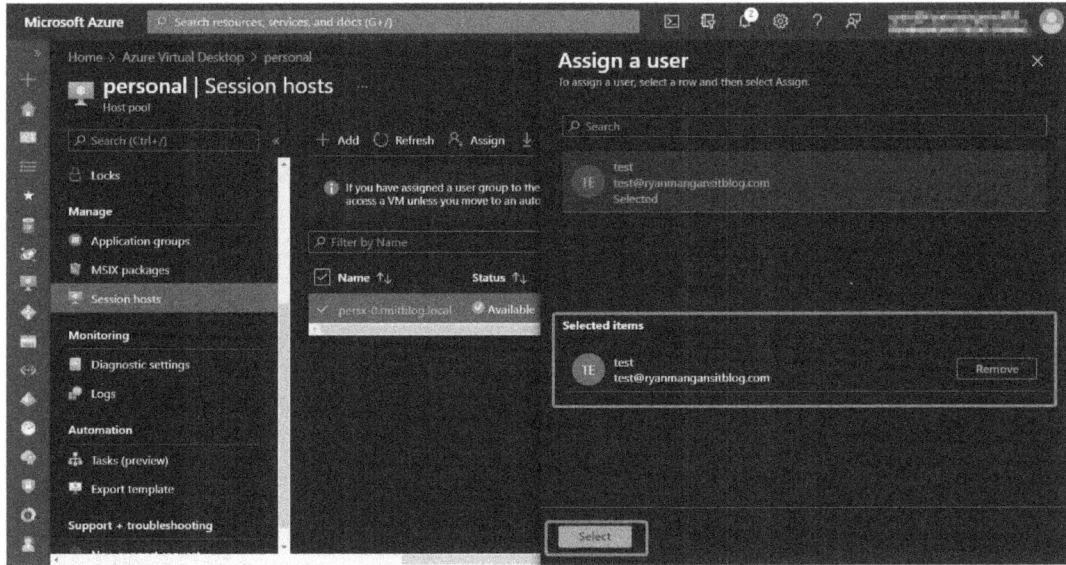

Figure 7.23 – Assign a user blade within the host pool

12. When you have finished, click **Select**.

There you have it – you have assigned a user to a session host:

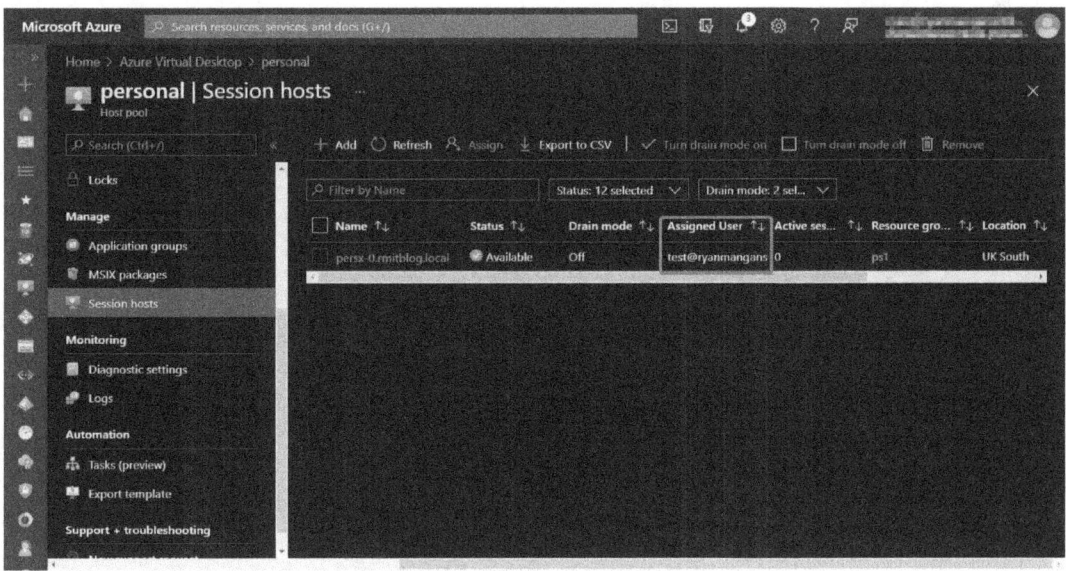

Figure 7.24 – User now assigned to a personal session host within the host pool

In this section, we looked at assigning users personal desktop hosts via the Azure portal UI. We will now take a look at how to do it via PowerShell.

Assigning users to host pools via PowerShell

Before you can use PowerShell to assign personal desktops, you will need to set up and load the PowerShell module for AVD as a prerequisite. I showed you how to do this in *Chapter 6*, *Creating Host Pools and Session Hosts*.

Configuring automatic assignment

In this subsection, we will look at configuring automatic assignments, which is helpful for users who do not need or require a specific session host.

Automatic assignment works by first assigning users to the desktop in the personal host pool. Then, when the user opens the RD client and clicks on the advertised resources in their feed, they will be allocated an available session host, which completes the automatic assignment process.

To enable automatic assignment, you would need to run the following PowerShell cmdlets:

```
Update-AzWvdHostPool -ResourceGroupName <resourcegroupname> -Name
<hostpoolname> -PersonalDesktopAssignmentType Automatic
```

The following example shows the assignment:

```
Update-AzWvdHostPool -ResourceGroupName ps1 -Name personal
-PersonalDesktopAssignmentType Automatic
```

Here's the output:

```
PS C:\WINDOWS\system32> Update-AzWvdHostPool -ResourceGroupName ps1 -Name personal -PersonalDesktop
AssignmentType Automatic

Location Name     Type
-------- ----     ----
uksouth  personal Microsoft.DesktopVirtualization/hostpools


PS C:\WINDOWS\system32>
```

Figure 7.25 – Cmdlets for setting the personal assignment to Automatic

To assign users to a personal desktop host pool, you would use the following cmdlets:

```
New-AzRoleAssignment -SignInName <userupn> -RoleDefinitionName
"Desktop Virtualization User" -ResourceName <appgroupname>
-ResourceGroupName <resourcegroupname> -ResourceType 'Microsoft.
DesktopVirtualization/applicationGroups'
```

The following provides an example of the usage:

```
New-AzRoleAssignment -SignInName testuser@ryanmangansitblog.com
-RoleDefinitionName "Desktop Virtualization User" -ResourceName
personal-DAG -ResourceGroupName ps1 -ResourceType 'Microsoft.
DesktopVirtualization/applicationGroups'
```

Here's the output:

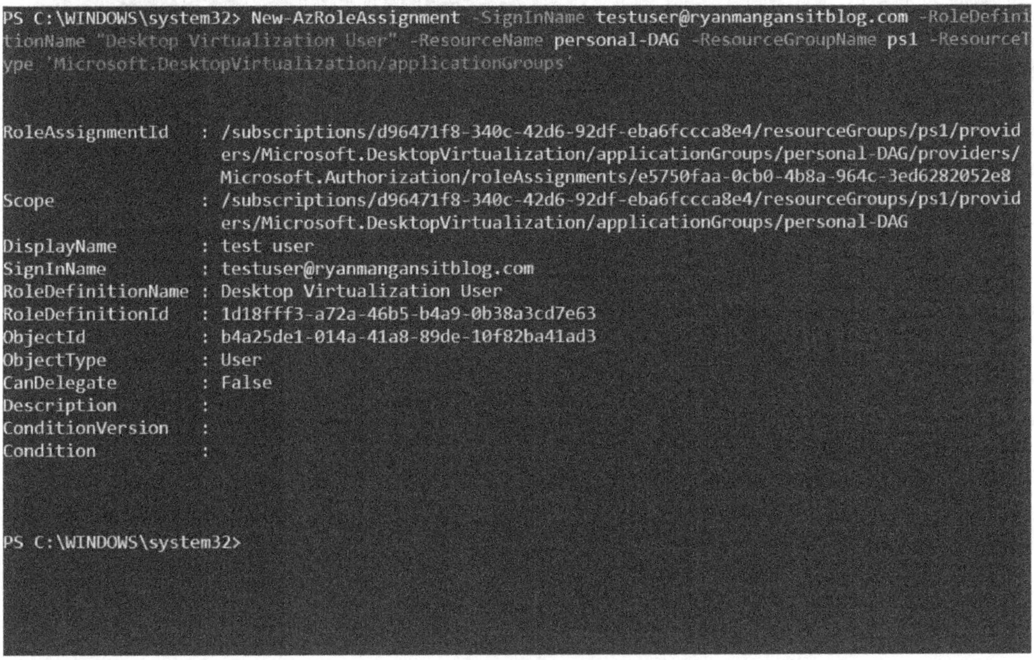

```
PS C:\WINDOWS\system32> New-AzRoleAssignment -SignInName testuser@ryanmangansitblog.com -RoleDefini
tionName "Desktop Virtualization User" -ResourceName personal-DAG -ResourceGroupName ps1 -ResourceT
ype 'Microsoft.DesktopVirtualization/applicationGroups'


RoleAssignmentId   : /subscriptions/d96471f8-340c-42d6-92df-eba6fccca8e4/resourceGroups/ps1/provid
                     ers/Microsoft.DesktopVirtualization/applicationGroups/personal-DAG/providers/
                     Microsoft.Authorization/roleAssignments/e5750faa-0cb0-4b8a-964c-3ed6282052e8
Scope              : /subscriptions/d96471f8-340c-42d6-92df-eba6fccca8e4/resourceGroups/ps1/provid
                     ers/Microsoft.DesktopVirtualization/applicationGroups/personal-DAG
DisplayName        : test user
SignInName         : testuser@ryanmangansitblog.com
RoleDefinitionName : Desktop Virtualization User
RoleDefinitionId   : 1d18fff3-a72a-46b5-b4a9-0b38a3cd7e63
ObjectId           : b4a25de1-014a-41a8-89de-10f82ba41ad3
ObjectType         : User
CanDelegate        : False
Description        :
ConditionVersion   :
Condition          :


PS C:\WINDOWS\system32>
```

Figure 7.26 – Assignment of a user to an app group

You can see in the following screenshot that the assignment worked:

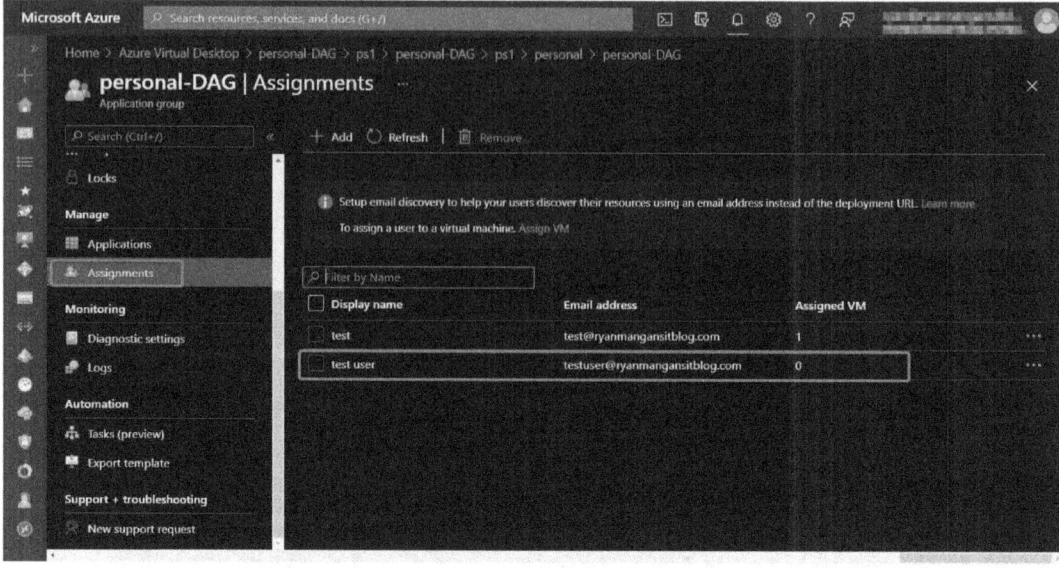

Figure 7.27 – The same user in the Azure portal confirming the assignment worked

As shown in the preceding screenshot, the user has been added to the app group.

> **Tip**
> You can also assign group object IDs to the app group rather than assigning individual users.

In this section, I showed you how to enable automatic assignment and assign users to an app group associated with a personal host pool. We will now look at direct assignments using PowerShell.

Re-assigning a personal desktop

As well as assigning users a desktop, we can also re-assign an existing desktop to a different user. This saves you from having to rebuild any existing desktops. However, if there is any personal data that is held on the VM, it is recommended to rebuild the VM.

To re-assign the VM in the Azure portal, select an existing VM in the host pool and then select **Assign to a different user**.

The VM will then be unassigned from the existing user and assigned to the new user that you specify.

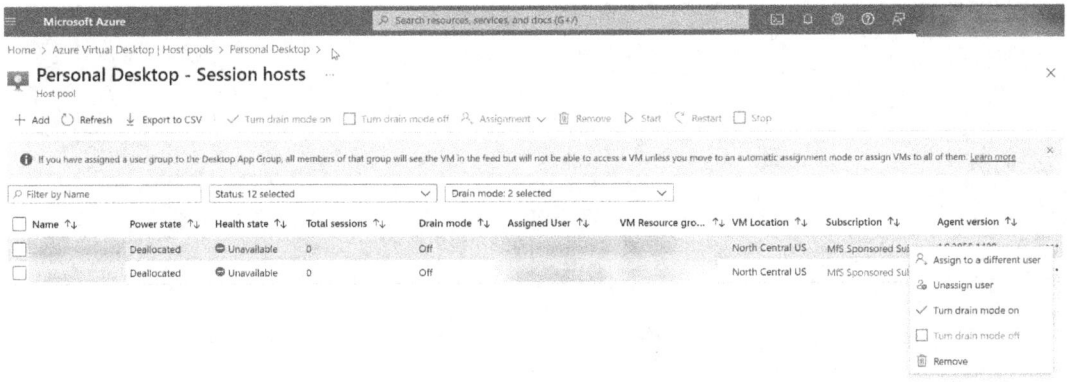

Figure 7.28 – Re-assigning a user to a session host in the Azure Portal

Configuring direct assignment using PowerShell

This section looks at assigning users to personal host pool session hosts using direct assignment. The key difference between direct assignment and automatic assignment is that with direct assignment, the user must be assigned to the personal session host before they can access the resources.

> **Remember**
>
> You need to follow the steps in *Chapter 6, Creating Host Pools and Session Hosts*, to configure PowerShell for AVD before proceeding.

The following steps walk you through the process of direct assignment using PowerShell:

1. The first step is to configure the host pool for the direct assignment of users to session hosts. Use the following cmdlets to complete this:

    ```
    Update-AzWvdHostPool -ResourceGroupName <resourcegroupname>
    -Name <hostpoolname> -PersonalDesktopAssignmentType Direct
    ```

    The following provides an example of the usage:

    ```
    Update-AzWvdHostPool -ResourceGroupName <resourcegroupname>
    -Name <hostpoolname> -PersonalDesktopAssignmentType Direct
    ```

The following screenshot shows the update of a host pool to direct assignment:

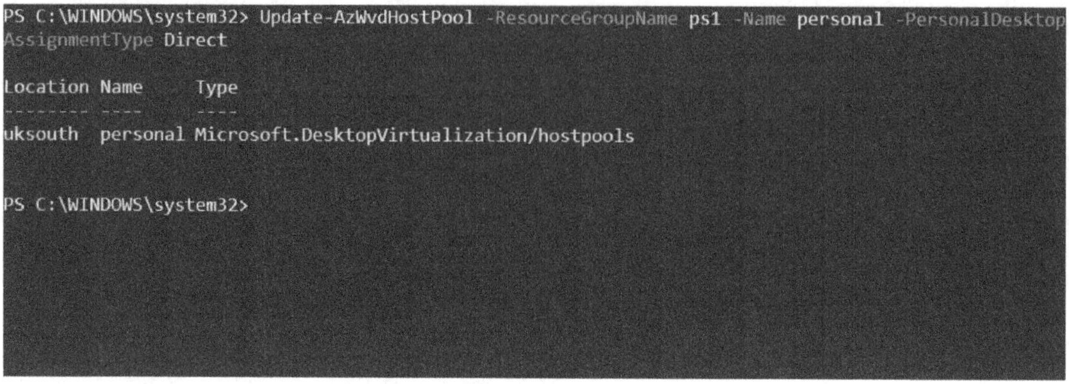

```
PS C:\WINDOWS\system32> Update-AzWvdHostPool -ResourceGroupName ps1 -Name personal -PersonalDesktop
AssignmentType Direct

Location Name     Type
-------- ----     ----
uksouth  personal Microsoft.DesktopVirtualization/hostpools


PS C:\WINDOWS\system32>
```

Figure 7.29 – Cmdlets for setting the direct assignment type

> **Tip**
> You can change between direct and automatic personal session host assignments. All you need to do is run the cmdlets for your chosen configuration.

2. Once we have set the host pool to Direct, as shown previously, we then need to assign a user to the personal desktop host pool:

   ```
   New-AzRoleAssignment -SignInName <userupn> -RoleDefinitionName
   "Desktop Virtualization User" -ResourceName <appgroupname>
   -ResourceGroupName <resourcegroupname> -ResourceType 'Microsoft.
   DesktopVirtualization/applicationGroups'
   ```

   As you will note from the previous section, the cmdlets are the same for assigning users to direct or automatic personal desktop host pools.

3. The final step is to assign the user to a specific session host; you would need to use the following cmdlets:

   ```
   Update-AzWvdSessionHost -HostPoolName <hostpoolname> -Name
   <sessionhostname> -ResourceGroupName <resourcegroupname>
   -AssignedUser <userupn>
   ```

   The following provides an example of the usage:

   ```
   Update-AzWvdSessionHost -HostPoolName personal -Name persx-1.
   rmitblog.local -ResourceGroupName ps1  -AssignedUser testuser@
   ryanmangansitblog.com
   ```

The following screenshot shows the cmdlets for assigning a user to a specific session host:

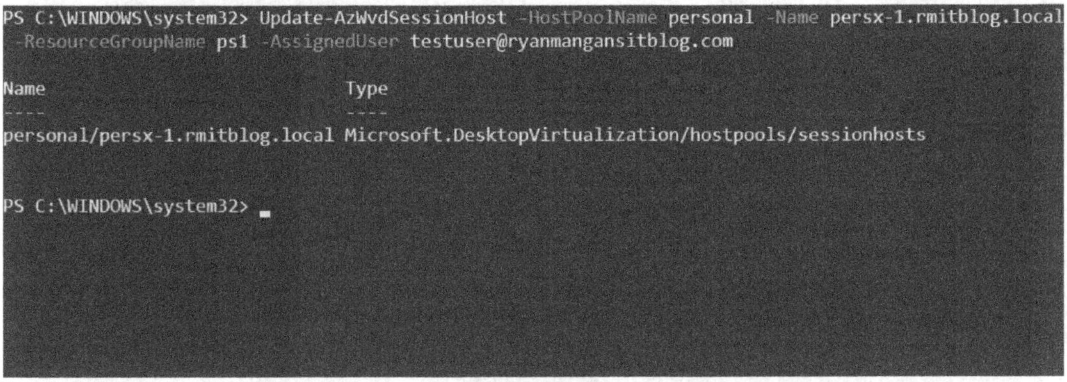

PS C:\WINDOWS\system32> Update-AzWvdSessionHost -HostPoolName personal -Name persx-1.rmitblog.local -ResourceGroupName ps1 -AssignedUser testuser@ryanmangansitblog.com
Name Type
---- ----
personal/persx-1.rmitblog.local Microsoft.DesktopVirtualization/hostpools/sessionhosts
PS C:\WINDOWS\system32>

Figure 7.30 – Assignment of a session host to a specific user

Once the cmdlets have been run successfully, you will then see the user assigned in the portal:

Microsoft Azure
Search resources, services, and docs (G+/)
Home > personal
personal | Session hosts
Host pool
Search (Ctrl+/)
Overview
Activity log
Access control (IAM)
Tags
Diagnose and solve problems
Settings
RDP Properties
Properties
Locks
Manage
Application groups
MSIX packages
Session hosts
Add
Refresh
Assign
Export to CSV
Turn drain mode on
Turn drain mode off
Remove
Filter by Name
Status: 12 selected
Drain mode: 2 sel...
Name
Status
Drain mode
Assigned User
Active ses...
Resource gro...
persx-1.rmitblog.local
Available
On
testuser@ryanmangansitblog.com
0
ps1

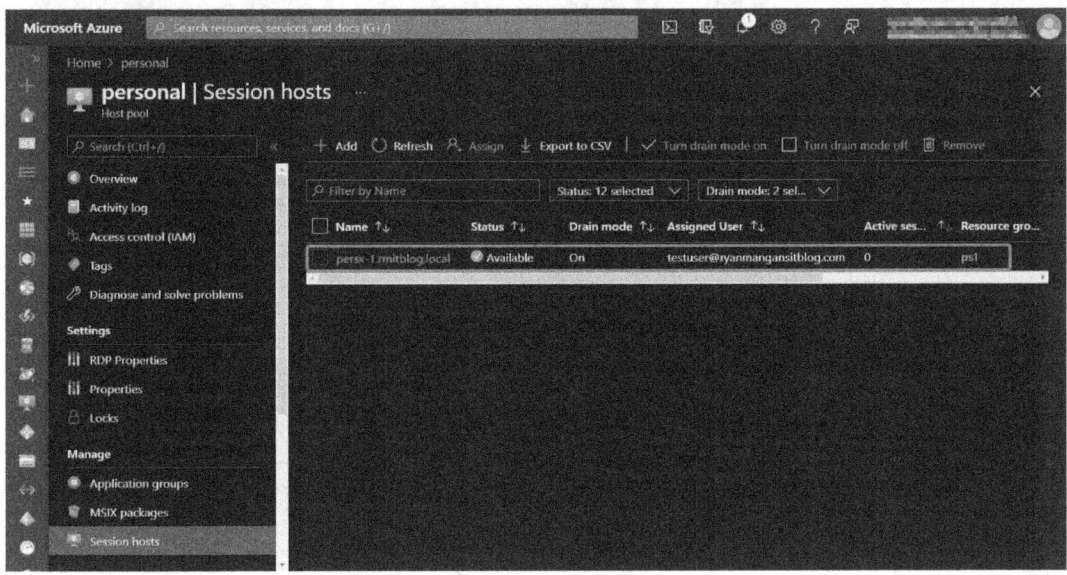

Figure 7.31 – Assignment in the Azure portal

There you have it – users assigned using direct assignment via PowerShell.

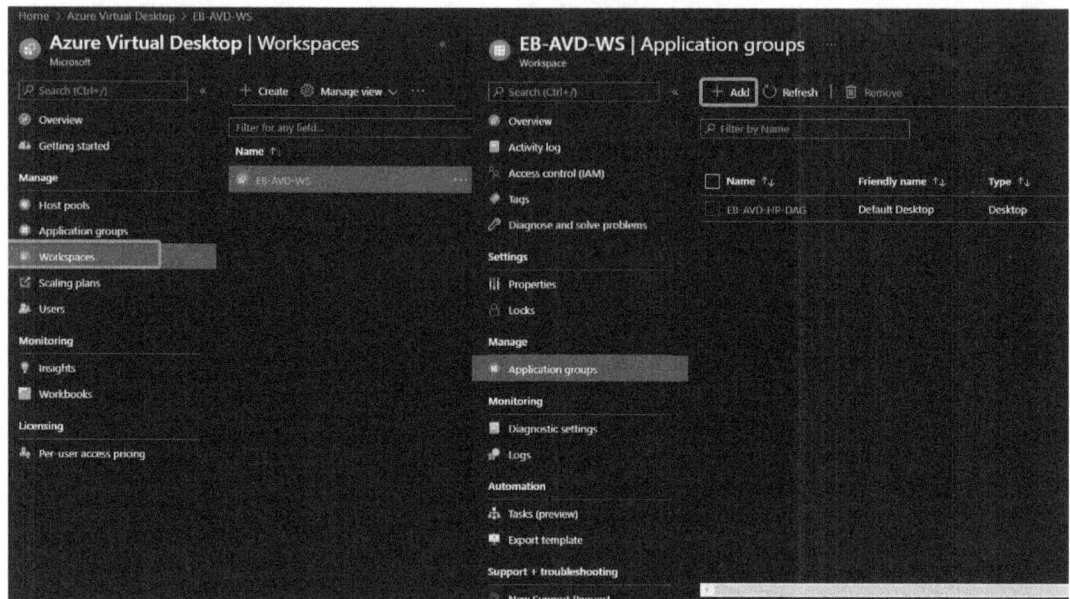

Figure 7.32 – Add an app group to a workspace in AVD

> **Important note**
> Don't forget to link your application group to your workspace. Workspaces are a logical grouping of app groups. Each app group must be linked with a workspace for users to see the desktops and remote apps published to them. See the following screenshot, showing how to add an app group under the **Workspaces** tab.

This section looked at the two options for personal session host assignment in AVD. We covered both automatic assignment and direct assignment.

We'll now move on to applying OS and application updates on a running AVD host.

Applying OS and application updates on an Azure Virtual Desktop host

When it comes to updating AVD personal desktops or multi-session, it's recommended that you use Microsoft Intune. However, you can also carry out manual update tasks on the image and then update the version within the Azure Compute Gallery.

> **Important note**
>
> The use of Azure Update Management is not supported for client Windows OSs. You can read more here: `https://docs.microsoft.com/azure/automation/update-management/operating-system-requirements#supported-operating-systems`

It is recommended that you use automatic enrollment to Microsoft Intune using Group Policy. It is also advised that you use device credentials for enrolment rather than user credentials. The steps for enrolling single-session and multi-session desktops to Microsoft Intune can be found here: `https://learn.microsoft.com/mem/intune/fundamentals/azure-virtual-desktop`

Configuring a validation pool

Validation pools are great for validating AVD service updates. Validation host pools are where service updates are applied in the first instance. This essentially allows you to monitor and test service updates before they are applied to your production or non-validation environment.

Validation host pools help you identify and discover issues from changes that are introduced from service updates. If you don't use a validation host pool, you may not discover any changes or errors introduced, which could impact your production environment.

> **Tip**
>
> It's recommended that you deploy a validation host pool to enable the testing of all future updates to the AVD service.

You can also change none validation host pools to a validation environment; here is how:

1. Sign in to the Azure portal UI using the following URL: `https://portal.azure.com`.
2. Search for and select **Azure Virtual Desktop** within the search bar located at the top of the web page.
3. On the **Azure Virtual Desktop** page, from the left-hand side menu, select **Host pools**.
4. Select the name of the host pool you want to edit.
5. Select **Properties**:

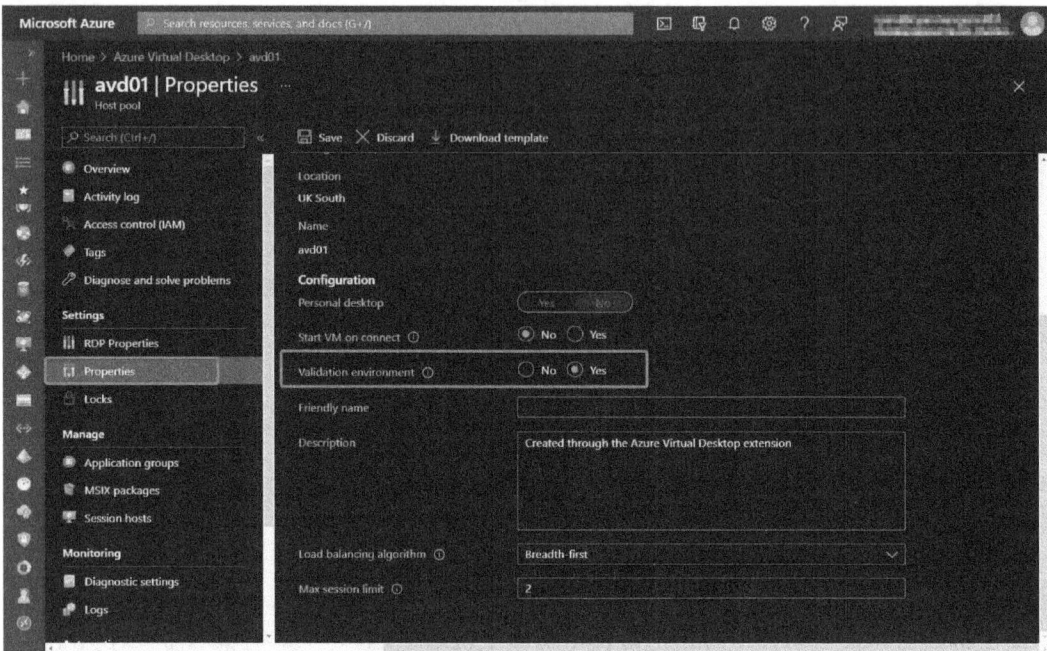

Figure 7.33 – Validation environment field within the properties of a host pool

6. Change **Validation environment** to **Yes** to enable the validation environment.
7. Select **Save**. Then, the settings will be applied.

To configure the validation environment using PowerShell, you would use the following cmdlets:

```
Update-AzWvdHostPool -ResourceGroupName <resourcegroupname> -Name
<hostpoolname> -ValidationEnvironment:$true
```

The following example shows you how to set an existing host pool to a validation environment; see the following example cmdlets:

```
Update-AzWvdHostPool -ResourceGroupName az140pw -Name pwdeployment
-ValidationEnvironment:$true
```

Here's the output:

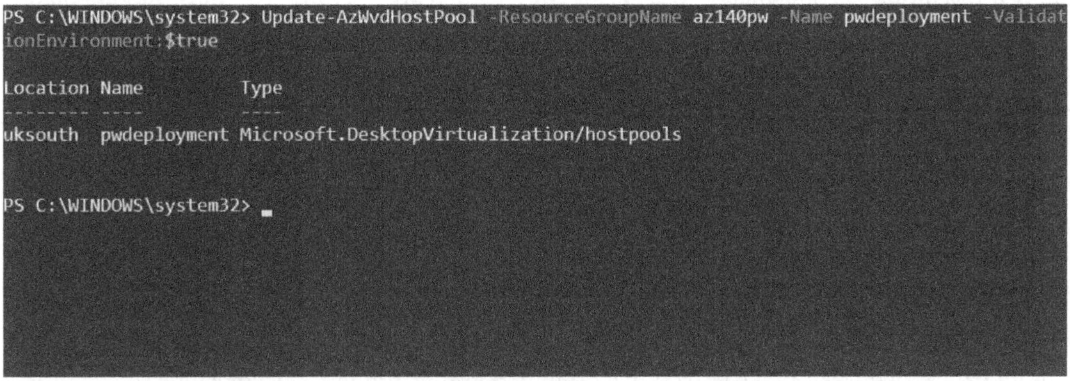

```
PS C:\WINDOWS\system32> Update-AzWvdHostPool -ResourceGroupName az140pw -Name pwdeployment -Validat
ionEnvironment:$true

Location Name          Type
-------- ----          ----
uksouth  pwdeployment  Microsoft.DesktopVirtualization/hostpools


PS C:\WINDOWS\system32>
```

Figure 7.34 – Set a host pool to be a validation environment via PowerShell

There you have it – you've changed an existing host pool to a validation pool using the Azure portal and PowerShell. We'll now move on to looking at security and compliance settings that you can apply to session hosts.

Applying security and compliance settings to session hosts

Session hosts are VMs that run within your Azure subscription, which are connected to an Azure virtual network. The overall security of an AVD deployment depends on the security controls you apply to your session hosts.

Here are some tips for session host security: https://learn.microsoft.com/azure/virtual-desktop/security-recommendations

Endpoint protection, endpoint detection and response, and threat and vulnerability management

It is advised that you enable a supported endpoint production product on your AVD session hosts. There is the option of using Microsoft Defender antivirus or third-party vendors.

> **Tip**
> When configuring antivirus for AVD, make sure you exclude FSLogix VHD/VHDX and MSIX app attach VHD/VHDX/CIMFS file extensions.

As security threats are becoming a daily occurrence and threats are becoming more complex, it is recommended that you deploy some form of **endpoint detection and response** (**EDR**) product to enable advanced detection and response capabilities. For example, you can use Microsoft Defender for Endpoint or others, including network virtual appliances that come with built-in threat detection capabilities that listen on the Azure virtual network.

You can use Azure Security Center to Identify vulnerabilities with applications and OSs. This helps you identify problems using vulnerability assessments. You can also use Microsoft Defender for Endpoint, which offers a level of vulnerability management for desktop OSs and web content filtering. You can read more about web content filtering here: `https://learn.microsoft.com/azure/virtual-desktop/security-recommendations`

As shown in the previous section, Azure helps you address vulnerabilities within your environment by allowing you to schedule updates and apply patches to your session hosts. These can be both personal and pooled host pools.

Session screen locks

As the world has made a significant shift to working from anywhere and working from home due to the global pandemic, COVID-19, you may want to consider configuring a machine screen lock during idle time, requiring the user to reauthenticate when they come back to the screen. This is to prevent unauthorized access to company devices. This not only applies to AVD but should apply to all devices deployed in a business. It's also important to not get confused between screen locks and session timeouts. Screen locks are useful for thin client devices where the device is static and usually installed within the organization's offices. Session timeouts are for remote connections where a user has not used the keyboard or mouse for a period of time and the session disconnects.

Configuring maximum inactive time and disconnection policies

A good security hardening technique for virtual desktop environments is configuring maximum inactive times and disconnection policies. The inactive time and disconnection policies can vary for each organization. However, 5 to 15 minutes is the time range you should consider when configuring these settings.

The following screenshot shows the user settings within a Group Policy object for the **Set time limit for active but idle Remote Desktop Services sessions** setting. This setting is used to control idle settings.

You can configure the policy setting using the **Computer Configuration | Administrative Templates | Windows Components | Remote Desktop Services | Remote Desktop Session Host | Session Time Limits** Group Policy path.

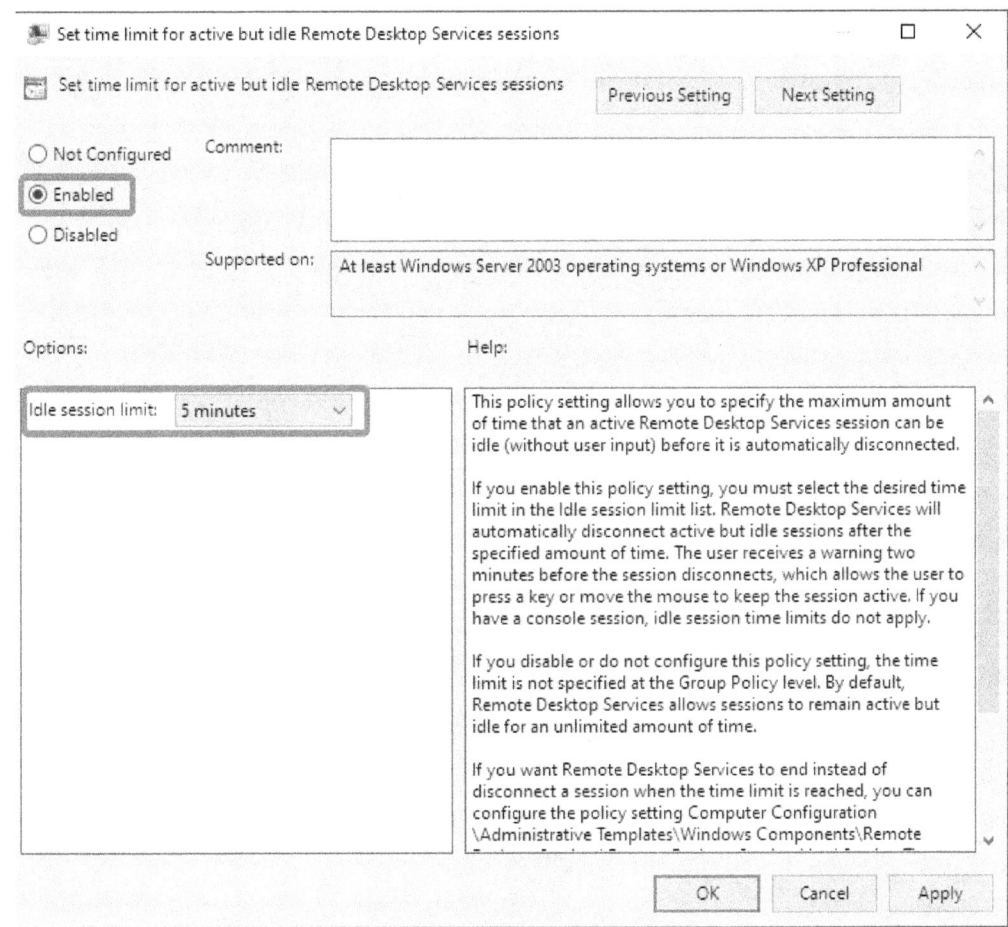

Figure 7.35 – Idle session limit set to 5 minutes

The following screenshot shows you the setting for **Set time limit for disconnected sessions**. This is used to remove any disconnected sessions from a session host after the specified time value set:

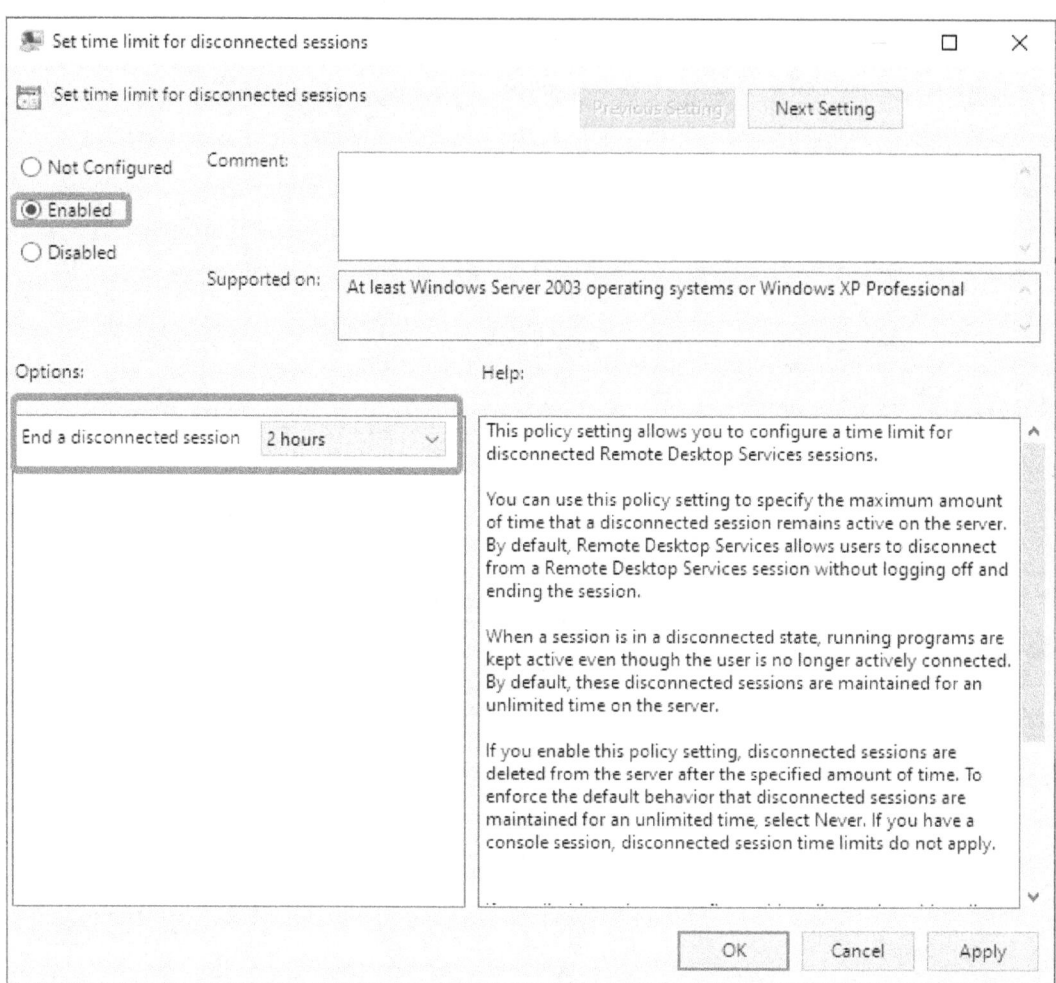

Figure 7.36 – Setting for Set time limit for disconnected sessions

The following screenshot shows all the policies you can configure within the `Session Time Limits` folder:

Configuring host pool settings   211

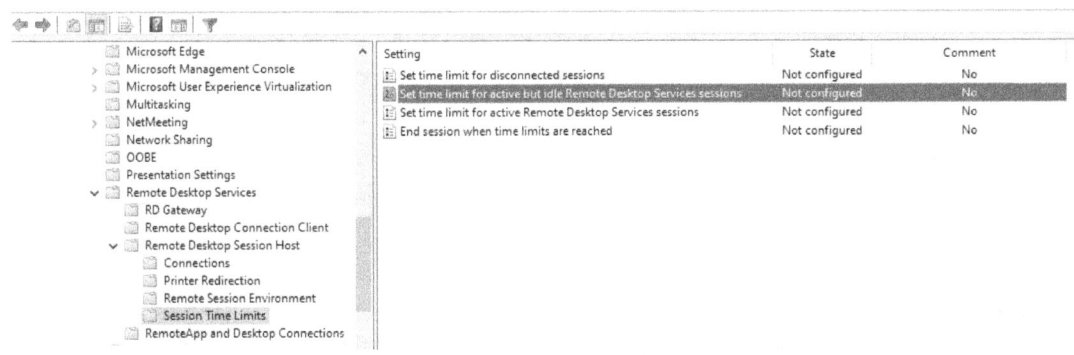

Figure 7.37 – Settings available under Session Time Limits

> **Important note**
> Please note that disconnecting idle tasks such as rendering or simulations can interrupt users. It's advised that you check before enabling such policies.

Application security within session hosts

It is important to secure the applications running within the session host. It is recommended that you use tools such as Microsoft Security Policy Advisor (https://learn.microsoft.com/deployoffice/admincenter/overview-cloud-policy) to help identify possible issues with Microsoft 365 apps for enterprise.

> **Tip**
> Remember to check other applications and apply the appropriate app locker and firewall rules to limit communication if not required for use within the organization.

Limiting OS capabilities

It is advised that you go through a security hardening exercise to ensure you have limited security exposure or reduce your security exposure. Some of the additional ways of achieving this would be to use Group Policy and RDP custom policies.

You can control device redirection by redirecting drives, USB devices, and printers to a user's local device remotely.

The following link details the different device redirections and the restrictions you can set:

https://learn.microsoft.com/azure/virtual-desktop/configure-device-redirections

You can find the full list of RDP settings here:

https://learn.microsoft.com/azure/virtual-desktop/rdp-properties

> **Recommendation**
> Microsoft recommends you assess/evaluate your security requirements and check which features you need and should disable.

A common VDI practice is to restrict Windows Explorer access by hiding local and remote drive mappings. This still applies to AVD. This prevents users from discovering unwanted information about systems and other users.

You should avoid direct RDP access to session hosts in your environment. If you need direct RDP access for administration or troubleshooting, enable just-in-time access, use Azure Bastion as covered in a previous chapter, or use Quick Assist if you are supporting a user.

Consider limiting user permissions when they access local and remote filesystems. You can restrict user permissions by ensuring that your local and remote filesystems use access control lists with the least privilege configured. This ensures that users can only access what they need and can't change or delete critical resources.

Prevent users from installing and running unwanted software on session hosts. As mentioned previously, you can enable AppLocker for additional security on session hosts, ensuring that only the apps you allow can run on the host. You can also use app masking for applications you don't want specific user groups to see.

In this section, we took a look at some high-level session host security tips. However, we look at security in more detail in *Chapter 11, Managing Security*.

Summary

In this chapter, we built on the previous chapter, where we deployed a session host.

In addition, this chapter looked at configuring session hosts within a host pool. We learned how to deploy and configure a Windows Server session host licensing server to support the use of the Windows Server OS within AVD. We then looked at the configuration of host pool settings, RDP properties, load-balancing methods, and assigning users to host pools. We then finished off the chapter by learning how to apply OS and application updates to a session host as well as applying security compliance settings to session hosts.

In the next chapter, we will take a look at creating and managing session host images.

Questions

Here are a few questions to test your understanding of this chapter:

1. When using server-based OSs for AVD, what resource do you require?

    When using server-based OSs, you will need an additional server to host the RD CALs that the server license will require.

2. What is the purpose of MaxSessionLimit within a host pool?

    MaxSessionLimit is used to limit the number of user sessions on a single session host.

3. What considerations are there to be taken into account when configuring antivirus for AVD?

    Ensure that you have set the exclusion rules for both FSLogix profiles and MSIX app attach on the appropriate file shares/storage.

4. Which Group Policy would you set to idle timeout sessions for user sessions in AVD?

    Set time limit for active but idle Remote Desktop Services sessions

8
Entra ID Join for Azure Virtual Desktop

In this chapter, we will look at **Entra ID** join for Azure Virtual Desktop. Using Entra ID join for Azure Virtual Desktop has many benefits for organizations, including **Single Sign-On** (**SSO**), virtual machines that use a single identity provider, and being able to avoid some of the complexities associated with having an Active Directory domain controller.

It is important to note that other services may still require an Active Directory Domain Services environment for access to applications and **Server Message Block** (**SMB**).

In this chapter, we will take a look at the following:

- The prerequisites for Entra ID join for Azure Virtual Desktop
- Deploying an Entra ID-joined host pool
- Enabling user access
- Configuring local admin access

Prerequisites

It is important to note that there are a few limitations when using Entra ID join for Azure Virtual Desktop at the time of writing. As you may know, many Microsoft services, third-party platforms, and others require access to an Active Directory environment for authentication and user/group permissions. Therefore, assessing your organization's current requirements is important to ensure that Entra ID join is a suitable solution. The requirements and prerequisites are as follows:

- The session hosts must use Windows 10 Enterprise Version 2004 or later
- Entra ID-joined virtual machines don't currently support external users
- FSLogix does not support cloud-only accounts right now

> **Important note**
> Entra ID join is different from an Active Directory Domain Services controller in that the session-host **Virtual Machines** (**VMs**) are automatically joined to the Entra ID tenant of the subscription that deploys them. There is no way to specify a different Entra ID tenant for the host VMs. This means you will need to ensure that the required Azure tenant is linked to the subscription you wish to deploy the Entra ID-joined VM to.

Let's now look at deploying an Entra ID-joined host pool.

Deploying an Entra ID-joined host pool

In this section, we will look at deploying a host pool using Entra ID join.

Before we get started, I want to cover the use of FSLogix profile containers with Entra ID join. When using Entra ID join, there are a few slight differences compared to the traditional way of using Active Directory Domain Services. The following link takes you to the Microsoft documentation that details how to configure FSLogix profile containers with Azure Files and Entra ID: https://docs.microsoft.com/azure/virtual-desktop/create-profile-container-azure-ad.

Let's now move on and see how to create an Entra ID-joined host pool:

1. Navigate to the Azure Virtual Desktop service.
2. Proceed to create a new host pool.
3. On the host pool creation screen, under the **Virtual Machine** tab in the **Domain to join** section, select **Microsoft Entra ID**:

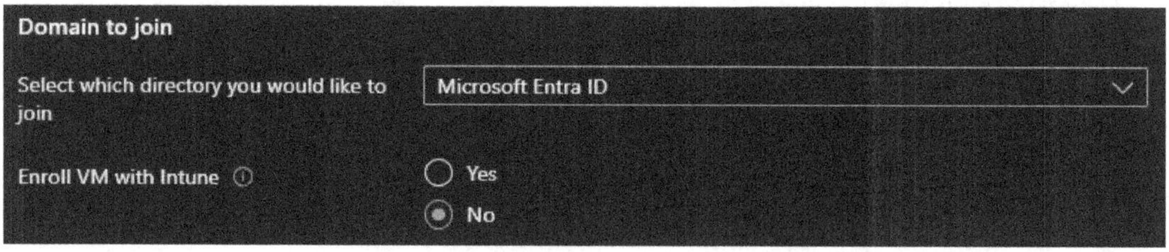

Figure 8.1 – The domain join options within the host pool creation wizard

4. When you select **Microsoft Entra ID**, you will see the **Enroll VM with Intune** option appear. You would choose the **Enroll VM with Intune** option if you wanted to use Intune to manage policies, distribute software, and generally manage VMs:

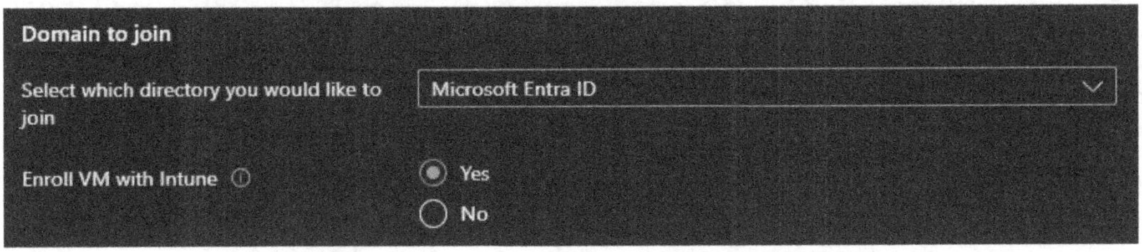

Figure 8.2 – A screenshot showing the Enroll VM with Intune field

You can read more about **Intune** here: https://learn.microsoft.com/mem/intune/.

> **Important Note**
> You need to ensure that you have set up Intune before using the **Enroll VM with Intune** feature; otherwise, the host deployments will fail.

5. Once the deployment has finished, you should see an extension called **AADLoginForWindows**, which is used to create the Entra ID join, and the **AADLoginForWindowsWithIntune** extension for Intune enrollment if you selected the **Enroll VM with Intune** option. If you choose not to use Intune enrollment, you will need to configure customizations and policies locally using a master image.

6. Once the deployment has been completed, you will see the device registered within Entra ID under **Devices | All devices**, as shown in the following screenshot:

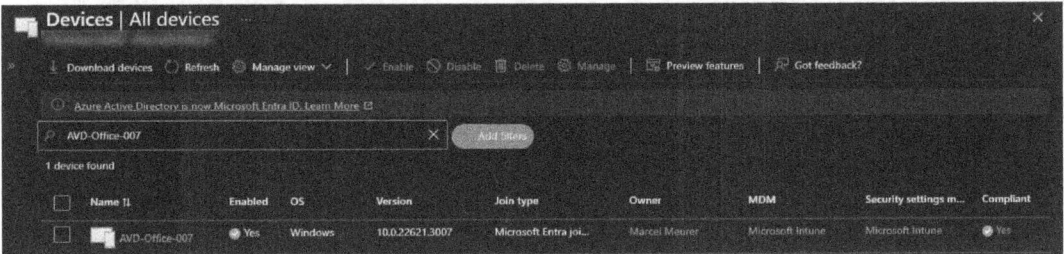

Figure 8.3 – A VM registered under devices within Entra ID

7. The device will also appear in **Microsoft. Endpoint Manager** if enrolled with Intune:

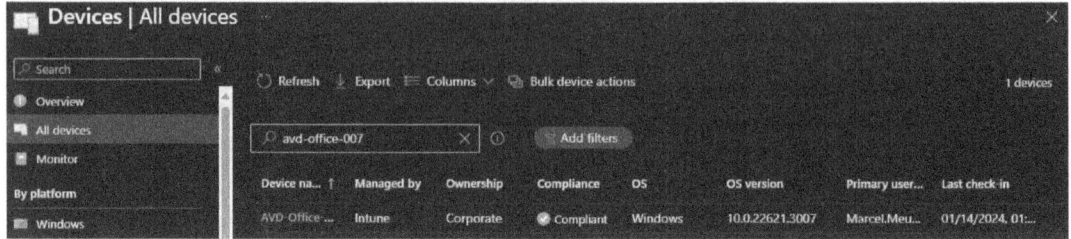

Figure 8.4 – The Intune Admin Center | All devices

8. You will also see the Entra ID registration within the **Audit logs** section when you navigate to **Entra ID | Devices | Audit logs**, as shown in the following screenshot:

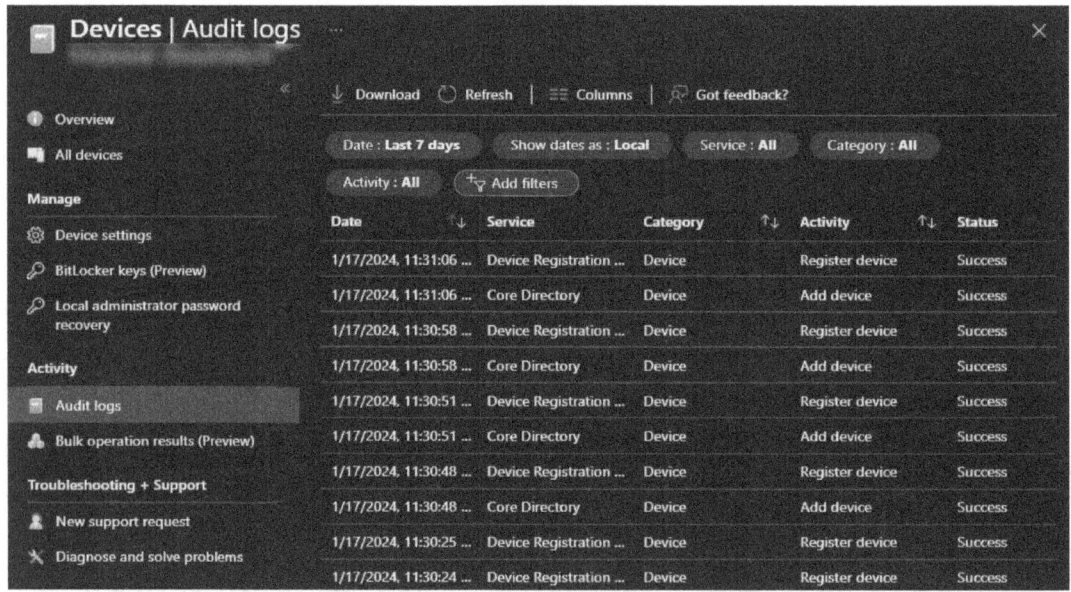

Figure 8.5 – Entra ID device registrations within the Audit logs section

9. If the VM does not appear to enroll or you want to confirm enrollment, you can log in locally to the VM and use the following command, using an elevated Command Prompt – `dsregcmd /status`:

Figure 8.6 – The dsregcmd /status command

10. Finally, you can also check the event logs using Event Viewer. The Entra ID registration logs are in the following section of Event Viewer – **Applications and Services Logs | Microsoft | Windows | User Device Registration | Admin**:

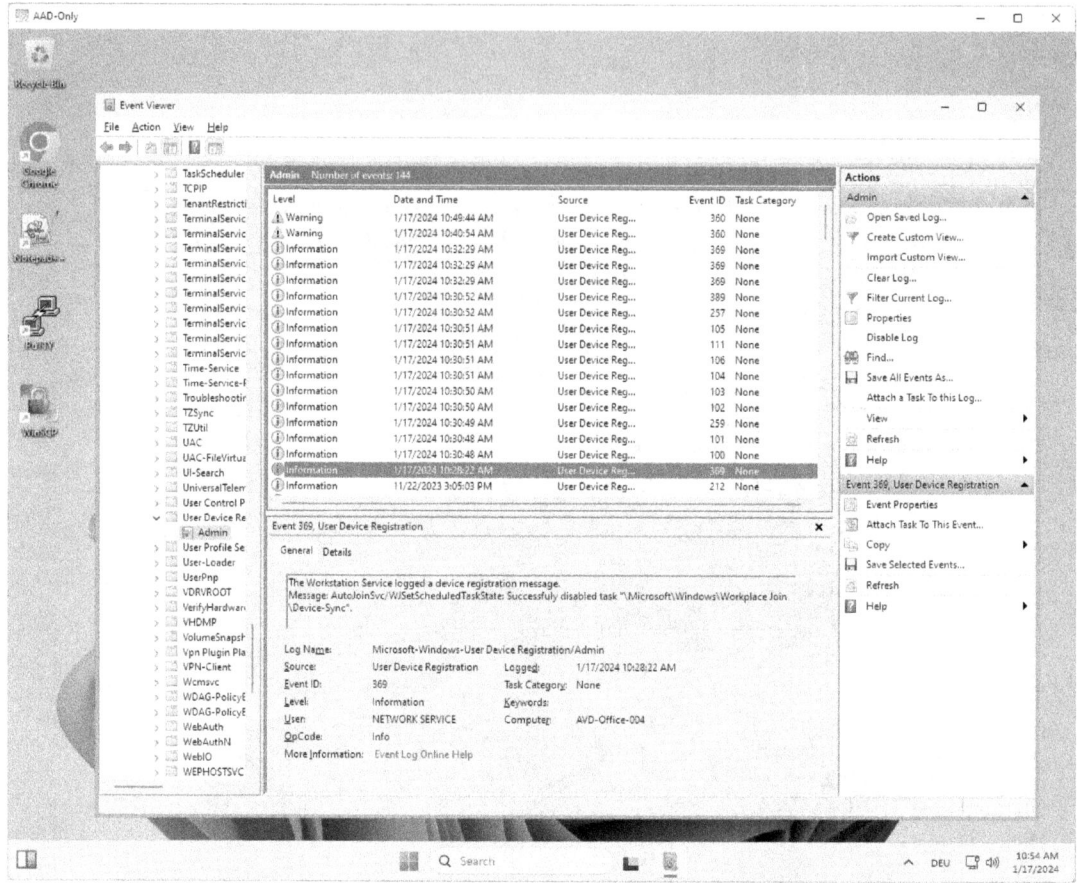

Figure 8.7 – Entra ID registration logs within Event Viewer

Now that we have deployed the Entra ID-joined host pool, we will take a look at enabling access for users in the next section.

> **Important note**
> If you roll out an Entra-joined VM, ensure that there is no device with the same name in Entra ID. If so, delete the orphan device. You cannot join a device to Entra ID with a name that still exists.

Enabling user access

Before users can sign in to the session hosts within the Entra ID-joined host pool, you must configure the required permission using **Role-Based Access Control** (**RBAC**). First, we need to add the required users and Entra ID groups to the host pool default desktop application group. We also need to add the **Virtual Machine User Login** RBAC role.

> **Important Note**
> The **Virtual Machine User Login** RBAC role is not an Azure Virtual Desktop role. This is required to enable access to sign in to a VM. The Azure role enables logon by applying the `DataAction` permission.

Depending on your requirements and host pool deployment, you should review the scope for this role. For example, assigning an Entra ID group at the resource-group level may make more sense than assigning the RBAC role for each user per VM.

> **Important Note**
> It is not advised to set the **Virtual Machine User Login** RBAC role at the subscription level; you would essentially give all assigned users the ability to sign in to all VMs within the subscription.

To assign the **Virtual Machine User Login** role, you will need to do the following:

1. Go to your host pool resource group in the Azure portal and select **Access control (IAM)**.
2. Select **+ Add**:

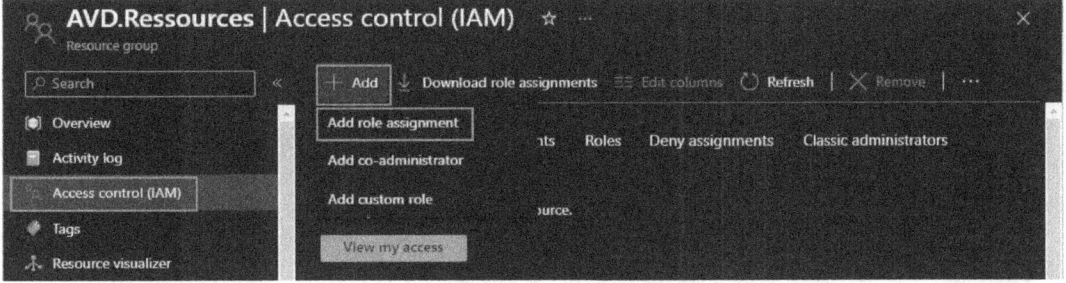

Figure 8.8 – The + Add button

3. On the **Add role assignment** page, select **Virtual Machine User Login**:

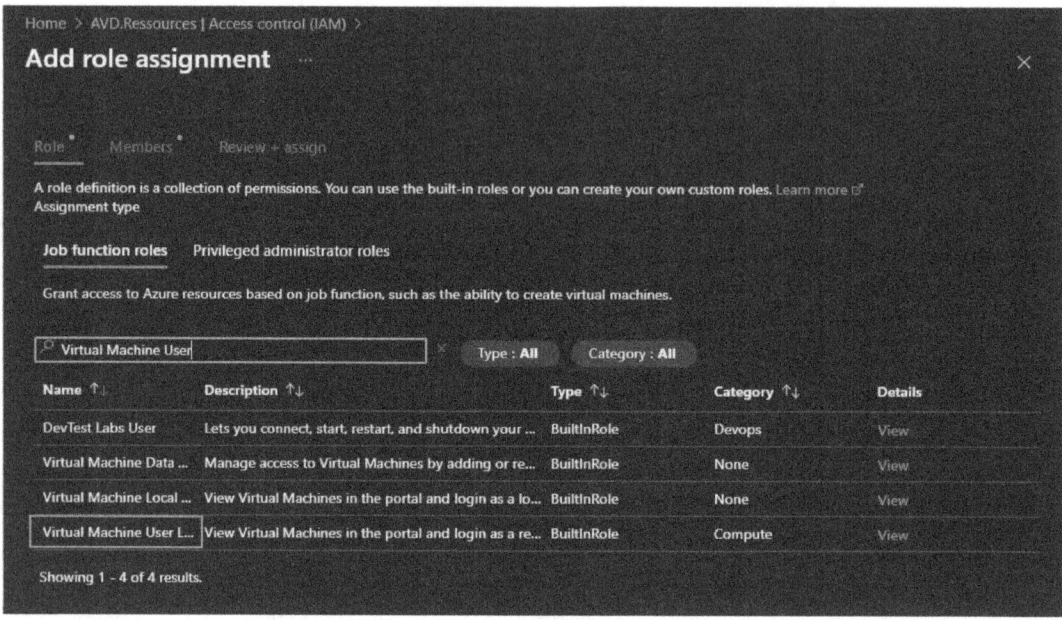

Figure 8.9 – Selecting the Virtual Machine User Login role

4. Under **Members**, click **select members**, and then select the required user group assigned to the Desktop application group; in this example, the user group is called `AVD-AADOnly-Users`:

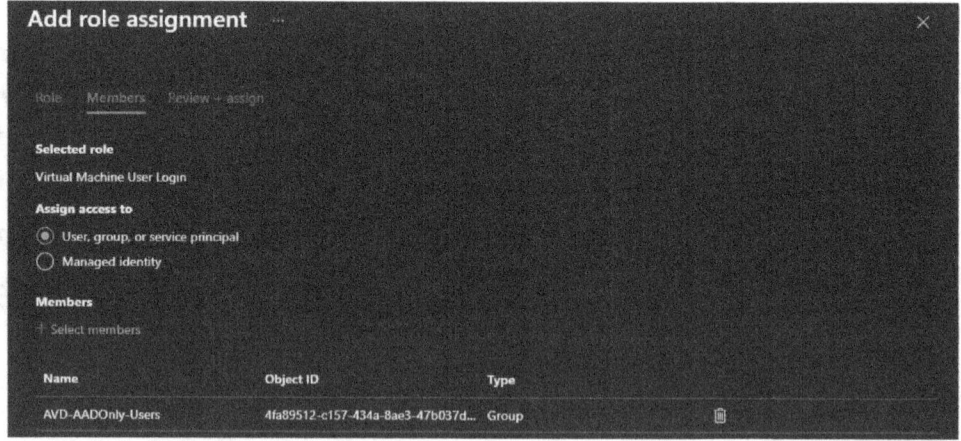

Figure 8.10 – The Add role assignment screen with the AVD-AADOnlyUsers group added as a member

5. Click **Review + assign**.

6. You should now see **Virtual Machine User Login** appear on the **Access Control (IAM)** page, as shown in the following screenshot:

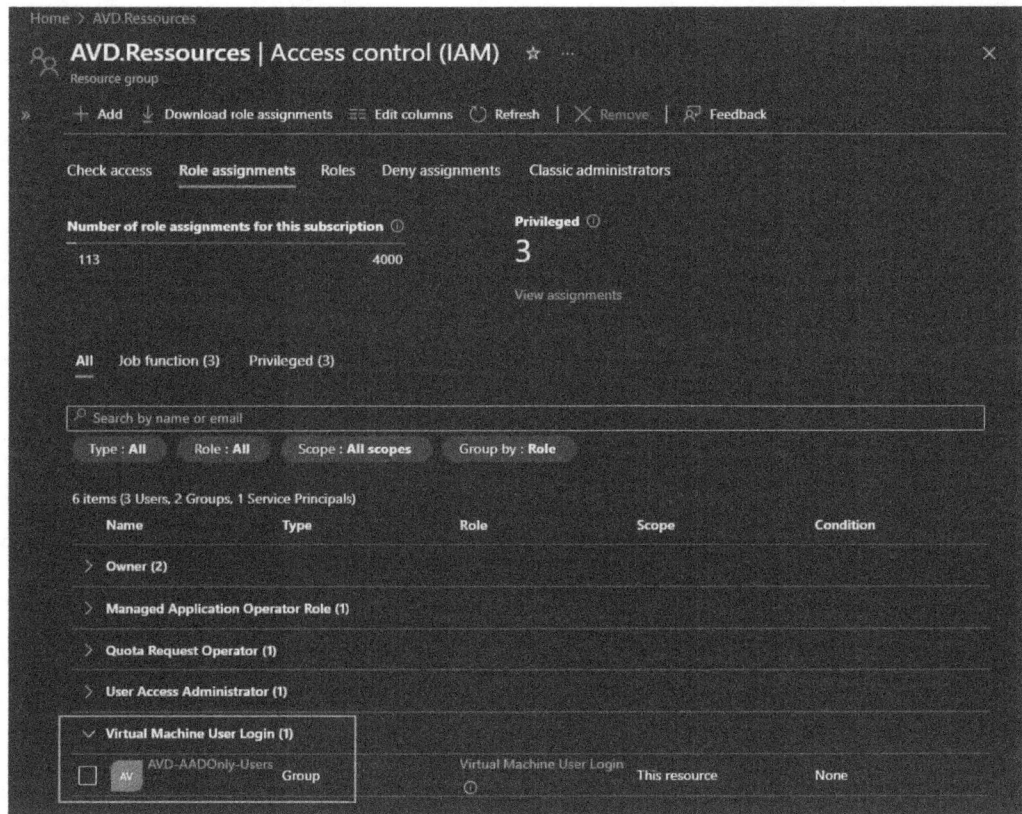

Figure 8.11 – The Entra ID group added to the Virtual Machine User Login role

This section looked at assigning the **Virtual Machine User Login** role to give Azure user accounts access to VMs. In the next section, we will look at connecting to session hosts using the Windows Desktop client.

Connecting to Entra ID-joined session hosts using the Remote Desktop client

To use Entra ID to log in to session hosts, we also need to configure the host pool correctly. If we open the Entra ID-joined host pool, we can click on **RDP Properties**. RDP properties can be used for different configurations of a host pool, such as the following:

- Connection information
- Session behavior

- Device redirection
- Display settings
- Advanced

The RDP properties are valid for all session hosts and users of the host pool. The **Connection information** configuration tab can be used to configure access via Microsoft Entra ID.

Configure **Connections will use Microsoft Entra authentication to provide single sign-on** and **RDP will use CredSSP if the operating system supports CredSSP**, as shown in the following screenshot:

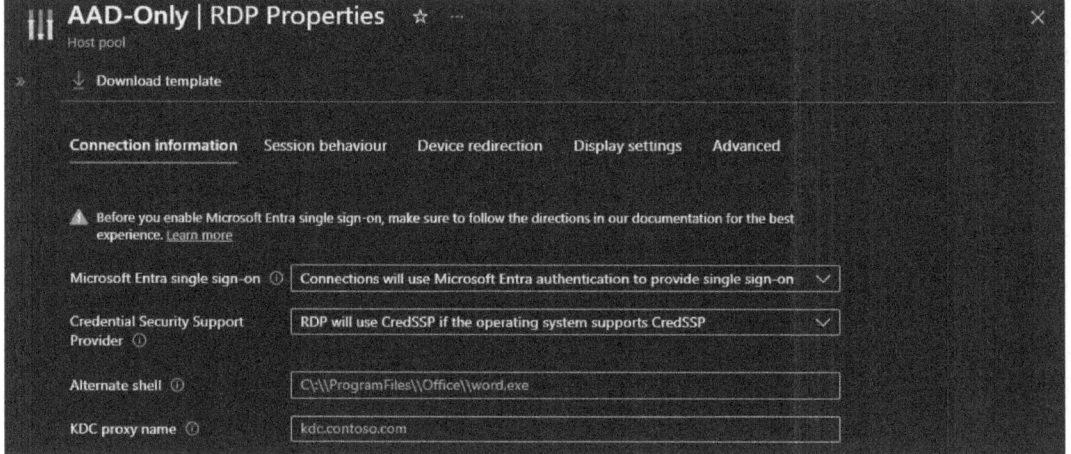

Figure 8.12 – RDP Properties

Single sign-on is only supported if the clients meet the following requirements:

- The local PC is Entra ID-joined to the same Entra ID tenant used for Azure Virtual Desktop.
- The local PC is hybrid AD-joined to the same Entra ID tenant used for Azure Virtual Desktop.
- You use Windows 10 build 2004 or later and Entra ID-registered with the same Entra ID tenant for Azure Virtual Desktop.

If you do not meet the preceding criteria, or users connect from other clients, such as the web, macOS, iOS, or Android, you can enable the legacy authentication. To learn more about Entra ID joined hosts, read.

```
https://learn.microsoft.com/azure/virtual-desktop/azure-ad-joined-
session-hosts
```

In **RDP Properties**, add the following string to the advanced configuration, separated by a semicolon - `targetisaddjoined:i:1`. The advanced settings reflect the complete settings:

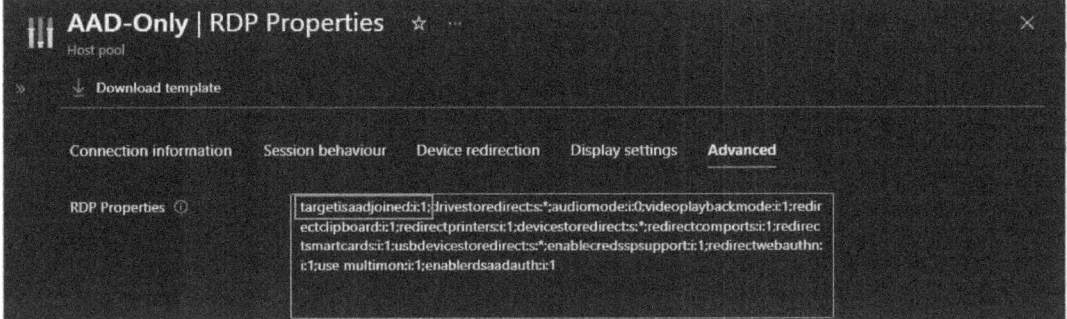

Figure 8.13 – The targetisaddjoined RDP property

After configuring the host pool to use Entra ID as authentication, we can explore configuring single-sign on with Entra authentication for RDP.

Enabling Microsoft Entra authentication for RDP

If you log on to an Entra-joined session host for the first time (or again after a while), you will be asked if you trust the specific session host:

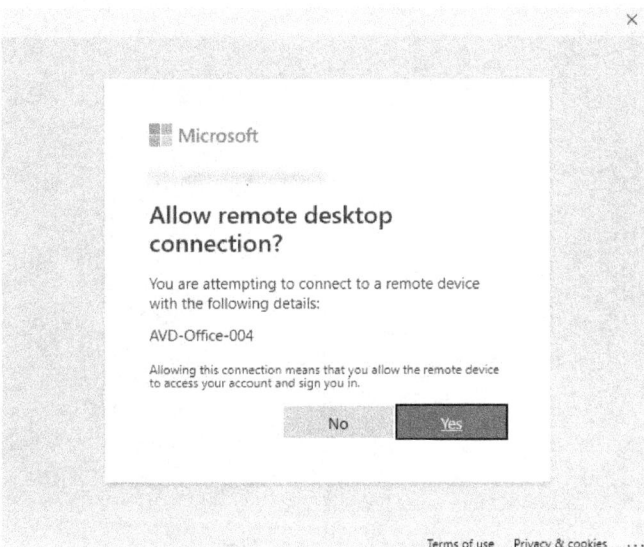

Figure 8.14 – Allow the connection to a specific host

Entra will remember the decision for 15 individual hosts for up to 30 days. This can be annoying in a multi-user pool, where a user gets another host each day. Fortunately, there is a solution to prevent this behavior.

We can create a dynamic device group containing our session hosts and configure this Entra to no longer ask this question to users.

We will start creating a group in Entra with the following settings:

Configuration	Value
Security type	Security
Group name	Name of the group (e.g., `AVD-Hosts-DynamicGroup`)
Group description	Description
Membership type	Dynamic device

Table 8.1 – Configuring a dynamic group

You can directly enter the settings in the **New Group** configuration:

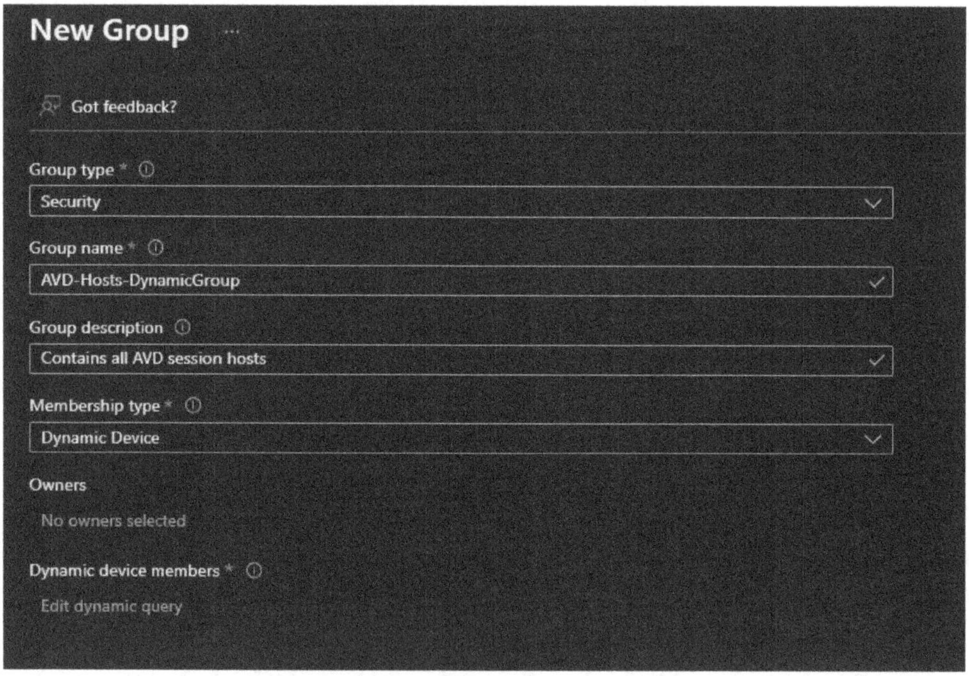

Figure 8.15 – New Group

Also, click on **Dynamic device members** and add a configuration. In my environment, all AVD host names start with **AVD**. Hence, I can configure the expression in the window, as shown in the following screenshot:

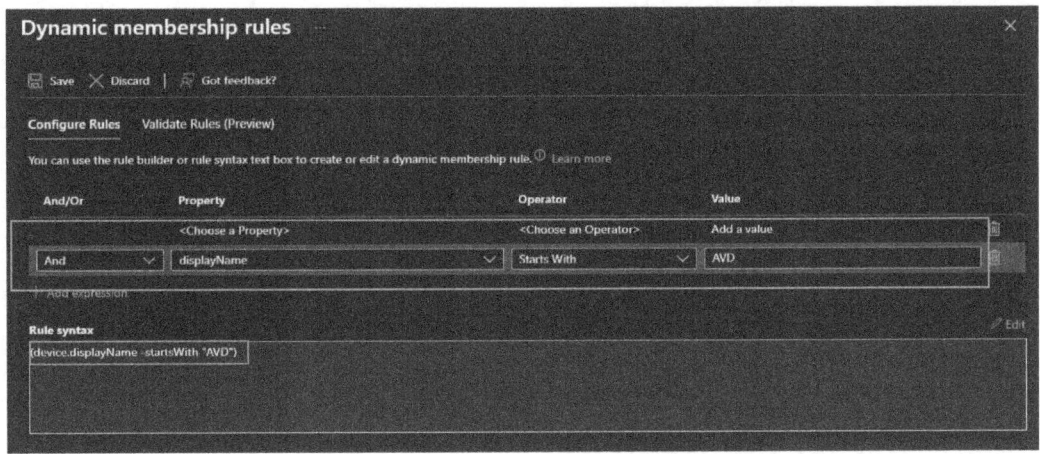

Figure 8.16 – Adding a rule for a dynamic group

Click on **Save** to save the full group configuration.

After a while, you will see the number of devices/hosts in the group on the **Overview** page:

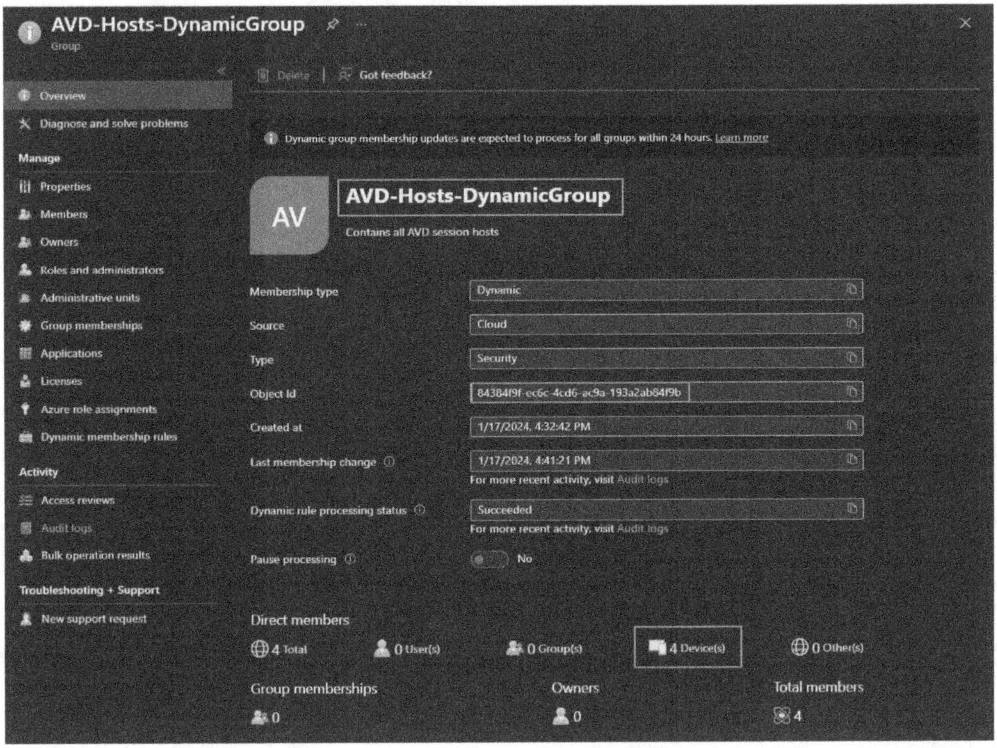

Figure 8.17 – The properties of a dynamic group

Copy the object ID and name of the group for the next step.

Open a PowerShell window and prepare the following script (modify lines 22 and 23 with your object ID and group name, respectively):

```
Import-Module Microsoft.Graph.Authentication
Import-Module Microsoft.Graph.Applications

Connect-MgGraph -Scopes "Application.Read.All","Application-RemoteDesktopConfig.ReadWrite.All"

$MSRDspId = (Get-MgServicePrincipal -Filter "AppId eq 'a4a365df-50f1-4397-bc59-1a1564b8bb9c'").Id
$WCLspId = (Get-MgServicePrincipal -Filter "AppId eq '270efc09-cd0d-444b-a71f-39af4910ec45'").Id

If ((Get-MgServicePrincipalRemoteDesktopSecurityConfiguration -ServicePrincipalId $MSRDspId) -ne $true) {
    Update-MgServicePrincipalRemoteDesktopSecurityConfiguration -ServicePrincipalId $MSRDspId -IsRemoteDesktopProtocolEnabled
}

If ((Get-MgServicePrincipalRemoteDesktopSecurityConfiguration -ServicePrincipalId $WCLspId) -ne $true) {
    Update-MgServicePrincipalRemoteDesktopSecurityConfiguration -ServicePrincipalId $WCLspId -IsRemoteDesktopProtocolEnabled
}

# Verify:
Get-MgServicePrincipalRemoteDesktopSecurityConfiguration -ServicePrincipalId $MSRDspId
Get-MgServicePrincipalRemoteDesktopSecurityConfiguration -ServicePrincipalId $WCLspId

$tdg = New-Object -TypeName Microsoft.Graph.PowerShell.Models.MicrosoftGraphTargetDeviceGroup
$tdg.Id = "84384f9f-ec6c-4cd6-ac9a-193a2ab84f9b"
$tdg.DisplayName = "AVD-Hosts-DynamicGroup"

New-MgServicePrincipalRemoteDesktopSecurityConfigurationTargetDeviceGroup -ServicePrincipalId $MSRDspId -BodyParameter $tdg
New-MgServicePrincipalRemoteDesktopSecurityConfigurationTargetDeviceGroup -ServicePrincipalId $WCLspId -BodyParameter $tdg
```

Here is the source of the preceding script: https://learn.microsoft.com/azure/virtual-desktop/configure-single-sign-on?WT.mc_id=Portal-Microsoft_Azure_WVD. It's also available at the Packt Publishing GitHub site: https://github.com/PacktPublishing/Mastering-Azure-Virtual-Desktop-2nd-Edition/blob/main/Enable%20Microsoft%20Entra%20authentication%20for%20RDP.ps1.

If everything worked as expected, you should see the following output:

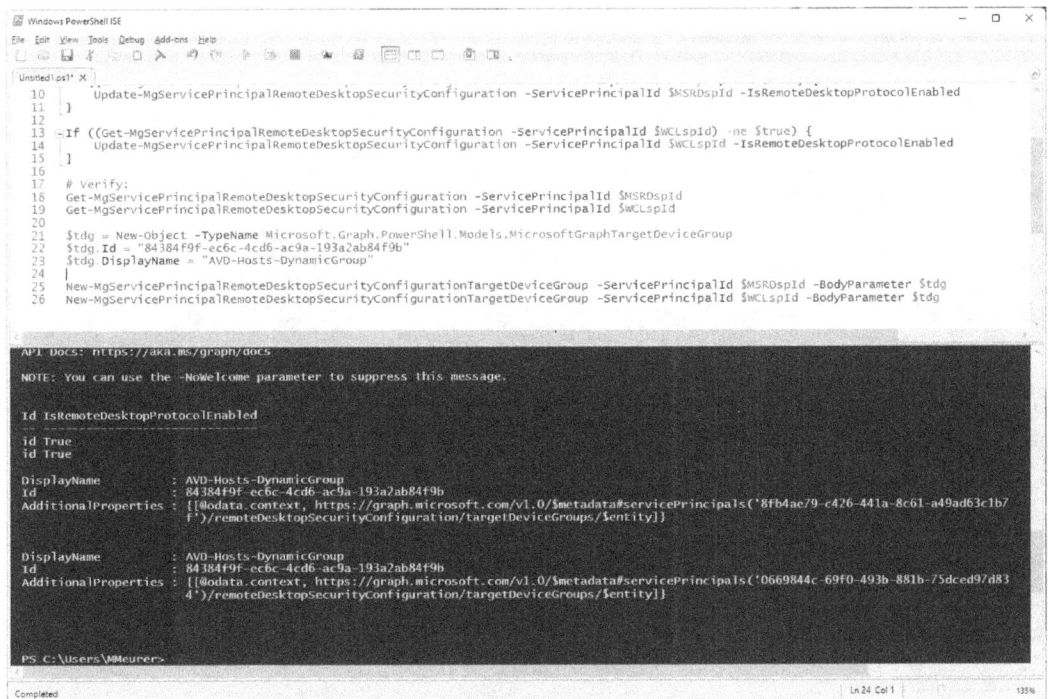

Figure 8.18 – The script to enable Entra authentication for RDP

To give administrators privileged permissions on the VM, we can configure local admin access.

Configuring local admin access

To give the user local admin access to a VM, you need to assign the **Virtual Machine Administrator Login** role to the VM by following the same process shown in the *Enabling user access* section.

Entra ID Join for Azure Virtual Desktop

> **Important note**
> It is recommended that you only assign users to the required VMs when assigning the **Virtual Machine User login** role. For example, if you assign this role to a group at the subscription level, all users within the group would have local admin rights to all the VMs. This is not a recommended approach.

You would add the required user to the **Virtual Machine Administrator Login** role, as shown in the following screenshot:

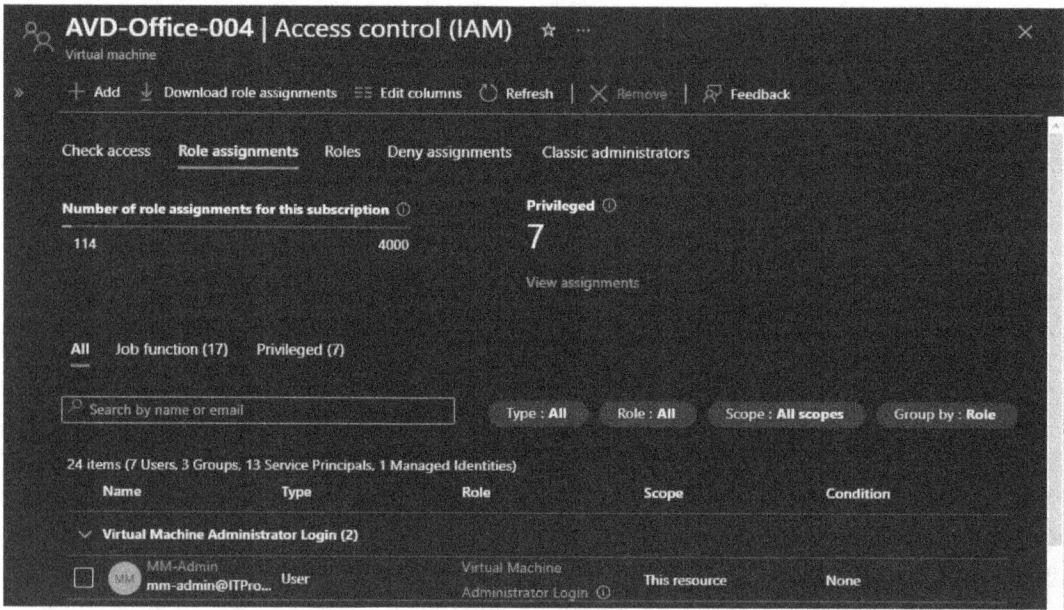

Figure 8.19 – The Virtual Machine Administrator Login role assigned to the VM

Once you have added the required permissions, you should see the user account now logged in as an administrator:

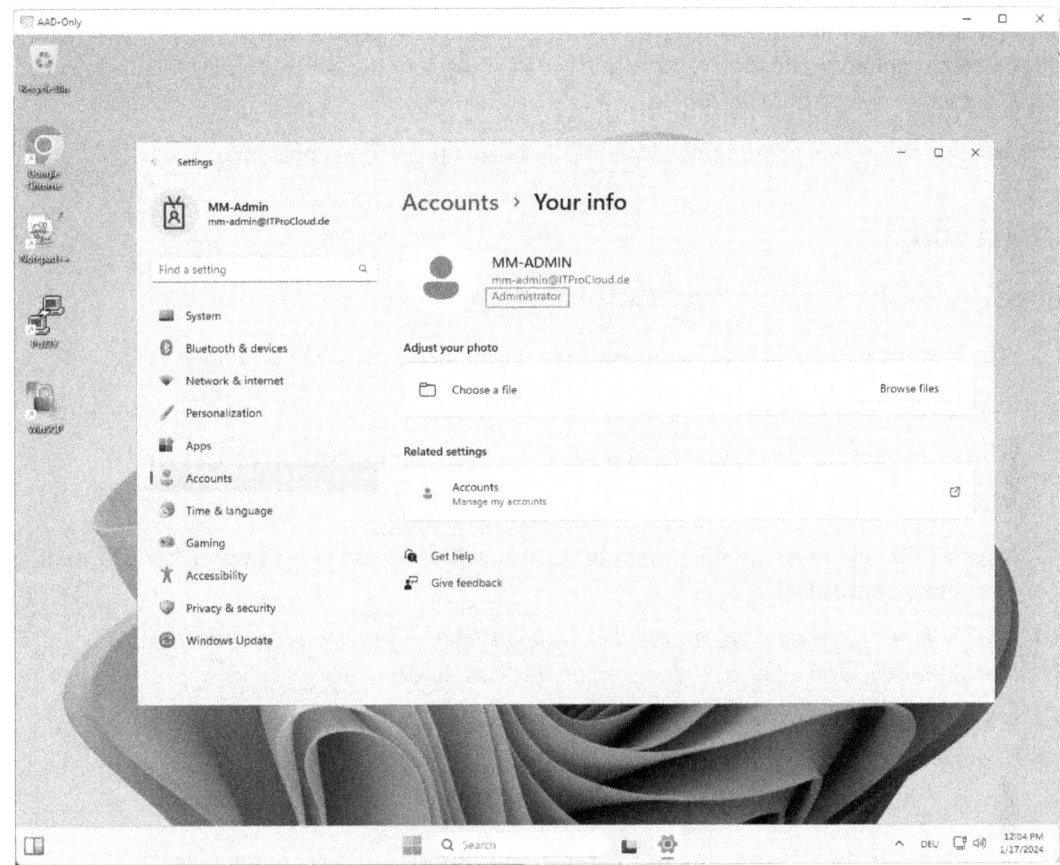

Figure 8.20 – The user logged in as an administrator

This brief section looked at assigning local admin rights to an Entra ID-joined host pool.

Summary

This chapter looked at the Entra ID join feature for Azure Virtual Desktop. First, we looked at the prerequisites, then we studied deploying an Entra ID-joined host pool, and we finished by looking at applying the required permissions and setting the RDP property for access on devices that are not Entra ID-joined or hybrid domain-joined.

In the next chapter, we will take a look at creating and managing session host images.

Questions

Here are a few questions to test your understanding of this chapter:

1. Which command would you use to check to see whether a session host is Entra ID-joined?

    `dsregcmd /status`

2. What is the minimum version of Windows 10 that you can use with Entra ID join?

    Windows 10 Enterprise build 2004

3. After deploying an Entra ID-joined host pool, what are the two things you need to do regarding permissions?

    Add the Entra ID group to the default desktop application group. Then, add the Virtual Machine User Login role for the Entra ID group within the host pool resource group.

4. What custom RDP property should you assign when using a macOS device with an Entra ID-joined host pool?

    `targetisaddjoined:i:1`

5. What must be done if you plan to roll out a new host in Entra ID as a replacement for an older host?

    Delete the existing device object in Entra.

9
Creating and Managing Session Host Images

In this chapter, we will look at creating a custom image that we can use with **Azure Virtual Desktop** (**AVD**) and some of the customization, updates, and publishing capabilities that are available in the **Azure Compute Gallery** (**ACG**). It is important to note that images for AVD are one of its core components as this is what the user will access. It's important to ensure that you configure and optimize them correctly so that you provide a good user experience. We will also take a look at troubleshooting some OS issues related to AVD.

In this chapter, we will cover the following topics:

- Creating a golden image
- Custom image templates
- Modifying a session host image
- Creating and using an ACG
- Troubleshooting OS issues related to AVD

Creating a golden image

In this section, we will look at provisioning a **virtual machine** (**VM**) in Microsoft Azure and preparing it for AVD.

Creating a VM

To create a standardized template image for AVD, you will need to spin up a VM. In this section, we will look at doing just that.

To get started, you will need to navigate to the Azure portal at https://portal.azure.com. From there, follow these steps:

1. Type `virtual machines` in the search bar located at the top of the page.
2. Under **Services**, select **Virtual machines**.
3. On the **Virtual machines** page, select **+ Create** and then **Virtual machine**:

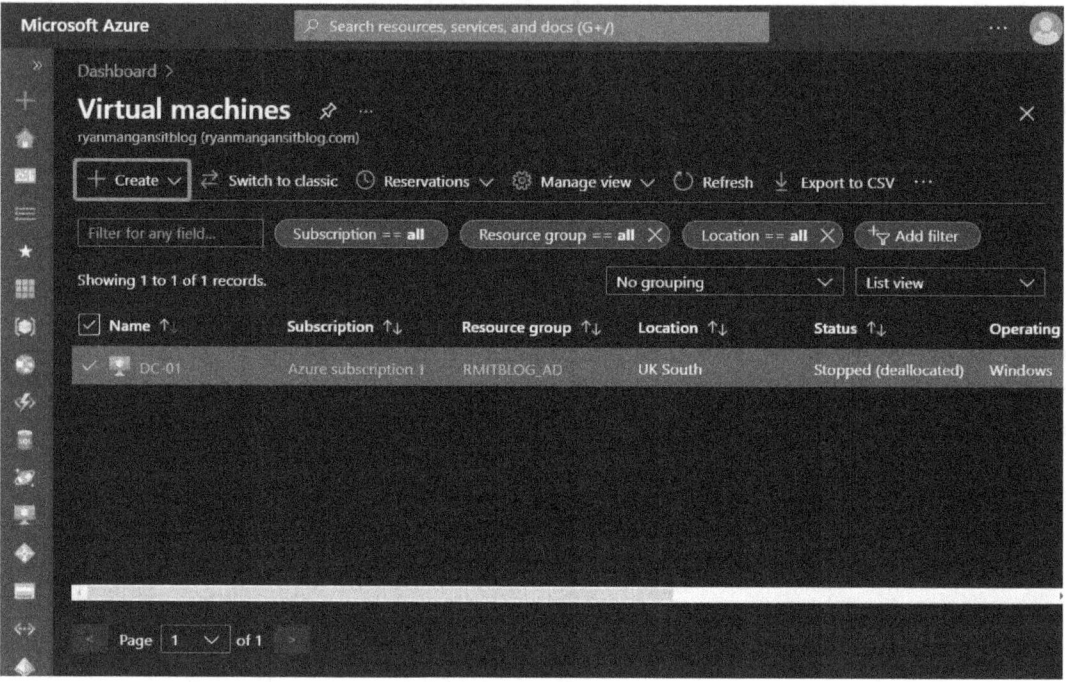

Figure 9.1 – The Virtual machines page

4. Under **Project details**, select the correct subscription and then choose to create a new resource group in the **Basics** tab. Type `myResourceGroup` for **Name**:

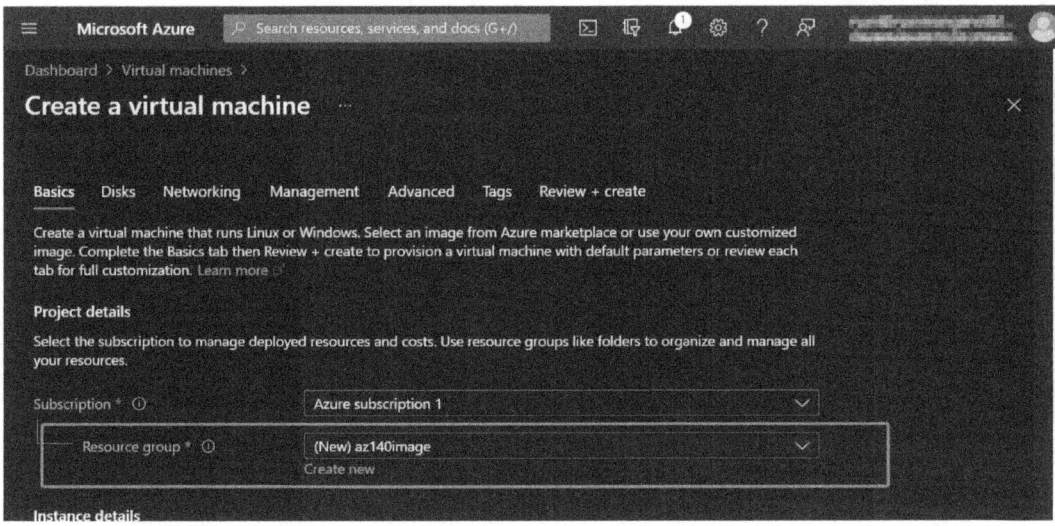

Figure 9.2 – The Create a virtual machine page

5. Under **Instance details**, enter a name under **Virtual machine name** and choose your required region. Next, choose **Windows 11 Enterprise multi-session** for **Image** and select the required VM size. Leave the other options as they are:

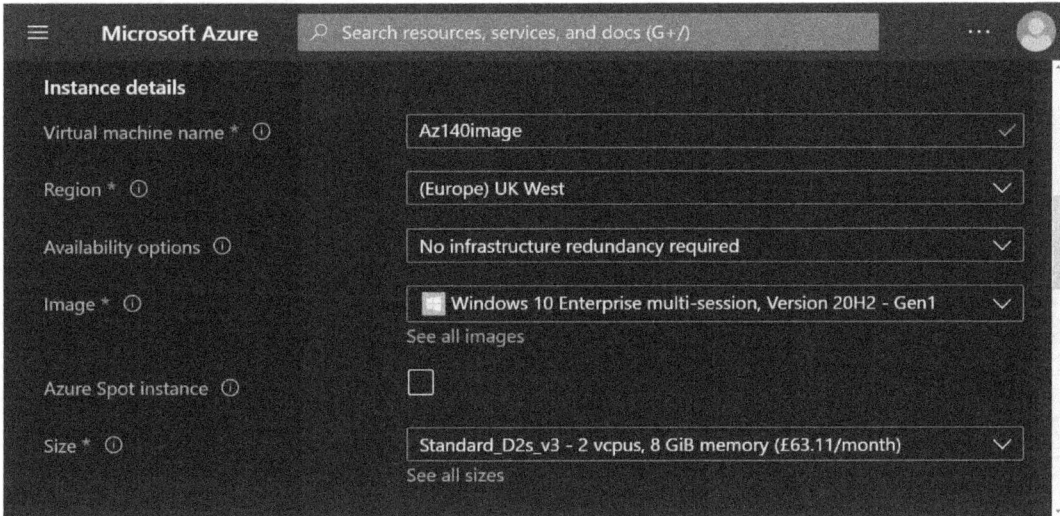

Figure 9.3 – Instance details when creating a VM

6. Under **Administrator account**, enter a username and a password:

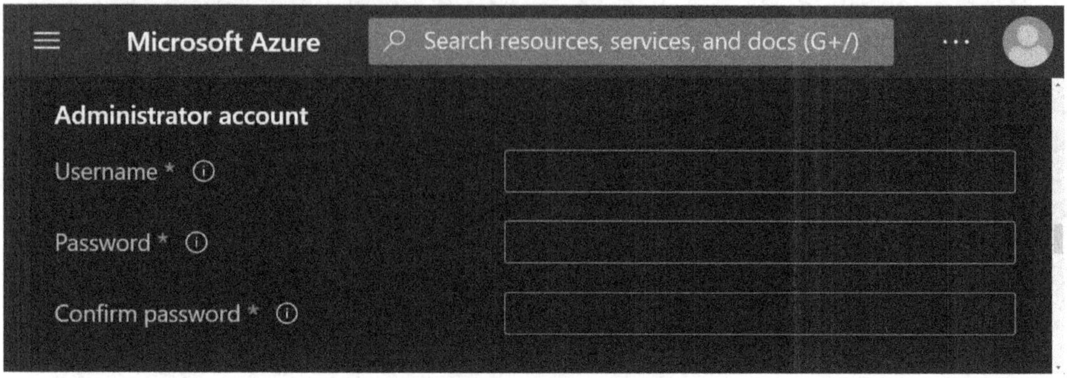

Figure 9.4 – The Administrator account section

> **Tip**
>
> Your password must be at least 12 characters long and meet the required complexity. For more information, go to https://docs.microsoft.com/azure/virtual-machines/windows/faq#what-are-the-password-requirements-when-creating-a-vm-.

7. Under **Inbound port rules**, choose **Allow selected ports** and then select **RDP (3389)** from the dropdown:

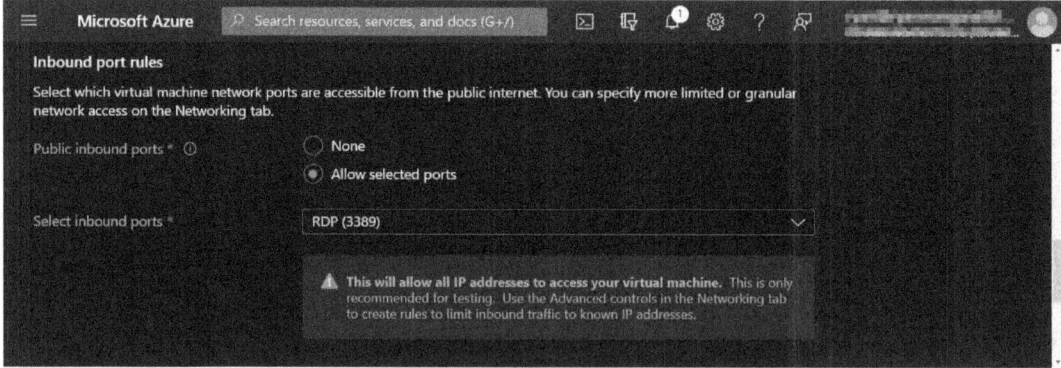

Figure 9.5 – Inbound port rules

8. Confirm the license eligibility for Windows 11 and click the **Review + create** button at the bottom of the page:

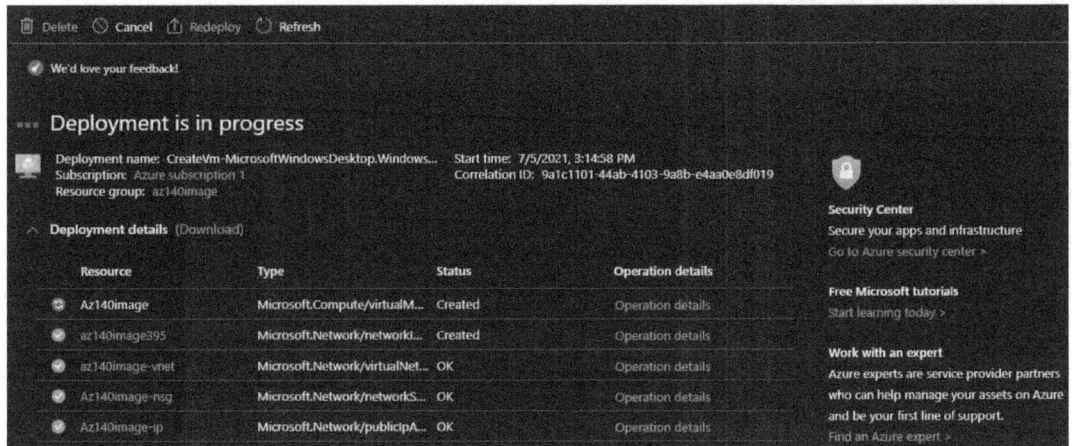

Figure 9.6 – Deployment is in progress

9. Once the deployment has finished, navigate to the VM by clicking **Go to the resource**.

Once you have created the VM, you are ready to connect to it.

Connecting to the VM

You can connect to an Azure VM using a remote desktop connection. You have two options – you can connect using a direct connection via a public IP address or connect using the local private IP address behind a VPN connection. Let's get started:

1. Navigate to the overview page of the VM you just deployed, select the **Connect** button, and then select **RDP**:

> **Important note**
> Even though this section shows you how to connect using RDP via a public IP address, it is recommended that you connect over a VPN connection or Azure Bastion.
>
> In *Chapter 4, Implementing and Managing Networking for Azure Virtual Desktop*, we learned how to configure Azure Bastion. It is recommended that you use Azure Bastion as RDP is insecure.

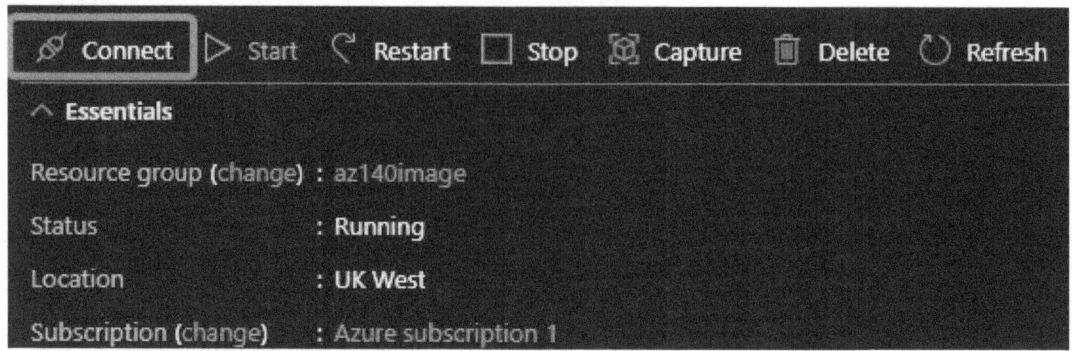

Figure 9.7 – Using the Connect button to download an RDP file for the VM

2. On the **Connect with RDP** page, keep the default options to connect by IP address, over port 3389, as-is, and click **Download RDP file**:

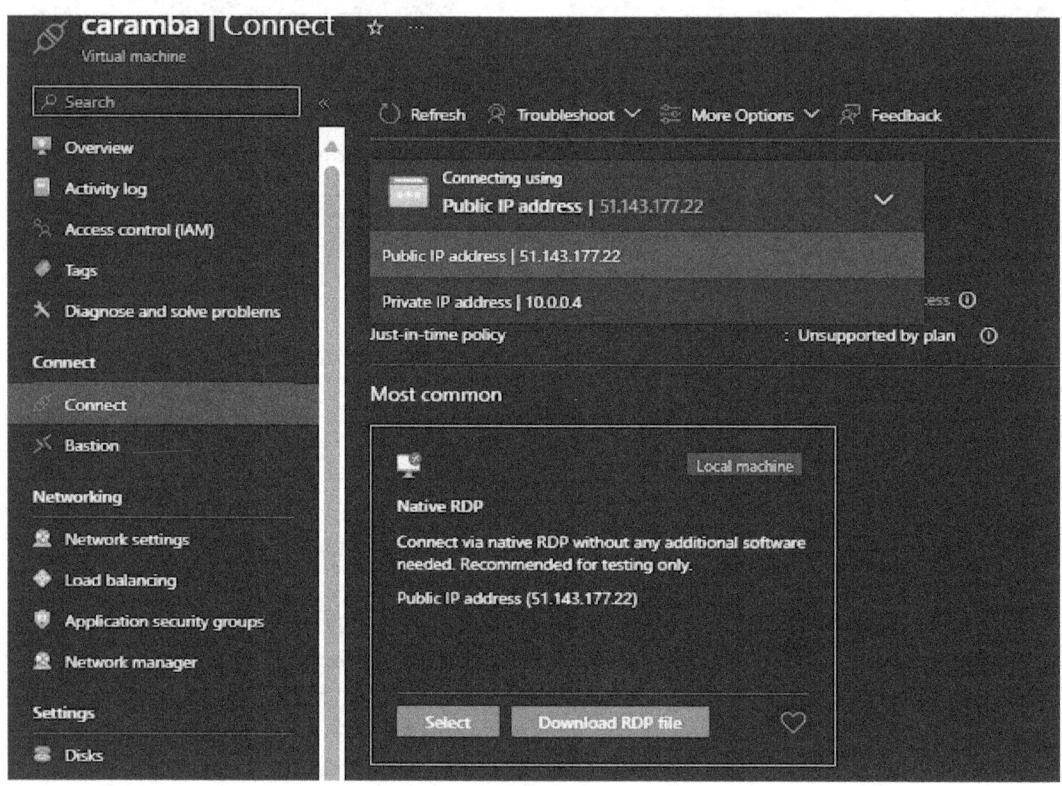

Figure 9.8 – The public and private IP address options for the RDP file

3. Open the downloaded RDP file and click **Connect** when prompted:

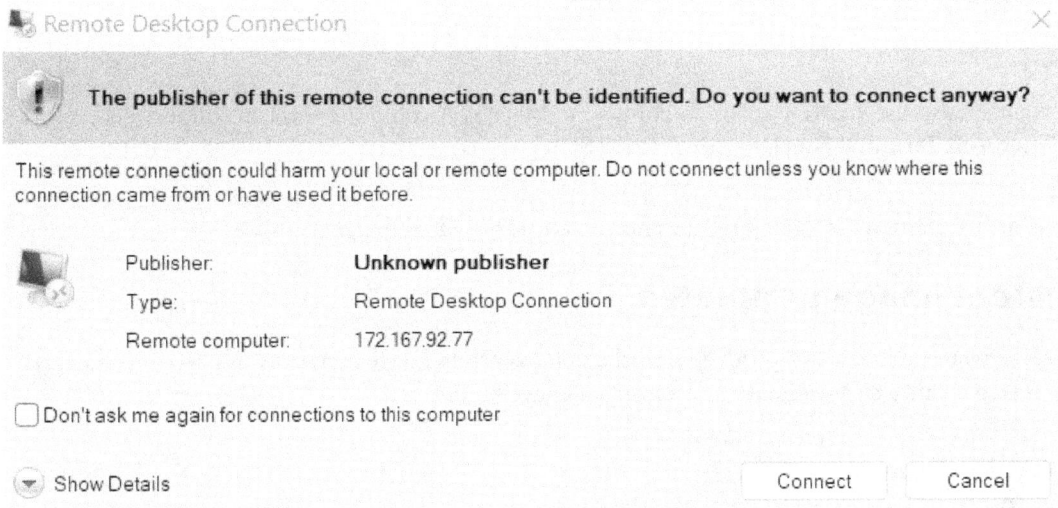

Figure 9.9 – RDP file connecting to the VM

Once connected to the RDP session, you will see the session appear, as shown in the following screenshot:

Figure 9.10 – VM connected via RDP

Once you've connected to the VM via a public or local IP address, you can then modify/customize the image based on your organization's requirements.

> **Tip**
> You can also use Azure Bastion to connect to the VM if you don't want to connect via the **Microsoft Terminal Services Client (MSTSC)**.

In the next section, we are going to learn how to create a custom image template.

Custom image templates

In this section, we provide you with an introduction to custom image templates and demonstrate how easy it is to create a custom image and automate image builds.

> **Important note**
> Custom images are built using the Azure VM Image Builder service. Read more here: `https://learn.microsoft.com/azure/virtual-machines/image-builder-overview`.

Creating your first custom image

To get started, you will need to navigate to the Azure portal at `https://portal.azure.com`. From there, follow these steps:

1. Type `Azure Virtual Desktop` in the search bar located at the top of the page.
2. Under **Manage**, select **Custom Image Templates**.

3. On the **Custom image templates** page, select **Add custom image template**:

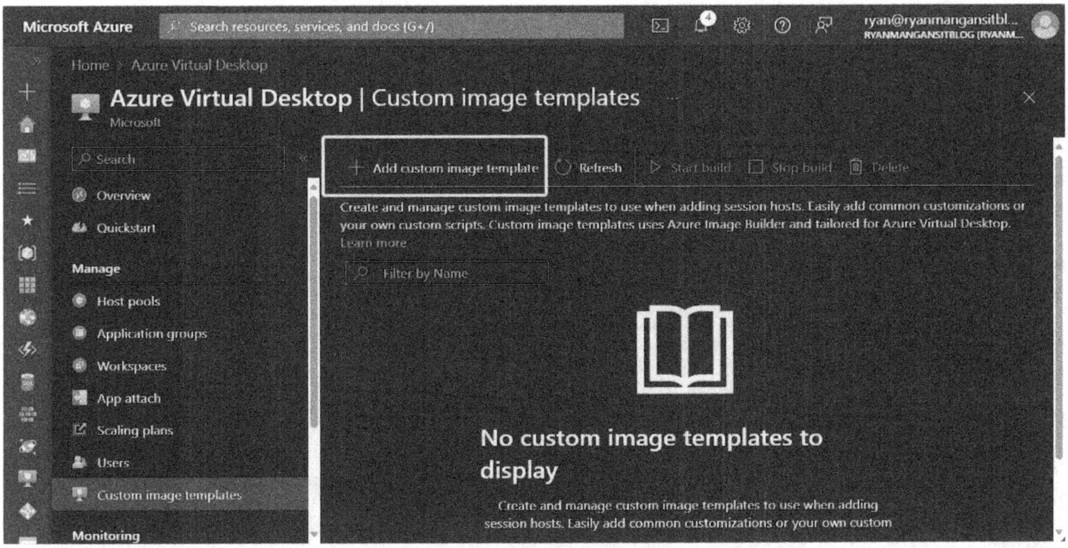

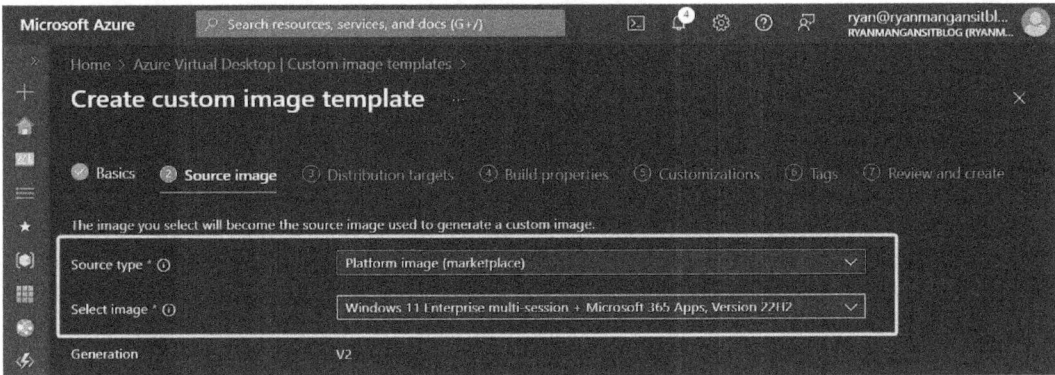

Figure 9.11 – AVD Custom image templates page

4. Under **Basics**, enter a name under **Template name**, select an existing resource group, and select a location (in this example, **UK South**) and also a managed identity. If you don't have a managed identity, you will need to create one.

5. Under **Source image**, choose a source type (in this example, we will use **Platform image (marketplace)**), and then select an image (in this example, we will select **Windows 11 Enterprise multi-session + Microsoft 365 Apps, Version 22H2**):

Figure 9.12 – The Source image tab within the Create custom image template wizard

6. In this example, we are going to create a managed image; however, in most cases, you would use an ACG. Within the **Distribution targets** tab, check the **Managed image** checkbox, choose a resource group, enter an image name, and provide a run output name:

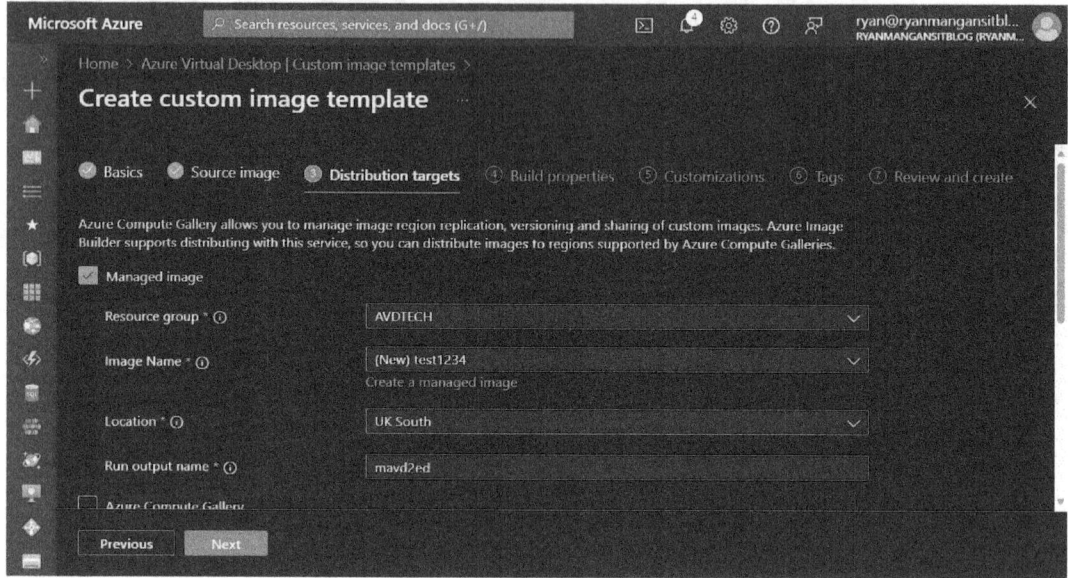

Figure 9.13 – The Distribution targets tab within the Create custom image template wizard

7. Within the **Build properties** tab, leave the **Build timeout (minutes)** box empty; at its default, it's set to 240. Select the required build VM size, adjust the **OS disk size (GB)** setting to the requirement, and provide a staging group (resource group) used for storing build resources and logs:

Custom image templates 243

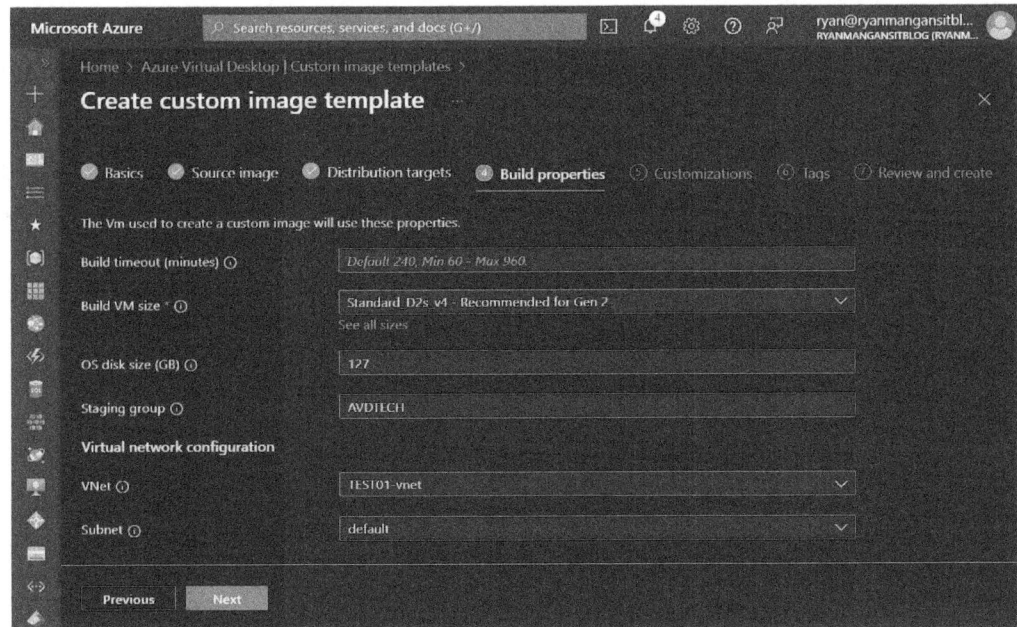

Figure 9.14 – The Build properties tab and the VM configuration

8. Within the **Customizations** tab, you can add both built-in scripts and your own. In this example, we will only use built-in scripts. Click the **Add built-in script** button:

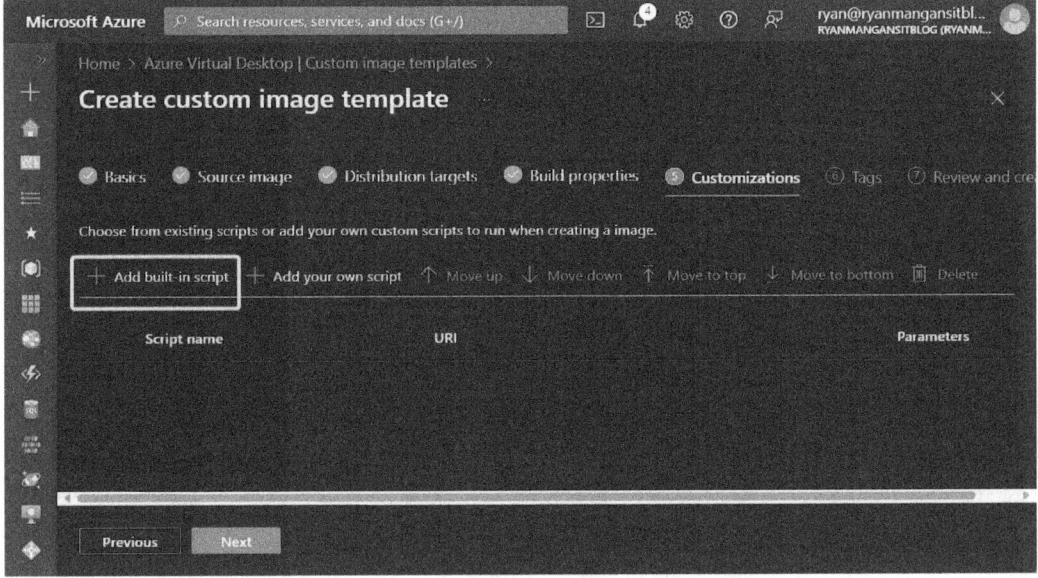

Figure 9.15 – The Customizations tab for creating a custom image template

9. Within the **Select built-in scripts** blade, there are a large number of customizations and configurations you can choose from. In this example, I have checked the **Install Languages** box, chosen **English (United Kingdom)**, and checked the **Install FSLogix and enable file containers** box:

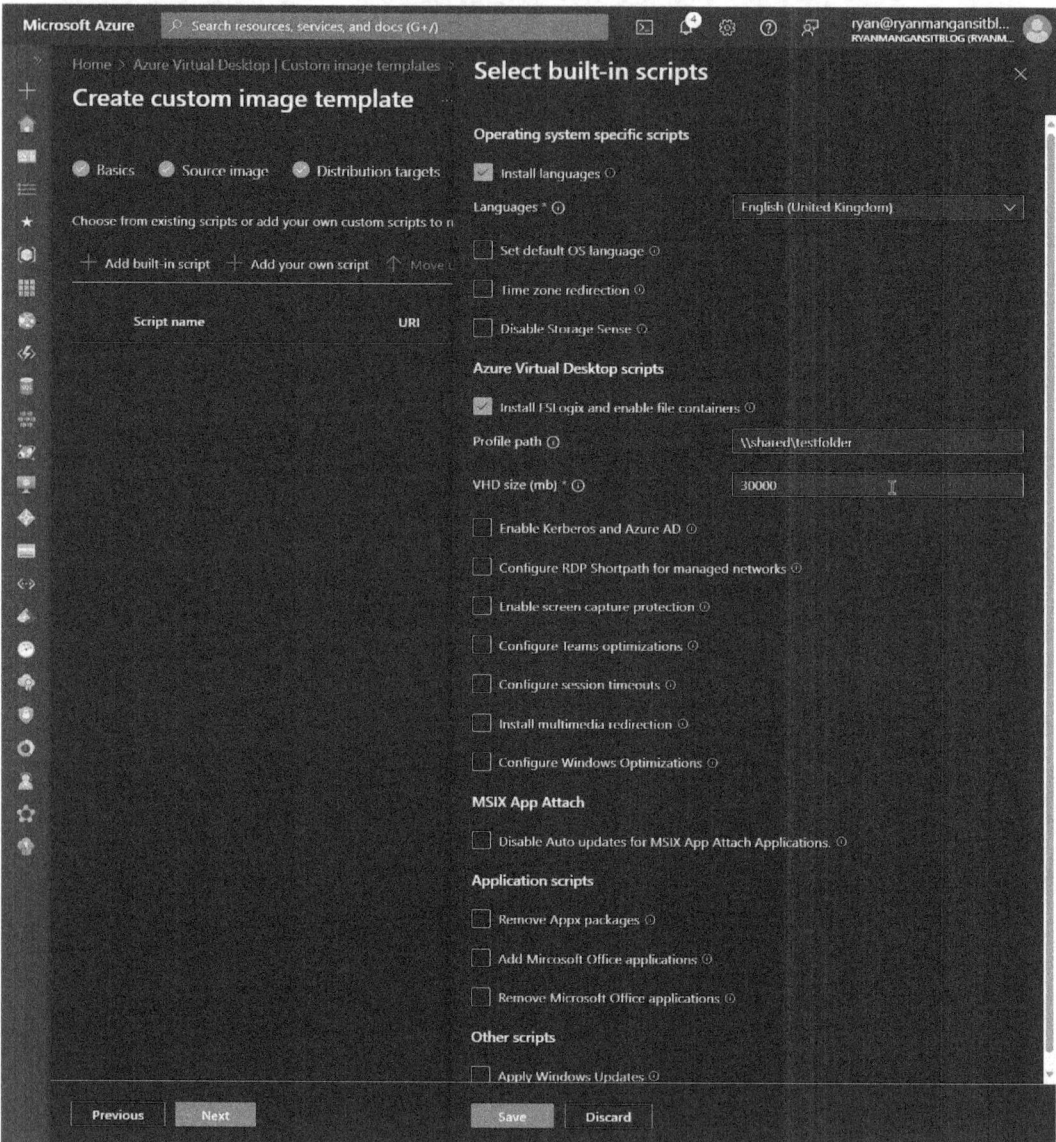

Figure 9.16 – The built-in scripts you can choose from

10. Once you click **Save**, these will appear in the **Customizations** blade, as shown in the following screenshot:

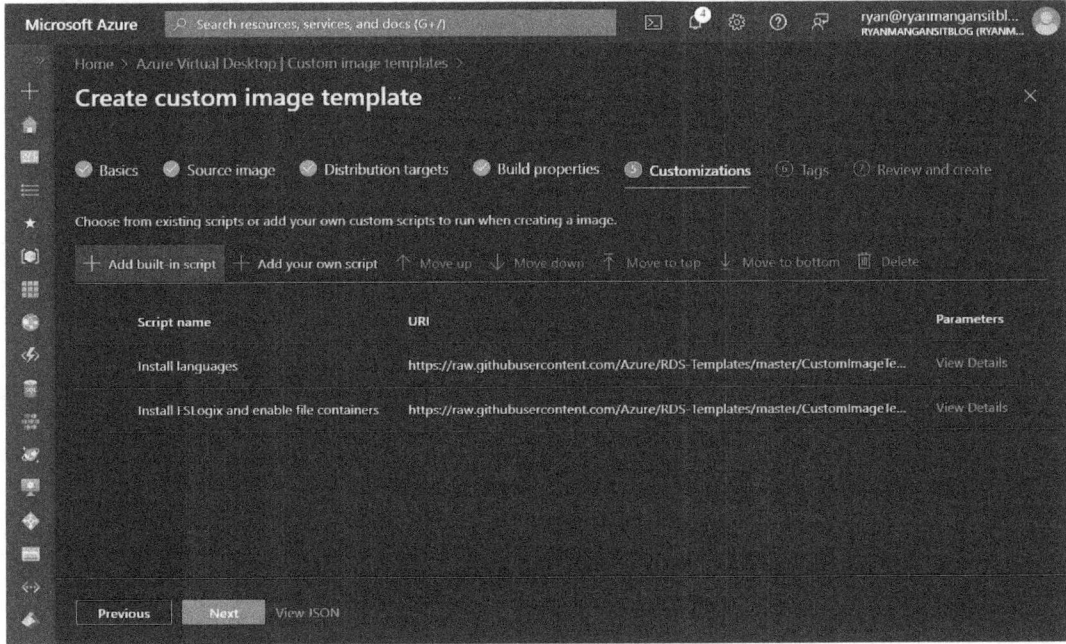

Figure 9.17 – Two scripts added to the Customizations section

11. Finally, proceed with adding tags if required, and under the **Review and create** tab, check the configuration; when happy, proceed with **Create**:

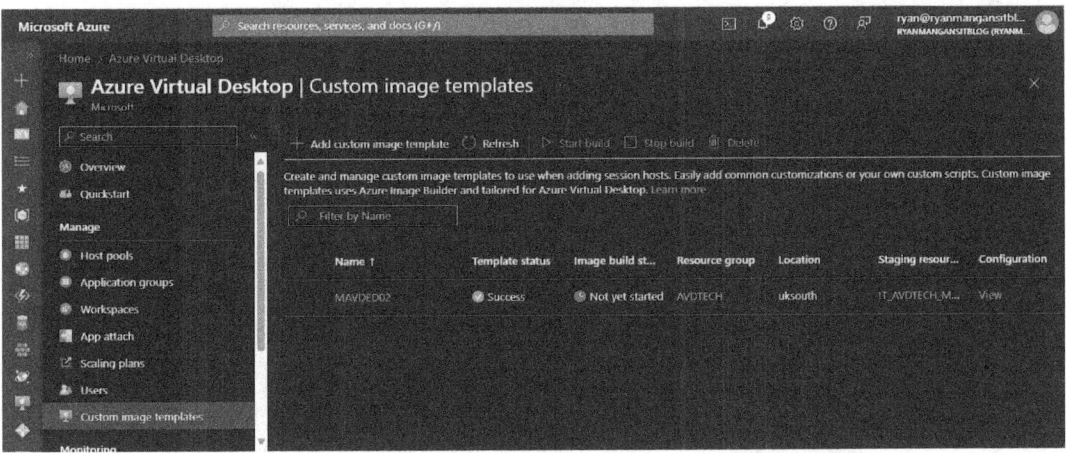

Figure 9.18 – The successful creation of a custom image template

12. You can now click **Start Build** within the template, as shown in the following screenshot:

Figure 9.19 – Showing where the Start build button is located within the template

There you have it – we have created our first simple custom image template.

In the next section, we are going to learn how to modify a session host image, including customizing and optimizing it.

Modifying a session host image

In this section, we will look at some of the *customizations/optimizations* you can apply to an image for AVD.

> **Tip**
> When you're using Windows 11 multi-session images from ACG, FSLogix profile containers come pre-installed. It's recommended that you use an ACG template when possible.

Disabling automatic updates

When you're using pooled desktops, images are typically deployed centrally. First, updates are completed on the master image, and then VMs are redeployed to the host pool on the next scheduled maintenance window.

You should consider disabling Windows updates for AVD images as these should be carried out on the master image during a maintenance window. This enables control over patch updates and consistency through the deployed virtual desktop estate.

You may also want to consider configuring a validation environment, as discussed in *Chapter 7, Configuring Azure Virtual Desktop Host Pools*.

You can disable automatic updates directly on the Windows image using Regedit:

```
reg add "HKLM\SOFTWARE\Policies\Microsoft\Windows\WindowsUpdate\AU" /v NoAutoUpdate /t REG_DWORD /d 1 /f
```

Alternatively, you can use a Group Policy to apply the same change, as detailed in the following steps:

1. Open **Local Group Policy Editor** | **Administrative Templates** | **Windows Components** | **Windows Update**.
2. Right-click **Configure Automatic Update** and set it to **Disabled**.

Now, let's install some language packs that we can use on the AVD image.

Installing language packs in AVD

To use multiple languages within AVD, you need to ensure that all the required languages are installed.

> **Tip**
> Windows 11 onward will only allow language packs to be distributed as `.cab` files, which can be used for imaging. **Language Interface Pack (LIP)** languages that aren't distributed as `.cab` files will only be available as `.appx` packages, which can be acquired through the **Settings** app after logging in.

Before you start customizing your image with multiple languages, you will need to download the required files for the language configuration; these can be found at the following links:

- Windows 11, version 21H2 Language and Optional Features ISO: `https://software-static.download.prss.microsoft.com/dbazure/988969d5-f34g-4e03-ac9d-1f9786c66749/22621.1.220506-1250.ni_release_amd64fre_CLIENT_LOF_PACKAGES_OEM.iso`

- Windows 11, version 22H2 and 23H2 Inbox Apps ISO: `https://software-download.microsoft.com/download/sg/19041.928.210407-2138.vb_release_svc_prod1_amd64fre_InboxApps.iso`

> **Important note**
>
> Please note that the inbox apps that are included in the ISO are not the latest versions of the pre-installed Windows apps. You will need to update the apps using the Windows Store app and perform a manual search for updates after installing the additional languages.
>
> You can find all the language ISO files here: `https://docs.microsoft.com/azure/virtual-desktop/language-packs#prerequisites`

When you're installing languages on Windows 10 version 2004, 20H2, and 21H1, it's recommended that you check the known issues to ensure you choose the correct ISO: `https://docs.microsoft.com/windows-hardware/manufacture/desktop/language-packs-known-issue`

> **Tip**
>
> You can read more about language packs here: `https://docs.microsoft.com/en-gb/azure/virtual-desktop/language-packs`

The following summary steps will show you how to create a custom Windows 11 Enterprise multi-session image and connect the repository as a drive letter:

1. Deploy an Azure VM, as shown previously in the *Creating a golden image* section.
2. Then, go to ACG and select the current version of Windows 11 Enterprise multi-session that you plan on using.

Once you have deployed the VM, connect to the VM using RDP as a local administrator.

Make sure that your VM has all the latest Windows updates installed. Then, download the updates and restart the VM if required.

Connect to the language package, features on-demand, and Inbox Apps file share repository and mount it to a letter drive (for example, the `Z:` drive). Note that you can use it on the root of `C:\` if you have a large OS drive.

> **Important note**
>
> Please note that the script should be customized to your needs and requirements. Running the entire script can take some time to complete.
>
> You can find the script for adding languages to run Windows images here: Link HERE

Once the script has finished running, you can check that the language packs have been installed correctly by navigating to **Start | Settings | Time & Language | Language**. If the language files have been installed, you will see them here.

Once you have confirmed that the licenses have been installed, you can install the inbox apps for each required language. Then, you can update the inbox apps by refreshing the pre-installed apps using the inbox app's ISO image. You can use the PowerShell script template available at the following link to automate this process and update only the installed versions for inbox apps with no internet access: `https://github.com/PacktPublishing/Mastering-Azure-Virtual-Desktop-2nd-Edition`.

Once you have finished, make sure that you disconnect the share you attached previously (for example, the `Z:` drive).

> **Tip**
> To ensure that modern apps can use the additional language packs, you should use the following PowerShell cmdlet, which is used to disable language pack cleanup:

```
Disable-ScheduledTask -TaskPath "\Microsoft\Windows\AppxDeploymentClient\" -TaskName "Pre-staged app cleanup"
```

> You can also find this within the example script provided here: `https://github.com/PacktPublishing/Remote-Productivity-with-Windows-Virtual-Desktop/blob/main/B17392_07/Languages_winimage.ps1/`

Before you `Sysprep` your image, after adding the language packs, you will need to run the following cleanup operations on the golden image using PowerShell:

```
##Cleanup to prepare sysprep##
Remove-AppxPackage -Package Microsoft.LanguageExperiencePackes-ES_22000.8.13.0_neutral__8wekyb3d8bbwe
Remove-AppxPackage -Package Microsoft.OneDriveSync_22000.8.13.0_neutral__8wekyb3d8bbwe
```

Now, let's learn how to optimize an image.

Optimizing an image

In this section, we will look at two ways to optimize AVD images. Optimizing an image can improve the user experience for AVD users and can also improve user density per session host.

> **Important Note**
> Remember – turning off/disabling features and services can impact the functionality and behavior of the user desktop and applications that are running. Make sure that you fully understand what you are optimizing before applying it.

Virtual desktop optimization tool

The second option would be to use the optimization tool and a set of PowerShell scripts that you can run on the template image to configure optimizations for the virtual desktop image.

You can access these scripts by going to the following GitHub repository: `https://github.com/The-Virtual-Desktop-Team/Virtual-Desktop-Optimization-Tool`

First, you will need to download the scripts from the aforementioned GitHub repository. Once you've downloaded them, you must find the build version of your operating system – for example, `C:\temp\optfiles\ConfigurationFiles` – and change the required settings to **Enabled** (default) or **Disabled**. The following screenshot provides an example of the JSON configuration file for `AppxPackages.json`:

```json
[
    {
        "AppxPackage": "Microsoft.OneConnect",
        "VDIState": "Disabled",
        "URL": "https://www.microsoft.com/en-us/p/mobile-plans/9nblggh5pnb1",
        "Description": "Microsoft Mobile Plans app"
    },
    {
        "AppxPackage": "Microsoft.MSPaint",
        "VDIState": "Enabled",
        "URL": "https://www.microsoft.com/en-us/p/paint-3d/9nblggh5fv99",
        "Description": "Paint 3D"
    },
    {
        "AppxPackage": "Microsoft.BingWeather",
        "VDIState": "Disabled",
        "URL": "https://www.microsoft.com/en-us/p/msn-weather/9wzdncrfj3q2",
        "Description": "MSN Weather app"
    },
```

Figure 9.20 – Snippet of the AppxPackages.json file

Once you have finished customizing the configuration files, you need to run the optimization script using PowerShell.

First, open PowerShell as an administrator and then set the execution policy to `Bypass`:

```
Set-ExecutionPolicy -ExecutionPolicy Bypass
```

The following screenshot shows the set execution policy's bypass cmdlets that have been run:

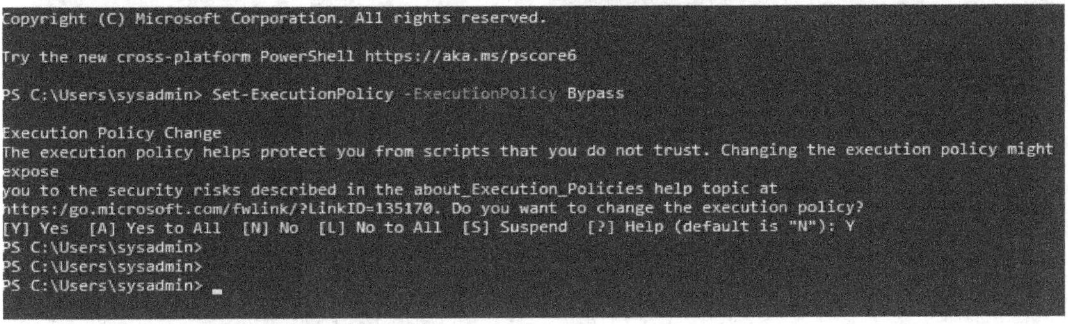

Figure 9.21 – The set-execution policy cmdlet being run

Once set, you need to run the script. Remember to use the required Windows version you are using:

```
.\Windows_VDOT.ps1 -Optimizations All -Verbose -AcceptEula
```

Usually, we must accept to run the script by pressing *R* and *Enter*.

The following screenshot shows the script being run and that the optimization process is in progress, as denoted by the progress bar:

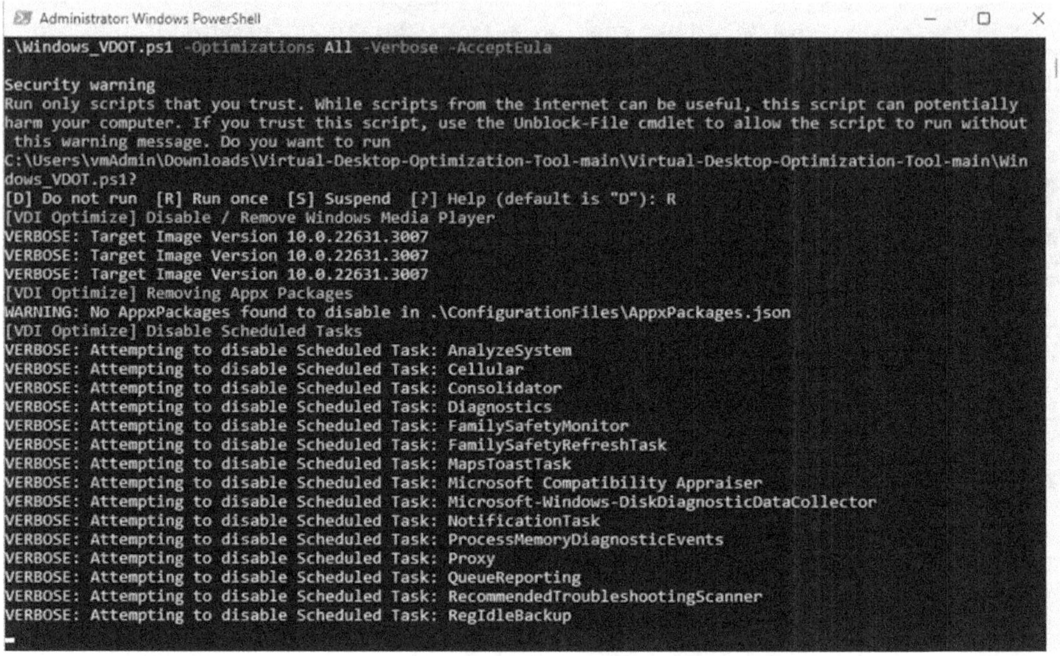

Figure 9.22 – Optimization script running

Once complete, the script will finish and prompt you to reboot.

Once you have rebooted, you may notice that the processes and threads may have been reduced within the **Task Manager** | **Performance** tab.

The following screenshot shows the number of processes and threads before optimization:

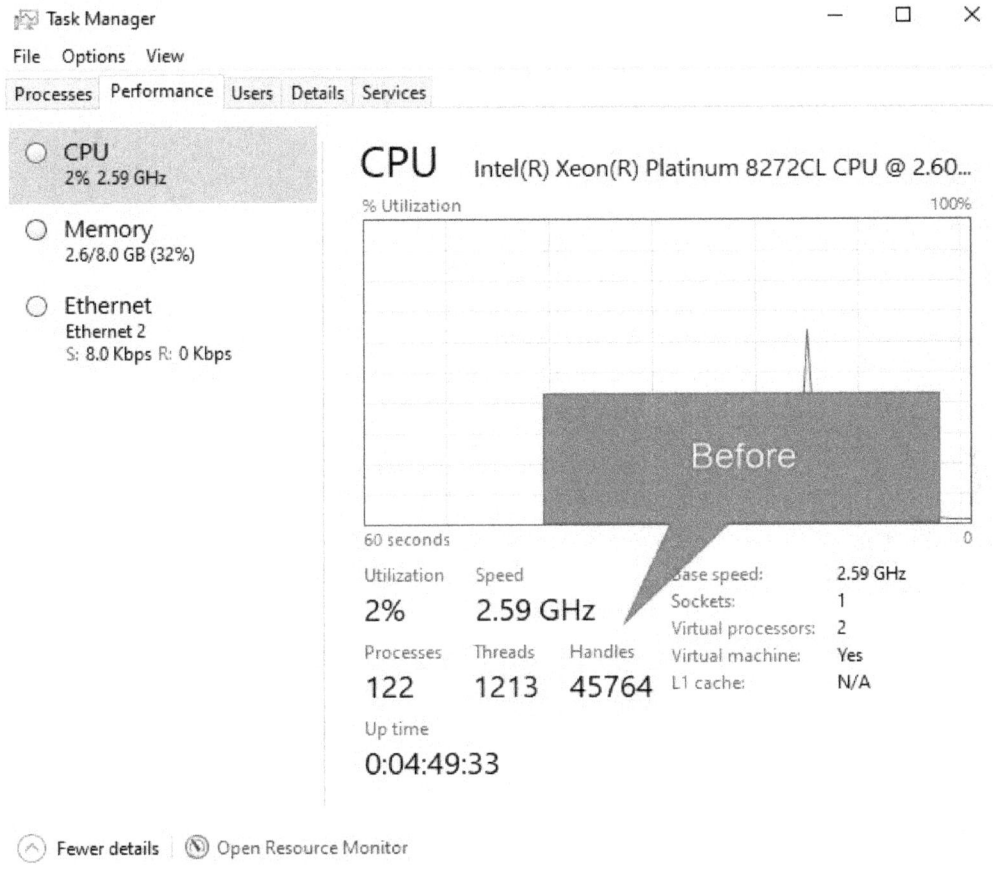

Figure 9.23 – The number of processes, threads, and handles before optimization

The following optimization screenshot shows a reduction in processes, threads, and handles:

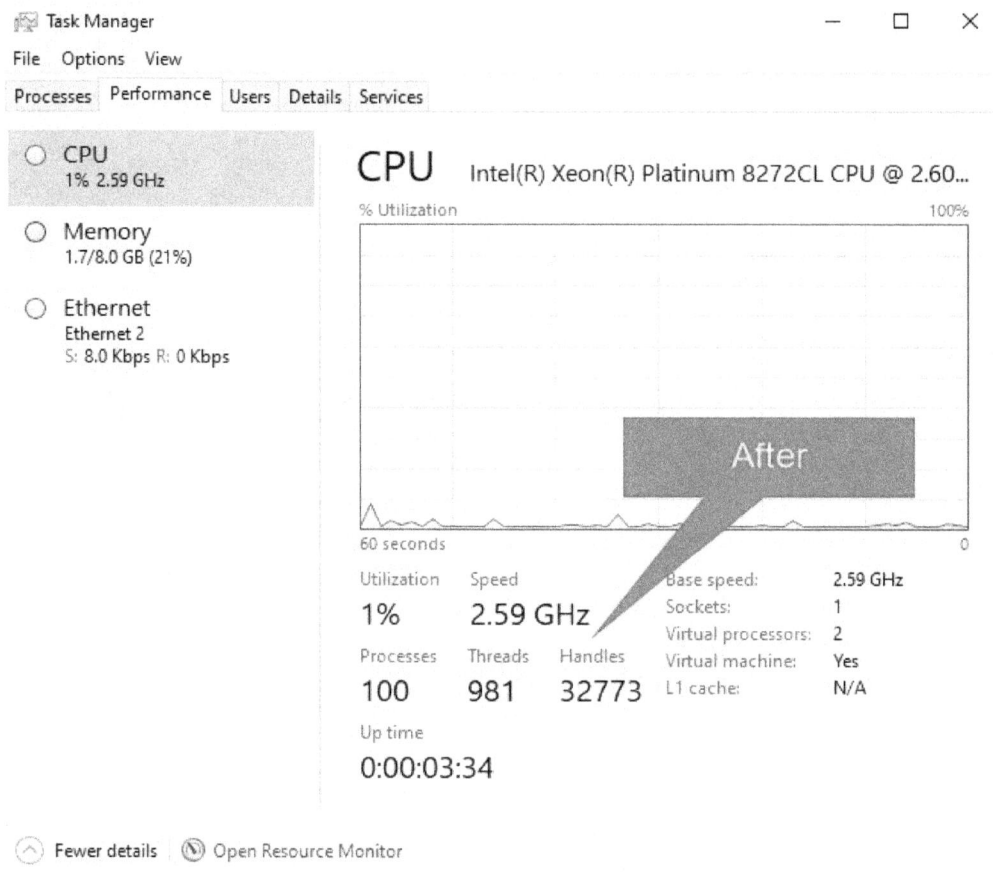

Figure 9.24 – The number of processes, threads, and handles after optimization

> **Important note**
> The AVD optimization script is community-driven and not supported by Microsoft.

Remember – optimizations have an impact on the user experience, and some can be difficult to restore. So, ensure you create a snapshot before applying such changes to your image as it is not as easy to roll back using the virtual desktop optimization script.

In the next section, we will look at capturing an image template.

Windows updates

While we are connected to the VM, we can manually install new Windows updates to our golden master. From the Start menu, start **Check for updates** and install the updates:

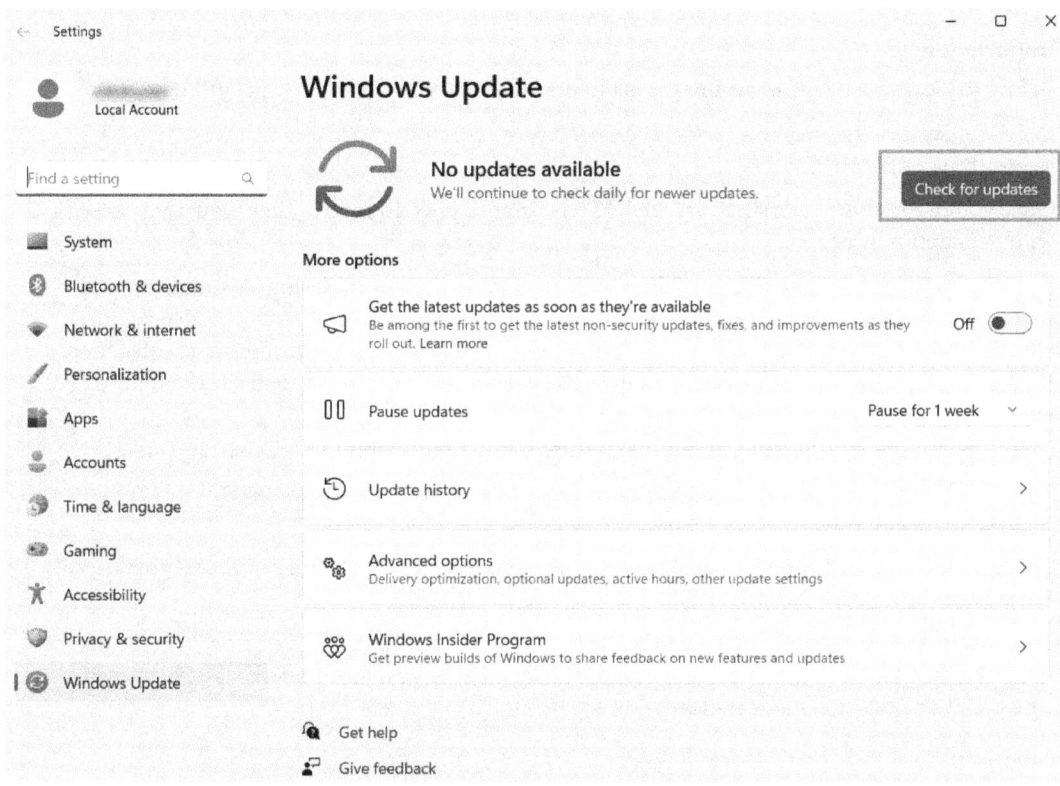

Figure 9.25 – Windows updates installation

After installing the updates, restart Windows to apply all updates. After logging in, verify that no new updates are pending.

Capturing an image template

In this section, we will look at capturing an image template so that you can distribute the same image across multiple hosts within a host pool or even multiple host pools.

> **Important note**
>
> Before you capture a VM in an image, make sure that you run the following command first:
>
> `C:\Windows\System32\Sysprep\Sysprep.exe /generalise /oobe /mode:vm /shutdown`
>
> Once you've run this, make sure that the VM has been stopped in Azure. Then, you can proceed with the capture process.
>
> It is also important to note that installing Microsoft Store apps or updating existing Store apps before generalizing a Windows image will cause `Sysprep` to fail. `Sysprep` also requires all apps to be provisioned for all users. When you update an app from the Microsoft Store, that application will become associated with the user account. You will then see the following error message:
>
> `<package name> was installed for a user, but not provisioned for all users. This package will not function properly in the Sysprep image.`
>
> This will be located within the `Sysprep` log file, here: `%WINDIR%\System32\Sysprep\Panther`
>
> You can read more about `Sysprep` here: `https://docs.microsoft.com/windows-hardware/manufacture/desktop/Sysprep--generalize--a-windows-installation?view=windows-11`

To capture a template, follow these steps:

1. First, you need to navigate to the Azure portal (`https://portal.azure.com/`) to manage the VM image. Then, search for and select **Virtual machines**.
2. Select your VM from the **Virtual Machine** list.

3. On the **Virtual machines** page for the VM, go to the top menu and click the **Capture** button:

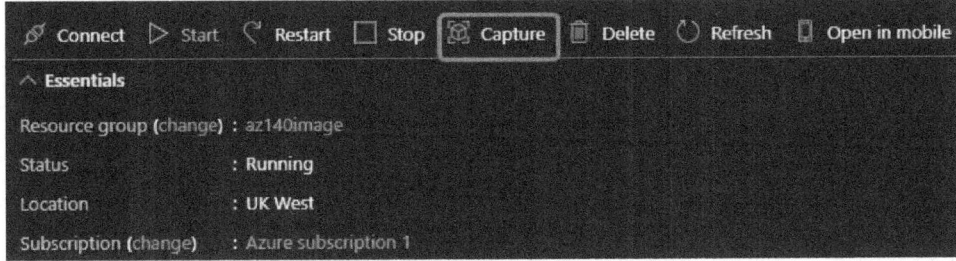

Figure 9.26 – The Capture button located within the Virtual machines page

4. A **Create image** page will then appear.
5. For **Name**, use the pre-populated name or enter a new name that you would like to use for the image:

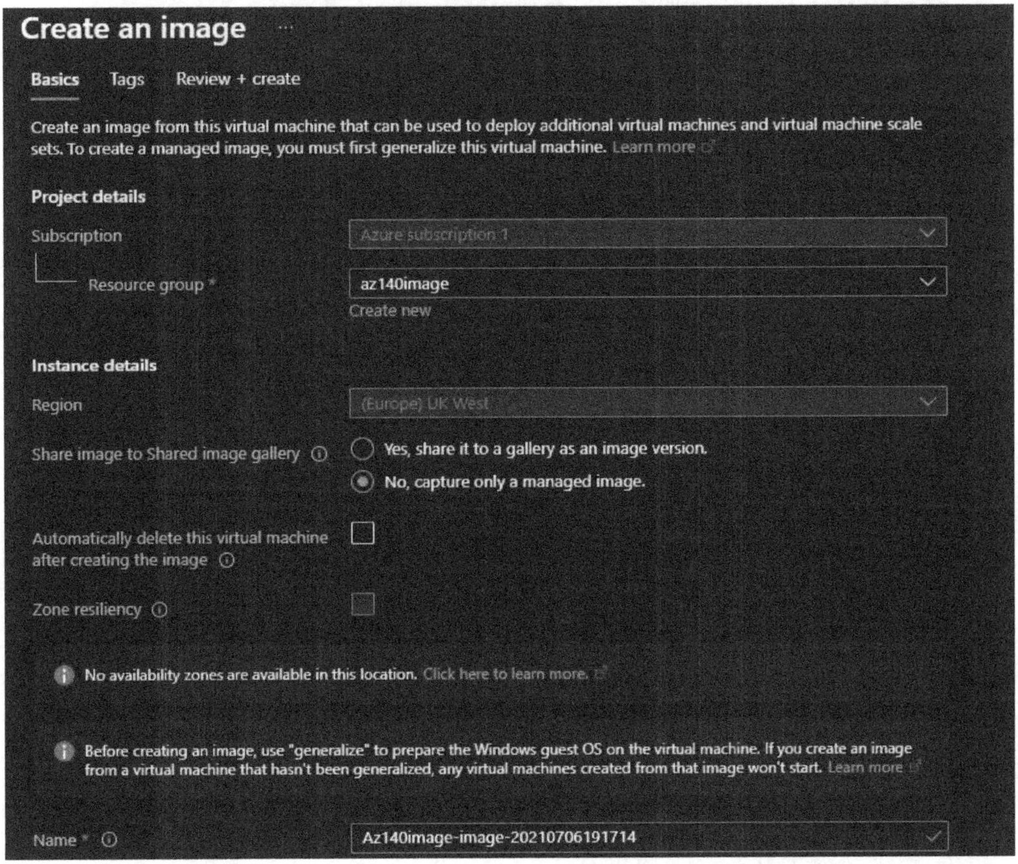

Figure 9.27 – The Create an image page within the Azure portal

6. For **Resource group**, use the existing one, select one from the dropdown, or create a new one.
7. For **Instance details**, ensure you select the **No, capture only a managed image.** option to create a managed image only.
8. To delete the source VM once the image has been created, select **Automatically delete this virtual machine after creating the image**.
9. If you want to use the image in any availability zone, check the **Zone resiliency** box.
10. Select **Create** to create the image; this will start the image creation process.
11. Once the image has been created, you can find it as an image resource in the list of resources in the resource group:

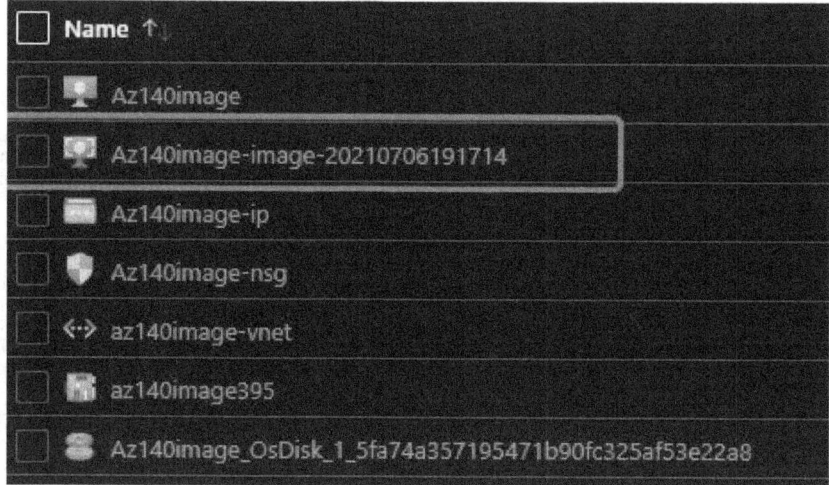

Figure 9.28 – Image created: post-creation process

In this section, we created an image from a VM and customized it. I will now show you how to create and use an ACG.

Creating and using an ACG

An ACG is a service that distributes images that can be shared across multiple regions and subscriptions within a **Microsoft Entra ID (MEID)** tenant. This is extremely useful for AVD as ACGs enable easy deployment of standard company desktop/server image templates across multiple Azure regions. One of the other benefits of **Shared Image Galleries (SIGs)** is that you can have different image versions and the newest can be referenced easily.

The following table details the different resource types within a SIG, now known as an ACG:

> **Tip**
> It is recommended that you become familiar with the terms within the following table before continuing.

Resource	Description
Image source	This is a resource that can be used to create an image version in an image gallery. An **image source** can be an existing Azure VM that is either generalized or specialized, a managed image, a snapshot, or an image version in another image gallery.
Image gallery	As with the Azure Marketplace, an **image gallery** is a repository for managing and sharing images, but you control who has access to it.
Image definition	**Image definitions** are created within a gallery and carry information about the image and the requirements for using it internally. This includes whether the image is for Windows or Linux, release notes, and the minimum and maximum memory requirements. It is a definition of a type of image.
Image version	An **image version** is what you use to create a VM when you're using a gallery. You can have multiple versions of an image as needed for your environment. As with a managed image, when you use an image version to create a VM, the image version creates new disks for the VM. Image versions can be used multiple times.

Table 9.1 – Different resource types

The preceding table was taken from Microsoft's documentation: `https://learn.microsoft.com/azure/virtual-machines/shared-image-galleries`

This section provided an overview of ACG and some of the specific terms that are used. In the next section, we will look at creating an ACG.

Creating your first ACG

In this section, you will learn how to create an ACG, which will help you get ready to capture an image inside it:

> **Important note**
> SIG has been renamed ACG. Note that some of the new names have not been fully updated on the Microsoft documentation site. You can read more here: `https://docs.microsoft.com/azure/virtual-machines/create-gallery`.

1. Sign in to the Azure portal by going to `https://portal.azure.com`.
2. Type `Azure Compute Gallery` in the search box and select **Azure Compute Gallery** from the results.
3. On the **Azure compute galleries** page, click the **Create** button:

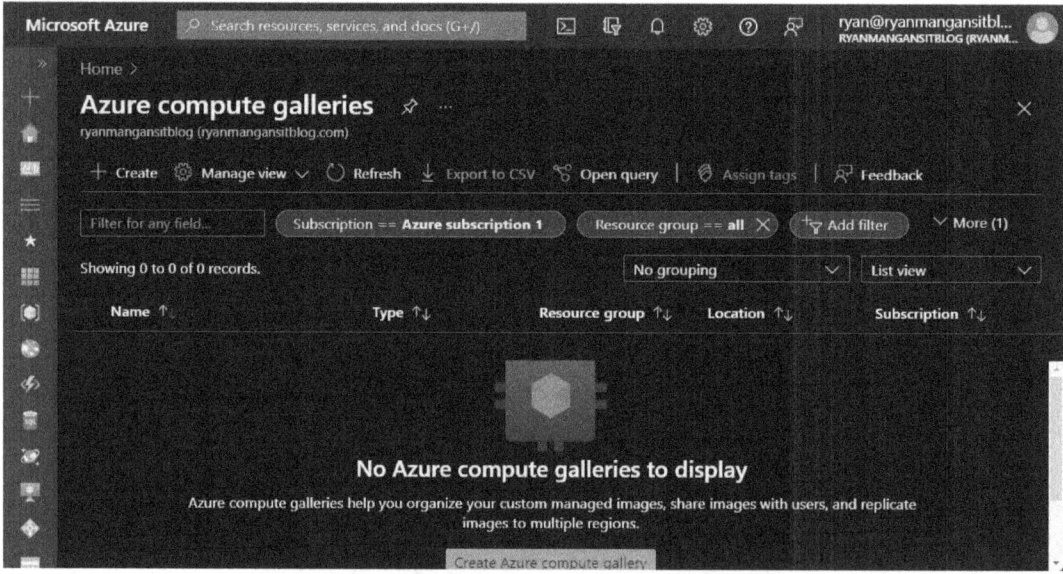

Figure 9.29 – The Azure compute galleries page in the Azure portal

4. On the **Create Azure compute gallery** page, ensure you select the correct subscription.
5. For **Resource group**, select **Create new** and enter the required resource group name.
6. For **Name**, provide a name for the name of the gallery.

7. Choose the required region.
8. You can enter a short description of the gallery, such as it being a desktop VM gallery, for testing purposes:

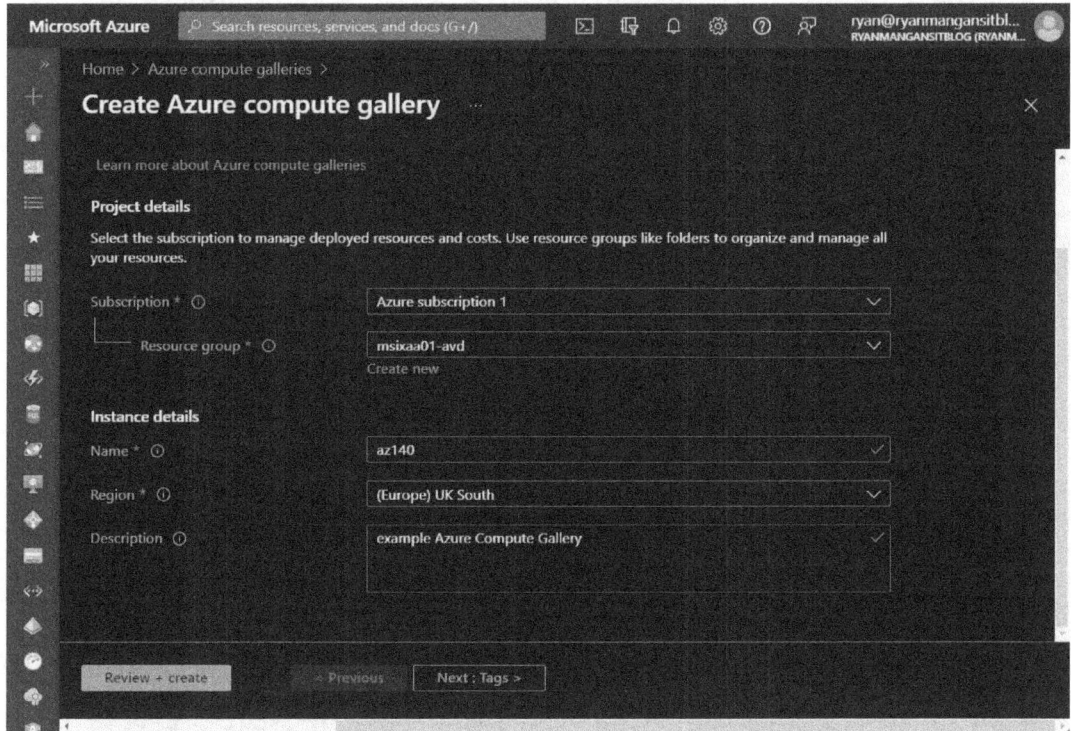

Figure 9.30 – The Create Azure compute gallery page

9. Then, click **Review + create**.
10. Once the validation process has been completed, select **Create**.

11. Once the deployment has finished, select **Go** to be taken to the resource. This will open ACG:

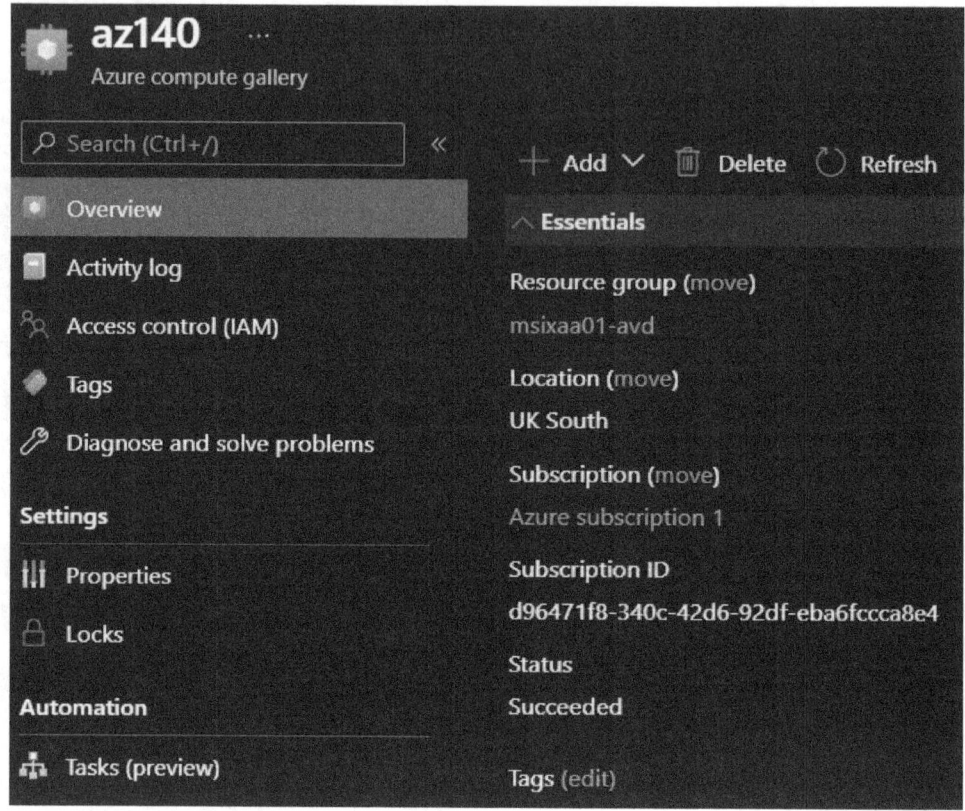

Figure 9.31 – ACG

There you have it; we have now created an ACG and are ready to add image definitions. In the next section, we will capture an image in an ACG and create an image definition and our first version.

Capturing an image in an ACG

In this section, we will capture an image in the ACG we created previously. Similar to the capture process shown in the previous section, when we captured a template, we can capture the image from a generalized VM within the **Virtual machines** page:

> **Important note**
> What does a generalized VM mean? Before you can deploy a Windows image to your AVD environment, you need to run a process called **Generalize**. This essentially removes computer-specific information such as the computer's **security identifier** (**SID**) and other items, such as drivers. This process enables you to turn the operating system into a deployable template.

1. Select your VM from the **Virtual Machine** list.
2. On the **Virtual machines** page for the VM, from the top menu, click the **Capture** button:

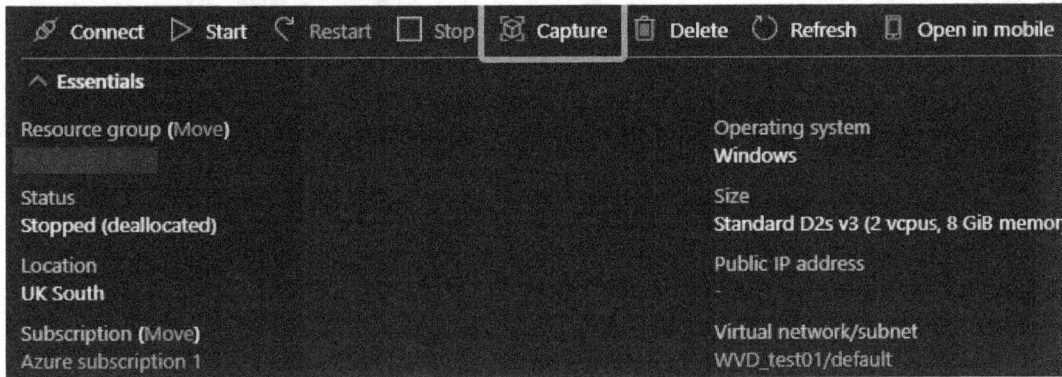

Figure 9.32 – The Capture button on the Virtual machines page in the Azure portal

3. A **Create image** page will appear.

4. Select the required resource group and ensure you select **Yes** to share the image to your ACG:

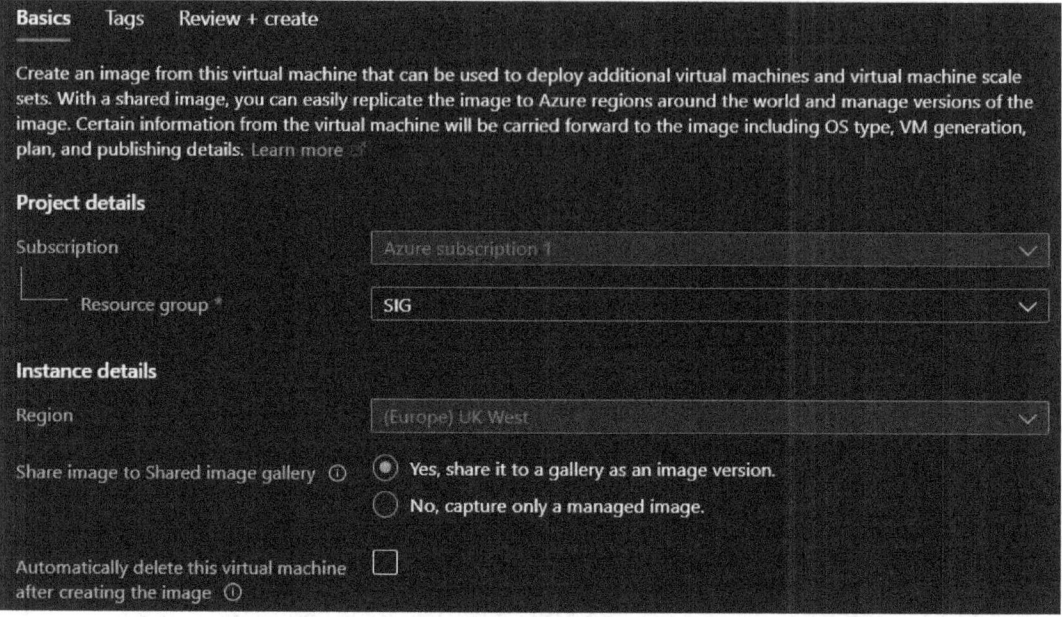

Figure 9.33 – The capture page for capturing an image in an ACG

Select the **Automatically delete this virtual machine after creating the image** checkbox if required.

Choose an availability zone if required (if your chosen region supports this).

5. Select the target image gallery we created previously and state if the image was generalized or specialized:

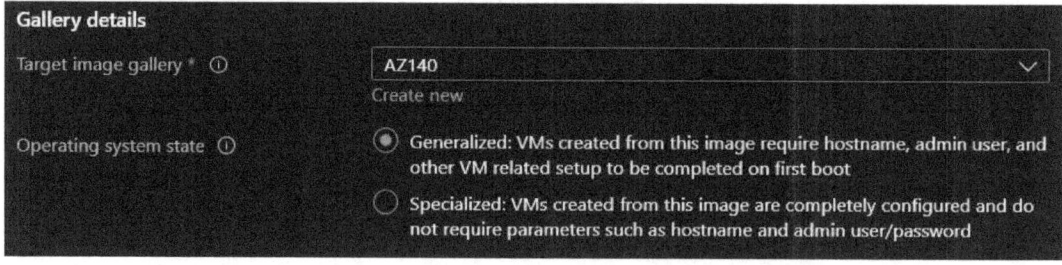

Figure 9.34 – The Target image gallery and Operating system state options

6. Create a new target image definition or use an existing one if you created one previously.

> **Important note**
>
> An image definition's purpose is to carry information about the image and its requirements. This information is for IT use, and you can include release notes, as well as the minimum/maximum memory. You can also enter a publisher, offer, and SKU code, which is typically used for Azure Marketplace deployments.
>
> The Azure Marketplace uses the following terminology:
>
> **Publisher**: The organization that created the image; for example, `MicrosoftWindowsServer`.
>
> **Offer**: The name of a group of related images that have been created by a publisher; for example, `WindowsServer`.
>
> **SKU**: An instance of an offer, such as the major release of a distribution; for example, `2019-Datacenter Version`, which is the version number of the image's SKU.

If you are not using the Azure Marketplace, you don't have to enter descriptive information, though it is recommended that you provide a description so that other IT admins can identify the specific image definition.

As shown in the following screenshot, I have provided the **Publisher**, **Offer**, and **SKU** fields with non-descriptive information. However, it is recommended that **Publisher** is set to your organization's name, **Offer** is set to the OS type (Windows 11), and **SKU** is set to the version (`MultiSession`):

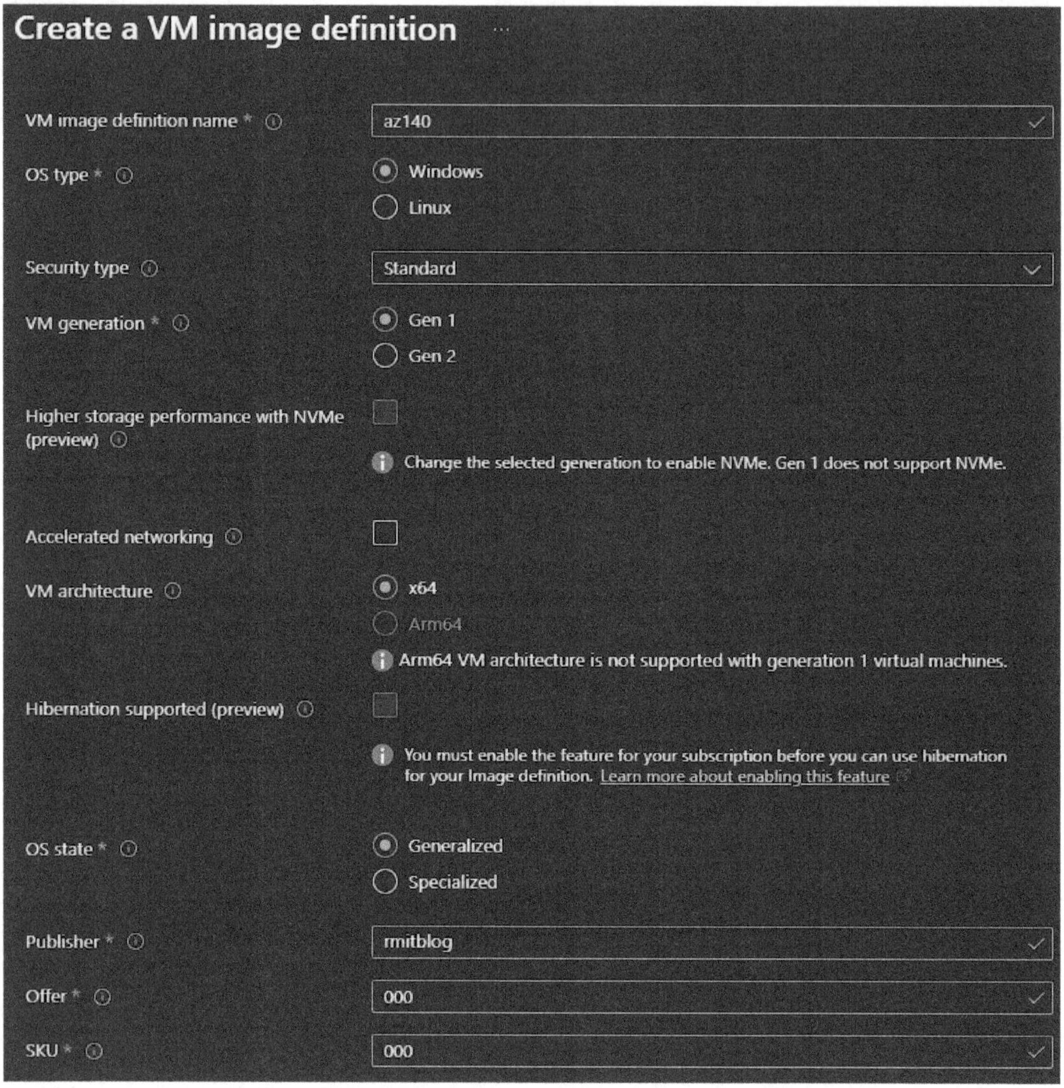

Figure 9.35 – The Create a VM image definition page

7. In this example, we created a new image definition:

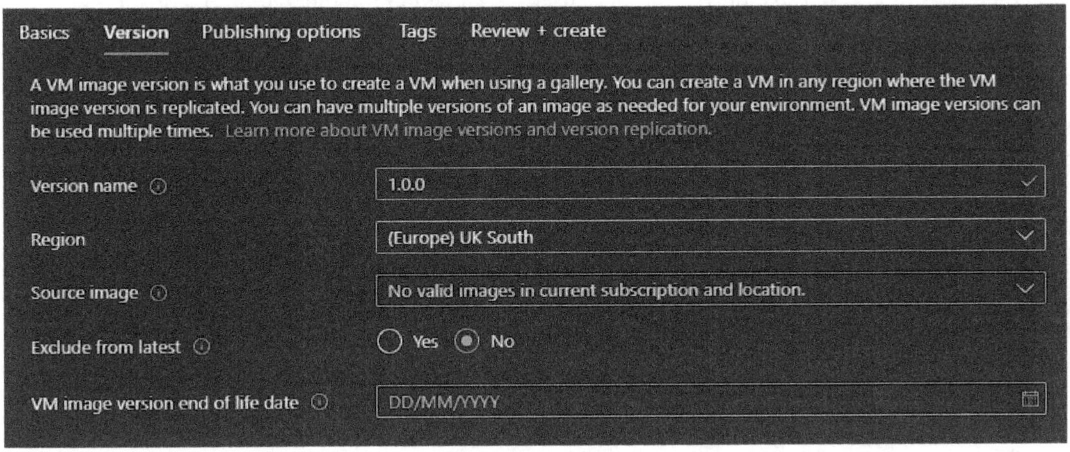

Figure 9.36 – Version details for an image definition

The following are the five fields within the **Version** section:

- **Version name**: The version number is used as the name of the image version. The format for this is `MajorVersion.MinorVersion.Patch`. When you specify to use **Latest** while creating a VM page or via PowerShell, the latest image is chosen based on the highest `MajorVersion`, then `MinorVersion`, then `Patch`.

- **Source image**: This can be a VM, managed disk, snapshot, managed image, or another image version.

- The **Exclude from latest** setting allows you to keep a version from being used as the latest image version.

- The **VM image version end of life date** field is used to indicate the **end-of-life** (EOL) date for the image version. EOL dates are for informational purposes only, and users can still create VMs from versions past the set EOL date.

8. You can use the **Replication** settings to configure the replication of the template across different regions by entering the required target regions, as shown in the following screenshot:

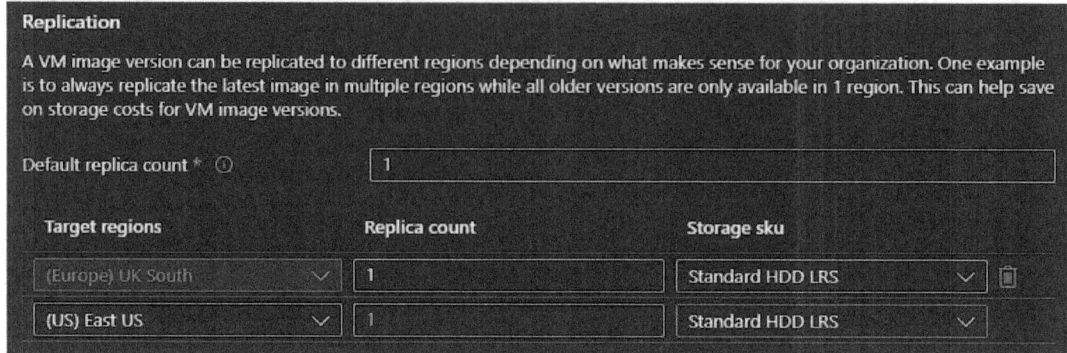

Figure 9.37 – The Replication section for specifying target regions that you would like to replicate the image to

9. Then, click on **Review + create**.

Once you've finished, click **Go to resource**. You will see the new image version:

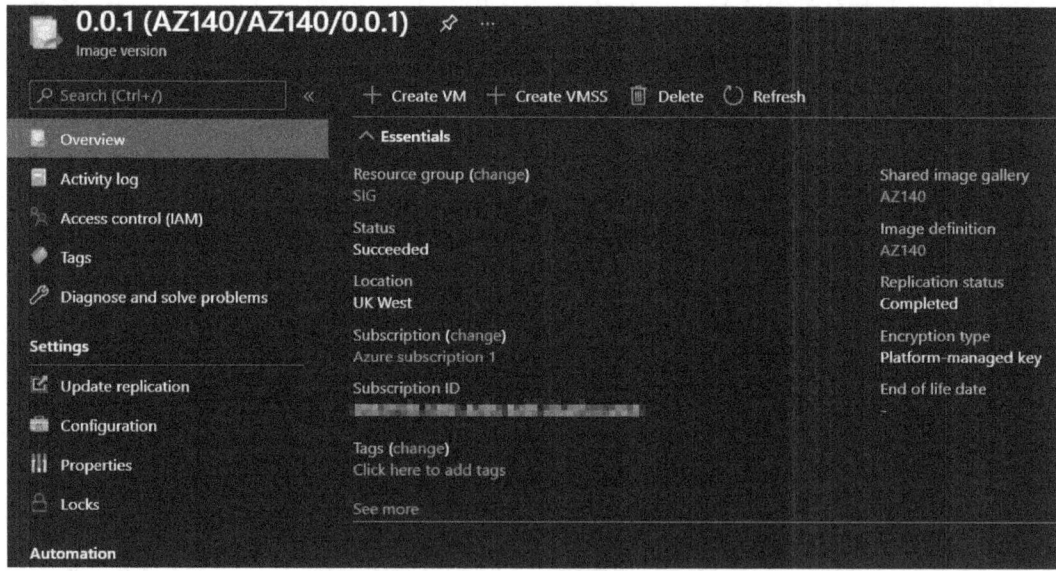

Figure 9.38 – Image version

With that, we have captured an image in the ACG; this has created both an image definition and an image version. Next, we will look at creating an image definition without using the capture feature shown previously.

Creating an image definition from the ACG

This section will show you how to create a new image definition from the ACG:

1. On the page for your recently created image gallery, select **Add a new image definition** at the top.
2. Under **Add new image definition to the Azure Compute Gallery**, for **Region**, select the required region:

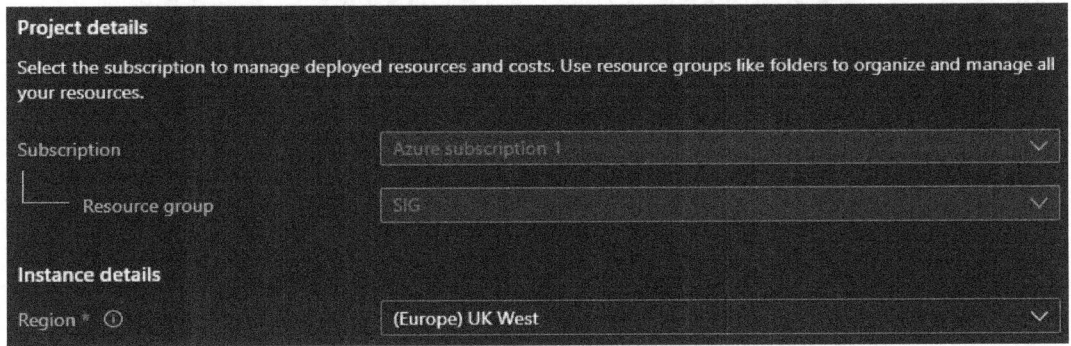

Figure 9.39 – The image definition page

3. For **VM image definition name**, enter a name.
4. For **OS type**, select your source VM.
5. For **OS state**, select an option based on your source VM.
6. For **VM generation**, select the option based on your source VM type. These are typically Gen 1, though you can build Gen 2 images.
7. For **Publisher**, enter an appropriate name.
8. For **Offer**, enter an appropriate name.

9. For **SKU**, enter an appropriate name:

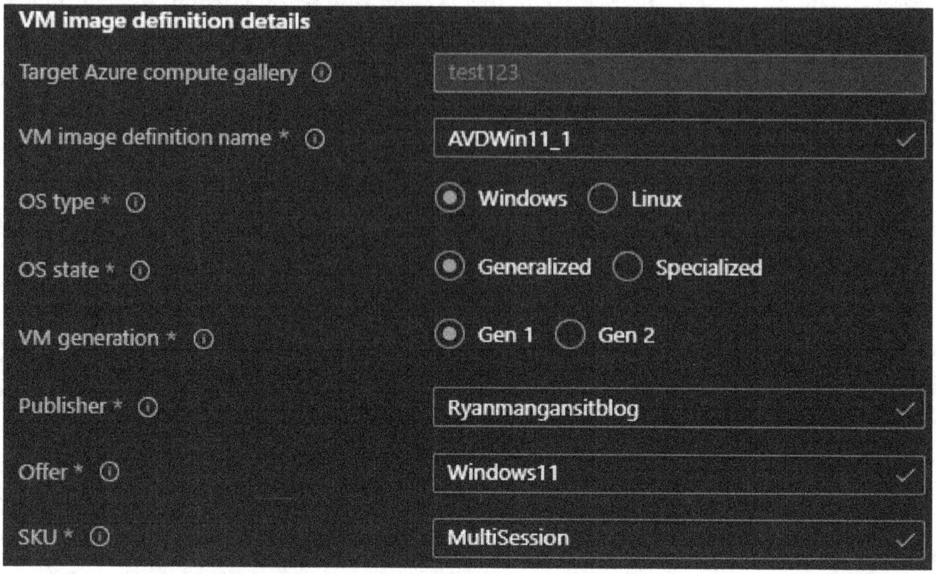

VM image definition details	
Target Azure compute gallery	test123
VM image definition name *	AVDWin11_1
OS type *	● Windows ○ Linux
OS state *	● Generalized ○ Specialized
VM generation *	● Gen 1 ○ Gen 2
Publisher *	Ryanmangansitblog
Offer *	Windows11
SKU *	MultiSession

Figure 9.40 – Required image definition details

10. When you're finished, click **Review + create**.
11. Once the image definition has been validated, click **Create**.
12. When the deployment has finished, click **Go to the resource**.

In this short section, we looked at creating an image definition from ACG. Now, let's learn how to create an image version.

Creating an image version

In this section, we will learn how to create an image version within the image definition:

> **Tip**
> Image versions are useful for updating images and their general management.

1. Within the page of your created image definition, select **Add version** from the top of the page:

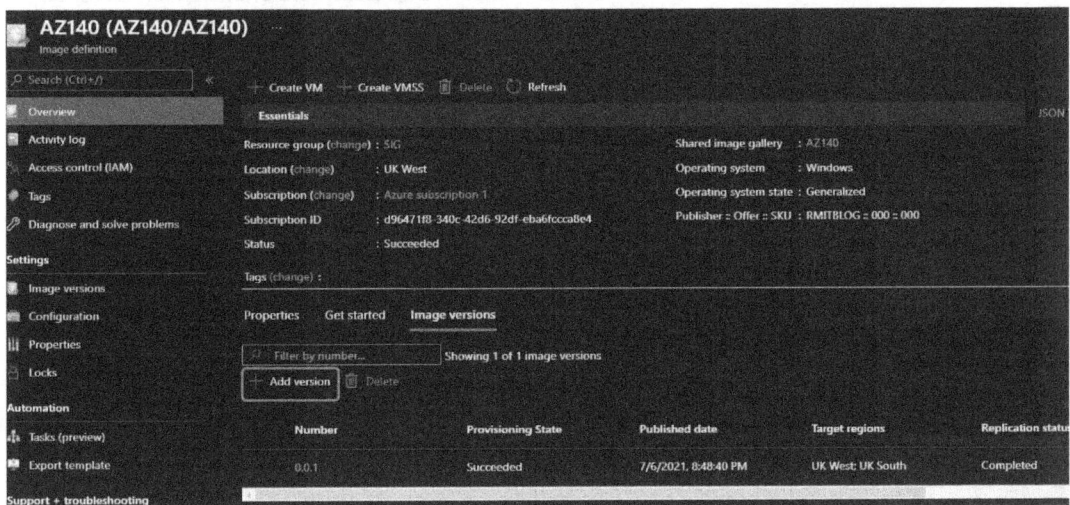

Figure 9.41 – The Add version button on the image definition

2. For **Region**, select the region where you want to create the image.
3. For **Version number**, enter a number, such as `1.0.0`. The image version name must follow the `MajorVersion.MinorVersion.Patch` format and use integers.

4. For **Source image**, select your source-managed image from the dropdown.

   The following table shows specific details for each source type:

Source	Other Fields
Disks or snapshots	- For the OS disk, select the required disk or snapshot from the dropdown - To add a data disk, type the logical unit number (LUN) and then select the data disk from the dropdown
Image version	- Select the source gallery from the dropdown - Select the correct image definition from the dropdown - Select the existing image version that you want to use from the dropdown
Managed image	Select the source image from the dropdown The managed image must be in the same region that you chose under Instance details
VHD in a storage account	Select Browse to choose a storage account for the VHD

   Table 9.2 – Source types

   The preceding table was taken from Microsoft's documentation: `https://docs.microsoft.com/azure/virtual-machines/linux/shared-images-portal`

1. For **Exclude from latest**, leave it set to **No**.
2. For **End-of-life date**, choose a date from the calendar. This can be a few weeks, months, or even years.

3. In the **Replication** tab, select the required storage type from the dropdown:

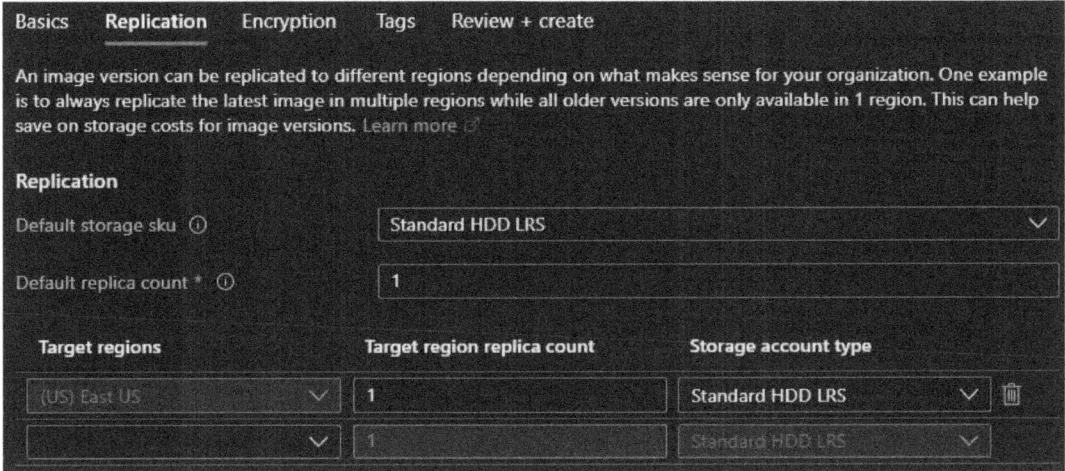

Figure 9.42 – The replication options for the image version

4. Enter a number for **Default replica count**. You can also override this for each Azure region you add.

5. You need to replicate to the source region. This means that the first replica in the list will be in the region where you created the image. You can add more replicas by selecting the required region from the dropdown and adjusting the replica count as necessary.

6. When you are done, click **Review + create**. Azure will validate your configuration.

7. When the image version passes validation, click **Create**.

8. When the deployment is finished, click **Go to the resource**.

9. It may take a while to replicate the image to all of the target regions.

In this section, we looked at creating an image version. Now, let's learn how to troubleshoot operating system image issues related to AVD.

Troubleshooting OS issues related to AVD

In this section, we will learn how to troubleshoot session host configuration issues. We will focus on the most common issues, including domain joins and communication between the AVD service and session host agent.

VMs are not joined to the domain

VMs not joining a domain typically occurs because the username and password that were entered during the host pool's setup/adding VMs were incorrect. Make sure that you check your password and use the full UPN of your Active Directory domain; for example, `domJoin.company.local`.

> **Important note**
> When you're using MEID-joined hosts, you would expect to see the hosts joined without the domain extension shown previously.

The other issue that may be preventing your VMs from joining the domain is a networking-related one, specifically **DNS**. Ensure that you have configured DNS correctly so that it points to your Entra ID DS infrastructure before trying to deploy session hosts.

> **Tip**
> Ensure that the account that you used for the domain join does not have **multi-factor authentication** (MFA) configured. You should use a service account with its password expiry set to **Disabled**. You should also make sure that the account has delegated permissions to make sure the account can join multiple devices to the domain.

You should also check that you have configured the correct permissions for the domain join account; otherwise, the domain join process will fail. As shown in the following screenshot, there is no check password feature in the form. It is advised that you write the password out in text form and copy and paste it into the form to ensure it's correct:

Figure 9.43 – Active Directory details for a domain join when adding VMs to a host pool

In this section, we looked at how to troubleshoot domain join issues when deploying session hosts to a host pool within AVD. We will now look at some AVD Agent and bootloader issues that may occur.

AVD Agent and AVD bootloader are not installed

For AVD to see the session hosts, you need to ensure that the session hosts have the AVD Agent and AVD bootloader installed. You can check this by reviewing the installed programs via **Control Panel | Programs | Programs and Features**. Alternatively, you can review the `scriptlog.log` file by navigating to the `c:\windows\temp\ScriptLog.log` file path. The log file will show error messages to help you diagnose the root cause.

> **Important note**
> The AVD Agent and AVD bootloader should not be installed on the master image. This is installed automatically through a VM deployment for a host pool or a manual process.

If the `ScriptLog.log` file is missing, this would indicate that the **Azure Resource Manager (ARM)** template had the incorrect permissions entered into it or that these credentials do not have the required permissions. Likewise, if PowerShell **Desired State Configuration (DSC)** was unable to start and run, this would indicate a permissions issue, such as that the hostname is incorrect or that MFA is enabled, causing the sign-in process to fail.

The next section will look at the AVD Agent not registering issues.

AVD Agent is not registering with the AVD service

If you encounter an issue where the session host is unavailable within the AVD portal, this is typically due to the agent not communicating correctly with the AVD service. It is advised that you check connectivity from the session host to the AVD service using Sysinternals tools such as PsPing.

You can download PsPing from `https://docs.microsoft.com/Sysinternals/downloads/psping/`.

To test your connectivity, run the following as an administrator within the command line:

> **Important note**
> For those who have already deployed an AVD environment, you can use the `WVDAgentUrlTool.exe` tool, which can be found in the `C:\Program Files\Microsoft RDInfra\RDAgent_*` folder.

`psping rdbroker.wvdselfhost.microsoft.com:443`

The following screenshot shows the use of PsPing to confirm that the host can communicate with the `rdbroker` for AVD:

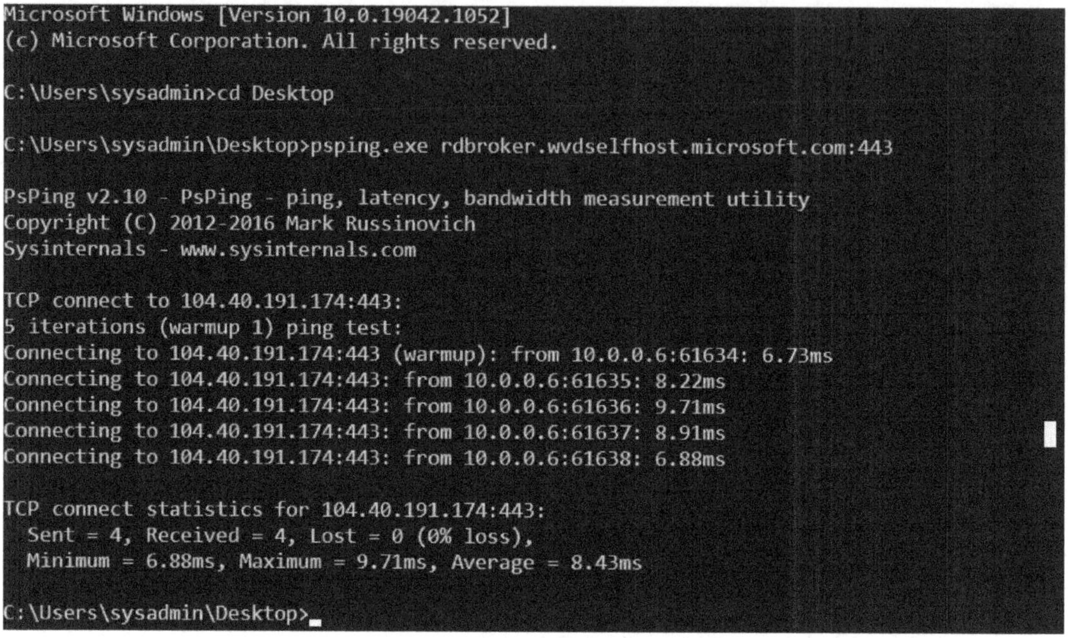

```
Microsoft Windows [Version 10.0.19042.1052]
(c) Microsoft Corporation. All rights reserved.

C:\Users\sysadmin>cd Desktop

C:\Users\sysadmin\Desktop>psping.exe rdbroker.wvdselfhost.microsoft.com:443

PsPing v2.10 - PsPing - ping, latency, bandwidth measurement utility
Copyright (C) 2012-2016 Mark Russinovich
Sysinternals - www.sysinternals.com

TCP connect to 104.40.191.174:443:
5 iterations (warmup 1) ping test:
Connecting to 104.40.191.174:443 (warmup): from 10.0.0.6:61634: 6.73ms
Connecting to 104.40.191.174:443: from 10.0.0.6:61635: 8.22ms
Connecting to 104.40.191.174:443: from 10.0.0.6:61636: 9.71ms
Connecting to 104.40.191.174:443: from 10.0.0.6:61637: 8.91ms
Connecting to 104.40.191.174:443: from 10.0.0.6:61638: 6.88ms

TCP connect statistics for 104.40.191.174:443:
  Sent = 4, Received = 4, Lost = 0 (0% loss),
  Minimum = 6.88ms, Maximum = 9.71ms, Average = 8.43ms

C:\Users\sysadmin\Desktop>
```

Figure 9.44 – The output of PsPing when testing communication with the broker

> **Tip**
> This test is also useful if the AVD Agent is not reporting a heartbeat when you run the `Get-AzWvdSessionHost` PowerShell cmdlet.

Once you have checked your network connectivity and confirmed that this is not the issue, you should follow these steps to update the agent manually:

1. Download a new version of the AVD Agent on the problematic session host VM.
2. Open Task Manager and, within the **Service** tab, stop the `RDAgentBootLoader` service from running.
3. Run the installer for the downloaded copy of the AVD Agent.
4. When you're prompted for the registration token, remove the `INVALID_TOKEN` entry and click **Next** (no token is required).
5. Complete the installation wizard and close it.
6. Open Task Manager and start the `RDAgentBootLoader` service.

7. If the AVD Agent registry entry called `IsRegistered` shows a value of 0, then the registration token has expired. You will need to generate a new registration token to fix this:

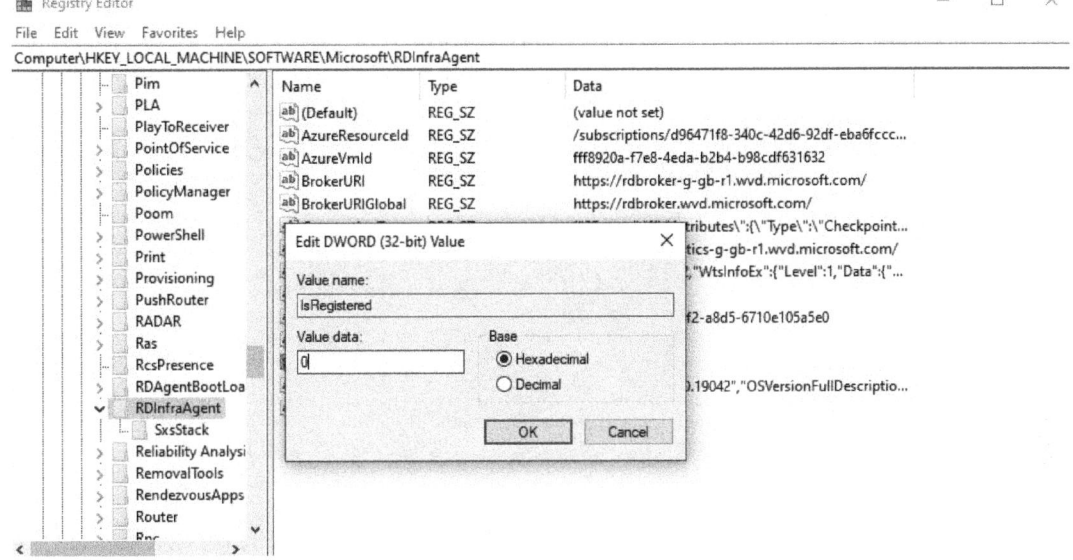

Figure 9.45 – The AVD Agent registry entry: IsRegistered

8. You can check this using a PowerShell cmdlet.
9. If there's already a registration token, remove the token using the following command:

    `Remove-AzWvdRegistrationInfo`

10. Run the `New-AzWvdRegistrationInfo` cmdlet to generate a new token.
11. Confirm that the `-ExpirationTime` parameter has been set to 3 days.

> **Further information**
>
> For more information on common VDA issues, you can find a complete list of errors and troubleshooting guidance at `https://docs.microsoft.com/azure/virtual-desktop/troubleshoot-agent`.

This section looked at some troubleshooting issues that may affect how the image communicates with AVD. We will now look at basic performance troubleshooting in AVD.

Basic performance troubleshooting in AVD

In this section, we will look at ways to identify and resolve common performance issues that may occur on session hosts within an AVD environment.

Four key resources can impact performance on a session host, as follows:

- **Central processing unit (CPU)**
- **Random-access memory (RAM)**
- **Storage (Disk)**
- **Network**

Typical performance issues you may experience include capacity, constraints, and overall performance degradation or lag.

First, we will look at troubleshooting CPU performance issues.

CPU troubleshooting

CPU constraints are a common issue with session-based desktops. Performance and high CPU utilization issues can occur for several reasons and not be due to a single app or service. The web browser can be one such suspect, especially if hardware rendering has been left set to **On** once the master image has been optimized. You can view your CPU usage by going to Task Manager and then going to **Azure monitoring features** within the Azure portal for specific VMs and using Sysinternals Process Explorer:

> **Tip**
> You can download Process Explorer here: `https://docs.microsoft.com/Sysinternals/downloads/process-explorer`

Troubleshooting OS issues related to AVD 279

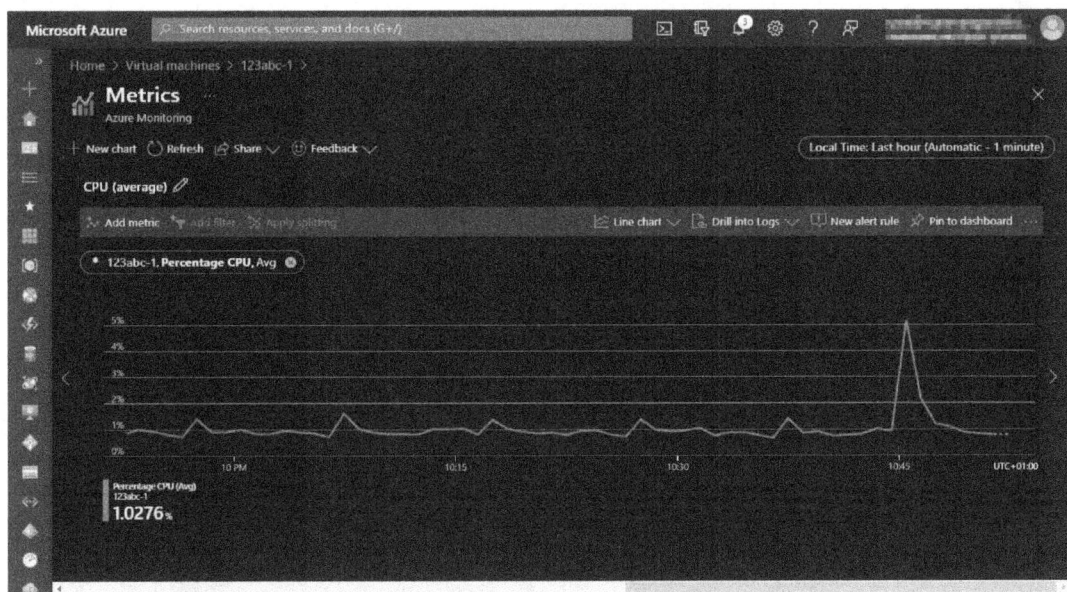

Figure 9.46 – The CPU utilization of a session host within Azure monitoring features

The following screenshot shows using Sysinternals Process Explorer to gauge the CPU usage when monitoring the performance of an operating system:

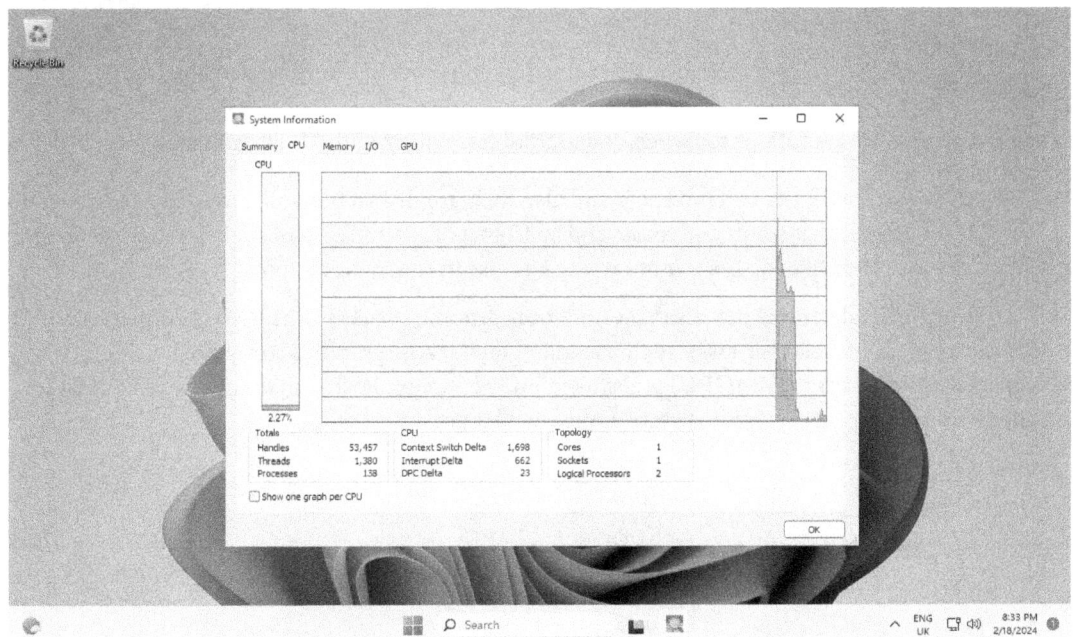

Figure 9.47 – CPU utilization of a session host using Sysinternals Process Explorer

The following screenshot shows the use of Sysinternals Process Explorer to monitor running processes to see which are consuming the most CPU and memory. This is a great tool for identifying processes that are consuming excessive resources:

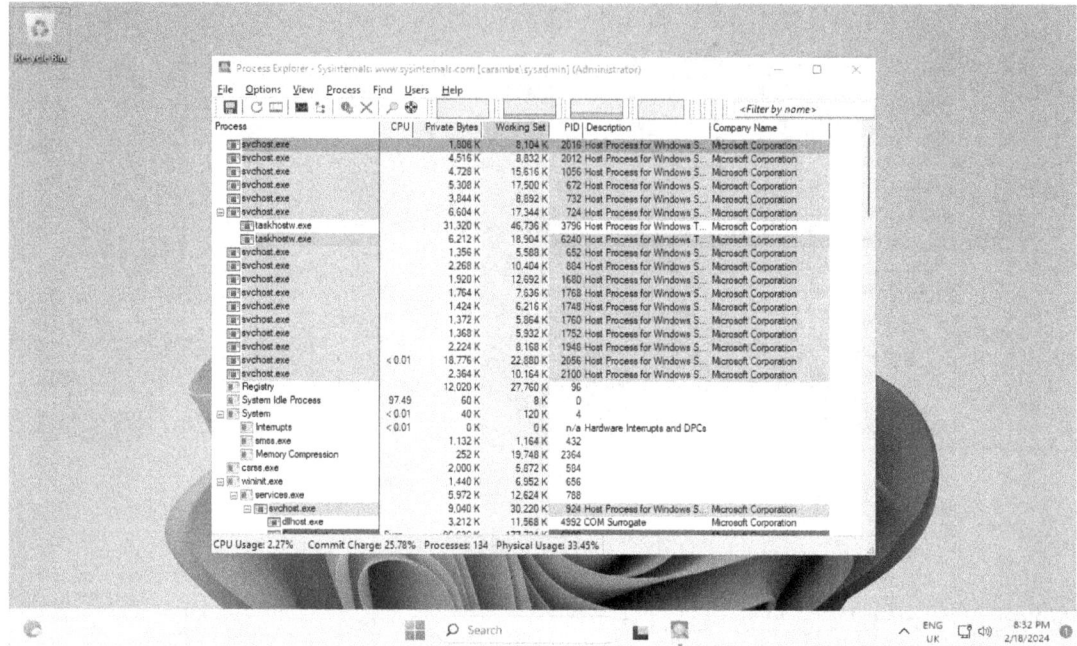

Figure 9.48 – Sysinternals Process Explorer (running processes) on a session host

The following are some useful tips for when you're troubleshooting CPU resource issues:

- Low RAM can cause the CPU to spike and also increase disk activity due to memory paging, which is a process that stores and retrieves data from a disk for use in the main memory. Paging typically uses the OS disk or a temporary disk to use in the main memory.

- If hardware rendering is not disabled on a non-graphic process unit VM, then more CPU resources will be used for tasks such as video playback or graphics software. However, if a **graphics processing unit** (**GPU**) is installed on the session host, you do not need to disable hardware rendering on web browsers, Office applications, and others.

Specific processes may be causing performance issues on the session host. You can use Sysinternals Process Explorer to view CPU time, allowing you to review your processes, the time they've run for, and if they are causing the system to hang. One good example of a process impacting the CPU is large Excel files with macros. Additionally, multiple users on the same session host using large Excel files or other resource-heavy applications may impact the session host. In this case, it's advised that you move these power users to higher-spec session hosts or personal pools.

> **Tip**
> It is recommended that you build more session hosts with fewer CPUs. This is advised because when you increase the number of cores, the system's synchronization overhead also increases. Smaller-resourced VMs would perform better than fewer and larger-specification VMs.

Here are a few examples of potential symptoms you may see when the CPU is under contention:

- Slow/lagging switching between windows
- The loading cursor appearing for long periods
- The logon and logoff process is slow, taking minutes to process
- Applications are consistently going into a **Not responding** state or locking/crashing
- Web pages are jumping, slow to load, or become unresponsive

Now, let's look at a few RAM performance issues and how to spot some of them.

RAM challenges

Modern applications consume more RAM. For example, when you combine the Microsoft Office suite and Microsoft Teams and have multiple other applications open simultaneously, this can impact RAM consumption. When RAM reaches high consumption or becomes full, it will revert to the page file as secondary storage for the main memory.

The challenge with page files is speed. For example, a **solid-state drive** (**SSD**) could have a typical write speed of 456 MB per second, whereas RAM writes at an estimated speed of 12,800 MB per second. When all the RAM has been consumed, the performance drops as secondary storage is being used. This performance drop will be seen by users and will most likely cause performance degradation for multiple users on a session host.

You can view your session host RAM utilization in **Azure Monitoring**, as shown in the following screenshot:

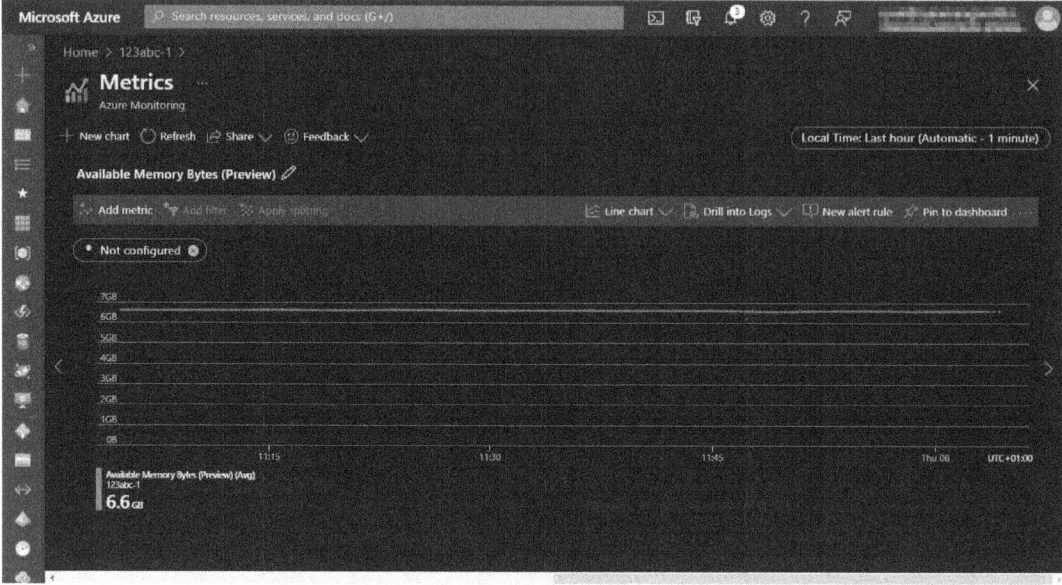

Figure 9.49 – Available RAM on a session host in Azure Monitoring

The following are some tips for troubleshooting RAM issues:

- Try to spread your users across multiple session hosts. You can manage this using the **Depth** and **Breadth** load-balancing modes.

- One of the ways to address issues with a session that has become saturated would be to enable **Stop drain** and ask a subset of users to log off and then back on. This will clear resources, free up RAM, and restore the expected user experience for the remaining users on that session host.

- Leaving users in an idle state and not rebooting session hosts can cause longer-term resource consumption issues. Therefore, it is advised that users log off and that the session hosts are restarted regularly.

Some of the symptoms you may see with RAM issues are as follows:

- Applications crashing
- Windows errors stating that there's low memory
- A slow logon and logoff experience, which will be visible
- Applications locked in a **Not responding** state

- Application launch times are slow
- Running applications causes performance degradation and they are slow to use

We now move on to the next section, where we look at disk performance and troubleshooting.

Disk performance troubleshooting

This is a common problem as it's easy to forget about the operating system disk when it comes to multi-session deployments. Within Microsoft Azure, different-sized disks have different performance outputs, which are measured in **input/output operations per second** (**IOPS**). To avoid any disk performance degradation, you should size the disk based on IOPS and factor in the expected total number of users per session host. P15 Azure managed disks or higher is recommended to ensure there are enough IOPS to serve all users and applications in use. Every environment can be different; it is advised to baseline your requirements for IOPS by using one or two users on a session host to generate IOPS consumption. You can then see the total usage of the disk over a set period from the test users carrying out their typical day-to-day tasks.

You can then use the data that's been collected to calculate the requirements and estimate the required IOPS based on a set number of users per session host.

See the following link for a list of the premium SSD disk sizes and their allocated IOPS: https://docs.microsoft.com/azure/virtual-machines/disks-types#premium-ssd-size

> **Tip**
> It is also important to note that choosing a VM that's the wrong size may limit the throughput that's allowed by the disk. So, you must check the VM specs as well as the disk IOPS output. The following link provides a table detailing the max IOPS per VM size: https://docs.microsoft.com/azure/virtual-machines/dav4-dasv4-series#dav4-series.

The following command can help you identify disk performance issues by spotting high disk queues:

```
typeperf -si 2 "\PhysicalDisk(*)\Avg. Disk Queue Length"
```

The following screenshot shows the output of the current disk queue length of the operating system. This is a great tool for identifying any issues with the disk performance on the operating system:

```
Microsoft Windows [Version 10.0.19042.1052]
(c) Microsoft Corporation. All rights reserved.

C:\Users\sysadmin>typeperf -si 2 "\PhysicalDisk(*)\Avg. Disk Queue Length"

"(PDH-CSV 4.0)","\\123abc-1\PhysicalDisk(0 C:)\Avg. Disk Queue Length","\\123abc-1\PhysicalDisk(1 D:)\Avg. Disk Queue
ngth","\\123abc-1\PhysicalDisk(_Total)\Avg. Disk Queue Length"
"07/07/2021 22:46:32.066","0.000178","0.000000","0.000178"
"07/07/2021 22:46:34.066","0.000517","0.000000","0.000517"
"07/07/2021 22:46:36.082","0.000870","0.000000","0.000870"
"07/07/2021 22:46:38.097","0.000071","0.000000","0.000071"
"07/07/2021 22:46:40.113","0.000286","0.000000","0.000286"
"07/07/2021 22:46:42.113","0.000000","0.000000","0.000000"
"07/07/2021 22:46:44.128","0.000441","0.000000","0.000441"
"07/07/2021 22:46:46.129","0.000238","0.000000","0.000238"
"07/07/2021 22:46:48.144","0.001872","0.000000","0.001872"
"07/07/2021 22:46:50.144","0.000497","0.000000","0.000497"
"07/07/2021 22:46:52.160","0.000000","0.000000","0.000000"
"07/07/2021 22:46:54.175","0.000210","0.000000","0.000210"
"07/07/2021 22:46:56.191","0.000974","0.000000","0.000974"
"07/07/2021 22:46:58.207","0.001537","0.000000","0.001537"
"07/07/2021 22:47:00.222","0.001956","0.000000","0.001956"
"07/07/2021 22:47:02.238","0.000101","0.000000","0.000101"
"07/07/2021 22:47:04.238","0.002081","0.000000","0.002081"
"07/07/2021 22:47:06.254","0.000314","0.000000","0.000314"
"07/07/2021 22:47:08.269","0.000090","0.000000","0.000090"
"07/07/2021 22:47:10.269","0.001163","0.000000","0.001163"
"07/07/2021 22:47:12.285","0.001131","0.000000","0.001131"
"07/07/2021 22:47:14.285","0.010351","0.000000","0.010351"
```

Figure 9.50 – Disk queue output from running typeperf

For a centralized view of potential disk performance issues, you can use Azure Monitor or Log Analytics. Azure Monitor and Log Analytics will be covered in *Chapter 18*, *Monitoring and Managing Performance and Health*.

The following are some key points regarding disk troubleshooting:

- You should not use standard disks for multi-session environments
- When you're deploying multi-session host pools, it is advised to use P-type SSD disks
- Prevent users from using streaming services such as radios and music as this can impact the disk

Now, let's take a look at the symptoms. The following are some disk performance symptoms:

- Files take a long time to open and save
- Slow switching between screens
- Logon and logoff times are long
- Applications stop responding or repeatedly show **Not responding** before becoming active again
- Applications are slow to launch

This section looked at disk performance issues, as well as some troubleshooting hints and tips that you can use to resolve some of the issues you may face.

Networking troubleshooting

Networking issues can occur in three areas: the user endpoint device, AVD's management service, and the session host itself. When you're reviewing network issues, it is recommended that you start with the client device and then look at the AVD management service and session host connectivity.

There are four typical network contention issues that you can experience:

- A jittery mouse or delayed typing
- Audio degradation or distortion
- The connection to AVD keeps dropping
- The screen goes blank and reappears frequently

These issues are usually found on the user side, all of which can be due to many different reasons. However, it is important to follow the necessary processes and check all three areas. There can be issues with the broker service, gateway communication, and even the session host agent service.

You can also review the connection information by right-clicking on the **SessionDesktop** title bar, as shown in the following screenshot:

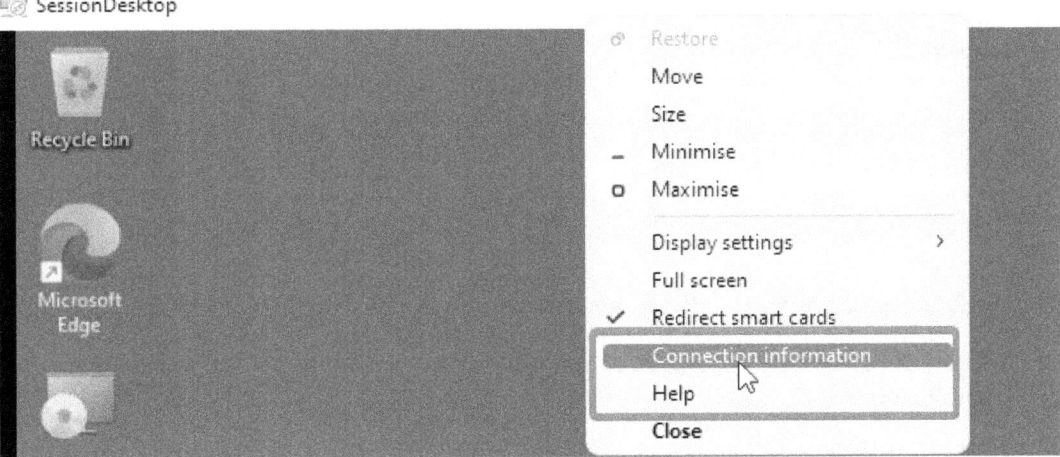

Figure 9.51 – The menu that appears after right-clicking the SessionDesktop title bar

Click the **Connection information** button to see a **Connection information** window appear. Click **Show details** to view the connection details for the client connection. Note that the **round-trip time (RTT)**, available bandwidth, and frame rate are shown here:

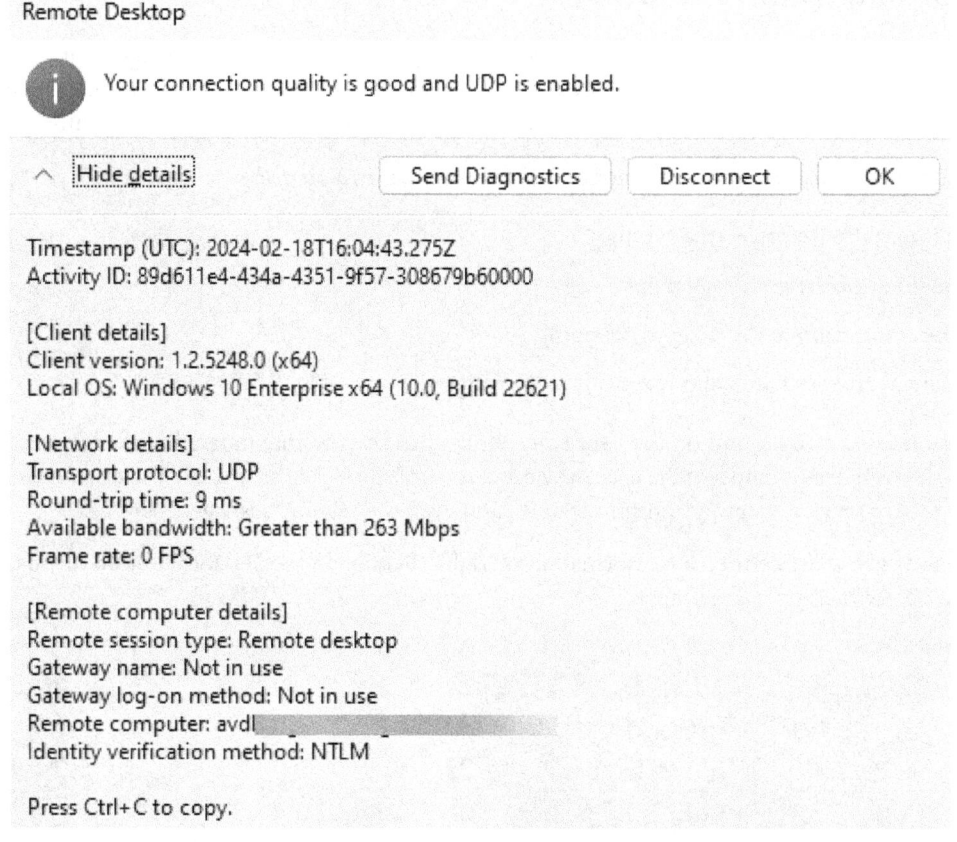

Figure 9.52 – The connection information for the Remote Desktop client on a Windows device

The following are some tips for identifying network issues:

- Conduct a broadband/ISP speed test to understand if the user has network connectivity issues to the public network
- Use tools such as PsPing to test connectivity to www.microsoft.com and rdweb.microsoft.com:443, as well as the broker service shown in the previous section
- Force session hosts to **redeploy**, which moves the VM within a specific region to another physical host

- Check the status of the session host using PowerShell
- Use AVD's **Insights** dashboard to view end-user latency and poor user bandwidth

Azure Monitor and AVD Insights will be covered in *Chapter 18*, *Monitoring and Managing Performance and Health*.

This section looked at some hints and tips for spotting possible issues and resource contentions that may impact or affect your session hosts and connecting users.

Summary

In this chapter, we looked at creating a golden image for AVD, as well as modifying the session image, including customizing it, optimizing it, and capturing it in an image template. Then, we created an ACG, including creating image definitions and image versions. Finally, we looked at troubleshooting the OS and diagnosing possible performance and resource contention issues.

In the next chapter, we will look at managing access and configuring IT admin and user security permissions.

Questions

Answer the following questions to test your knowledge of this chapter:

1. When you're configuring a custom image, you need to ensure that modern apps can use the additional language packs. Which PowerShell cmdlet would you use to disable pre-stage app cleanup?

    ```
    Disable-ScheduledTask -TaskPath "\Microsoft\Windows\
    AppxDeploymentClient\" -TaskName "Pre-staged app cleanup"
    ```

2. You have created a VM to be used as the image master. You install applications on the image and apply the necessary optimizations and configurations. What should you do before making the image distributable as a template?

    At Command Prompt, run the Sysprep command

3. You need to replicate the custom session host image to multiple Azure regions. Which Azure service would you use to complete this task?

    ACG

4. To automate the image creation of AVD session hosts, which tool can you use to complete this task?

    Custom image templates

Part 4: Managing Access and Security

This section covers the two key topics of managing user access and security for Azure Virtual Desktop. This part of the book comprises the following chapters:

- *Chapter 10, Managing Access*
- *Chapter 11, Managing Security*

10
Managing Access

In this chapter, we will journey through the process of managing access within **Azure Virtual Desktop** (**AVD**). First, we'll look at planning, managing, and restricting access to AVD resources. This consists of exploring how **role-based access control** (**RBAC**) works in combination with the permissions that must be synchronized with the Active Directory domain. We will also briefly look at Group Policy and the value it brings to a virtual desktop environment.

In this chapter, we will cover the following topics:

- Introduction to Azure RBAC
- Planning and implementing Azure role-based controls
- The delegated access model
- Assigning RBAC roles to IT admins
- Creating a custom role using rights assignments
- Configuring user restrictions by using Entra ID groups

Introduction to Azure RBAC

Azure RBAC stands for **Azure role-based access control**. This feature allows you to apply access management to your Azure management groups/subscriptions. It enables you to configure granular control with access to Azure resources and specify which user/admin actions can be taken with the resources you have granted access to.

Here are some examples of the access control conditions you can set in Azure RBAC:

- Allow one user to manage **virtual machines** (**VMs**) in a specific subscription and another user to only manage networks. This is an example of separating the network controls from the VM controls.
- Provide access to a specific user so that they can only manage all of the resources within a specific resource group.
- Grant access to an application so that it can access specific resources or a resource group.

Now, let's take a look at the components of an RBAC item:

- Security principal
- Role definition
- Scope

Let's understand these elements in more detail:

- **Security principal**: A security principal is an object that represents a user, group, service principals, or managed identity. You can assign a role to any of these security principals:

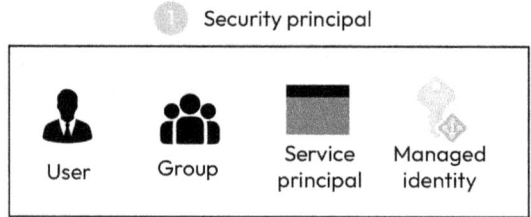

Figure 10.1 – Different security principals

- **Role definition**: A role definition is a collection of permissions you can assign for different operations. These can then be assigned to the security principal. Azure does include several built-in roles, so you don't have to configure custom roles, but you can if you want to. A good example is the VM contributor role, which allows users to create and manage VMs.

- **Scope**: The scope refers to the resources you apply to grant access to. There are four levels within the scope that you can specify – that is, the management group, subscription, resource group, and the specific resources themselves:

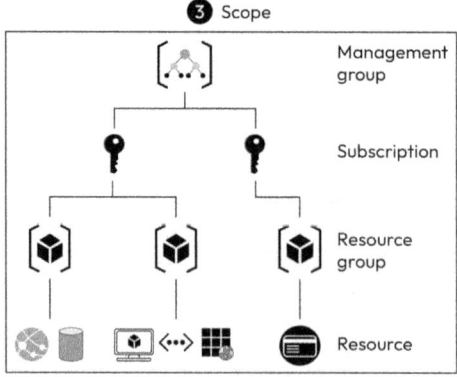

Figure 10.2 – The different scope levels

Now that we've reviewed the three core elements of RBAC in more detail, let's move on to role assignments.

Essentially, a role assignment is the process of assigning a role definition to a security principal at the required scope level (that is, applying RBAC permissions to users and groups).

> **Tip**
> You can find a complete list of Azure's built-in roles at `https://docs.microsoft.com/azure/role-based-access-control/built-in-roles`.

In this section, we learned about the basics of Azure RBAC. Next, we'll look at the built-in roles of AVD and learn how to use RBAC with AVD.

Planning and implementing Azure roles and RBAC for AVD

As with all Azure resources, AVD uses Azure RBAC to assign roles and permissions to users and IT admins, as mentioned earlier. In this section, we will look at the RBAC roles that can be used within AVD, and I will show you how to apply these role assignments and create custom roles.

> **Tip**
> RBAC roles that are specific to AVD do appear in the AZ140 exam. You are advised to pay attention to these different roles as you might see a question on them in the exam.

AVD has many built-in management roles that you can use for host pools, app groups, and workspaces. This gives you more granular control over administrative tasks, which can be extremely useful in larger organizations. It is recommended that all organizations leverage RBAC roles to use the least-privilege model, ensuring access to systems is carefully controlled.

Additionally, it is important to note that the roles are named in compliance with Azure's least-privilege methodology and standard role naming conventions.

> **Tip**
> AVD does not have a specific owner role. However, you can use the Azure standard owner role for service objects. It is recommended that you follow the least-privilege methodology when assigning admin permissions to AVD administrators.

The following screenshot shows the specific desktop virtualization roles that are available in RBAC:

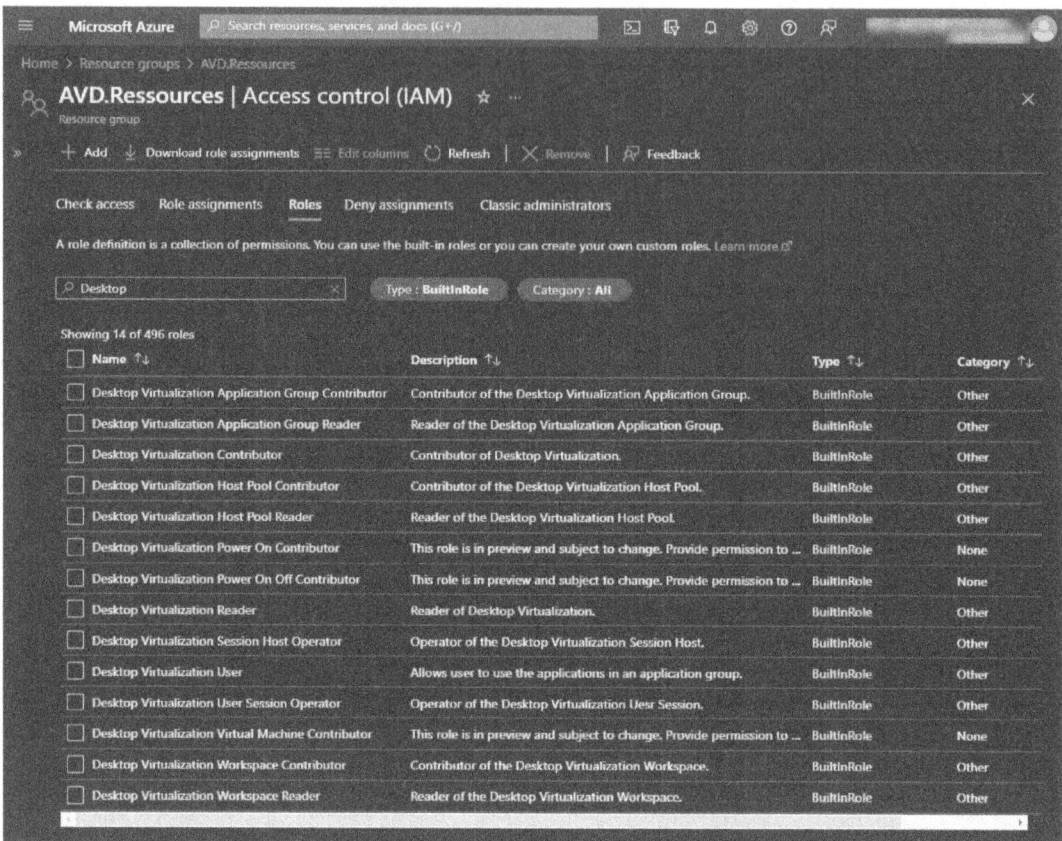

Figure 10.3 – The Access control (IAM) menu within an Azure resource

The following table details the built-in roles for AVD that you can use to separate management roles from host pools, app groups, and workspaces. Note that these roles are for managing AVD, not user access:

Role	Description
Desktop Virtualization Contributor	The Desktop Virtualization Contributor role enables you to manage all aspects of the deployment. However, it doesn't grant you access to compute resources. Note that you will also need the User Access Administrator role to publish app groups to users or user groups.
Desktop Virtualization Reader	The Desktop Virtualization Reader role allows you to view everything in the deployment but doesn't let you make any changes.

Role	Description
Desktop Virtualization Host Pool Contributor	The Host Pool Contributor role allows you to manage all aspects of host pools, including access to resources. You do need an extra contributor role, Virtual Machine Contributor, to create VMs. Additionally, you will need the AppGroup and Workspace contributor roles to create a host pool using the portal or to use the Desktop Virtualization Contributor role.
Desktop Virtualization Host Pool Reader	The Host Pool Reader role enables you to view everything within the host pool but does not allow you to make any changes.
Desktop Virtualization Application Group Contributor	The Application Group Contributor role allows you to manage all aspects of app groups. If you want to publish app groups to users or user groups, you will also need to add the User Access Administrator role.
Desktop Virtualization Application Group Reader	The Application Group Reader role enables you to view everything in the app group and will not allow you to make any changes.
Desktop Virtualization Workspace Contributor	The Workspace Contributor role allows you to manage all aspects of your workspaces. To get information on applications that have been added to the app groups, you will also need to assign the Application Group Reader role.
Desktop Virtualization Workspace Reader	The Workspace Reader role enables you to view everything in the workspace but doesn't allow you to make any changes.
Desktop Virtualization User Session Operator	The User Session Operator role enables you to send messages, disconnect sessions, and use the `logoff` function to sign sessions out of the session host. However, this role doesn't enable you to perform session host management such as removing the session host, changing the drain mode setting, and more. This role can view assignments, but you cannot modify admins. It is recommended that you assign this role to specific host pools. If you give this permission at the resource group level, the admin will have read permissions on all host pools under a resource group.

Role	Description
Desktop Virtualization Session Host Operator	The Session Host Operator role enables you to view and remove session hosts and change the drain mode. The session host operator can't add session hosts using the Azure portal because they don't have write permissions for host pool objects. If the registration token is valid (generated and not expired), you can also use this role to add session hosts to the host pool outside of the Azure portal. This is subject to ensuring the admin has compute permissions through the Virtual Machine Contributor role.
Desktop Virtualization Virtual Machine Contributor	This extends the Desktop Virtualization Session Host Operator role with the permission to create, delete, update, start, and stop VMs.
Desktop Virtualization Power On Contributor	Permission to start VMs. Typically, this is used to give Microsoft's AVD Service principal permission to power on VMs upon connecting.
Desktop Virtualization Power On Off Contributor	Permission to start and stop VMs. Typically, this is used to give Microsoft's AVD service principal permission for autoscaling.

Table 10.1 – Desktop virtualization-related built-in roles

The following link provides an up-to-date list of RBAC roles for AVD: `https://docs.microsoft.com/azure/virtual-desktop/rbac`.

Now, we will take a look at the delegated access model for AVD.

The delegated access model

Delegated access in AVD lets you specify the level and total amount of access a particular user is allowed. This can be done by assigning a role that can be built-in or custom.

> **Important note**
> The delegated access model is based on the Azure RBAC model. Essentially, you can customize roles with granular controls, ensuring the least-privilege methodology is followed as per security best practices. Also, note that the `Desktop Virtualization User` role is the lowest role and is required for user access to the AVD environment.

You can learn more about AVD delegated access at https://docs.microsoft.com/azure/virtual-desktop/delegated-access-virtual-desktop.

This section examined different built-in RBAC roles for AVD and briefly covered the delegated access model. Next, we'll learn how to configure RBAC through the Azure portal.

Assigning RBAC roles to IT admins

This section looks at assigning RBAC roles to specific resources and resource groups for AVD.

Within a subscription, resource group, or a specific resource, you will see **Access control (IAM)** in the menu options:

Figure 10.4 – The Access control (IAM) menu button is shown in each Azure resource

This is where you can assign roles to users, groups, and service principals. As mentioned earlier, these are called security principals.

To assign a security principal to a role within scope, you can click on the **Access control** button; as shown in this example, this is at the subscription level. Click **Add | Add role assignment**:

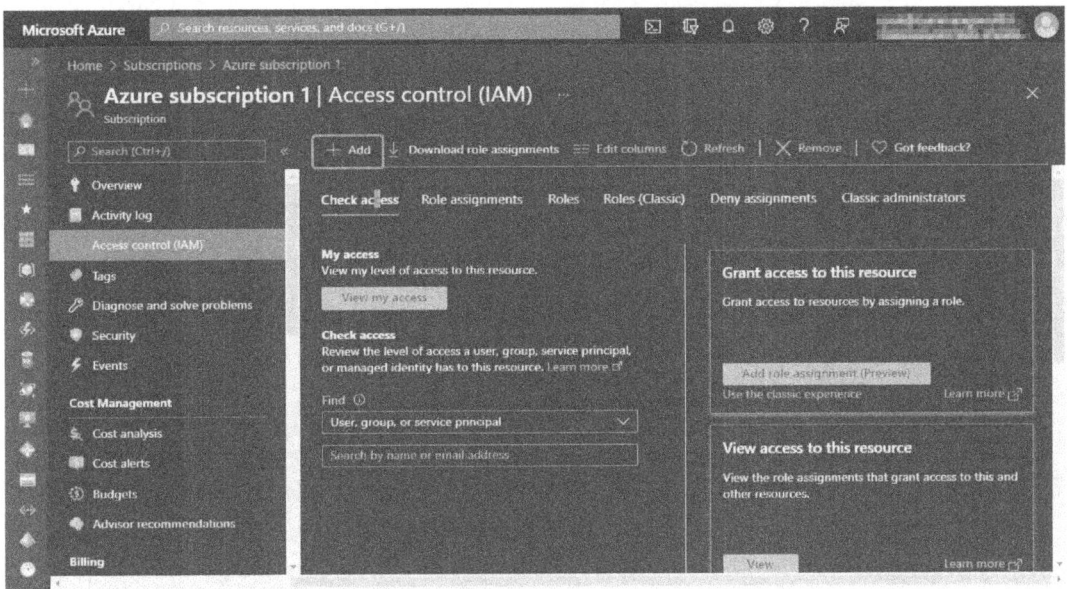

Figure 10.5 – The buttons for adding role assignments

Then, you should see the **Add role assignment** tab appear. Select the role you require, select the security principal, and click **Save**:

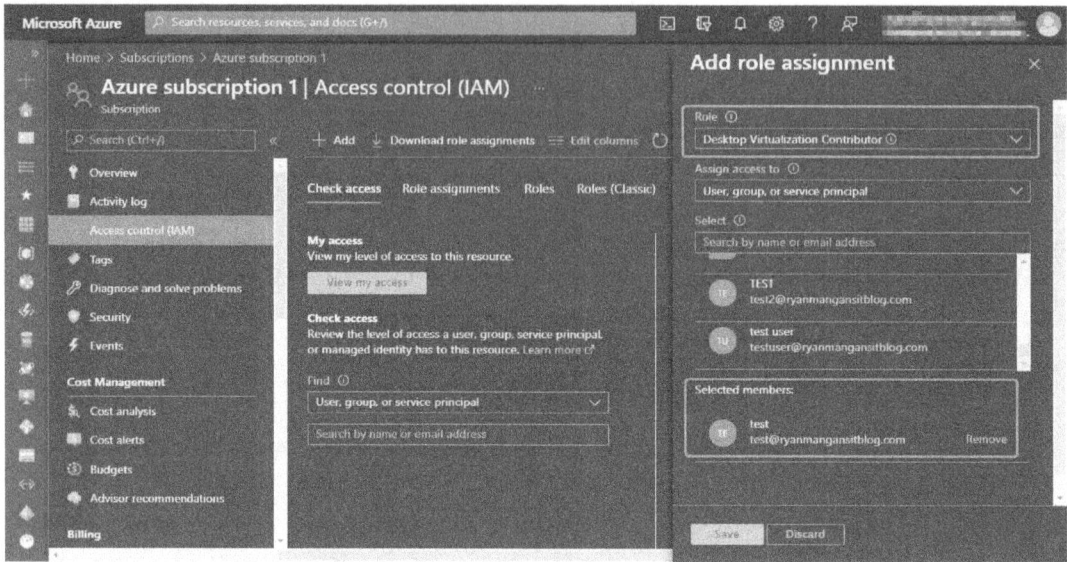

Figure 10.6 – Selecting a role and user when adding a role assignment

Now, you will see the **Desktop Virtualization Contributor** role with the user account assigned in the bottom section of *Figure 10.7*:

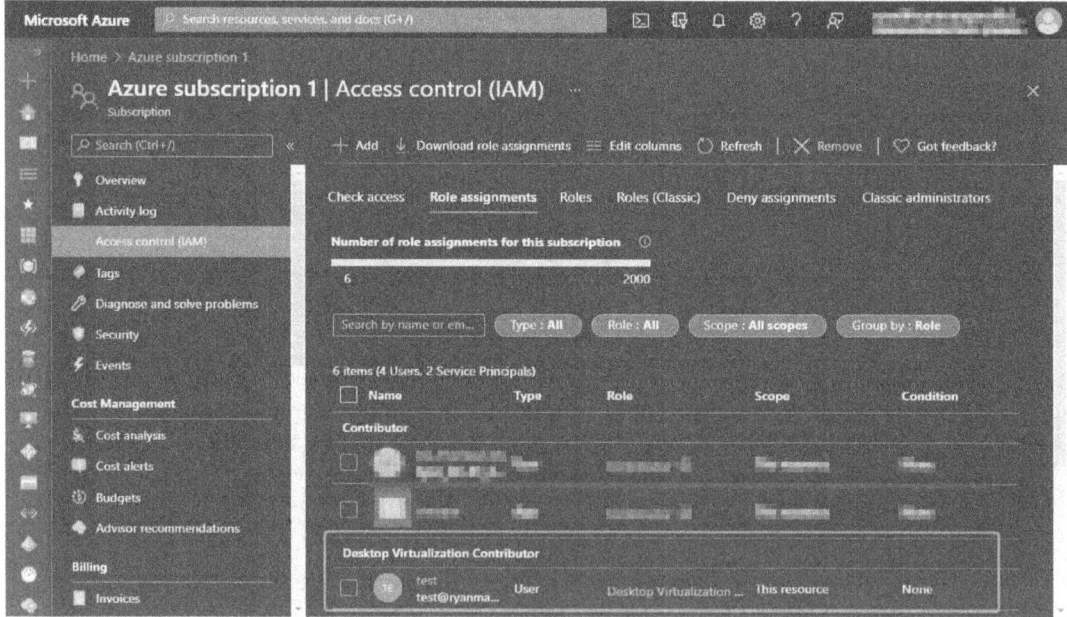

Figure 10.7 – The new role assignment added to the resource

There you have it; you have just assigned a user account to the **Desktop Virtualization Contributor** RBAC role at the subscription level. Additionally, as a best practice, it is recommended that you assign roles to Entra ID groups that offer more flexibility and security rather than specific users.

> **Important note**
> Deny assignments are created and managed by Microsoft Azure. You cannot directly create your own denied assignments.

Now, we'll learn how to do the same using PowerShell.

The PowerShell way to assign role assignments

Before you can start assigning roles via PowerShell, you must ensure that the AVD PowerShell model is set up. Please refer to the *Setting up PowerShell for Azure VD* section in *Chapter 6, Creating Host Pools and Session Hosts*.

To assign a user to an app group using PowerShell, you can use the following cmdlets:

```
New-AzRoleAssignment -SignInName <userupn> -RoleDefinitionName
"Desktop Virtualization User" -ResourceName <appgroupname>
-ResourceGroupName <resourcegroupname> -ResourceType 'Microsoft.
DesktopVirtualization/applicationGroups'
```

Here's the result of running the preceding command:

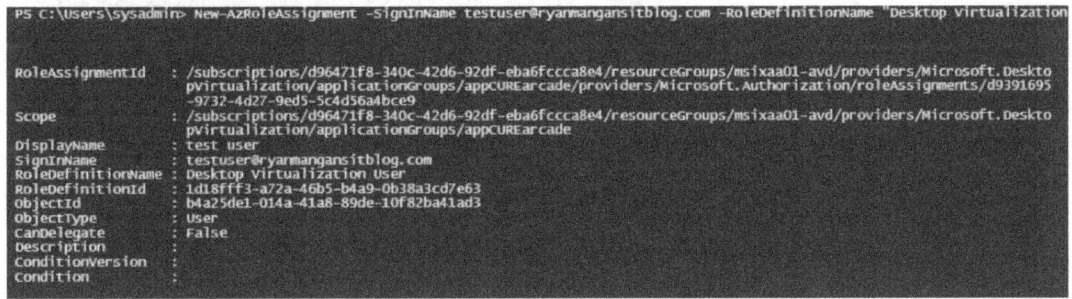

Figure 10.8 – Using New-AzRoleAssignement

To assign a group, you can run the following command:

```
New-AzRoleAssignment -ObjectId <usergroupobjectid> -RoleDefinitionName
"Desktop Virtualization User" -ResourceName <appgroupname>
-ResourceGroupName <resourcegroupname> -ResourceType 'Microsoft.
DesktopVirtualization/applicationGroups'
```

When using a group, you need to use the object ID, which can be found on the Entra ID page within the Azure portal. Within groups, select the required group. You will see the object ID that's required:

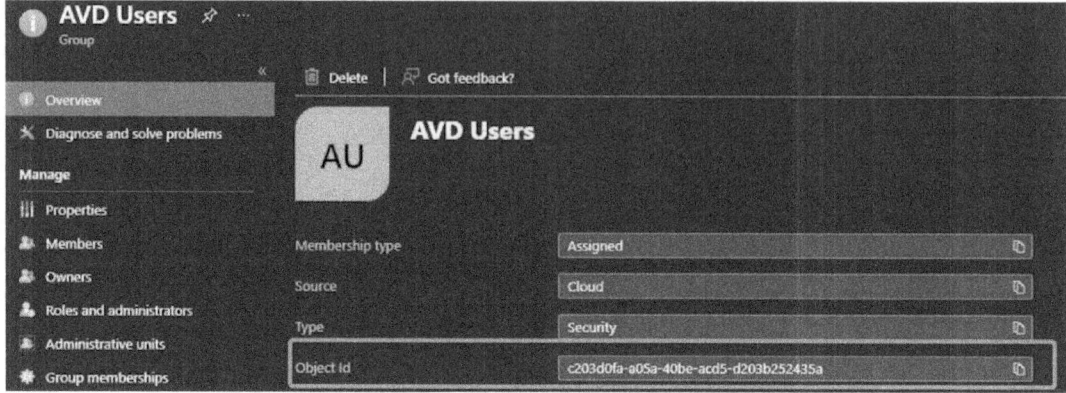

Figure 10.9 – An example group and its associated object ID

Once you have the required object ID for the associated group, you are ready to assign the desktop virtualization user role definition:

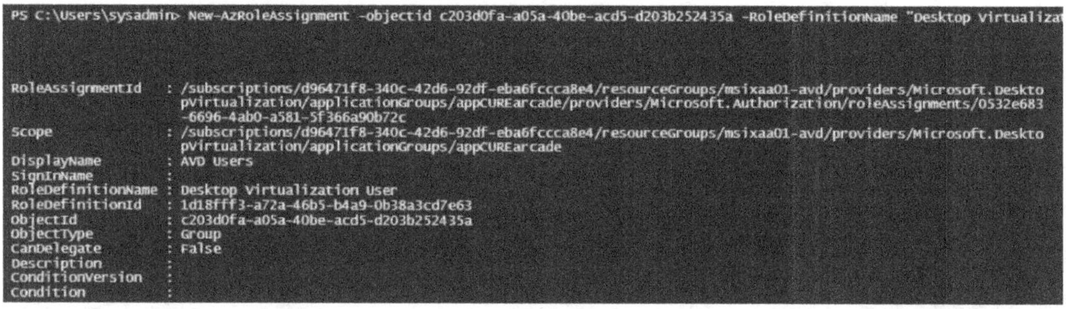

Figure 10.10 – An example of assigning a role assignment to a group using PowerShell

You can read more about assigning roles using PowerShell at https://docs.microsoft.com/azure/role-based-access-control/role-assignments-powershell.

In the next section, we'll look at how to create a custom role using the Azure portal.

Creating a custom role using the Azure portal

This section provides high-level steps regarding how to create a custom role assignment. Custom role assignments are useful when you want to grant very specific actions to admins and users. In larger organizations, you might wish to create more granular roles to suit your IT department's support structures or other use cases:

The following screenshot shows the **Access control (IAM)** menu within a resource. You can see the **Add custom role** option within the drop-down menu:

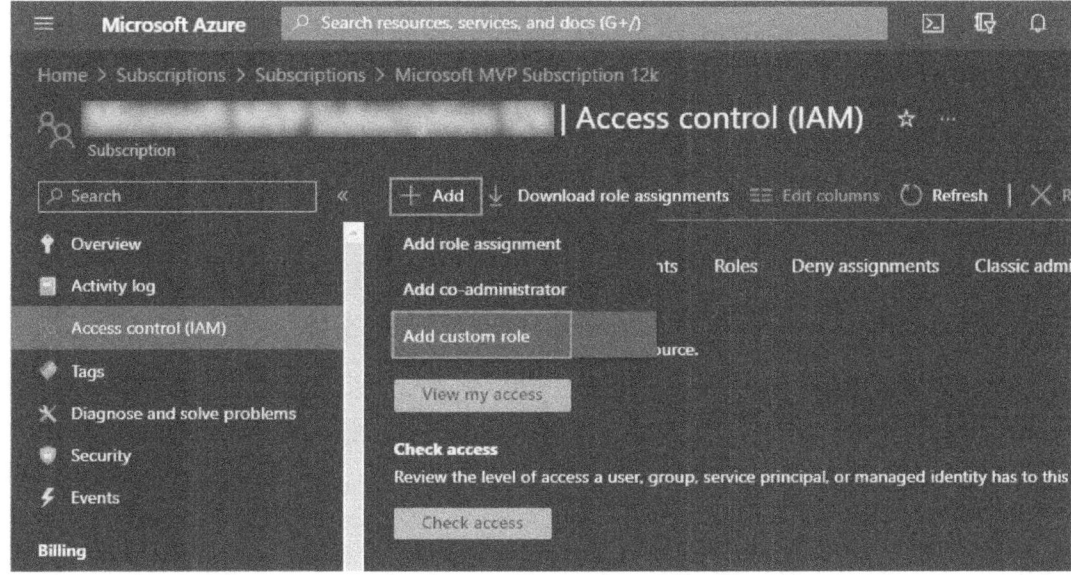

Figure 10.11 – The + Add function of a custom role

When clicking on the **Add custom role** option, you will see the **Create a custom role** page appear. Enter a role name and description and choose the base permissions. It is recommended that you clone a role rather than start from scratch. You can also start from JSON if you wish:

1. As shown in the following screenshot, we are cloning a role. In this case, we will use the **Desktop Virtualization User** role. Once you have filled out the form, click **Next**:

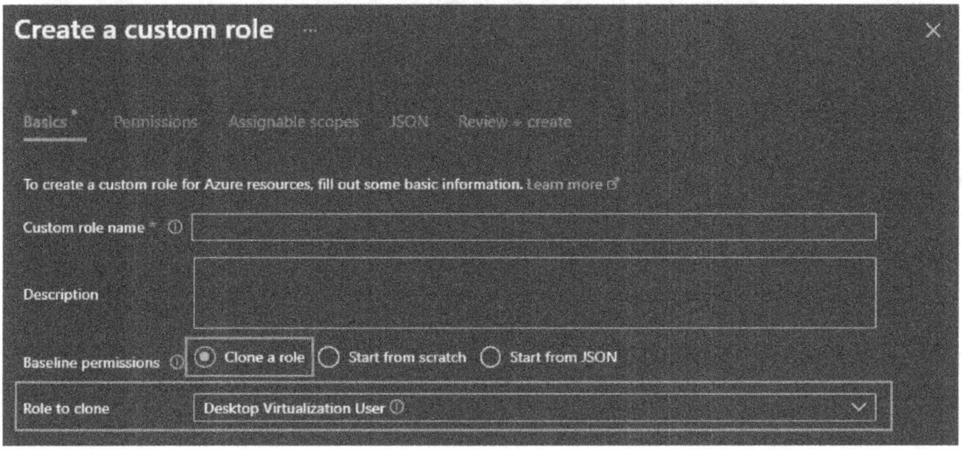

Figure 10.12 – The Create a custom role page

2. On the **Permissions** tab, you will see two buttons: **Add permissions** and **Exclude permissions**. To add permissions, click on the **Add permissions** button:

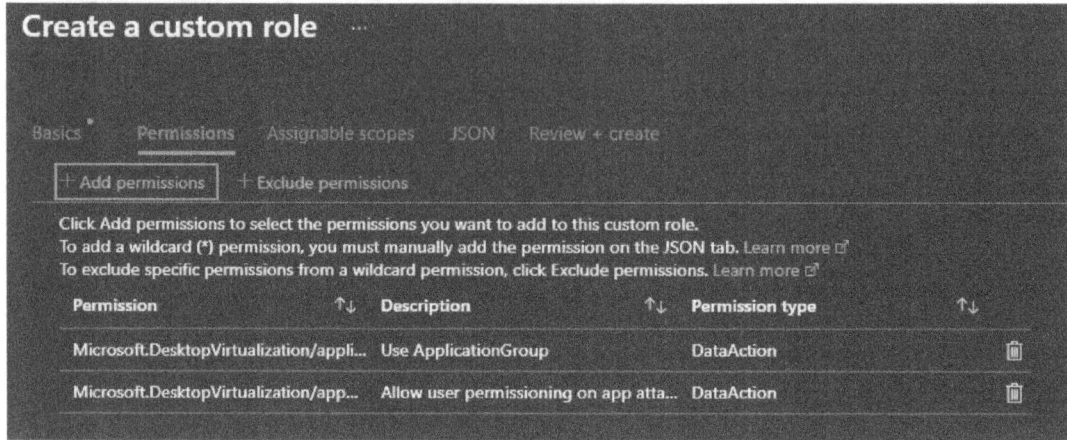

Figure 10.13 – The Add permissions button within the Create a custom role page

3. You will see the **Add permissions** page appear. You can use the search bar to search for specific permissions, as shown in the following screenshot:

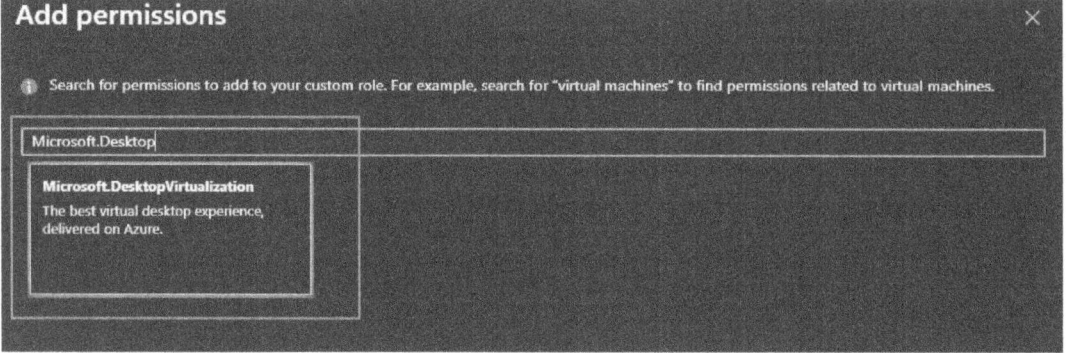

Figure 10.14 – Searching for Microsoft.DesktopVirtualization permissions

4. In this example, I will select the **Microsoft.DesktopVirtualisation** permissions. Once selected, you will be presented with a new page, as shown in the following screenshot. Select the permissions you want; these can be either **Actions** or **Data Actions**:

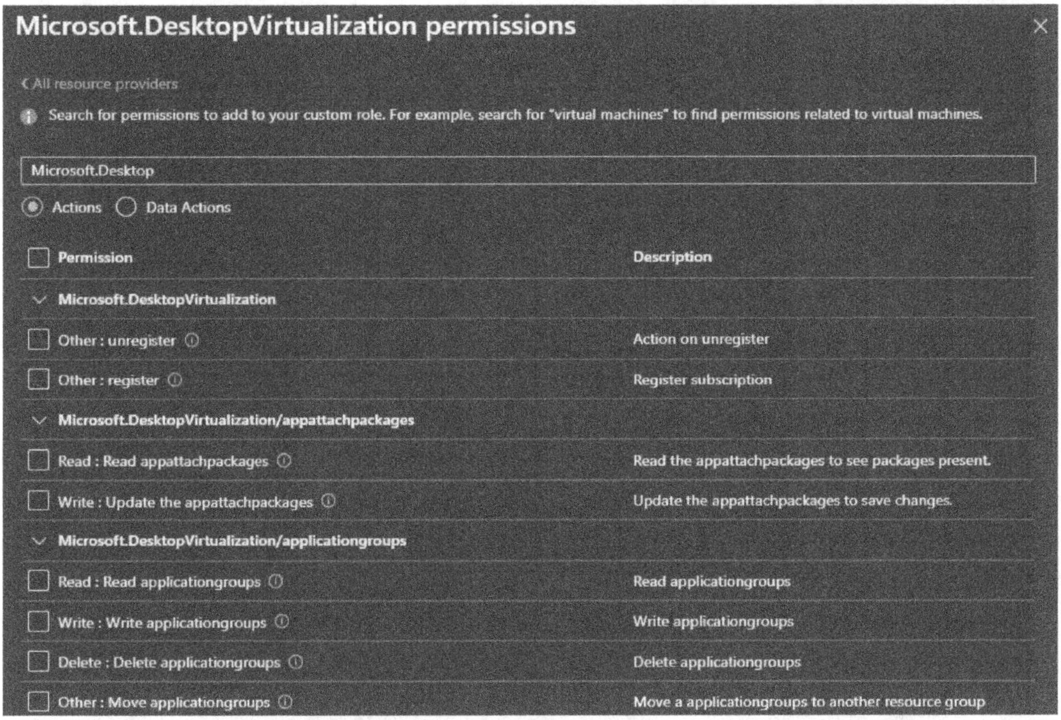

Figure 10.15 – The Microsoft.DesktopVirtualization permissions page with the add custom role

> **Tip**
>
> **Actions** are an array of strings that specify the management operation that the role allows. **DataActions** refers to an array of strings that specifies the data operations the role allows to be performed on the data within that object. You can read more at `https://docs.microsot.com/azure/role-based-access-control/custom-roles#how-to-determine-the-permissions-you-need`.

5. Once you have selected the required permissions, click **Add**.

6. You will then see the added permission appear on the **Create a custom role** page:

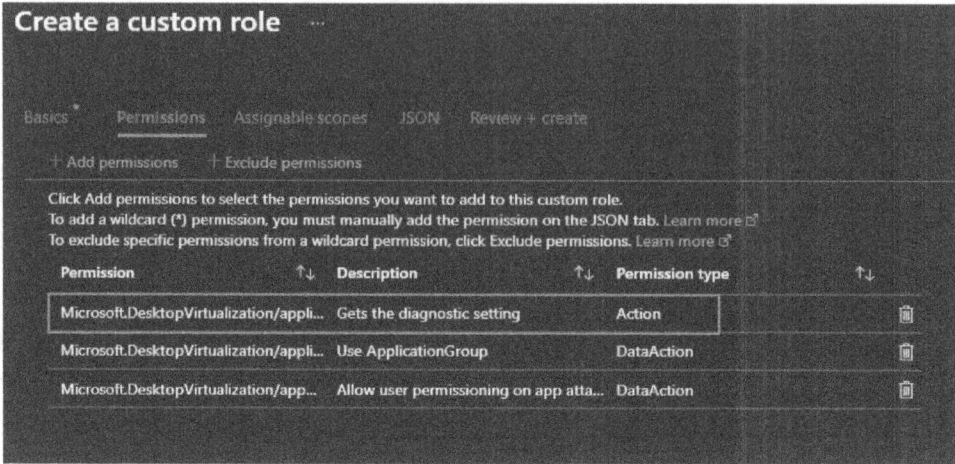

Figure 10.16 – The added role of getting diagnostic settings

7. Once you have finished adding permissions, click on the **Review + create** button, as shown in the preceding screenshot.
8. Check what you have configured and, once complete, proceed by clicking on the **Create** button:

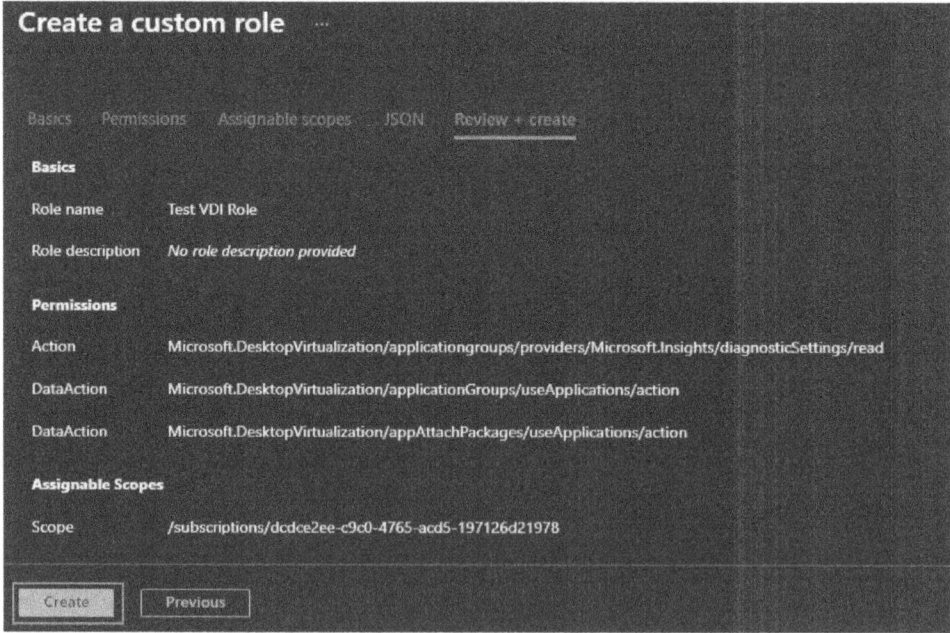

Figure 10.17 – Configuring the custom role before creating it

9. Once created, you can search for the customer role underneath the **Roles** tab within the **Access control (IAM)** page:

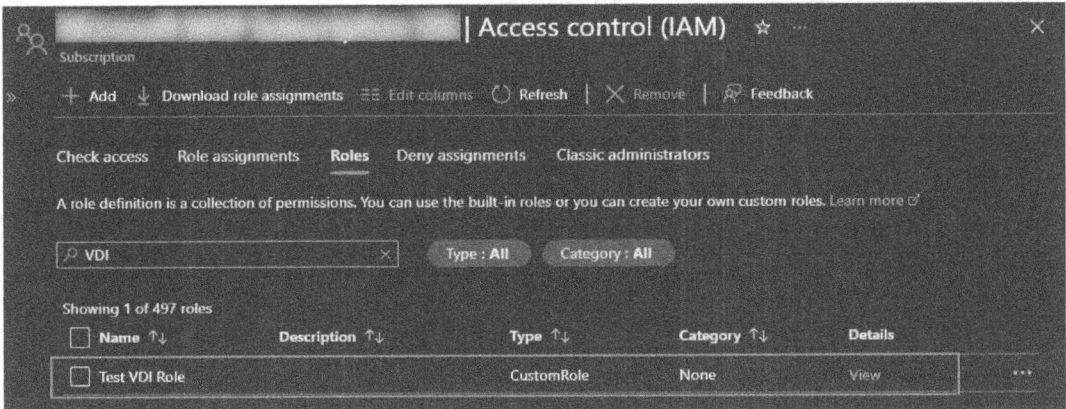

Figure 10.18 – The newly created custom role

With that, you've learned how to create a custom role. In the next section, we will look at how to manage local roles, groups, and right assignments on session hosts.

Managing local roles, groups, and rights assignments on AVD session hosts

In this section, we will take a look at managing local roles, groups, and rights assignments. When working in a pooled desktop environment, it's important to ensure that each session host is configured identically within the pool so that they have the same user experience. When configuring local roles, groups, and rights assignments, you can configure these on the gold master image (image template) or configure Group Policy to apply to single or multiple host pools centrally.

> **Tip**
> Using Group Policy is great when you're working with multiple host pools as you don't need to worry about customizing the image with roles, groups, and rights assignments. In Entra ID-only or hybrid deployments, you can also use Intune policies.

The use of local roles and groups might be required for specific requirements such as applications or allowing additional access permissions to the local operating system.

For example, if you're using FSLogix profile containers, you might wish to have a specific group of users assigned to the FSLogix profile containers.

Let's suppose you need to apply local groups to all session hosts within a pool. In that case, you can use the restricted groups setting via Group Policy. This allows you to configure local groups and apply them to your session hosts within an organizational unit.

The following screenshot shows us going to **Group Policy Management Editor | Computer Configuration | Policies | Security Settings | Restricted Groups**.

The **Restricted Groups** setting is a great way to automatically configure local groups to session hosts.

The following screenshot shows the **Restricted Groups** setting within the **Computer Configuration** tab, underneath **Windows Settings**:

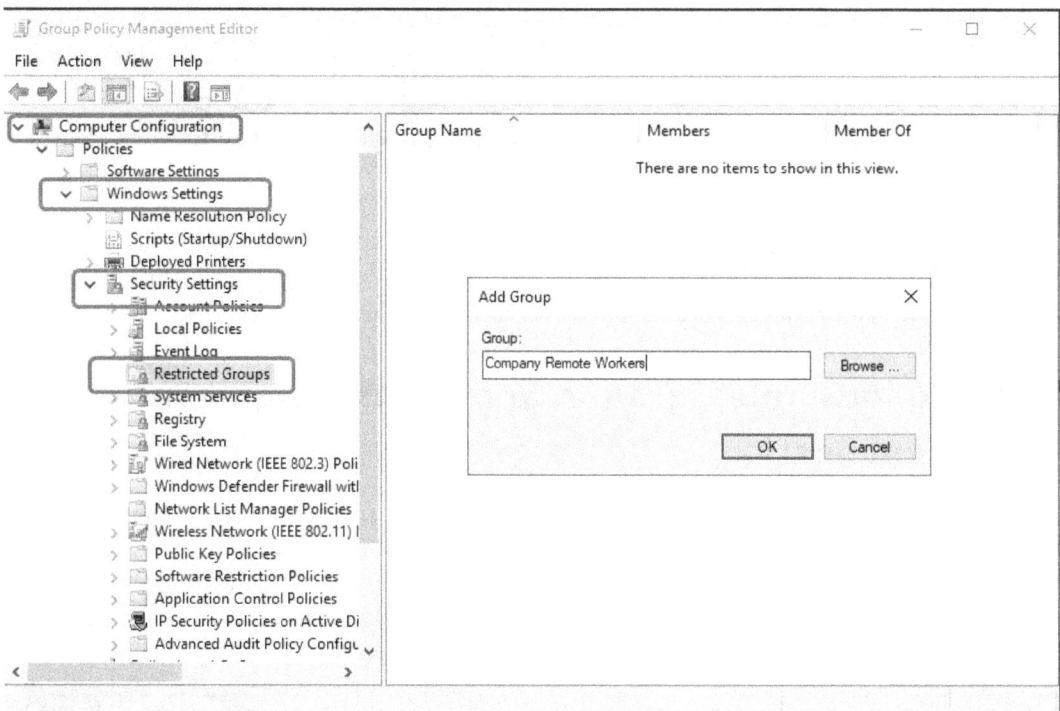

Figure 10.19 – Creating a group within the Restricted Groups setting

The following screenshot shows how to add members to a group that can be either users or groups. Additionally, you can link this group to a member of another group. For example, here, we can see that the **Company Remote Workers** group has been added and configured with a domain security group:

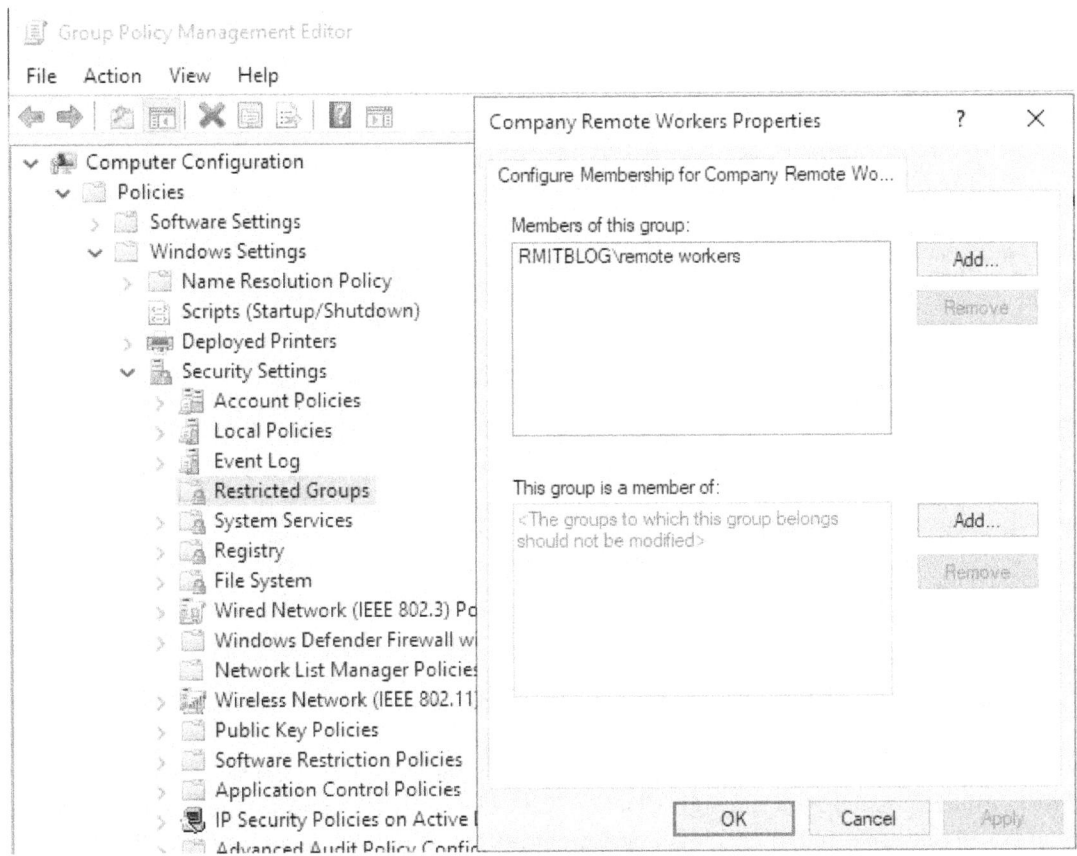

Figure 10.20 – Configuring membership for the new group

Now that we have covered the **Restricted group** settings, let's take a look at **User Rights Assignments**.

User Rights Assignments are used to customize specific items, such as preventing a time zone from being changed or making changes to a specific application; you can even configure who can shut down the session host. Again, as stated earlier, the configuration and customization of user rights assignments are unique to the organization, and it is recommended that rights assignments are carefully considered and tested.

The following screenshot shows the Group Policy setting for rights assignments under **Computer Configuration** | **Policies** | **Windows Settings** | **Security Settings** | **User Rights Assignment**:

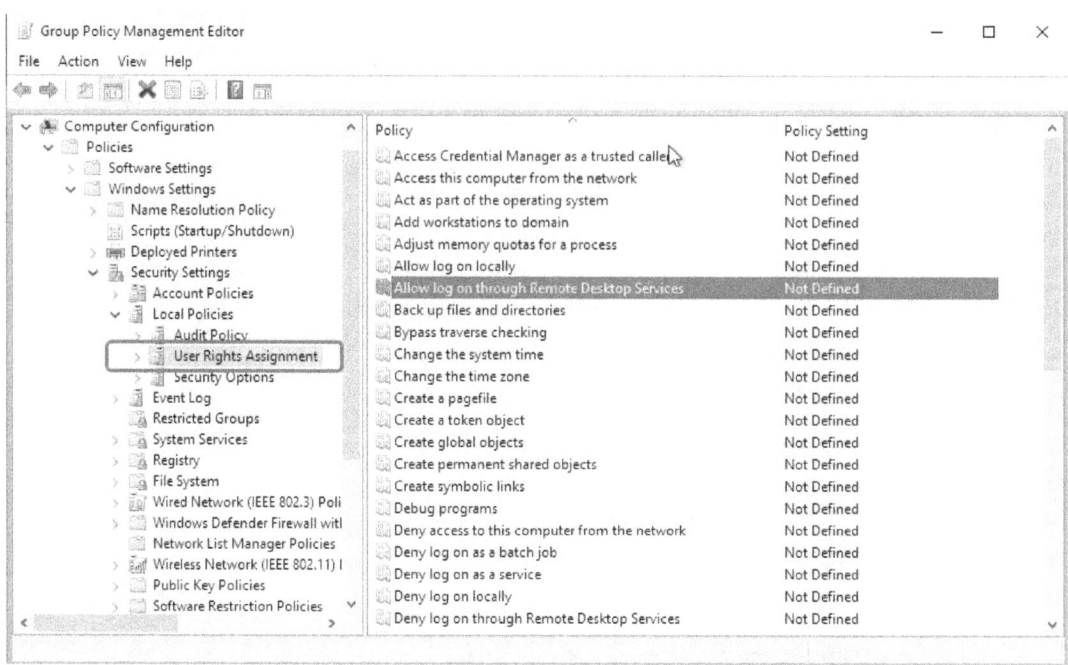

Figure 10.21 – The User Rights Assignment settings you can configure within Group Policy

You can use Group Policy to customize a wide range of optimizations/customizations and controls for AVD. Where possible, you should use Group Policy; however, it's important to note that some Group Policy settings could impact login performance compared to configuring them directly on the image template.

> **Group Policy tip**
>
> **LSDOU**: This is a really good acronym to remember when it comes to Group Policy. It means that local Group Policy objects are applied first (**LS**), then domain (**D**), and, finally, organizational unit policies (**OU**).
>
> **Block inheritance**: You can use this if you want to ignore group policies above the organizational unit your session hosts reside in. It is important to note that any enforced GPOs still apply when block inheritance is set.
>
> **Loopback**: This is useful for desktop virtualization environments as you can take the user settings and apply them to a group of session hosts (user settings linked to computer OU will be applied). Remember to ensure that both user and computer objects are added to the **Scope** tab of the Group Policy object for this to work and ensure the GPO status is set to enabled.
>
> Please note that loopback is known to slow down Group Policy processing.

The next section looks at configuring user restrictions by using Entra ID group policies and AD policies.

Configuring user restrictions by using Entra ID Domain Services group policies

Using Group Policy to restrict visibility and access to items such as the operating system disk and removing the restart button from the start menu are typically done via Group Policy. It is recommended, where possible, to use Group Policy rather than making direct configurations on the image template.

If you are using Entra ID group policies, you will need to build a Windows Server management VM that is joined to Entra ID Domain Services. Additionally, you will need a user account that is a member of the AAD DC Administrators group in your Entra ID tenant. You can read more about the requirements and setup process at `https://docs.microsoft.com/azure/active-directory-domain-services/manage-group-policy`.

> **Tip**
> You can now use AVD with Intune to configure policies and controls. For more information, please refer to `https://docs.microsoft.com/mem/intune/fundamentals/azure-virtual-desktop`. Please refer to `https://learn.microsoft.com/enus/mem/intune/fundamentals/azure-virtual-desktop-to` for multi-session OS.

Summary

This chapter looked at Azure roles and the specific RBAC for AVD resources. Hopefully, I refreshed your memory of Active Directory and then discussed some useful tips for managing local roles, groups, and rights assignments. We finished by looking at the requirements for Entra ID Domain Services to be able to use group policies.

In the next chapter, we will look at security and compliance.

Questions

Answer the following questions to test your knowledge of this chapter:

1. You have a Group Policy object named `user_personalization` that only contains user settings. The policy is linked to the AVD Session host OU. What should you configure for the settings within the policy so that you can apply them to users when they sign in to the session hosts?

    Loopback processing.

2. Which Desktop Virtualization role allows you to manage all aspects of the AVD service?

    Desktop Virtualization Contributor.

3. What are the three key elements of RBAC?

    Security principal, role definition, and scope.

11
Managing Security

In this chapter, we'll take a look at security and compliance settings for **Azure Virtual Desktop** (**AVD**). First, we'll look at planning and implementing **multi-factor authentication** (**MFA**) and Conditional Access policies for AVD. Next, we'll look at Microsoft Defender for Cloud and the benefits of turning this feature on and enabling Azure Defender. To finish this chapter, we'll look at Microsoft Defender Antivirus and additional configurations you can apply to streamline the security signature updates to session hosts.

This chapter covers the following topics:

- Planning and implementing MFA
- Managing security by using Microsoft Defender for Cloud
- Using Microsoft Defender for Cloud for AVD
- Enabling Azure Defender for AVD
- Configuring Microsoft Defender Antivirus for AVD

Introduction to MFA

MFA is an authentication layer you can add to the sign-in process to improve sign-in security. The user must provide additional identity verification when accessing corporate accounts, apps, or other services. This additional verification can be scanning a fingerprint or entering a code received by a phone or token-generating device.

> **Important note**
>
> The security threat landscape is consistently changing, with new threats appearing daily. As a best practice, it's advised that organizations use MFA as a standard practice to harden the sign-in process to protect users and corporate data.

How does Entra MFA Work?

Entra ID MFA works by requiring the user to use two or more authentication methods to complete a sign-in process. The first method is typically a password. Trusted devices such as a phone or hardware key or biometrics such as a fingerprint or face scan can be used as a second method.

> **Important note**
>
> Entra ID MFA also offers a feature known as secure password reset. This can be enabled when users register for Entra ID MFA, which is an additional step.

You can use the following forms of authentication when using Entra ID:

- Microsoft Authenticator app
- OATH hardware token (preview)
- OATH software token
- SMS
- Voice call
- FIDO2 security key
- Windows Hello for Business

The verification, when using Entra MFA, looks similar to what's shown in the following screenshot:

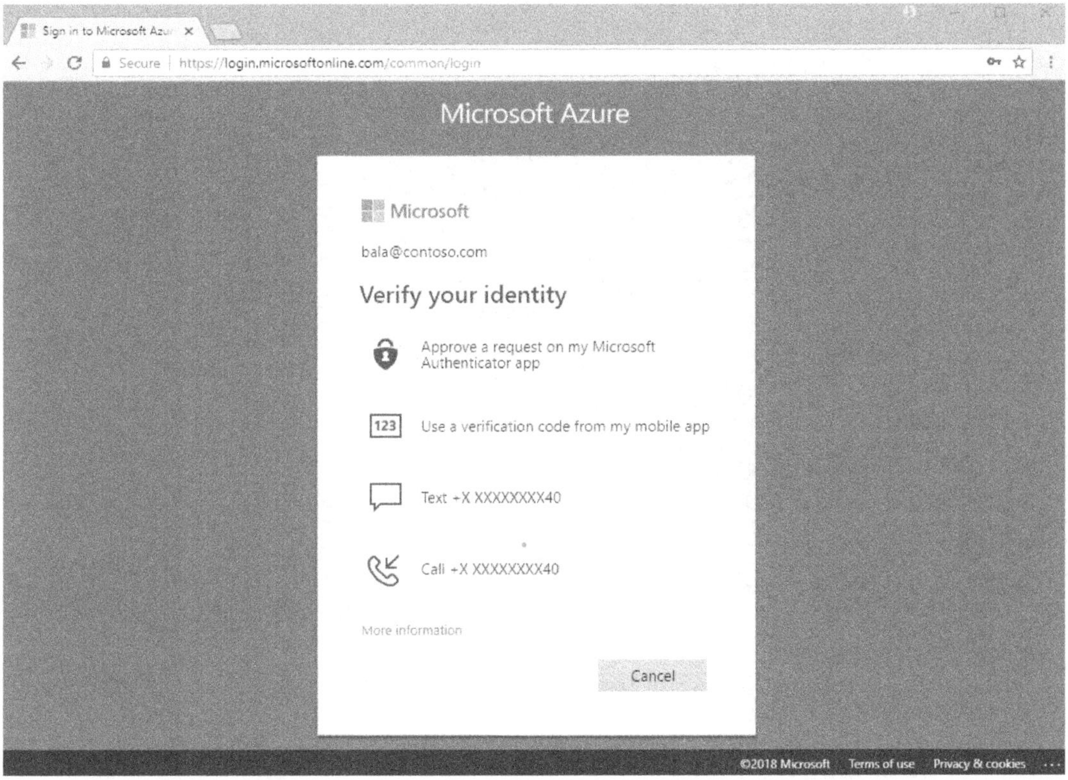

Figure 11.1 – Azure MFA prompt during a user sign-in process

You have the option of configuring the security defaults to enable Authenticator for all users or choosing Conditional Access policies to control specific events and applications. For example, you can configure Conditional Access policies to allow regular sign-in or to include a prompt for additional verification when a user is remote or on a personal device.

Now, let's look at the security defaults available to you on Azure.

Security defaults

Security defaults are features that help simplify security hardening when applying MFA to your organization's Azure tenant. When using preconfigured security settings, you essentially set the following:

- Requiring all users to register for Entra ID MFA
- Requiring administrators to do MFA
- Blocking legacy authentication protocols

- Requiring users to do MFA when necessary
- Protecting privileged activities such as access to the Azure portal:

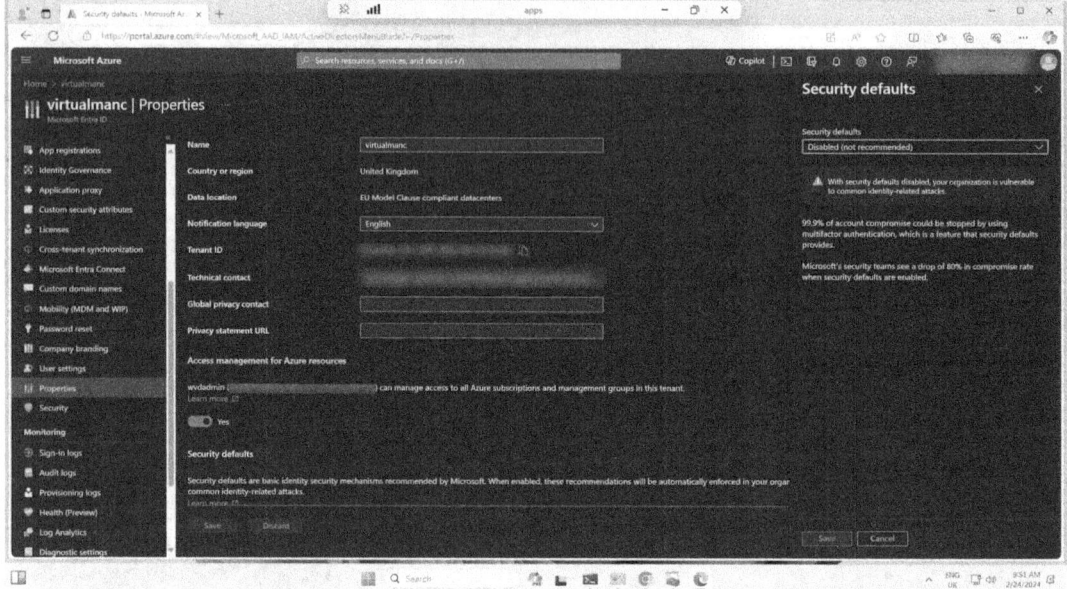

Figure 11.2 – The Security defaults section

You can read more on security defaults here: `https://learn.microsoft.com/entra/fundamentals/security-defaults`.

Conditional Access

As the technology ecosystem continues to evolve and change daily, the way people work and access corporate resources changes. This can also be described as the *modern security perimeter*, which refers to users and device identities that access corporate data and network resources outside the corporate network.

When considering conditional access, we must first understand its three core principles: signals, decisions, and enforcements.

> **Important note**
>
> To use Conditional Access policies, you need the Entra ID Premium P1 license. You can read more here: `https://www.microsoft.com/security/business/microsoft-entra-pricing`.

Let's take a look at these three components, all of which are required for Conditional Access organizational policies:

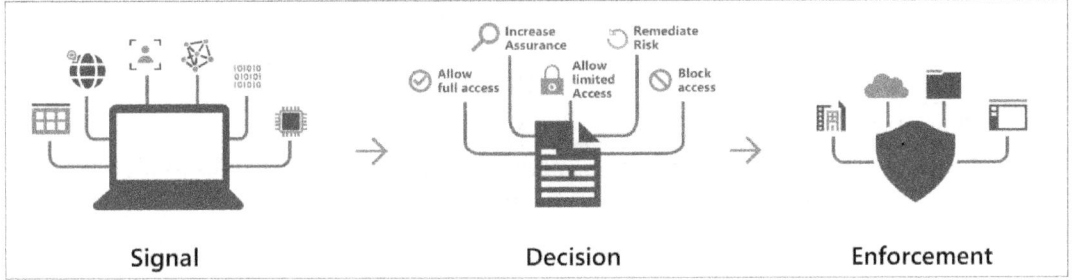

Figure 11.3 – The three components of Conditional Access policies

Signals

You must consider the following when making policy decisions using Conditional Access:

Signal	Description
User and group membership	Policies can target specific users or group memberships, enabling granular control over access to organizational resources.
IP location information	Trusted locations are common. If a user needs to pass through an organization's physical security controls, such as security cards or biometrics, they may not need to use MFA at their desk.
Device	You can control access to specific devices or types of devices.
Application	You can control the required access controls based on the application the user is trying to connect to.
Real-time calculated risk detection	You can use advanced signals such as real-time or calculated risk detection based on identity risk sign-in behavior. This means that if the behavior of an identity is deemed unusual, then access can be blocked until the admin takes manual intervention.
Microsoft Defender for Cloud apps	Application access and sessions can be monitored and controlled in real time, providing improved security controls as well as improved visibility of activities within your cloud environment.

Table 11.1 – Key Signals for Conditional Access Policies

Decisions

The following table details the two decisions and the options that are available when you select **Grant Access**:

Decision	Description
Block Access	This is the most restrictive decision.
Grant Access	You can configure one or many of the following controls with the **Grant Access** decision: • Require MFA • Require authentication strength • Require the device to be marked as compliant • Require a Microsoft Entra Hybrid-joined device • Require approved client app • Require an app protection policy • Require a password change • Require terms of use

Table 11.2 – Key decisions for Conditional Access Policies

Enforcements

The following is a list of examples of some of the applied policies you can set:

- Blocking risky sign-in behaviors
- Requiring MFA for Azure management tasks
- Blocking sign-ins for users attempting to use legacy authentication protocols
- Requiring trusted locations for Entra ID MFA registration
- Requiring MFA for users with administrative roles
- Blocking or granting access from specific locations
- Requiring organization-managed devices for specific applications

You can read more about the three components of Conditional Access here: https://learn.microsoft.com/entra/identity/conditional-access/overview.

The following diagram shows the use of the three components in actions to enforce Conditional Access for the required apps and data for your organizations:

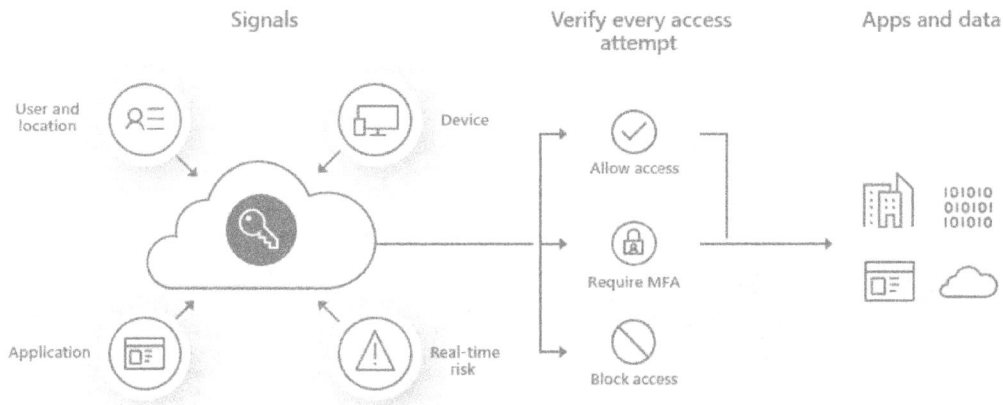

Figure 11.4 – Logical diagram of how Conditional Access works

Next, we'll look at how to plan and implement MFA.

Planning and implementing MFA

This section goes into detail on how to implement MFA for **Azure Virtual Desktop** (**AVD**). We will navigate the process step by step. The benefit of MFA is that it provides an extra layer of security for users, and only the user with access to the token can log in, reducing the risk of unauthorized access to the network and IT resources.

The prerequisites for getting started are as follows:

- First, you need to assign a license to users that include Entra ID Premium P1 or P2.
- You also need to create a new Entra ID group for MFA and ensure that the users you want to assign MFA to are included.
- You must ensure that you assign users a license that includes an Entra ID Premium P1 or P2 license.
- Ensure you enable Azure MFA for all required users.

Managing Security

For more information on the prerequisites, please see the following link: `https://learn.microsoft.com/entra/identity/authentication/howto-mfa-getstarted#prerequisites-for-deploying-microsoft-entra-multifactor-authentication`.

You also need to ensure that your users are configured to use MFA. You can do this by following the steps I have summarized here:

1. First, sign in to the Azure portal using an administrator account.
2. Select **Microsoft Entra ID** from the left-hand menu, then select **Users | All users**.
3. Select **Per-user MFA**. The following screenshot shows the **All users** section within Entra ID:

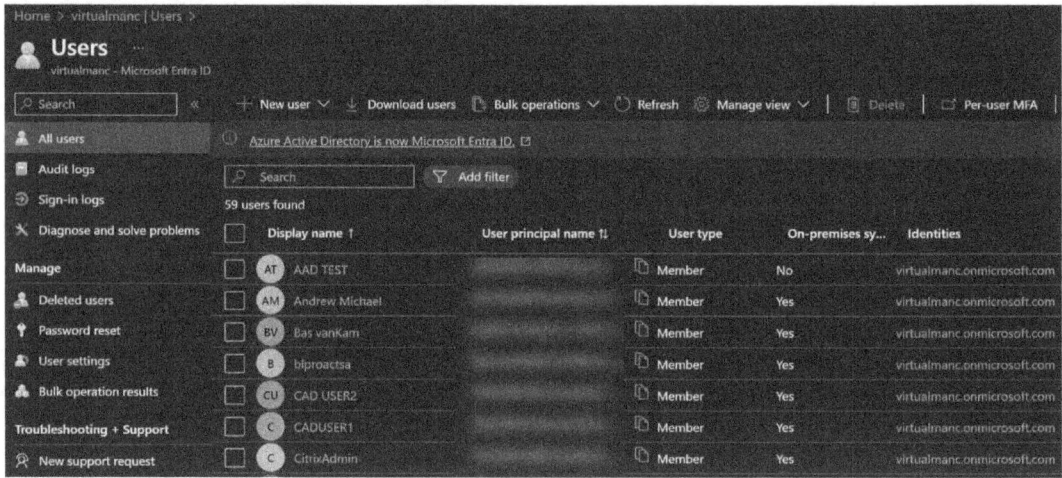

Figure 11.5 – The All users page within Entra ID

4. The following screenshot shows the **Enable** button for enabling a user to use MFA:

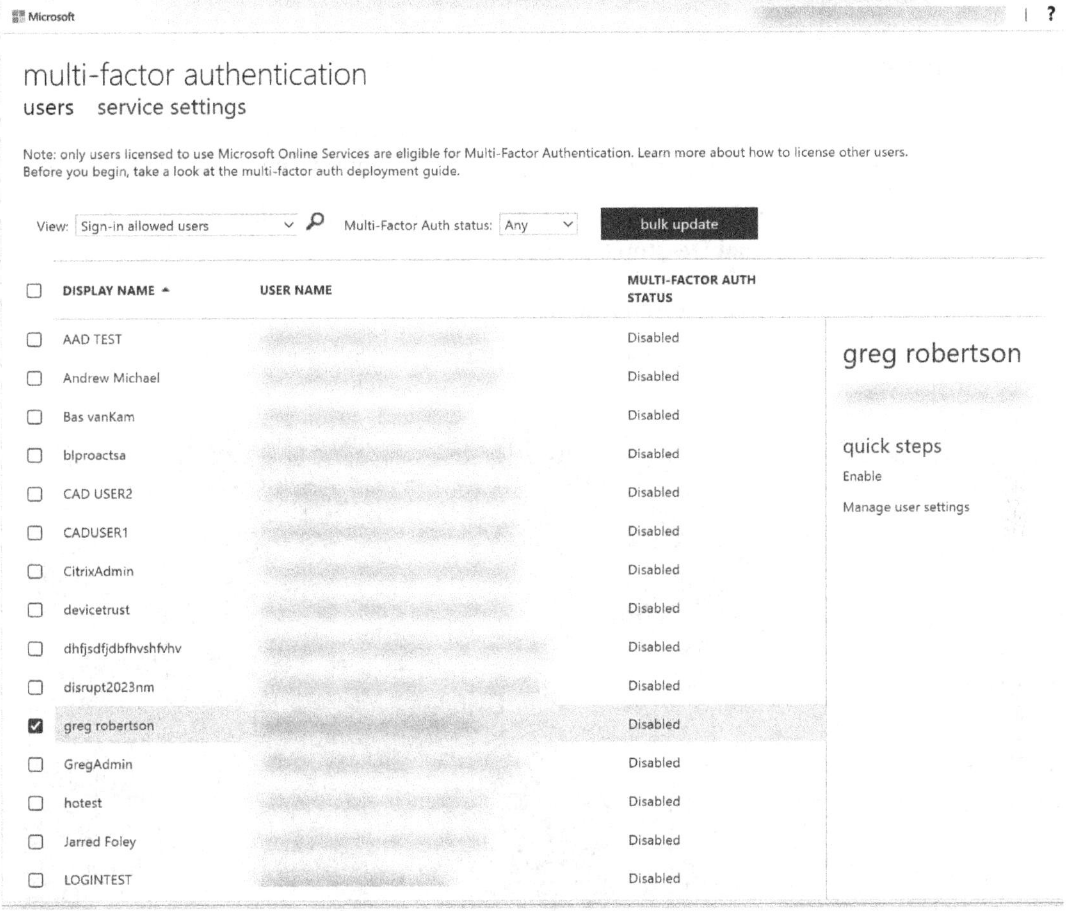

Figure 11.6 – The multi-factor authentication page for enabling MFA for users found in the Office.com portal

To learn how to configure multi-factor user states, see the following link: `https://learn.microsoft.com/en-gb/entra/identity/authentication/howto-mfa-userstates`.

> **Tip**
> It is recommended that you do not manually change the user state to **Enforced** unless the user is already registered or if the user understands that connections to legacy authentication protocols will be interrupted.

In the following subsection, we'll learn how to configure the required Conditional Access policy for AVD to enforce MFA.

Creating a Conditional Access policy for MFA

In the previous section, we discussed Conditional Access policies and their three components: signals, decisions, and enforcements. Now, we'll examine the process of creating a Conditional Access policy.

The following steps will guide you through creating a Conditional Access policy that requires MFA when connecting to **Azure Virtual Desktop (AVD)**:

1. Log in to the Azure portal as an administrator.
2. Navigate to **Entra ID | Security | Conditional Access**. The following screenshot shows the **Security** menu within the **Entra ID** menu in the Azure portal:

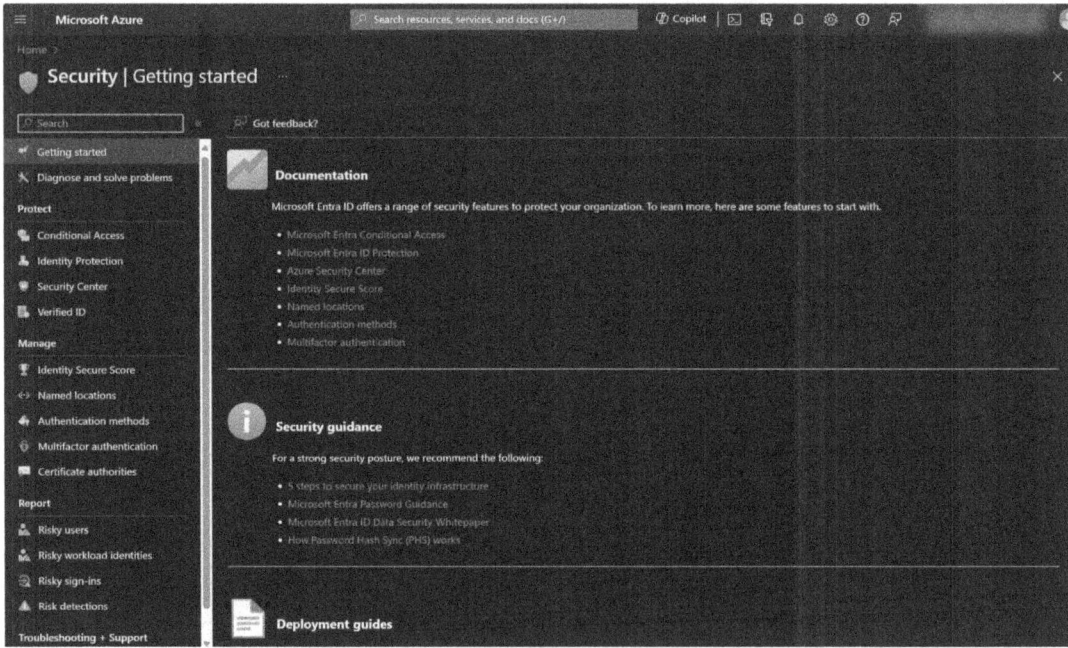

Figure 11.7 – The Conditional Access option on the Entra ID page

3. Select **+ Create new policy**. The following screenshot shows the button to add a new Conditional Access policy:

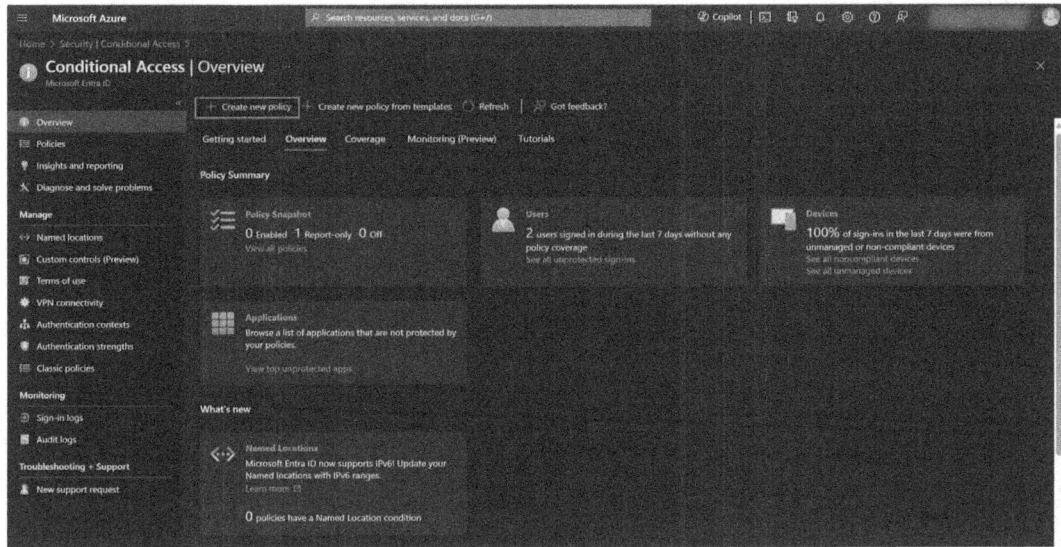

Figure 11.8 – The + Create new policy button

4. Enter a name for your policy.

5. Under the **Assignments** section, select **users and groups**. Under **Include**, click **select users and groups** | **users and groups** and choose the group you created in the prerequisites stage. Then, click **Done**.

6. Under **Target Resources** | **Include**, click **Select apps**. Based on the version of AVD you're using, select one of the following apps:

> **Tip**
> Please note that you may find that the name has not changed from Windows Virtual Desktop, and it is advised that you check for both.

- If you're using AVD, choose the **Azure Virtual Desktop** app (app ID `9cdead84-a844-4324-93f2-b2e6bb768d07`)

- If you're using AVD with single-on enabled, choose **Microsoft Remote Desktop** (app ID `a4a365df-50f1-4397-bc59-1a1564b8bb9c`) and **Windows Cloud Login** (app ID `270efc09-cd0d-444b-a71f-39af4910ec45`)

The following screenshot shows the details filled in for creating a new Conditional Access policy. We will apply these to AVD, as specified in the **Cloud apps | Include** section:

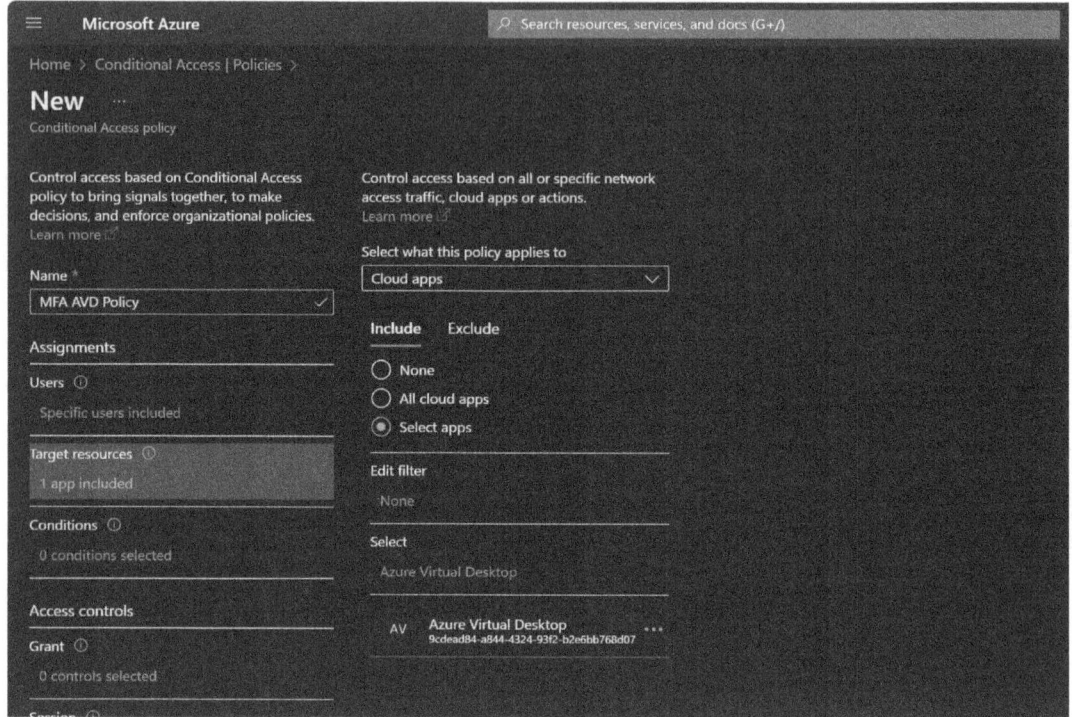

Figure 11.9 – Cloud apps AVD

> **Tip**
> To find the app ID of the app you want to select, navigate to **Enterprise Applications** and select **Microsoft Applications** from the **Application type** drop-down menu. You can read more here: https://learn.microsoft.com/en-gb/entra/identity/enterprise-apps/view-applications-portal?tabs=azure-portal#search-for-an-application.

7. Next, navigate to **Conditions | Client apps**. In **Configure**, select **Yes**, and then select where you wish to apply the policy:

   - If you want the policy to apply to the web client, select **Browser**
   - If you want to apply the policy to other clients, select **Mobile apps and desktop clients**
   - If you want to apply the policy to all clients, select both checkboxes

   The following screenshot shows the conditions that have been set for this Conditional Access policy. Note that both **Browser** and **Mobile apps and desktop clients** have been set for this policy:

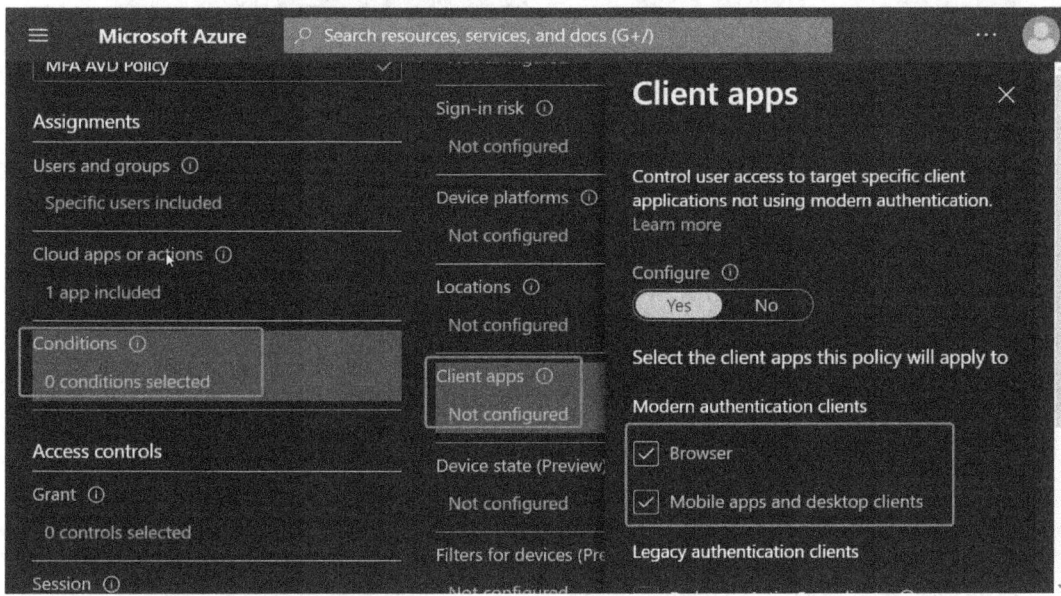

Figure 11.10 – Conditions being set for browser and mobile apps and desktop clients

8. Once you've selected the required client apps shown in *Figure 11.10*, click **Select**, then **Done**.
9. Under **Access controls | Grant**, select **Grant access | Require multi-factor authentication**, then **Select**.

The following screenshot shows **Access controls** configured, specifically the **Grant access** control. By setting **Require multi-factor authentication**, you force the user to complete two-factor verification to access the resources that have been configured within the Conditional Access policy:

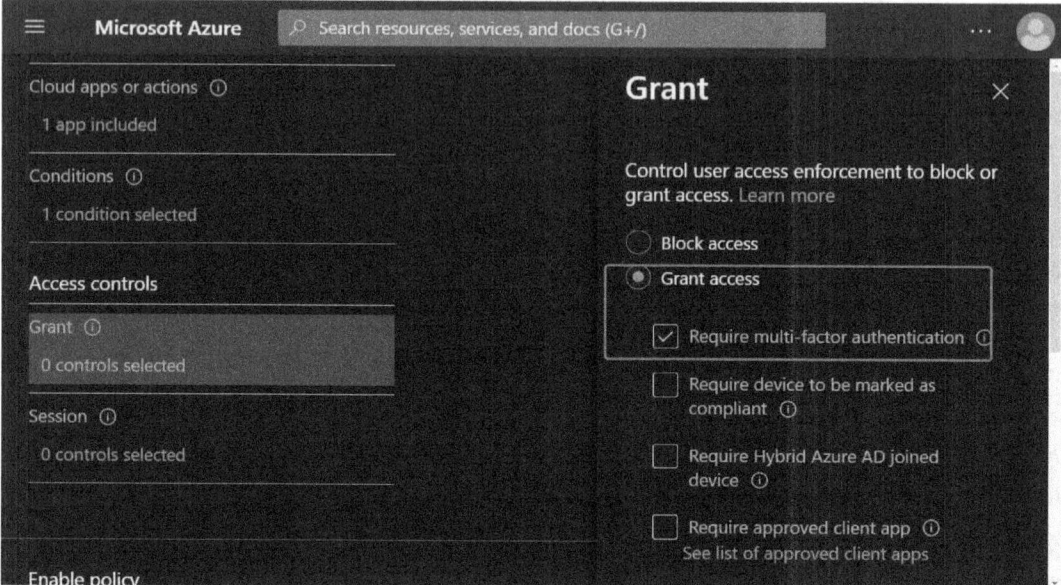

Figure 11.11 – Control for enforcing MFA for granting access to the cloud app

10. Under **Access controls | Session**, select **Sign-in frequency**, set the time between MFA prompts to the desired value, and then click **Select**. For example, setting the value to **4** and the unit to **Hours** will require MFA if a connection is launched 4 hours after the last.

The following screenshot shows how to customize the **Sign-in frequency** setting. You can set a specific time before a user needs to reauthenticate:

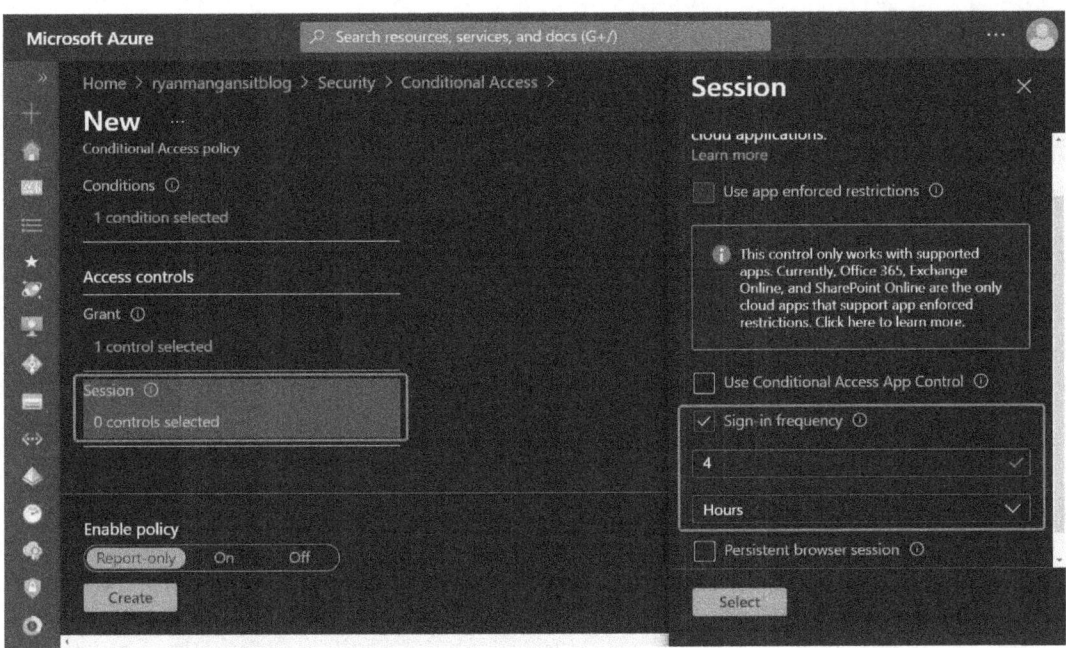

Figure 11.12 – The Sign-frequency setting for enforcing re-authentication

11. Confirm the settings and set **Enable policy** to **On**. The following screenshot shows that **Enable policy** has been turned on for the Conditional Access policy:

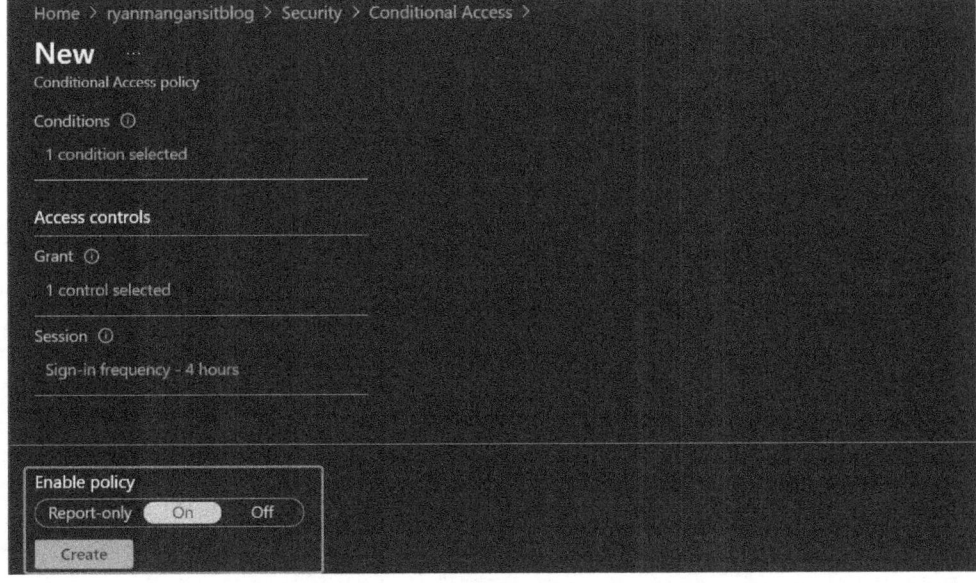

Figure 11.13 – Setting the policy to On

12. Click **Create** to enable your policy.

There you have it; you have enabled MFA and configured the required Conditional Access policy for AVD.

> **Tip**
> It is advised that you use **Report-only** before introducing this to a production environment. **Report-only** allows you to identify any issues and ensure the configured Conditional Access policy is functioning correctly.

This section looked at enabling MFA for users and then configuring an AVD Conditional Access policy. The following section will look at managing security by using Microsoft Defender for Cloud.

Managing security by using Microsoft Defender for Cloud

Microsoft Defender for Cloud was previously known as Azure Security Center and Azure Defender. I want to set some context around the reasoning and detail of the responsibilities that are split between Microsoft and the customer.

We previously spoke about advanced security features, such as reverse connect, which reduces the risk of exposing Virtual Desktop resources directly to the public network. We'll now look at the security responsibilities and some of the Azure security best practices available.

Here are the security areas you're responsible for in your **Azure Virtual Desktop** (**AVD**) deployment. Note that the value under the **Customer responsibility** column is **Yes** if the customer is responsible and **No** if Microsoft is responsible:

Security areas	Customer Responsibility
Identity	Yes
User devices (mobile and PC)	Yes
App security	Yes
Session host operating system	Yes
Deployment configuration	Yes
Network controls	Yes
Virtualization control plane	No
Physical hosts	No
Physical network	No
Physical data center	No

Table 11.3 – Customer Responsibilities in Azure Virtual Desktop (AVD) Deployment

This table was adapted from Microsoft:

`https://learn.microsoft.com/en-gb/azure/virtual-desktop/security-recommendations#security-responsibilities`.

As detailed in the table, Microsoft takes care of the physical aspects of the cloud infrastructure and the virtualization control plane. The customer is responsible for everything else. This is why it makes sense to use Microsoft Defender for Cloud to assist with security hardening all the required components for your AVD environment.

> **Important note**
>
> Microsoft Defender for Cloud is an essential security posture manager that has two offerings, the first being a free version. The second option, known as enhanced security, offers several security features and tools to help you harden your environment.
>
> Microsoft Defender for Cloud represents several security services specific to different workloads, such as databases, storage accounts, containers, and key vaults.

Microsoft Defender for Cloud helps you harden your resources, as well as map your current security posture, track future changes to help protect against cyberattacks, and streamline your IT security. As Microsoft Defender for Cloud is natively integrated, it provides a simple and easy way to deploy Defender so that you can secure your resources by default.

The following figure shows the three core needs when managing security with Microsoft Defender for Cloud:

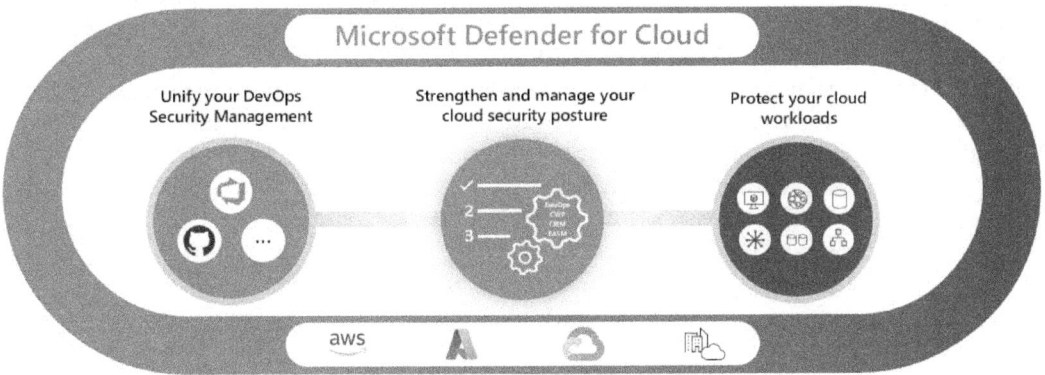

Figure 11.14 – The three core needs when managing the security of your resources in the cloud or on-premises

For a detailed breakdown of Microsoft Defender for Cloud, go to https://learn.microsoft.com/azure/defender-for-cloud/defender-for-cloud-introduction.

The following table describes the three core security requirements that are used within Defender for Cloud Core security requirements:

Security Requirement	Defender for Cloud Solution
DevSecOps	A Development Security Operations solution that unifies security management at the code level across multi-cloud and multiple pipeline environments
Cloud Security Posture Management (CSPM)	A CSPM solution that surfaces actions that you can take to prevent breaches
Cloud Workload Protection Platform (CWPP)	Specific protection for servers, containers, storage, databases, and other workloads, including AVD

Table 11.4 –

This table has been adapted from the Microsoft documentation site: https://learn.microsoft.com/azure/defender-for-cloud/defender-for-cloud-introduction.

> **Important note**
>
> Please note that when using custom/third-party technologies such as **network virtual appliances (NVAs)**, you may get false positive alerts regarding best practices from Microsoft Defender for Cloud. These alerts or recommendations become a false positive because you are effectively bypassing the default Azure configurations with a third-party security feature/technology. One good example is the use of port forwarding, which is typically enabled for an NVA. This would flag an alert in Microsoft Defender for Cloud. However, port forwarding must pass traffic through the NVA and thus be enabled on the NVA.

Microsoft Defender for Cloud provides a security score, which is essentially a set of recommendations and best practices for improving your AVD environment:

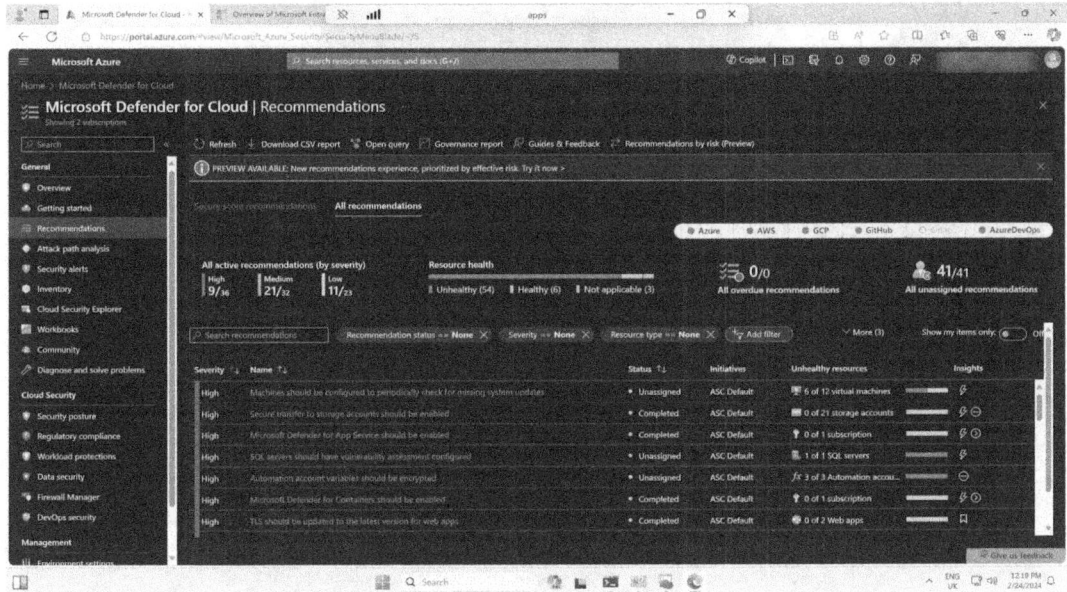

Figure 11.15 – Security score and list of recommendations below

The good news is that recommendations are prioritized to help you select the most important ones. There is also the **Fix** option to help you quickly identify and address any vulnerabilities. This is important to note. The **Fix** button helps with some issues but does not provide full coverage. It's advised that you conduct internal security reviews to ensure that you meet the requirements for your organization's security posture:

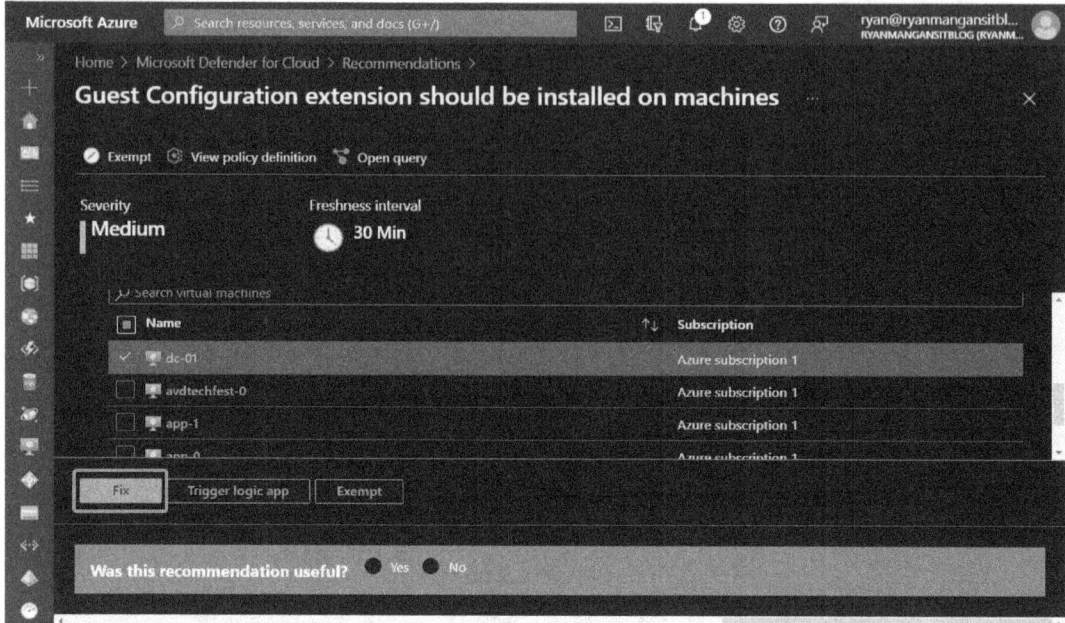

Figure 11.16 – The Fix option within the Recommendations section of Microsoft Defender for Cloud

> **Important note**
> The recommendations will be updated when changes in the IT ecosystem occur, meaning that new recommendations will be provided when the security landscape changes, such as when new vulnerabilities occur and new/better ways to maintain your AVD environment's security are developed.

The next section examines how to secure your AVD security environment and enable enhanced security within Microsoft Defender for Cloud.

Securing AVD using Microsoft Defender for Cloud

As summarized in the introduction to this section, the customer is responsible for the following areas under the shared responsibility model:

- Network
- Deployment configuration
- Session host operating system
- Application security
- Identity

Security posture refers to an organization's overall cybersecurity strength. It can also be used to predict, prevent, and respond to ever-changing threats. Therefore, it is advised that you examine both the required level of threat protection and the security posture for your AVD environment.

Misconfiguring the network and/or virtual machines can increase the attack surface or possibly compromise an endpoint.

> **Important note**
> You must ensure all management ports are closed on your AVD virtual machines. Direct access to session hosts from the public network is not required. If you want direct access to virtual machines, using Azure Bastion or connecting over a VPN is advised.

We'll cover endpoint protection in the next section; however, the following security controls must be mentioned to protect users from browsing malicious sites or connecting to malicious devices.

Here is a list of benefits Microsoft Defender for Cloud offers for improving security posture and threat protection for AVD when enabling Azure Defender:

- Secure configuration assessment and Secure Score
- Industry-tested vulnerability assessment
- Host-level detections
- Agentless cloud network micro-segmentation and detection
- File integrity monitoring
- Just-in-time virtual machine access
- Adaptive application controls

The following table shows the different security areas and what Microsoft Defender for Cloud offers in terms of capabilities:

AVD Security Area	Azure Security Center Security Posture Enhancement Capabilities	Microsoft Defender for Cloud Threat Protection Capabilities
Network security	Secure configuration assessment and security score Just-in-time virtual machine access	Agentless cloud network micro-segmentation and detections
Deployment configuration	Secure configuration assessment and security score Session host operating system	Not available

AVD Security Area	Azure Security Center Security Posture Enhancement Capabilities	Microsoft Defender for Cloud Threat Protection Capabilities
Session host operating system	Industry-tested vulnerability assessment	Host-level detections
Application security	Industry-tested vulnerability assessment File integrity monitoring Adaptive application controls	Host-level detections
Identity	Secure configuration assessment and security score	Agentless cloud network micro-segmentation and detections

Table 11.5 – Security Areas and Capabilities Offered by Microsoft Defender for Cloud

This section provided a high-level overview of Microsoft's and the customer's security responsibilities. It also introduced Microsoft Defender for Cloud to set the scene for the following sections of this chapter. We'll now move on to examine using Microsoft Defender for Cloud and Azure Defender for AVD.

Using Microsoft Defender for Cloud and AVD

Microsoft Defender for Cloud provides continuous assessments, security recommendations, fixes, and Azure security scores, all of which can be used to gauge your security posture.

Enabling Azure Defender offers additional features, including just-in-time virtual machine access, adaptive application controls/network hardening, compliance dashboards/reports, and threat protection for Azure virtual machines and non-Azure servers.

> **Important note**
> It is important to note that Microsoft Defender for Cloud is a **security posture manager** (SPM).

The following screenshot shows the differences between Microsoft Defender for Cloud being switched on and off:

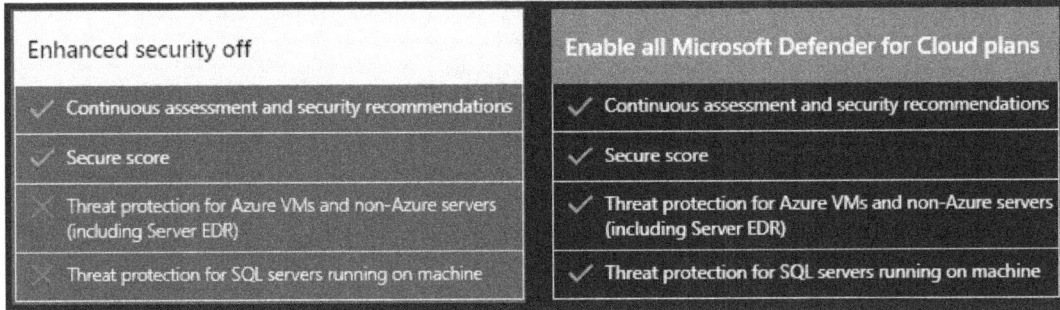

Figure 11.17 – The different features that are available when Microsoft Defender for Cloud is on and off

To access Microsoft Defender for Cloud, you can click the Security Center icon in the main window of the Azure portal. When you click this, you will be taken to the **Overview** window of Microsoft Defender for Cloud, as shown in the following screenshot:

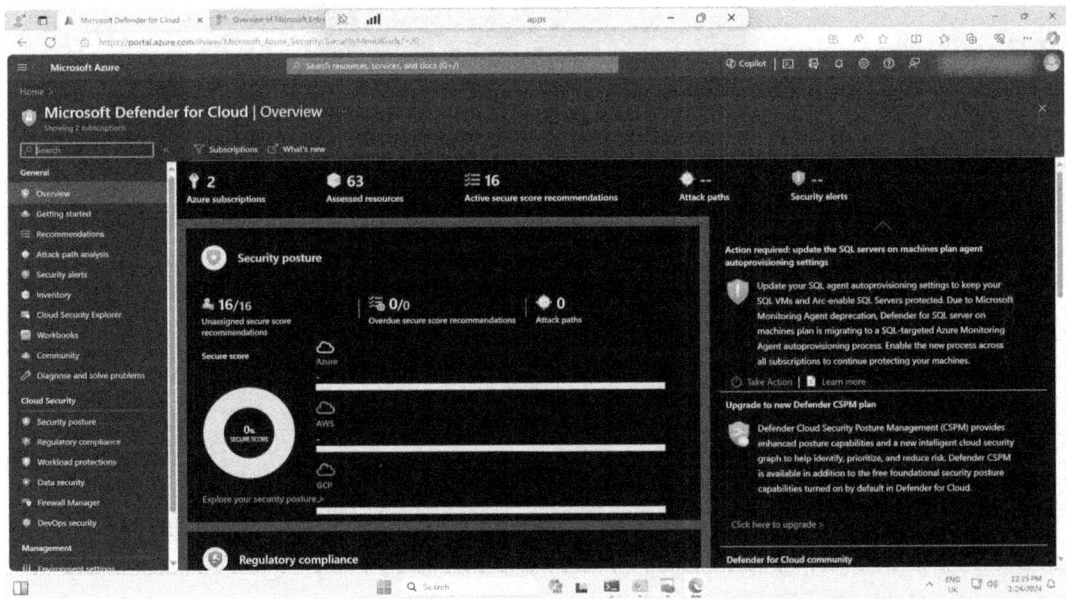

Figure 11.18 – The Microsoft Defender for Cloud Overview page

Within Microsoft Defender for Cloud, you can review and configure various security controls/policies and review best practices. A lot of the content within Microsoft Defender for Cloud is outside the scope of this book; however, we will take a brief look at what you can use for AVD to improve your desktop virtualization security posture.

As highlighted in the following screenshot, you can view your resources via the **Inventory** page. This lets you see which resources have been configured with monitoring agents, turn Azure Defender on/off, and see recommendations.

The **Inventory** page of Microsoft Defender for Cloud is used for reviewing all resources, including total, unhealthy, unmonitored, and unregistered subscriptions:

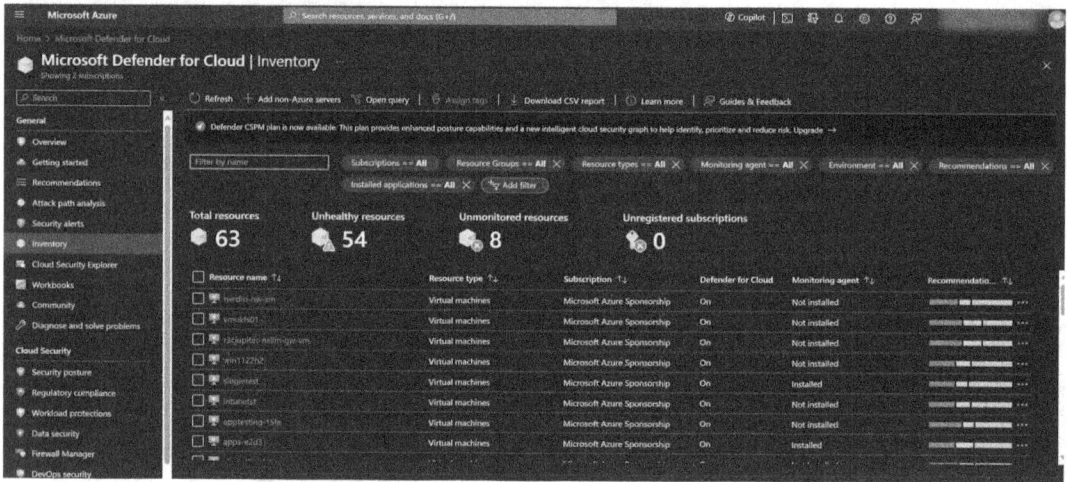

Figure 11.19 – The Microsoft Defender for Cloud Inventory page

One final part I wanted to cover before we move on to enabling Microsoft Defender for Cloud is the **Recommendations** page. This provides a centralized list of recommendations to improve your Azure security score and gauge your current state and future security score.

> **Tip**
> Did you know that the regulatory compliance feature within Microsoft Defender for Cloud is part of the free module?

The following screenshot shows the list of recommendations for Azure Security Center based on the current score and configuration of the Azure subscription:

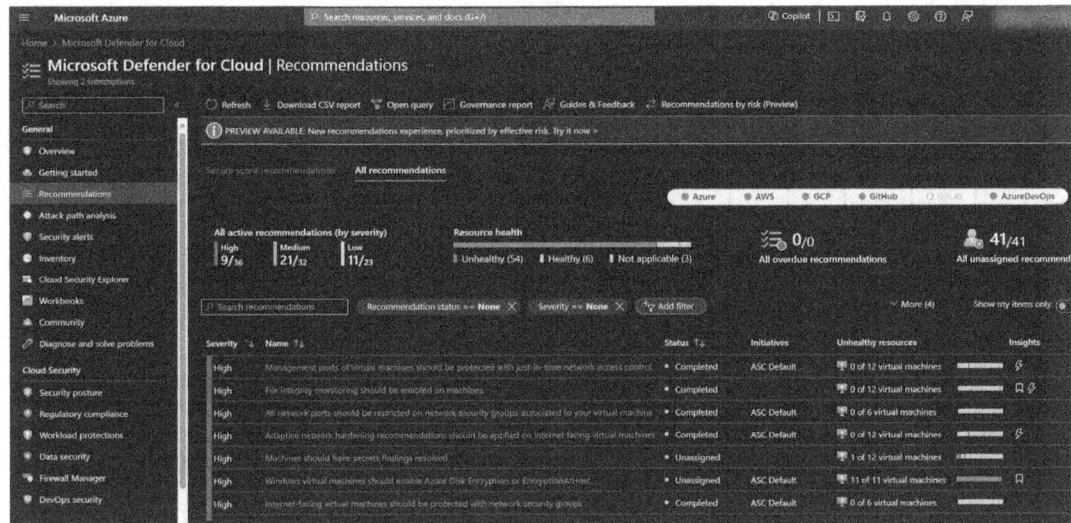

Figure 11.20 – The Recommendations page within Azure Security Center

In this section, we examined Microsoft Defender for Cloud and how it can improve your AVD's security posture. The next section examines enabling Azure Defender for AVD.

Enabling enhanced security for AVD

This section summarizes the basic steps for enabling enhanced security for Microsoft Defender for Cloud on your Azure subscription. This will allow you to use the more advanced features of the Security Center at a cost.

You can find pricing information here: `https://azure.microsoft.com/en-gb/pricing/details/defender-for-cloud/`.

> **Important note**
> You must enable enhanced security for Microsoft Defender for Cloud for each subscription you use.

The basic steps for enabling Azure Defender on your Azure subscription are as follows:

1. Navigate to the Security Center, which is located in the left-hand menu. Within the **Microsoft Defender for Cloud** menu, select **Environment settings**. The following screenshot shows the **Environment settings** menu option, which lists the subscriptions:

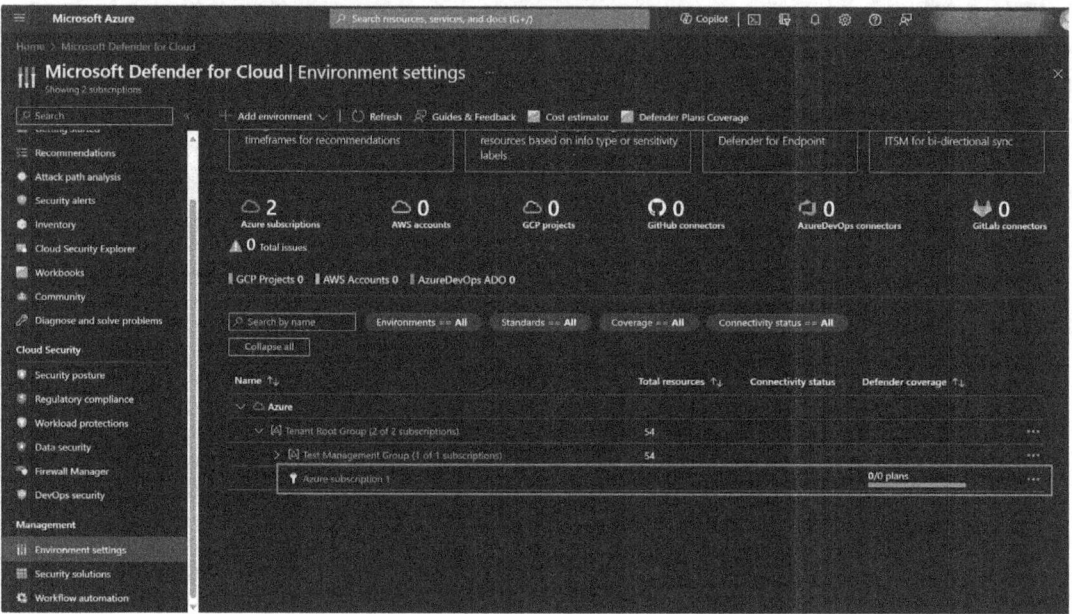

Figure 11.21 – The Environment settings page in Microsoft Defender for Cloud

2. Click on the required Azure subscription.
3. Select **Enable all Microsoft Defender for Cloud plans**.

Enabling enhanced security for AVD

> **Important note**
> It is important to note that if you select the **Enable all Microsoft Defender for Cloud plans** option, it will onboard all resources within the subscription. If you want to onboard only a subset, you will need to manually onboard the specific required resources.

The following screenshot shows the option to turn Microsoft Defender for Cloud on and off:

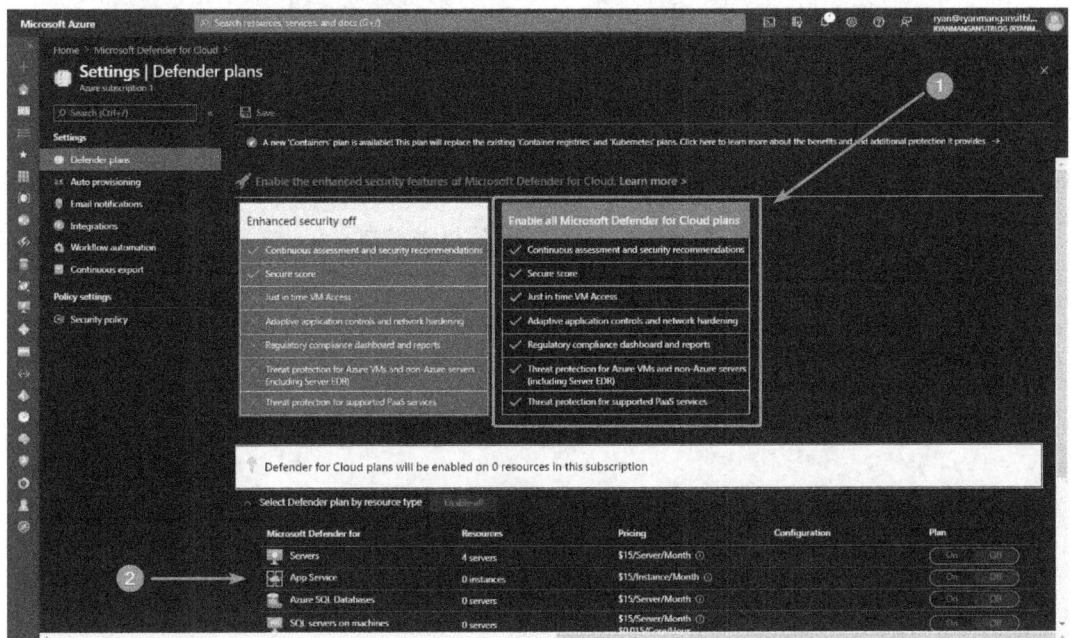

Figure 11.22 – Selecting Enable Microsoft Defender for Cloud plans on the Defender plans page

1. Select the resource types that you would like to enable on the **Defender plans** page. The following screenshot shows the different plans you can configure when you want to enable enhanced security:

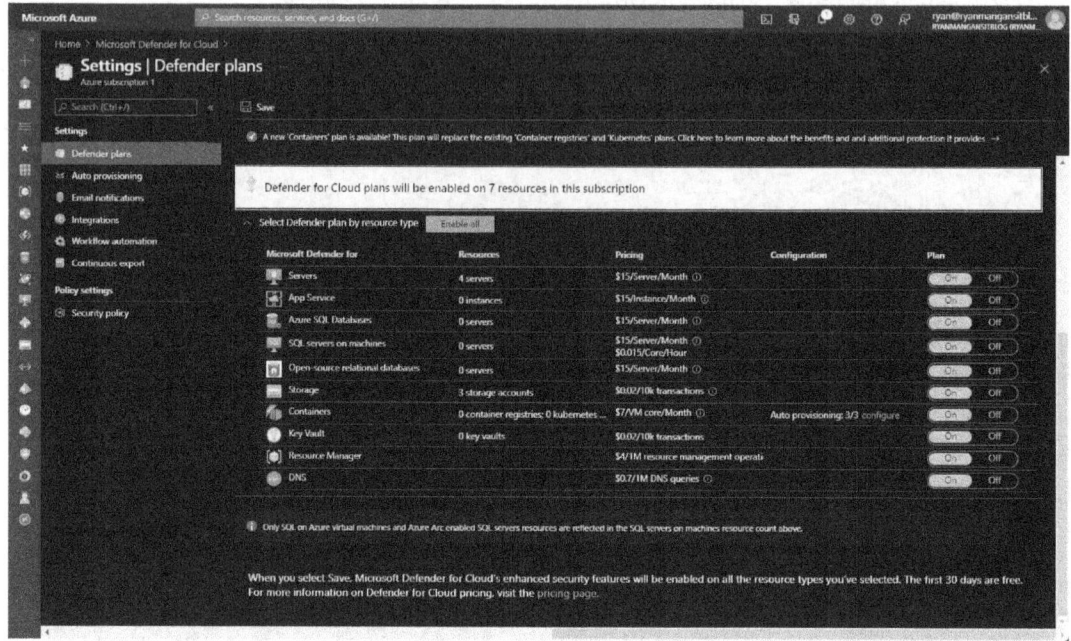

Figure 11.23 – Selected plans and saving the chosen plans

2. Once you have chosen the resource types (plans) you require, click **Save**.
3. Then, navigate back to the main Microsoft Defender for Cloud page and click **Workload protections** from the left-hand menu. The following screenshot shows the **Workload protections** page within Microsoft Defender for Cloud:

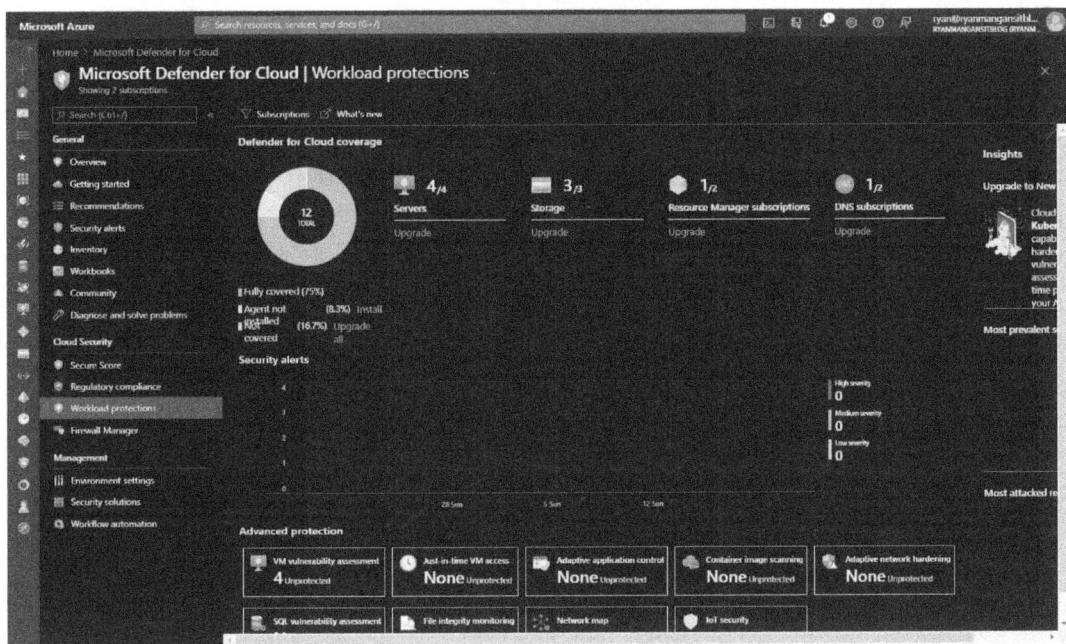

Figure 11.24 – The Workload protections page detailing the coverage graphically

As shown in the preceding screenshot, you can now review Azure Defender's coverage. With that, you've learned how to enable enhanced security within Microsoft Defender for Cloud for AVD.

The following link provides the planning and operations guide for Microsoft Defender for Cloud: https://learn.microsoft.com/azure/defender-for-cloud/defender-for-cloud-planning-and-operations-guide.

In the next section, we'll learn how to configure Microsoft Defender Antivirus for session hosts and useful configurations for ensuring that antimalware signatures are constantly updated.

Configuring Microsoft Defender Antivirus for session hosts

This section examines Microsoft Defender Antivirus for session hosts. Before we discuss scans and prevent notifications, I want to examine offloading security intelligence updates onto a host machine.

The benefit of doing this is that you can reduce the impact of CPU, disk, and memory resources on session hosts when security intelligence updates are processed. You can manage Microsoft Defender Antivirus using Group Policy; however, you can also use System Center Configuration Manager, Intune, and other third-party **mobile device management** (**MDM**) platforms.

See the following link from Microsoft on deploying Microsoft Defender Antivirus: `https://learn.microsoft.com/microsoft-365/security/defender-endpoint/deployment-vdi-microsoft-defender-antivirus?view=o365-worldwide`.

What's the difference between Microsoft Defender Antivirus and Microsoft Defender for Endpoint?

Microsoft Defender for Endpoint is an additional license you can purchase that essentially offers an extra layer of security for your endpoints. It is an enterprise endpoint security platform that provides other features for antivirus offerings, including advanced threat detection.

The following table details why you should consider both Microsoft Defender Antivirus and Microsoft Defender for Endpoint together:

#	Advantage	Why It Matters
1	Antivirus signal sharing	Microsoft applications and services share signals across your enterprise organization, providing a stronger single platform. See `https://www.microsoft.com/security/blog/2018/12/03/insights-from-the-mitre-attack-based-evaluation-of-windows-defender-atp/`.
2	Threat analytics and your score for devices	Microsoft Defender Antivirus collects underlying system data used by threat analytics and Microsoft Secure Score for Devices. This provides your organization's security team with more meaningful information, such as recommendations and opportunities to improve your organization's security posture.
3	Performance	Microsoft Defender for Endpoint is designed to work with Microsoft Defender Antivirus, so you get better performance when you use these offerings together.
4	Details about blocked malware	More details and actions for blocked malware are available with Microsoft Defender Antivirus and Microsoft Defender for Endpoint.
5	Network protection	Your organization's security team can protect your network by blocking specific URLs and IP addresses.
6	File blocking	Your organization's security team can block specific files.
7	Attack surface reduction	Your organization's security team can reduce your vulnerabilities (attack surfaces), giving attackers fewer ways to perform attacks. Attack surface reduction uses cloud protection for several rules.
8	Auditing events	Auditing event signals are available in Endpoint Detection and Response capabilities. (These signals are not available with non-Microsoft antivirus solutions.)

#	Advantage	Why It Matters
9	Geographic data	Compliant with ISO 270001 and data retention, geographic data is provided according to your organization's selected geographic sovereignty.
10	File recovery via OneDrive	If you are using Microsoft Defender Antivirus together with Office 365, and your device is attacked by ransomware, your files are protected and recoverable.
11	Technical support	By using Microsoft Defender for Endpoint together with Microsoft Defender Antivirus, you have one company to call for technical support.

Table 11.6 – Benefits of Using Microsoft Defender Antivirus and Microsoft Defender for Endpoint Together

The preceding table was adapted from the following link from Microsoft: `https://learn.microsoft.com/en-gb/microsoft-365/security/defender-endpoint/why-use-microsoft-defender-antivirus?view=o365-worldwide`.

Now, let's take a look at configuring some of the Microsoft Defender Antivirus features.

In this example, you will use Group Policy to enable the Microsoft shared security intelligence feature:

> **Important note**
> The shared security intelligence feature is used to offload the processing that's required by an endpoint in terms of unpackaging and installing security intelligence updates. Using a network or local path reduces the resource utilization of a client when security intelligence updates are applied.

1. On the management machine with Group Policy installed, open the Group Policy management console, right-click the Group Policy object you want to configure, and then click **Edit**.
2. In **Group Policy Management Editor**, navigate to **Computer configuration**.
3. Click the **Administrative** templates.
4. Expand the tree to **Windows components | Microsoft Defender Antivirus | Security Intelligence Updates**.

5. The following screenshot shows the **Security Intelligence Updates | Define security intelligence location for VDI clients** policy's location:

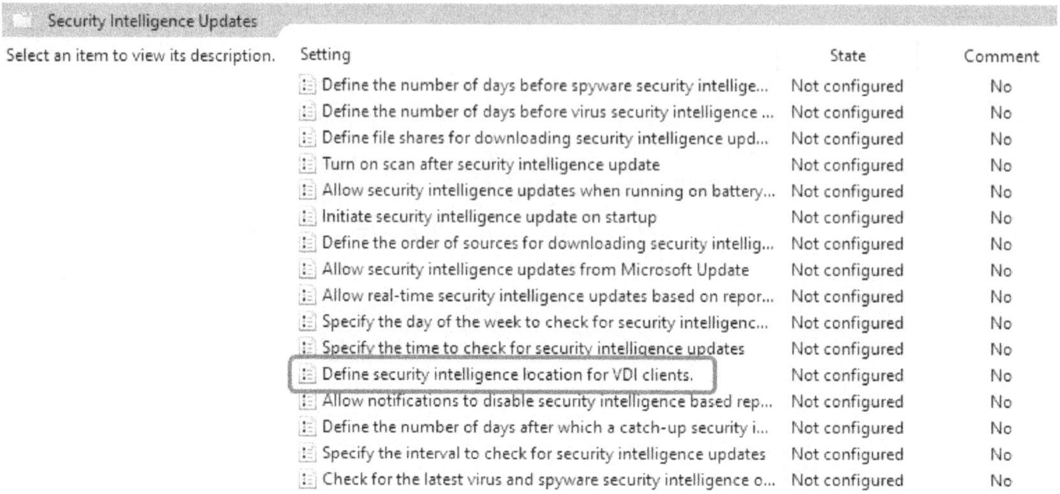
Security Intelligence Updates
Select an item to view its description.
Setting
State
Comment
Define the number of days before spyware security intellige... Not configured No
Define the number of days before virus security intelligence ... Not configured No
Define file shares for downloading security intelligence upd... Not configured No
Turn on scan after security intelligence update Not configured No
Allow security intelligence updates when running on battery... Not configured No
Initiate security intelligence update on startup Not configured No
Define the order of sources for downloading security intellig... Not configured No
Allow security intelligence updates from Microsoft Update Not configured No
Allow real-time security intelligence updates based on repor... Not configured No
Specify the day of the week to check for security intelligenc... Not configured No
Specify the time to check for security intelligence updates Not configured No
Define security intelligence location for VDI clients. Not configured No
Allow notifications to disable security intelligence based rep... Not configured No
Define the number of days after which a catch-up security i... Not configured No
Specify the interval to check for security intelligence updates Not configured No
Check for the latest virus and spyware security intelligence o... Not configured No

Figure 11.25 – The Define security intelligence location for VDI clients policy within the Security Intelligence Updates policy folder

6. Double-click **Define security intelligence location for VDI clients** and then set the option to **Enabled** within the form. A field should automatically appear:

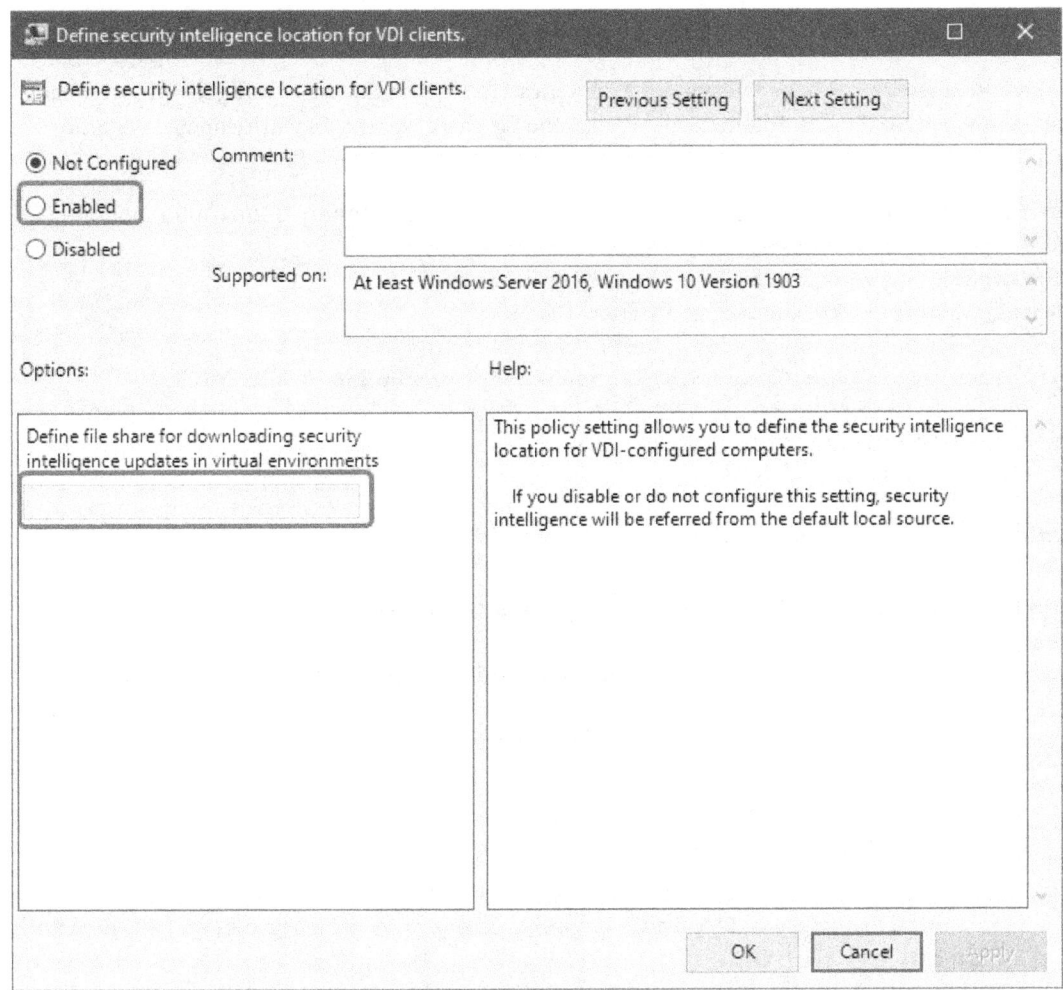

Figure 11.26 – Configuration form for the Define security intelligence location for VDI clients policy

7. Enter \\<fileshare\>\av-update (we will look at how to download these updates shortly).
8. Click **OK** to close the form with the new configuration.
9. Deploy the Group Policy object to the virtual machines you want to test.

> **Tip**
> You can also use PowerShell to enable the feature using the `Set-MpPreference -SharedSignaturesPath \\<fileshare>\av-update` cmdlet. You can deliver these on each machine or using Custom Script Extension.

Getting the latest updates

To download and unpack the latest security updates, it's advised that you configure a PowerShell script as a scheduled task to automatically update the file share with new security update definitions when they are released:

```
$vdmspathbase = "$env:systemdrive\av-update\{00000000-0000-0000-0000-"
$vdmspathtime = Get-Date -format "yMMddHHmmss"
$vdmspath = $vdmspathbase + $vdmspathtime + '}'
$vdmspackage = $vdmspath + '\mpam-fe.exe'

New-Item -ItemType Directory -Force -Path $vdmspath | Out-Null

Invoke-WebRequest -Uri 'https://go.microsoft.com/
fwlink/?LinkID=121721&arch=x64' -OutFile $vdmspackage

cmd /c "cd $vdmpath & c: & mpam-fe.exe /x"
```

Setting the scheduled task to run the PowerShell script

The following steps will guide you through setting a scheduled task to run the PowerShell script:

1. On your chosen management machine, open the *Start* menu and type `Task Scheduler`. Open and select **Create task...** on the side panel.

2. Enter the name as `Security intelligence unpacker` or another of your choosing. Go to the **Trigger** tab. Select **New...** | **Daily** and click **OK**.

3. Go to the **Actions** tab. Select **New....** Enter **PowerShell** in the **Program/Script** field. Enter `-ExecutionPolicy Bypass c:\av-update\vdmdlunpack.ps1` in the **Add arguments** field. Click **OK**.

4. You can also choose to configure additional settings if you require.

5. Click **OK** to save the configured scheduled task.

You can start the update manually by right-clicking on the task and clicking **Run**.

Manually downloading and unpacking

If you would prefer to configure manually, follow these steps to replicate the script's behavior:

1. Create a new folder on the machine root called `av_update` to store intelligence updates; for example, create the `c:\av_update` folder.

2. Create a subfolder under `av_update` with a GUID name, such as `{00000000-0000-0000-0000-000000000000}`.

Here's an example: `c:\av_update\{00000000-0000-0000-0000-000000000000}`.

> **Important note**
> In the *Getting the latest updates* section, you will note that the script includes the date, month, and year within the GUID so that a new folder is created for each update. This can be changed so that files are downloaded to the same folder each time.

1. Download a security intelligence package from `https://www.microsoft.com/wdsi/definitions` into the GUID folder. The file should be named `mpam-fe.exe`.
2. Open a Command Prompt window and navigate to the GUID folder you created previously. Use the `/X` extraction command to extract the files – for example, `mpam-fe.exe /X`.

> **Tip**
> The session host virtual machines will pick up the updated package when a new GUID folder is created with an updated package or whenever the existing folder is updated with new packages.

The following section looks at configuring quick scans for AVD session hosts.

Configuring quick scans

This section quickly looks at configuring Group Policy for specifying the scan type. In this example, we will be configuring a quick scan:

1. In **Group Policy Editor**, go to **Administrative templates | Windows components | Microsoft Defender Antivirus | Scan**:

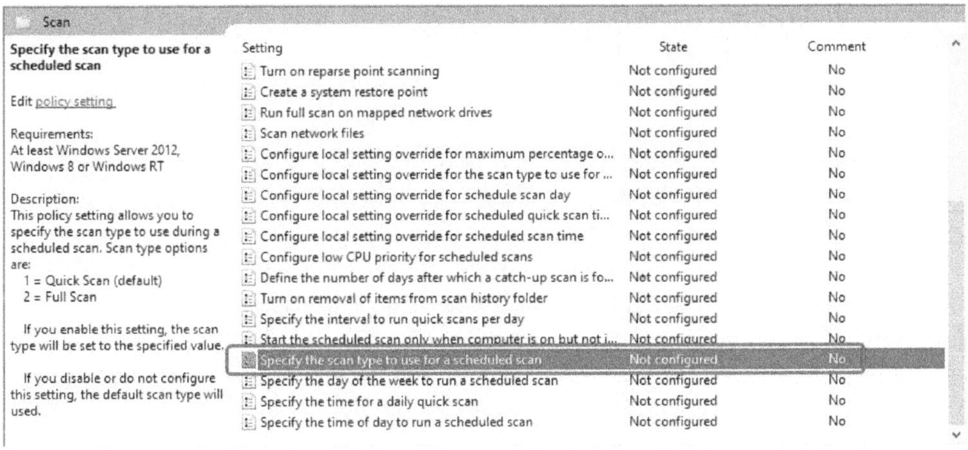

Figure 11.27 – The Specify the scan type to use for a scheduled scan policy in the Scan policy folder

2. Select **Specify the scan type to use for a scheduled scan** and then edit the policy setting. The following screenshot shows the **Specify the scan type to use for a scheduled scan** form set to **Enabled** and the scan type set to **Quick scan**:

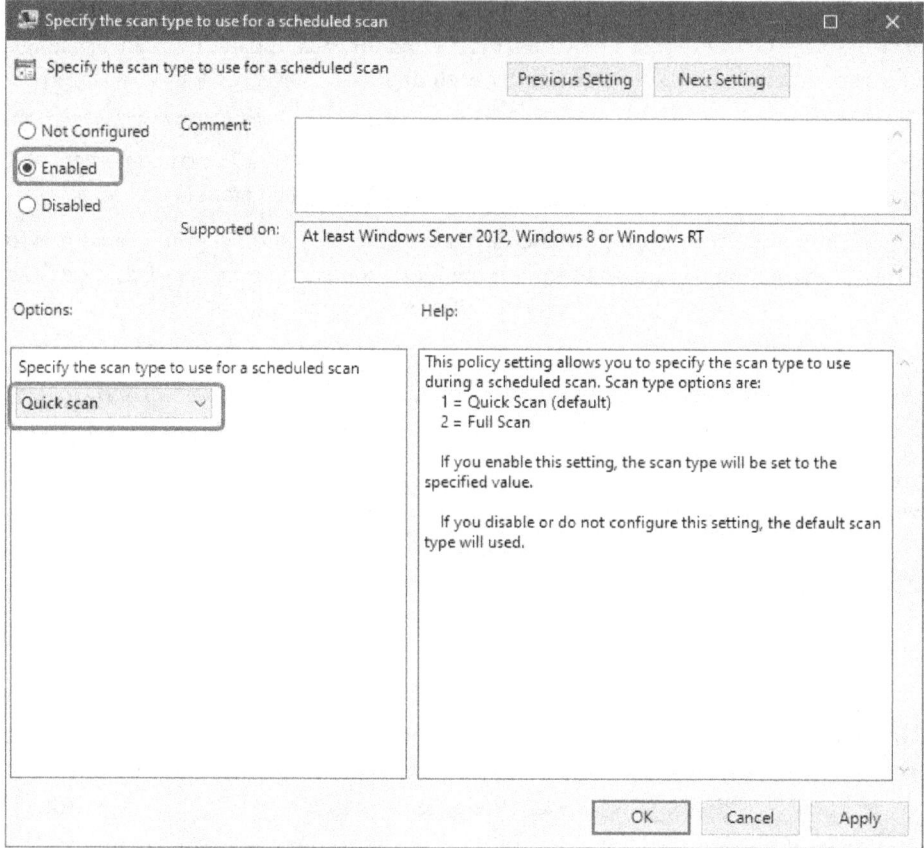

Figure 11.28 – Policy form for configuring the scan type

3. Set the policy to **Enabled**. Then, under **Options**, select **Quick scan**.
4. Click **OK**.
5. Deploy your Group Policy object.

This section showed you how to configure a quick scan for AVD session hosts. Next, we'll learn how to suppress notifications for Microsoft Defender Antivirus.

Suppressing notifications

This section explains how to suppress Microsoft Defender Antivirus notifications. Follow these steps to configure this:

1. In **Group Policy Editor**, go to **Windows components | Microsoft Defender Antivirus | Client Interface**. The following screenshot shows the **Microsoft Defender Antivirus | Client Interface | Suppress all notifications** policy:

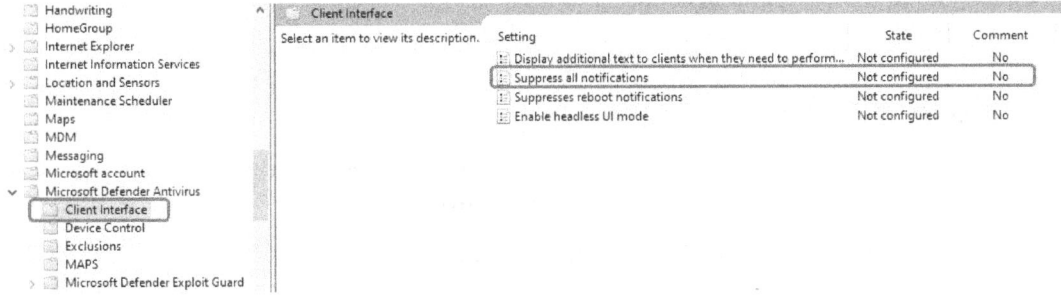

Figure 11.29 – Suppress all notifications in Group Policy

2. Select **Suppress all notifications** and then edit the policy settings. The following screenshot shows the **Suppress all notifications** policy set to **Enabled**:

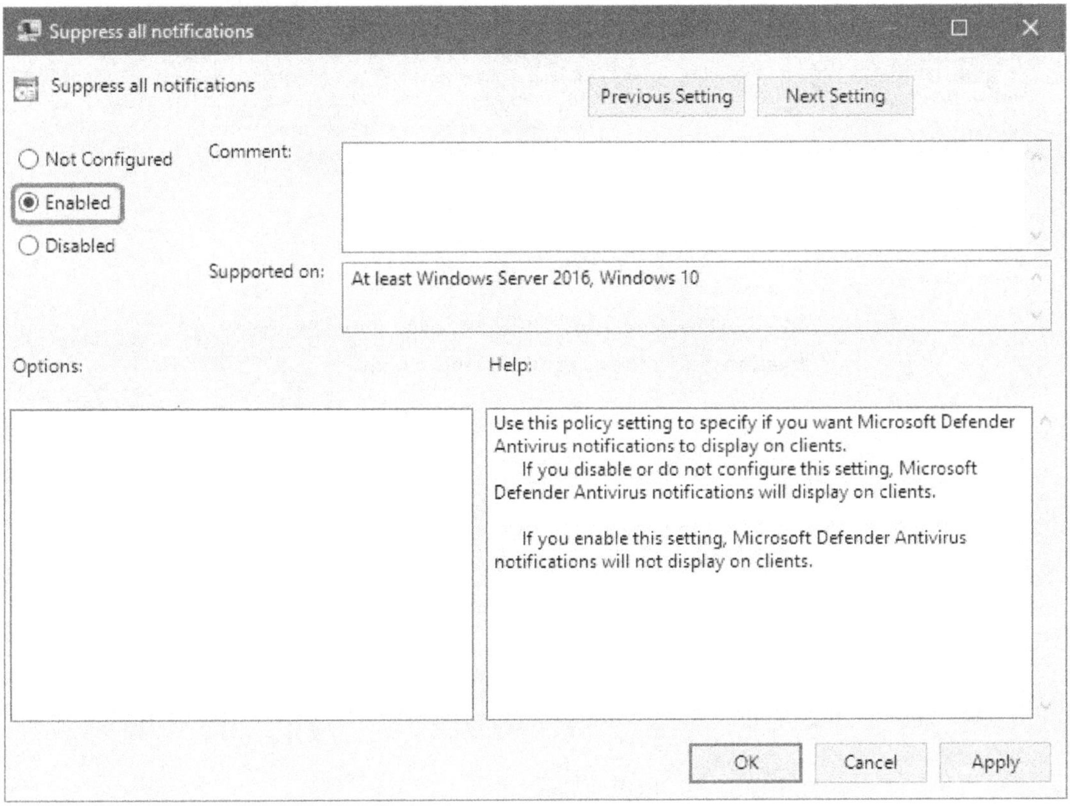

Figure 11.30 – The Suppress all notifications policy form

3. Set the policy to **Enabled**, then click **OK**.
4. Deploy your Group Policy object.

This section looked at suppressing all notifications for Microsoft Defender Antivirus. Next, we'll look at enabling headless UI mode, which essentially hides the UI from the user.

Enabling headless UI mode

Headless UI mode is a great feature for AVD as it hides the UI from the end user. This means that the IT admin is in full control and schedules scans when required.

The following steps detail how to configure headless UI mode:

1. In **Group Policy Editor**, navigate to **Windows components | Microsoft Defender Antivirus | Client Interface**.
2. Select **Enable headless UI mode** and edit the policy. The following screenshot shows the **Microsoft Defender Antivirus | Client Interface | Enable headless UI mode** policy:

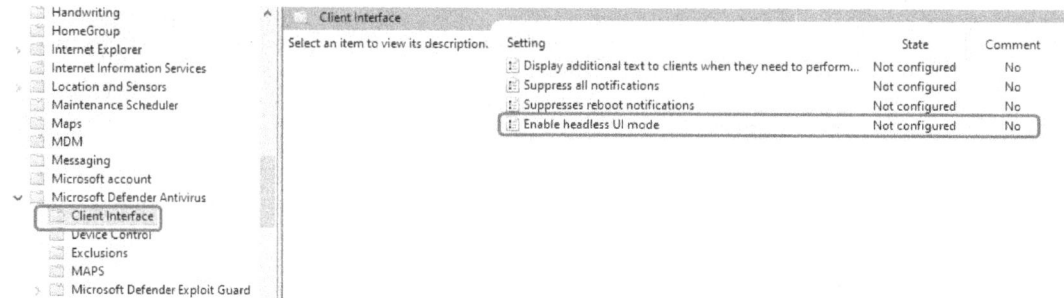

Figure 11.31 – The Enable headless UI mode policy within the Client Interface folder for Microsoft Defender Antivirus

3. Set the policy to **Enabled**. The following screenshot shows headless UI mode enabled:

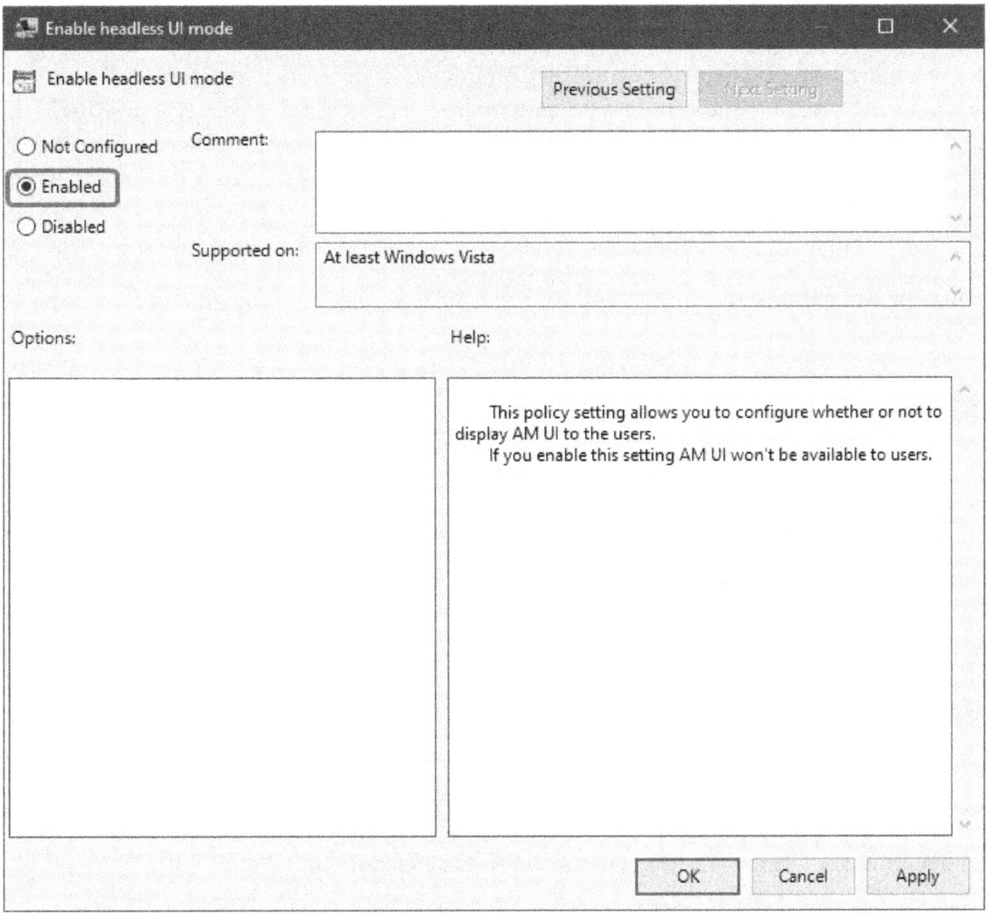

Figure 11.32 – The Enable headless UI mode policy form

4. Click **OK**.
5. Deploy the Group Policy object.

In this section, we looked at enabling headless UI mode to hide Microsoft Defender Antivirus from the users' view.

Summary

This chapter provided insights into Microsoft Defender for Cloud with a focus on AVD. We started by looking at enabling MFA and then configuring a Conditional Access policy to enforce MFA on AVD. Then, we looked at security responsibilities for both Microsoft and the customers. After, we dived into Microsoft Defender for Cloud, the value it offers Azure customers, and how you can use it to improve your AVD security posture as well as wider Azure resources running within your subscription(s). To finish off this chapter, we looked at Microsoft Defender Antivirus at a high level, focusing on some of the features you may want to configure for AVD.

In the next chapter, we'll change topics and look at implementing and managing FSLogix profile containers in AVD.

Questions

Answer the following questions to test your knowledge of this chapter:

1. What is the difference between Microsoft Defender Antivirus and Microsoft Defender for Endpoint?

   Microsoft Defender Antivirus comes native with the operating system. Microsoft Defender for Endpoint is an additional service for which you require a license.

2. What are the three core principles that are required for setting up a Conditional Access policy?

   Signals, decisions, and enforcements.

3. What does the security defaults policy do in regards to Azure MFA?

   Applies a default set of preconfigured security settings.

4. What are the two options when configuring Azure Defender for Cloud?

   Enhanced security off and enhanced security on.

Part 5: Managing User Environments and Apps

The section of the book looks at the management and configuration of user environments and apps for Azure Virtual Desktop.

This part of the book comprises the following chapters:

- *Chapter 12, Implementing and Managing FSLogix*
- *Chapter 13, Configuring User Experience Settings*
- *Chapter 14, MSIX App Attach*
- *Chapter 15, Configuring Apps on a Session Host*

12
Implementing and Managing FSLogix

This chapter examines FSLogix and its associated benefits in more depth. In *Chapter 5, Implementing and Managing Storage for Azure Virtual Desktop*, we covered some of the planning requirements for FSLogix and storage, and now we'll move on to its implementation and management.

FSLogix profile containers enable you to roam user data between computing session hosts, removing the user's dependency on a specific device. This reduces the user's sign-in time, as they don't have to create a new profile with each logon, and provides the flexibility of connecting to different business desktops/remote apps with the same user profile experience.

To summarize, FSLogix uses both a filesystem driver and a registry filter driver. The latter handles any filesystem or registry requests and allows user profiles to be redirected.

Here are the topics that will be covered in this chapter:

- Installing and configuring FSLogix
- Configuring antivirus exclusions
- Configuring profile containers
- Configuring Cloud Cache
- Microsoft Teams integration
- FSLogix profile container best practices

Installing and configuring FSLogix

The following diagram shows how FSLogix works within the Windows operating system:

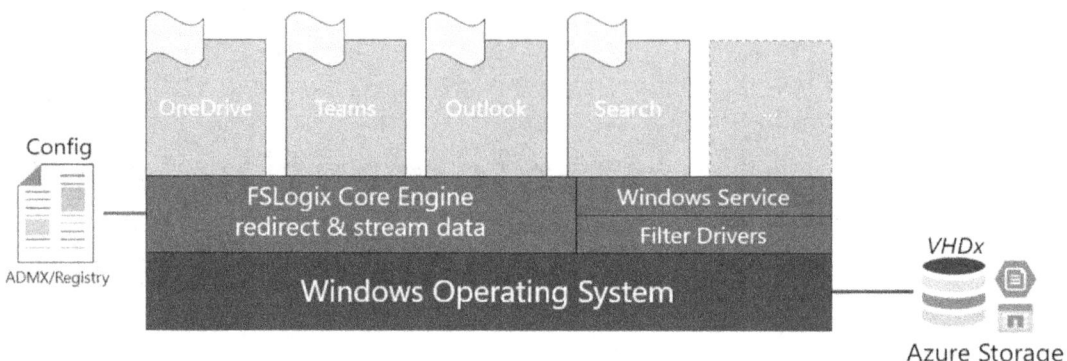

Figure 12.1 – Architecture diagram of FSLogix

The preceding architecture diagram shows how the FSLogix filter driver sits within an operating system and some services it redirects. The diagram depicts the redirection of user data using filter drivers to a remote storage solution.

Now, let's briefly examine the license requirements for FSLogix profile containers.

License requirements for FSLogix profile containers

To use FSLogix profile containers, you must have one of the following licenses:

- Microsoft 365 E3/E5
- Microsoft 365 A3/A5/Student Use Benefits
- Microsoft 365 F1/F3
- Microsoft 365 Business
- Windows 11 Enterprise E3/E5
- Windows 11 Education A3/A5
- Windows 11 VDA per User

- **Remote Desktop Services (RDS) Client Access License (CAL)**
- RDS **Subscriber Access License (SAL)**
- **Azure Virtual Desktop (AVD)** per-user access license

You will note that these license requirements are the same as those for accessing the AVD service.

Let's now take a quick look at the key capabilities of FSLogix profile containers.

FSLogix key capabilities

We now take a look at the key capabilities/benefits of using FSLogix profile containers:

- Redirect user profiles to a network storage location. Mounting the profile as a VHD(X) and using the profile over the network eliminates delays often associated with other solutions that copy profiles to and from the network location. In addition, this provides a seamless experience for the end user. When using FSLogix, the experience is similar to native local profiles where the operating system looks for the data in `C:\users`, which has been redirected using FSLogix.
- FSLogix also offers the ability to redirect only the required portion of the profile that contains Office data by using FSLogix Office Container. Office Container allows your organization, which is already using an alternate profile solution, to enable Office in a pooled virtual desktop environment.
- FSLogix Profile Container is used with Cloud Cache to create resilient and highly available user profile environments. Cloud Cache places a portion of the profile VHD(X) on the local hard drive and allows the administrator to specify multiple remote profile locations. With multiple remote profile containers, local cache insulates users from network and storage failures, making this an excellent solution for enterprise and **disaster recovery (DR)** scenarios.

Now that we have covered the basics of what FSLogix is, let's look at installing FSLogix components for profile containers.

FSLogix installation and configuration

In this section, I will provide instructions on downloading and installing FSLogix.

Before we start, I want to set out the steps for deploying FSLogix profile containers at a high level. The following diagram shows the steps for configuring FSLogix profile containers:

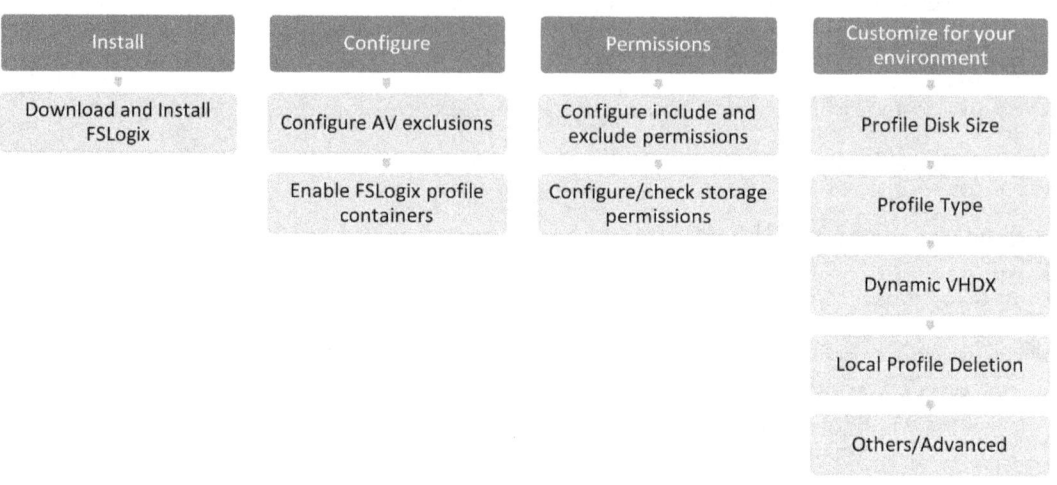

Figure 12.2 – Process diagram detailing the high-level steps for deploying FSLogix

Now we will look at the four areas of deployment set out in the preceding diagram. You may refer back to this to ensure you have configured all the required steps.

Getting started

> **Important note**
> To obtain the latest copy of FSLogix, you can download it from the following link: https://aka.ms/fslogix/download

To configure both profile containers and Office containers, you must download and install the FSLogix apps on the session host. FSLogix apps install all the required core drivers and components for FSLogix to work.

> **Tip**
> When using the Azure Marketplace templates, FSLogix will come preinstalled on the Windows 11 Multi-Session + Microsoft 365 app image template.

Once you have downloaded and unzipped the files, follow these steps:

1. Navigate to the `FSLogix_Apps_2.9.8784.63912\x64\Release` directory and run the `FSLogixAppsSetup.exe` installer.
2. Click **Install**, as shown on the installer:

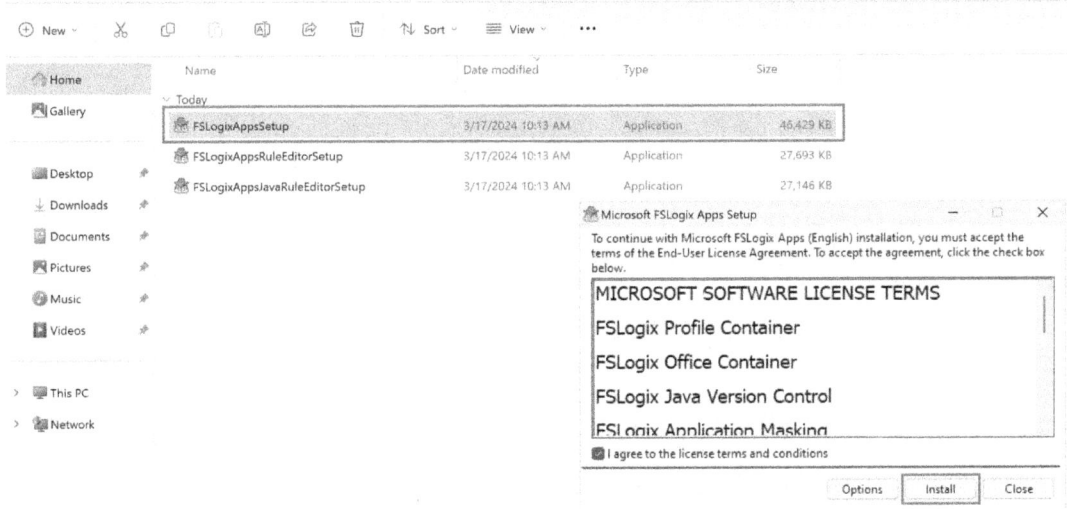

Figure 12.3 – Screenshot showing the installation of FSLogixAppsSetup

3. Once installed, you should see a **Setup Successful** message from the installer:

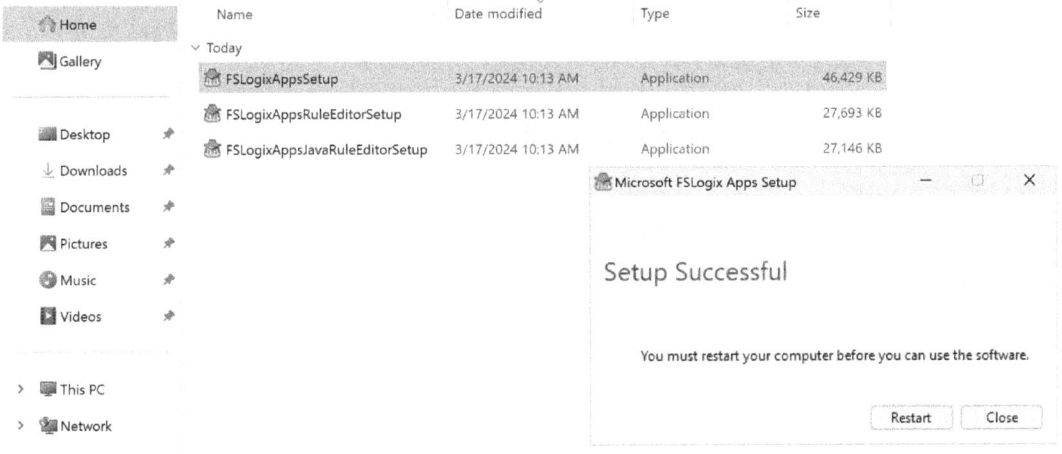

Figure 12.4 – Screenshot showing the FSLogix app setup complete

4. You can also check to ensure that the files are present as FSLogix is installed on the `C:\Program Files\FSLogix\Apps` path:

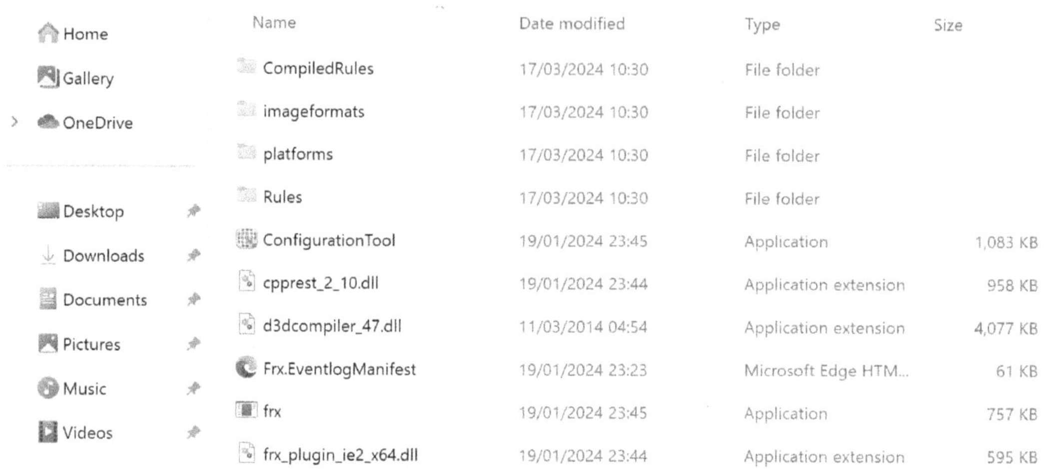

Figure 12.5 – Screenshot showing the location of the installed FSLogix files

You should now see the FSLogix files within the program files directory.

> **Tip**
> To install silently, use the following switches: `"C:\temp\Fslogix\Win32\Release\FSLogixAppsSetup.exe" /install /quiet /norestart`

We now move on to configuring the required antivirus exclusions for FSLogix profile containers.

Configuring antivirus exclusions

This section looks at the required antivirus file and process exclusions as necessary for FSLogix. Even if you are using Microsoft Defender Antivirus, it is recommended that you configure the recommended exclusions.

The following list shows the files, folders, and Cloud Cache file exclusions you should configure in your AVD environment:

- `%TEMP%\*\*.VHD`
- `%TEMP%\*\*.VHDX`
- `%Windir%\TEMP\*\*.VHD`
- `%Windir%\TEMP\*\*.VHDX`

- `\\server-name\share-name\*\*.VHD`
- `\\server-name\share-name\*\*.VHD.lock`
- `\\server-name\share-name\*\*.VHD.meta`
- `\\server-name\share-name\*\*.VHD.metadata`
- `\\server-name\share-name\*\*.VHDX`
- `\\server-name\share-name\*\*.VHDX.lock`
- `\\server-name\share-name\*\*.VHDX.meta`
- `\\server-name\share-name\*\*.VHDX.metadata`

Cloud Cache-specific exclusions include the following:

- `%ProgramData%\FSLogix\Cache\*` (folder and files)
- `%ProgramData%\FSLogix\Proxy\*` (folder and files)

This list was taken from https://learn.microsoft.com/fslogix/overview-prerequisites#configure-antivirus-file-and-folder-exclusions, so please check for any updates.

Configuring exclusions using PowerShell

This script is designed to set up your template image by incorporating necessary exclusions for FSLogix profile containers. It is specifically tailored for Microsoft Defender. The script's primary function is to include exclusions for the virtual disks of profile containers, encompassing both VHD and VHDX formats:

```
# A140 FSLOGIX AV Exc
#Defender Exclusions for FSLogix
  $Cloudcache = $false             # Ensure you Set for true if using cloud cache
  $StorageAcct = "
  storageacct"  # Enter a Storage Account Name

  $filelist = ` # Files to be excluded
  "%ProgramFiles%\FSLogix\Apps\frxdrv.sys", `
  "%ProgramFiles%\FSLogix\Apps\frxdrvvt.sys", `
  "%ProgramFiles%\FSLogix\Apps\frxccd.sys", `
  "%TEMP%\*.VHD", `
  "%TEMP%\*.VHDX", `
  "%Windir%\TEMP\*.VHD", `
  "%Windir%\TEMP\*.VHDX", `
```

```
    "\\$Storageacct.file.core.windows.net\share\*.VHD", `
    "\\$Storageacct.file.core.windows.net\share\*.VHDX"

    $processlist = ` # processes to be excluded
    "%ProgramFiles%\FSLogix\Apps\frxccd.exe", `
    "%ProgramFiles%\FSLogix\Apps\frxccds.exe", `
    "%ProgramFiles%\FSLogix\Apps\frxsvc.exe"

    Foreach($item in $filelist){
        Add-MpPreference -ExclusionPath $item}
    Foreach($item in $processlist){
        Add-MpPreference -ExclusionProcess $item}

    If ($Cloudcache){
        Add-MpPreference -ExclusionPath "%ProgramData%\FSLogix\Cache\*.VHD"
        Add-MpPreference -ExclusionPath "%ProgramData%\FSLogix\Cache\*.VHDX"
        Add-MpPreference -ExclusionPath "%ProgramData%\FSLogix\Proxy\*.VHD"
        Add-MpPreference -ExclusionPath "%ProgramData%\FSLogix\Proxy\*.VHDX"}
```

Now that we have installed FSLogix on the session host/session host template, we can proceed with configuring profile containers.

Configuring profile containers

In this section, we run through the configuration of FSLogix profile containers. The configuration of profile containers to redirect user profiles is relatively straightforward as this consists of several registry settings you can add to the master image template or rollout using Group Policy.

The two areas we will look at are the configuration of storage location and the setup of the `include` and `exclude` user groups.

> **Important note**
> Ensure that you exclude your VHD(X) files from any antivirus software.

Before you can configure Group Policy, you need to ensure that the FSLogix ADM/ADMX files are copied into the correct policy definition folder or central store. These files can be found in the downloaded ZIP file I showed in the previous section:

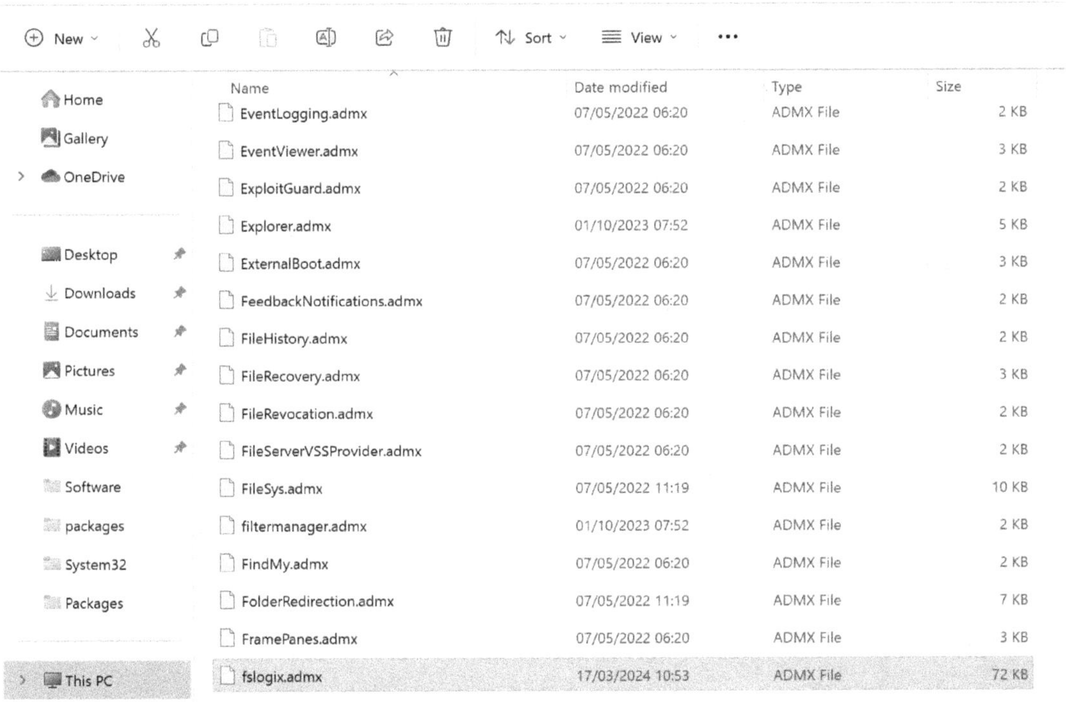

Configuring antivirus exclusions

Name	Date modified	Type	Size
EventLogging.admx	07/05/2022 06:20	ADMX File	2 KB
EventViewer.admx	07/05/2022 06:20	ADMX File	3 KB
ExploitGuard.admx	07/05/2022 06:20	ADMX File	2 KB
Explorer.admx	01/10/2023 07:52	ADMX File	5 KB
ExternalBoot.admx	07/05/2022 06:20	ADMX File	3 KB
FeedbackNotifications.admx	07/05/2022 06:20	ADMX File	2 KB
FileHistory.admx	07/05/2022 06:20	ADMX File	2 KB
FileRecovery.admx	07/05/2022 06:20	ADMX File	3 KB
FileRevocation.admx	07/05/2022 06:20	ADMX File	2 KB
FileServerVSSProvider.admx	07/05/2022 06:20	ADMX File	2 KB
FileSys.admx	07/05/2022 11:19	ADMX File	10 KB
filtermanager.admx	01/10/2023 07:52	ADMX File	2 KB
FindMy.admx	07/05/2022 06:20	ADMX File	2 KB
FolderRedirection.admx	07/05/2022 11:19	ADMX File	7 KB
FramePanes.admx	07/05/2022 06:20	ADMX File	3 KB
fslogix.admx	17/03/2024 10:53	ADMX File	72 KB

Figure 12.6 – Screenshot showing the FSLogix ADMX file added to the PolicyDefinitions folder

Once FSLogix is added to the `PolicyDefinitions` folder/`Group Policy Central Store`, you will see the folder appear within `Computer Configuration | Administrative Templates | FSLogix`:

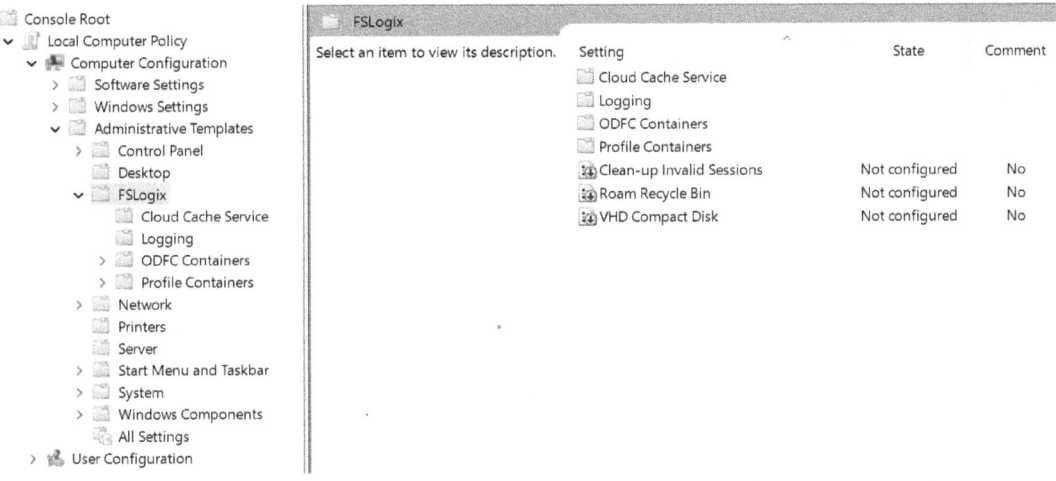

Figure 12.7 – Screenshot showing the FSLogix folder within Group Policy

Two required settings must be applied for FSLogix profile containers to work. These are (1) the setting to enable profile containers, and (2) specifying the `VHDLocations` storage location.

To configure these, perform the following steps:

1. Create a **Group Policy Object** (**GPO**) within Active Directory.
2. Navigate to the FSLogix policies within `Computer Configuration | Administrative Templates | FSLogix`.
3. Select the `Profile Containers` folder:

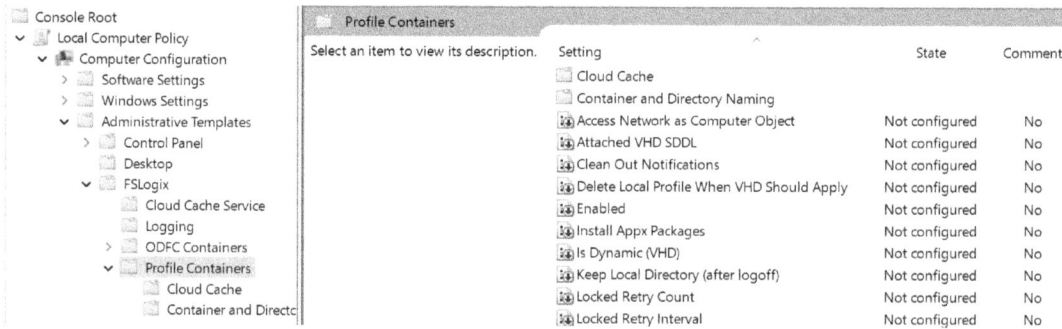

Figure 12.8 – Screenshot showing Profile Containers in Group Policy

4. Open the `Enabled` policy, set it to **Enabled**, and then click **OK**:

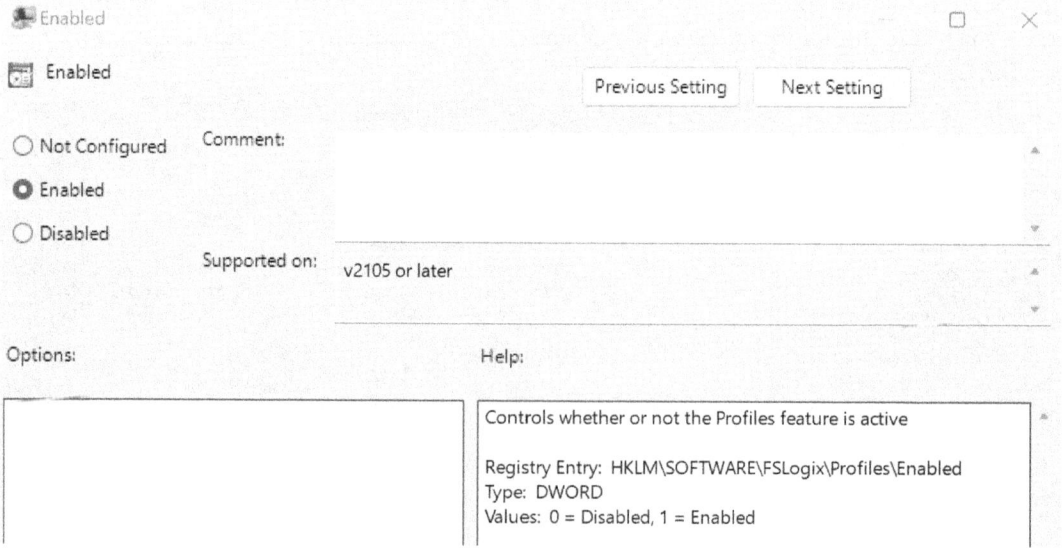

Figure 12.9 – Screenshot showing the FSLogix profile containers' Enabled policy

5. Then, open the VHD location policy and enter a file path to store your user's roaming profile containers. Then, click **OK**:

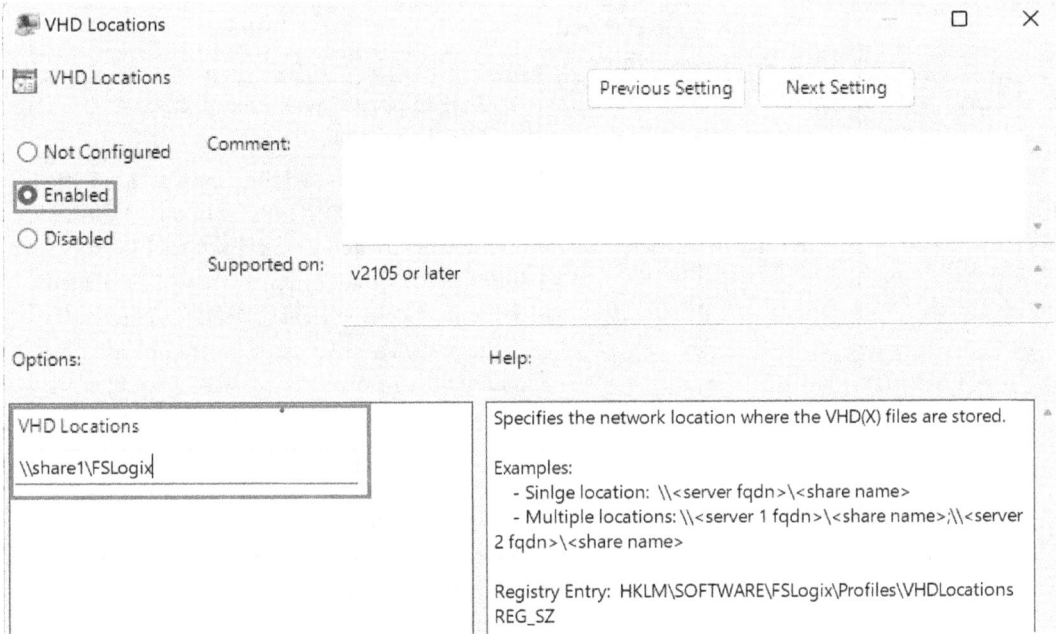

Figure 12.10 – Screenshot showing the VHD location path for storing FSLogix profiles

> **Tip**
> It is recommended that you use Azure files for smaller deployments and Azure Netapp Files for enterprise FSLogix deployments.

Let's now look at configuring the same thing using the registry rather than Group Policy.

Configuring using the registry

You can also configure FSLogix profile containers within the registry. For example, you would enable FSLogix profile containers and set the path configuration settings with the following registry path: HKLM\SOFTWARE\FSLogix\Profiles.

The following table lists two mandatory settings that must be configured for FSLogix profile containers to work:

Value	Type	Configured Value	Description
`Enabled` (Mandatory setting)	DWORD	1	`0`: Profile containers disabled. `1`: Profile containers enabled.
`VHDLocations` (Mandatory setting)	MULTI_SZ or REG_SZ		A list of filesystem locations to search for the user's profile VHD(X) file. If one isn't found, one will be created in the first listed location. If the VHD path doesn't exist, it will be created before it checks whether a VHD(X) file exists in the path. These values can contain variables that will be resolved. Supported variables are `%username%`, `%userdomain%`, `%sid%`, `%osmajor%`, `%osminor%`, `%osbuild%`, `%osservicepack%`, `%profileversion`, and any environment variable available at the time of use. When specified as a `REG_SZ` value, multiple locations can be separated by a semicolon.

Table 12.1 – Configuration settings for profile containers

The table is taken from the Microsoft documentation: `https://learn.microsoft.com/fslogix/tutorial-configure-profilecontainers#configure-profile-container-registry-settings`

The following screenshot shows you what the configuration looks like within the registry:

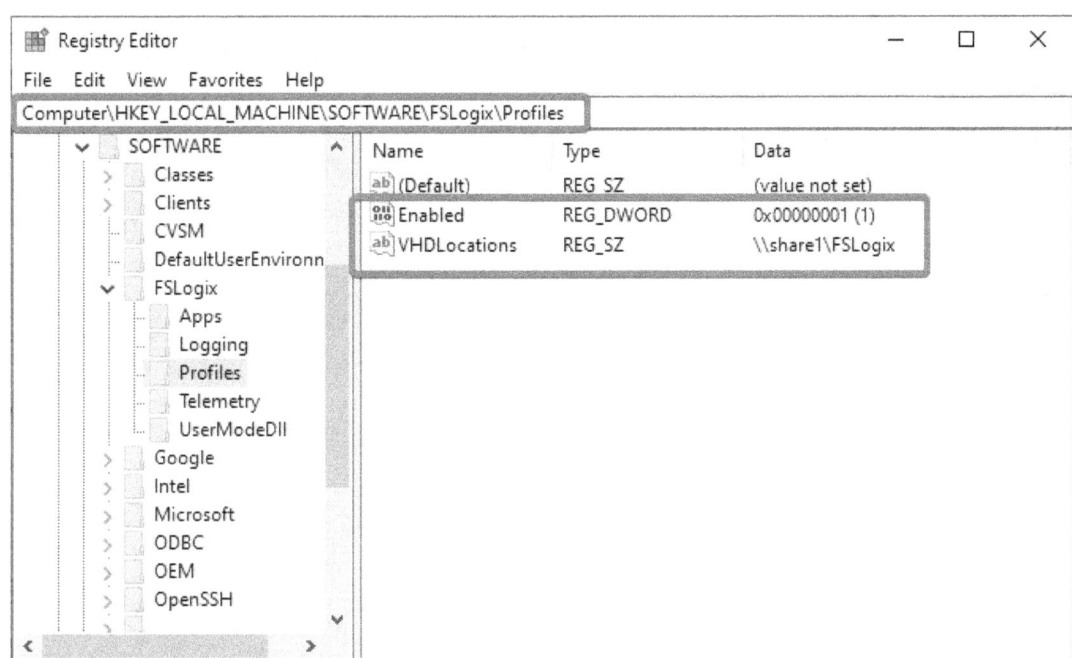

Figure 12.11 – Screenshot of registry settings for enabling and configuring the storage path for FSLogix

This configures the basics for getting started with FSLogix profile containers.

Configuring using Microsoft Intune

As more companies move to devices joined to Entra ID, we cannot use Group Policy to manage them. Microsoft has given the ability to manage the FSLogix configuration settings using Microsoft Intune.

To configure the Intune settings, we have to navigate to https://intune.microsoft.com. Next, go to **Devices | Manage Devices | Configuration**. We then need to create a new profile while choosing **Windows 10 and later** as the platform and selecting **Settings catalog** as the profile type:

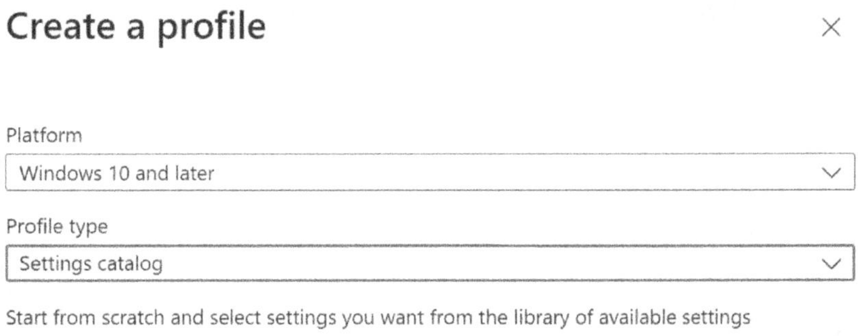

Figure 12.12 – Screenshot of Create a profile settings

In the **Settings picker** section, you will see an **FSLogix** category where we can configure the relevant settings that we need:

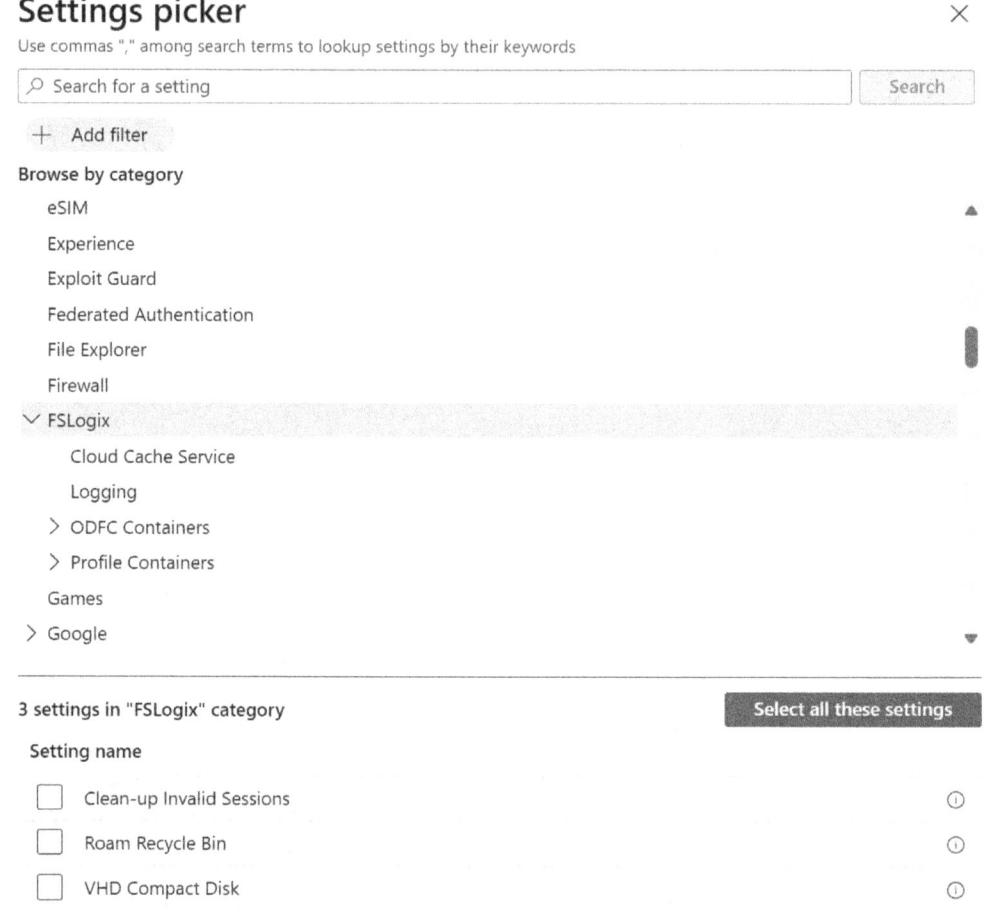

Figure 12.13 – Screenshot of Settings picker settings

Once the policy has been created, we can then assign the policy settings to the relevant Intune-enabled devices.

FSLogix exclude/include lists

We will now take a look at FSLogix profile exclude and include lists.

> **Important note**
> By default, the **Everyone** group is added to the **FSLogix Profile Include List** group.

When configuring profile containers, you need to consider the local administrators and other user accounts that should remain as local profiles. The way to do this would be to enter users or groups into the FSLogix profile exclude list.

You can do this by navigating to **Local Users and Groups** > **Select Groups**, and you will see the **FSLogix Profile Exclude List** group. Within this group, add users and groups you would like to exclude:

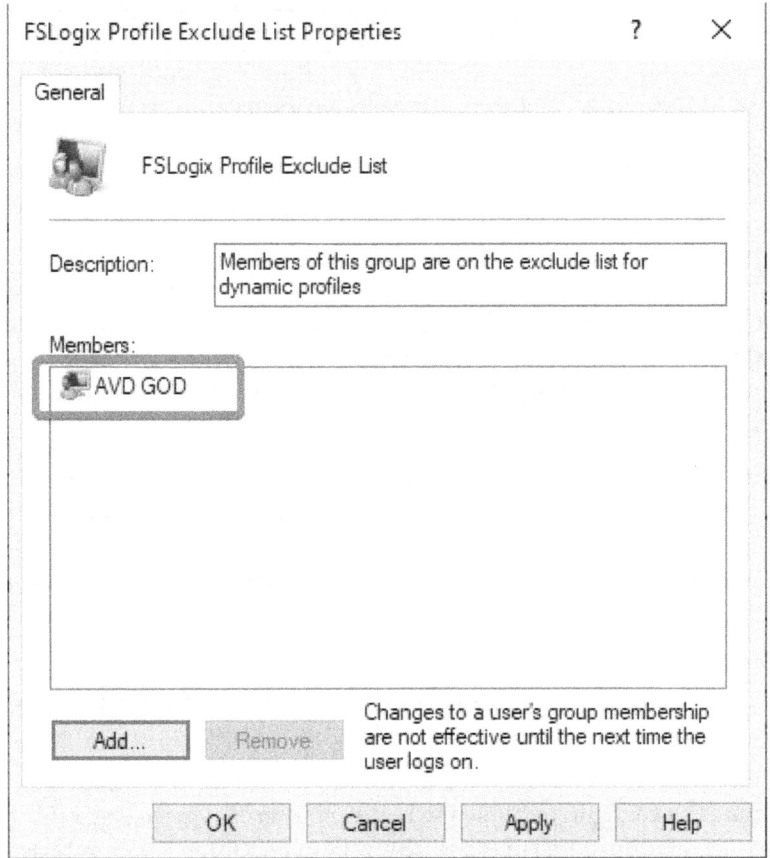

Figure 12.14 – Screenshot showing the exclude list for FSLogix profile containers

There may be cases where you want to remove everyone's configuration from the included group; however, this does not need to be changed in most deployments.

The following table details the permissions you should configure on the storage for FSLogix profile containers:

User Account	Folder	Permissions
Users	This Folder Only	Modify
Creator/Owner	Subfolders and Files Only	Modify
Administrator (optional)	This Folder, Subfolders, and Files	Full Control

Table 12.2 – User account permissions for folder access

In the next section, we will take a look at Cloud Cache and its benefits.

Cloud Cache

Cloud Cache is a technology (a part of FSLogix) that enables you to use multiple remote locations continually updated during the user session. One of the use cases for Cloud Cache is **disaster recovery** (**DR**) and **business continuity** (**BC**) as it provides real-time, active-active redundancy for a profile container.

If the main or storage provider becomes unavailable, the Cloud Cache will continue to function/operate with the other storage providers. If an unavailable storage provider becomes available within the user session, it will be updated with the local cache. If the provider does not become available until after the user has signed out of their session, then the storage provider will be updated during the next session.

> **Important note**
> When using Cloud Cache, all profile data writes go through the local cache first, and then the remote storage locations.

It is important to note that configuring Cloud Cache differs from configuring typical FSLogix profile containers. The settings are configured in the same way as in the Group Policy or registry; however, you need to ensure that you remove any settings in the `VHDLocations` policy as Cloud Cache uses `CCDLocations`.

You should also note that the value when configuring Cloud Cache is also different from profile containers. Instead of entering the **Universal Naming Convention** (**UNC**) path, you will need to use the value described next. Finally, you will note that I have detailed both **Server Message Block** (**SMB**) shares and the use of Azure Page Blob storage:

> **Important note**
> To use Azure Blob storage, you must generate an access key within the chosen Azure Storage account.

Cloud Cache

Cloud Cache Storage Configuration Types	
Type	**Value Example**
SMB share	`type=smb,connectionString=\\Location1\Folder1;type=smb,connectionString=\\Location2\folder2`
Azure Page Blobs	`type=azure,connectionString="DefaultEndpointsProtocol=https;AccountName=;AccountKey=;EndpointSuffix="`

Table 12.3 – Cloud Cache storage configuration types

We now take a closer look at configuring Cloud Cache and the associated steps.

Configuring Cloud Cache

The following diagram shows the steps for configuring Cloud Cache; you will note that the diagram covers all the bases for existing FSLogix profile container implementations and greenfield deployments:

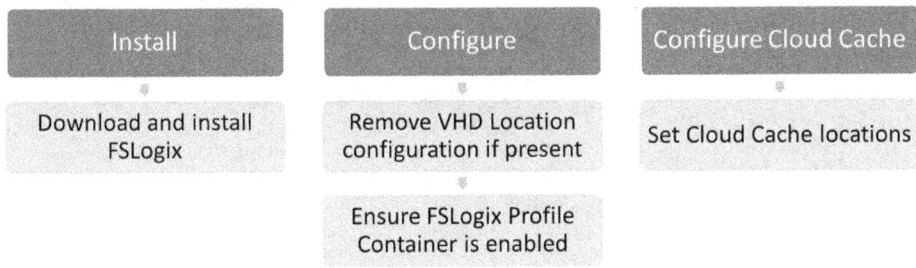

Figure 12.15 – Process diagram for FSLogix Cloud Cache

We will skip the installation step as this was shown in a previous section, *Installing and configuring FSLogix*.

To configure Cloud Cache, we need to enable and configure the GPO Cloud Cache locations:

1. Within `Group Policy`, navigate to `Computer Configuration > FSLogix > Profile Containers > Cloud Cache > CCD Locations`.

2. Set the policy to **Enabled** and enter the storage paths described in the policy description and the **Cloud Cache Storage Configuration types** table, as shown in the following screenshot:

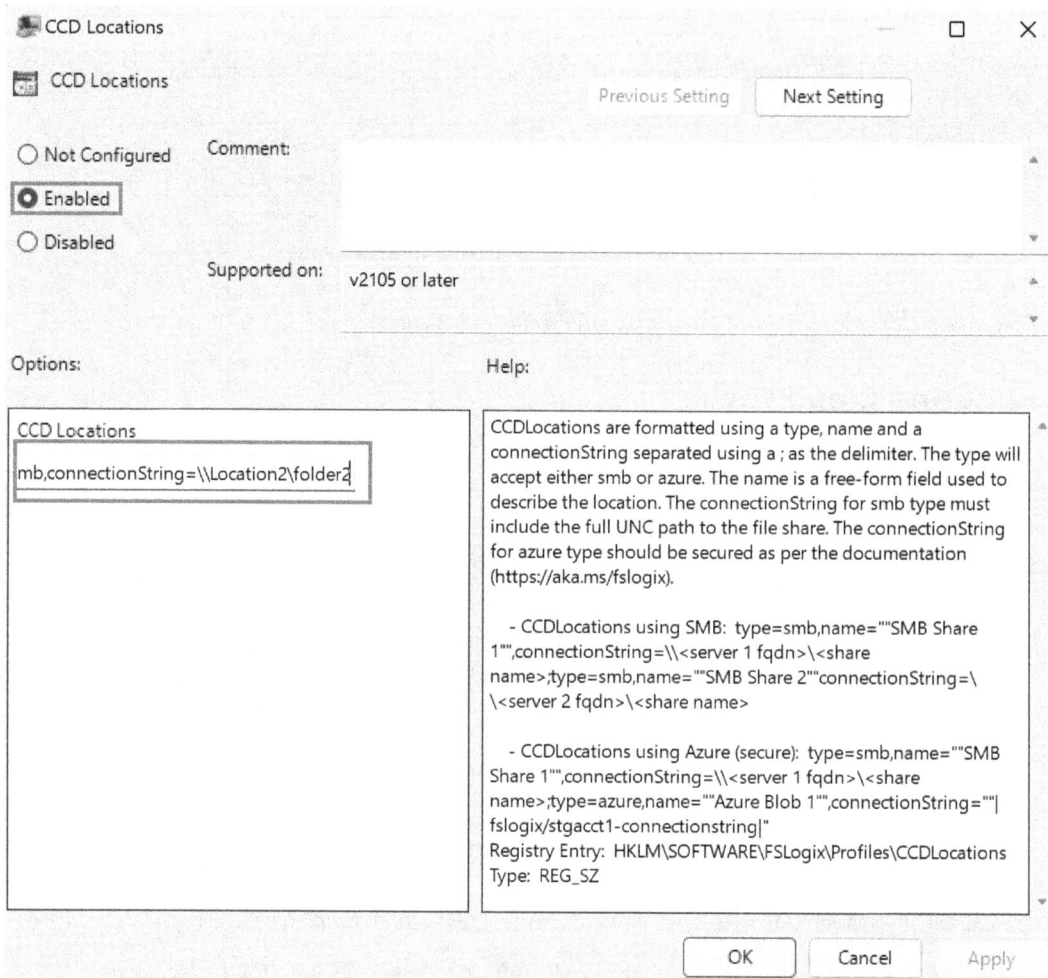

Figure 12.16 – Screenshot showing the Cloud Cache locations policy within Group Policy

3. Once you have set the policy to **Enabled** and entered the required Cloud Cache locations, click **OK**.

Cloud Cache

You can also configure the same thing using the registry on the image template by using the following:

Registry Value	Type	Value
CCDLocations	REG_SZ / MULTI_SZ	`type=smb,connectionString=<\Location1\Folder1>;type=smb,connectionString=<\Location2\folder2>`
Enabled	DWORD	1

Table 12.4 – Registry values for configuration

The preceding table was adapted from the following Microsoft page: https://learn.microsoft.com/en-gb/fslogix/tutorial-cloud-cache-containers#configuring-cloud-cache-for-office-container

As shown in the following screenshot, you can see that `CCDLocations` has been set with the storage provider locations and that FSLogix is enabled:

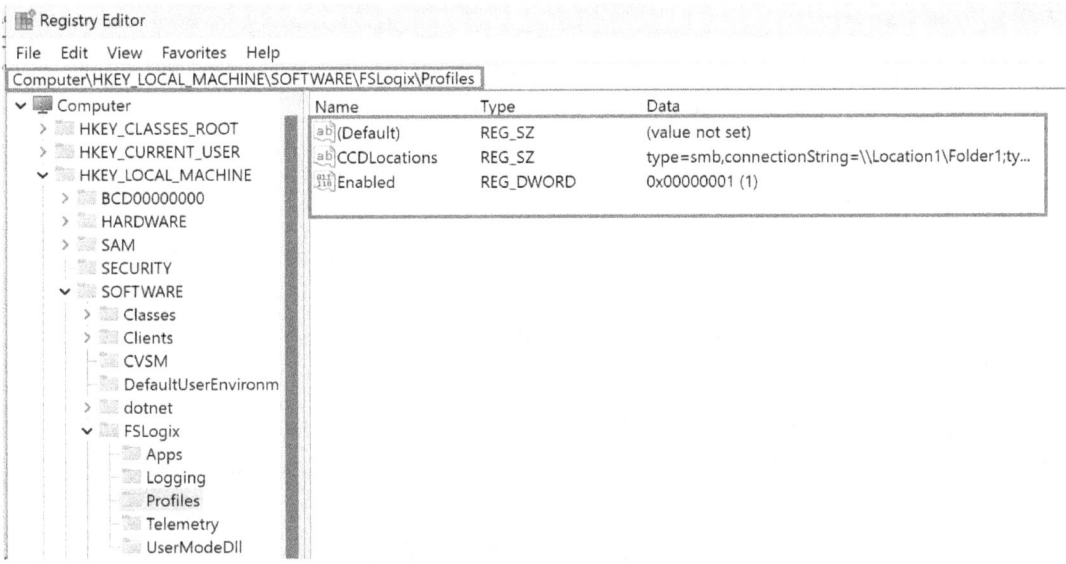

Figure 12.17 – Screenshot showing the registry way of configuring Cloud Cache

> **Note**
> There are several additional configurations for FSLogix Cloud Cache; however, they are not covered within this book.

In this section, we looked at how to configure Cloud Cache for user-profile roaming. In the next section, we will take a look at Microsoft Teams exclusions for FSLogix.

Microsoft Teams integration

In October 2023, Microsoft released a new version of Microsoft Teams, which included many improvements over the previous version. The new version of Teams brought many enhancements to the VDI version. For a full list of supported features, please visit https://learn.microsoft.com/azure/virtual-desktop/teams-supported-features.

To use the latest version of Teams with FSLogix, we must ensure that we are using **FSLogix 2210 Update 4** as the minimum version, which is 2.9.8884.27471.

The reason for this change is that the new version of Teams stores the user/data cache in the %LocalAppData%\Packages\MSTeams_8wekyb3d8bbwe\LocalCache folder, which was not previously persisted in the FSLogix profile prior to FSLogix Hotfix 3.

Teams exclusions

The following folders are recommended to be excluded from the user profile for the newer version of Teams to save disk space inside the user FSLogix profile:

- AppData\Local\Packages\MSTeams_8wekyb3d8bbwe\LocalCache\Microsoft\MSTeams\Logs
- AppData\Local\Packages\MSTeams_8wekyb3d8bbwe\LocalCache\Microsoft\MSTeams\PerfLogs
- AppData\Local\Packages\MSTeams_8wekyb3d8bbwe\LocalCache\Microsoft\MSTeams\EBWebView\WV2Profile_tfw\WebStorage

For a detailed description of what these folders are used for and for additional potential exclusions, please visit https://learn.microsoft.com/microsoftteams/new-teams-vdi-requirements-deploy.

For instructions on how to implement the redirections.xml file, which would exclude these folders, please visit https://learn.microsoft.com/fslogix/tutorial-redirections-xml.

FSLogix profile container best practices

This short section details the best practices for enterprises. There are a few settings you can apply to your FSLogix deployment to improve management and the cleanup of profiles.

The following table shows some of the best practice settings when configuring FSLogix profile containers within an enterprise:

Setting	Value	Reason
DeleteLocalProfileWhenVHDShouldApply	1	The benefits of this are to prevent errors when logging in and also clean up any local profiles that may appear following a storage issue.
SizeInMBs	30000	Set the size of the profile disk.
VolumeType	VHDx	VHD disks offer limited management capability; however, VHDX disks do offer this. It is recommended that you use VHDX when configuring FSLogix profile containers.
FlipFlopProfileDirectoryName	1	This setting is recommended to make it easier to search for the specific profile container user folder on the network share. When enabled, a SID folder is created as %username%%sid% instead of the default %sid%%username.

Table 12.5 – FSLogix configuration settings best practices

In this section, we briefly took a look at some of the best practices/recommended configurations you should apply when configuring FSLogix profile containers.

Summary

In this chapter, we ran through the implementation of FSLogix profile containers and Cloud Cache. We started with the basics of installing the software on a session host and the configuration of antivirus exclusions, and then we moved on to what is required to set up profile containers. We then discussed the need to remove the profile containers' VHDLocation path when configuring Cloud Cache and some of your other options. We then took a look at creating profile exclusions for Microsoft Teams and finished off the chapter by looking at a few enterprise best practices you can configure for FSLogix.

In the next chapter, we will take a look at configuring user experience settings. We will talk about universal printing, Group Policy configurations, and troubleshooting both user-profile issues and AVD clients.

Questions

Here are a few questions to test your understanding of this chapter:

1. What are the maximum concurrent handles you can have in Azure Files?

   2,000 handles.

2. When configuring the storage path for FSLogix Cloud Cache, what is the correct value for the storage location?

   CCDLocations.

3. What is the reason for configuring Microsoft Teams exclusions for FSLogix profile containers?

   Microsoft Teams has a caching folder that can cause bloat on your user profile disk.

4. What do you need to do before you can start configuring and using FSLogix profile containers?

   Ensure that you download and install the FSLogix apps. Remember that if you are using a Microsoft gallery image, this is usually installed as part of the image. Always check beforehand.

13
Configuring User Experience Settings

In this chapter, we will look at user experience. We will start by looking at the powerful capabilities of Universal Print before looking at Microsoft Intune. Then, we will look at Start VM on Connect, which is very useful for those who want to reduce the cost and control the startup of **virtual machines** (**VMs**). After that, we will cover screen capture protection and watermarking for protecting corporate data, FSLogix profile troubleshooting, and provide some useful information on Remote Desktop client connection issues.

The following topics will be covered in this chapter:

- Configuring Universal Print
- Configuring user settings using Microsoft Intune
- Start VM on Connect
- Enabling screen capture protection for AVD
- Enabling watermarking
- Troubleshooting FSLogix profile issues
- Troubleshooting AVD client issues

> **Important note**
> For more information on configuring persistent and non-persistent desktops and configuring **Remote Desktop Protocol** (**RDP**) properties for a host pool, please refer back to *Chapter 6, Creating Host Pools and Session Hosts*.

We will kick off this chapter by looking at Universal Print and the fundamentals of how the product works.

Configuring Universal Print

What is **Universal Print**? This is a cloud-managed print service that's provided by Microsoft through Microsoft Azure. Universal Print runs solely on Microsoft Azure. So, when it's deployed with Universal Print-compatible printers, you do not require any on-premises infrastructure to use the service.

The service is essentially a Microsoft 365 subscription-based service that you can use to centralize print management through the Universal Print portal. It's important to note that this service is fully integrated into Entra ID and supports **single sign-on** (**SSO**) scenarios.

This section will look at Universal Print and how you can use this Azure service with **Azure Virtual Desktop** (**AVD**).

Let's take a quick look at its architecture:

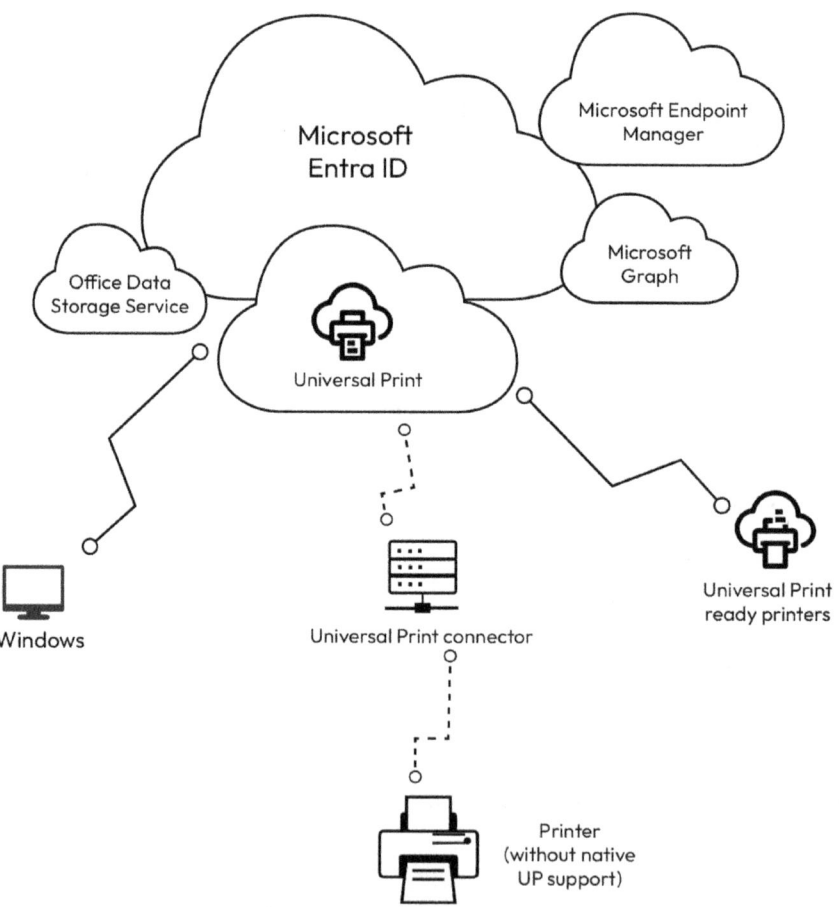

Figure 13.1 – The component architecture of Universal Print

The Universal Print Service leverages the following components:

Component	Description
Universal Print	Cloud print service
Entra ID	User and device identity and authorization service
Office Data Storage Service	Print queue data storage service
Microsoft Intune	Client device printer provisioning policy service
Microsoft Graph	Printer management API
Universal Print connector	A component that handles communication between printers and the Universal Print service
Universal Print ready printer	A printer that has built-in support for communicating with Universal Print
Printer (without native Universal Print support)	A printer that needs to be registered using the Universal Print connector to communicate with Universal Print

Table 13.1 – Description of components related to Universal Print

The preceding table was taken from the following Microsoft site: https://learn.microsoft.com/en-gb/universal-print/fundamentals/universal-print-whatis#architecture

Now, let's look at the required licensing and Universal Print's prerequisites.

Before we look at Universal Print, we must look at the required licensing for using the service. The following subscriptions include Universal Print:

- Microsoft 365 Enterprise F3, E3, E5, A3, and A5
- Windows 10 Enterprise E3, E5, A3, and A5
- Microsoft 365 Business Premium

If you have issues accessing the Universal Print service with the correct licenses, you need to ensure that you have checked the **Universal Print** service plan. To check this, follow these steps:

1. Within the Azure portal, go to Entra ID | Licenses | All products.
2. Select the product from the list shown.
3. Navigate to the **Service plan details** section in the left-hand menu.

4. Check if **Universal Print** is in the service plan list:

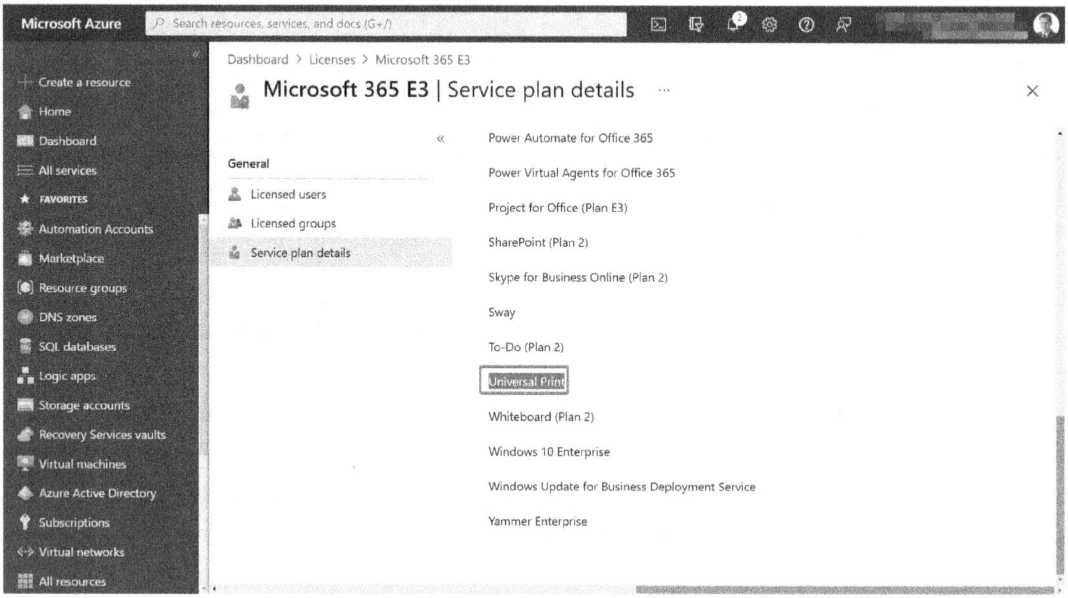

Figure 13.2 – Universal Print listed in the service plan for Microsoft 365 E3

5. If this is included in one or more product licenses, you need to ensure that the required licenses are assigned to those users who require the Universal Print service.

Next, let's look at the requirements we need to have in place for configuring Universal Print.

> **Important note**
> Where is print data stored? Universal Print stores all print queues in its office data storage. This is the same storage that's used to store Office 365 mailboxes and OneDrive files. A job can be queued for a few days. If the job is not claimed by a printer within 3 days, the job gets marked as aborted. You may see jobs stay within Universal Print for up to 10 days. Please note that when print jobs are sent using Universal Print, they are cloud encrypted.

Prerequisites for Universal Print

In this section, we will look at the prerequisites for Universal Print:

- You will need a Universal Print (eligible) license that has been assigned to the user by a Global Administrator.
- To configure and manage Universal Print, the IT administrator must have a Universal Print-eligible license assigned.

- To configure and manage Universal Print, an administrator must be assigned either of the following two Entra ID roles: Printer Administrator or Global Administrator.
- To install and print from Universal Print, you need a client device running a Windows client OS that's at version 1903 or beyond. Where possible, use the latest operating system for the best user experience. For a list of partner integrations that you can use for Universal Print, please go to https://learn.microsoft.com/universal-print/fundamentals/universal-print-partner-integrations.
- An internet connection.

Ensure that the following firewall rules have been applied to the device that's been chosen to host the connector. If you are using a Windows client device, remember that you will need to disable the hibernation/sleep controls and that you also need to ensure that the following firewall rules are set at the perimeter and on the client device:

- *.print.Microsoft.com
- *.microsoftonline.com
- *.azure.com
- *.msftauth.net
- go.microsoft.com
- aka.ms

> **Important note**
> Make sure that both TCP 443 and 445 are open on the firewall to ensure you don't experience any issues when using Universal Print.

Universal Print administrator roles

There are two designated limited Azure administrator roles that you can use to manage Universal Print. The following table details these two roles:

Name	Description
Printer Administrator	Users with this role have full access to manage all aspects of printers in Universal Print.
Printer Technician	Users in this role can register and un-register printers and set the printer's status.

Table 13.2 – Roles and descriptions in Universal Print

The preceding table was taken from the following Microsoft resource: `https://learn.microsoft.com/en-gb/universal-print/fundamentals/universal-print-administrator-roles`

Now, let's learn how to set up Universal Print.

Setting up Universal Print

When setting up Universal Print, we first need to deploy the Universal Print connector, which enables printers to communicate with the Universal Print service.

> **Important note**
>
> Some in-market printers do not support the required Universal Print protocols. It is most likely that printer manufacturers will offer printer firmware upgrades that add Universal Print support directly to the printer; however, you need to use the Universal Print connector for those that do not.

The connector's purpose, as its name suggests, is to ensure that a wide range of printers can connect to and communicate with the Universal Print service. Printers with the Universal Print protocol within their firmware don't require the Universal Print connector.

The key functions of the connector are as follows:

- The connector enables IT admins to register printers with the Universal Print service
- It reports a print job's status and the printer's status to the Universal Print service
- The connector collects print jobs from the Universal Print service and delivers them to their target printer

> **Important note**
>
> The print connector passes jobs to the print spooler without locally storing the files the user intends to print. However, depending on their size, the connector may need to store the file to ensure it is submitted successfully to the spooler. In some cases, the deletion may not be successful and IT admin intervention will be required to clear these no-longer-needed files.

Connector prerequisites

Ensure that the following firewall rules are applied to the device chosen to host the connector. If you are using a Windows client device, remember that you will need to disable hibernation/sleep controls and that you also need to ensure the required firewall rules are set as detailed in the *Prerequisites for Universal Print* section.

The recommended operating system is Windows 10 64-bit Pro or Enterprise on build 1809 onward. If you are using a server, you will need a minimum of Windows Server 2016 64-bit. Both Windows Server 2019 and 2022 are supported.

> **Important note**
> When you create a print connector, a device object is created in Entra ID with an object ID.

Installing the connector

Follow these steps to install the connector:

1. First, you will need to download the connector file, which you can download from the following link: `https://aka.ms/UPConnector`.

> **Important note**
> For those using proxy services, you can use `bitsadmin` to set the proxy. You can find more information on this here: `https://learn.microsoft.com/en-gb/windows-server/administration/windows-commands/bitsadmin-util-and-setieproxy`.

2. Once you have downloaded the installation files, run the EXE file and follow the steps to install it, as guided by the wizard:

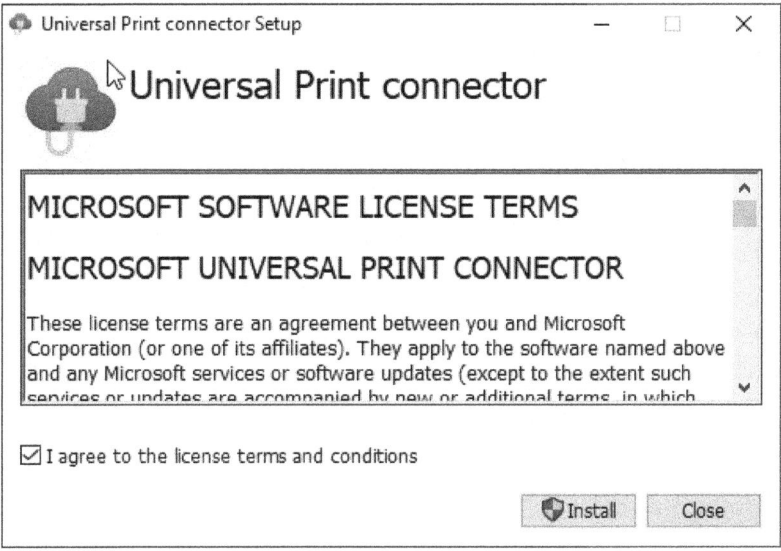

Figure 13.3 – The Universal Print connector installer form

3. Once installed, you will be prompted to launch the connector. Click the **Launch** button, as shown in the following screenshot:

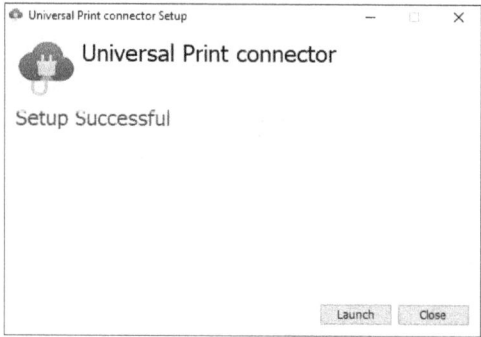

Figure 13.4 – The Universal Print connector has been installed successfully

Once the Universal Print connector has been launched, you will see an option to sign in. Go ahead and sign in using the administrator account that's been assigned to the Universal Print license:

> **Important note**
> Make sure that the user account that's used to configure the Universal Print connector's configuration has either the role of **Printer Administrator** or **Global Administrator**.

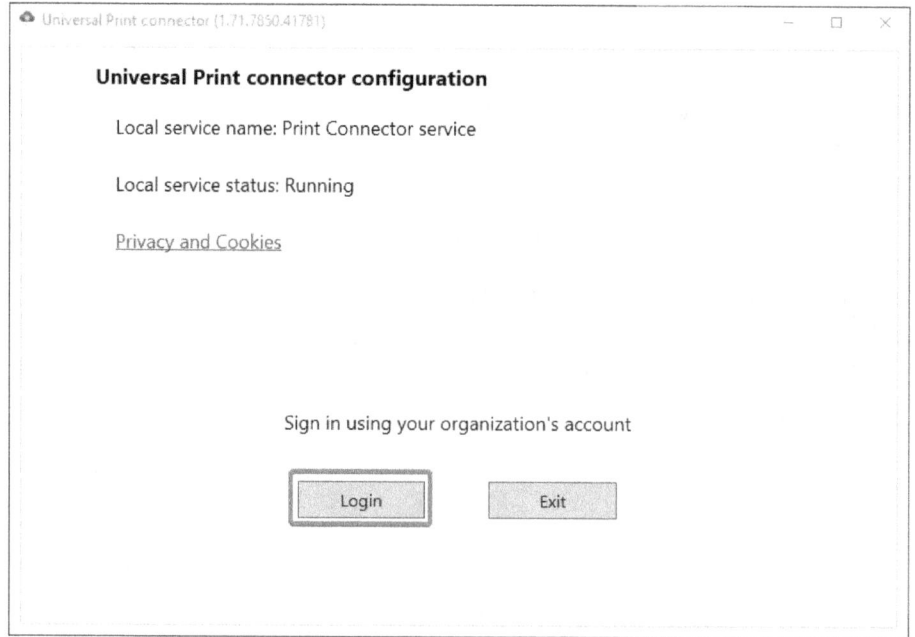

Figure 13.5 – Universal Print connector configuration form

1. Once you've signed in, you will see a box appear within the Universal Print connector form showing the connector name label and textbox for registration. Enter a name and register:

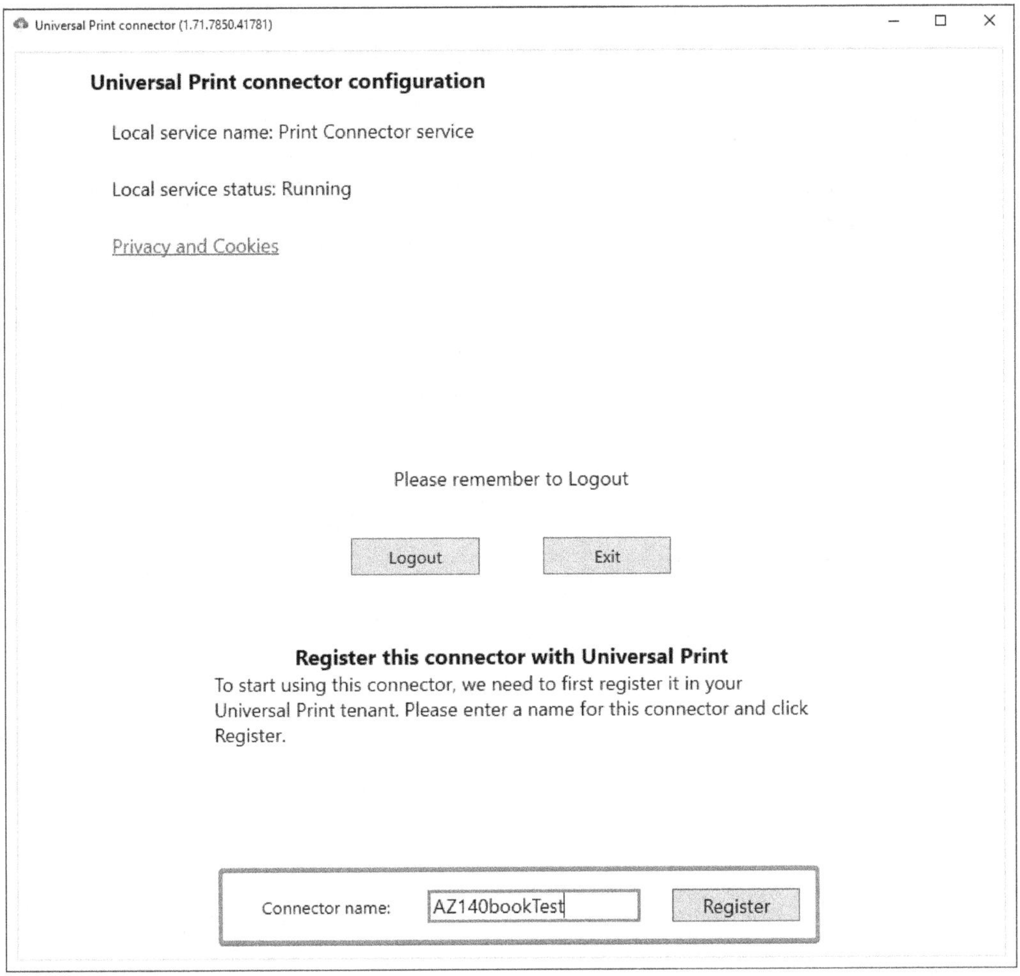

![Universal Print connector configuration window showing Local service name: Print Connector service, Local service status: Running, Privacy and Cookies link, Logout and Exit buttons, Register this connector with Universal Print section, and Connector name field with "AZ140bookTest" entered and Register button]

Figure 13.6 – Successfully signing in to the Universal Print connector with the registration field shown

2. At this point, you will see the registration loading screen:

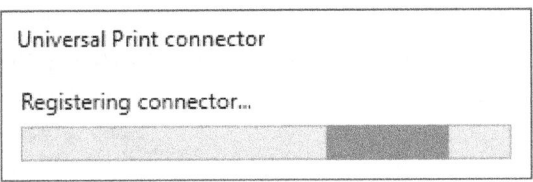

Figure 13.7 – Registration progress message/progress bar

3. Once you've registered successfully, you will see the configuration page:

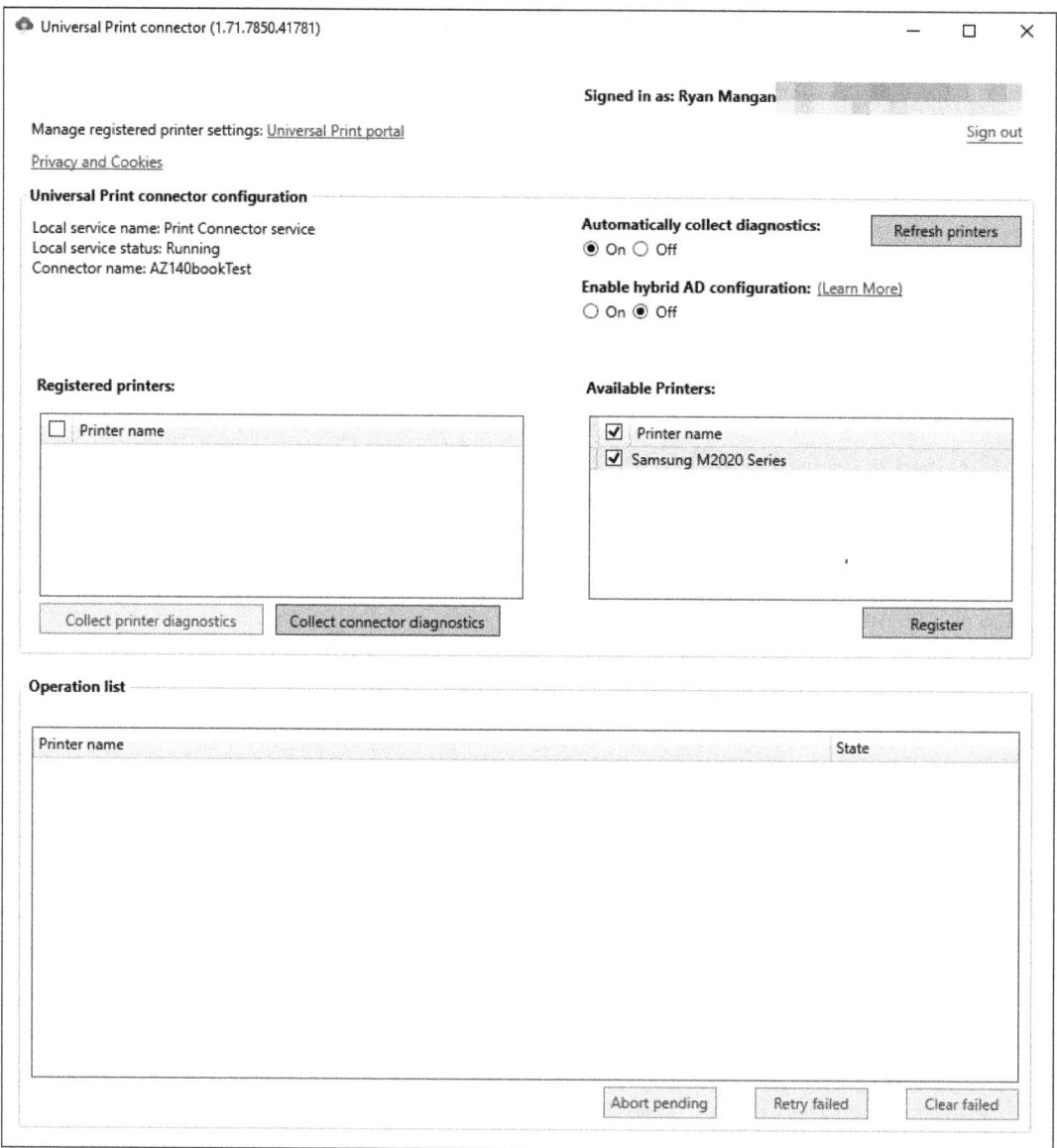

Figure 13.8 – Registered Universal Print connector listing the available printers to register

Within the Universal Print portal, you will see the connector under **Universal Print | Connectors**:

Dashboard > Universal Print

Universal Print | Connectors

- Search (Ctrl+/)
- Refresh
- Overview

Manage

↑↓ Name

- Printers
- Printer Shares
- Connectors
- Document conversion

AZ140bookTest

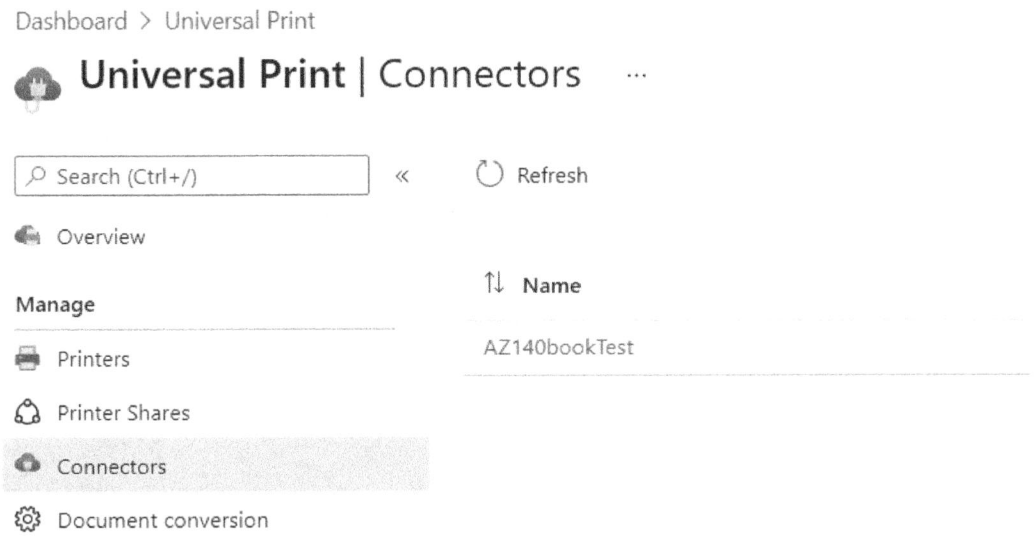

Figure 13.9 – Registered Universal Print connectors

In this subsection, we looked at the Universal Print connector's prerequisites and installed the connector on a device. Now that we have installed the Universal Print connector, we can register the printer.

Registering printers using the Universal Print connector

In this section, we'll learn how to register printers with Universal Print using the Universal Print connector we installed in the previous section.

> **Important note**
> Ensure that you have configured the correct external URLs through the perimeter security and localhost firewall.

There are essentially two steps to registering a printer with Universal Print:

1. Select the printers you want to register from the available list and click the **Register** button. The printers will appear in the **Available Printers** list if the registration process is successful:

Figure 13.10 – Available printers that can be registered

The registration process can take between 10 to 30 seconds on a typical internet connection to register. It may take longer, so please be patient.

> **Important note**
> The option to set **Enable Hybrid AD Configuration** to **Enable** is used for those organizations that use both Active Directory as well as Entra ID. In this type of setup, the user account exists in both directory services.

You can read more about Hybrid AD Configuration here: `https://learn.microsoft.com/universal-print/fundamentals/universal-print-hybrid-ad-aad-environment-setup`

2. Once the registration process is complete, you will see the printer(s) appear in the **Registered printers** section, as shown in the following screenshot:

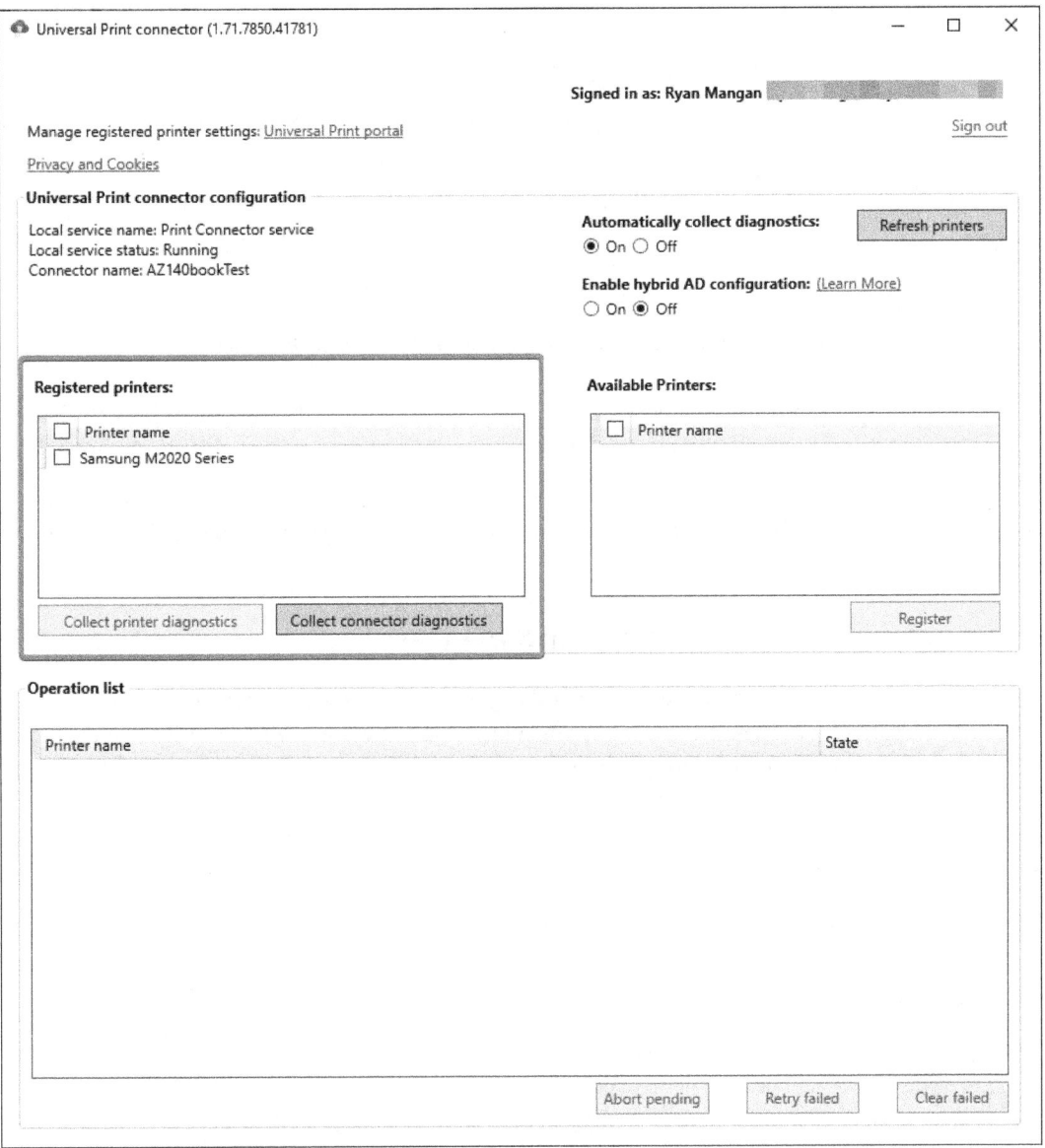

Figure 13.11 – The printer is now registered

Now that we have registered the printer, you should see the newly registered printer within **Azure portal | Universal Print | Printers**:

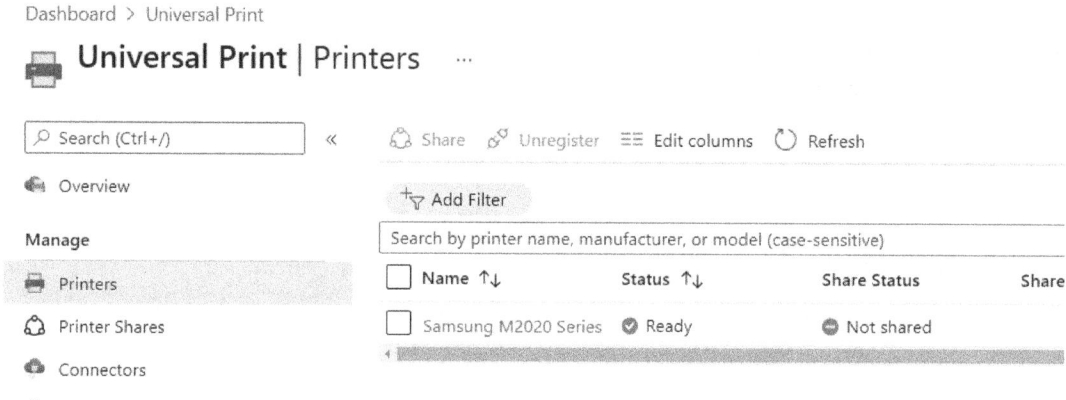

Figure 13.12 – Printers registered within the Azure Universal Print page

If the printer registration process fails, you will note that it will remain in the operations list and show a failure in the **Status** section. You can retry the failed registration process and clear it, as shown in the preceding screenshot.

Now that we have registered a printer with Universal Print, we can assign permissions to the registered printers and share them.

Assigning permissions and sharing printers

Now that we have installed the print connector and registered the printers, the next stage is to share the printers. This means making the printers accessible to users. Before a user can print to a printer, the printer must be shared, thus granting access.

The quickest way to share a printer is to navigate to **Printer Shares** and click the **Add** button (https://portal.azure.com/#blade/Universal_Print/MainMenuBlade/PrinterShares):

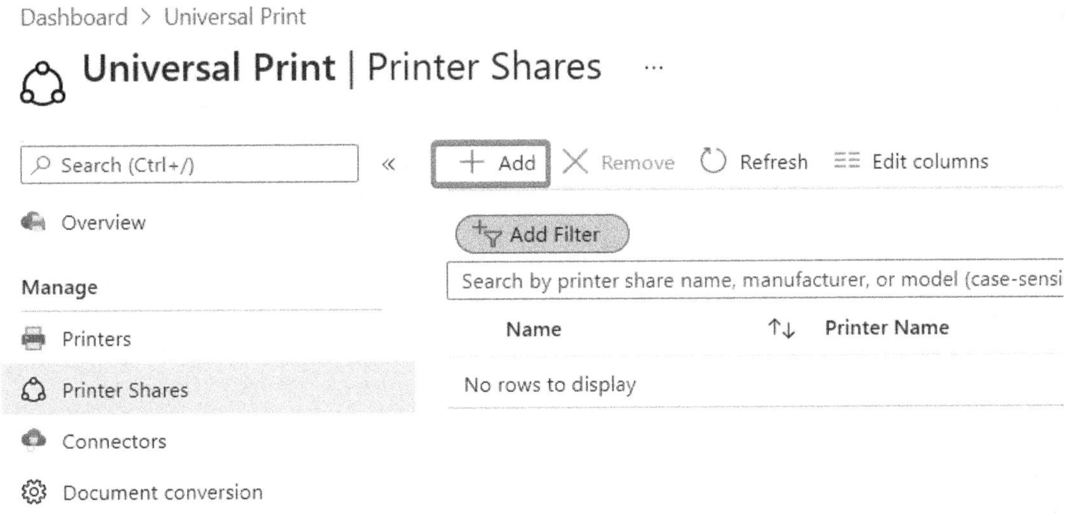

Figure 13.13 – The Printer Shares page with the Printer Shares function highlighted

Once you have clicked the **Add** button, a **Create printer share** blade will appear. Enter the share name, the printer/printers, and the selected Azure users/Azure groups, or allow access to everyone in your organization. Once you have entered the correct information for the share, click **Share Printer**:

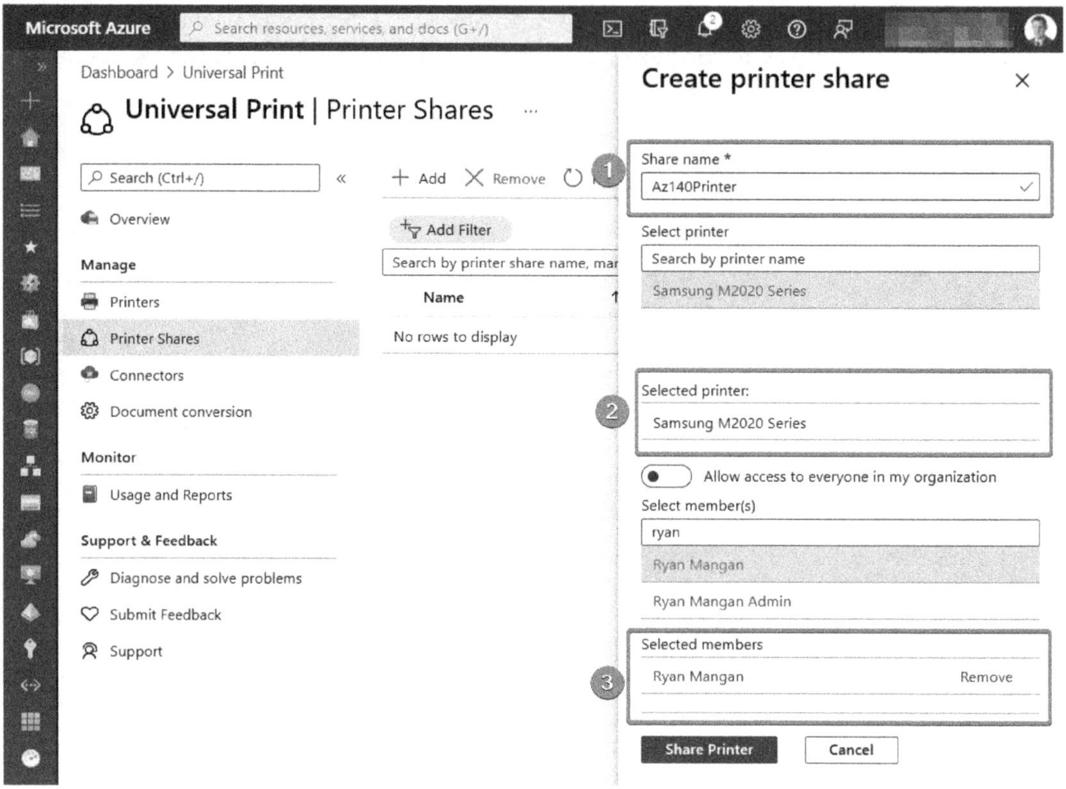

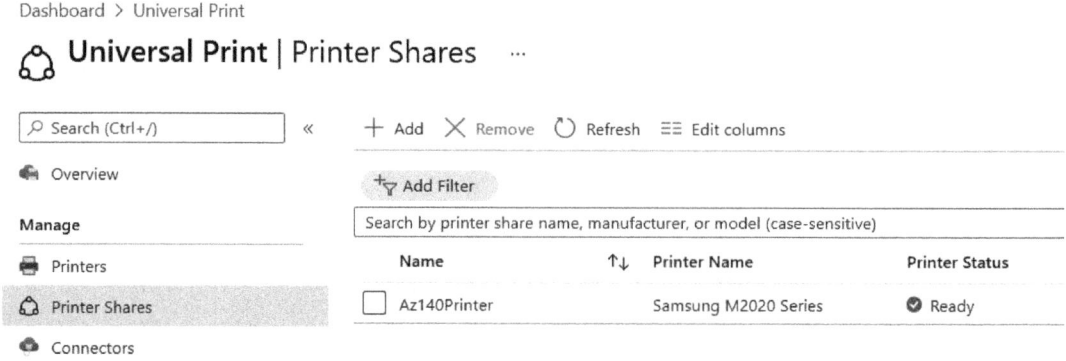

Figure 13.14 – The Create printer share blade

Once created, you will see the printer share appear on the **Printer Shares** page within Universal Print, as per the following screenshot:

Figure 13.15 – Universal Print: the Printer Shares page

When you click on **Printer Shares**, you will see options for managing access control, which is where you can add more users to the printer share, have the opportunity to delete the printer share, and swap the printer, as shown in the following screenshot:

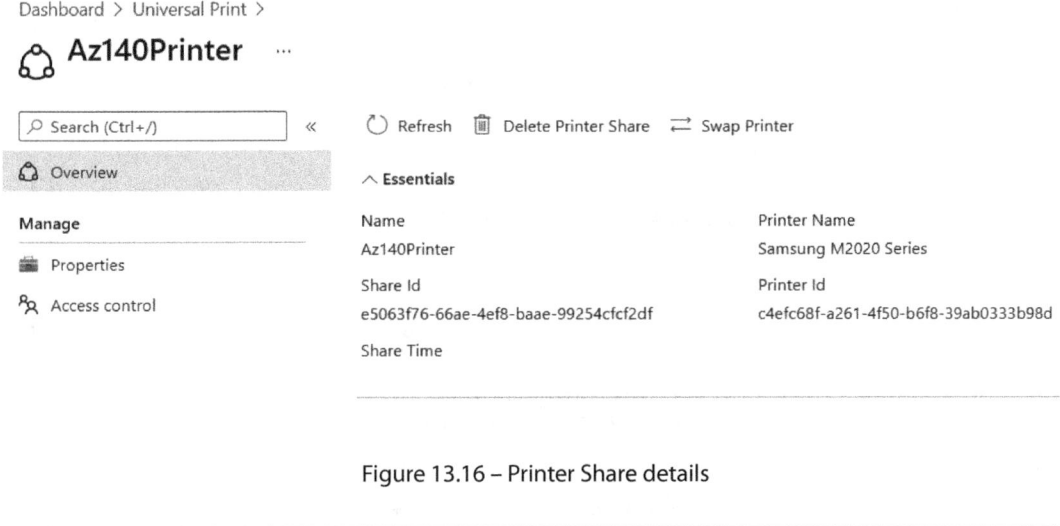

Figure 13.16 – Printer Share details

> **Tip**
> When a printer needs to be replaced, you can use the **Swap Printer** button to choose another printer that has been registered with Universal Print.

In this section, we learned how to assign printers to a printer share and how to assign user (member) permissions to use the Universal Print service. Now that we have configured printer sharing, we must add the Universal Print printer to a Windows device.

Adding a Universal Print printer to a Windows device

In this final section on configuring Universal Print, we will assign a printer to a Windows device. The following are the prerequisites we must have before we can add the printer to the user's device:

- The user's device needs to be Entra ID joined, Entra ID registered, or hybrid domain joined
- The Universal Print printer must have been shared, and the user must have been assigned the appropriate permissions to use the printer
- The user must have the appropriate license for Universal Print

392　Configuring User Experience Settings

There are typically three steps to add a Universal Print printer to a device:

1. Navigate to **Settings** on the Windows client and search for **Printers & scanners**:

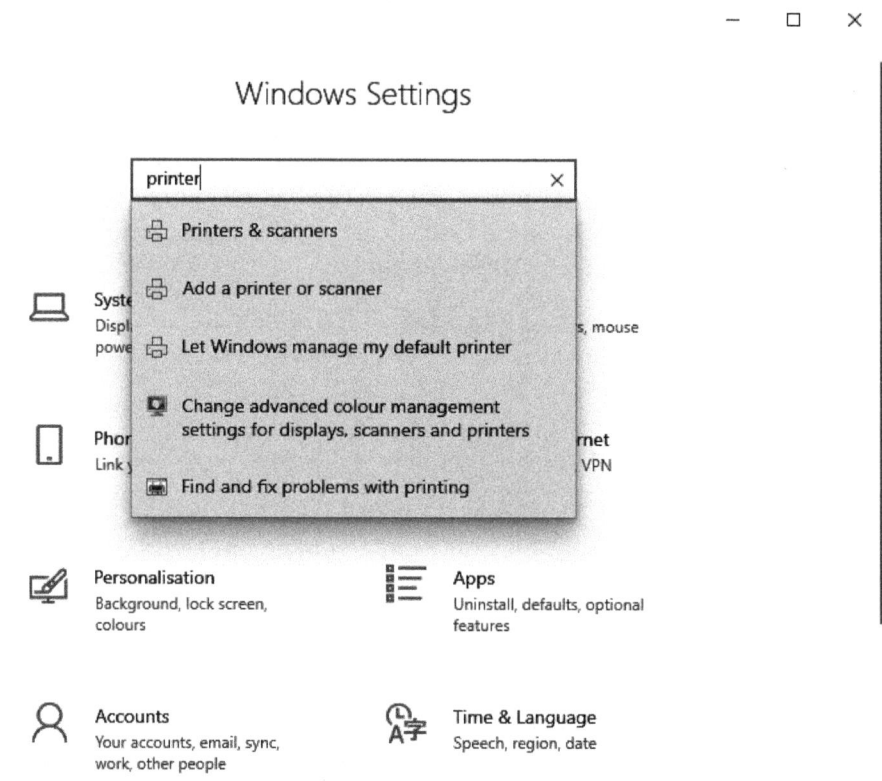

Figure 13.17 – The Windows Settings page

2. Click **Add a printer or scanner** to start scanning for printers. You should then see the Universal Print printer appear, as shown in the following screenshot, as **Az140Printer**:

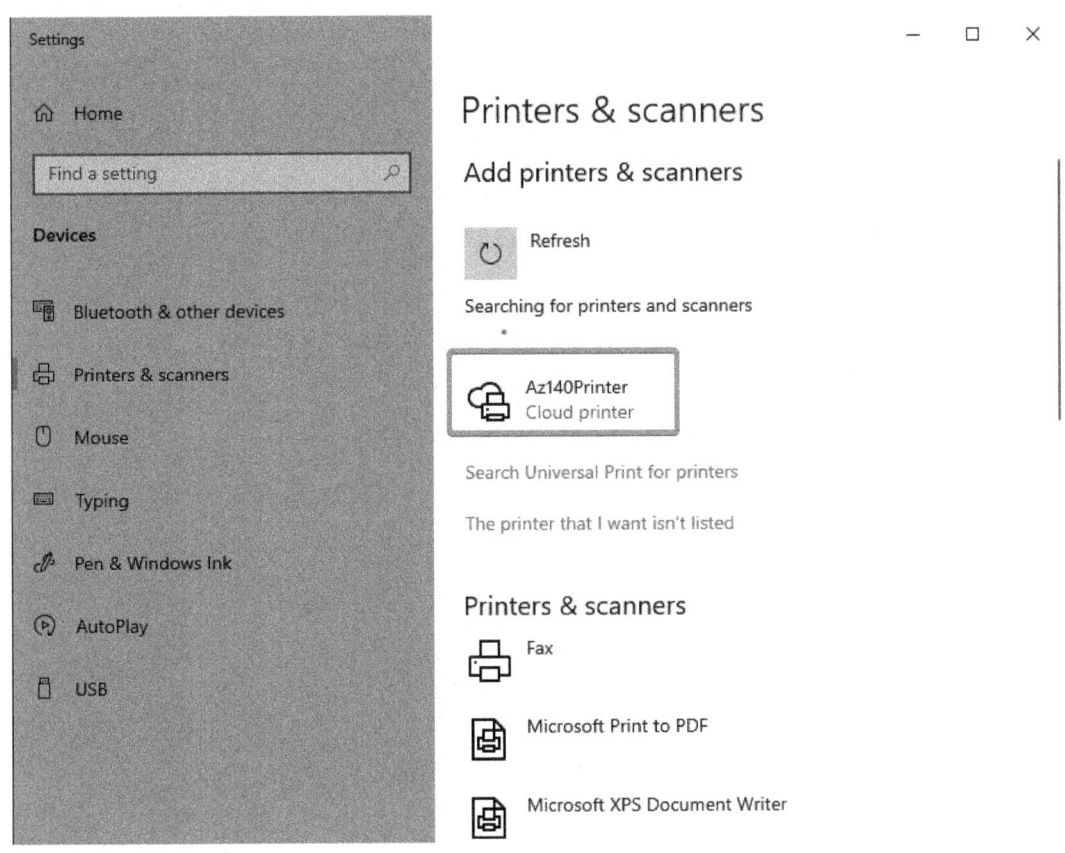

Figure 13.18 – The Printers & scanners page

3. Once the printer has been installed, you will see it appear as ready and that it's in the **Printers & scanners** list:

> **Important note**
> It is advised that you don't change the driver for Universal Print printers as this could cause the printer to stop printing.

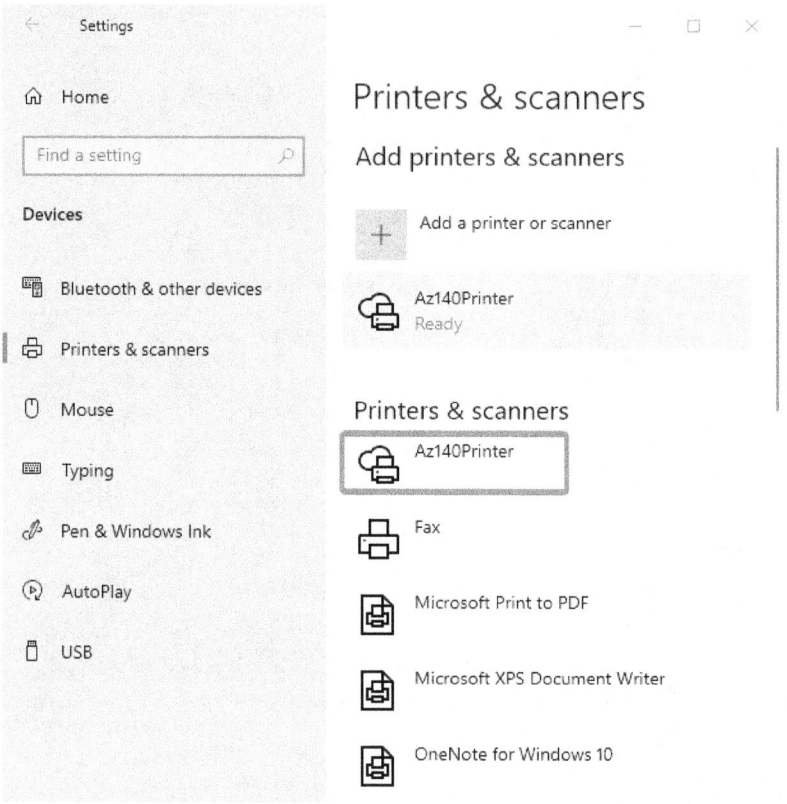

Figure 13.19 – Universal Printer added to the client device

There you have it – with that, we have deployed Universal Print for AVD. The key takeaway from this section is that you need to install a connector on a device on-premises to register printers within the Azure Universal Print portal page. Then, unless you have a printer that supports the Universal Print protocols out of the box, you need to share the printer and assign the necessary permissions. Once it has been shared, you can add the printer to user devices that have the required license to use Universal Print.

In the next section, we will look at user settings in Group Policy and Microsoft Intune.

Configuring user settings using Microsoft Intune

This section will briefly examine ways to configure and manage user settings on AVD. You can use local policies and registry entries, Group Policy settings, and more, all through Microsoft Intune.

Microsoft Intune is a cloud-based platform that focuses on both mobile and device management. This type of service offering is **Mobile Device Management** (**MDM**) or **Mobile Application Management** (**MAM**).

Microsoft Intune enables you to control and manage your organization's devices and how they are used. This is the same for AVD and Microsoft's latest offering, Windows 365 or, as some call it, Cloud PC.

In the **Devices** | **Windows** section, there are several Windows policies you can configure. This includes compliance, configuration, the use of PowerShell scripts, and Windows updates/feature updates.

Let's take a brief look at creating a configuration profile for AVD users:

> **Important note**
> Intune supports both Windows 10/11 Enterprise machines and Windows 10/11 Enterprise Multi-Session for AVD.

1. To create a profile, navigate to **Devices** | **Windows** | **Configuration profiles**. Then, click **Create**:

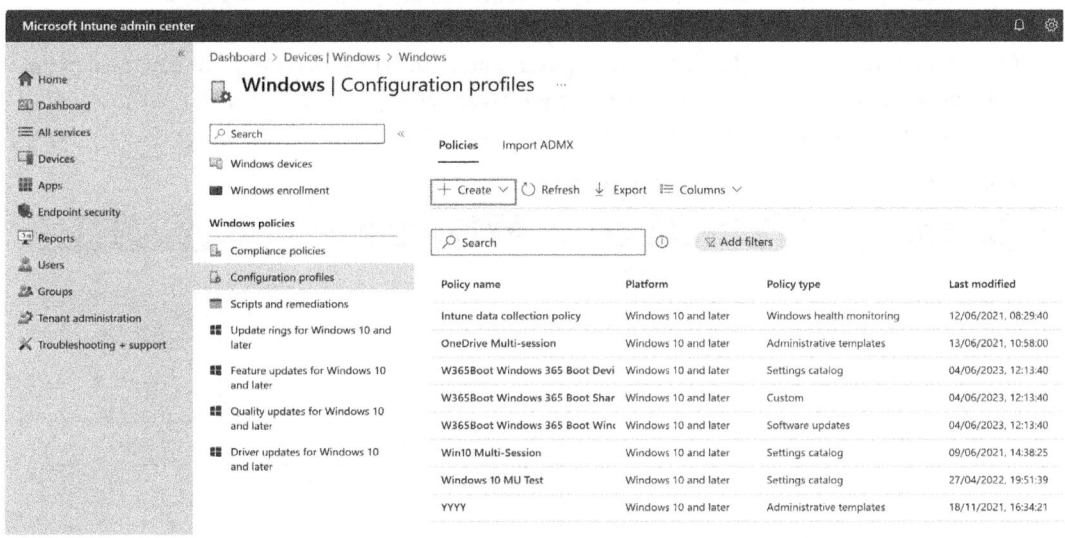

Figure 13.20 – The Windows Configuration profiles page within Microsoft Endpoint Manager

2. Once you click **Create** | **New Policy**, a blade will appear on the right-hand side, where you can choose your platform and the profile type. In this example, we have chosen **Windows 10 and later** as the platform and **Templates** as the profile type. We have also selected administrative templates that you would find in a Group Policy:

Configuring User Experience Settings

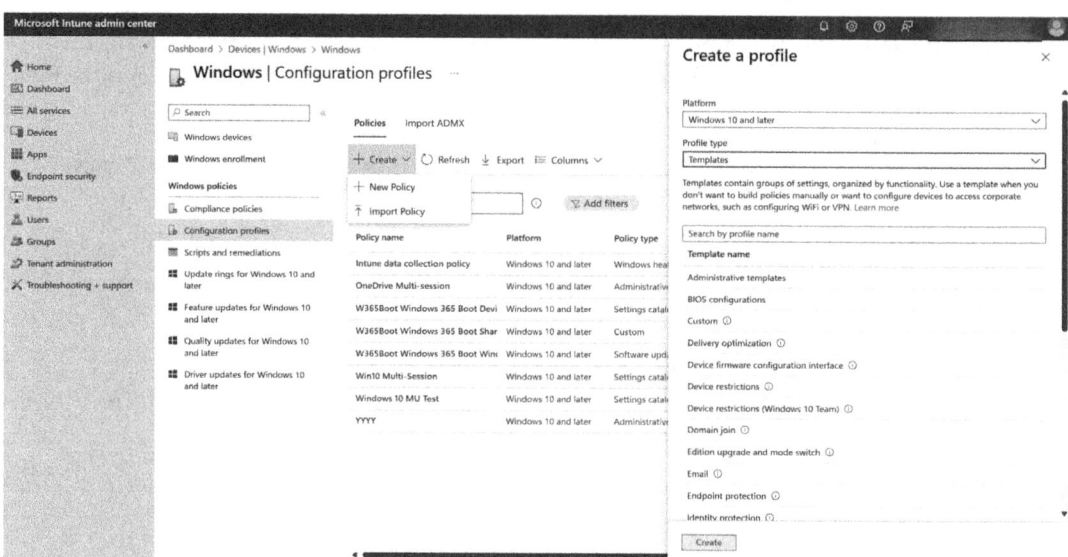

Figure 13.21 – The Create a profile blade within the Configuration profiles page

3. Once you have finished selecting your requirements, click the **Create** button. A wizard form will appear with four sections: **Basics**, **Configuration settings**, **Scope tags**, and **Assignments**. You can click **Review + create** to check the configurations you have specified.

Enter a name and description for the configuration policy:

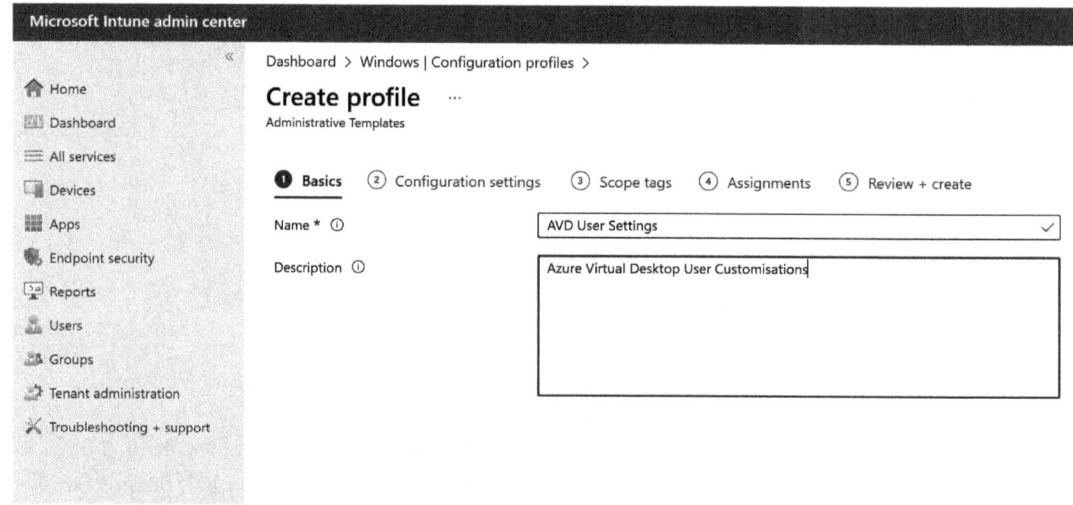

Figure 13.22 – The Create profile page when creating a new configuration profile

4. Click **Next** to move on to the **Configuration settings** section of the **Create profile** wizard form. Within the **Configuration settings** page, you can select both user and computer settings. In this example, I am going to choose a few user settings to lock down the desktop:

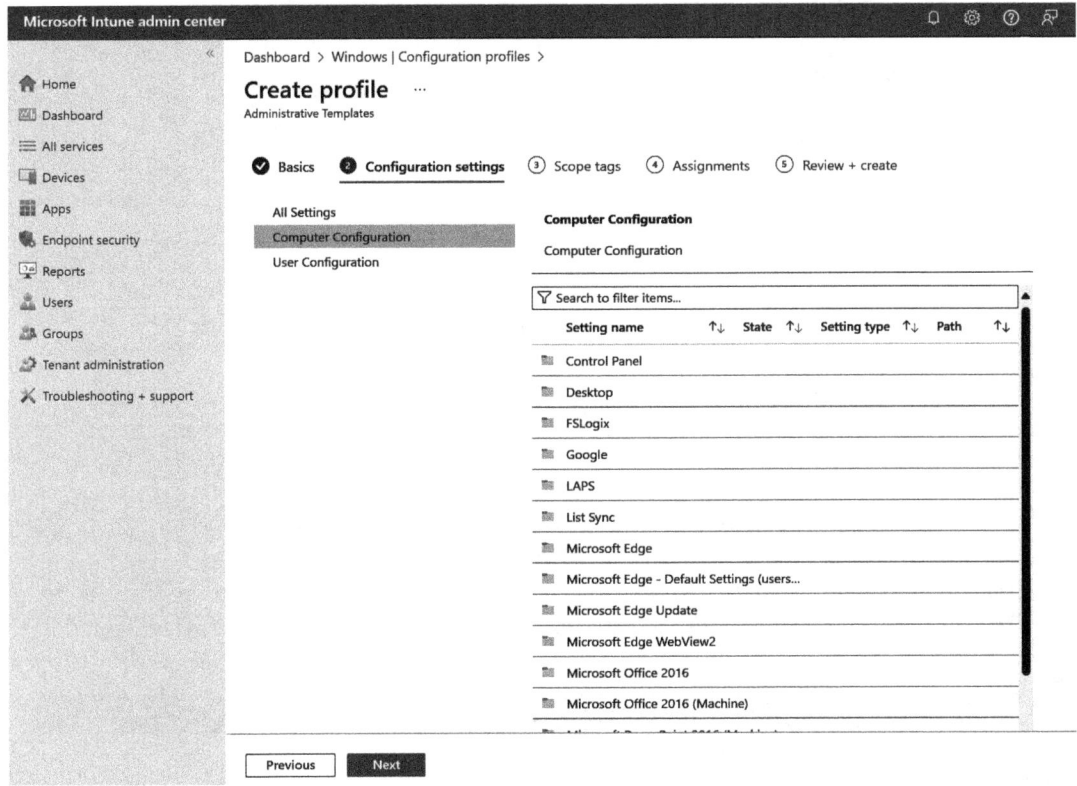

Figure 13.23 – Configuration options that are available for the computer and user settings

As you can see, there are several different policies you can enable to lock down a desktop. These are the same ones that you would find in a group policy or the local policy on the session host itself.

5. Choose the policies you require and, once complete, click **Next**:

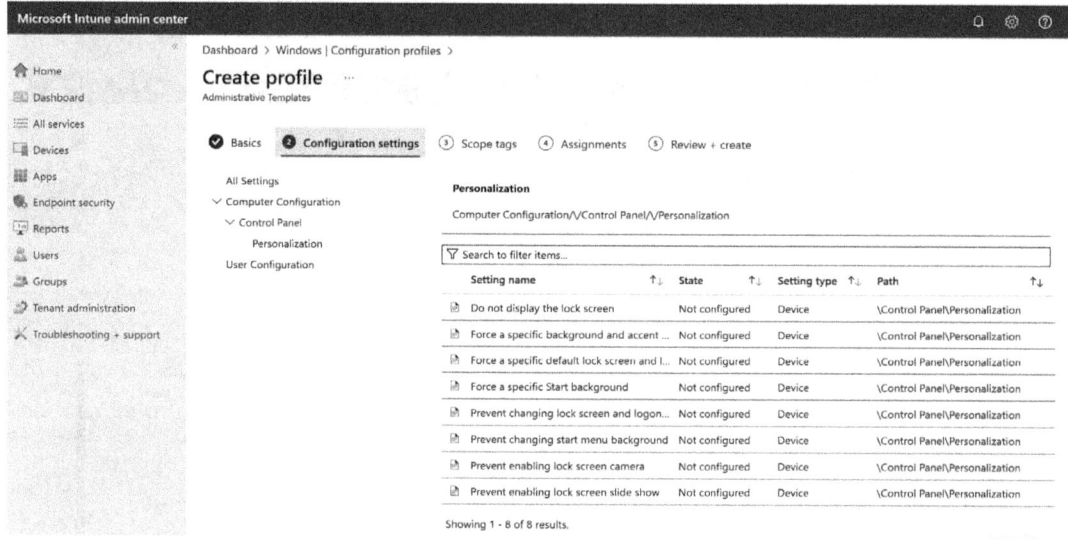

Figure 13.24 – Some of the policies you can add to the configuration profile

6. You should now see a **Scope tags** section. A scope tag is like a virtual version of an Active Directory **organizational unit** (**OU**) that can be used to assign/group devices to configuration profiles and compliance policies:

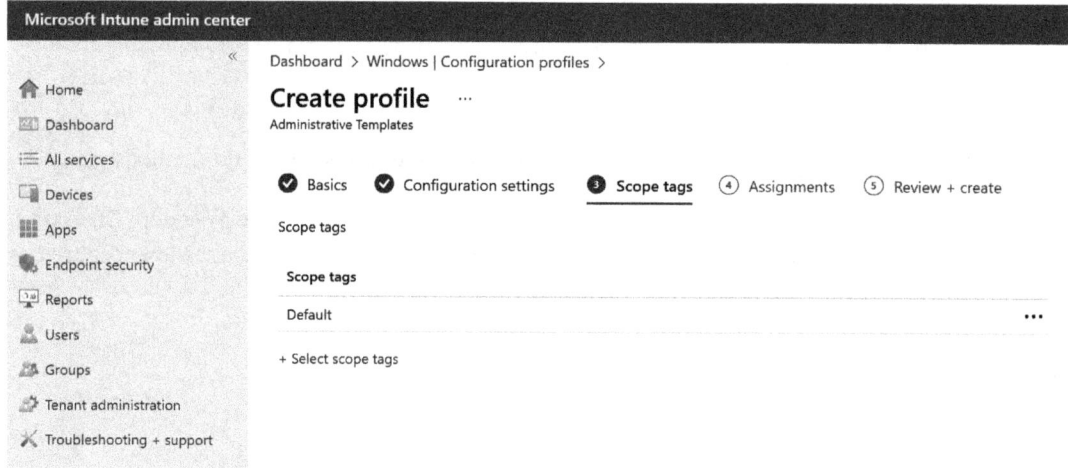

Figure 13.25 – The Scope tags setting

7. Once you have finished adding your scope tags, you can start assigning users. In this example, we will assign all users. You can add users or groups; however, you cannot mix user and device groups across included and excluded groups:

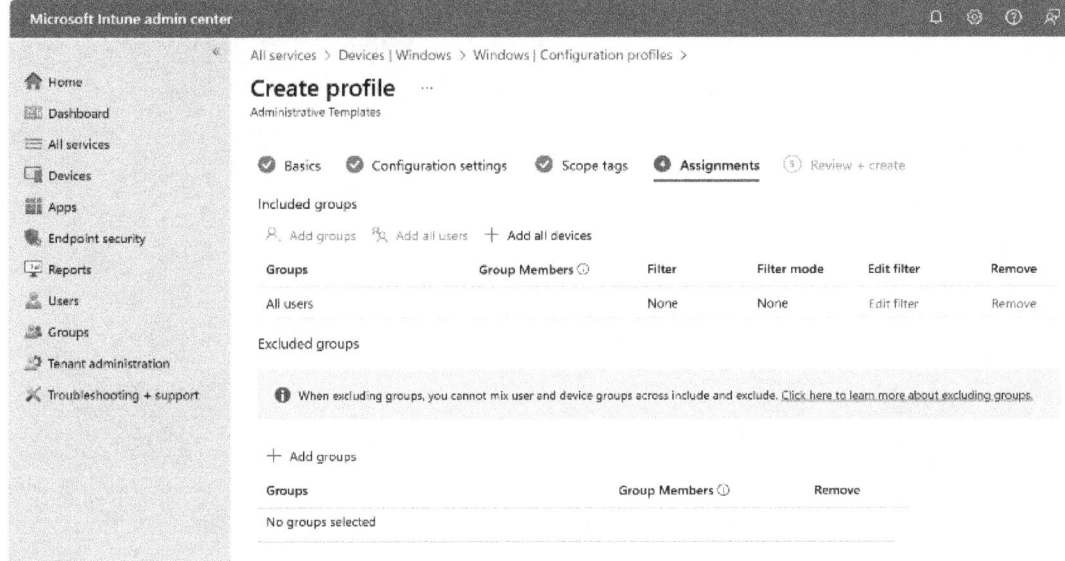

Figure 13.26 – The Assignments tab of the configuration policy allows you to select included and excluded groups

8. Once you have finished configuring the assignment, click **Review + create**.

   As shown in the following screenshot, I have selected three user configuration items in this policy that will be rolled out to the AVD session hosts.

9. Once you are happy with the settings/configuration, click **Create** to create the policy:

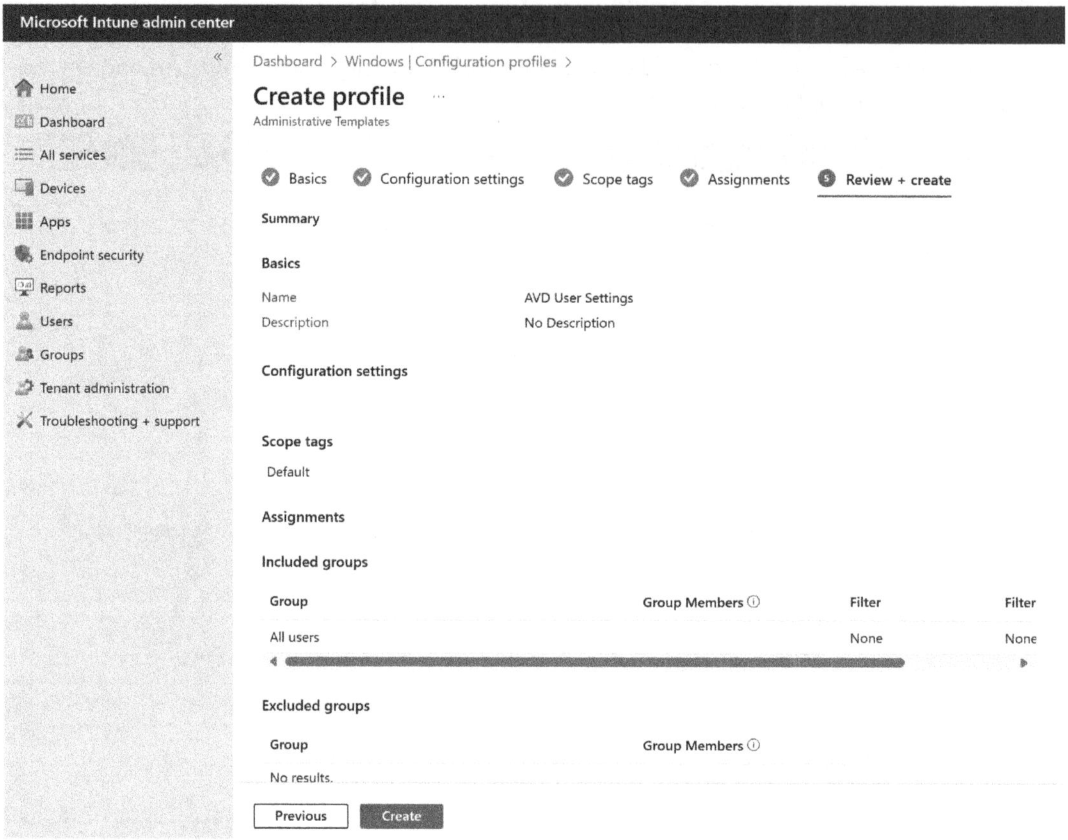

400 Configuring User Experience Settings

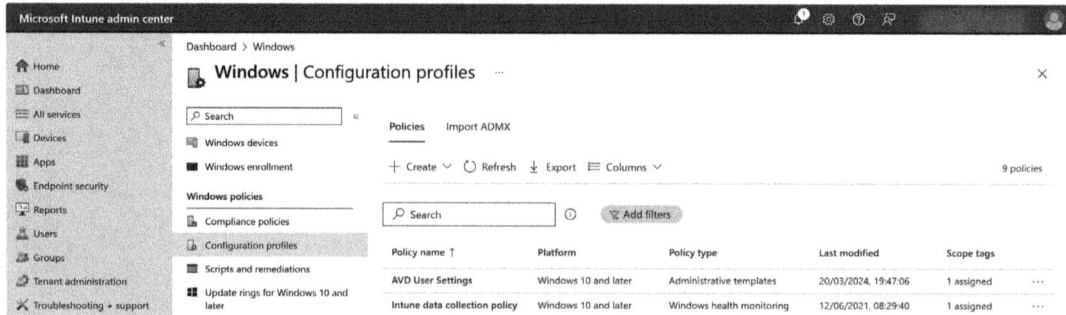

Figure 13.27 – The Review + create section of the Create profile page

Once created, you will see the new policy on the **Configuration policies** page:

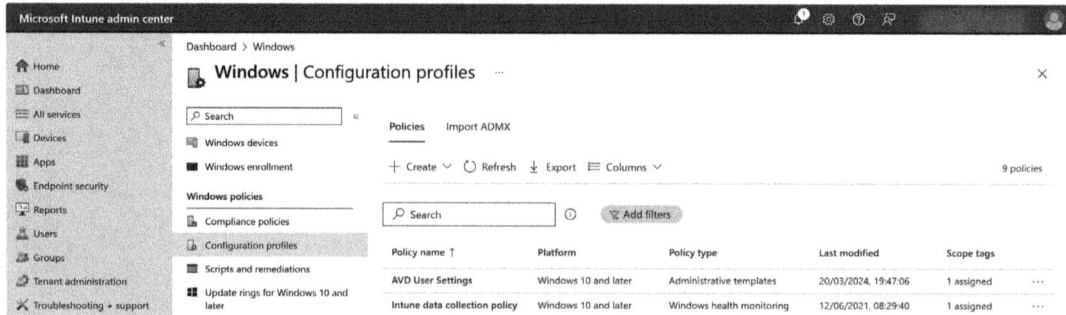

Figure 13.28 – The created configuration profile

There you have it – we have created our first configuration policy for AVD using Microsoft Intune! You can learn more about Microsoft Intune by reading *Mastering Microsoft Intune – Second Edition*, by *Christiaan Brinkhoff, Per Larsen, Packt Publishing*.

This section looked at how we can assign configuration policies to AVD session hosts using Microsoft Intune. In the next section, we will look at configuring Start VM on Connect.

Start VM on Connect

In this section, we will look at the **Start VM on Connect** feature. This offers a cost-saving mechanism for organizations as Start VM on Connect essentially allows you to turn VMs on when required; when they are not needed, the user can turn them off. Start VM on Connect is a great way for you to boot VMs on demand. This can be useful for personal desktops or resources that shut down every evening, as scheduled by the IT department.

The following screenshot shows Start VM on Connect in action:

Figure 13.29 – Start VM on Connect

> **Important note**
> Start VM on Connect is available for personal and pooled host pools and uses the Azure portal and PowerShell.

We now take a look at using the Azure portal to configure Start VM on Connect.

Configuring with the Azure portal

Before we enable Start VM on Connect, we need to assign the *Desktop Virtualization Power On Contributor* RBAC role to the AVD service principal in the Azure subscription, which contains the AVD session hosts you want to use Start VM on Connect with:

1. Go to the Azure portal and then go to **Subscriptions**.
2. Select the subscription that contains the AVD session hosts you want to use Start VM on Connect with.

3. Select **Access Control (IAM)**.
4. Select **Add Role Assignment** and select **Next**:

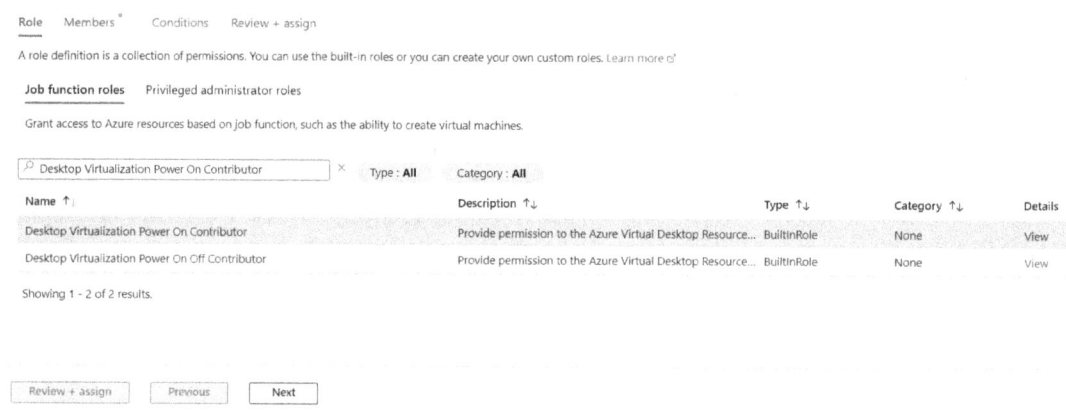

Figure 13.30 – Add role assignment settings

5. Select **User, group, or service principal** and click on **Select members**. Add **Azure Virtual Desktop** and click on **Select**:

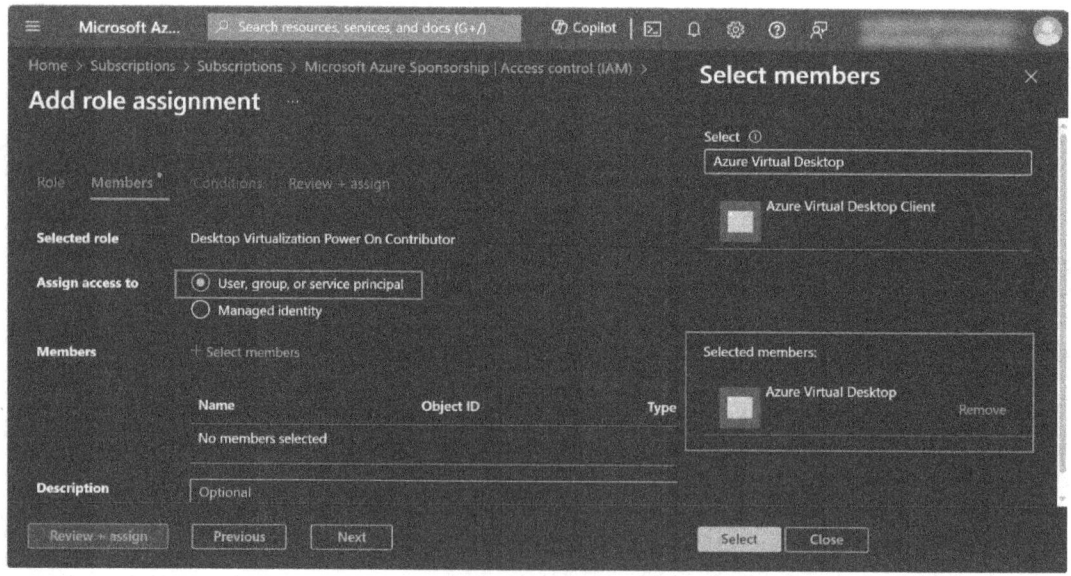

Figure 13.31 – Adding Azure Virtual Desktop service principal

6. Confirm that the **Azure Virtual Desktop** service principal is added and select **Next**:

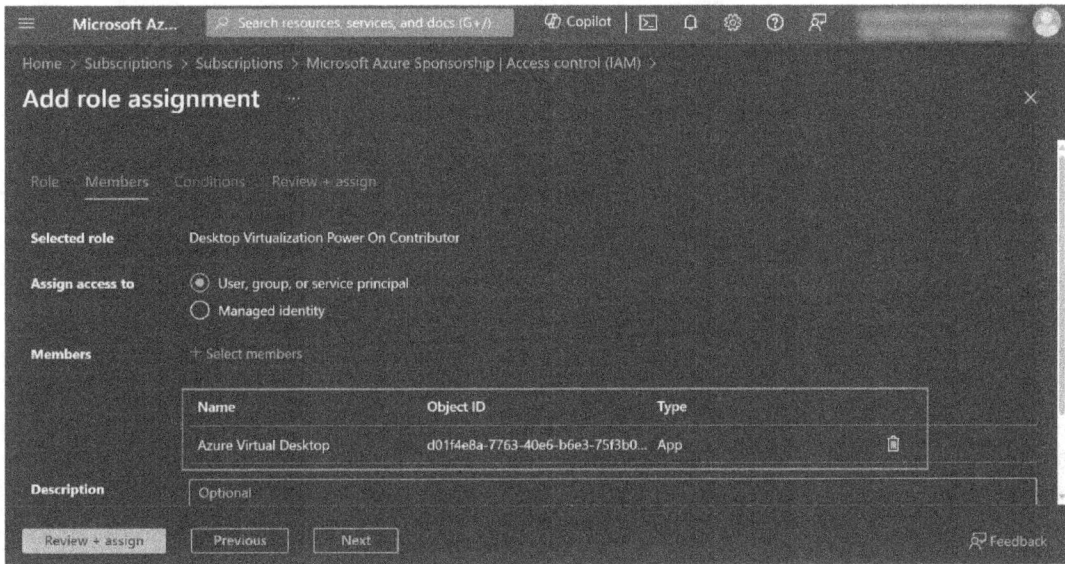

Figure 13.32 – Confirming settings

7. Review the settings and select **Review + assign**:

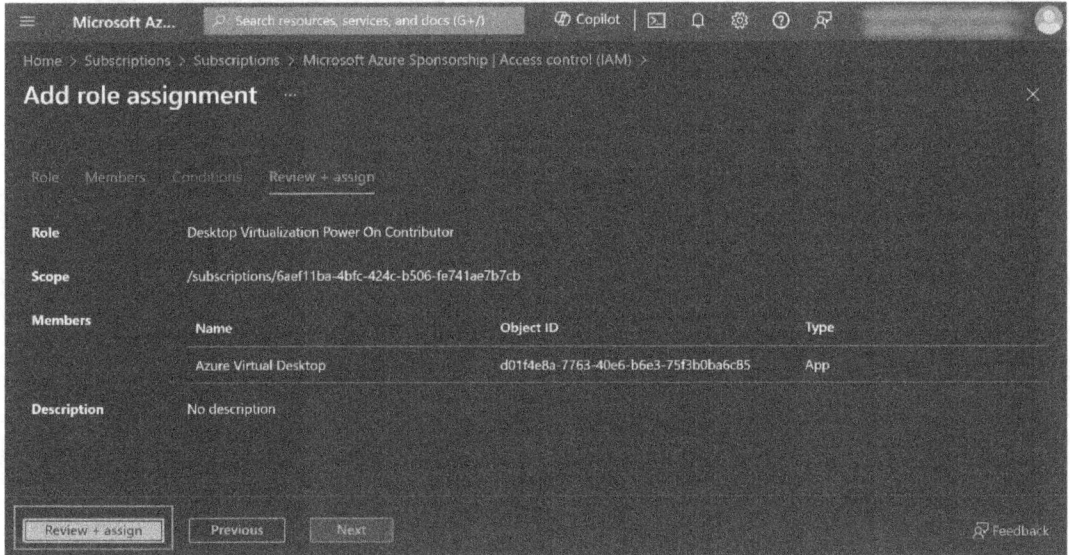

Figure 13.33 – Review + assign settings

Configuring User Experience Settings

Once you have completed this step, you will be ready to enable Start VM on Connect within the required host pool.

Enabling Start VM on Connect

Now that we have configured Start VM on Connect permissions, we can enable it on the host pool:

1. From the Azure portal, go to **Azure Virtual Desktop | Host pools** and select the host pool you want to enable it for.
2. Go to **Properties | Start VM on connect** and toggle the setting to **On**:

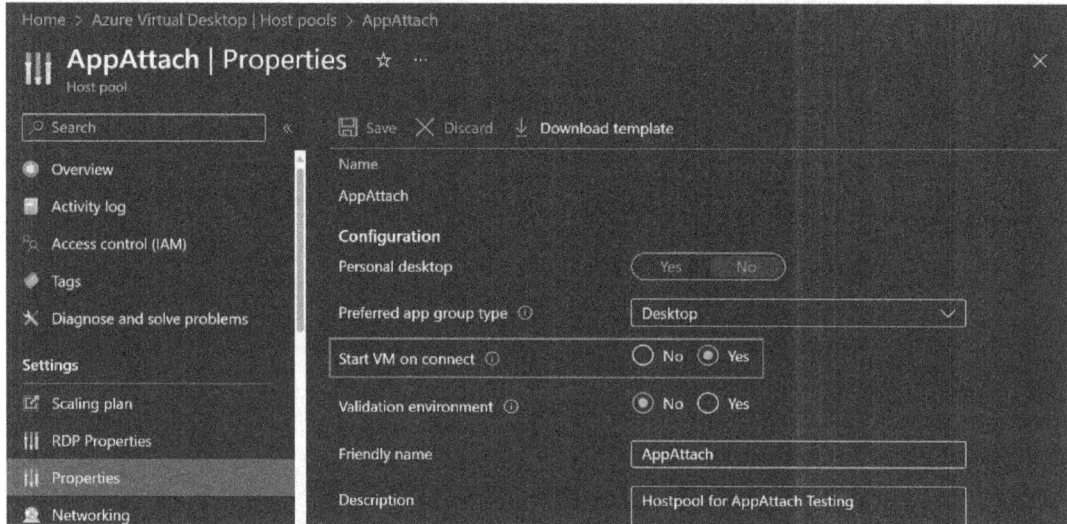

Figure 13.34 – Configuring Start VM on Connect

In the next section, we will cover hibernation mode, which allows us to put the VM into a hibernated state instead of shutting it down.

Hibernation mode

Microsoft has recently announced the general availability of Hibernation mode for Azure Virtual Machines. Hibernation allows you to pause VMs that aren't being used and save on compute costs.

When hibernating a VM, the memory contents are stored on the OS disk. The VM is then deallocated and powered off. When the VM is started again, the memory contents are transferred from the OS disk back into memory.

From an AVD perspective, this is useful for single-user desktop scenarios as the desktop users do not need to log out of the desktop and save their work; they can resume where they left off.

Supported VM sizes

Hibernation is currently limited to the following SKUs only:

- DasV5-series
- Dadsv5-series
- Dsv5-series
- Ddsv5-series

For an up-to-date list, please visit https://learn.microsoft.com/azure/virtual-machines/hibernate-resume.

Prerequisites for enabling hibernation mode

The following is a list of prerequisites for using hibernation mode for AVD:

- The hibernation feature must be enabled in the Azure subscription
- The OS disk must be large enough to store the contents of the RAM, operating system, and all applications running on the VM at the time of hibernation
- The VM SKU and OS support hibernation
- The Azure VM Agent is installed
- Hibernation is enabled upon VM creation
- If the VM is being created from an Azure Compute Gallery image, hibernation support must be enabled on the Gallery image definition

Integrating hibernation into AVD

Hibernation is actually a feature of Azure Compute, but we can integrate it into AVD to enhance the user experience and also reduce the cost of personal desktops.

We can integrate it into AVD by configuring the AVD scaling policy to put the VM into hibernation mode instead of powering off the desktop when the user goes into a disconnected state.

To configure this, we must do the following:

1. Go to the AVD portal.
2. Go to **Scaling Plans** and edit the scaling plan you want to enable the hibernation mode on.
3. Go to **Schedules**, edit the required schedule, and select **Next**.
4. Under the **Ramp-up** configuration settings, we can configure either of the following:

    - **Disconnect settings**: The behavior when a session goes into a disconnected state
    - **Log off settings**: The behavior when the user logs off a session.

When configured for **Disconnect settings**, if **Hibernate** is selected, the session will go into a hibernated state so that when the user re-connects to a session, they can resume where they left off:

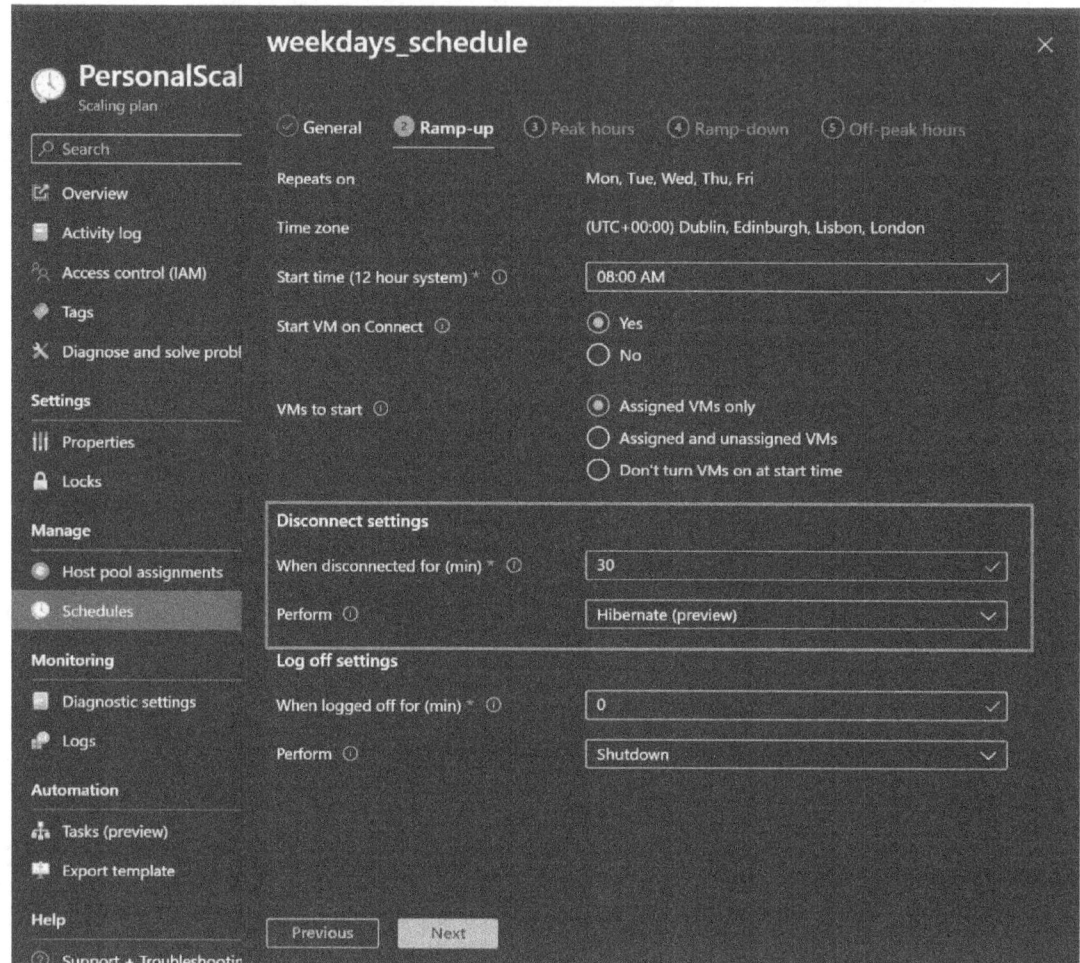

Figure 13.35 – Configuring hibernation autoscale settings

The next section will look at screen capture protection for AVD.

Enabling screen capture protection for AVD

The screen capture protection feature can be used as a data leak prevention tool to prevent sensitive information from being captured on endpoint clients. When this feature is enabled, remote content will be automatically blocked or hidden in screenshots and screen shares.

> **Important note**
> The Remote Desktop client hides content from any malicious software that may be capturing the screen.

Prerequisites

Screen capture protection is supported on both Desktop sessions and RemoteApp sessions. The following table shows the clients and versions that are supported:

Client	Client version	Desktop session	RemoteApp session
Remote Desktop client for Windows	1.2.1672 or later	Yes	Yes. Client device OS must be Windows 11, version 22H2 or later.
Azure Virtual Desktop Store app	Any	Yes	Yes. Client device OS must be Windows 11, version 22H2 or later.
Remote Desktop client for macOS	10.7.0 or later	Yes	Yes

Table 13.3 – Client compatibility for different operating systems

This table was referenced from https://learn.microsoft.com/azure/virtual-desktop/screen-capture-protection#prerequisites.

Configuring screen capture protection

In this subsection, we will configure screen capture protection:

> **Tip**
> This example shows you how to configure the session host itself. However, you can use the Group Policy Central Store within Active Directory to do this.

1. To configure screen capture protection, you will need to download a set of administrative templates and add them to the session host/session host template.
2. You can copy the latest ADMX files from https://aka.ms/avdgpo.
3. Then, you need to copy the terminalserver-avd.admx file to the %windir%\policyDefinitions folder.

4. A similar process must be followed for the ADML file – copy it to the `%windir%\policyDefninitions\en-us` folder.

5. Once both files have been copied to their required locations, you can check that the policies have been added correctly by opening the local Group Policy editor and navigating to `Computer Configuration|Administrative Templates|Windows Components|Remote Desktop Services|Remote Desktop Session Host|Azure Virtual Desktop`.

6. You should then see the **Enable screen capture protection** policy:

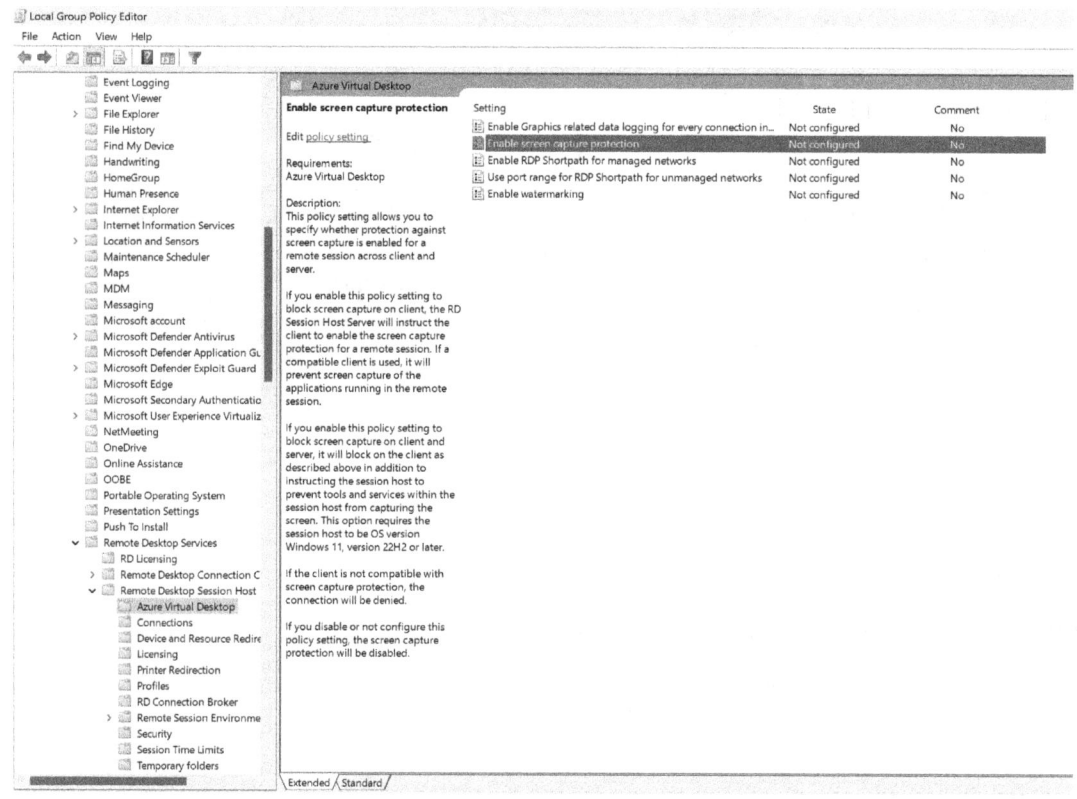

Figure 13.36 – Enable screen capture protection within the local computer policy of the session host

7. Set the policy to **Enabled**:

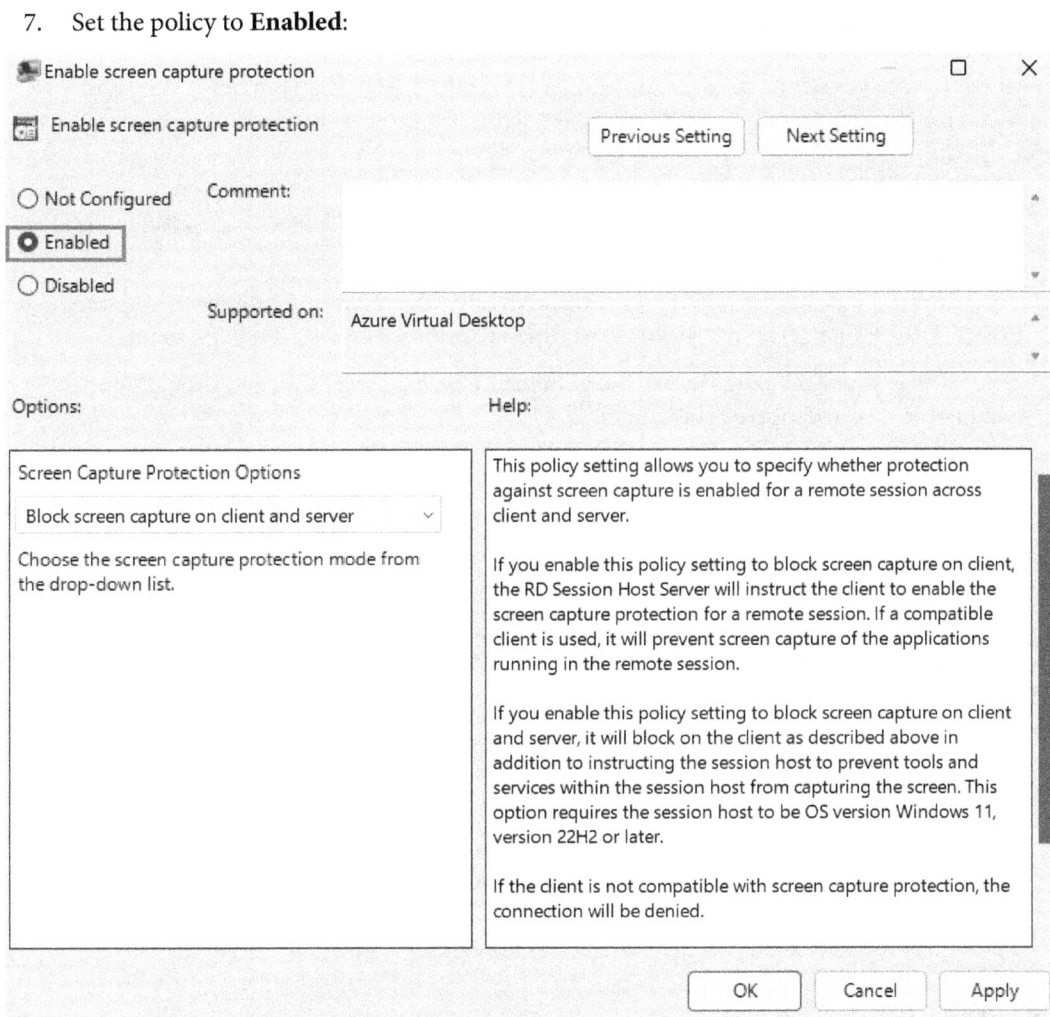

Figure 13.37 – The Enable screen capture protection page

8. Once enabled, you will need to reboot the host if you're testing.

> **Important note**
> There is no guarantee that the feature will fully restrict protected content, and it is recommended that you test it before rolling it out to a production environment. It is also recommended that you consider restricting access to items such as the clipboard, drive, and printer redirection and using screen capture protection. It is also important to understand that users cannot use local collaboration software such as Microsoft Teams when the screen capture protection feature is enabled.

Enabling screen capture protection via Intune

For devices that are not joined to **Microsoft Entra ID Domain Services (Entra ID DS)** or that we want to manage through Intune, we can also enable screen capture protection via Intune.

1. Go to the Intune page, which is `https://intune.microsoft.com`.
2. Create or edit a configuration profile for **Windows 10 and later devices** and **Settings Catalog** profile types.
3. In the **Settings picker** section, go to **Administrative Templates | Windows Components | Remote Desktop Services | Remote Desktop Session Host | Azure Virtual Desktop**.
4. The settings you can configure are shown in the following screenshot and should be enabled via **Enable screen capture protection**:

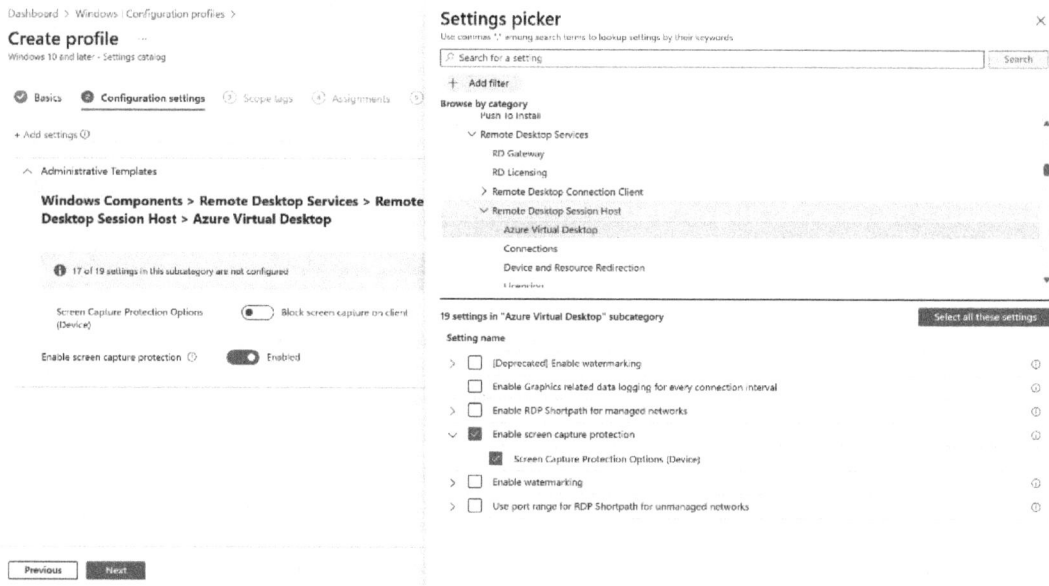

Figure 13.38 – Configuring screen capture protection via Intune

In this section, we looked at how to configure and enable screen capture protection for AVD. Now, let's learn how to troubleshoot FSLogix profile containers.

Enabling watermarking

Watermarking is a feature in AVD that places a watermark on the Desktop session.

This assists in preventing sensitive information from being captured on the client endpoints as a watermark appears as a QR code on the screen.

The QR code contains the connection ID of the remote session that admins can use to trace session details:

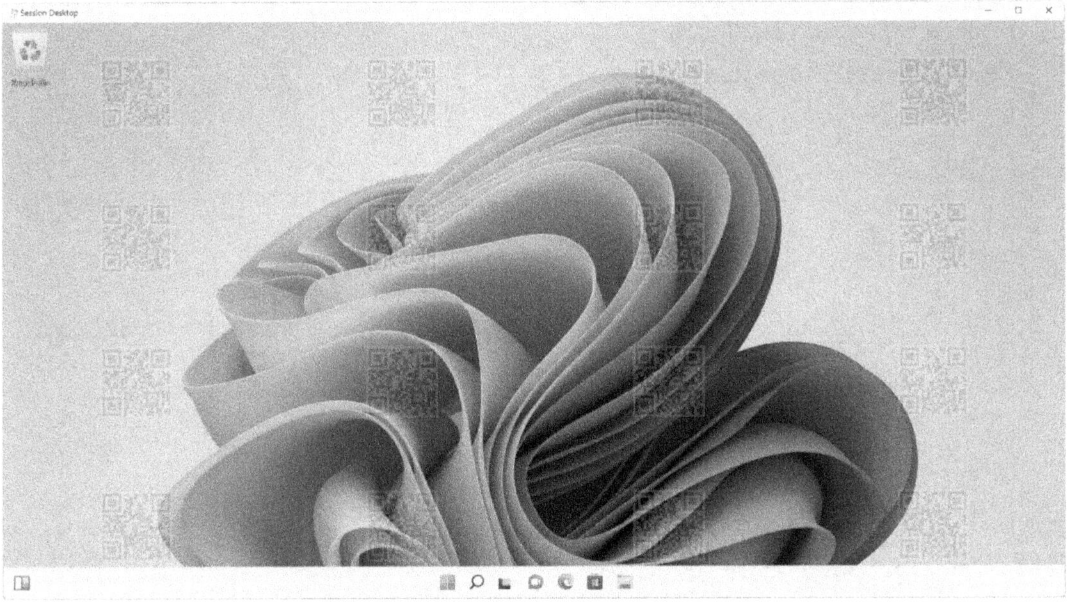

Figure 13.39 – Watermarking in action

Prerequisites for watermarking

Before we can implement watermarking, there are a few prerequisites that need to be met. For an up-to-date list, please visit `https://learn.microsoft.com/azure/virtualdesktop/watermarking#prerequisites`, but here's a summary of the prerequisites at the time of writing this book:

- Windows 10 or later
- macOS client v10.9.5 or later
- iOS client v10.5.4 or later
- AVD Insights configured

Enabling watermarking using Group Policy

We can also control the watermarking configuration settings via Group Policy. Please follow the next steps to achieve this:

1. Download the ADMX files used in the previous step to enable screen capture protection.

2. Navigate to `Computer Configuration|Administrative Templates|Windows Components|Remote Desktop Services|Remote Desktop Session Host|Azure Virtual Desktop`.

3. You should see **Enable watermarking**:

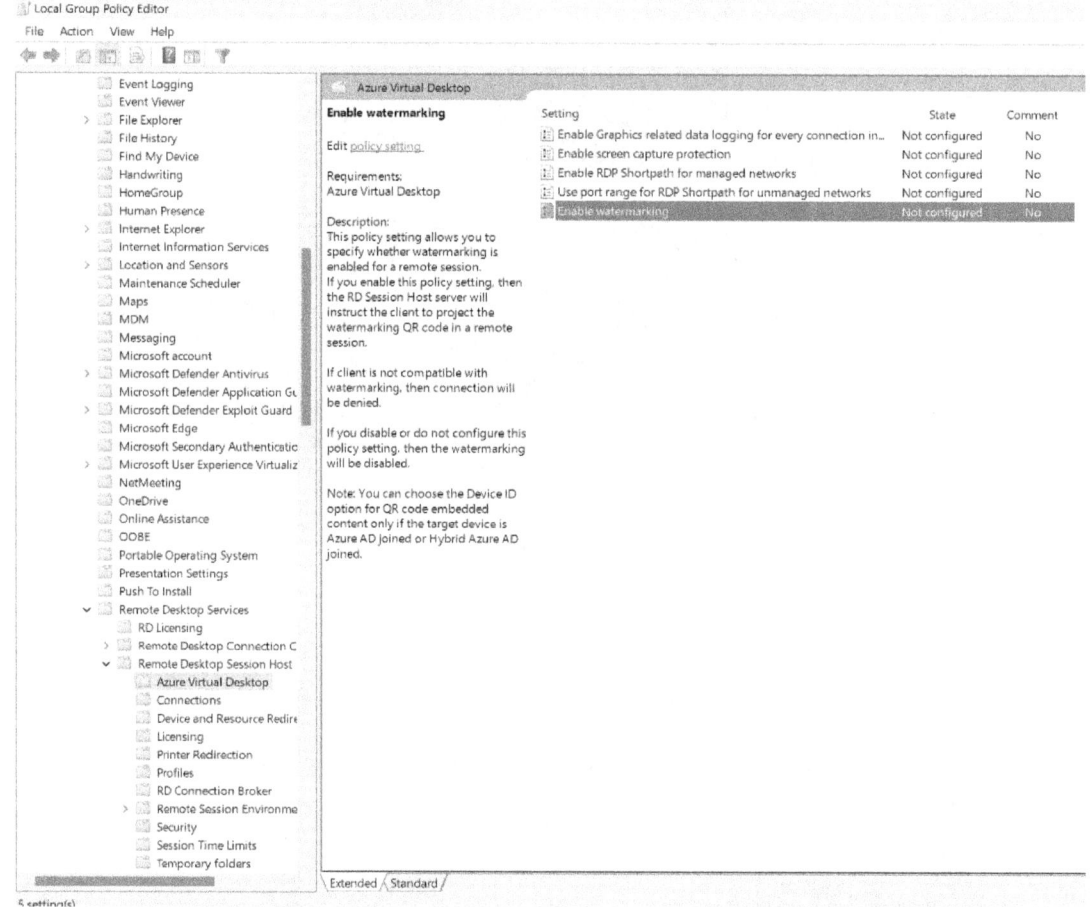

Figure 13.40 – Configuring watermarking via Group Policy

4. There are a few configuration settings that can be applied to set the width, height, and opacity of watermarking on the user's session:

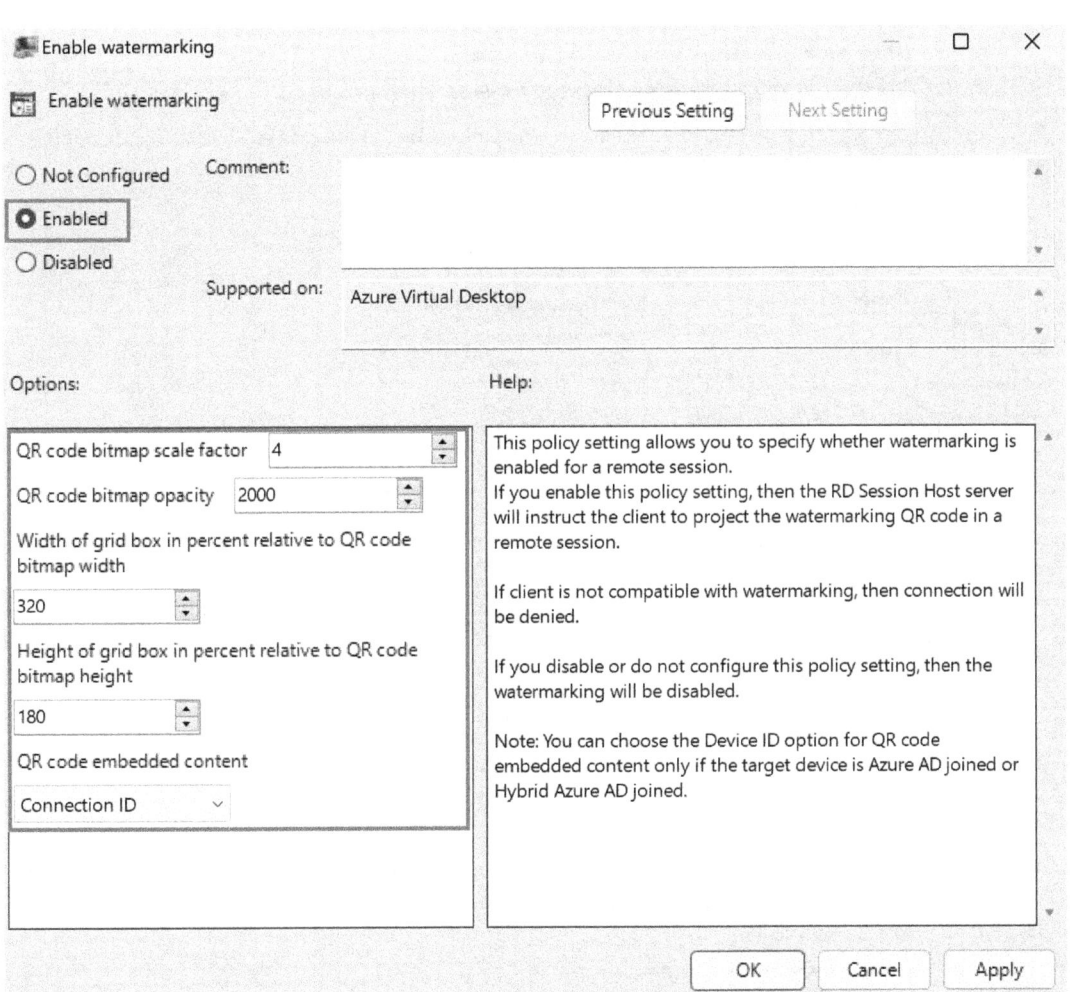

Figure 13.41 – Configuring watermarking configuration settings

Enabling watermarking via Intune

For devices that are not joined to Entra ID DS or that we want to manage through Intune, we can also enable watermarking via Intune:

1. Go to the Intune page, which is https://intune.microsoft.com.
2. Create or edit a configuration profile for **Windows 10 and later devices** and **Settings Catalog** profile types.

3. In the **Settings picker** section, go to **Administrative Templates | Windows Components | Remote Desktop Services | Remote Desktop Session Host | Azure Virtual Desktop**.
4. The settings you can configure are shown in the following screenshot and should be enabled via **Enable watermarking**:

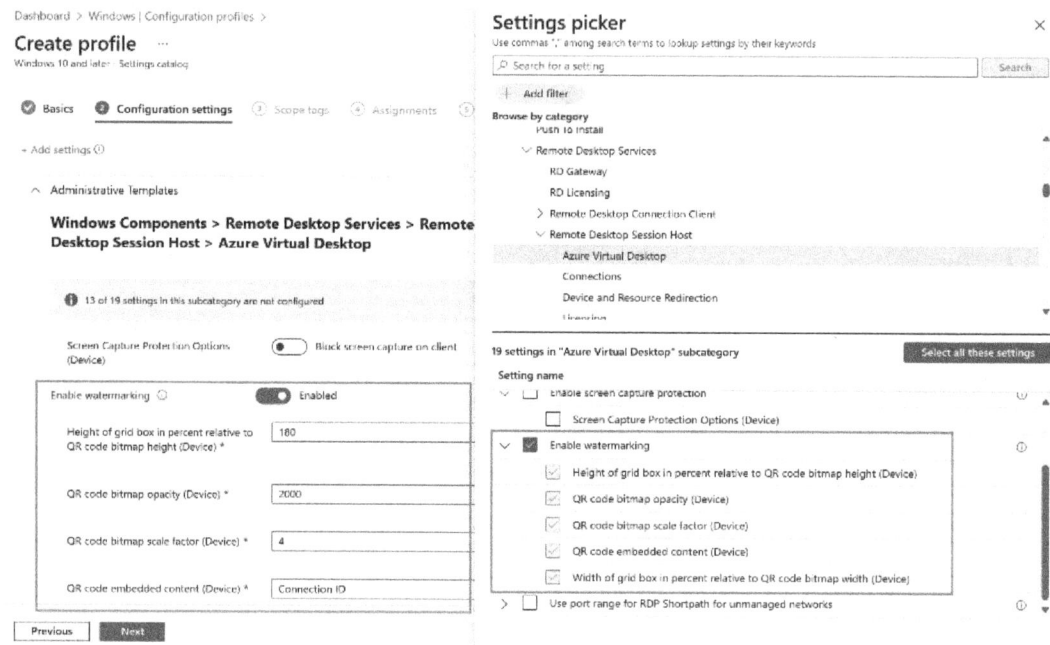

Figure 13.42 – Configuring watermarking via Intune

Let us now look at troubleshooting FSLogix profile issues.

Troubleshooting FSLogix profile issues

This section will provide an overview of troubleshooting FSLogix profile container issues. We will provide a few pointers to help you when you're diagnosing profile-related issues.

The quickest way to get insight into FSLogix profile issues is to review the logs using the FSLogix Profile Status utility. This enables you to view both administrative and operational profile-related events. You can find the Profile Status utility here: `C:\Program Files\FSLogix\Apps\frxtray.exe`.

> **Tip**
> You can review the logs for FSLogix remotely via `%ProgramData%\FSLogix\Logs`.

Using the FSLogix Profile Status utility and cross-referencing the status codes will help you quickly diagnose an issue. You can find a list of status codes on the Microsoft documentation site at `https://docs.microsoft.com/fslogix/fslogix-error-codes-reference`:

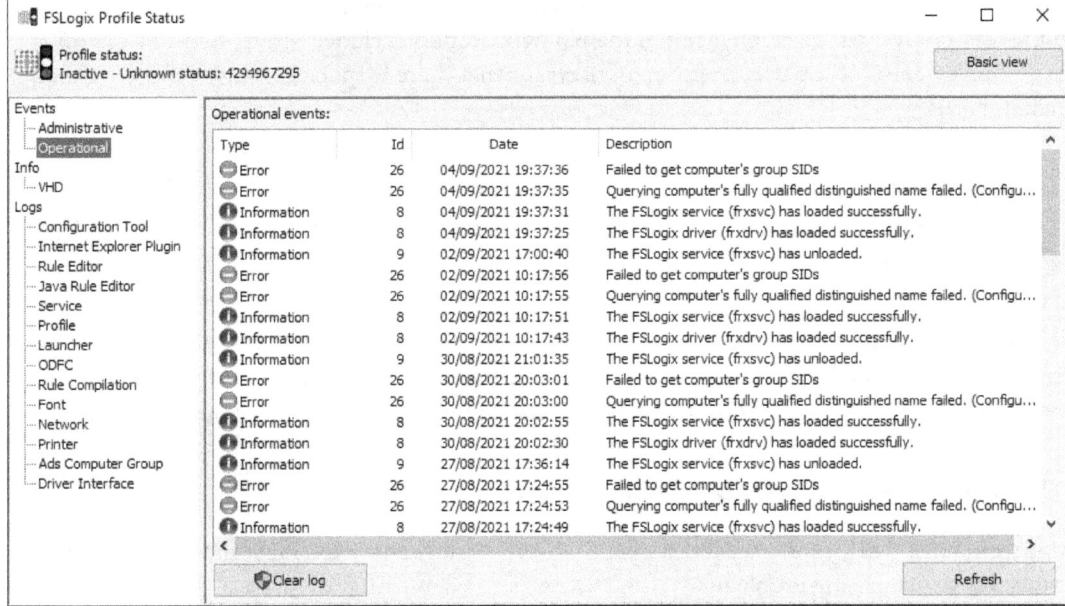

Figure 13.43 – The FSLogix Profile Status utility tool's Operational events page

You can also review VHD/VHD(X) disk usage and size using the FSLogix Profile Status utility tool and view logs that are specific to a particular area, such as the profile or service.

> **Important note**
> You can review the FSLogix logs via Event Viewer by going to **EventViewer - Applications and Services Logs** | **Microsoft** | **FSLogix**.

You can read more about FSLogix profile logs by going to the Microsoft document site: `https://learn.microsoft.com/en-gb/fslogix/troubleshooting-events-logs-diagnostics`.

One of the more typical issues related to FSLogix profile containers is ensuring that you have enabled the service on the session host or via Group Policy central management. Also, ensure that you have specified a valid VHD location (`VHDLocations`) and that permissions have been set correctly for the profile to be mounted for the user trying to log on.

> **Important note**
> When you're using the FSLogix Cloud Cache, `VHDLocations` is replaced with `CCDLocations`.

A couple of final closing issues to watch out for when troubleshooting FSLogix profile containers are that the user or group in question is not in the FSLogix profile's excluded group, that a local profile doesn't already exist for the user or group of users, or that there is space available on the selected profile storage.

> **Important note**
> For more information on configuring FSLogix profile containers, please refer back to *Chapter 12, Implementing and Managing FSLogix*.

In this section, we looked at troubleshooting FSLogix profile container issues within AVD. We provided a high-level summary and discussed the FSLogix Profile Status utility, as well as some common problems to watch out for. In the next section, we will look at troubleshooting AVD client issues.

Troubleshooting AVD client issues

In this section, we will look at troubleshooting AVD client issues and some hints and tips to help you on your way to diagnosing problems.

You need to ensure that your user can communicate with the AVD service and on corporate devices; you should add the correct firewall rules to do so.

The following table details the required URLs that the client should be able to access:

Address	Outbound TCP Port	Purpose	Client(s)	Azure Gov
`*.wvd.microsoft.com`	443	Service traffic	All	`*.wvd.microsoft.us`
`*.servicebus.windows.net`	443	Troubleshooting data	All	`*.servicebus.usgovcloudapi.net`
`go.microsoft.com`	443	Microsoft FWLinks	All	None
`aka.ms`	443	Microsoft URL shortener	All	None

Address	Outbound TCP Port	Purpose	Client(s)	Azure Gov
`docs.microsoft.com`	443	Documentation	All	None
`privacy.microsoft.com`	443	Privacy statement	All	None
`query.prod.cms.rt.microsoft.com`	443	Client updates	Windows Desktop	None

Table 13.4 – Outbound URLs required for AVD

The preceding table was adapted from `https://learn.microsoft.com/en-gb/azure/virtual-desktop/required-fqdn-endpoint?tabs=azure#remote-desktop-clients`.

In the next section, we will look at testing connectivity to help you troubleshoot any issues.

Testing connectivity

This subsection will look at how to test that the client can communicate correctly with AVD. We will look at two tests – one that uses PsPing and another that uses `nslookup`.

You can complete client tests using Sysinternals PsPing, which allows you to test connectivity from the client to the AVD service.

You can download PsPing from `https://learn.microsoft.com/sysinternals/downloads/psping`.

The test that was run in the following screenshot is essentially pinging the RDWeb service using port 443; that is, `"psping64.exe" -t rdweb.wvd.microsoft.com:443`.

As you can see, the device can communicate with the service:

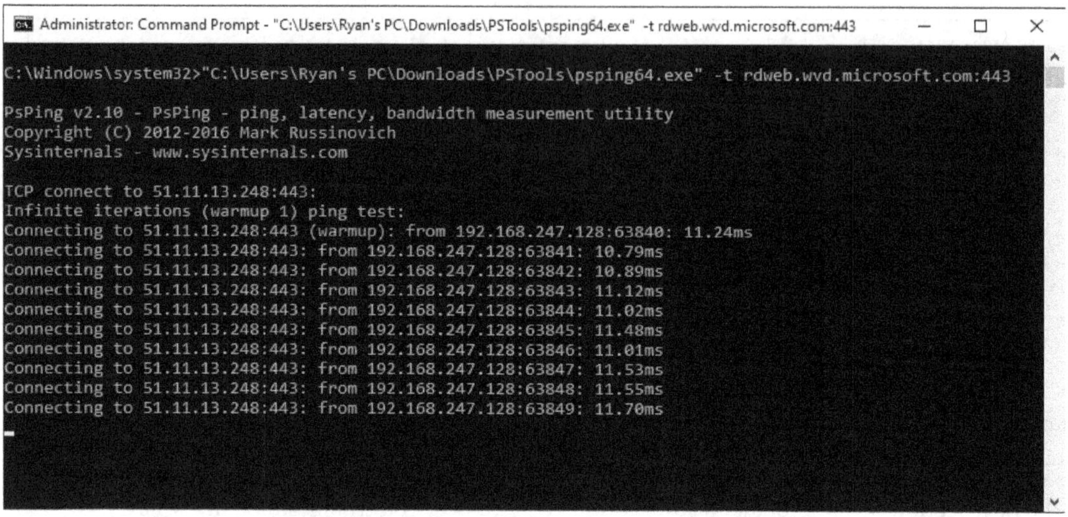

Figure 13.44 – PsPing testing port 443's connectivity to the AVD RDWeb service

You can also use `nslookup` to ensure that DNS is working as expected:

```
nslookup rdweb.wvd.microsoft.com
```

You should see the following output. If DNS is working, it should resolve with IP addresses and other information, such as aliases:

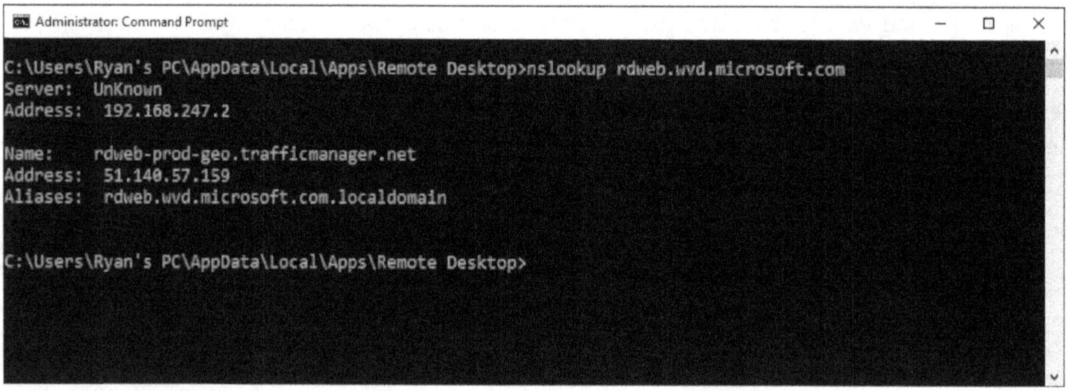

Figure 13.45 – PsPing testing DNS connectivity to the AVD RDWeb service

Now that we have looked at how to test client connectivity to the AVD service, let's learn how to reset the client.

Resetting the Remote Desktop client

If you find that a user's Remote Desktop client stops responding, cannot be opened, or you receive messages such as certificate errors, you may want to try resetting the client as a way to resolve the issue:

```
"%userprofile%\appdata\local\apps\Remote Desktop\msrdcw.exe" /reset
```

The result of the preceding query is shown in the following screenshot:

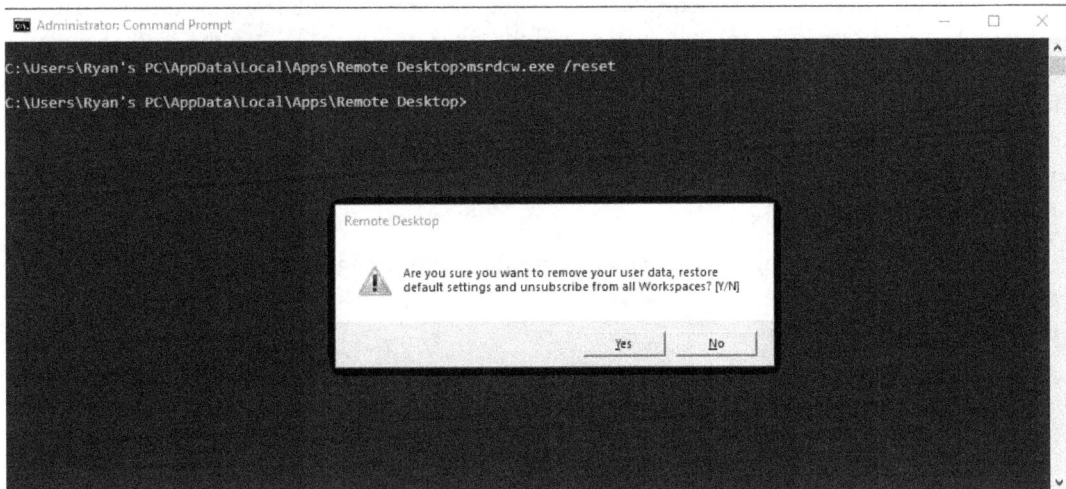

Figure 13.46 – Command line for resetting the Remote Desktop client

You can also add the [/f] switch to force the reset without receiving a pop-up message. This is useful if you want to automate the process of resetting multiple devices using an endpoint manager or other tool.

To reset with the force switch, you can use the following command:

```
"%userprofile%\appdata\local\apps\Remote Desktop\msrdcw.exe" /reset /f
```

Now that we have covered resetting the Remote Desktop client, let's move on to the next section, where we will learn what to do if the Remote Desktop client is showing no resources.

The Remote Desktop client is showing no resources

If your client is showing no resources, this is usually because the user has been taken out of the app group. If there has been a resource move between resource groups, this can impact the configuration of AVD. If no resources exist, check the app groups first. It is also advised that the user logs out of the client and re-authenticates to see if the issue persists.

In this section, we looked at troubleshooting AVD clients. We looked at troubleshooting and confirming connectivity, including testing connectivity to the AVD service using PsPing from Sysinternals and `nslookup` to ensure that DNS was working correctly.

Summary

In this chapter, we started by learning about Universal Print, which offers a modern, flexible print solution for organizations and complements AVD as both services are built inside Microsoft Azure. We then looked at Microsoft Endpoint Manager and the benefits it offers AVD in the sense that you can centralize and configure configuration policies for users using scope tags. Next, we looked at a relatively new feature to the AVD portfolio known as screen capture protection, which offers better protection for sensitive workspaces. We looked at how to add the required policies and configure them. Finally, we looked at troubleshooting FSLogix profile containers and troubleshooting the Remote Desktop client.

The next chapter is very exciting as we will look at installing and configuring apps on a session host.

Further reading

Step-by-Step: Configure and manage Microsoft Universal Print: `https://techcommunity.microsoft.com/t5/itops-talk-blog/step-by-step-configure-and-manage-microsoft-universal-print/ba-p/2227224`.

Questions

Answer the following questions to test your knowledge of this chapter:

1. Does screen capture protection support both remote apps and full desktops when you're using a Windows client?

    Windows clients only support full desktop screen capture protection at the time of writing.

2. Name the two Universal Print roles within AVD.

    Printer Administrator and Printer Technician.

3. Where can you retrieve the FSLogix logs from a session host?

    %ProgramData%\FSLogix\Logs.

4. When you're troubleshooting a Remote Desktop client on a Windows operating system, what command you would use to remove your user data, restore the default settings, and unsubscribe from all workspaces?

    "%userprofile%\appdata\local\apps\Remote Desktop\msrdcw.exe" /reset.

14
MSIX App Attach

In this chapter, we take a deep dive into MSIX app attach and the various terms and features associated with it. We will look at how to create an MSIX package and MSIX image. Then, we will progress to configuring MSIX app attach and publishing a remote app using MSIX app attach. To finish off the chapter, we will cover high-level troubleshooting for common issues.

In this chapter, we will look at the following:

- Configuring dynamic application delivery by using MSIX app attach
- What is MSIX?
- What does it look like inside MSIX?
- What is MSIX app attach?
- MSIX app attach terminology
- An overview of how MSIX app attach works
- Prerequisites
- Creating an MSIX package
- Creating an MSIX image
- Configuring Azure Files for MSIX app attach
- Importing the code-signed certificate
- Uploading MSIX images to Azure Files
- Configuring MSIX app attach
- Publishing an MSIX app to a RemoteApp application group
- Troubleshooting MSIX app attach

Configuring dynamic application delivery by using MSIX app attach

In this section, we'll look at **MSIX app attach** – a dynamic application delivery capability for Azure Virtual Desktop. This is a relatively new technology to the application delivery market that offers the ability to deliver applications to users either in session or as they log on. Let's learn more about it in detail.

What is MSIX?

MSIX is a modern application packaging format and development framework positioned as the future application packaging format. When combined with MSIX app attach, it provides a modern and dynamic application delivery mechanism.

MSIX offers a container-based packaging solution for Windows applications, simplifying the application installation process for IT admins and users. Applications packaged using MSIX run in a lightweight application container. The MSIX app and child processes run inside this container and are isolated using the filesystem and registry virtualization rather than running natively on the operating system.

MSIX applications can read the global registry. Additionally, an MSIX app writes to its own virtual registry and application data folder. This app data is then deleted when the app is uninstalled or reset using the apps and features settings page. Other applications do not have access to the virtual registry or virtual filesystem of an MSIX application.

> **Tip**
> Services work slightly differently from typical applications where the service runs outside of the container. If an application has a service, the service component does not run within the container.

Existing applications can be repackaged or converted to MSIX using the Microsoft MSIX packaging tool or via third-party packaging products such as **TMEditX** (https://www.tmurgent.com/appV/en/buy/tmeditx/tmeditx-documentation). This helps improve the success rates for MSIX.

Here are some of the benefits available when using MSIX as a packaging format:

- **Clean removal**: When you remove MSIX apps, you can delete all associated application data. As a result, no data remains in the registry or file system. However, there is the caveat that you must ensure the MSIX package has not written any configurations outside the container.
- It uses a container-type technology that isolates the application from the rest of the OS for security reasons.
- It offers predictable and secure deployment.

- The MSIX packaging format removes the deduplication of files across applications, and Windows manages the shared files across different applications. The applications are independent of each other, so updates will not impact other applications that share the file. A clean uninstall is guaranteed even if the platform manages shared files across applications.

Let's now explore the inner workings of MSIX and how the container architecture works.

What does it look like inside MSIX?

To recap the previous section, we looked briefly at how applications are packaged in an MSIX lightweight container. The MSIX package writes to its own virtual registry and application data folder. It is important to note that the application processes run inside that container.

Applications installed in the MSIX packaging format are located in the `C:\program files\WindowsApps` folder. Each package folder contains a set of standard files required for the MSIX application to communicate with the Windows operating system API.

The following figure shows the structure of a typical MSIX package:

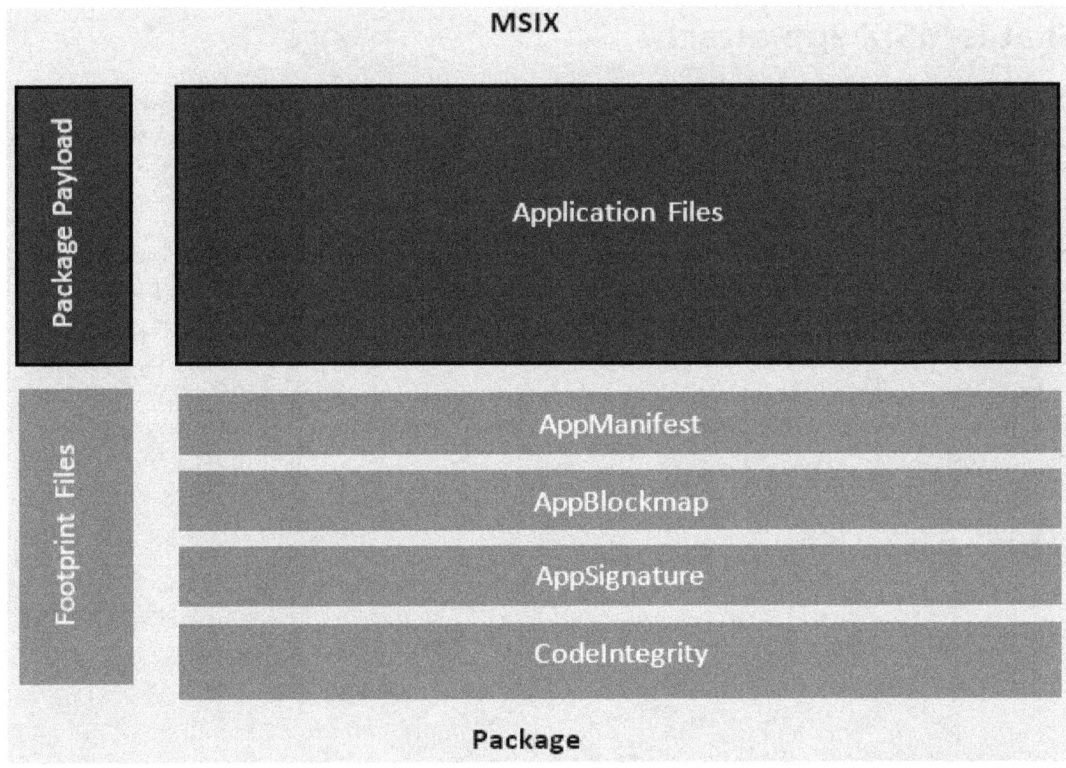

Figure 14.1 – Structure of an MSIX package

The following table details the core contents of an MSIX package:

File	Description
App payload	Contains the app code files and assets
`AppxBlockMapp.xml`	Contains a verified and secure list of all the files within the package
`AppxManifest.xml`	Essentially the configuration file for the MSIX package and contains the identity of the package and its dependencies
`AppxSignature.p7x`	Contains the signature of the package that the OS must trust before the application can be installed

Table 14.1 – Table showing the core contents of an MSIX package

Now that we have finished looking at the fundamentals of MSIX, we will move to looking at MSIX app attach.

What is MSIX app attach?

MSIX app attach is a relatively recent technology advancement and an addition to the Azure Virtual Desktop feature portfolio. This technology facilitates the delivery of applications to a user session within Azure Virtual Desktop by dynamically attaching the applications in an MSIX image. This new capability delivers applications upon logon and/or in-session, hence the term dynamic.

One of the critical benefits of MSIX app attach over other application delivery technologies is that it has been designed to deliver applications that do not impact or delay user login times. It also offers reduced management that IT needs to perform as applications are not installed natively on the Azure Virtual Desktop image. This allows organizations to modernize/enhance their virtual desktop estate and move from dedicated/specific virtual machines for applications. When using MSIX app attach, applications are delivered dynamically to any VM that a user may log in to. This saves on complexity, time, and costs.

The following figure depicts some of the benefits of MSIX app attach:

Single Instance Storage
- MSIX app attach uses one instance of the MSIX application to deliver to all hosts without consuming extra space.

Resistance to tampering
- After an MSIX package has been expanded into an MSIX image, the latter is read-only and locked down for modifcation by the OS.

Improvied use experience
- Providing the ability to attach applications in session or at logon.

Simplified gold image Management
- As department/group applications

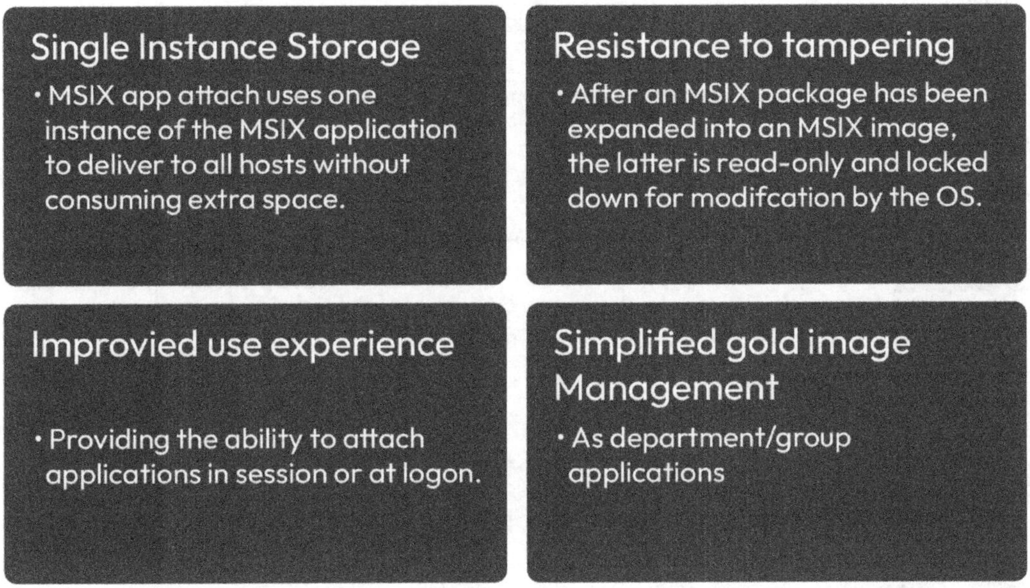

Figure 14.2 – The four key benefits of MSIX app attach

> **Important note**
> It is important to note that MSIX app attach requires that an MSIX package is in an expanded state, also known as an MSIX image (a virtual disk or **Composite File System** (**CimFS**) image). Expanding is the process term for taking the MSIX package, unzipping it, and applying the appropriate system permissions to the file structure inside the chosen virtual disk or **Composite Image File System** (**CimFS**) image.

The following figure illustrates both FSLogix profile containers and MSIX app attach in Azure Virtual Desktop:

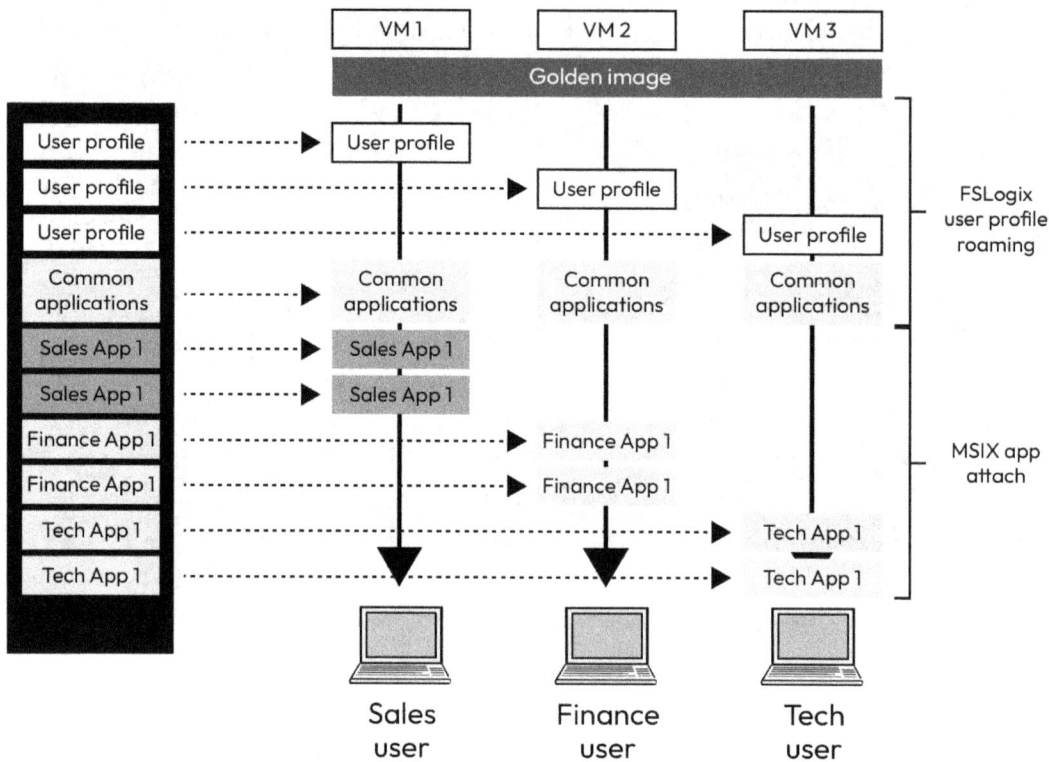

Figure 14.3 – MSIX app attach and FSLogix profile containers working together

From the preceding diagram, you will note that typical applications are installed within the operating system image. The user profiles are attached on logon, and departmental/group applications are dynamically delivered to the user on logon.

> **Important note**
> The term *dynamically* refers to the process of attaching a virtual disk or CimFS image to the operating system rather than the traditional method of installing an application natively.

MSIX app attach terminology

There are five different MSIX app attach process stages to be aware of:

Term	Definition
Stage	Azure Virtual Desktop notifies the operating system that an application is available. This means the virtual disk containing the MSIX package (also known as the MSIX image) is mounted.
Register	MSIX app attach uses a per-user process to make the application available to the user.
Delayed or deferred registration	Complete registration of the application is delayed until the user decides to run the application.
Deregistration	The application is no longer available to you after you sign out.
Destage	The application is no longer available from the VM following the shutting down or restarting of the VM.

Table 14.2 – Table showing the different stages of the MSIX app attach process

We'll now move on to looking at how MSIX app attach works within Azure Virtual Desktop.

An overview on how MSIX app attach works

There are typically five steps to delivering applications using Azure Virtual Desktop and MSIX app attach:

1. You would sign in and select the host pool you should have access to from the Azure Virtual Desktop client. The process is similar to opening published desktops or RemoteApp applications from the Azure Virtual Desktop environment.
2. You're assigned a virtual machine within the host pool from which a RemoteApp or Remote Desktop session is created. Then, the Azure Virtual Desktop client interacts with the session.
3. If the user profile is configured, the FSLogix agent on the session host provides the user profile from the file share. The file share can be via Azure Files, Azure NetApp Files, or an **Infrastructure-as-a-Service (IaaS)** file server.
4. Applications that are assigned to you are read from the Azure Virtual Desktop service.
5. MSIX app attach applications are registered to the operating system for you from the attached MSIX virtual disk. For example, the virtual disk might be on an IaaS file share or Azure NetApp Files.

The following diagram highlights the preceding five key steps:

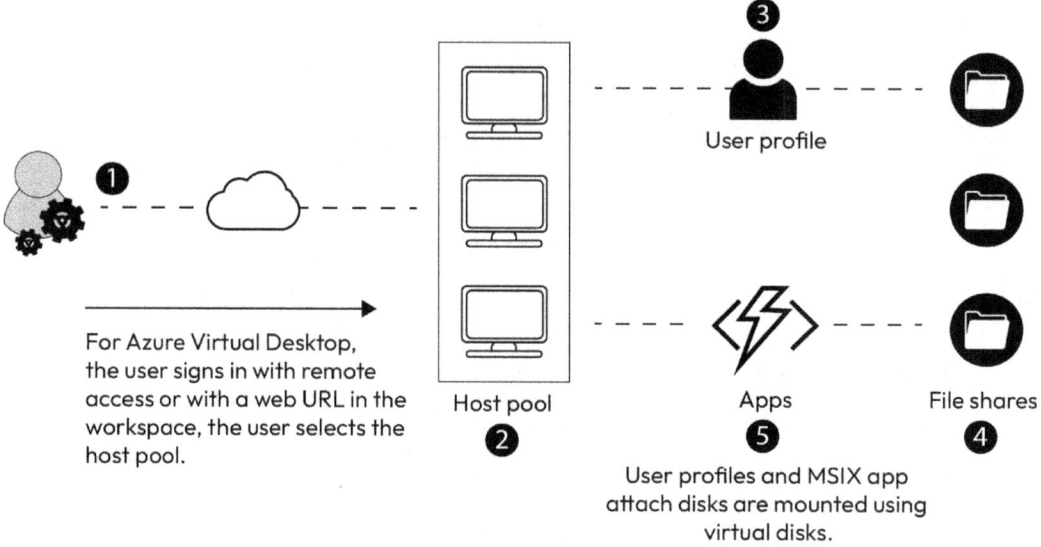

Figure 14.4 – Five steps for delivering MSIX app attach to Azure Virtual Desktop users

Now that we have covered the basics of MSIX app attach, we'll move on to the prerequisites and getting started with MSIX app attach.

Prerequisites

Before we can start provisioning MSIX images and applications to users, we first need to package some applications in the MSIX app attach format.

> **Important note**
> You will need a code-signing certificate to create MSIX packages. The certificate can either be a public, domain, or self-signed certificate.

The five steps to prepare your package for MSIX applications are as follows:

Step	Description
Create an MSIX package	The first step is to create an MSIX package containing the application you want to deliver dynamically.
Create an MSIX image	Create an MSIX image using the MSIX package you created in the previous step.
Configure Azure Files for MSIX app attach (It's recommended that Azure NetApp Files is used for enterprise deployments)	Configure Azure Files, including the required permissions for the file share.
Upload the MSIX images	Upload your prepared MSIX image.
Code-signed certificate	Install the required certificate (self-assigned or otherwise) on all the required session hosts for use with MSIX app attach.

Table 14.3 – Table showing the key steps for preparing for MSIX app attach in Azure Virtual Desktop

It is important to note that you cannot use Entra ID Domain Services with MSIX app attach.

> **Tip**
> Azure Files has a maximum concurrent handle limit of 2,000. It is advised that enterprises use Azure NetApp Files when planning to use MSIX app attach. Read more on this limitation here: `https://docs.microsoft.com/azure/storage/files/storage-files-scale-targets#file-scale-targets`

In the next section, we look at creating an MSIX package.

Creating an MSIX package

Before we can create an MSIX image, we need the application in the correct format, which is an MSIX package. You will first need to package your application with the MSIX packaging tool or a third-party tool such as **TMEditX**.

> **Important note**
>
> MSIX app attach does not support automatic application updates. Therefore, you must disable automatic updates on the operating system. Otherwise, you could end up with unusable applications for the end user(s) if updates occur. It is also important to note that you may need to turn off auto-updating within the application if it supports such a feature.

To turn off automatic updates on the operating system image, you need to run the following script within Command Prompt:

```
rem Disable Store auto-update:
reg add HKLM\Software\Policies\Microsoft\WindowsStore /v AutoDownload /t REG_DWORD /d 0 /f
Schtasks /Change /Tn "\Microsoft\Windows\WindowsUpdate\Automatic app update" /Disable
Schtasks /Change /Tn "\Microsoft\Windows\WindowsUpdate\Scheduled Start" /Disable
rem Disable Content Delivery auto download apps that they want to promote to users:
reg add HKCU\Software\Microsoft\Windows\CurrentVersion\ContentDeliveryManager /v PreInstalledAppsEnabled /t REG_DWORD /d 0 /f
reg add HKLM\SOFTWARE\Microsoft\Windows\CurrentVersion\ContentDeliveryManager\Debug /v ContentDeliveryAllowedOverride /t REG_DWORD /d 0x2 /f
```

> **Important note**
>
> You can create your own MSIX packages or you can download them from the **Integrated Software Vendor** (**ISV**). More and more ISVs are turning to MSIX as the packaging format of choice. A good example is Mozilla Firefox. You can read more here: http://gecko-docs.mozilla.org-l1.s3.us-west-2.amazonaws.com/browser/installer/windows/installer/MSIX.html

To get started with creating your first MSIX package, you will need to download the MSIX package tool from the following URL:

https://www.microsoft.com/p/msix-packaging-tool/9n5lw3jbcxkf

The following screenshot shows the main UI page of the MSIX packaging tool:

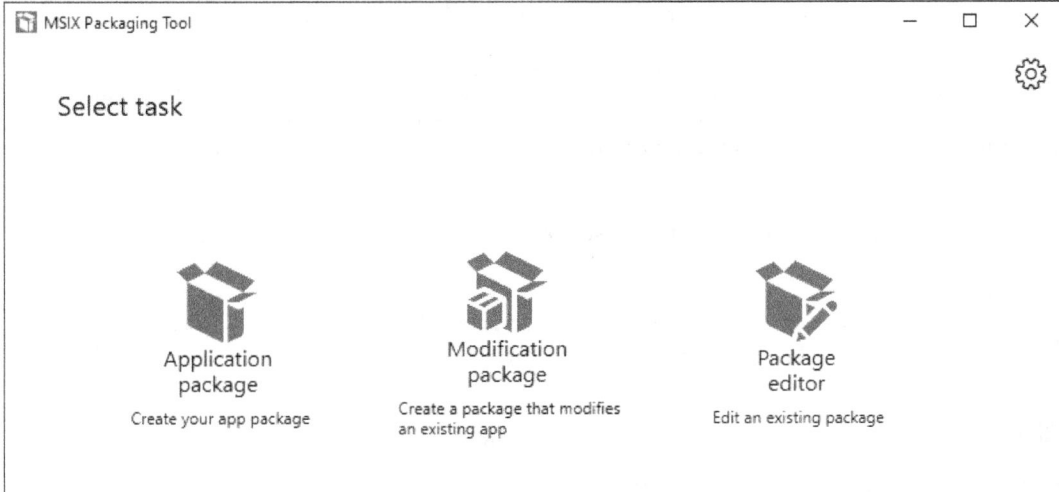

Figure 14.5 – The main page of the MSIX packaging tool

We'll now move on to look at packaging a simple application in the MSIX packaging format.

Packaging a simple application in an MSIX container

Before we take a look at creating a package, I want to briefly cover some of the options available to you when you create an MSIX package. Firstly, even though it's common, you don't have to package one app per MSIX package. What this means is that you can package multiple packages in the same MSIX package if required.

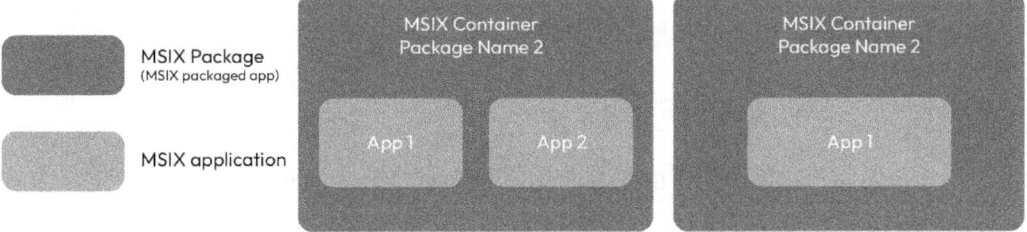

Figure 14.6 – Two different options when creating an MSIX package

Before we create an MSIX package, we need a certificate. As an example, I will create a self-signed certificate.

To create a self-signed certificate, you can use the following PowerShell script:

```
New-SelfSignedCertificate -Type Custom -Subject "CN=RMSITBLOG,
O=Ryanmangansitblog, C=GB" -KeyUsage DigitalSignature -FriendlyName
"Your friendly name goes here" -CertStoreLocation "cert:\
CurrentUser\ my" -TextExtension @("2.5.29.37={text}1.3.6.1.5.5.7.3.3",
"2.5.29.19={text}")
```

Once you have the CER certificate, you can use the following PowerShell cmdlets to convert it into a PFX certificate:

```
$password = ConvertTo-SecureString -String -Force -AsPlainText Export-
PfxCertificate -cert "Cert:\CurrentUser\My\" -FilePath .pfx -Password
$password
```

We'll now move on to looking at creating an MSIX package:

1. First, download and install the MSIX Packaging Tool.
2. Launch the application and you should see the following screen:

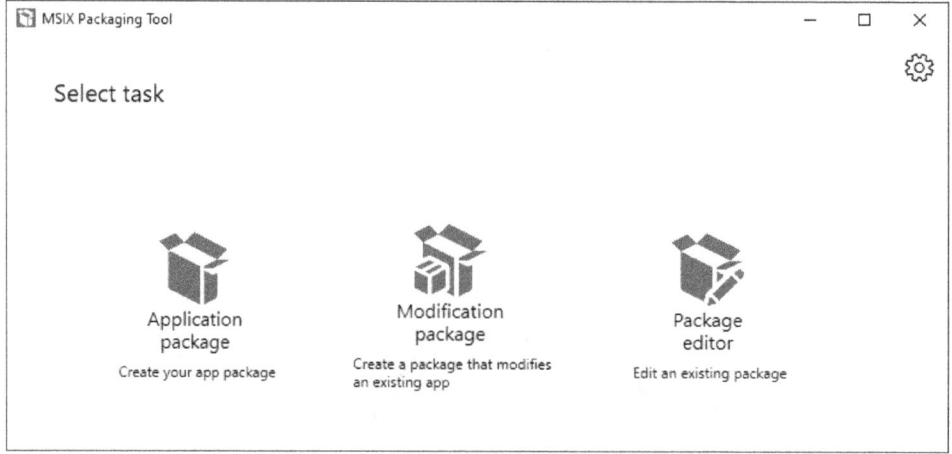

Figure 14.7 – The home page of the MSIX Packaging Tool

3. Click the cog located in the top-right corner of the home page.

4. Scroll down to the certificate section:

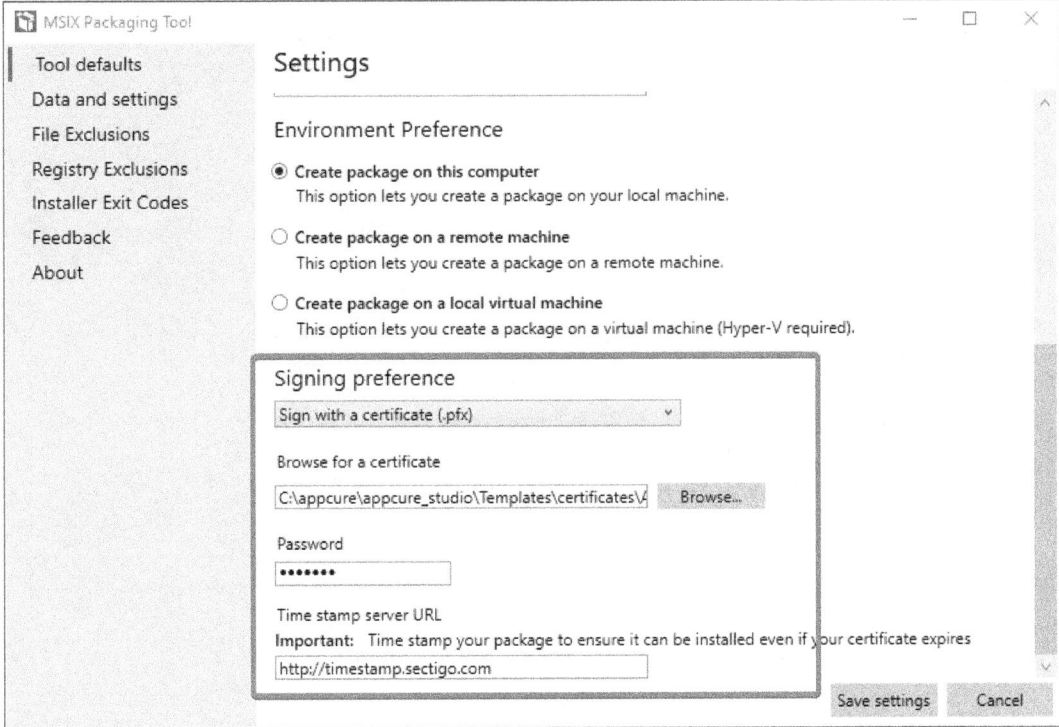

Figure 14.8 – Save the certificate settings for application packaging

5. Select **Sign with a Certificate (.pfx)** and enter the path of the certificate and password.
6. Optionally (however, somewhat important), enter a time stamp URL so that your packages do not stop working when the certificate expires.
7. Click **Save settings**.

We'll now move on to creating the MSIX package:

1. On the home page of the MSIX Packaging Tool, click **Application package**:

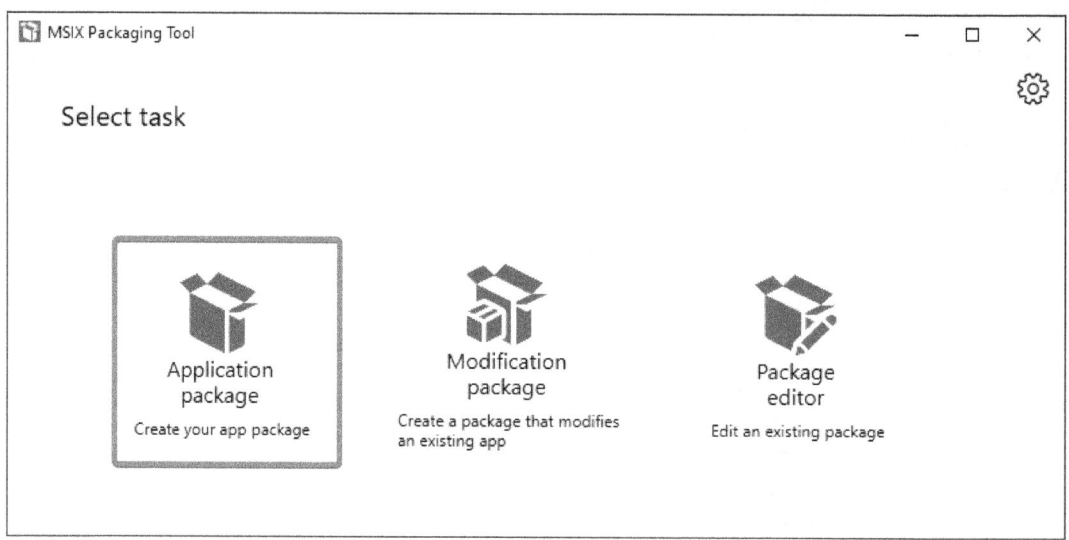

Figure 14.9 – Application package icon highlighted

2. Select **Create package on this computer**:

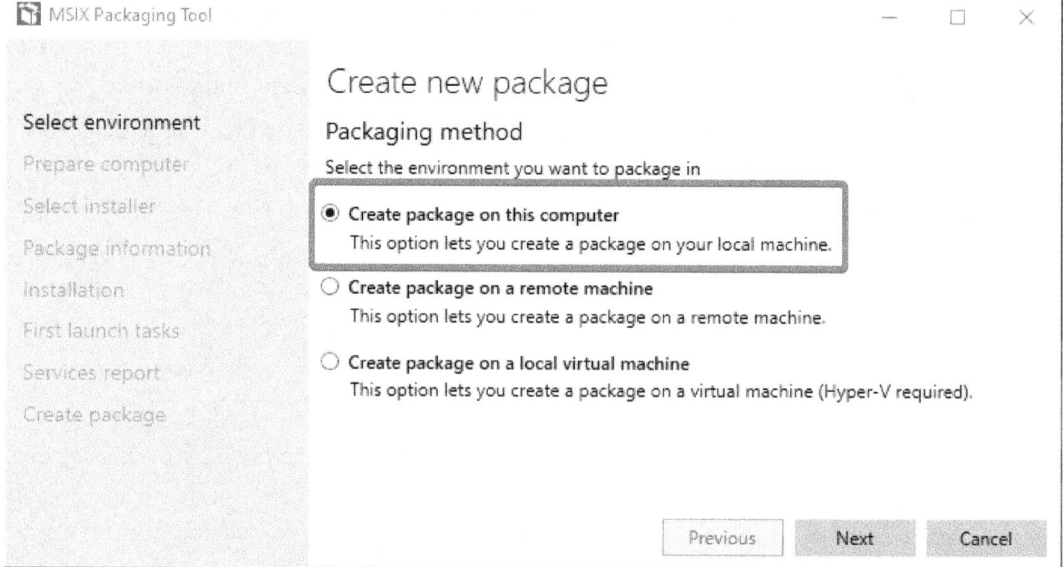

Figure 14.10 – Create package on this computer

3. Click **Next**.
4. Within the **Create new package** screen, you will see two sections, **Additional preparations** and **Recommended action items**:

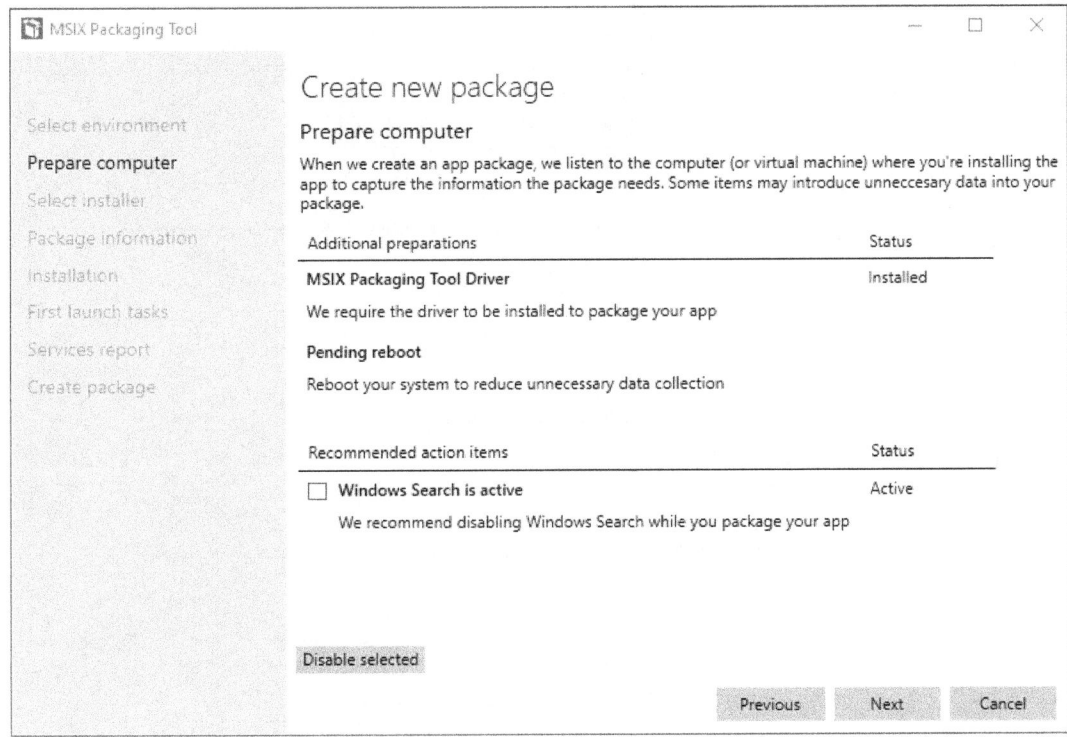

Figure 14.11 – Prepare computer page

5. Check the box **Windows Search is active** and click **Disable selected**.

6. Then click **Next**:

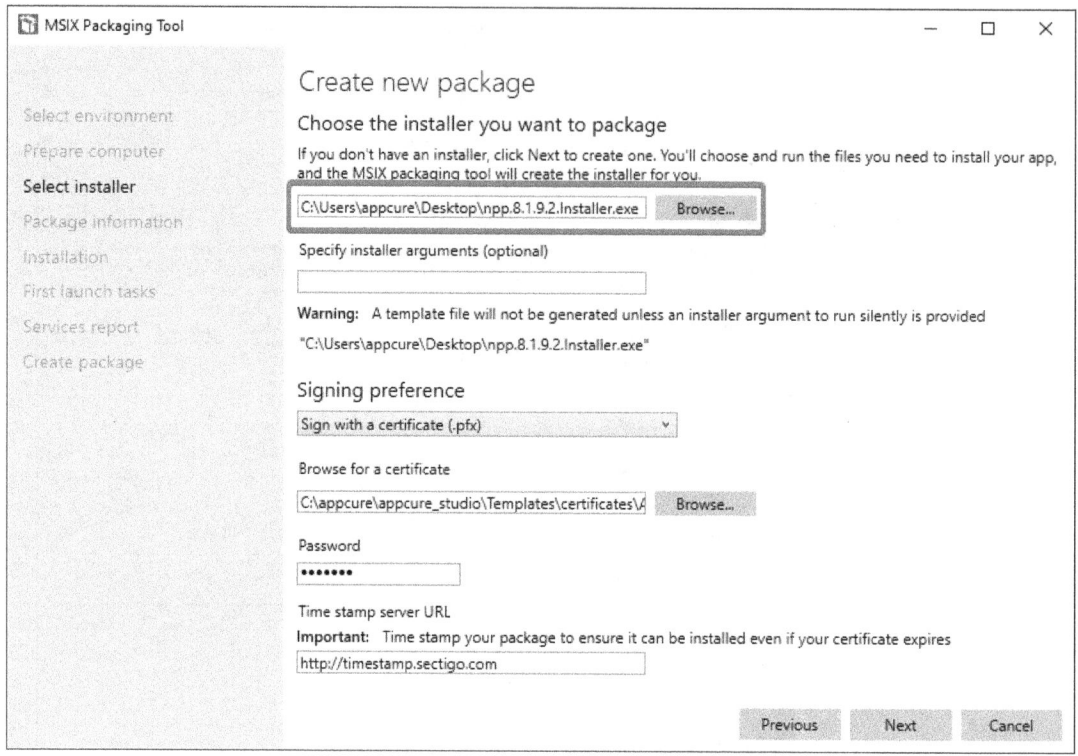

Figure 14.12 – Installer selection page

> **Important note**
> Please note that if you have not set a time stamp server URL, then when the certificate expires, you will no longer be able to use the app. This will require the application to be signed again with an in-date certificate.

7. Select the application installer you want to use. In this example, we are using Notepad++.
8. You will also note that the certificate settings we previously configured are now shown within the **Signing preference** section.
9. Click **Next** to continue.
10. You will now be presented with the **Package information** page.

11. All fields marked with an asterisk are mandatory:

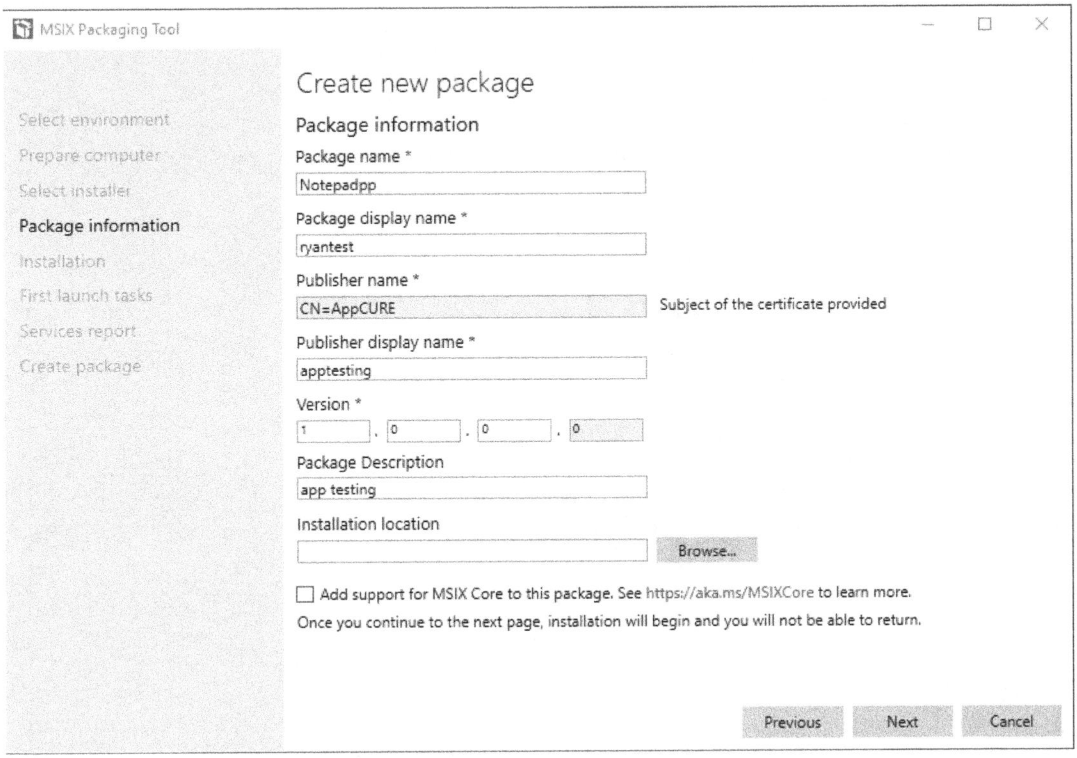

Figure 14.13 – Package information page

> **Type**
> It's important that the private key and CN (organization's name) matches the package for the code-signing process to work when the package is created. You should see the notification alongside the publisher name field stating **Subject of the certificate provided**. If you do not see this, then you may want to check your certificate.

12. Once you have filled out all the application information, you can proceed to the next page, **Installation**. Click **Next**.

13. You will then see the **Installation** page and the application installer will launch:

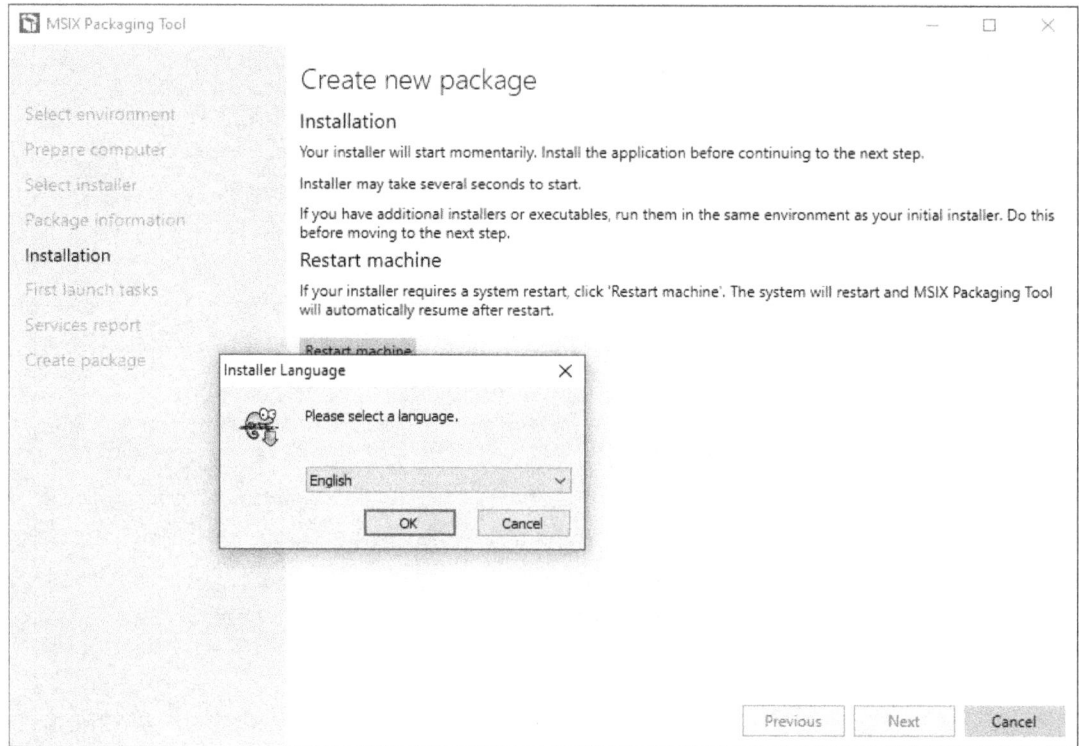

Figure 14.14 – Installation page

14. Follow the steps to install the package and launch it on install completion.
15. Once installed and launched, make the required setting changes within the app and, when complete, close the application.
16. Click **Next** within the MSIX Packaging Tool:

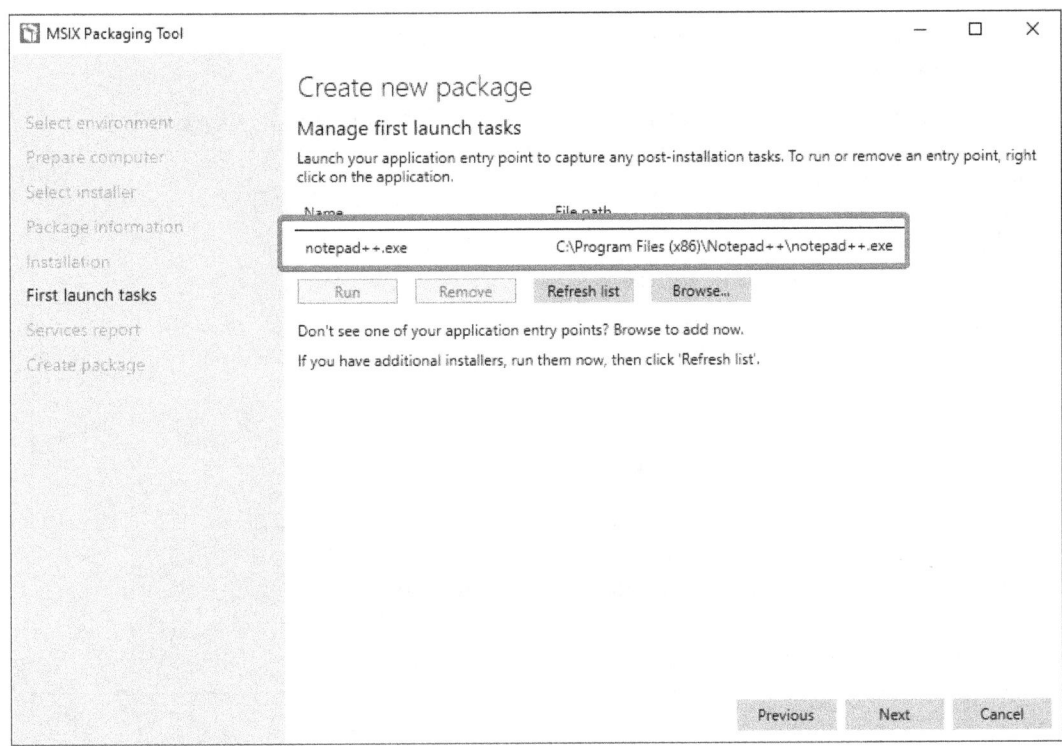

Figure 14.15 – Entry points for the application

17. Ensure you have set the required entry points on the **Manage first launch tasks** page, then click **Next**. You will then be prompted with the following:

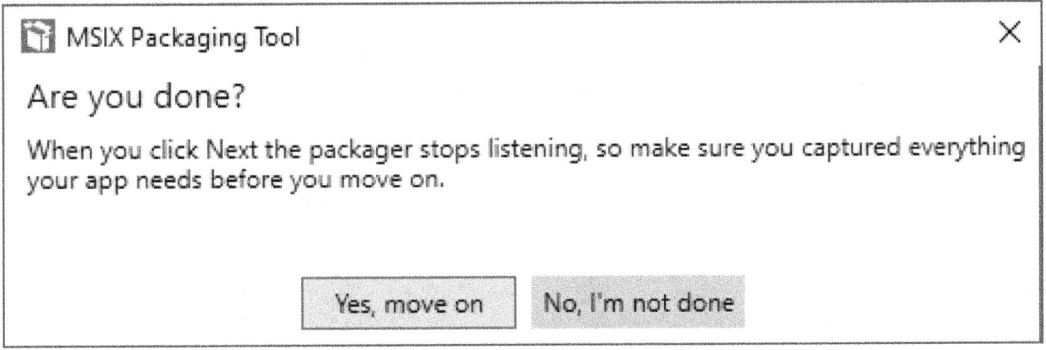

Figure 14.16 – Prompt

18. Click **Yes, move on** if you have finished installing apps.

19. On the **Services report** page, exclude any services if required. In this example, there are none. Click **Next** to continue.

20. On the final page, **Create package**, select the save location and click **Create package**:

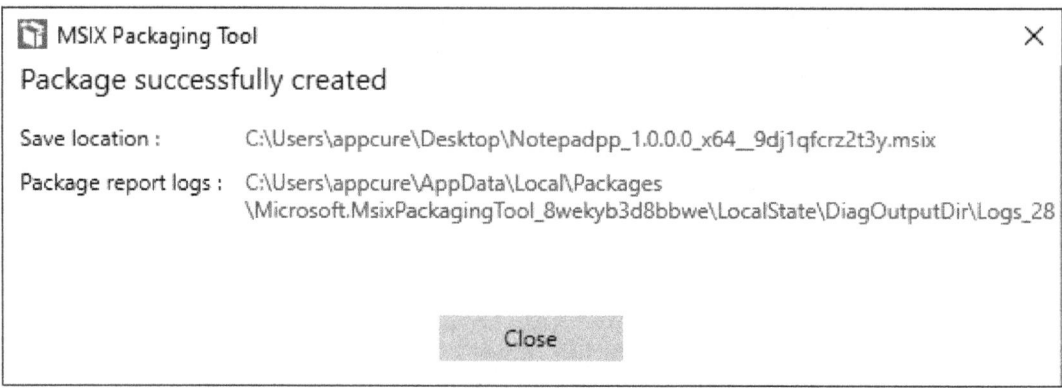

Figure 14.17 – Package successfully created

21. The final step is to launch the app to make sure it's working:

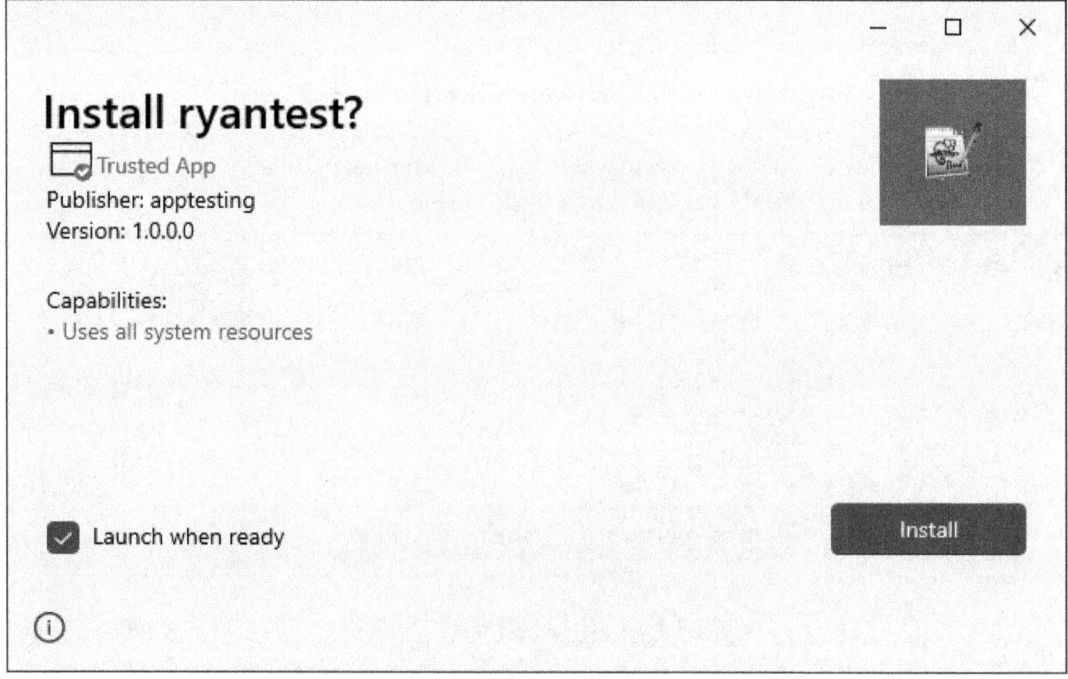

Figure 14.18 – MSIX package ready to install on the operating system

There you have it – we just walked through the steps of creating an MSIX package.

The final point to cover in this section is the Package editor feature within the MSIX Packaging Tool.

The Package editor feature offers the ability to configure app capability declarations for the MSIX package, which are essentially used for specifying specific access to areas of the Windows APIs and resources such as pictures, music, and devices such as a microphone or camera:

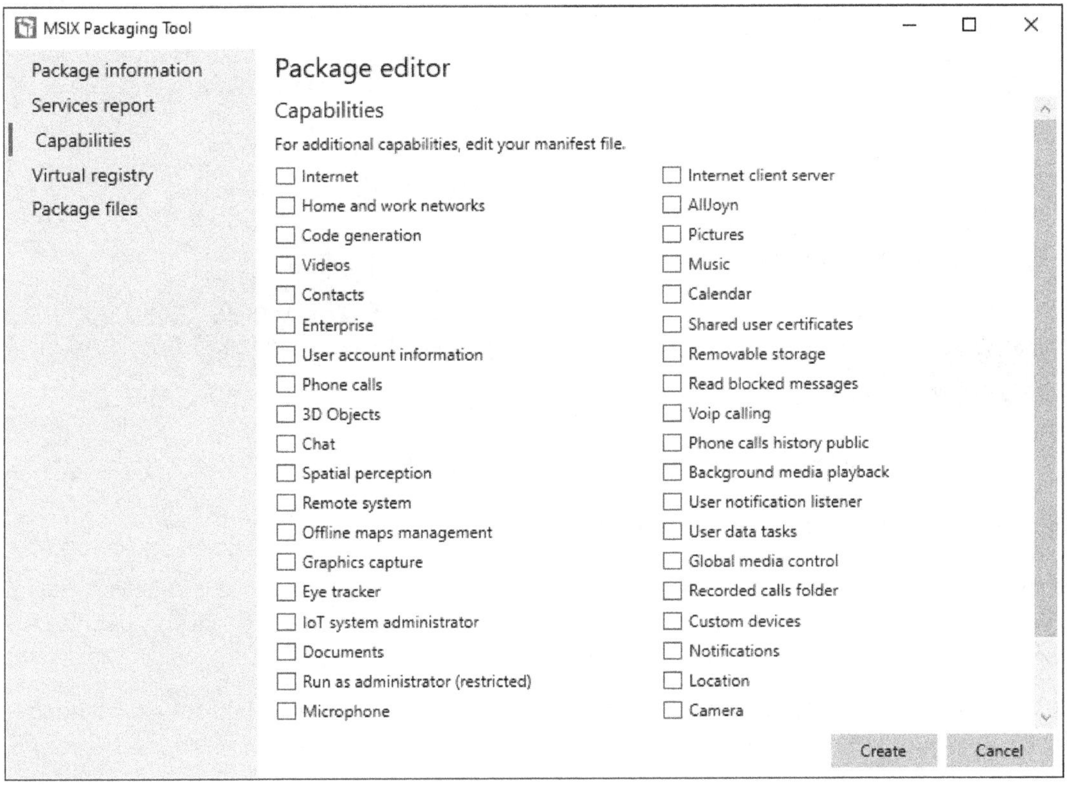

Figure 14.19 – Package editor

The Package editor also enables the ability to add and remove files and modify the virtual registry, which includes the importing and exporting of registry settings.

Let's now move on to taking a look at creating an MSIX image.

Creating an MSIX image

An **MSIX image** is essentially one of three different disk types: VHD, VHDX, or a CIM image. These virtual disks/images are mounted or attached to the session host, and the content is then registered using the native operating system APIs.

The following figure shows the typical structure of an MSIX image. As you can see from the key elements, three layers make up the structure of the MSIX image:

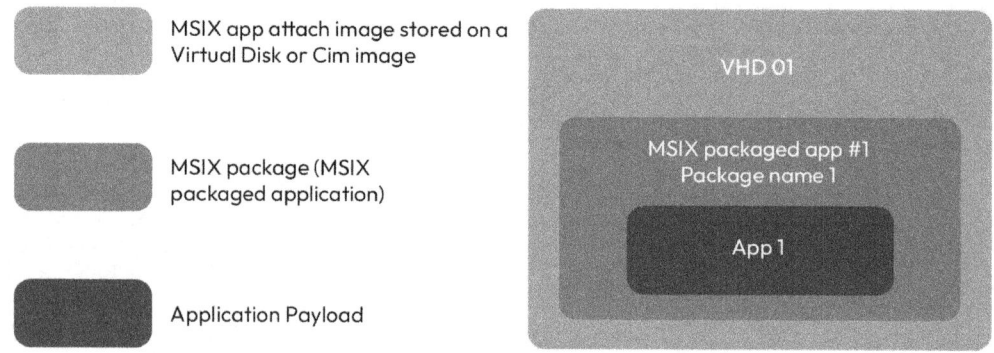

Figure 14.20 – An illustration of the structure of an MSIX image

What is the Composite File System (CimFS)?

The **CIM** image is essentially a file-backed image format that is a similar concept to WIM. Read more on WIM here: https://docs.microsoft.com/windows-hardware/manufacture/desktop/wim-vs-ffu-image-file-formats.

The term *composite* is used as it can contain multiple file system volumes that can be mounted individually but still share the same data files.

The following figure shows you the three types of files that you should expect to see when creating a CIM image:

Name	Date modified	Type	Size
objectid_fc0b3d45-5988-4edc-8a38-361...	24/02/2021 16:38	File	33 KB
objectid_fc0b3d45-5988-4edc-8a38-361...	24/02/2021 16:38	File	227 KB
objectid_fc0b3d45-5988-4edc-8a38-361...	24/02/2021 16:38	File	12 KB
region_fc0b3d45-5988-4edc-8a38-3611...	24/02/2021 16:38	File	1,056 KB
region_fc0b3d45-5988-4edc-8a38-3611...	24/02/2021 16:38	File	1,251,348 ...
region_fc0b3d45-5988-4edc-8a38-3611...	24/02/2021 16:38	File	170 KB
testapp.cim	24/02/2021 16:38	CIM File	1 KB

Figure 14.21 – The structure of a composite image (CIM)

When you create a CIM image, you will see three file types in the folder: `ObjectID`, region files containing the data, and the composite image file (`.cim`) that contains the metadata.

> **Tip**
>
> When creating a CIM image, it's advised that you create one per folder. If you create multiple CIM images in the same folder it will be very difficult for you to separate the package files. This is something I call *CIM sprawl*.
>
> The following link will help you address this issue: https://ryanmangansitblog.com/2021/02/11/msix-app-attach-how-to-manage-cimfs-file-sprawl-tips-and-tricks/

CimFS does offer performance benefits compared to traditional virtual disk types (VHD and VHDX). You can read more on the performance benefits here: https://docs.microsoft.com/azure/virtual-desktop/app-attach-glossary#cim.

Creating an MSIX image

To create an MSIX image, you will need to download the MSIXMGR tool used for creating MSIX images. You can download the MSIXMGR tool from https://aka.ms/msixmgr.

The following steps detail the process to expand an MSIX file:

1. Download the MSIXMGR tool if you haven't already.
2. Unzip `MSIXMGR.zip` into a local folder.
3. Open Command Prompt in elevated mode.
4. Find the local folder from *step 2*.
5. Run the following command in Command Prompt to create an MSIX image:

    ```
    msixmgr.exe -Unpack -packagePath -destination [-applyacls]
    [-create] [-vhdSize ] [-filetype ] [-rootDirectory ]
    ```

Here's an example of the use of the MSIXMGR tool:

```
msixmgr.exe -Unpack -packagePath "C:\apps\notepadpp_1.0.0.0_x64__
ekey3h7rct2nj.msix" -destination "C:\apps\testapp.vhdx" -applyacls
-create -vhdSize 2048 -filetype VHDX -rootDirectory MSIX
```

The output should look similar to the following screenshot:

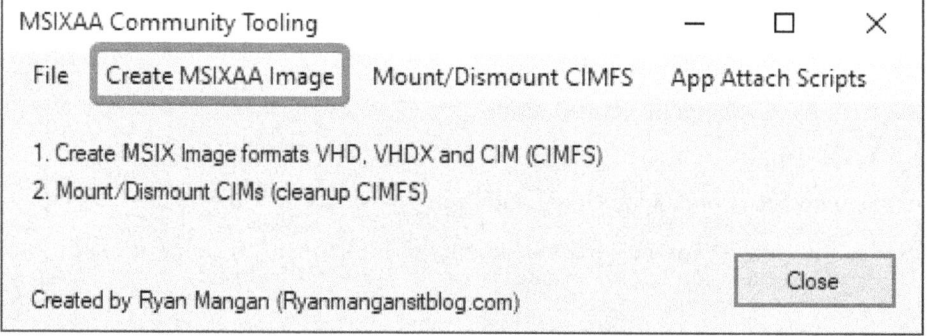

Figure 14.22 – The output of running the msixmgr.exe cmdlet

Once you have created your MSIX image, we need to progress to configuring Azure Files for MSIX app attach:

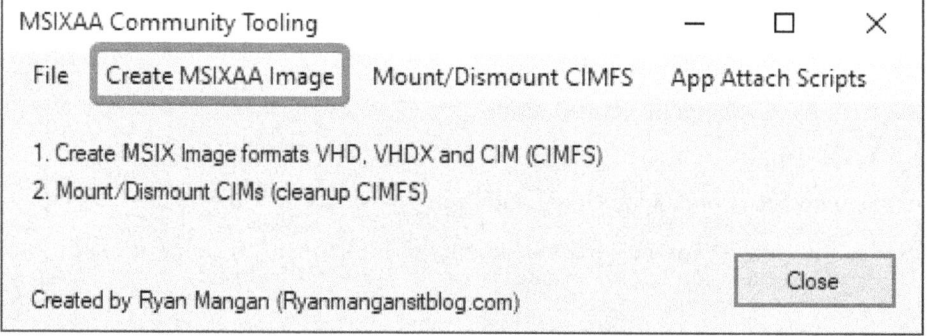

Figure 14.23 – MSIXAA community application main page

Select the **Create MSIXAA Image** button on the main form. This will then load the MSIXAA image creation form.

Within the form shown in the following screenshot, you have the option to create an MSIX image of one of the three types of MSIX images. Once you have selected the MSIX package you want to convert into an MSIX image and completed the required fields of the form, you can proceed to create an MSIX image by clicking **Create MSIX Image**:

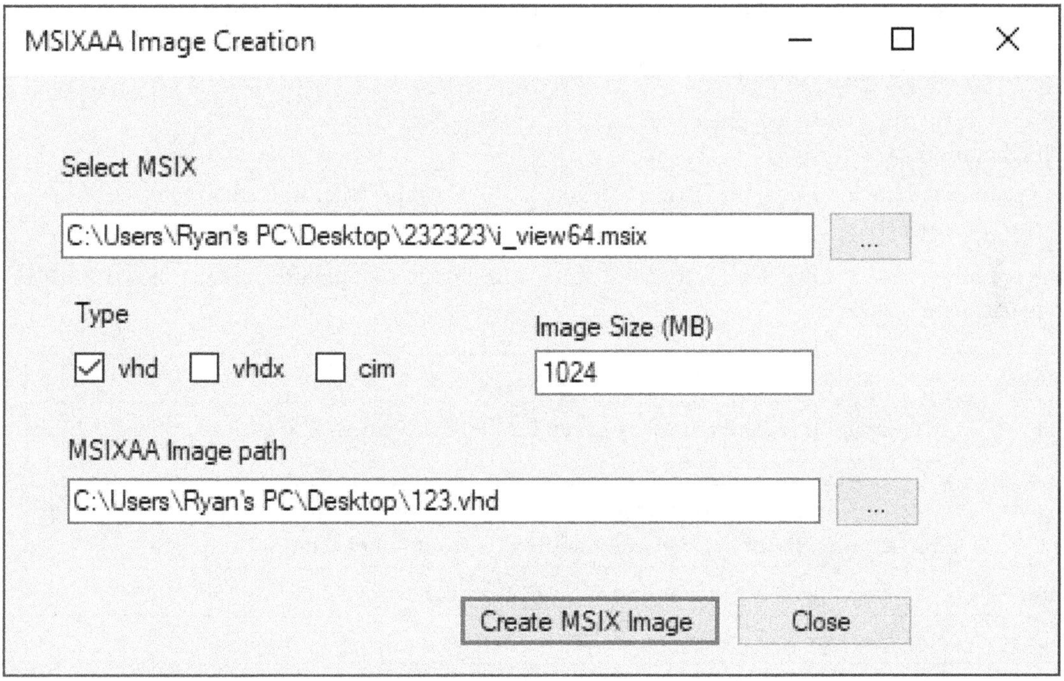

Figure 14.24 – MSIXAA community tooling app that can be used to create MSIX images

You can also use the MSIXAA Community Tooling application found on GitHub: `https://github.com/RMITBLOG/MSIX_APP_ATTACH/releases/download/2.1/MSIXAA_27052021.msi`

For more information on using the Microsoft MSIXMGR tool, you can visit this link: `https://docs.microsoft.com/azure/virtual-desktop/app-attach-msixmgr`

Configuring Azure Files for MSIX app attach

In this section, we take a look at configuring Azure Files for MSIX app attach. This is an important step as incorrectly configuring the storage and permissions will impact users when applications are trying to attach and register within user sessions. The file share is a process similar to FSLogix in terms of setting up. However, one significant difference is that you need to ensure that the specific permissions are assigned correctly.

> **Important note**
> Make sure the storage used has a latency of less than 400 ms as per Microsoft recommendations.

There are nine steps required for configuring Azure Files for MSIX app attach. The following table summarizes these steps:

Step	Description
One	Create an Entra ID DS security group for the session hosts you wish to use with MSIX app attach.
Two	Add the computer accounts for all session host VMs as members of the created group. Ensure you reboot the VMs after adding the computer accounts to the group.
Three	You then need to Sync the Entra ID DS group to Entra ID.
Four	Create a storage account.
Five	You will then need to create a file share under the storage account by following the instructions for creating an Azure file share.
Six	Join the storage account to Entra ID DS.
Seven	Add the storage file data SMB share contributor role to the synced Entra ID DS group that contains the computer object of the session hosts.
Eight	To be able to manage the file share, you need to grant NTFS permissions on the file share to the computer object's Entra ID DS group.
Nine	Grant NTFS permission for the user or user group containing the user accounts, sourced from Entra ID DS.

Table 14.5 – 9 important steps to configure Azure Files for MSIX app attach

Let's now take a look at these steps in more detail:

1. We first need to create the security group within Active Directory Domain Services. To do this, we create a security group in Active Directory as shown in the following screenshot:

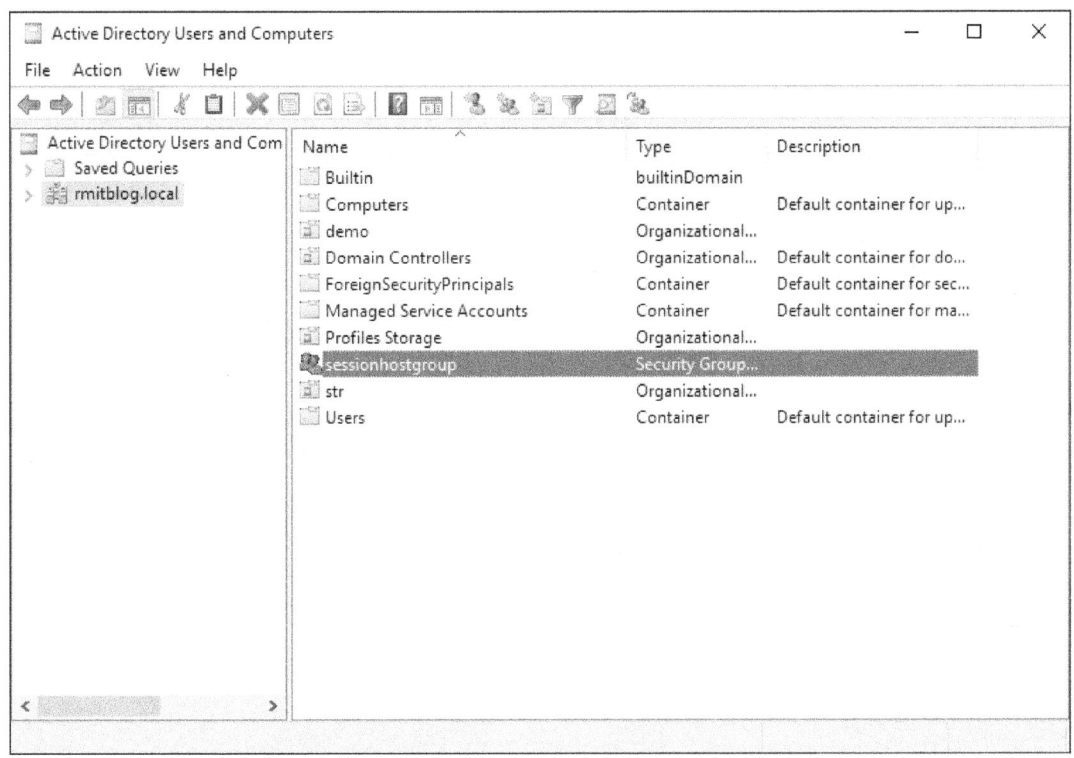

Figure 14.25 – The session host group being created

2. Once the group has been created, we then add the computer accounts to the group as shown in the following screenshot:

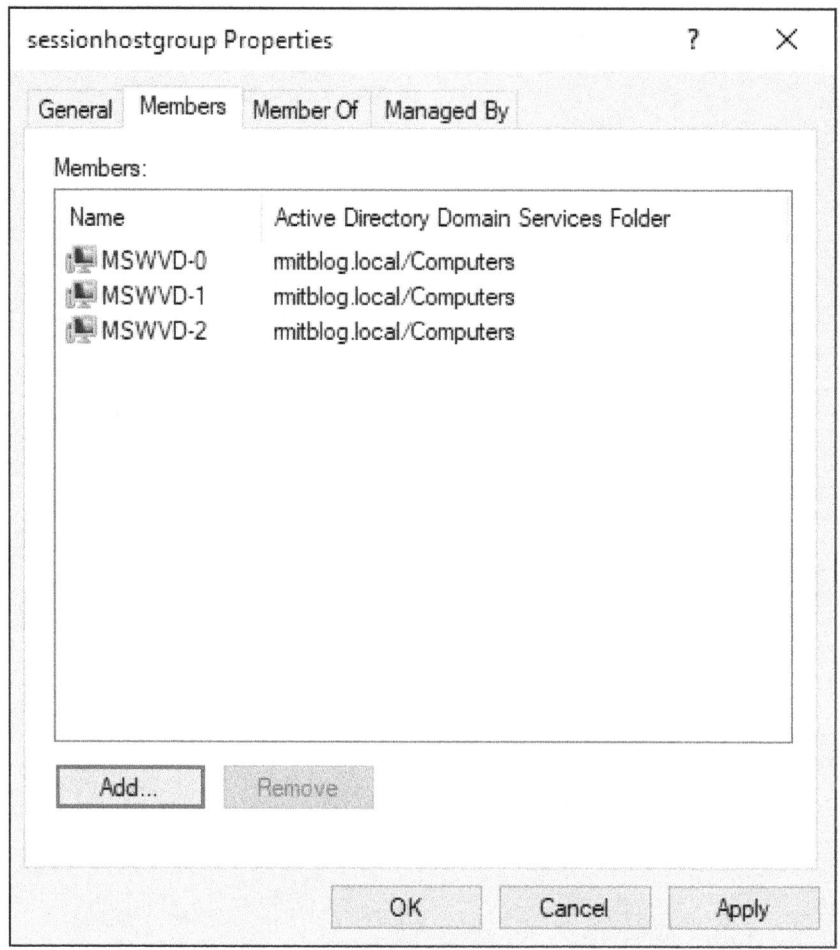

Figure 14.26 – Session host computer accounts added to the synchronized AD group

3. Once complete, you then need to ensure the security group has been synchronized with Entra ID to ensure that the computer accounts are synced:

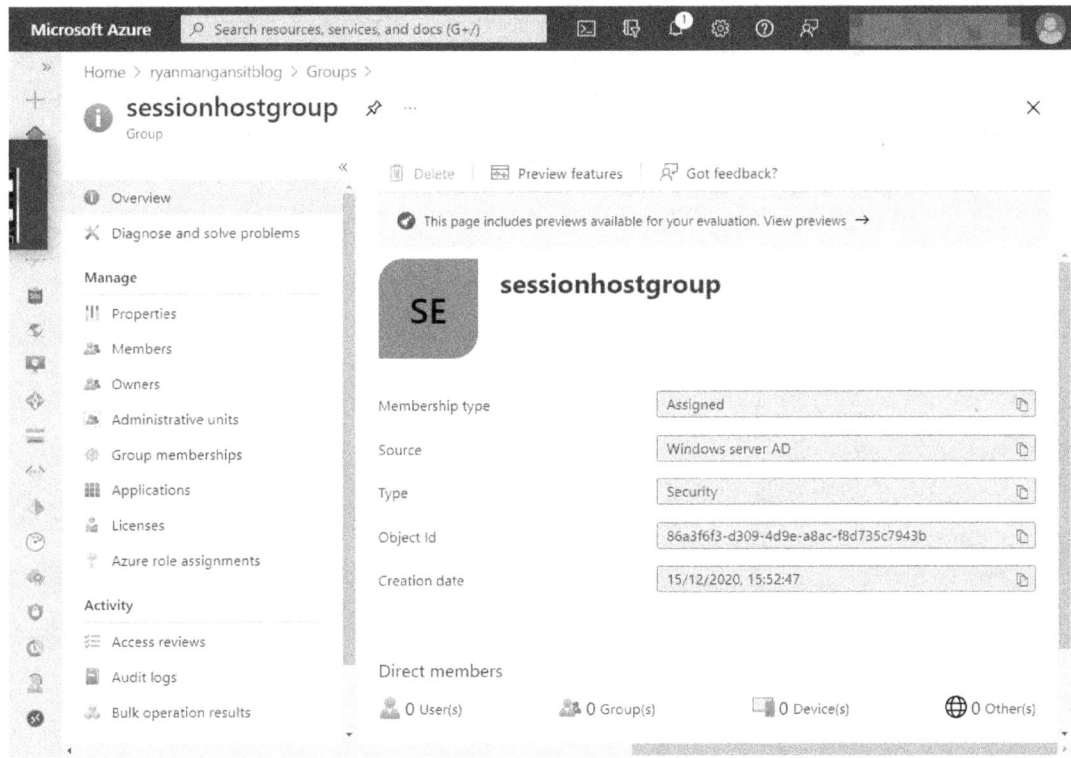

Figure 14.27 – Active Directory Domain Services group synchronized with Entra ID

4. Create an Azure Storage account as shown in *Chapter 5, Implementing and Managing Storage for Azure Virtual Desktop*, in the *Configure storage accounts* section.

5. Create a file share within the storage account as shown in *Chapter 5, Implementing and Managing Storage for Azure Virtual Desktop* in the *Configuring file shares* section.

6. Join the storage account to Active Directory Domain Services. To complete these steps, you will need to follow the guidance in the Microsoft documentation detailed here: `https://docs.microsoft.com/azure/storage/files/storage-files-identity-auth-active-directory-enable`

Once configured, you should see the **Active Directory** field shows **Configured** on the **File shares** page under **File share settings** as shown in the following screenshot:

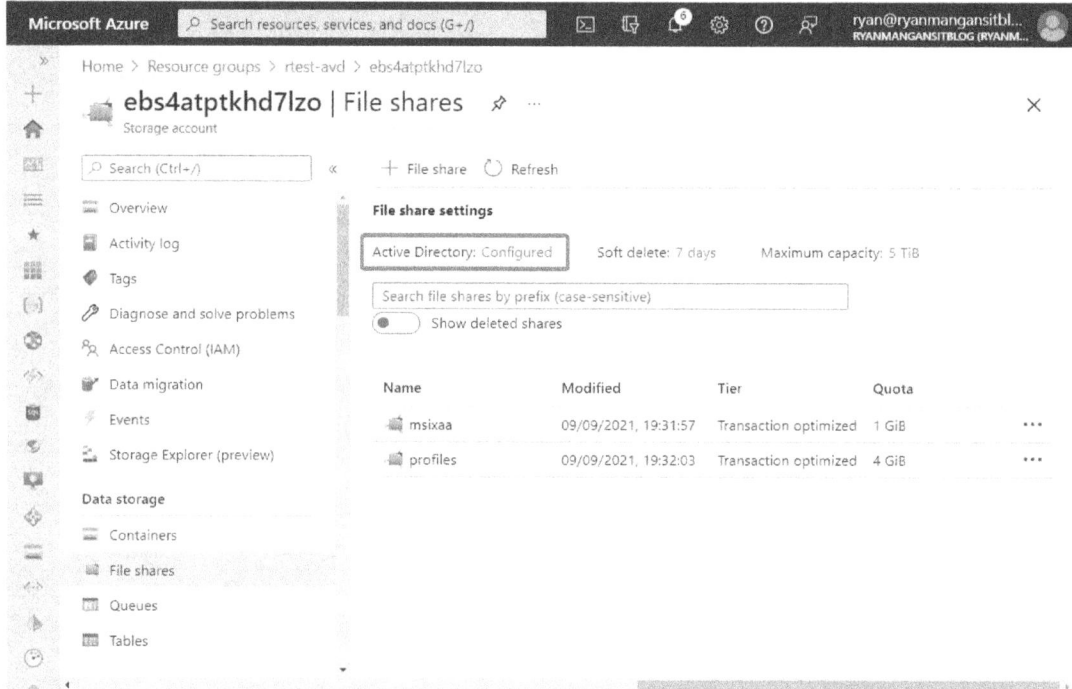

Figure 14.28 – Active Directory is configured for Azure Files

7. Add the storage file data SMB share contributor role to the synced Entra ID DS group that contains the computer object of the session hosts:

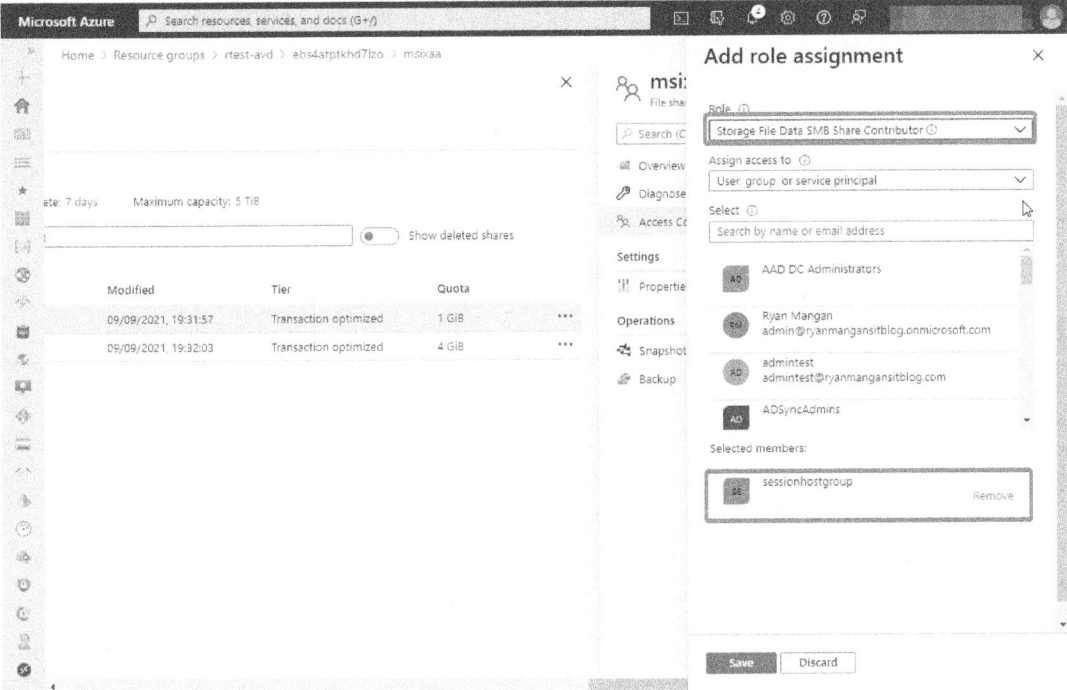

Figure 14.29 – Role assignment for the storage file data SMB share contributor role assigned to the session host group

8. Grant NTFS permissions on the file share to the computer object's Entra ID DS group:

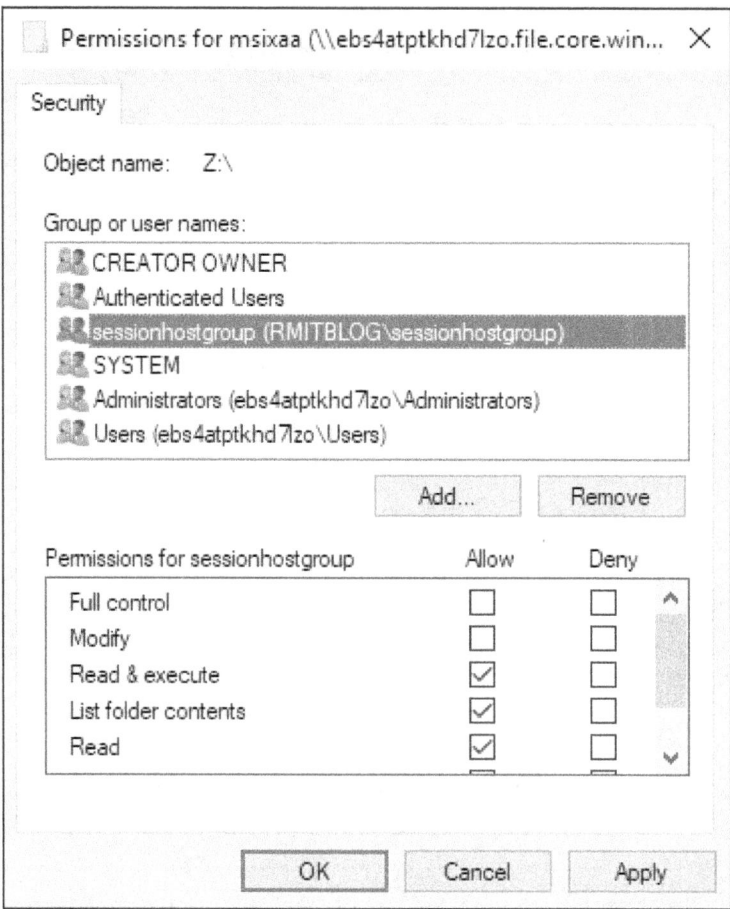

Figure 14.30 – Adding the sessionhostgroup group to grant NTFS permissions for the session hosts on the file share

9. Grant NTFS permission for the user or user group containing the user accounts, sourced from Entra ID DS:

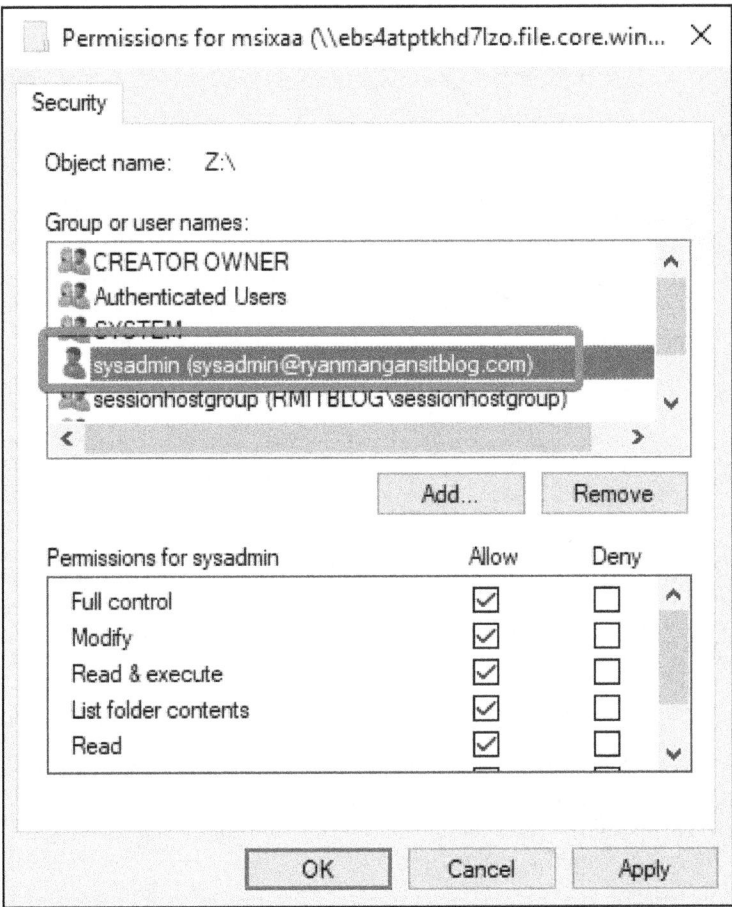

Figure 14.31 – Admin user account added for accessing and managing the file share

> **Important note**
> It is recommended that you reboot the session hosts once the configurations are complete, as access may fail until reboots have been completed.

Once all steps are complete and you have rebooted the session hosts, you should be able to navigate to the share path using the Active Directory user account you configured in *step 9* to test connectivity to the file share:

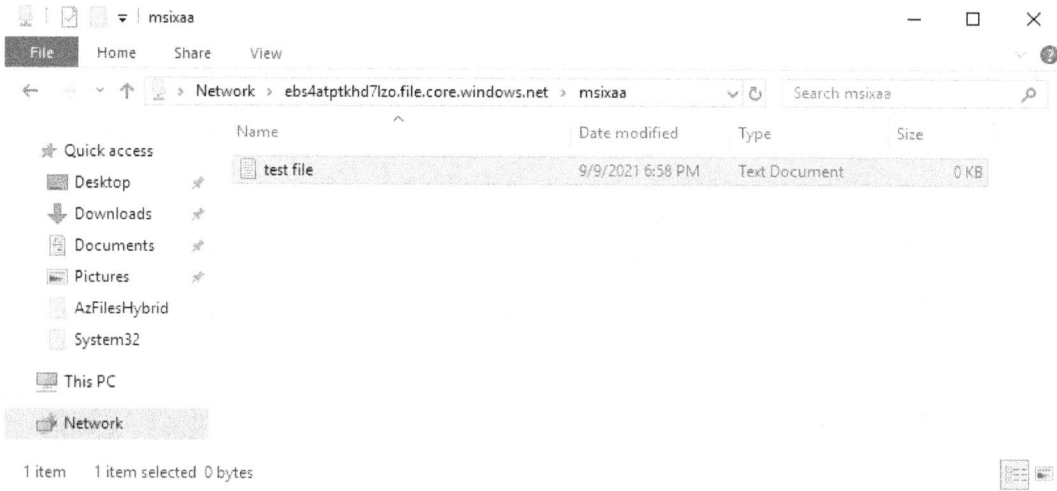

Figure 14.32 – Active Directory access to the Azure Files file share and a test file created to prove permissions are configured correctly

> **Important note**
> You can also use private endpoints when using Azure Storage to improve security. This enables you to configure network traffic between clients on the VNet and storage account to traverse over the VNet via a private link rather than publicly. This essentially means using the Microsoft backbone network rather than the public internet.

This section provided a high-level overview of configuring the Azure Files file share for MSIX app attach, including the nine steps required for Active Directory authentication. In the next section, we look at how to install the certificate on all the required session hosts using MSIX app attach.

Importing the code-signed certificate

Before we upload MSIX images to the file share, we'll take a look at installing the code-signed certificate on all required session hosts.

You can essentially use three types of code-signed certificates with MSIX app attach:

- Self-signed certificates
- Public code signed certificates
- Internal certificate authority certificates

> **Tip**
> Enterprises should use public code-signed certificates or an internal certificate authority. It's not recommended that you use a self-signed certificate for production. It is also important to note that you should time stamp your application packages because if the certificate expires and you have not signed the MSIX package, then the app will stop working.

To install the certificate on a session host, open a new MMC snap-in for the local computer. Then, add the **Certificates (Local Computer)** snap-in and click **OK**:

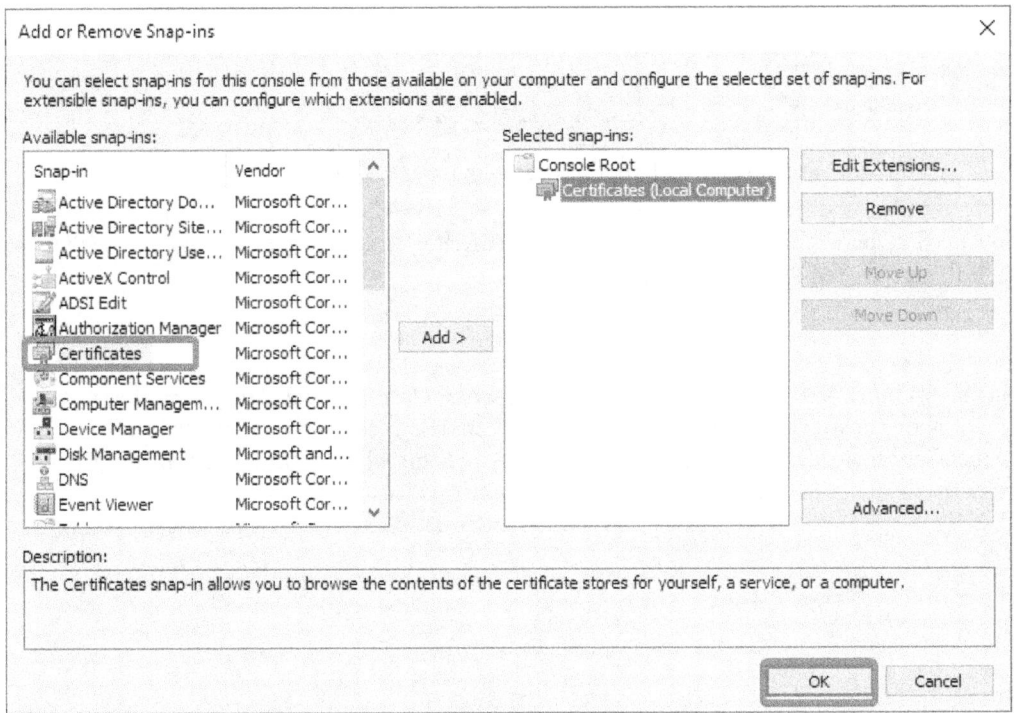

Figure 14.33 – MMC snap-in and adding the Certificates snap-in for the local computer

Once you are in the **Certificates (Local Computer)** snap-in, right-click within the Certificates window, typically the middle panel within the MMC console, and select **All Tasks | Import**:

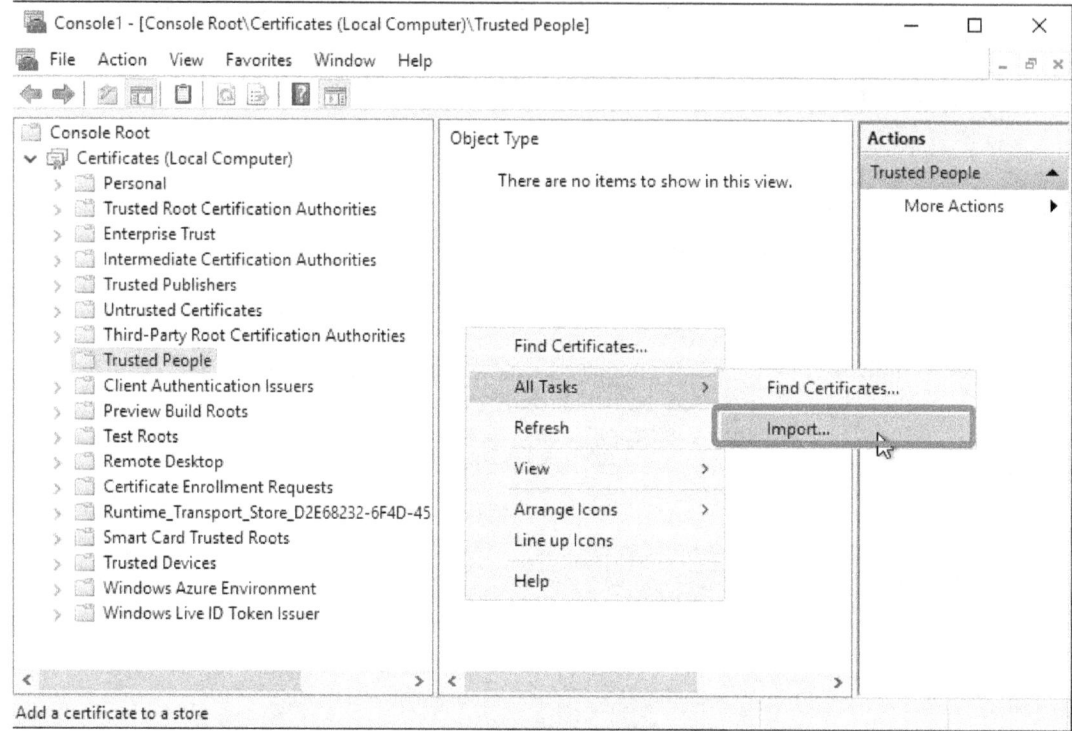

Figure 14.34 – Import button within the MMC snap-in

You now need to follow the wizard by selecting the certificate you want to import and ensure that you import it to the **Trusted People** certificate store:

> **Tip**
>
> You have a choice of importing the certificate as **Current User** or **Local Computer**. There is no right or wrong answer as it's dependent on your chosen design. Some may argue that controlling the installation of a certificate per user has security benefits whereas other organizations may prefer the local machine certificate installation option.

Importing the code-signed certificate 457

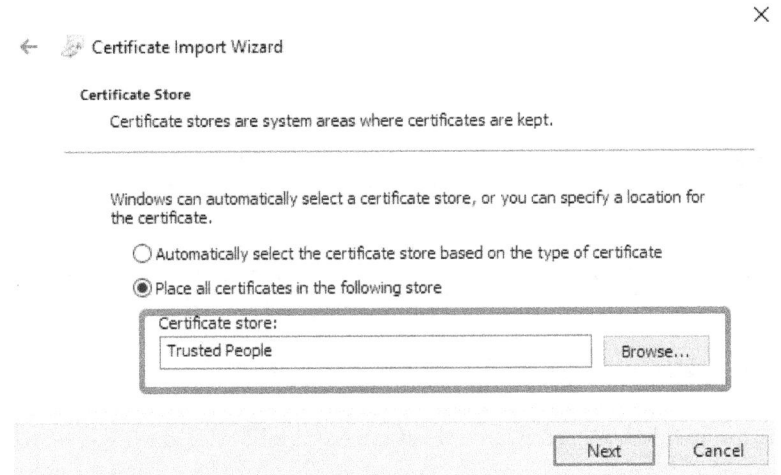

Figure 14.35 – Trusted People

Once imported, you will see the certificate inside the `Trusted People` folder as can be seen in the following screenshot:

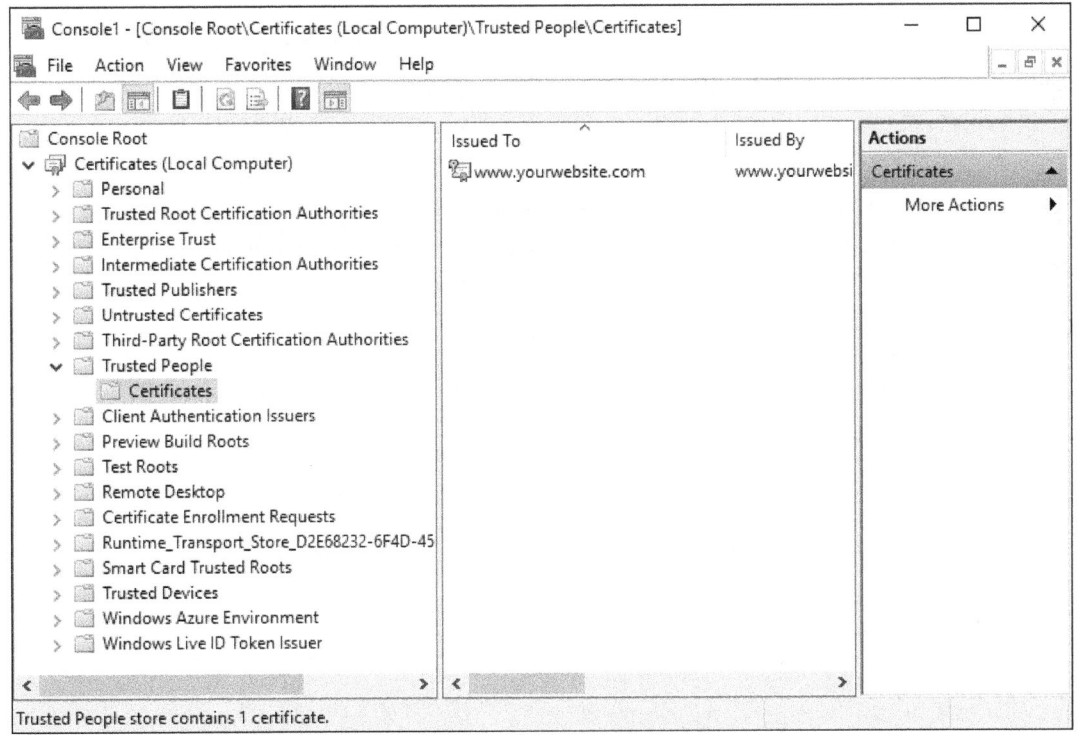

Figure 14.36 – The certificate has now been imported into the Trusted People store

You have now completed the steps for installing a code-signing certificate on a session host.

> **Important note**
> It is recommended that you deploy the certificate to a template image or use custom script extensions to automate the deployment of certificates to your session hosts.

Now that we have covered creating and installing certificates, we'll move on to looking at uploading MSIX images to Azure Files.

Uploading MSIX images to Azure Files

In this section, we will briefly look at how to upload the required MSIX images to Azure Files.

> **Tip**
> In larger environments, it is recommended that you package your applications on an isolated session host to keep file transfer times down. You will appreciate this as it's much faster to copy files from a packaging virtual machine within Azure than it would be from on-premises to Microsoft Azure.

The most common method for transferring MSIX image files to Azure Files is by connecting to the SMB share directly on your network. However, there are a few other tools and methods you can use, such as the following:

- **Azure portal**: You can upload files within the Azure portal using your web browser. You simply click the upload icon within the share as shown in the following screenshot:

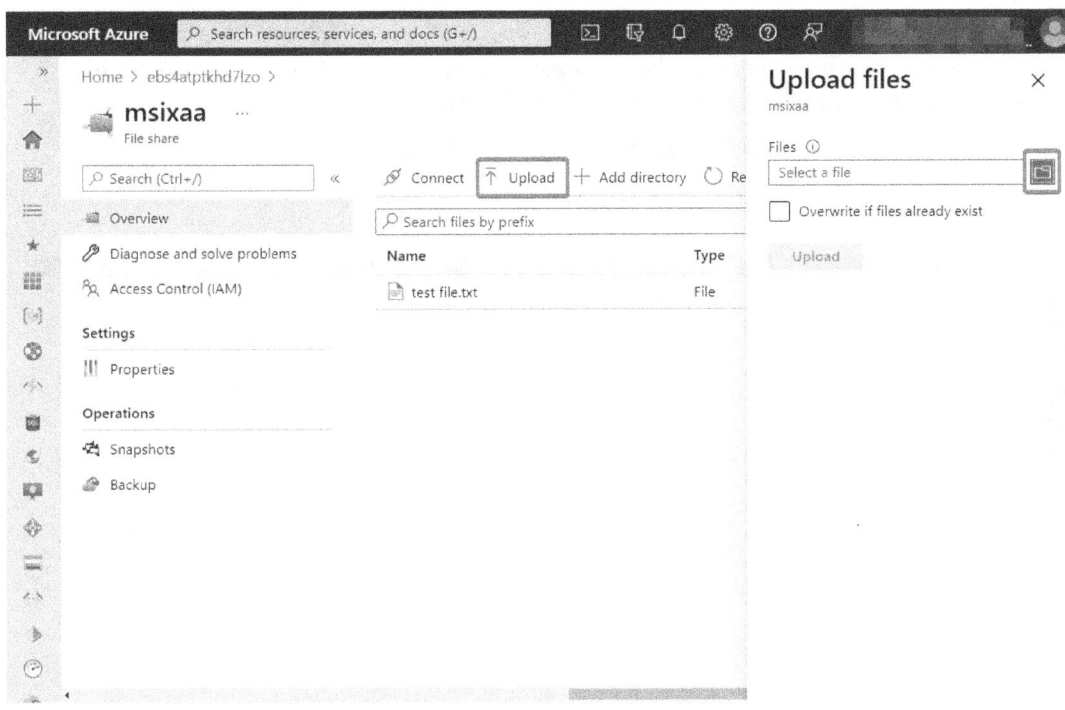

Figure 14.37 – The feature to upload files to Azure Files within the Azure portal

- **AzCopy**: You can also use AzCopy, which is a command-line transfer tool. You can download this from the following URL: https://docs.microsoft.com/azure/storage/common/storage-use-azcopy-v10

AzCopy also offers the capability to preserve file permissions during copying, similar to how Robocopy works if you have ever used it.

- **Storage Explorer**: The final tool I wanted to cover is **Storage Explorer**. This is a standalone application that you can run on the packaging machine. It simplifies the transfer of files to Azure Files and other storage offerings within Microsoft Azure.

    You can download Storage Explorer here: `https://www.storageexplorer.com/`.

    The following screenshot shows the Microsoft Azure Storage Explorer application:

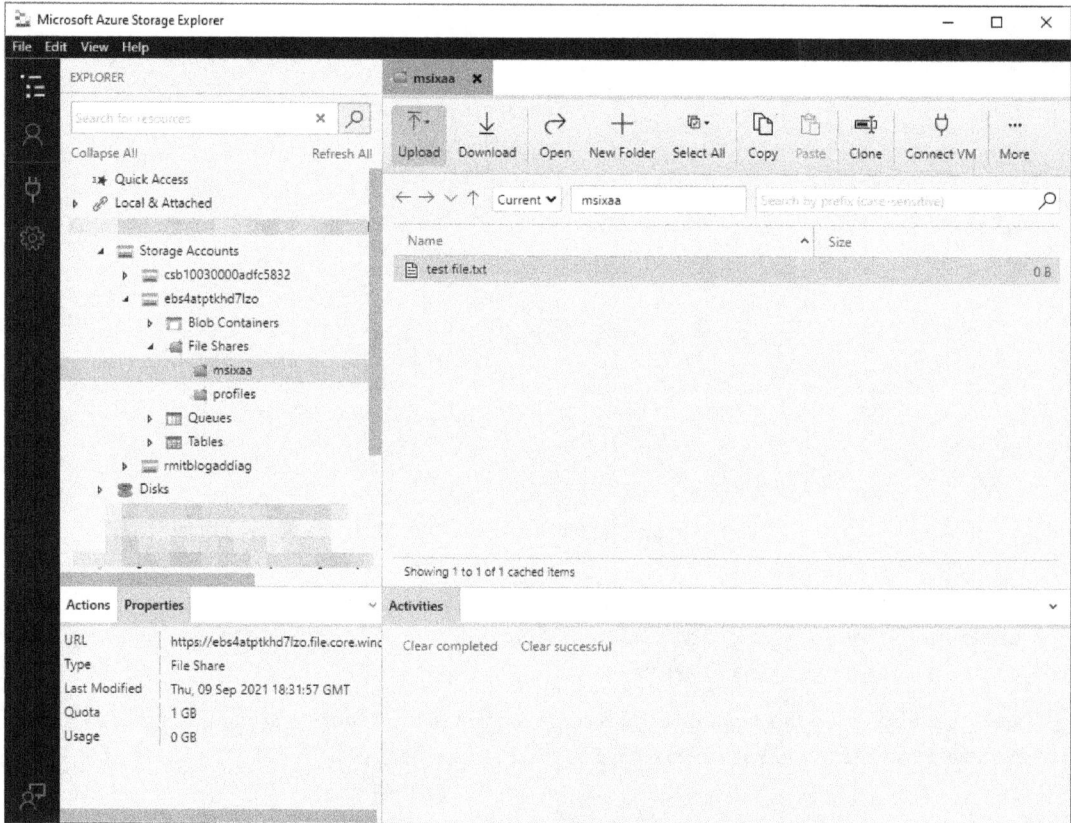

Figure 14.38 – Storage Explorer

This section looked at some of the options available to you for uploading files to Azure Files. Now that we have covered all the pre-MSIX app attach configuration tasks, we'll move on to look at configuring MSIX app attach.

Configuring MSIX app attach

> **Important**
>
> MSIX app attach will eventually change to the new version, app attach, which is currently in preview. 'MSIX App Attach' and 'App Attach'

In this section, we'll look at adding MSIX packages to a host pool within the Azure portal:

MSIX app attach: Delivery of applications occurs via RemoteApp or within a desktop session. Permissions are regulated through assignment to application groups. Additionally, all MSIX app-attached applications are visible to desktop users within the desktop application group.

> **Tip**
>
> Ensure you have installed the required code-signed certificates before proceeding with the steps detailed in this section. The configuration will fail on certificate validation if not installed on each session host within the host pool.

1. To add MSIX packages, navigate to the **MSIX packages** blade located within the host pool. You will see the icon and the name **MSIX packages** under the **Manage** section in the menu on the left side:

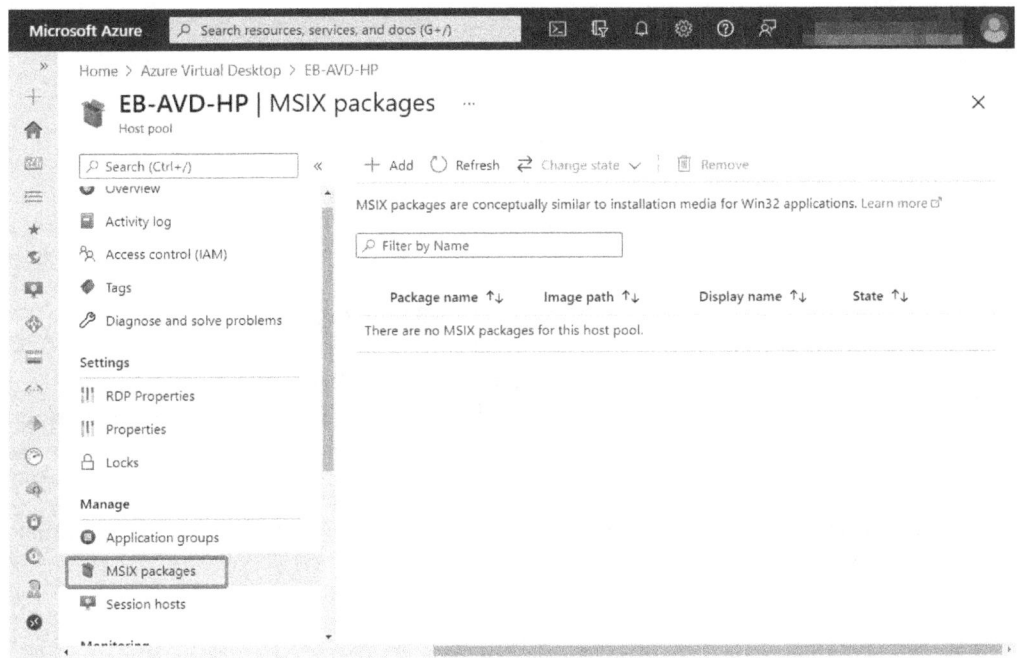

Figure 14.39 – MSIX packages page within the host pool

2. To add a package, simply click the **Add** button located within the menu bar.

The following steps detail how to add MSIX packages to your host pool. In the **Add MSIX package** tab, enter the following values:

1. For **MSIX image path**, enter a valid UNC path pointing to the MSIX image on the file share (for example, `\\storageaccount.file.core.windows.net\msixshare\appfolder\MSIXimage.vhd`). Select **Add** to query the MSIX container to check whether the path is valid when you're done:

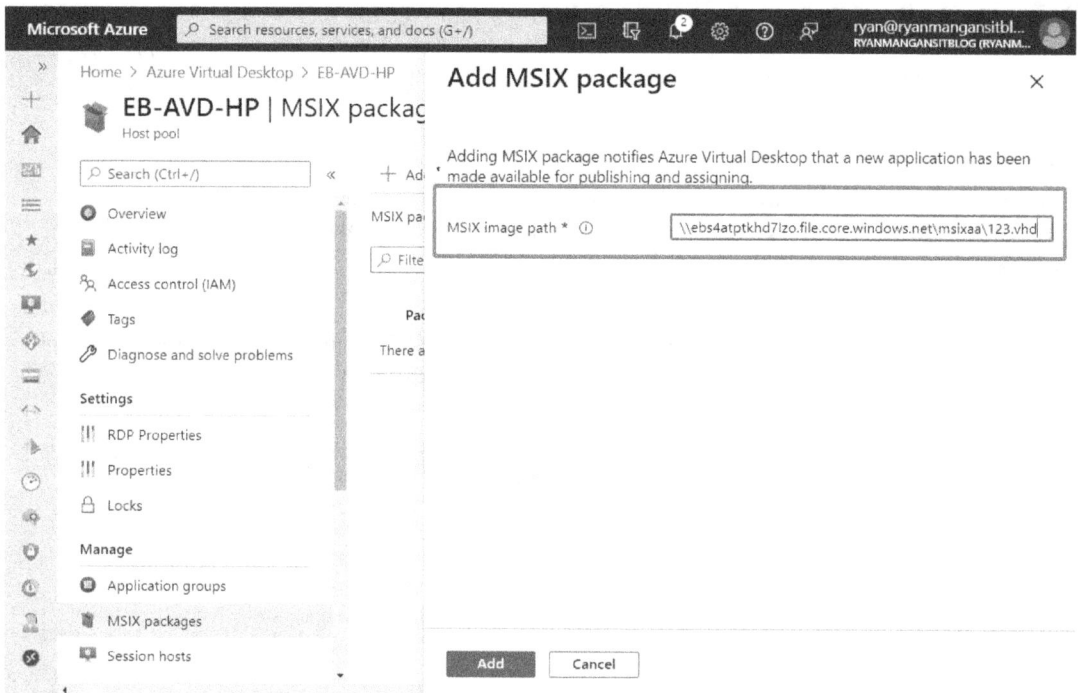

Figure 14.40 – Add MSIX package blade

2. To configure the MSIX package, select the required MSIX package name from the drop-down menu in the **MSIX package** field. This menu will only be populated if you've entered a valid image path in the **MSIX image path** textbox. If there is an error, you will see an error message appear onscreen in message format.

3. For package applications, make sure the list contains all MSIX applications you want to be available to users.

4. Alternatively, enter a display name if you want your package to be more user-friendly when used in your user deployments:

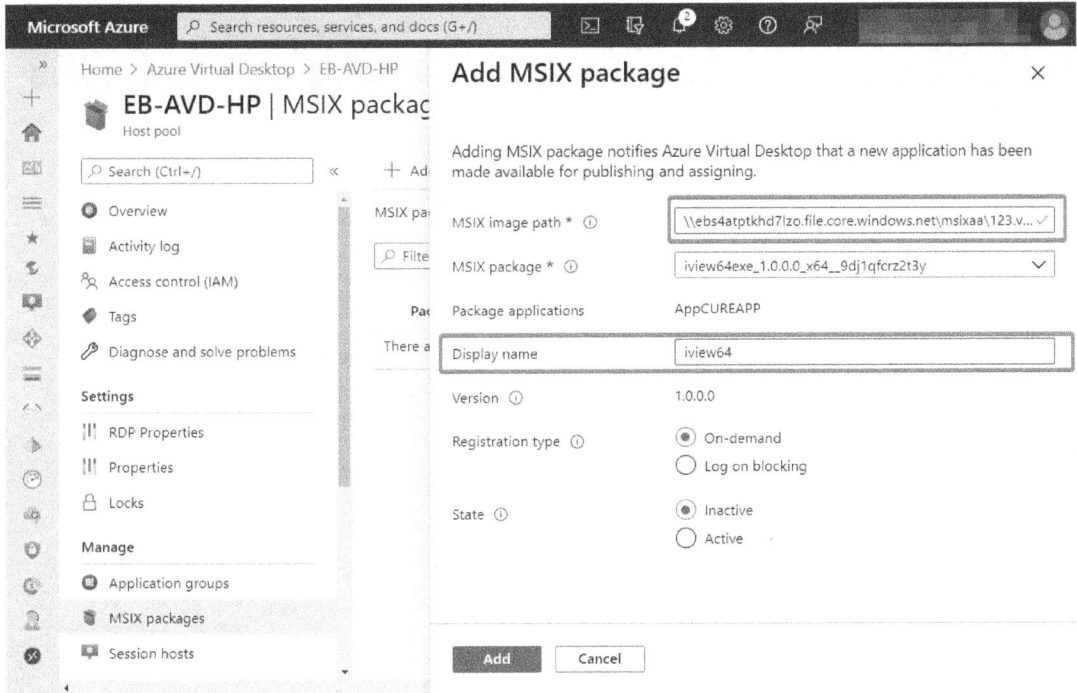

Figure 14.41 – The Add MSIX package blade and the package we are adding in this example

5. Ensure the version number is correct.

6. Select the **Registration type** option that you want to use. This will depend on your needs:

   - **On-demand** registration essentially postpones the complete registration of the MSIX application until the user has launched the application. This is the registration type Microsoft recommends.

   - **Log on blocking** is used to only register apps while the user is signing in. This is not recommended for most deployments as it can increase sign-in times for users.

7. For the **State** field, select your preferred state:

   - The **Active** status is used to enable users to interact with the package

   - The **Inactive** status instructs Azure Virtual Desktop to ignore the package and not deliver the package to users

8. When you're done, select **Add**.
9. Once complete, you will see the package appear on the **MSIX packages** page as shown in the following screenshot:

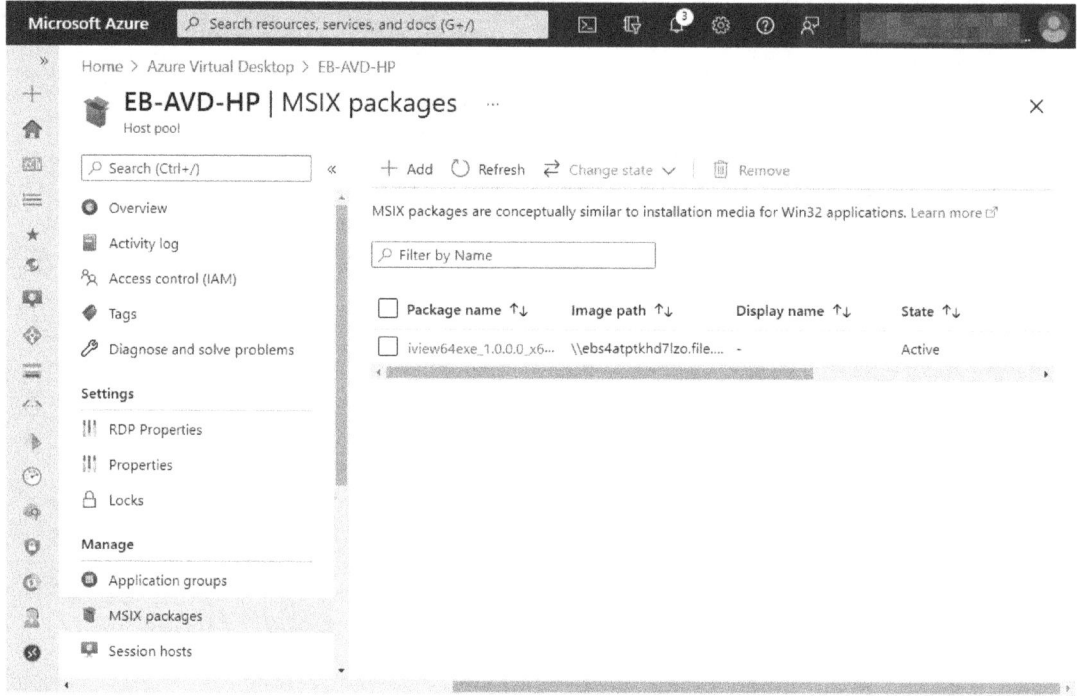

Figure 14.42 – The added package on the MSIX packages page

Now that we have added an MSIX package to the host pool, we'll now move on to assigning packages to an application group, also known as publishing an MSIX app to an application group.

Publishing an MSIX app to a RemoteApp application group

This section looks at assigning MSIX apps to a RemoteApp application group.

The following steps will guide you through publishing MSIX apps to application groups as remote apps:

> **Important note**
> You can deliver MSIX applications to both RemoteApp and desktop app groups. In this example, we are creating a remote app group for MSIX packages. To add applications to a desktop app group, you would navigate to the desktop app group and add the applications you require.

1. Navigate to the **Azure Virtual Desktop** page within the Azure portal, then select the **Applications group** button.
2. Click + **Add**.

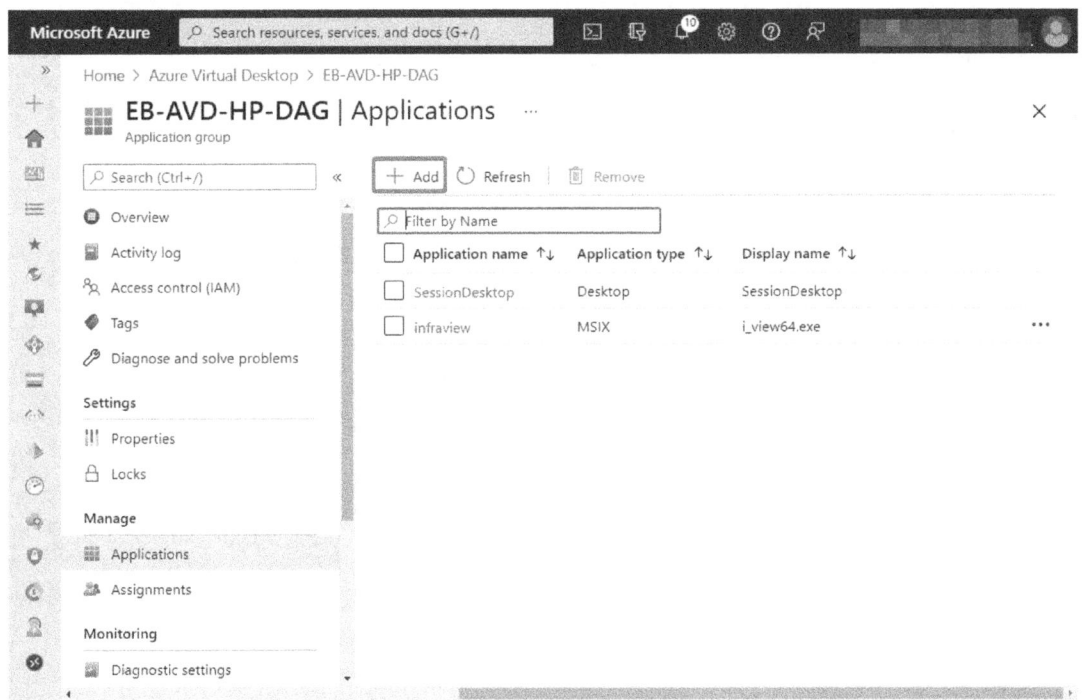

Figure 14.43 – The Add button to add applications to a desktop application group

3. You will then be presented with the **Create an application group** page:

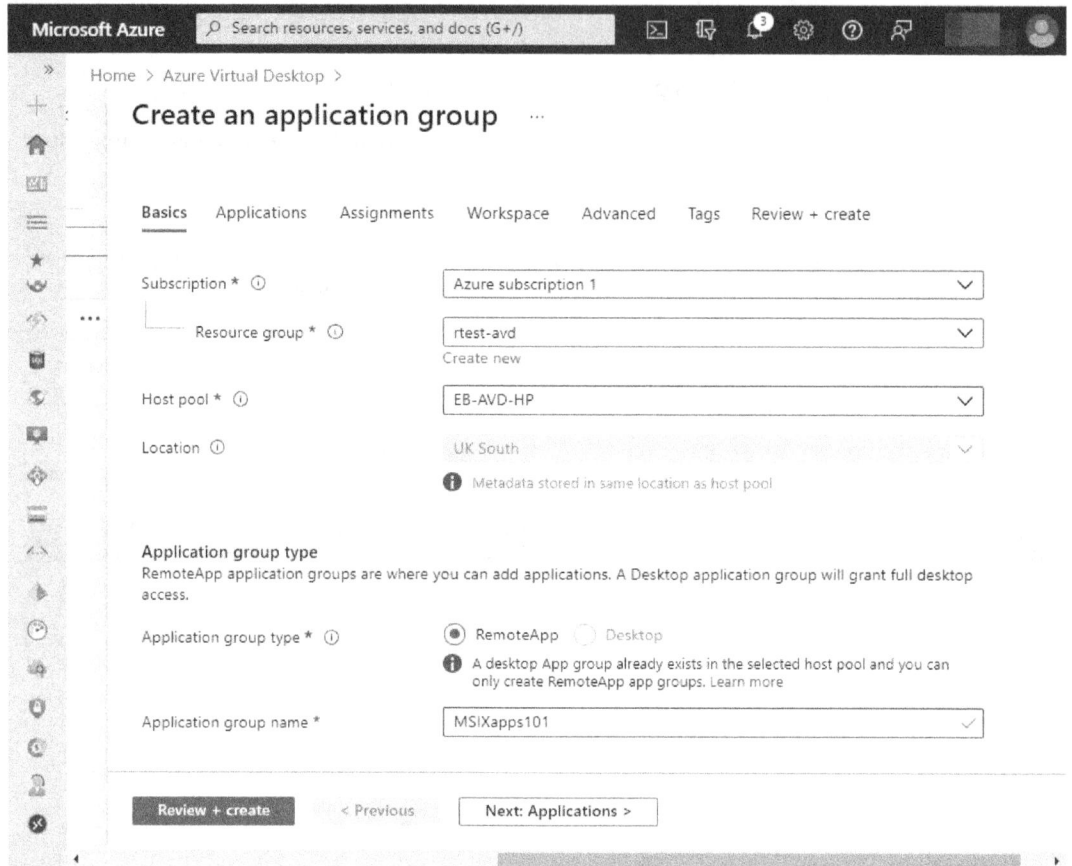

Figure 14.44 – Create an application group page

4. Within the **Basics** tab, select the **Host pool** and provide an application group name, then click **Next**.

5. You will then be presented with the **Applications** tab, where you can add applications by clicking **Add applications**. This will open the **Add application** blade as shown in the following screenshot:

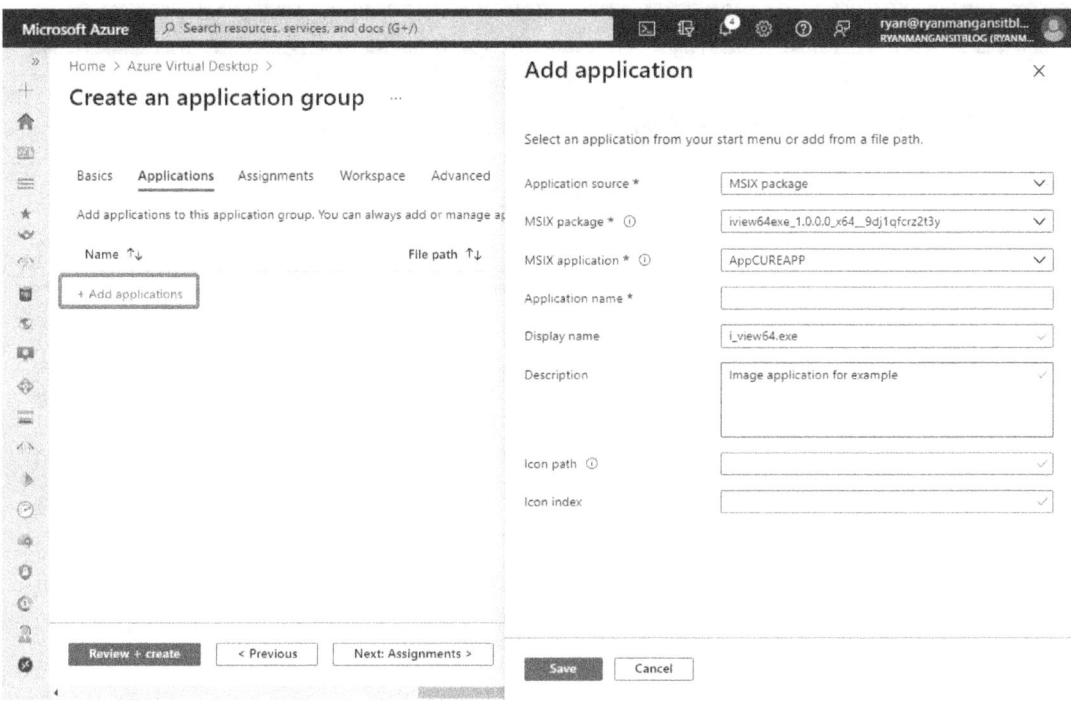

Figure 14.45 – Add application

6. Once you have added the application(s), you can proceed to the next tab, **Assignments**. This is where you select the users and/or groups you would like to access the published applications. Once complete, click **Next**, which will show the **Workspace** tab.

7. Within the **Workspace** tab, set the **Register Application Group** to **Yes**, then click **Next**.

8. You can enable diagnostic settings and add **Tags**. If you don't need these two features, skip to **Review + create** and check your configuration:

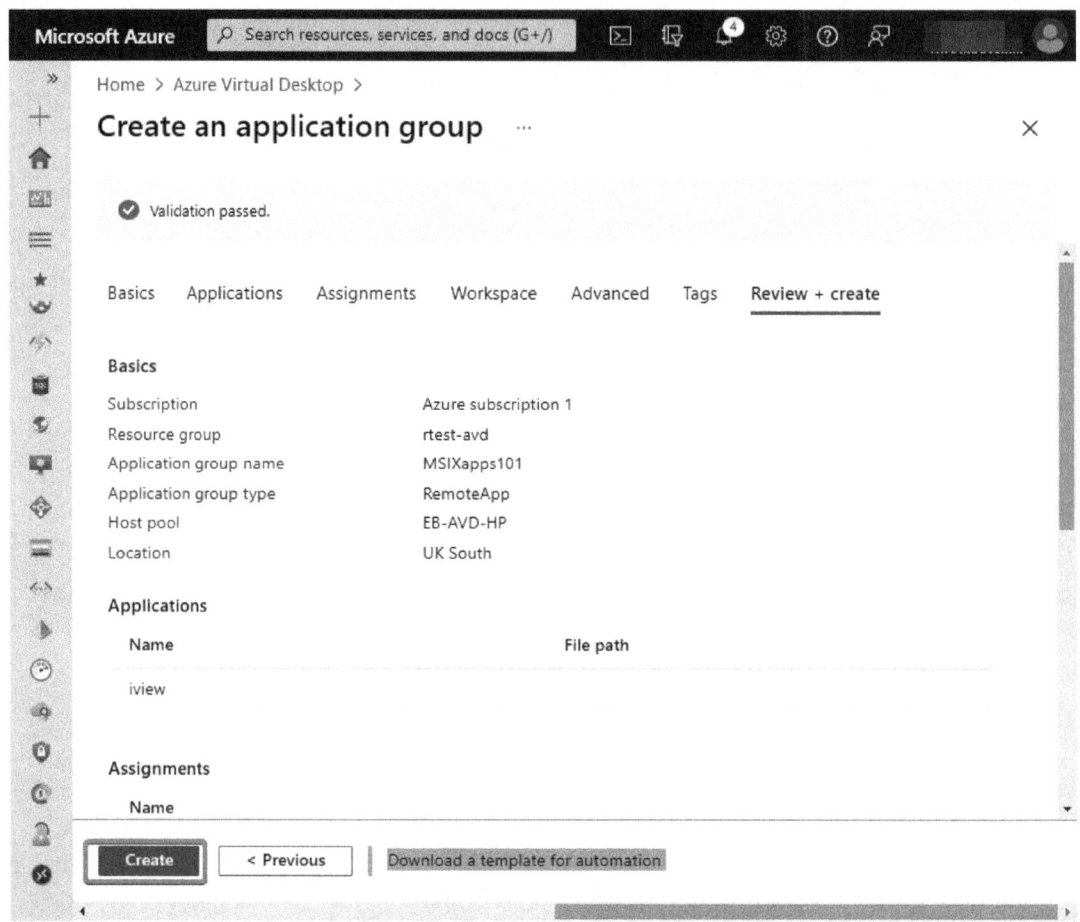

Figure 14.46 – The Review + create section of Create an application group

7. **Application assignment and access**:

   - **MSIX app attach**: Applications are assigned to a host pool and added to a desktop or RemoteApp application group. Users see these applications in their Start menu if they are part of the relevant application group.

   - **App attach**: Provides finer control over application access, allowing application assignment to individual users or groups within the same host pool or session host.

Creating an app attach package

This section provides a high-level overview of creating an app attach package.

> **Important**
> Similar to MSIX app attach, you will need to make sure all the prerequisites are configured as per those set out in the *Configuring Azure Files for MSIX app attach, Importing the code-signed certificate,* and *Uploading MSIX images to Azure Files* sections.

1. First, you need to navigate to the Azure Virtual Desktop blade and select **App attach**, as shown in the screenshot below.

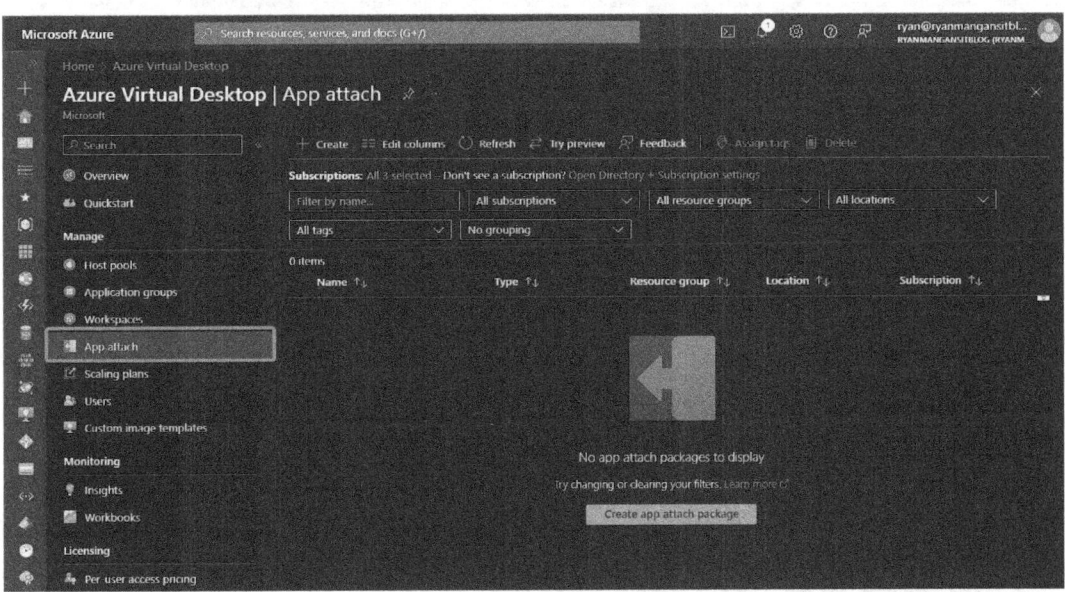

Figure 14.53 – App attach blade

2. Click the + **Create** button (**1** in the following screenshot) or the **Create App attach package** button (**2**).

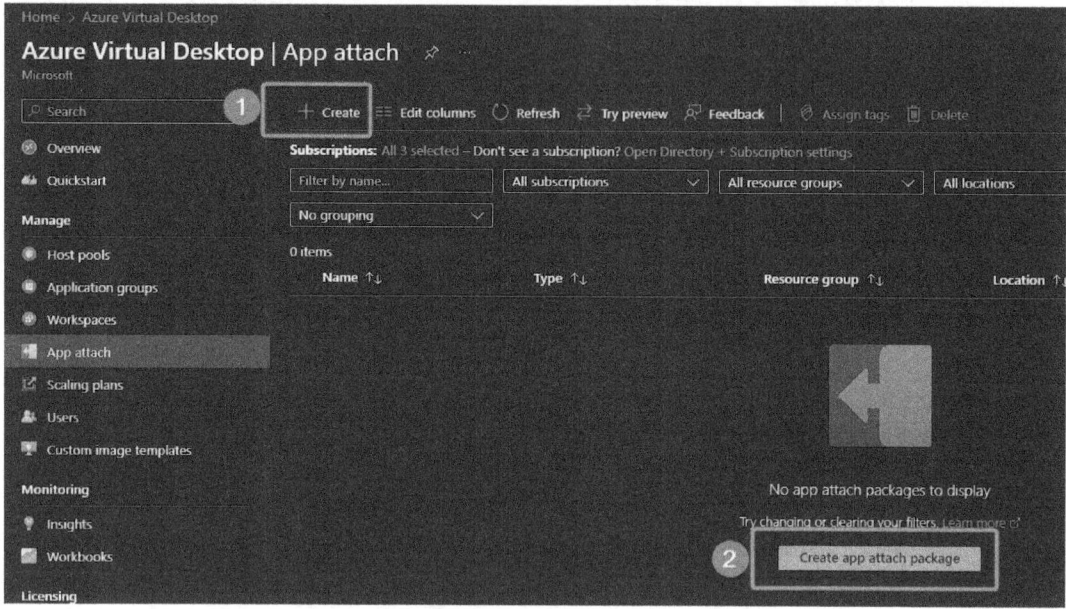

Figure 14.54 – The buttons to create an app attach package

3. You will then enter into the **Create App Attach** wizard. You need to complete the wizard by selecting a subscription, the resource group where the host pool resides, and the location.

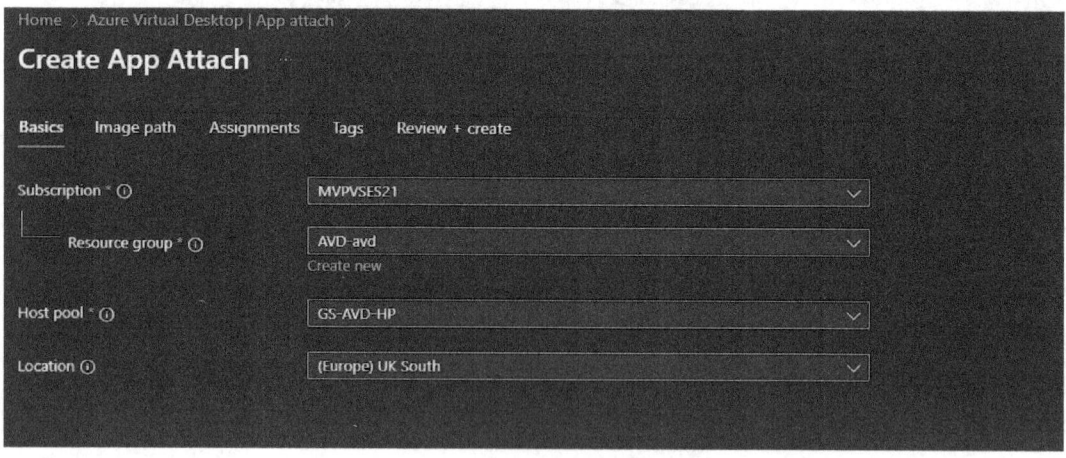

Figure 14.55 – The Basics tab of the Create App Attach blade

4. Once you have entered the required information in the **Basics** tab, proceed to the **Image path** tab.
5. Select either a **storage account** or input a **UNC path**. In this example, we will use a storage path.
6. Navigate through the storage browser and select the package you want to add.

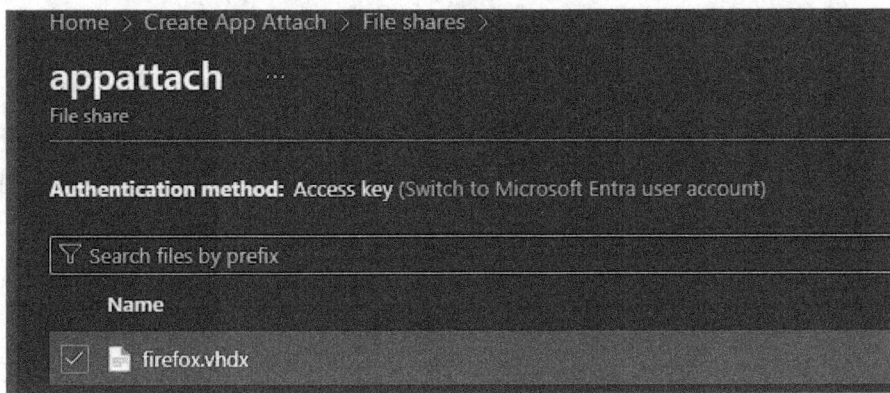

Figure 14.56 – The Image path tab of the Create App Attach blade

7. If you have not configured all the prerequisites detailed in this exercise, you will see the following message when you try to add a package.

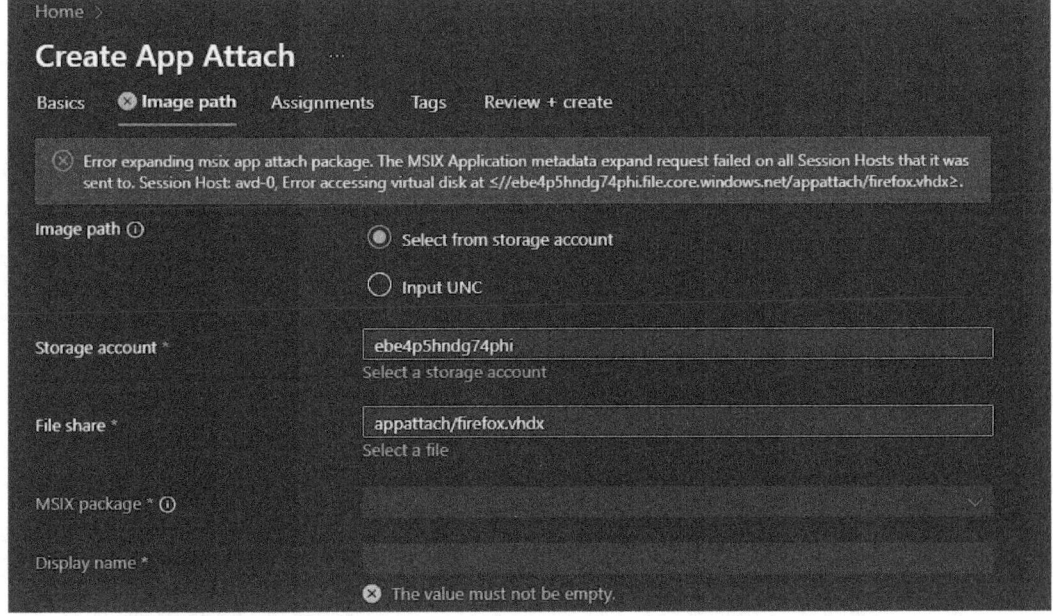

Figure 14.57 – The expected error if you have not configured the prerequisites

8. Once you have successfully added a package, you can proceed to select the host pools and add users and groups under the **Assignments** tab.

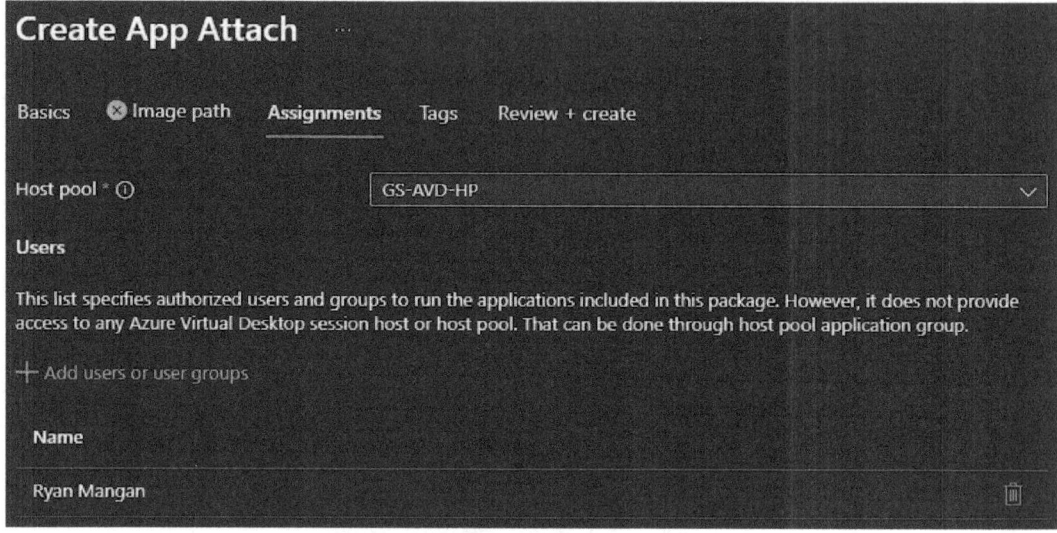

Figure 14.58 – Shows the assignments page for Create App Attach

9. Once configured, add any tags, click the **Review + create** tab, review the settings, and create the app attach package.

You will now see your app on the main **App attach** page.

Summary

In this chapter, we took a detailed look at MSIX app attach and learned how to configure it for Azure Virtual Desktop. We discussed the creation of MSIX packages and MSIX images, the configuration of Azure Files, and the configuration and publication of MSIX packages using MSIX app attach in Azure Virtual Desktop. Finally, we finished off by looking at the app attach preview feature and the differences between it and MSIX app attach.

In the next chapter, we will take a look at some of the other capabilities Azure Virtual Desktop has to offer, including FSLogix application masking, deploying applications as remote apps, and Microsoft Teams AV redirection.

Further reading

You'll find a list of approved Microsoft third-party packing partners here: `https://docs.microsoft.com/windows/msix/desktop/desktop-to-uwp-third-party-installer`

Questions

Here are a few questions to test your understanding of this chapter:

1. Which certificate store would you use for MSIX and MSIX app attach?

    The Trusted People store.

2. What is the difference between MSIX and an MSIX image?

    MSIX: An application package format for distributing and managing Windows applications.

    MSIX Image: An expanded MSIX package in a container format such as VHD, VHDX, or CIMFS, allowing deployment from a virtual disk.

3. What are three different code-signed certificate types you can use with MSIX and MSIX app attach?

    Self-signed, domain, and public certificates.

4. What is the registry key for setting the MSIX app attach package check interval for Azure Virtual Desktop?

    [HKEY_LOCAL_MACHINE\SOFTWARE\Microsoft\RDInfraAgent\MSIXAppAttach]
    "PackageListCheckIntervalMinutes"=dword:00000001

15
Configuring Apps on a Session Host

In this chapter, we will take a look at configuration applications for Azure Virtual Desktop. We will start by looking at application masking, which is used to hide applications from users who do not require those specific applications. We will then look at deploying an application as a RemoteApp application. As we progress through the chapter, we will also implement OneDrive for Business on a multi-session environment and configure Microsoft Teams AV redirection. Finally, we will provide troubleshooting advice and guidance on application issues relating to Azure Virtual Desktop.

In this chapter, we look at the following:

- Configuring application masking
- Deploying RemoteApp applications
- Configuring and managing OneDrive for Business for a multi-session environment
- Configuring and managing Microsoft Teams AV redirection
- Implementing and managing multimedia redirection
- Managing internet access for Azure Virtual Desktop sessions
- What are VM applications?

Application masking

Application masking is part of the FSLogix portfolio that can manage access to applications, fonts, and other items based on criteria. **Apps RuleEditor** is used to create app masking rules, edit rules, manage user and group assignments, and test created rule sets. This is incredibly useful for controlling application access via user groups.

To get started with application masking, you need to install the `FSLogixAppRuleEditorSetup.exe` application, which can be downloaded with the other FSLogix products here: `https://aka.ms/fslogix/download`.

Rule types available

Application masking supports the following three rule types:

- **Specify value rule**: Used to assign a value for the specified item
- **Redirection rule**: Used to redirect applications or app data to a specified item
- **Hiding rule**: Used to redirect applications or app data to specified criteria

The following figure shows **FSLogix Apps RuleEditor** in the Windows 11 **Start** menu:

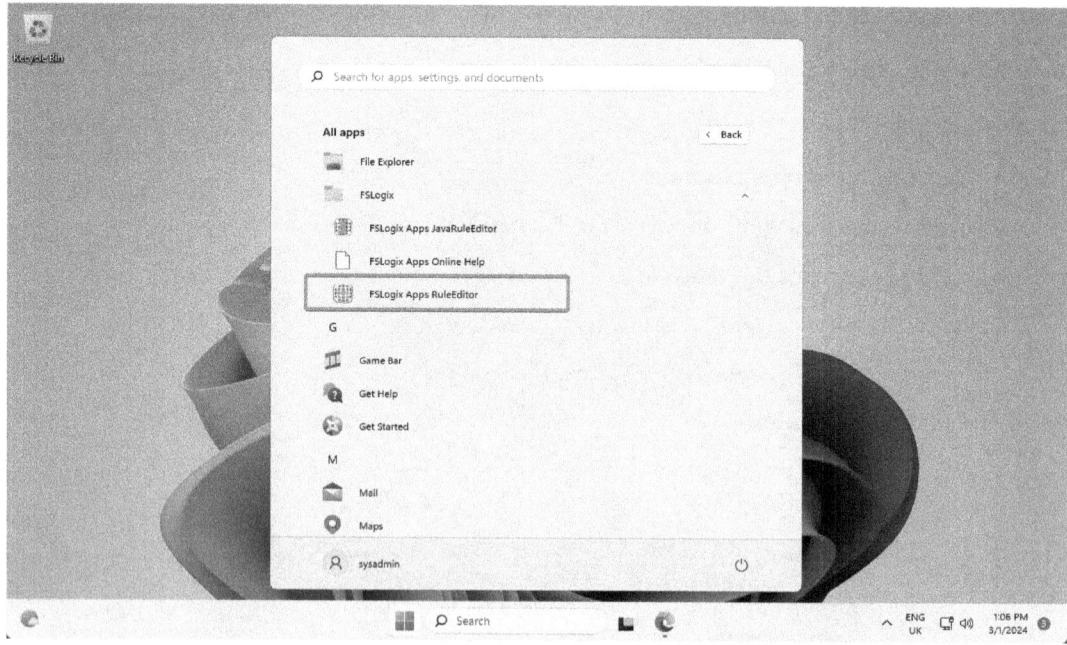

Figure 15.1 – The FSLogix Apps RuleEditor icon in the Start menu

Application masking

> **Important**
>
> App container rules have been deprecated as of August 22, 2023, and was retired in March 2024.

Now, let's create our first rule:

1. Once **FSlogix Apps RuleEditor** has been launched, click **File** and then **New** to create a new rule:

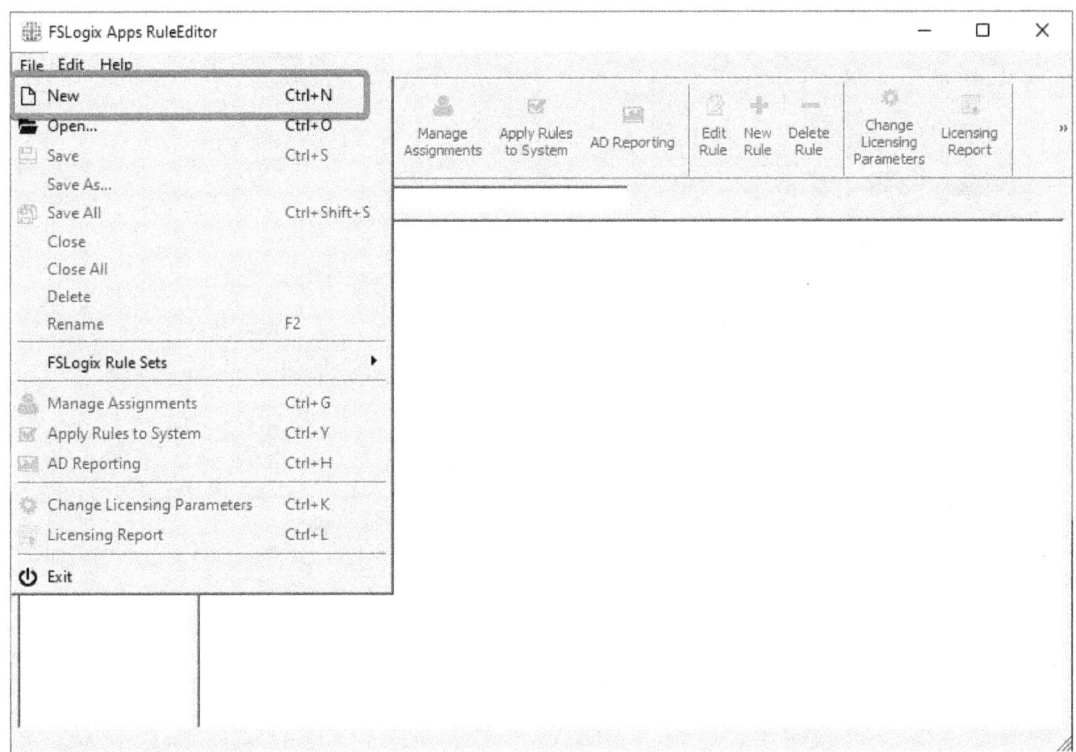

Figure 15.2 – Creating a new rule in FSLogix Apps RuleEditor

486 Configuring Apps on a Session Host

2. You will now need to enter a **File name** value for the rule we are about to create:

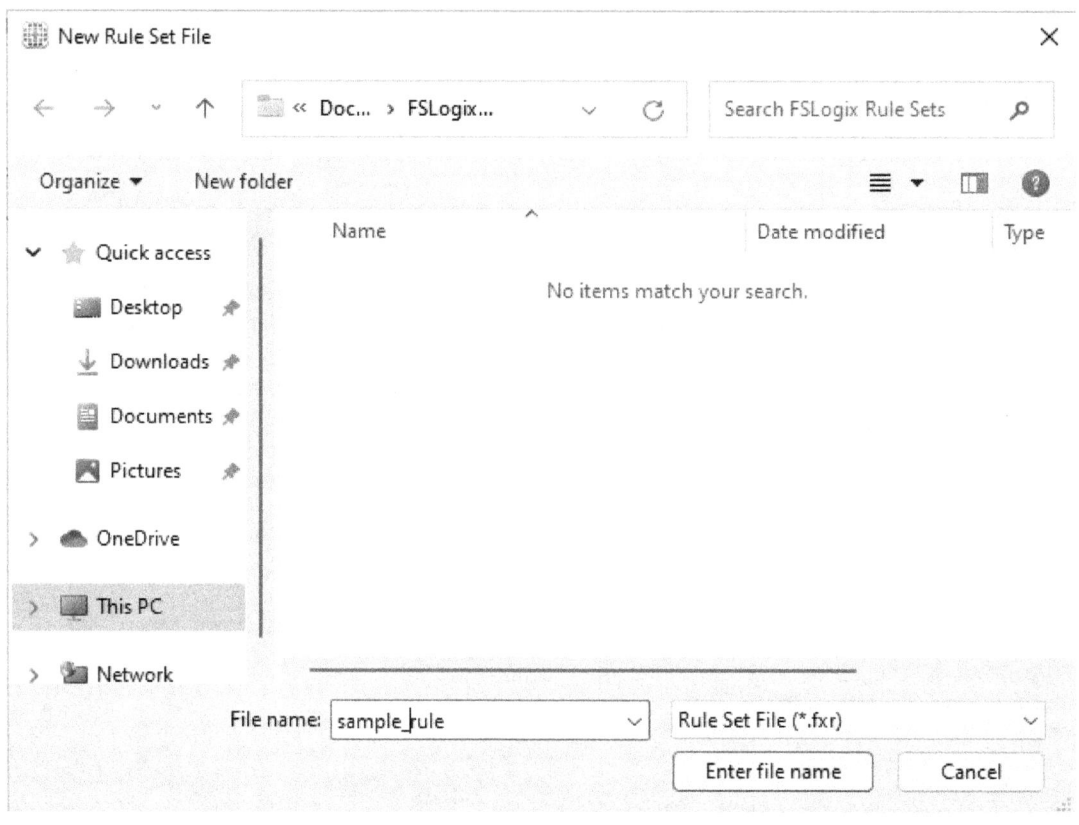

Figure 15.3 – Dialog box for saving the new rule file

3. Once you have added a name and clicked **Enter file name**, you will see the following **Rule Set** form appear:

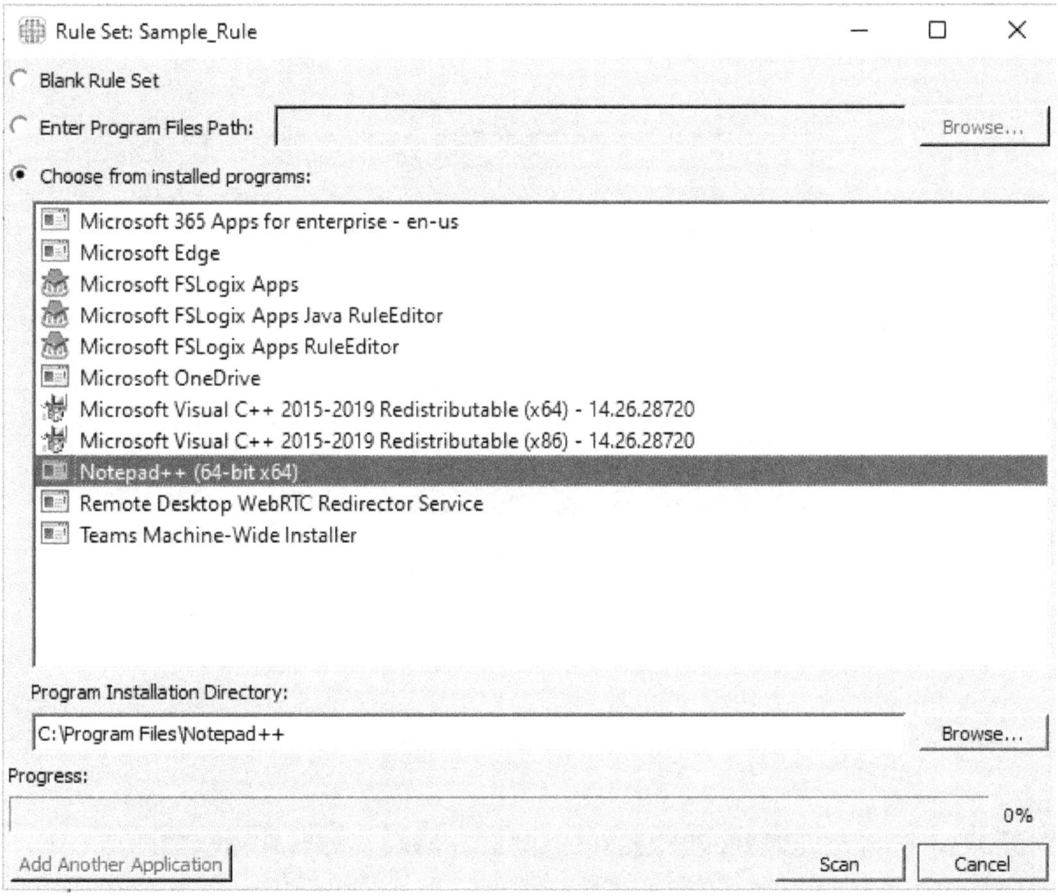

Figure 15.4 – The Rule Set form where we select an application from the installed programs list

4. We will choose a program from the installed programs for this example. Select **Notepad++** and click **Scan**. Once the scan has been completed, you can click the **Ok** button, which has changed from **Scan** to **Ok**:

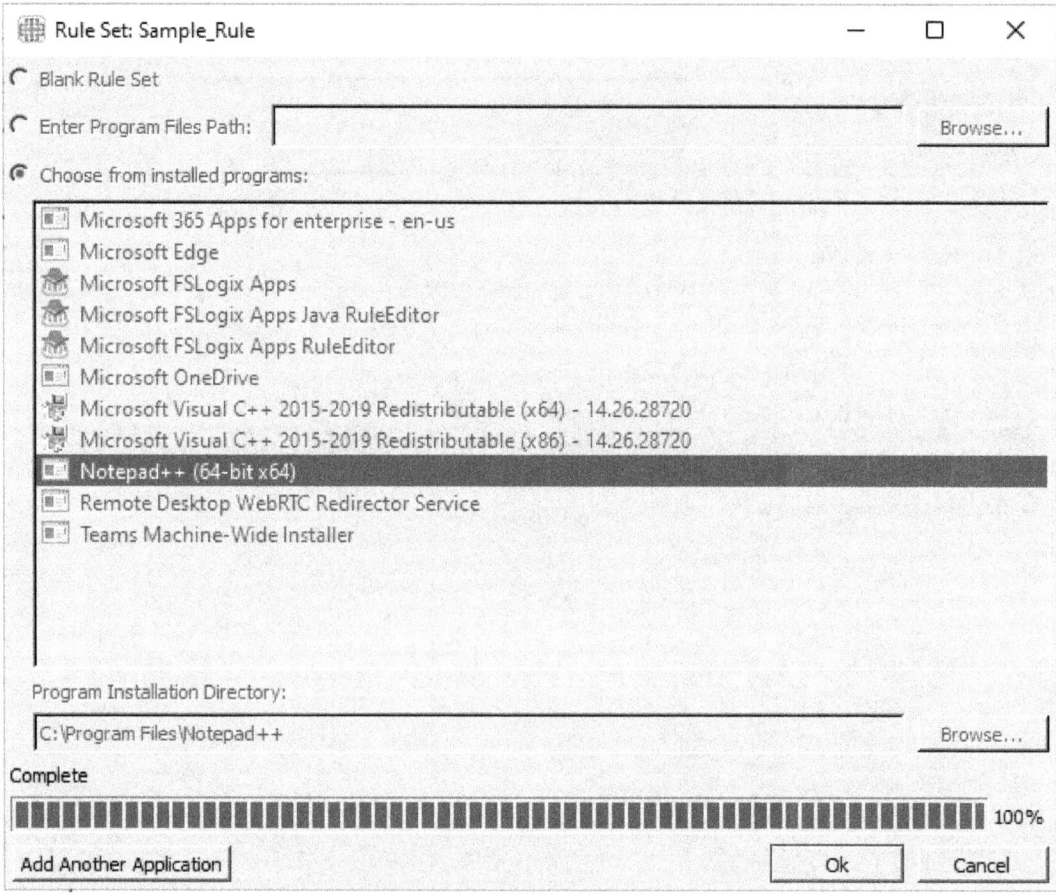

Figure 15.5 – The scan has been completed and you can now click Ok

5. You should now see the hiding rule for **Notepad++**:

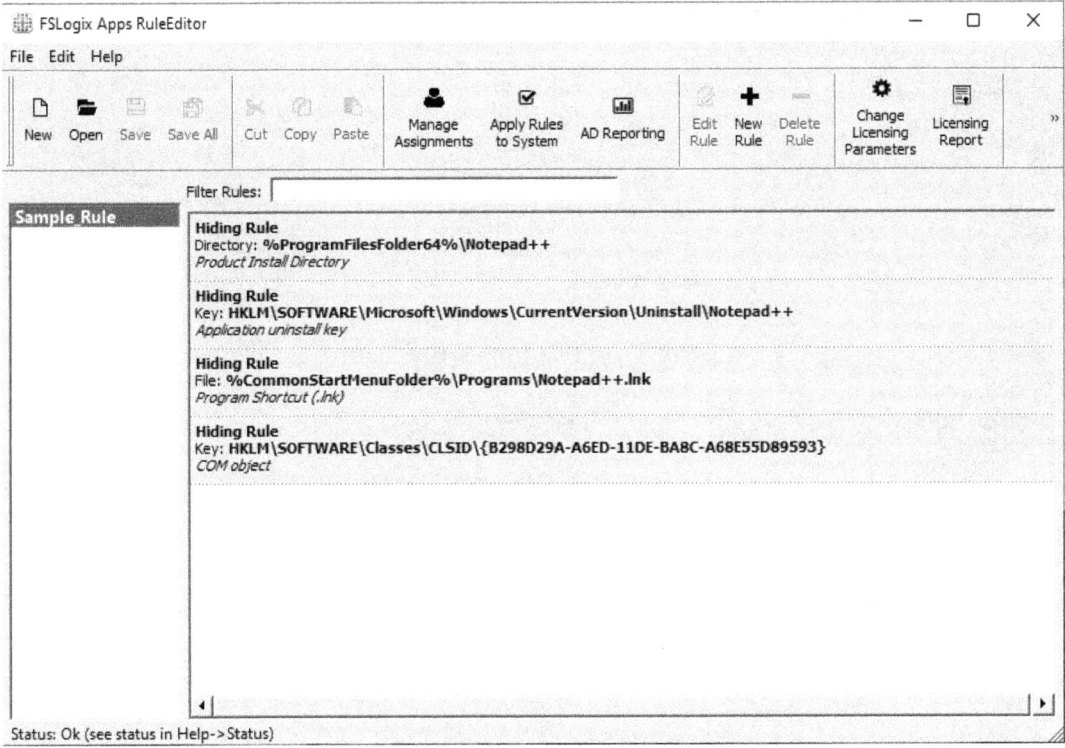

Figure 15.6 – A hiding rule has been created

6. To manage assignments, right-click on **Sample_Rule**, as shown in the following screenshot, and click **Manage Assignments**. This will load the **Assignments** page:

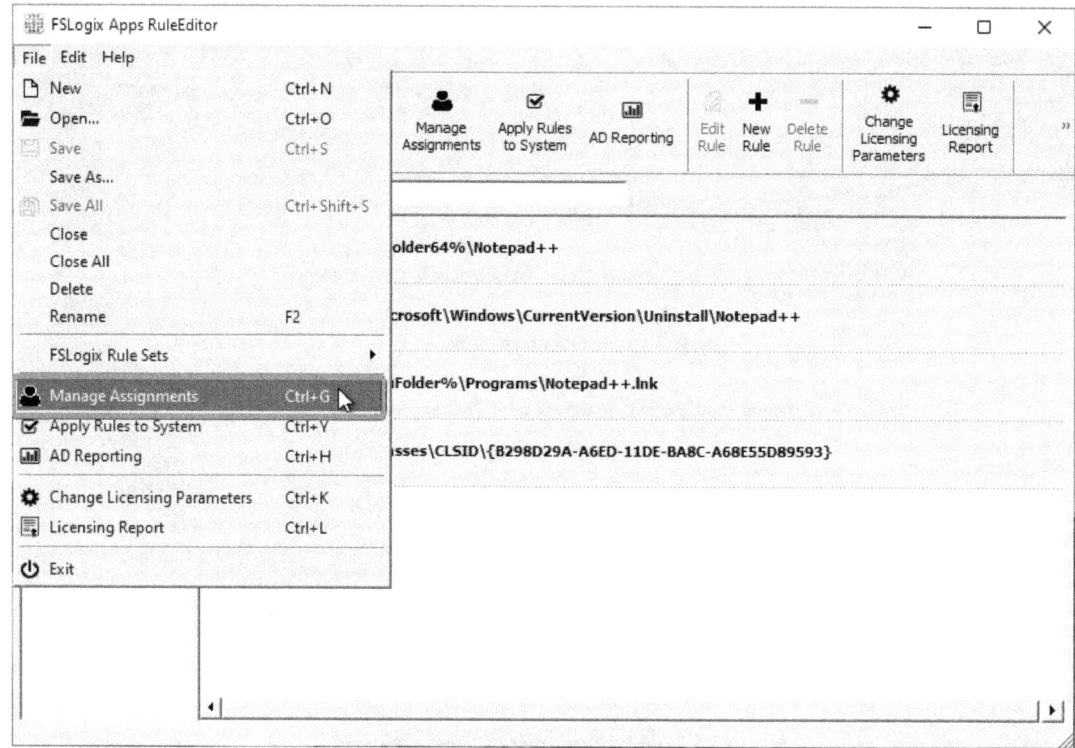

Figure 15.7 – The Manage Assignments option

7. Within the **Assignments** page, you can add the required user/group permissions for this rule.

> **Important note**
> There are several options available to you when it comes to assignments. You have a choice of not only user and group but also additional options including process, network location, computer, directory container, and environment variable.

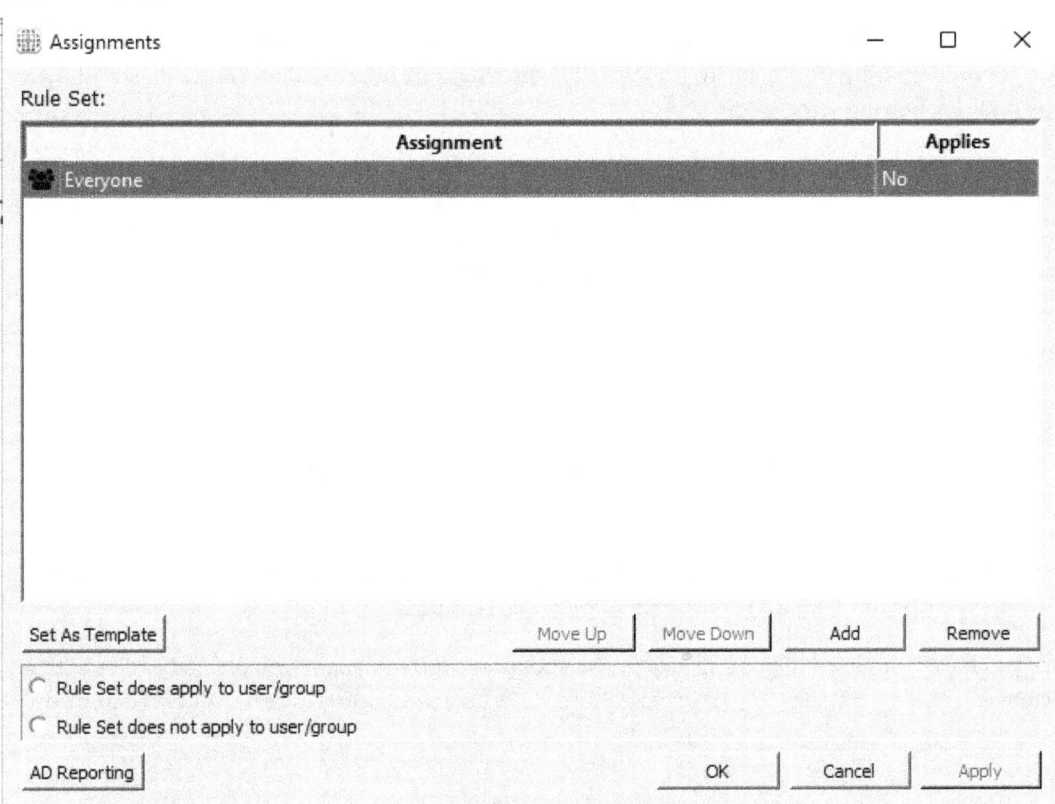

Figure 15.8 – The Assignments form where you can set different assignments such as users and groups

> **Important note**
> Good practice would be to use Active Directory groups to break out applications into department groups or groups relating to specific requirements. This then essentially enables granular access and better IT administrator control for user access to applications using FSLogix application masking.

8. Now that we have finished configuring the rule, you might want to test it before we deploy it to a production environment. To do that, click the checkbox in the main form taskbar, as shown in the following screenshot:

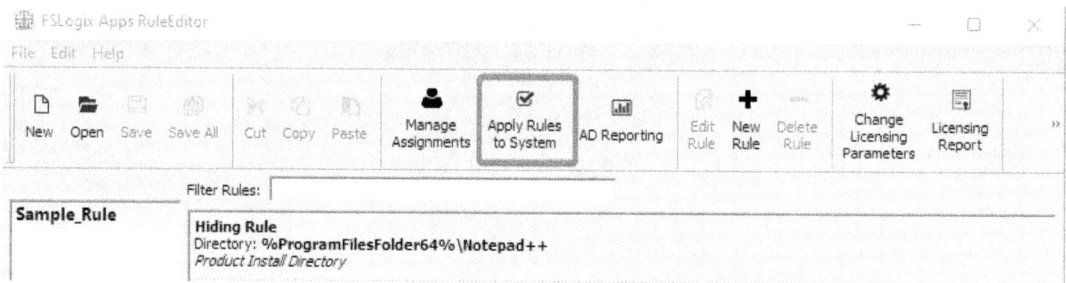

Figure 15.9 – The test check button on the taskbar

9. Once you have finished testing using the built-in test capability, the next step is to publish these rules to the session hosts. To do this, you would simply need to copy the created rule to the `c:\Program Files\FSLogix\Apps\Rules` path.

Once you have completed the task in *step 9*, you should see the rule name you created earlier with a different extension, `.fxc`, in `C:\Program Files\FSLogix\Apps\Rules\Compiledrules`:

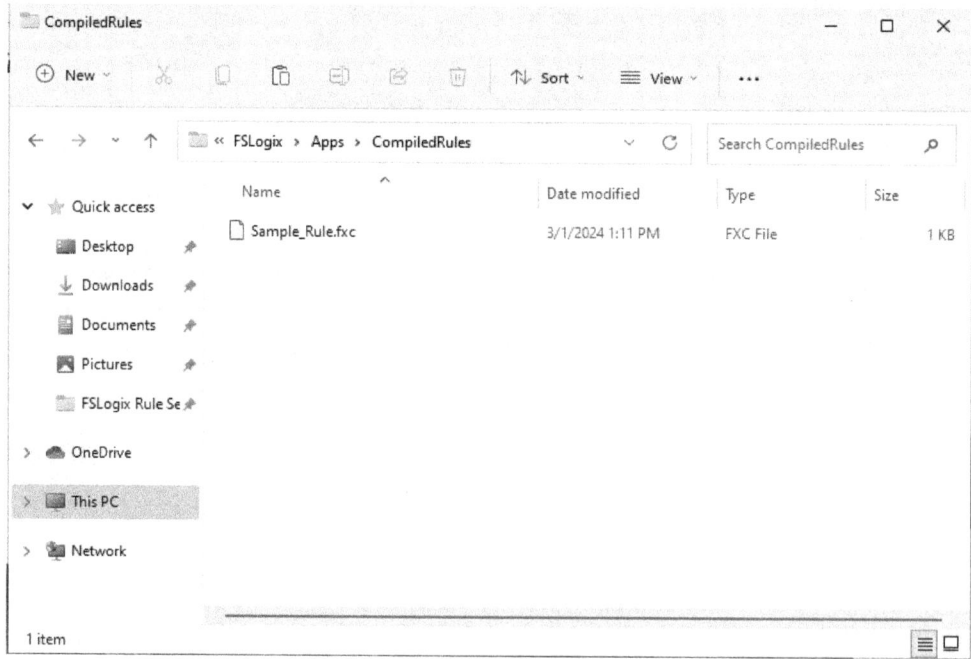

Figure 15.10 – The compiled rule once you've added the rule to c:\Program Files\FSLogix\Apps\Rules

> **Important note**
> You can use traditional file transfer technologies or apply the rules within a master image by copying the rule file to the required location. You can also use custom script extensions to distribute the rules.

To summarize, we have created a simple hiding rule from the installed programs. After that, you can apply many custom/granular configurations and controls to your rule sets, including redirection rules, specify value rules, and app container (VHD) rules.

Deploying an application as a RemoteApp application

Please note that before proceeding with this section, you will need to ensure that applications have been installed on the session host first.

In this section, we take a look at deploying a RemoteApp application within Azure Virtual Desktop. RemoteApp applications are essentially configured within the application groups. As mentioned in previous chapters, there are two types of application groups – one being a desktop app group and the second being a RemoteApp group.

> **Reminder**
> You can use **MSIX app attach** to deliver applications to RemoteApp user sessions.

Here are the steps to create a RemoteApp group:

1. Sign in to the Azure portal and search for `Azure Virtual Desktop`.
2. On the **Azure Virtual Desktop** page, select **Host pools** and then select **Application groups**.

494 Configuring Apps on a Session Host

3. Click **Create** to create a new application group:

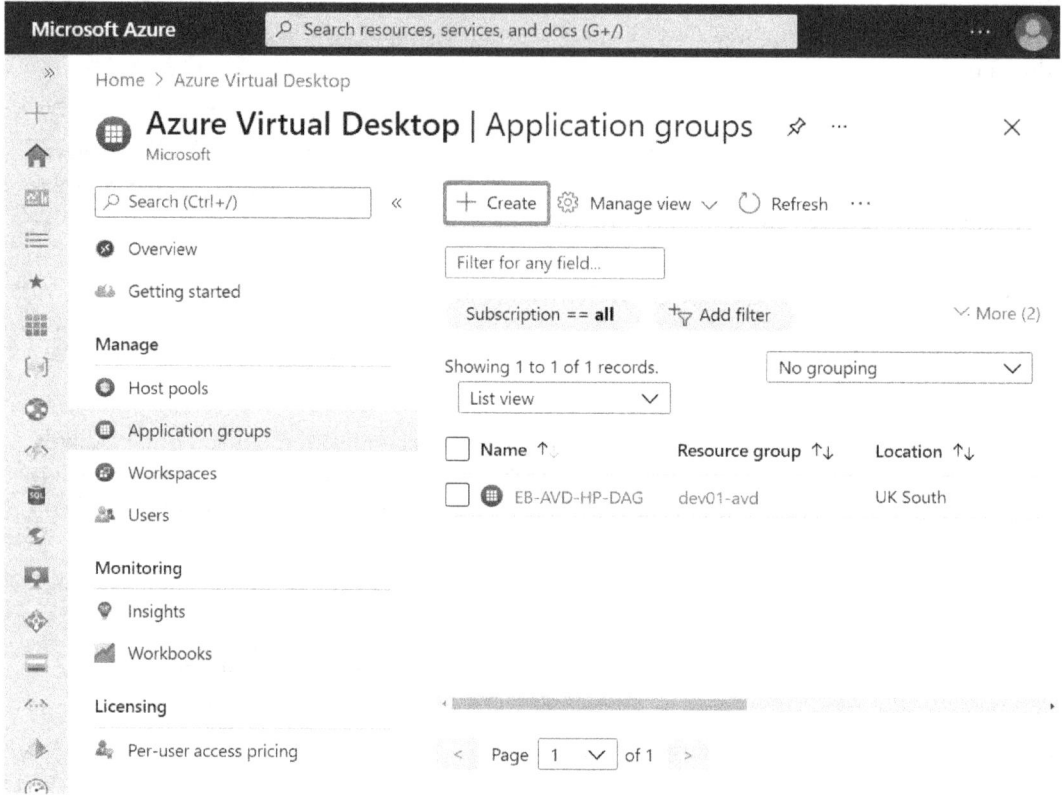

Figure 15.11 – The Application groups page within Azure Virtual Desktop

4. Within the **Basics** tab, select the required **Subscription**, **Resource group**, and **Host pool** options:

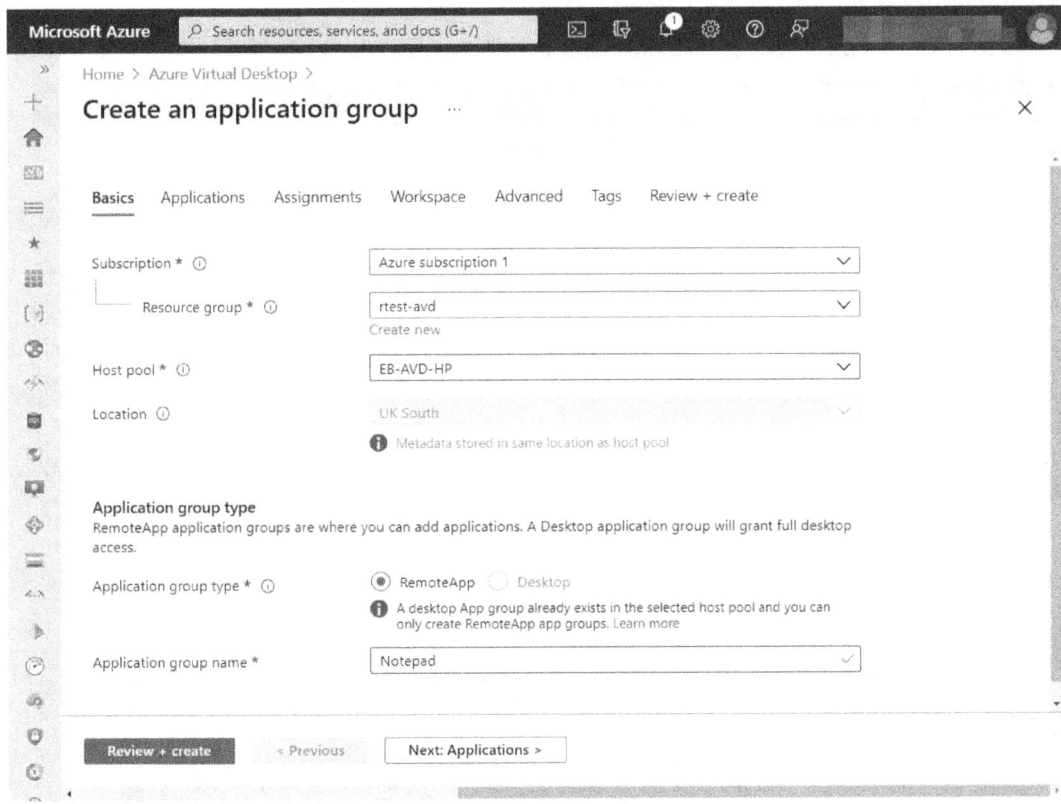

Figure 15.12 – The Create an application group page within Application groups

5. Ensure that **Application group type** is set to **RemoteApp** and enter a name:

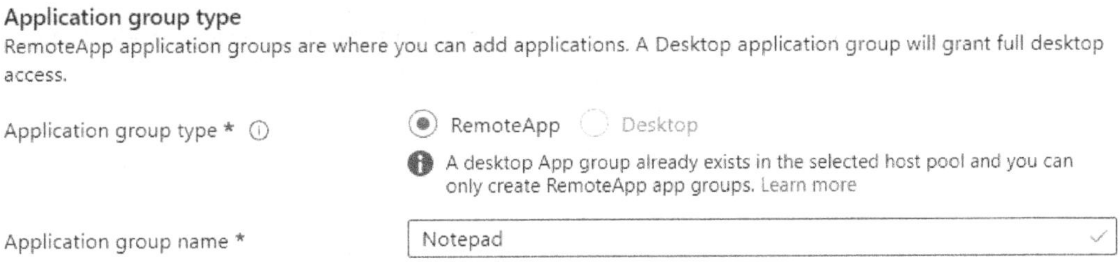

Figure 15.13 – The Application group type section

6. Once you have filled out the required fields, click **Next: Applications** to move on to the **Applications** tab.
7. Select **Start menu** as the application source and **Notepad** as the application in this example:

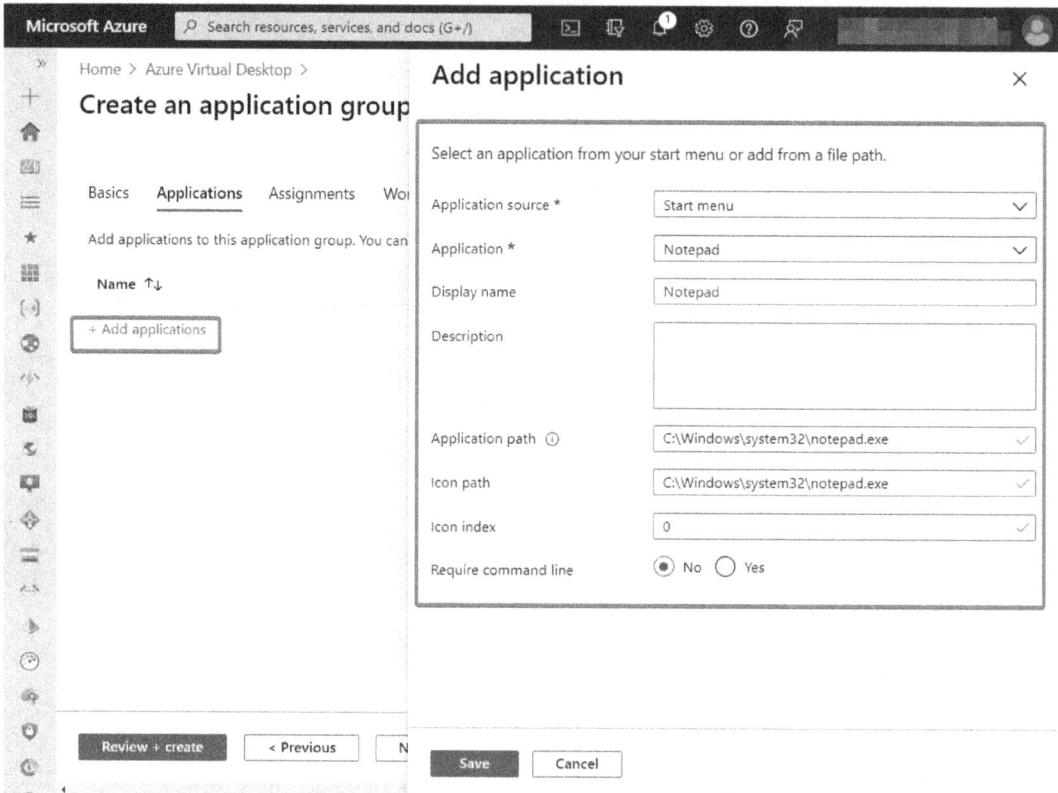

Figure 15.14 – The Add application blade within an application group

8. Enter a **Display name** value for the application and leave the other options as is, then click **Save**.

9. Once you have finished adding applications, click **Next: Assignments** to move on to the **Assignments** tab:

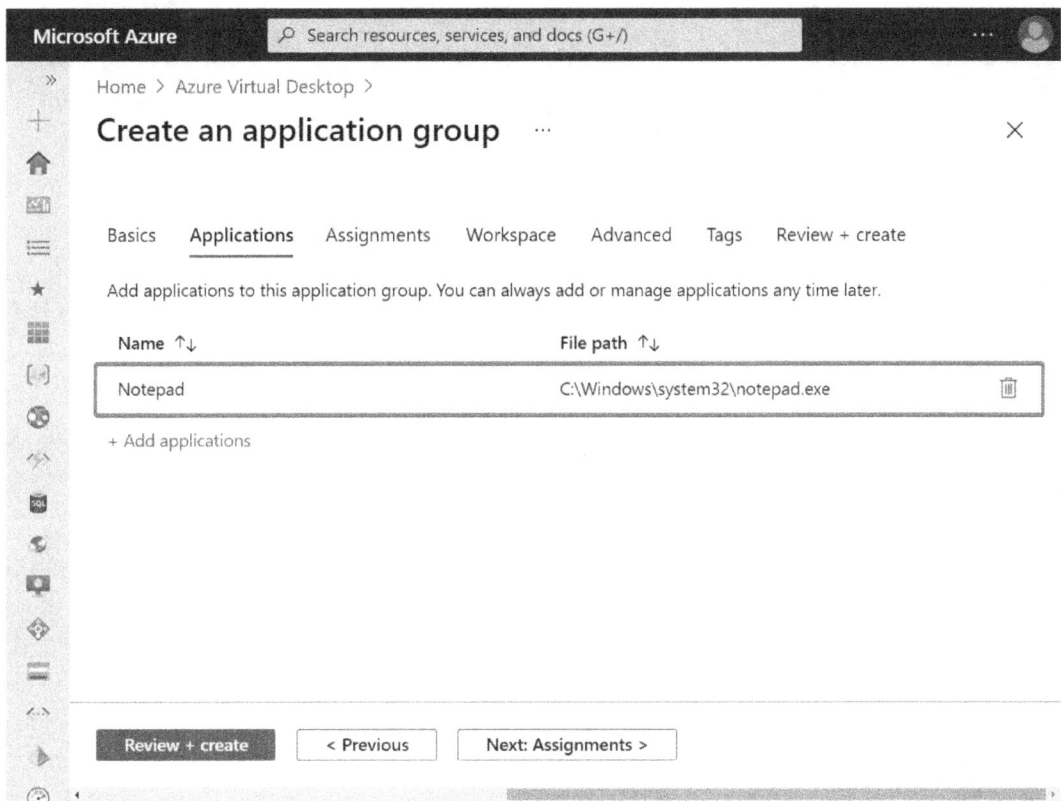

Figure 15.15 – Application added to the Create an application group page

10. Within **Assignments**, add the users/groups you want to have access to the remote application. Once complete, click **Next: Workspace** to move on to the **Workspace** tab:

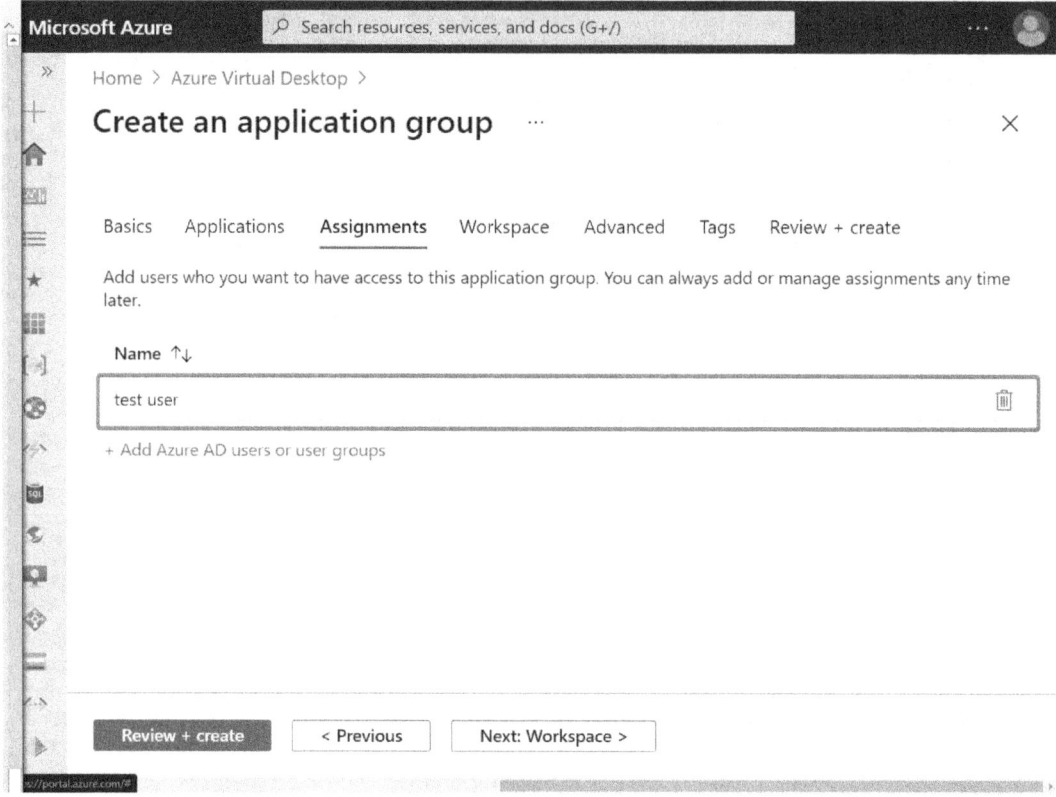

Figure 15.16 – A user account added to the Assignments tab

11. If you want to register the application group to a workspace, select **Yes** to register the application group. If you want to do this later, select **No**. If you have an existing workspace, you can register your application group to it:

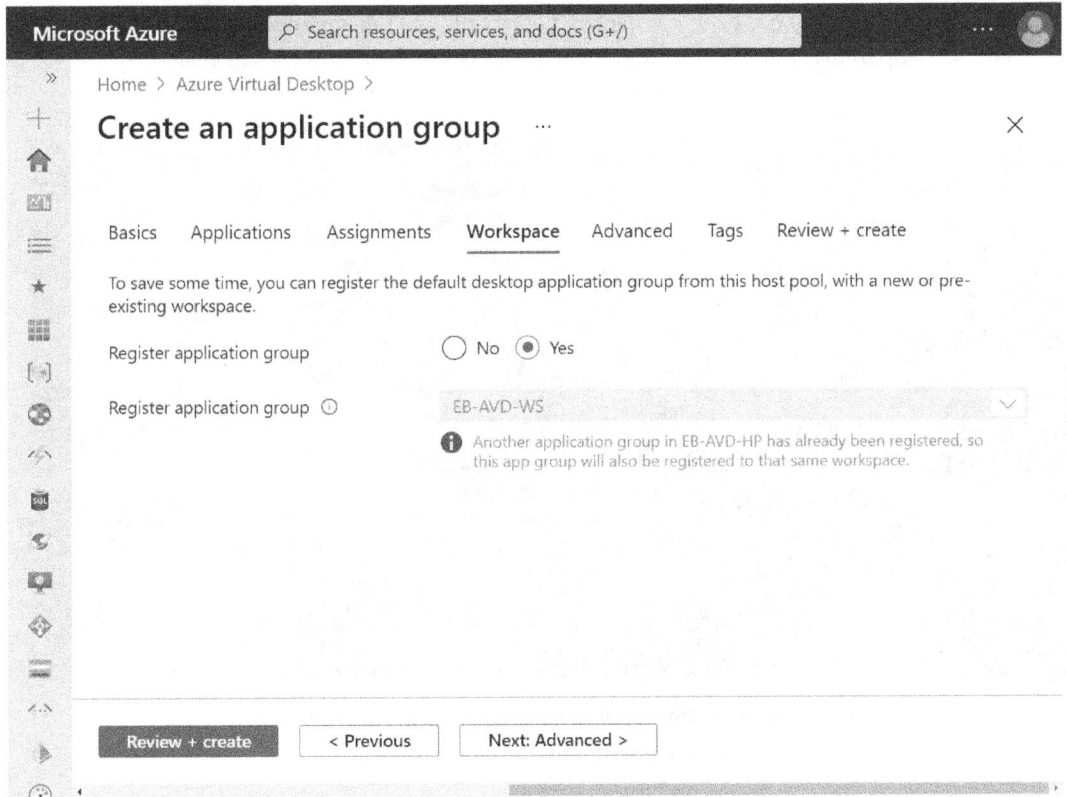

Figure 15.17 – The Workspace tab within the Create an application group page

> **Important note**
> Application groups can only be registered to workspaces created in the same region as the host pool. If you've previously registered another app group from the same host pool as your new app group to a workspace, it will be selected, and you won't be able to edit it, as shown in the preceding screenshot. All application groups from a host pool must be registered to the same workspace.

12. Click **Next: Advanced** and configure **Diagnostic** and **Tags**. If tags and diagnostics are not required, then you can click **Review + create** instead.
13. Within the **Review + create** tab, check your configurations, then click **Create**.
14. Once the deployment has been completed, you should see the **Notepad** application appear as a remote app in the **Remote Desktop** client, as shown in the following screenshot:

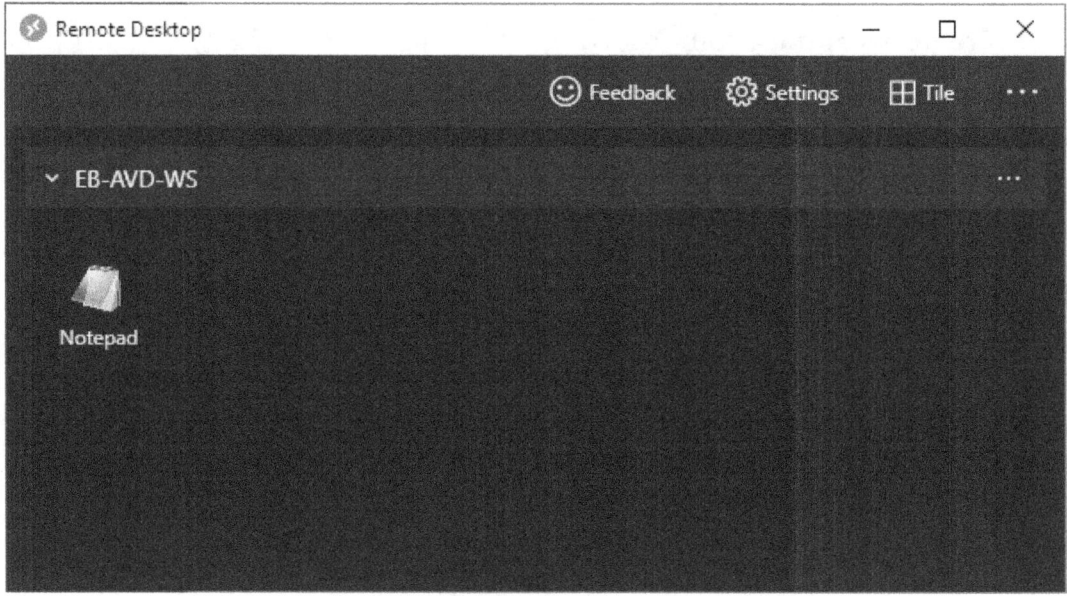

Figure 15.18 – The Remote Desktop client, displaying the Notepad application we just published as a remote app

This concludes the steps to deploy a RemoteApp application. We'll now look at how to implement and manage OneDrive for Business for a multi-session environment.

Implementing and managing OneDrive for Business for a multi-session environment

In this section, we will look at OneDrive for Business and how to deploy it for a multi-session environment. A typical OneDrive installation installs per user; this means that the OneDrive client is installed under the `%localappdata%` folder. When deploying OneDrive (sync app) for a multi-session/VDI environment, you need to install the per-machine installation option. This installs OneDrive under the `Program Files (X86)` or `Program Files` directory, depending on the operating system architecture. The reason is that if you install per user, you will need to run the OneDrive setup for each user on the session host, which causes issues.

> **Important note**
>
> To ensure that you can apply sync app updates, you need to ensure that computers in your environment can communicate with the following URLs: `oneclient.sfx.ms` and `g.live.com`. Make sure you don't block these URLs.

To Install OneDrive per machine, please follow these steps:

1. You can download OneDrive from the Microsoft site. The following link is for version 24.20.128.3 (February 22, 2024): `https://go.microsoft.com/fwlink/?linkid=844652`.

2. To deploy OneDrive per machine, you would use the `OneDriveSetup.exe /allusers` cmdlet, as shown in the following screenshot:

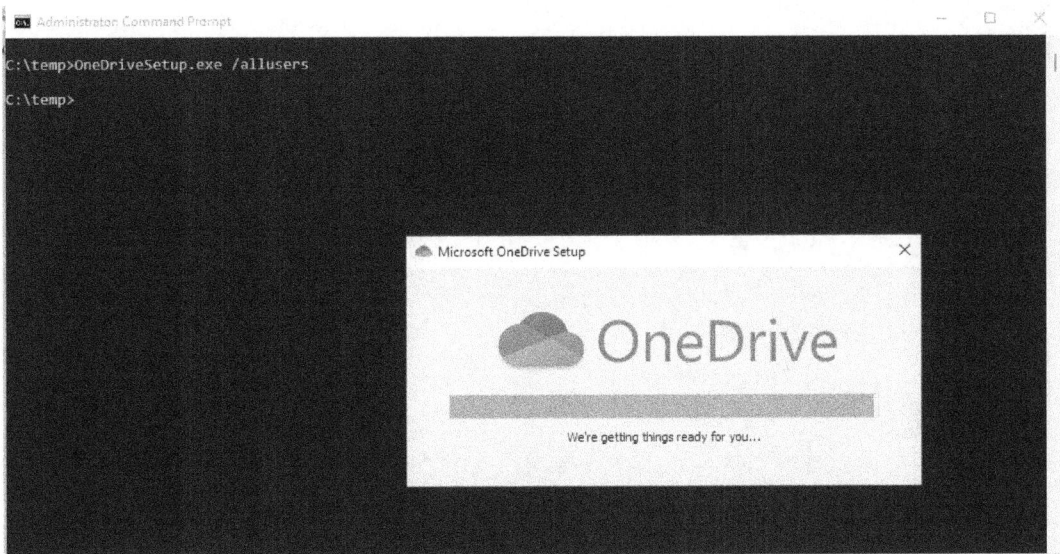

Figure 15.19 – Installation of OneDrive using the per-machine option

3. Add the following registry key to start the sign-in for all users:

   ```
   REG ADD "HKLM\Software\Microsoft\Windows\CurrentVersion\Run"
   /v OneDrive /t REG_SZ /d "C:\Program Files\Microsoft OneDrive\
   OneDrive.exe /background" /f
   ```

4. You then need to enable **Silent Configure User Account** with the following command:

   ```
   REG ADD "HKLM\SOFTWARE\Policies\Microsoft\OneDrive" /v
   "SilentAccountConfig" /t REG_DWORD /d 1 /f
   ```

5. You also need to redirect the Windows known folders to OneDrive using this command:

```
REG ADD "HKLM\SOFTWARE\Policies\Microsoft\OneDrive" /v
"KFMSilentOptIn" /t REG_SZ /d "<your-AzureAdTenantId>" /f
```

Once you have installed OneDrive (sync app), you should see the application within `Program Files (x86)` or `Program Files`, as shown in the following screenshot:

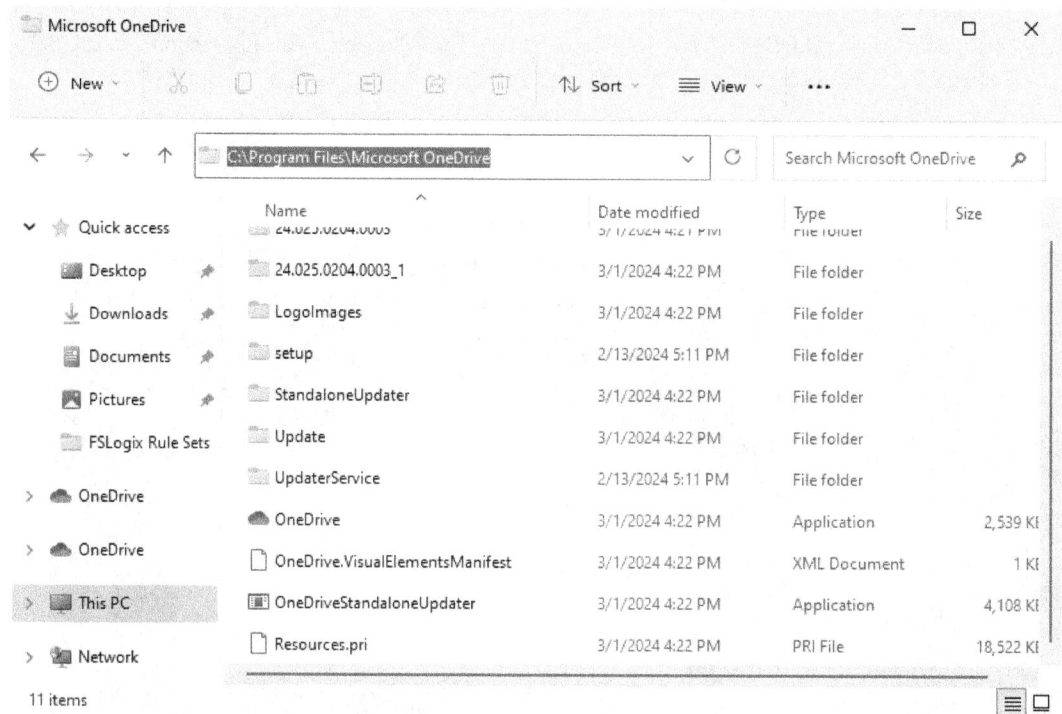

Figure 15.20 – OneDrive program files after installing using the per-machine option

> **Important note**
>
> Using the images provided by Microsoft can save you a lot of time as configurations have already been done by Microsoft.
>
> For more information on OneDrive and best practices for controlling its behavior, please go to the following link, which provides details on the settings you can configure for OneDrive with Azure Virtual Desktop: https://docs.microsoft.com/onedrive/use-group-policy#manage-onedrive-using-group-policy.

In this section, we looked at how to deploy OneDrive for multi-session deployments. In the next section, we take a look at implementing and managing Microsoft Teams AV redirection.

Implementing and managing Microsoft Teams AV redirection

This section looks at the specific configurations required for Microsoft Teams AV redirection for Azure Virtual Desktop. AV redirection is essentially an optimization for Microsoft Teams. It works by enabling the Windows desktop client to handle audio and video locally for Teams calls and meetings.

You can experience high CPU and poor performance when the session host handles audio and video. By redirecting to the local client, you reduce the resources used on the session host virtual machine and improve the overall experience as audio and video are handled by the local client device. Follow these steps:

1. The first step is to ensure that you have installed the Teams desktop application on the session host template. This must be installed per device and not per user. If you use the images managed by Microsoft, then everything is configured for you.

2. You then need to add the following registry key to HKEY_LOCAL_MACHINE\SOFTWARE\Microsoft\Teams, as shown in the following table:

Name	Type	Data/Value
IsAVDEnvironment	DWORD	1

   Table 15.1 – Registry key notifying Teams that the environment is Azure Virtual Desktop

   You can add the registry key using the following command within Command Prompt run as an administrator:

   ```
   reg add"HKLM\SOFTWARE\Microsoft\Teams" /v IsWVDEnvironment /t REG_DWORD /d 1 /f
   ```

   Or you can use PowerShell:

   ```
   New-Item -Path "HKLM:\SOFTWARE\Microsoft\Teams" -Force
   New-ItemProperty -Path "HKLM:\SOFTWARE\Microsoft\Teams" -Name IsWVDEnvironment -PropertyType DWORD -Value 1 -Force
   ```

3. Install the Teams WebSocket service on the virtual machine image. You can download it from here: https://query.prod.cms.rt.microsoft.com/cms/api/am/binary/RWFYsj. If you experience installation issues, you will need to install the latest Microsoft Visual C++ Redistributable download. You can find it at this URL: https://support.microsoft.com/help/2977003/the-latest-supported-visual-c-downloads.

4. Once you have installed the required components, you will need to reboot the session host image/template.

Configuring Apps on a Session Host

> **Important note**
>
> The configuration for AV redirection is already set up for you when using an image from the gallery, specifically the Windows 11 images with *Microsoft 365 Apps* in the title.

The following screenshot shows the **WebRTC Redirector Service** program installed:

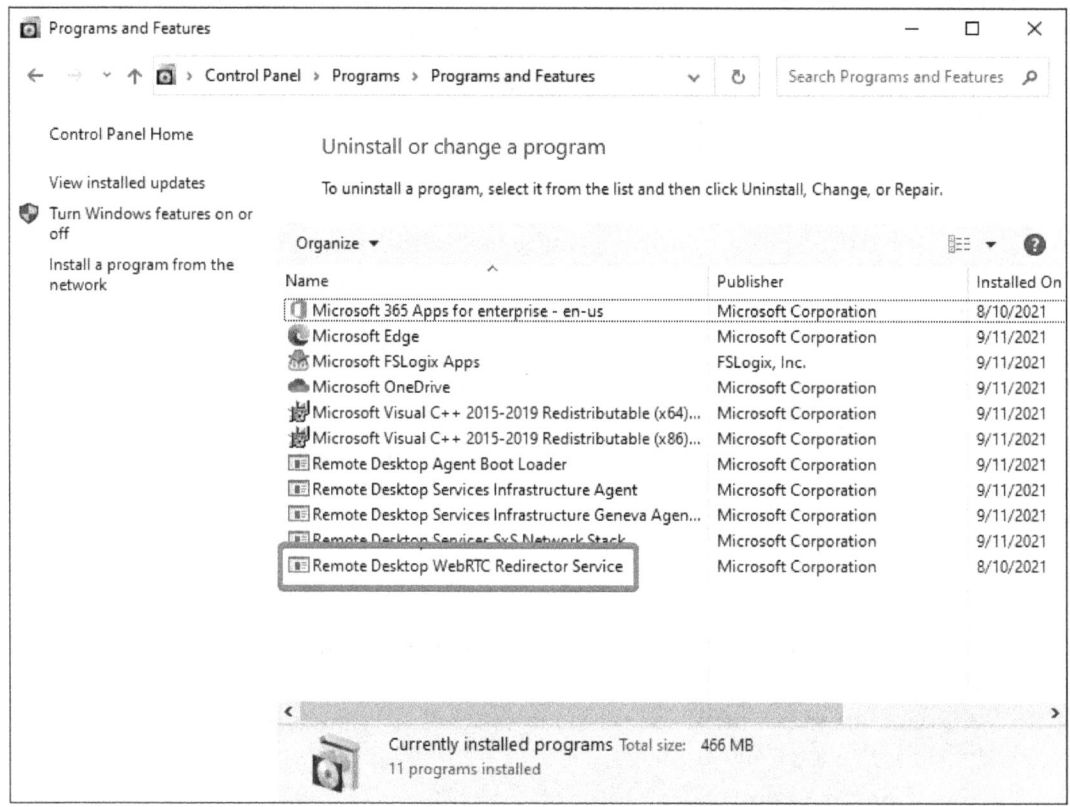

Figure 15.21 – Remote desktop WebRTC Redirector Service installed on a templated image

You can also see that the required registry key has been pre-added, as shown in the following screenshot:

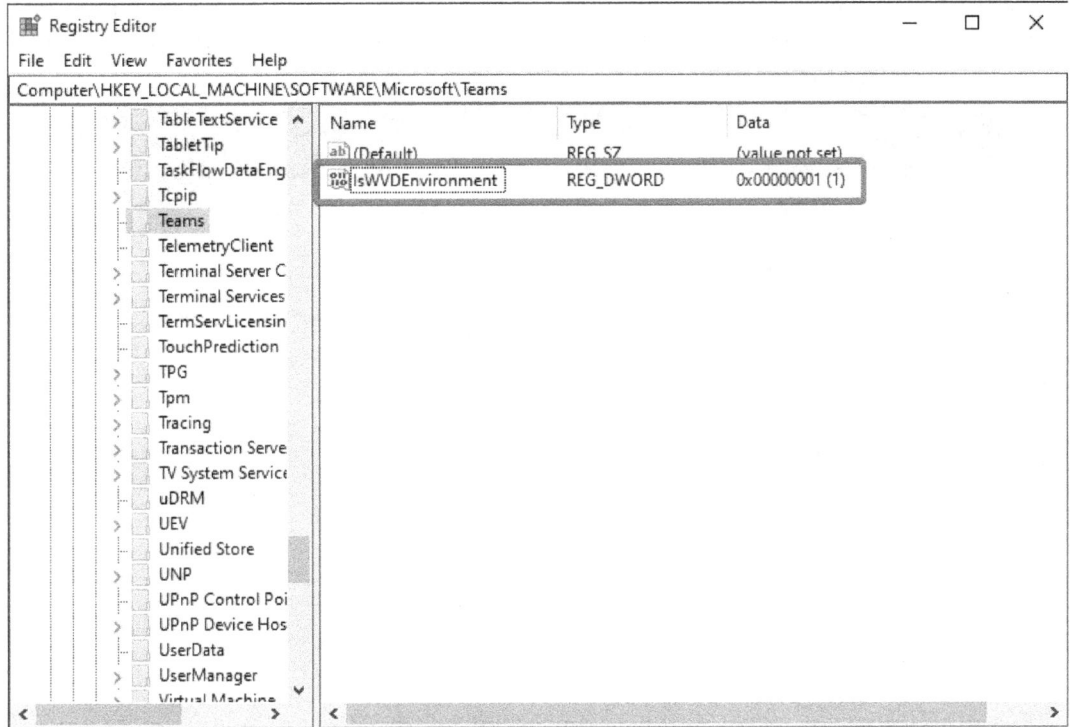

Figure 15.22 – Required registry setting included within the image template

Now that we have configured Teams AV redirection, let's look at the next steps of verifying that the configuration is working.

Verifying that media optimizations are loaded

We have installed the required components for AV redirection. We now need to verify that Teams is configured for AVD redirection. You can check this by navigating to the **Version** button, as shown in the steps in the following screenshot:

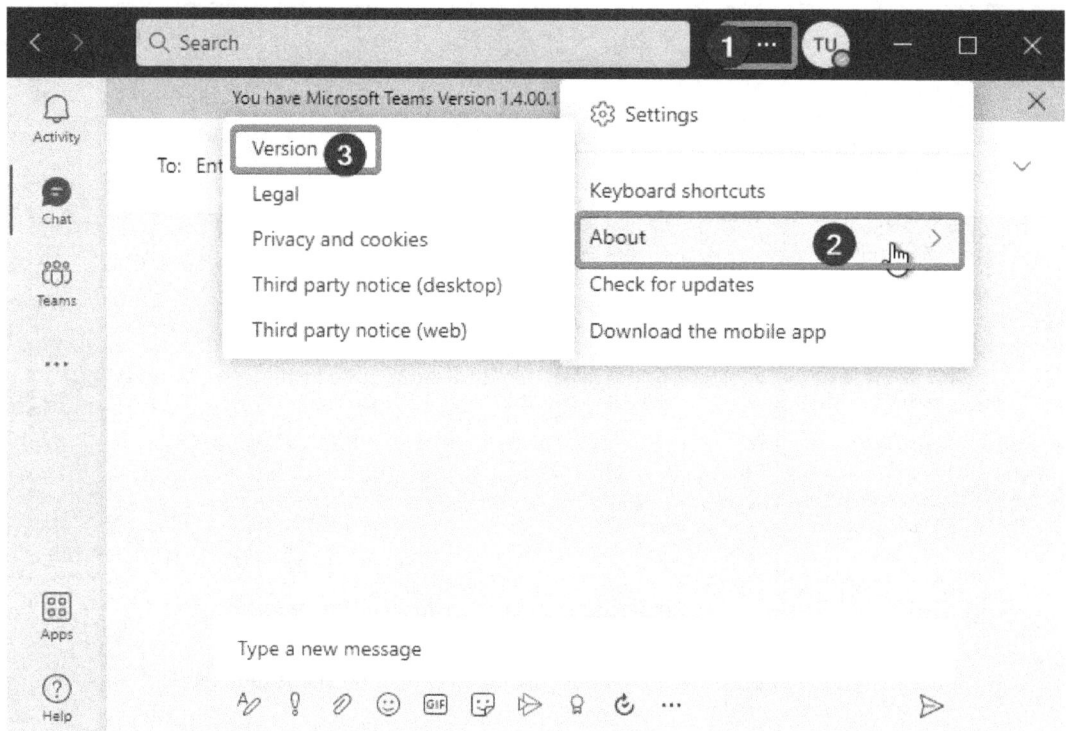

Figure 15.23 – Displaying the Microsoft Teams version

Once you have clicked **Version**, you should see the version appear as a banner within the Teams app, as shown in the following screenshot:

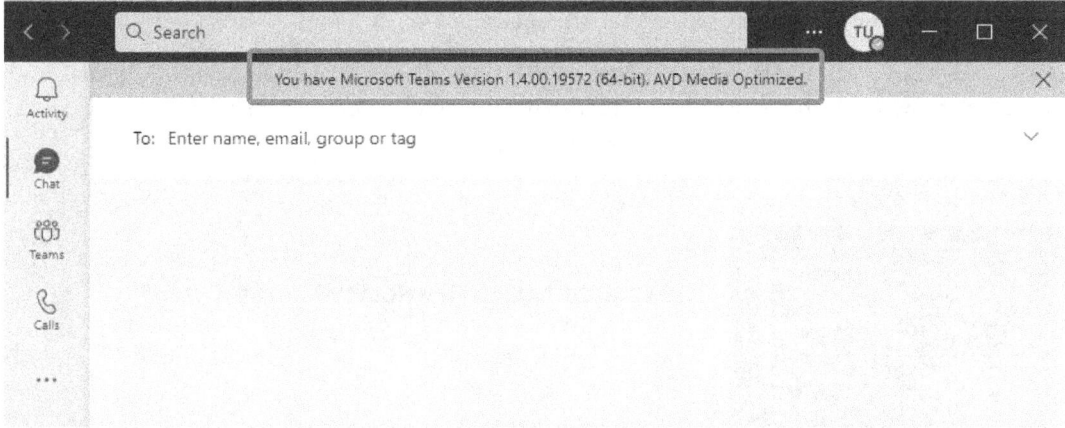

Figure 15.24 – Version output confirming that AVD Media Optimized is configured

As shown in the preceding screenshot, you can see that we now have confirmed that Microsoft Teams is configured with Microsoft Teams AV redirection. Additionally, the version will state **AVD Media Optimized** when configured correctly.

In this section, we looked at how to configure AV redirection for Microsoft Teams within Azure Virtual Desktop. This included how to check whether AV redirection is configured. In the next section, we will look at multimedia redirection.

Implementing and managing multimedia redirection

In this section, we will take a look at **multimedia redirection** (**MMR**) for Azure Virtual Desktop, which, at the time of writing, is in preview. MMR provides smooth video playback through the Microsoft Edge and Google Chrome web browsers. When MMR is enabled, the media element is remoted to the Remote Desktop client (local machine), reducing resource utilization, specifically the CPU on a session host. This is a feature similar to AV redirection for Microsoft Teams.

> **Important note**
>
> To use MMR, you need to make sure you use the Remote Desktop client version 1.2.2222 or later. Additionally, you can check whether MMR is supported as `MSMmrDVCPlugin.dll` should be present in the Remote Desktop client application installation path.

Getting started with MMR

In this section, we will configure MMR for use with Microsoft Edge and allow `youtube.com` to use MMR. You can add multiple sites or all of them if required.

You first need to download and install the **MsMmrHostMsi** installer, which installs the MMR extensions for your web browser on the session host. You can download it here: `https://query.prod.cms.rt.microsoft.com/cms/api/am/binary/RWIzIk`.

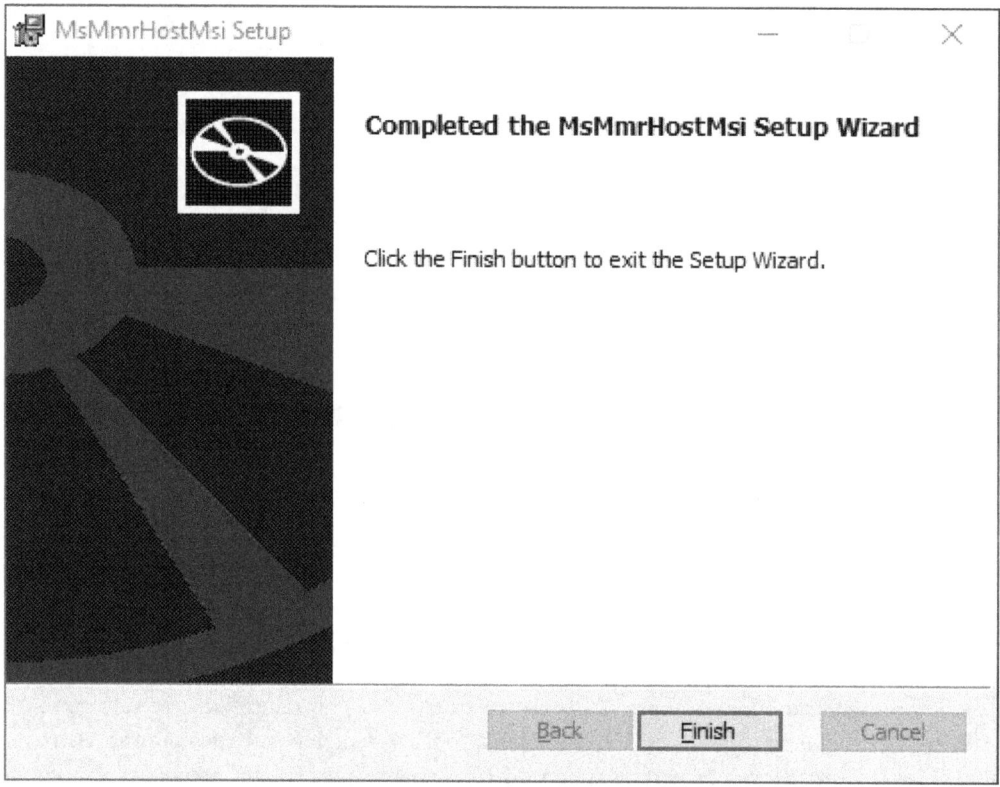

Figure 15.25 – The MsMmrhostMsi installation has been completed

Now that we have installed MSMmrHostMsi, we can proceed to the next part of this section, which looks at controlling which websites can use MMR.

Restricting which websites can use MMR

We'll now take a look at how to allow and block websites and configure Microsoft Edge or Google Chrome for MMR. This example looks at Microsoft Edge only. You can use Group Policy to configure MMR. Before we can do this, we need to download and install the MS Edge Group Policy administrative template: `https://aka.ms/EdgeEnterprise`.

1. Once you've downloaded the CAB file, extract it to a suitable location on your session host such as `c:\temp`.
2. Copy the ADM and ADMX files to the `PolicyDefinitions` folder. If using the **Group Policy Central Store**, copy to the policy definition path:

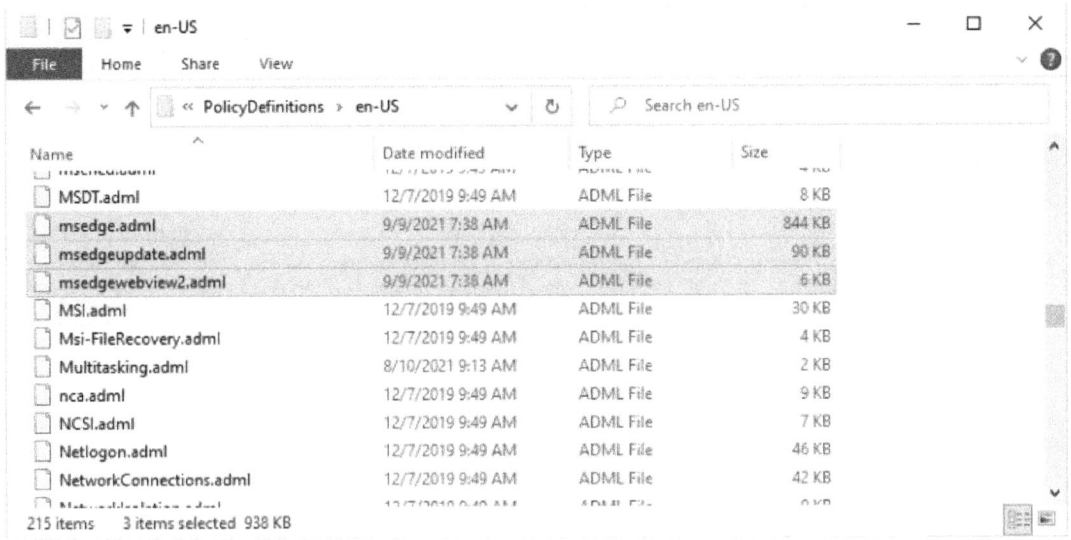

Figure 15.26 – ADML files that have been copied to the PolicyDefinitions folder

3. Create a new policy within **Group Policy Manager | User configuration | Administrative Templates | Microsoft Edge | Extensions | Configure extension management settings**.
4. Set the policy as enabled and enter the following code in the **Configure extension management settings** policy:

```
{ "joeclbldhdmoijbaagobkhlpfjglcihd": { "installation_mode":
"force_installed", "runtime_allowed_hosts": [ "*://*.youtube.
com" ], "runtime_blocked_hosts": [ "*://*" ], "update_url":
"https://edge.microsoft.com/extensionwebstorebase/v1/crx" } }
```

You can customize the runtime allowed/blocked hosts. In this example, we are allowing `youtube.com`:

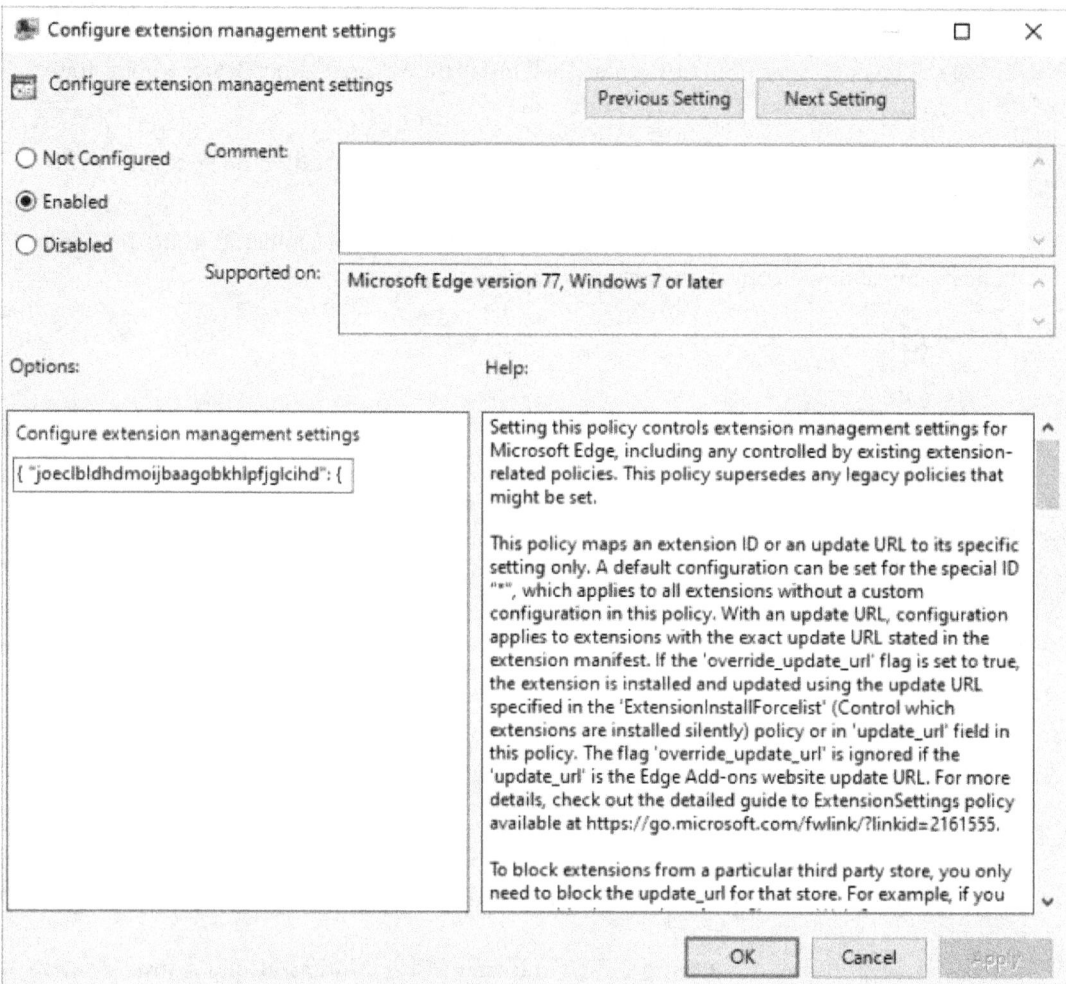

Figure 15.27 – The Configure extension management settings policy for Microsoft Edge

5. Once you have applied the policy, it is recommended that you reboot the session host once you've finished configuring.

We'll now move on to look at MMR in action.

Testing MMR

When the session host has rebooted, open Microsoft Edge, and you will see a Remote Desktop client icon in the menu bar, as shown in the following screenshot. Additionally, when loading an allowed/supported multimedia site, the MMR icon will show a green checkbox if configured correctly, including communicating correctly with the Remote Desktop client:

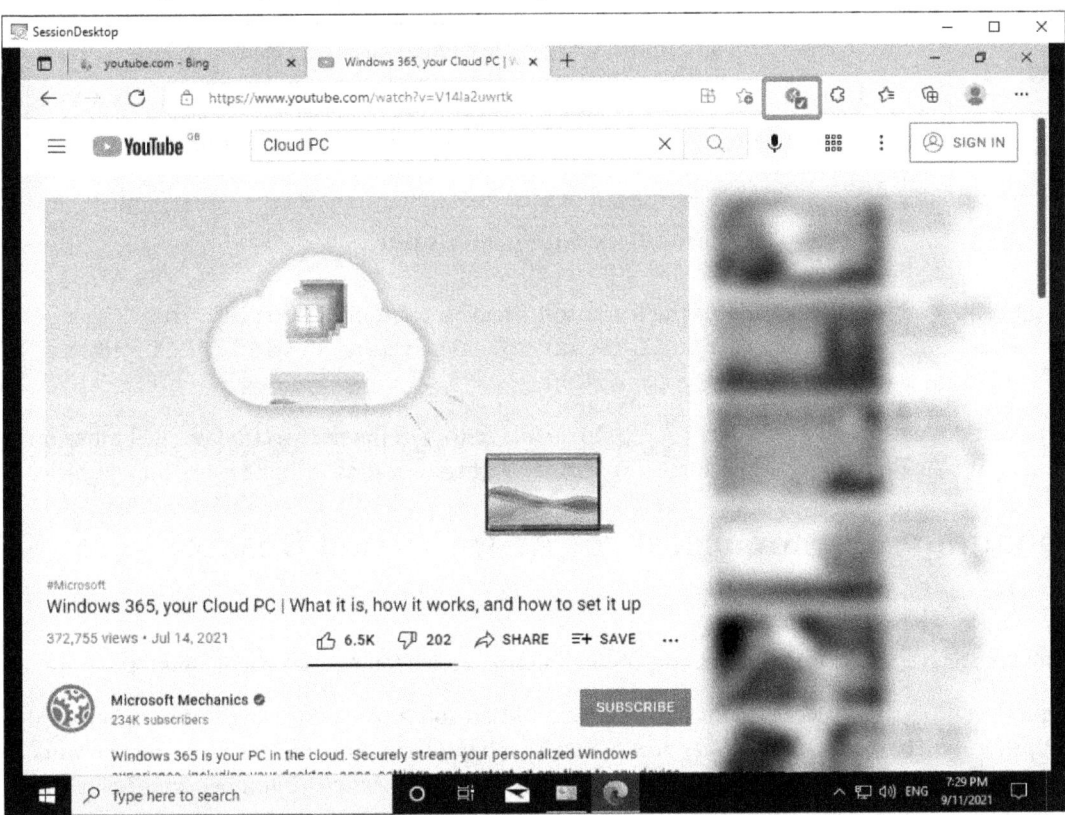

Figure 15.28 – MMR enabled on a Remote Desktop client session

The following table shows the three different icon states, which will help you if you need to troubleshoot any issues:

Icon State	Definition
	The default icon appearance with no status applied
	The red square with an *X* indicates that the client couldn't connect to MMR
	You will see the green square with a check mark when the client has successfully connected to MMR

Table 15.2 – Different states of MMR

The preceding table was taken from the Microsoft site; you can find it here: https://docs.microsoft.com/azure/virtual-desktop/multimedia-redirection#the-multimedia-redirection-status-icon.

In this section, we looked at MMR for Azure Virtual Desktop. In the next section, we will move on to look at managing internet access for Azure Virtual Desktop sessions.

Managing internet access for Azure Virtual Desktop sessions

In this section, we take a brief look at controlling internet access for Azure Virtual Desktop.

A typical way to manage access to the internet within Microsoft Azure is by using **Network Security Groups** (**NSGs**). NSGs are used to filter network traffic inbound and outbound from a virtual network subnet. You can filter traffic by IP address, port, and protocol. To restrict internet access for Azure Virtual Desktop users, you can use an NSG to block web traffic.

To block internet access, you should complete the following steps:

1. Navigate to the session host subnet's NSG.
2. Within the **Settings** menu located on the left-hand side of the NSG, click **Outbound security rules**.
3. Add a new rule.

4. Set **Source** as **Any**; the source port ranges should be the default *. Set **Destination** as **Service Tag** and **Destination service tag** as **Internet**. Then, specify **Service** as **HTTPS** and **Action** as **Deny**. You also need to specify a priority and ensure that the priority does not impact other services on the virtual network. Please note that you will need to repeat these steps for HTTP traffic to change the service to **HTTP**. The following screenshot shows an example:

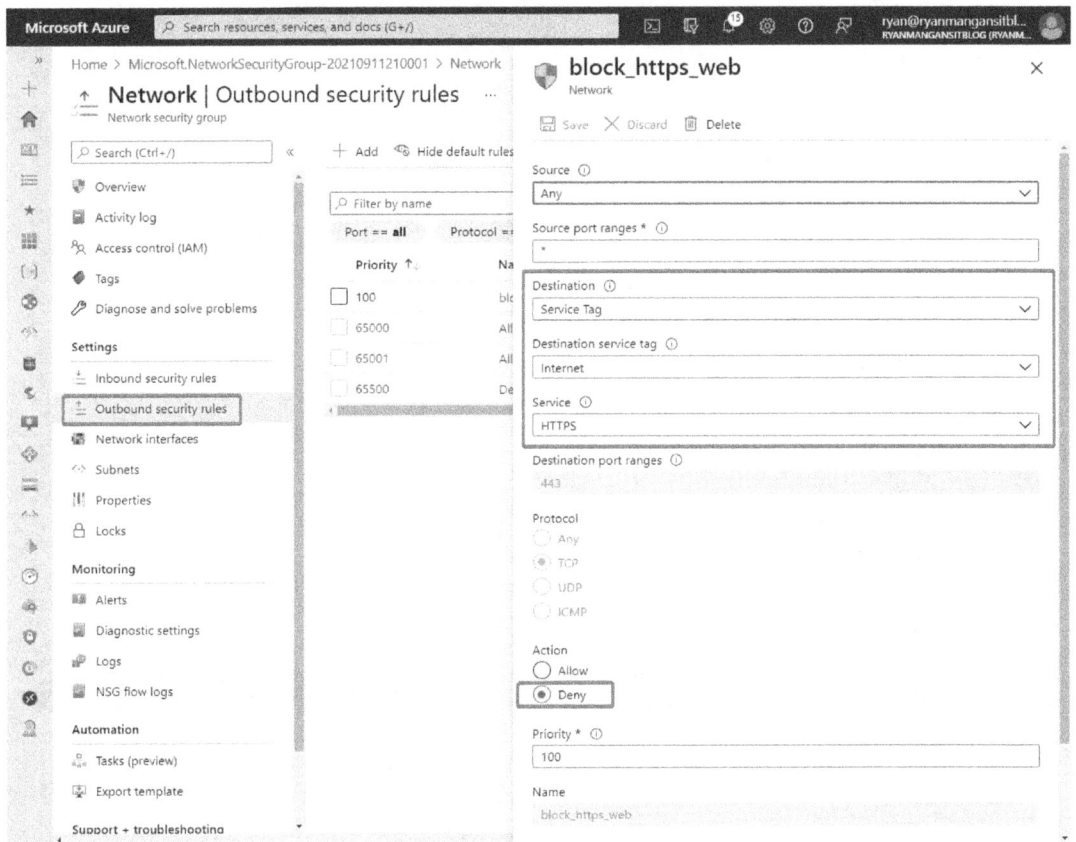

Figure 15.29 – The Outbound security rules page of an NSG and the blade for adding a security rule to restrict HTTPS access to the internet

5. Once the policy is set, you should see the message **ERR_CONNECTION_TIMED_OUT** when opening Microsoft Edge, as shown in the following screenshot:

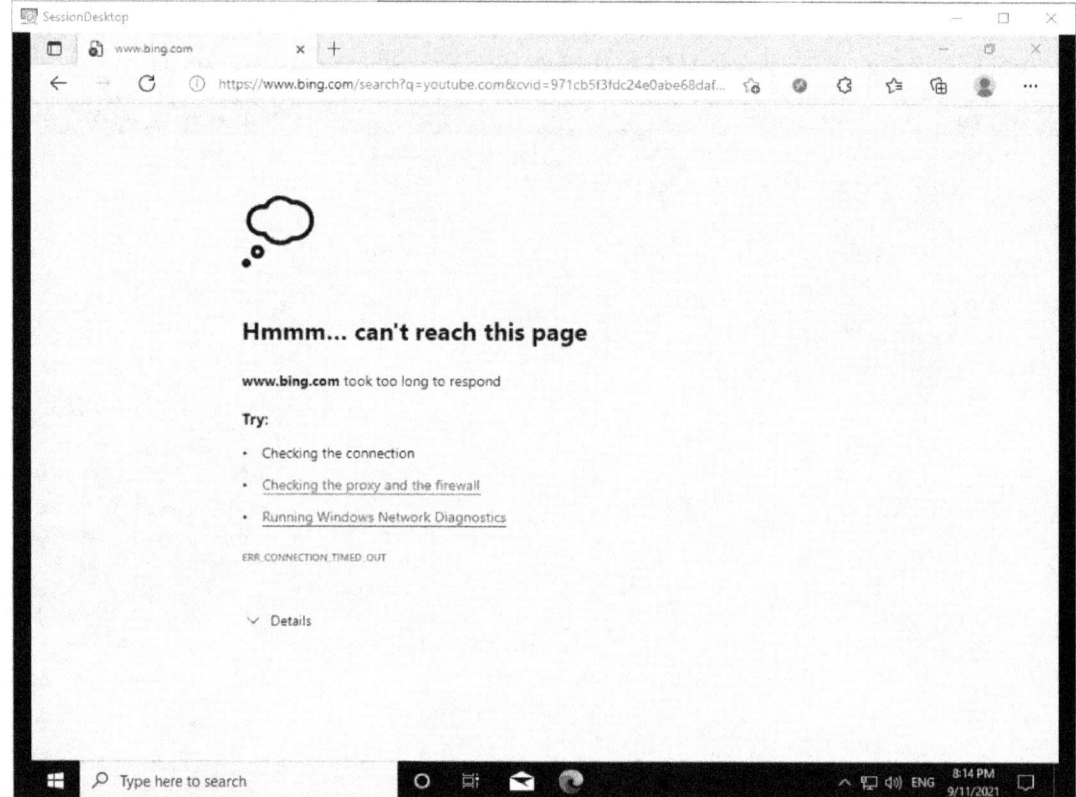

Figure 15.30 – The screen after applying the security rule

> **Tip**
> For more granular control of internet access from the session host, you should look at a third-party **Network Virtual Appliance** (**NVA**) or Azure Firewall. Using an NVA offers advanced third-party features, including content filtering and specific application control policies that take effect at the perimeter.

This section took a quick look at restricting internet access to session hosts without impacting other services and granular control access to specific services using an NSG.

What are VM applications?

VM applications in Azure Compute Gallery offer an application management solution for up to 25 different applications deployed on a virtual machine. This provides a simple but effective method for deploying, updating, and managing applications separately from VM images. This feature also enables global distribution, version control, and Azure RBAC integration without the need for frequent image updates. This could be particularly useful for persistent desktop deployments.

You can distribute applications in the `.msi` installer format, `.exe`, or even zipped files.

> **Tip**
>
> Here are a few tips to consider when you are using VM applications:
>
> **Tip 1**: The download directory for apps is `C:\Packages\Plugins\Microsoft.CPlat.Core.VMApplicationManagerWindows\1.0.9\Downloads\<appname>\<app version>`.
>
> **Tip 2**: When the application file is downloaded to the VM, it is renamed `MYvmApp` with no file extension. This is because the VM is unaware of the file's original extension or filename. You can use `PackageFileName` and corresponding `ConfigFileName` to rename the file.
>
> **Tip 3**: There is no additional cost for using VM applications apart from storage or network egress fees.

You can read more here:

https://learn.microsoft.com/azure/virtual-machines/vm-applications?

And you can read about deployment here:

https://learn.microsoft.com/azure/virtual-machines/vm-applications-how-to?

Summary

We started this chapter by looking at application masking and hiding applications from users who are not within the correct security groups, and other assignment options. We then moved on to take a look at the benefits of application file containers and how to redirect application files to a VHD or VHDX. Next, we looked at how to deploy and configure a RemoteApp application, and in the example used, we published Notepad as a RemoteApp application to the Remote Desktop client.

We then looked at deploying OneDrive for multiple sessions using per-machine installation. Next, we covered both Microsoft Teams AV redirection and MMR. Finally, to finish the chapter, we looked at how to restrict session host internet access using an NSG.

I hope you found this chapter interesting and are looking forward to the next, where we take a look at planning and implementing business continuity and disaster recovery.

Questions

Here are a few questions to test your understanding of this chapter:

1. What are the three non-deprecated application masking rules you can configure?

    Specify value rule, Redirection rule, and Hiding rule

2. Can application groups be registered to workspaces in different regions from the host pool?

    No

3. Which DLL should you check for to ensure that MMR has been installed correctly?

    `MSMmrDVCPlugin.dll`

4. What is the limit of applications you can deploy using VM applications?

    25 different applications

Part 6: Monitoring and Maintaining an Azure Virtual Desktop Infrastructure

This section of the book looks at how you design and plan a disaster recovery solution for Azure Virtual Desktop, as well as automate repeat admin tasks, monitor and manage environment health and performance, and use the *Getting started* feature to quickly deploy a full Azure Virtual Desktop environment quickly.

This part of the book comprises the following chapters:

- *Chapter 16, Planning and Implementing Business Continuity and Disaster Recovery*
- *Chapter 17, Automating Azure Virtual Desktop Management Tasks*
- *Chapter 18, Monitoring and Managing Performance and Health*
- *Chapter 19, Azure Virtual Desktop's Quickstart Feature*
- *Appendix, Microsoft Resources and Microsoft Learn*

16
Planning and Implementing Business Continuity and Disaster Recovery

This chapter examines planning and implementing a business continuity and disaster recovery solution for Azure Virtual Desktop. We first review the requirements before configuring a business continuity and disaster recovery solution, and then we explore the available options.

In this chapter, we examine the following topics

- Designing a backup strategy for Azure Virtual Desktop
- Planning and implementing a disaster recovery plan for Azure Virtual Desktop
- Configuring backup and restore for FSLogix user profiles and personal virtual desktops
- Infrastructure and golden images

Designing a backup strategy for Azure Virtual Desktop

Before we get started, I want to remind you of the shared responsibilities of using Azure Virtual Desktop. The following illustration details the responsibilities managed by Microsoft and what the customer controls. Within this chapter, we will look at all of the items within the customer section for which you are responsible.

The following screenshot shows the different responsibilities of both the customer and Microsoft. You will note that Microsoft takes care of the management plane, while the customer is responsible for everything inside the subscription.

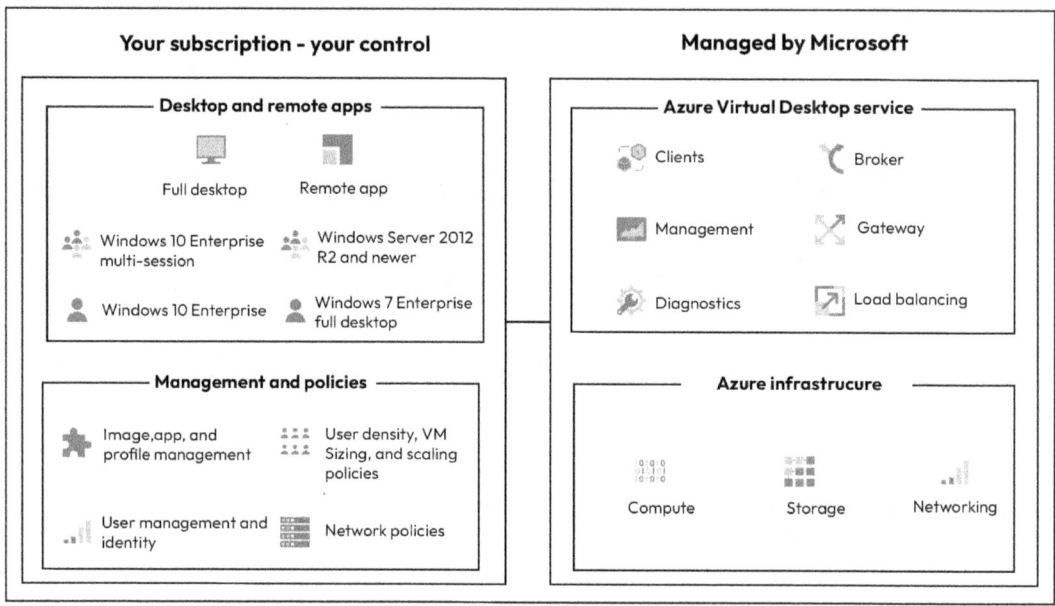

Figure 16.1 – Shared responsibilities of Microsoft and the customer for Azure Virtual Desktop

A robust business continuity and disaster recovery strategy keeps your applications and workloads running during unplanned and planned service outages. For example, when an outage occurs in an Azure region, the Azure Virtual Desktop service infrastructure components will fail over to a secondary location and continue to function.

> **Important note**
> Where required, ensure that your business applications that rely on data in the primary region can fail over to ensure that all services function correctly.

We'll look at the two types of host pool because each type will have a different backup strategy due to the design and use cases associated with the host pool type:

- **Personal desktops**: These are typically used for those who need admin permissions to the local machine or are completing specific tasks requiring a dedicated virtual machine. A personal desktop commonly has specific user data and operating system customizations that need to be protected.

- **Pooled desktops**: In contrast, pooled desktops are designed to provide users with any desktop in the pool. All the data is typically stored in a central profile store, such as FSLogix profile containers. Furthermore, in most environments, a pooled Desktop Host pool is treated as disposable. This is because personal data is not stored on the virtual machine, and the operating system is typically delivered from an image template. Pooled desktops are not usually backed up as they can be easily re-created by re-deploying the session hosts from the image.

Depending on your chosen host pool type, you will need to consider the required backup strategy. When using both host pool types, you should design a strategy for both.

The five crucial areas to consider for Azure Virtual Desktop business continuity and disaster recovery strategies are detailed in the following table:

	Azure component	Description of component
1	**Virtual network**	You should consider the design of your network connectivity during an outage.
2	**Virtual machines**	Depending on the host pool type and requirement, you may wish to replicate your virtual machines to the second region or replicate an image template using Azure Compute Galleries to deploy new session hosts within the second region. Availability zones should also be considered within a local region. Read more here: `https://learn.microsoft.com/azure/reliability/availability-zones-overview?tabs=azure-cli`.
3	**User and app data**	For user profile containers, you should set up data replication in the second region and consider features such as MSIX app attach.
4	**User identities**	Ensure you can access your user identities within the second region.
5	**Application dependencies**	Any line-of-business applications will need to be failed over to the secondary region in the event of an outage.

Table 16.1 – Key Azure components and their strategic roles

This table outlines key Azure components and their strategic roles in ensuring business continuity during outages. It emphasizes the importance of network design, virtual machine replication, data redundancy, user identity access, and application dependency management across multiple regions.

We will take a more detailed look at these five critical components in the next section of this chapter.

Planning and implementing a disaster recovery plan for Azure Virtual Desktop

This section looks at the five critical components for implementing a disaster recovery plan for Azure Virtual Desktop. First, we will look at virtual networks, which underpin the Azure Virtual Desktop environment.

Virtual network

In this section, we take a look at the network requirements for disaster recovery in an Azure Virtual Desktop environment.

Connectivity is the first consideration when planning and implementing a disaster recovery plan for Azure Virtual Desktop. Before resources can communicate in a secondary region, you need to ensure that VNet has been set up in your secondary region/location. You may also need to consider connectivity between on-premises and the secondary region.

The most common connectivity type to use between on-premises and the Azure virtual network would be a VPN gateway. The following diagram depicts a simple VPN gateway connecting to an on-premises site:

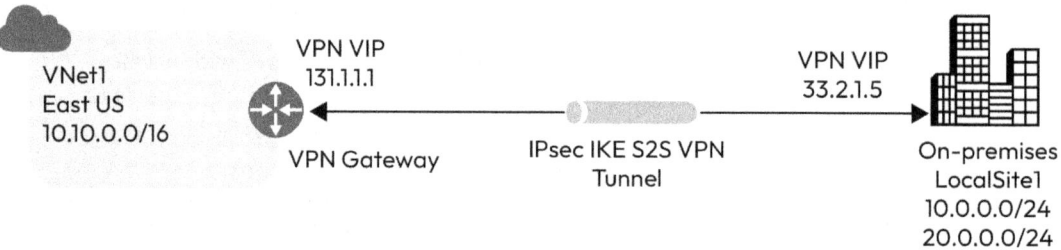

Figure 16.2 – A simple site-to-site VPN connecting an on-premises site to Azure

For smaller Azure Virtual Desktop deployments, you can use **Azure Site Recovery (ASR)** as this can be configured to set up the VNet in a secondary (failover) region. ASR can preserve your primary settings and does not require network peering.

For larger environments, customers use **ExpressRoute**. ExpressRoute goes through a connectivity provider at a colocation facility so it can be much faster and more secure than a site-to-site VPN as the traffic does not go through the public internet. However, it is more expensive and more complicated to implement than a site-to-site VPN.

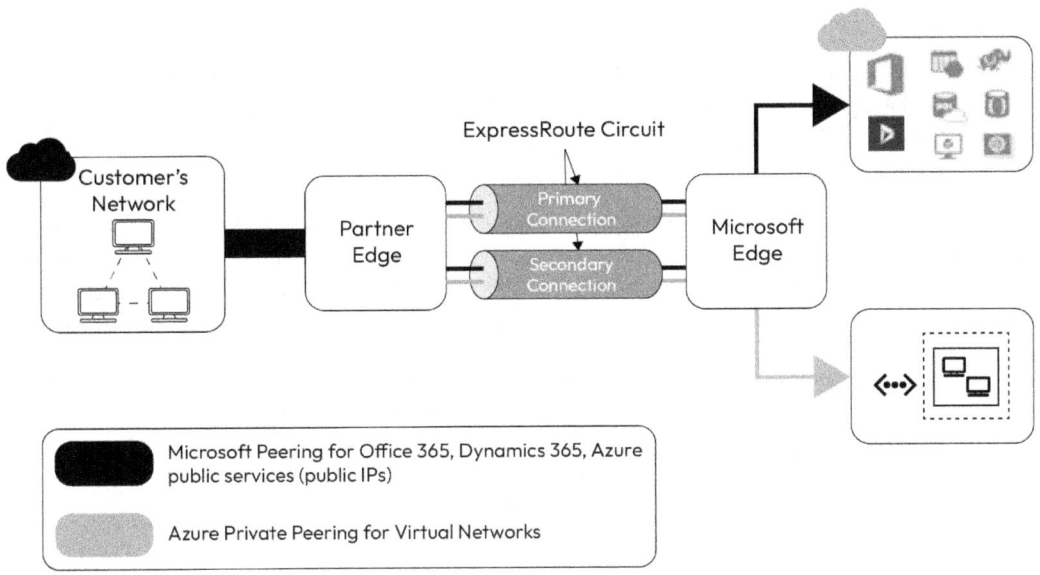

Figure 16.3 – Disaster recovery plan for Azure Virtual Desktop using ExpressRoute

One of the most important things to remember when working with Azure Virtual Desktop and multiple regions is **DNS**. A common problem customers face in Azure Virtual Desktop is DNS configuration issues relating to the virtual network. Ensure that you have configured the VNet's DNS to communicate with the required domain controller.

One final point within this section is the fact that you need to ensure that all the required URLs are included within the **Network Security Group** or other security technology you may have deployed on the virtual network. For more information on the URL list for Azure Virtual Desktop, please refer to *Chapter 4, Implementing and Managing Networking for Azure Virtual Desktop*.

Virtual machines

When we look at business continuity and disaster recovery options for virtual machines (session hosts) for Azure Virtual Desktop, two specific options are available. The first is **active-active**, and the second is **active-passive**.

Let's first take a look at the active-active option.

The active-active option

When using the active-active option, you can have a single host pool that is stretched across multiple Azure regions. This means creating a single host pool and deploying the virtual machines across a primary and secondary region. You need to use **FSLogix Cloud Cache** to handle profiles to replicate user data between the regions.

> **Important Note**
> This option should only be considered for pooled desktop host pools. Also, when deploying into secondary regions, you should always test your backend infrastructure to ensure that you do not need any additional firewall or NSG rules implemented.

If you choose active-active, you can protect against short storage outages without the user having to re-login. This also means that you can adopt continuous testing of disaster recovery locations as you can use both regions in normal operating conditions. This option does not provide any cost savings and has no benefits in terms of performance; however, as stated, you can continually test disaster recovery.

You can only manage connections using this option via the session host drain stop; otherwise, incoming user connections will be equal. What this means is that user connections could be directed to either the primary region or the secondary region.

We now move on to take a look at the active-passive option.

The active-passive option

The active-passive option should be considered for personal desktop host pools where you need to replicate the dedicated virtual machines over to the second region. You can use the active-passive option for pooled desktops as well if required.

Similar to the active-active option, you can use a single host pool. This is the recommended approach for simplicity. You can also deploy a new host pool in the secondary region with all associated resources turned off. There is still a cost associated with this and you would need to regularly power these virtual machines on to prevent the machine account passwords from expiring. In the event of an outage, you can turn on those resources for the organization to access as a disaster recovery solution. For this method, as with all new host pool configurations, you would need to set up application groups for the required users within the failover region. You can also use an ASR plan to turn on the resources orchestrating the overall process.

When using the active-passive option for pooled host pools, it is common to create hosts using the secondary option when they are needed. Using scripting and automation, you should be able to create around 100 multi-session hosts in about 1 hour.

> **Important note**
> It is recommended that you use ASR for personal desktop host pools.

Virtual machine availability sets versus availability zones

The default resiliency option for a host pool deployment is an **availability set**. The resiliency provided by an availability set is at the single Azure data center level. Microsoft provides a 99.5% SLA on an availability set.

The second option would be to use something called an availability zone. This is where virtual machines within a host pool are distributed across different data centers in a region. Again, this provides high resiliency, and Microsoft provides a 99.9% SLA on availability zones.

It is recommended that you consider multi-region resilience over availability sets/zones for better coverage. The following link details some of the challenges you may face in large organizations relating to availability sets and allocation failures: https://learn.microsoft.com/troubleshoot/azure/virtual-machines/windows/allocation-failure#allocation-failures-for-large-deployments-more-than-500-cores.

One common mistake we see is that customers will deploy all their VMs into a single availability zone. To ensure the smallest impact when an Azure Zone has an outage, it is recommended that when you create your session hosts, you spread them evenly across multiple zones. In the following diagram, we can see that there are multiple availability zones contained within a single region, so it would make sense to spread your workload across those zones:

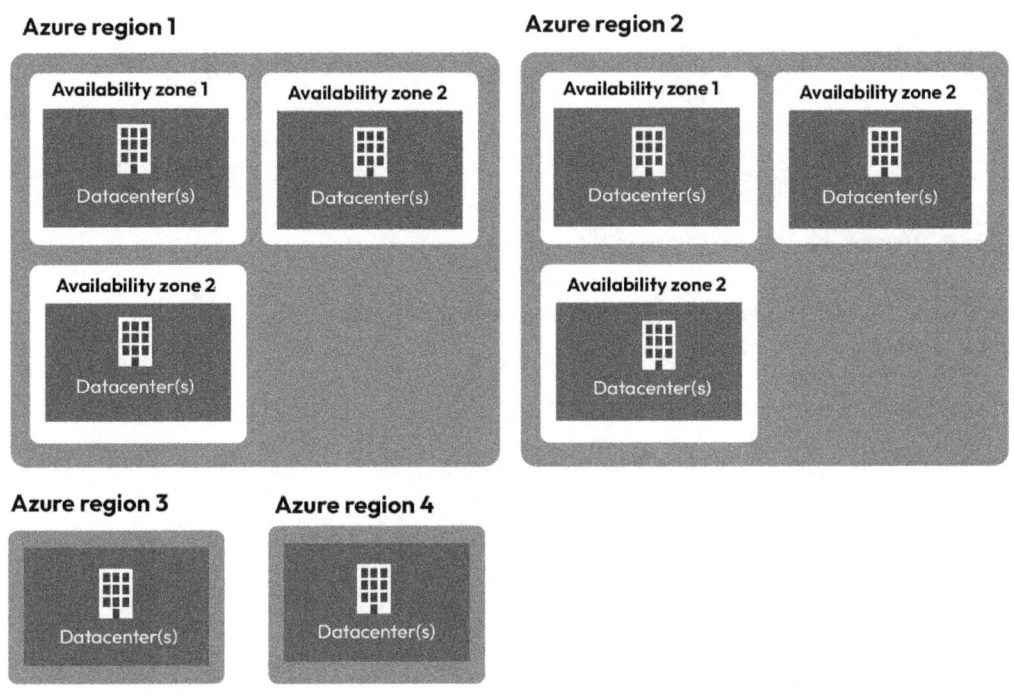

Figure 16.4 – differences between availability zones and Azure regions

We will now move on to take a look at managing user identities.

Managing user identities

In this section, we take a look at managing user identities and some of the different options available to you. The first thing you need to think about when managing user identities is that when you fail over to a secondary region, you need to ensure that the domain controller is available so your virtual machines can communicate with it.

I have listed three different options for enabling communication with the domain controller in the secondary region.

Deploying a domain controller in the failover region

Using Virtual Network peering, you can configure both **Region A** and **Region B** virtual networks to communicate using network peering. The following diagram shows how network peering facilitates the communication of domain controllers between the two regions:

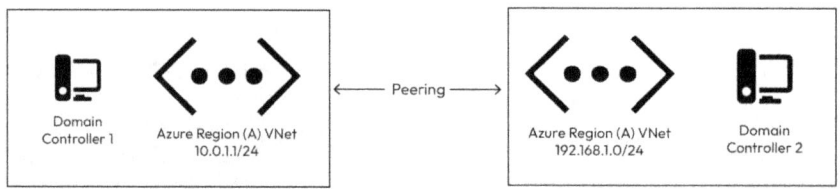

Figure 16.5 – Network peering used to enable domain controllers to communicate across Azure regions

Now that we have looked at deploying a domain controller in the failover region, let's look at how you would use an on-premises domain controller.

Using an on-premises domain controller

Another option would be to use a **virtual network gateway** to connect multiple virtual networks and an on-premises site enabling you to use the on-premises domain controller for multiple regions, as you can see in the following diagram:

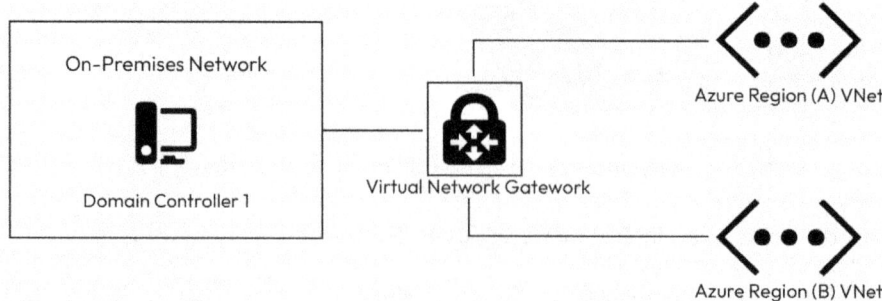

Figure 16.6 – This diagram shows multiple regions connecting via a VPN gateway to an on-premises site

> **Tip**
> Make sure you configure the DNS settings on the VNet to point to the domain controller. This is quite a common issue.

Let's now take a look at how you would logically use ASR to replicate your domain controller between regions.

Using ASR to replicate your domain controller

One of the most common approaches for ensuring that Azure Virtual Desktop resources can access the domain controller in the failover region is to use ASR in a similar manner to what we discussed earlier in the virtual machine part of this section.

The following diagram shows the use of ASR replication to replicate the domain controller from Azure Region A to Azure Region B:

Figure 16.7 – This diagram shows the replication of the domain controller from Region A to Region B

Now that we have taken a look at user identity options, let's move on to learn how best to configure user and app data.

Bring the simplicity of Entra ID

One of the modern authentication methods is to use Entra ID for authentication. If our Azure Virtual Desktop session hosts are joined to Entra ID instead of Active Directory it means that we do not need line-of-sight to the domain controllers.

This means that it would be a much simpler design because we would not need connectivity to our domain controllers for authentication, making the design simpler and highly resilient. However, if any backend services needed IP connectivity, then we would still need to have the required network infrastructure in place.

Configuring user and app data

In this section, we take a look at profiles and user data. When using local profiles, it's recommended that you use ASR to replicate user data and the session hosts to the failover region. This is a common approach for those who are using personal host pools.

FSLogix offers the ability to separate the user profile and office container disks. This also enables you to split user data components into different storage locations. In a typical environment, the office container would consume more disk space than a profile disk. The backup, replication, and user profile disk would be significantly quicker than using office containers as part of the solution. In most cases, it is typical to expect organizations to use user profile containers only.

The three recommended storage options for storing FSLogix profile containers are as follows:

- Azure Files
- Azure NetApp Files
- FSLogix Cloud Cache

There are multiple options available in terms of storing your FSLogix profile containers. To find out more about the options available to you and their limitations, please refer to *Chapter 3, Designing for User Identities and Profiles*.

> **Important note**
> Microsoft recommends storing FSLogix profile containers on Azure Files and NetApp Files for the majority of scenarios.

When looking to configure disaster recovery for user profiles, you have the following options:

- Use Azure replication (for example, Azure NetApp Replication or Azure Files Sync for file servers).
- Configure FSLogix Cloud Cache for both application and user data.
- Configure app data for disaster recovery only. What this means is that your users would have new user profiles, and the first-time sign-in experience would occur.
- OneDrive is also an option to consider because it can redirect well-known folders, including Desktop, Documents, and Pictures. Again, this provides a level of resilience without any specific business continuity or disaster recovery considerations.

> **Important note**
> When using Cloud Cache, make sure that your session hosts are configured using premium SSDs for the local cache file to prevent data loss.

For more information on Cloud Cache, please refer to the *Configuring Cloud Cache* section of *Chapter 12, Implementing and Managing FSLogix*.

Disaster recovery considerations for MSIX app attach

MSIX app attach is a dynamic application delivery feature in Azure Virtual Desktop. MSIX app attach delivers MSIX applications using a virtual disk or CimFS image to pooled and personal desktop host pools within Azure Virtual Desktop.

When implementing a disaster recovery plan, there are two areas of consideration for MSIX app attach. The first is *storage*; you need to ensure that the MSIX images are accessible in the failover region. Similar to FSLogix, MSIX app attach requires network storage for the application disks/images.

The second consideration relates to the *configuration* of MSIX app attach. If the storage path changes in a disaster recovery scenario, you will need to change all the MSIX image paths or reconfigure all the applications configured within a host pool.

To avoid the requirement to reconfigure MSIX app attach in a disaster recovery scenario, it is recommended that you use one of the following options:

- Create a separate host pool for the secondary region pre-configured for MSIX app attach
- Use Azure Files with geo-redundant storage
- Implement Azure NetApp Files cross-region replication

> **Tip**
> For enterprise deployments, it is recommended that you configure Azure NetApp Files cross-region replication.

To learn more about MSIX app attach, please refer to *Chapter 14, MSIX App Attach*. In the next section, we will take a high-level look at application dependencies.

Application dependencies

When an outage occurs, it may not just impact your Azure Virtual Desktop environment but also any critical business applications that rely on data located in the primary region. This could be web services, SQL databases, and more.

You should also consider any specific settings required for the applications and any additional configurations for these services once you have set up replication or high availability.

You can use ASR plans to model application dependencies, and you could consider configuring the app to use the second region as its default configuration or as part of the failover process.

Please note the following:

- You need to ensure that users can access on-premises applications in the event of a failover. See the *Virtual network* section for more details.
- You need to ensure that you review all dependent applications and any other resources to ensure availability in the disaster recovery location.

In this section, we looked at planning and implementing a disaster recovery plan for Azure Virtual Desktop, looking at the five critical areas: *virtual network*, *virtual machines*, *managing user identities*, *configuring user and app data*, and *application dependencies*.

In the next section, we look at configuring backup and restore for FSLogix user profiles.

Configuring backup and restore for FSLogix user profiles, personal virtual desktop infrastructures (VDIs), and golden images

In this section, we look at backing up and restoring Azure Virtual Desktop components. We will first take a look at backup and restore.

Virtual machine backup and restore

When managing personal host pools, you may want to back up the session hosts when local profiles are used. In this section, we'll look at how to take a backup of a session host virtual machine.

We start by creating an Azure Recovery Vault for Azure Virtual Desktop, the steps for which are as follows:

1. First, navigate to the Azure portal.
2. Search for `Backup center` in the Azure portal and navigate to the **Backup center** dashboard.
3. Click the **Vault** icon located in the main menu of the page.

Configuring backup and restore for FSLogix user profiles, personal virtual desktop infrastructures (VDIs), and golden images   531

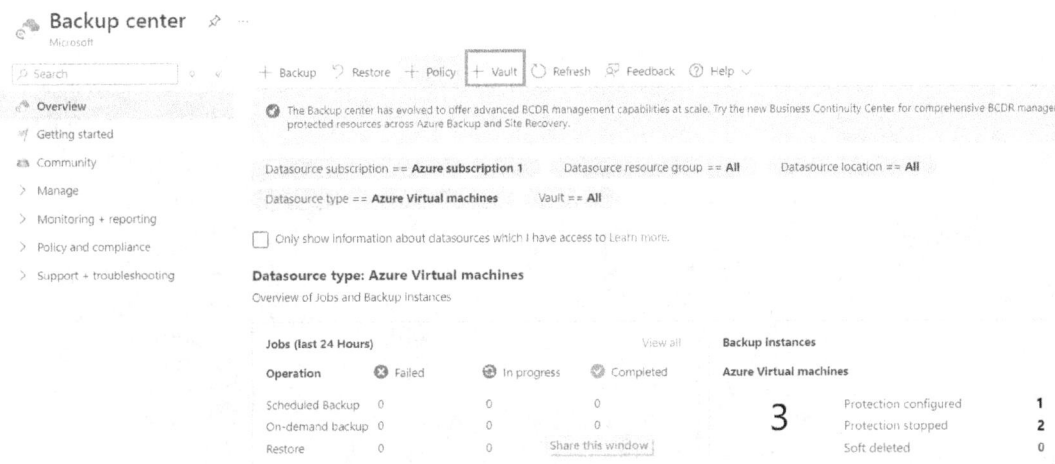

Figure 16.8 – Azure Backup center page

4. Within the **Create Vault** page, select **Recovery Services vault** and then click **Continue**:

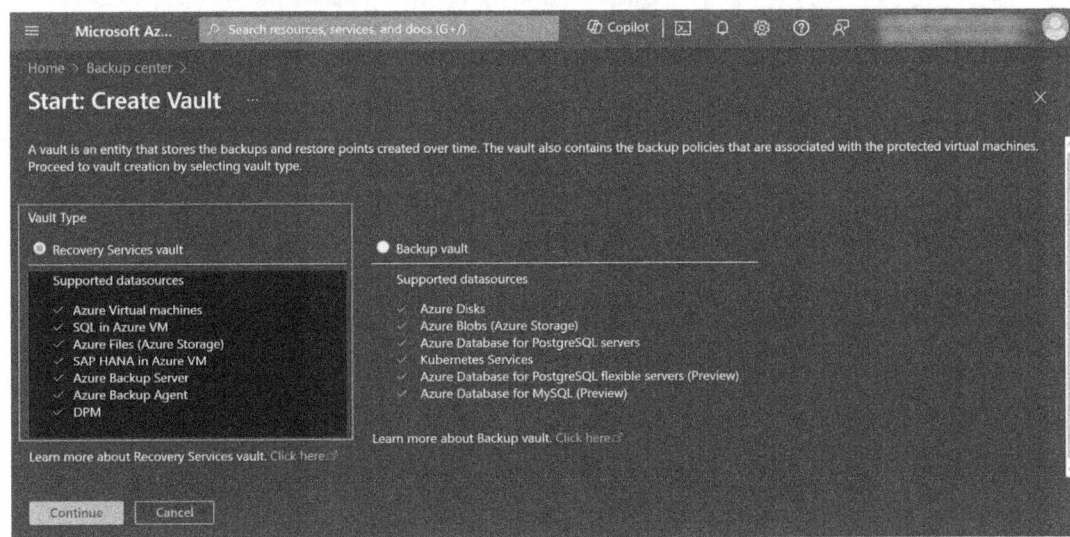

Figure 16.9 – Create Vault page within Backup center

5. You will then see the **Create Recovery Services vault** page.

6. Select the **Resource group** option or create a new one and specify a vault name and region.

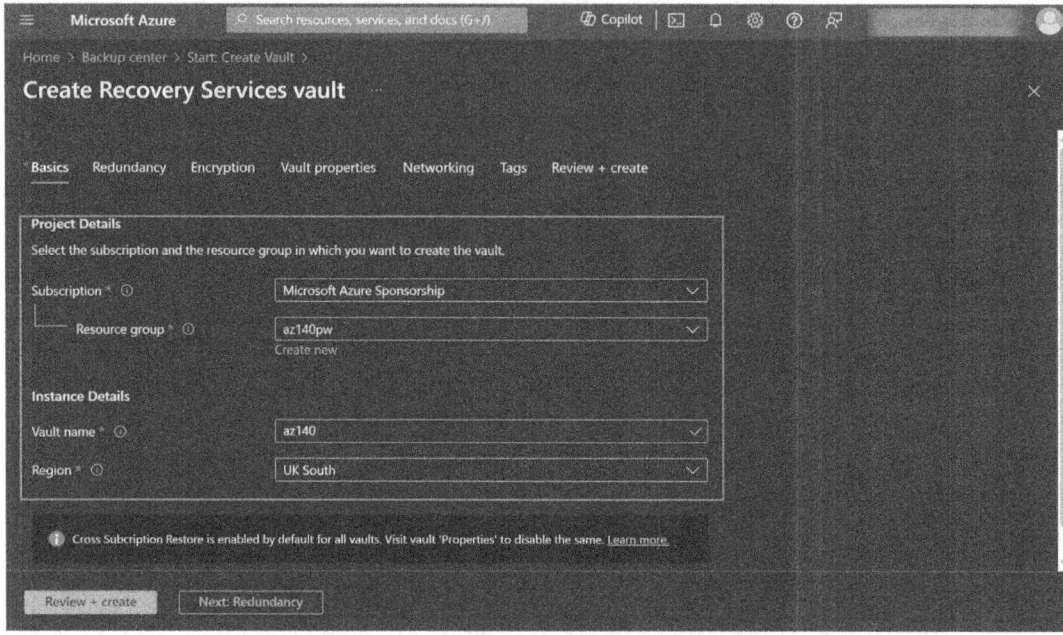

Figure 16.10 – Create Recovery Services vault page

7. Next, we need to decide whether the backup storage is locally redundant, zone redundant, or geo redundant. For a description of the options, please visit https://docs.microsoft.com/azure/backup/backup-vault-overview.

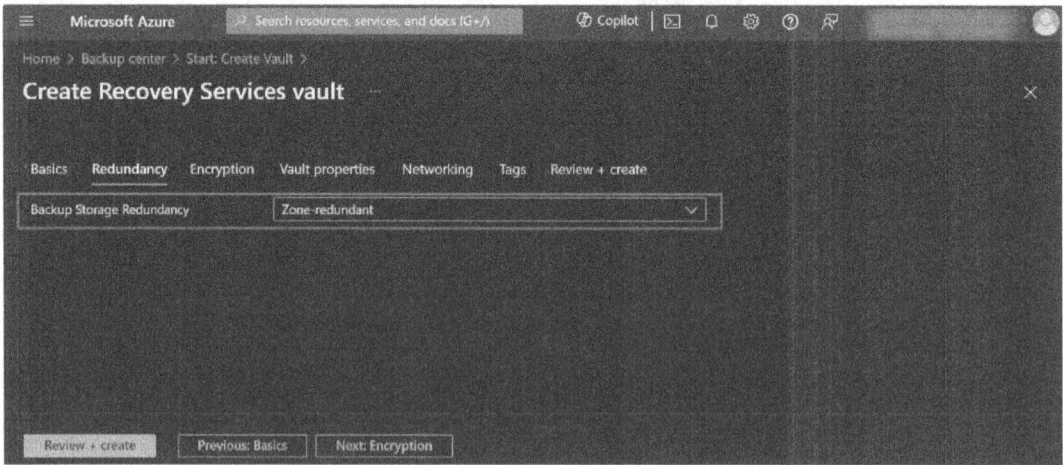

Figure 16.11 – Create Recovery Services vault page

8. Next up is Encryption options. Azure Backup now allows you to use your own customer-managed keys, if required. If not, then you default to **Use Microsoft Managed key** and let Azure manage the encryption keys for you.

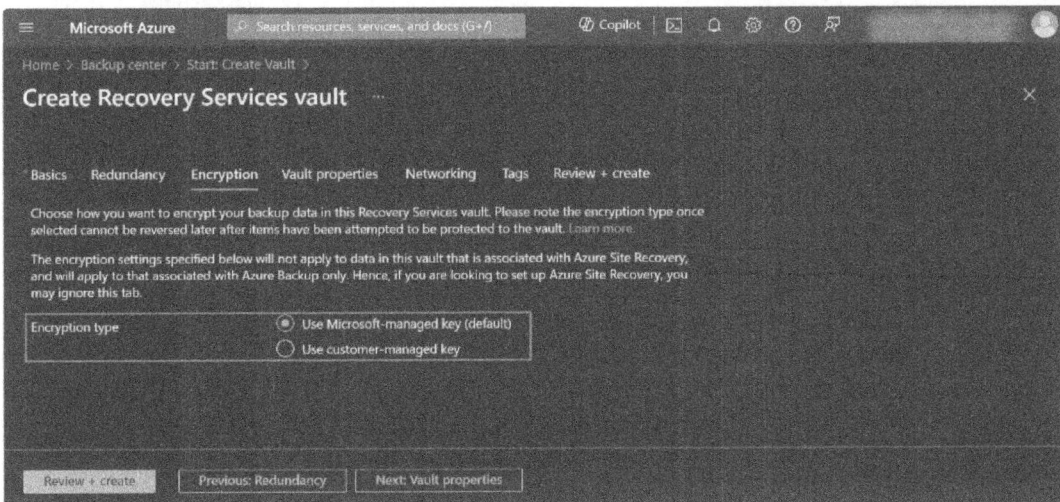

Figure 16.12 – Configure Recovery Services vault encryption

9. Next, we need to decide whether to enable immutability or not. Immutability protects your backup data by blocking operations that would remove your recovery points. We can also lock the settings to prevent any malicious actors, such as ransomware hackers, from disabling immutability and deleting your backups. More information is available at https://learn.microsoft.com/azure/backup/backup-azure-immutable-vault-concept?tabs=recovery-services-vault.

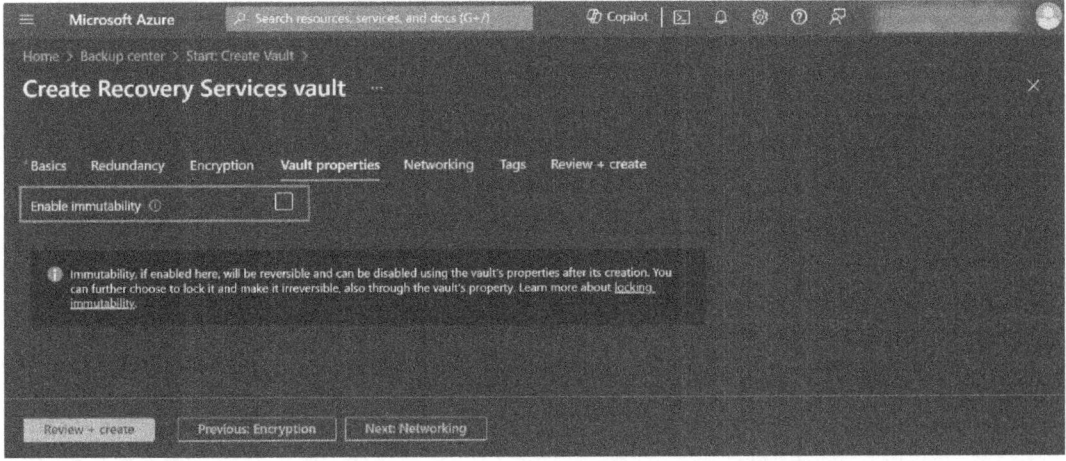

Figure 16.13 – Configuring immutability for Recovery Service vault

10. Now, we need to decide whether to allow public access to our backup vaults. It is highly advisable to create private endpoints for your backup vaults to ensure their security.

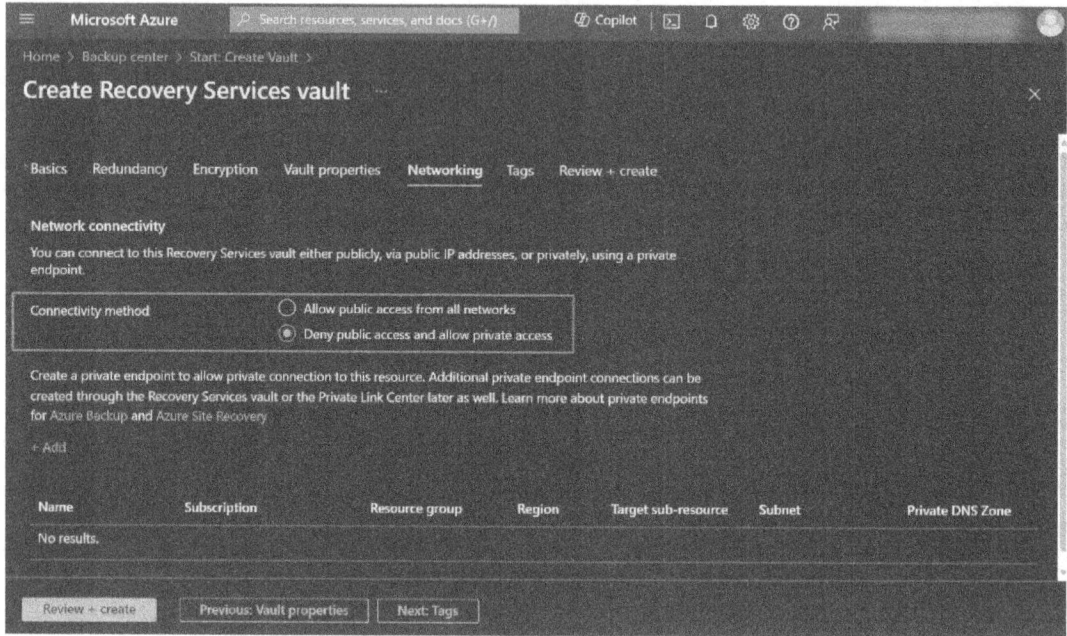

Figure 16.14 – Configuring Recovery Services Vault connectivity

11. The last step in the process is to create the tags.

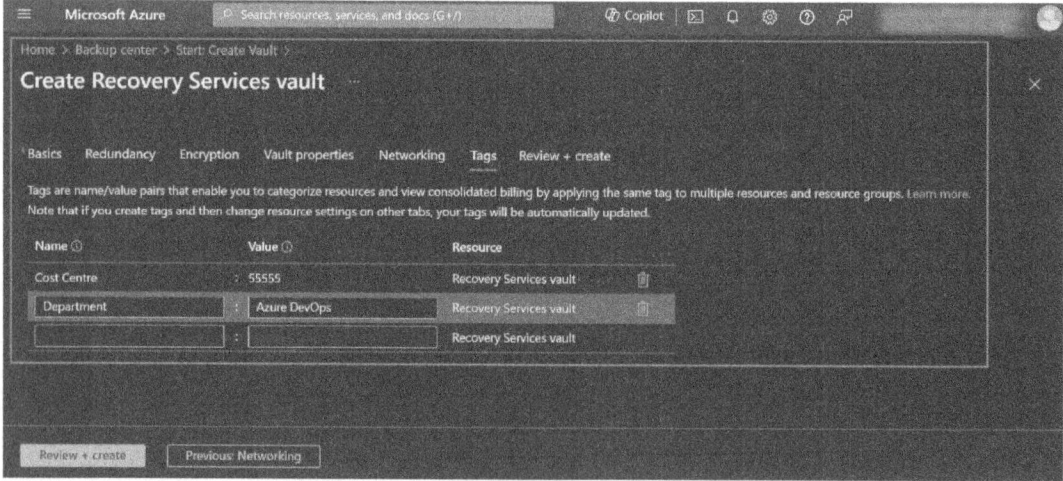

Figure 16.15 – Configuring Recovery Services vault tags

12. Once tags are configured, proceed to **Review + create**
13. You will now see the new **Recovery Services vault** in Backup center.

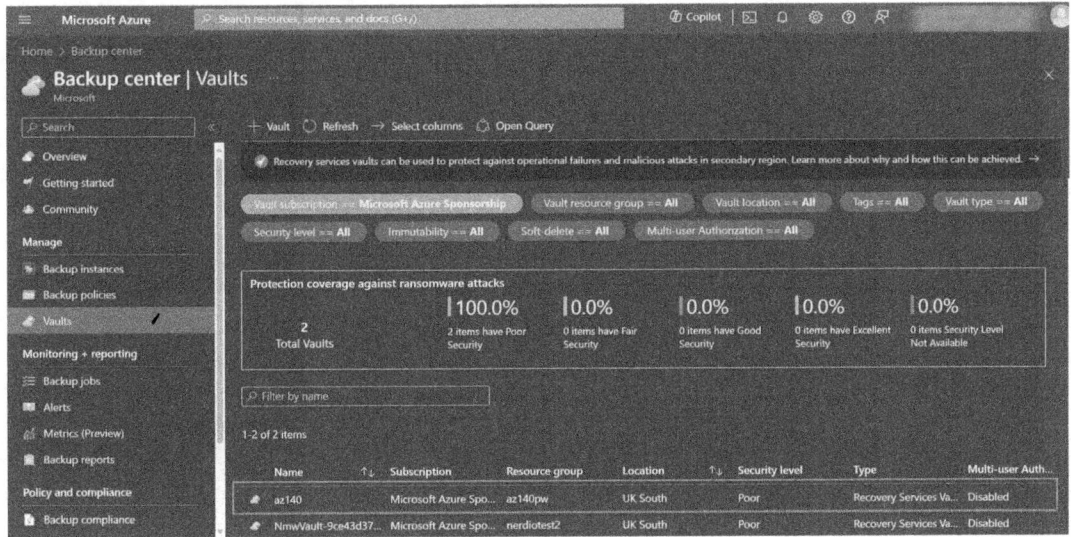

Figure 16.16 – New Recovery Services vaults in Backup center

Now that we have created a recovery services vault, we can progress with backing up a session host virtual machine.

14. Within Backup center, select the recently created recovery services vault.

15. Within the recovery services vault, select **Backup Instances**, located on the left-hand side under **Manage**.

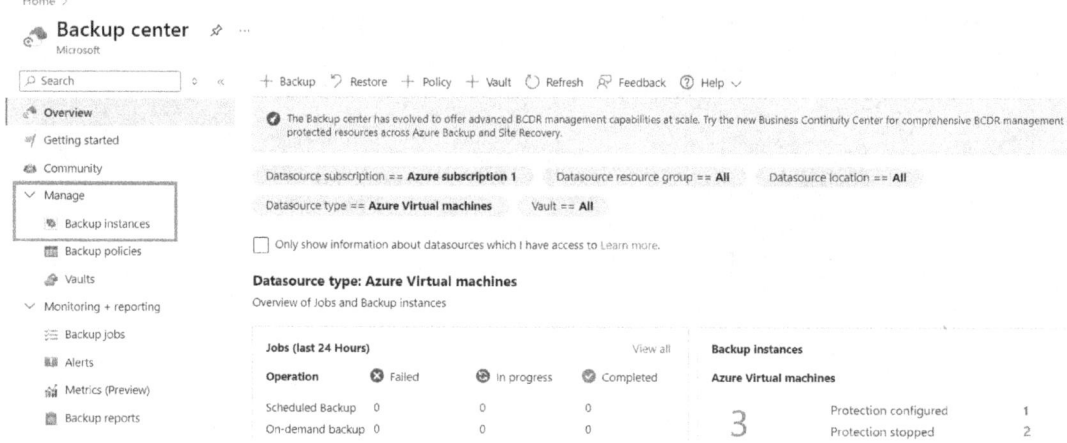

Figure 16.17 – Backup highlighted in the Getting started section

16. Once **Backup instances** is selected, the **Backup** blade will appear. Select the **Backup** button.

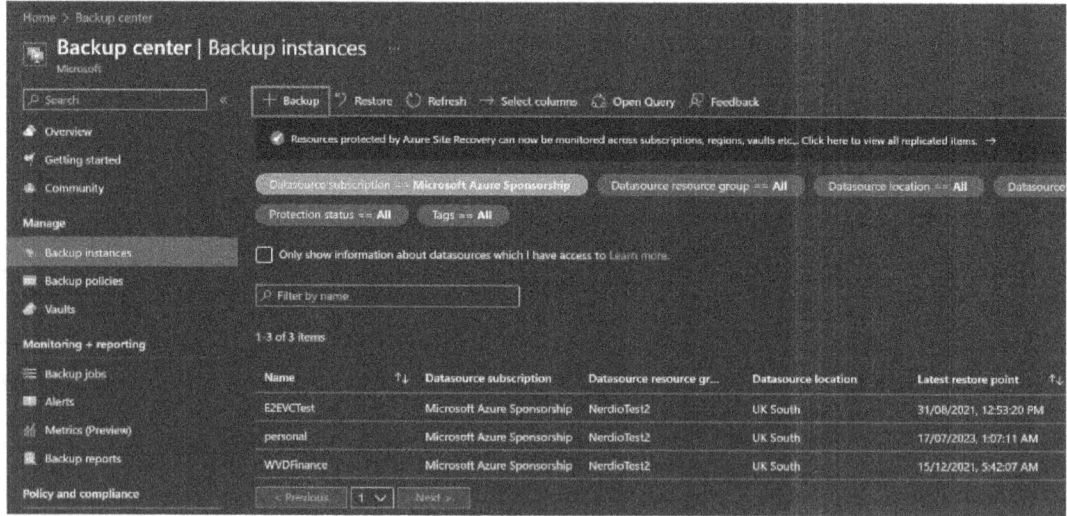

Figure 16.18 – Backup highlighted in Backup instances

17. Select **Azure Virtual Machines** as the **Datasource type** and ensure that the vault that you created earlier is selected.

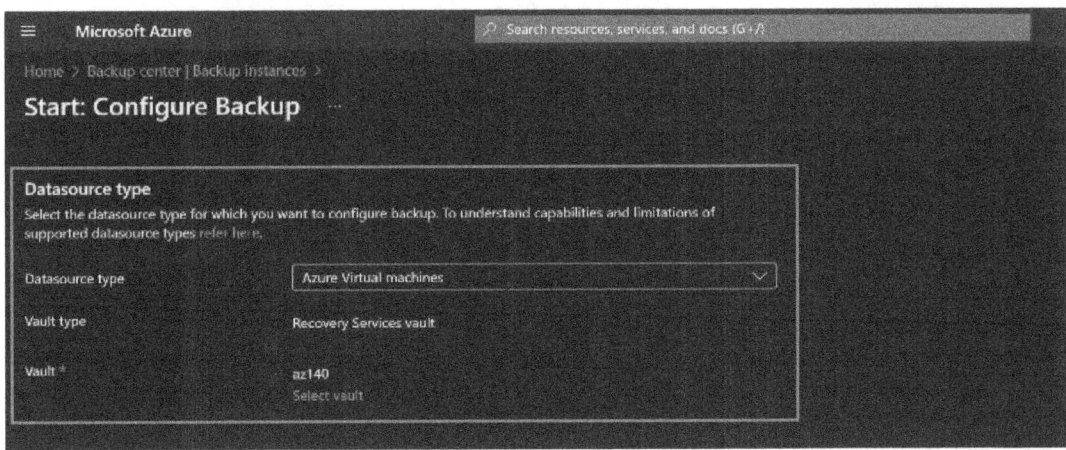

Figure 16.19 – Configure Backup Datasource type

18. Next, we decide what policy we want. You can choose one of two default policies, **Standard** or **Enhanced**. If they don't meet your requirements, you can create your own policy. For this example, we will select the **Standard** policy, which backs the VM up once a day and keeps the backup for 5 days.

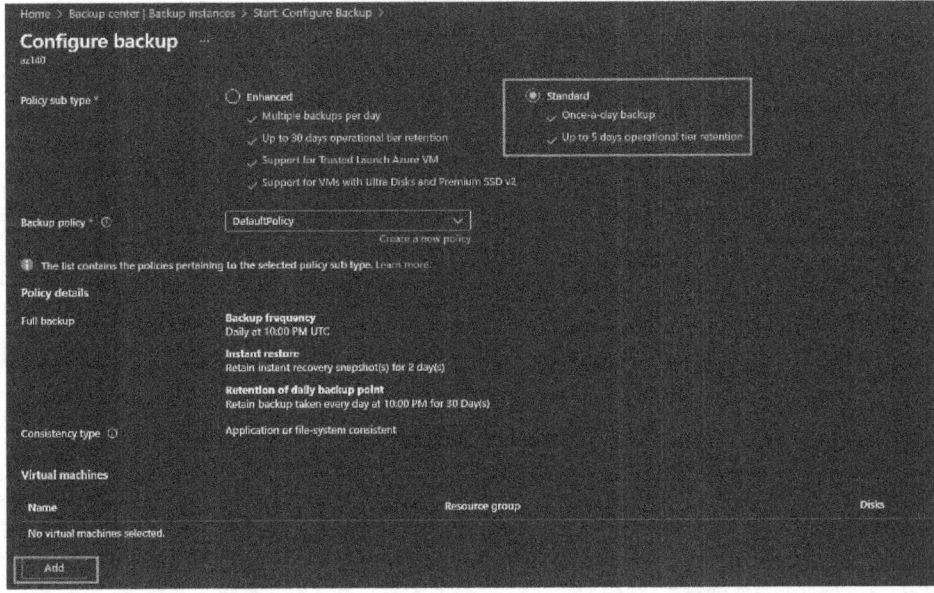

Figure 16.20 – Configure backup policy selection

19. Click the **Add** button in the **Virtual Machines** section, choose the virtual machines you wish to back up, and then click **OK**.

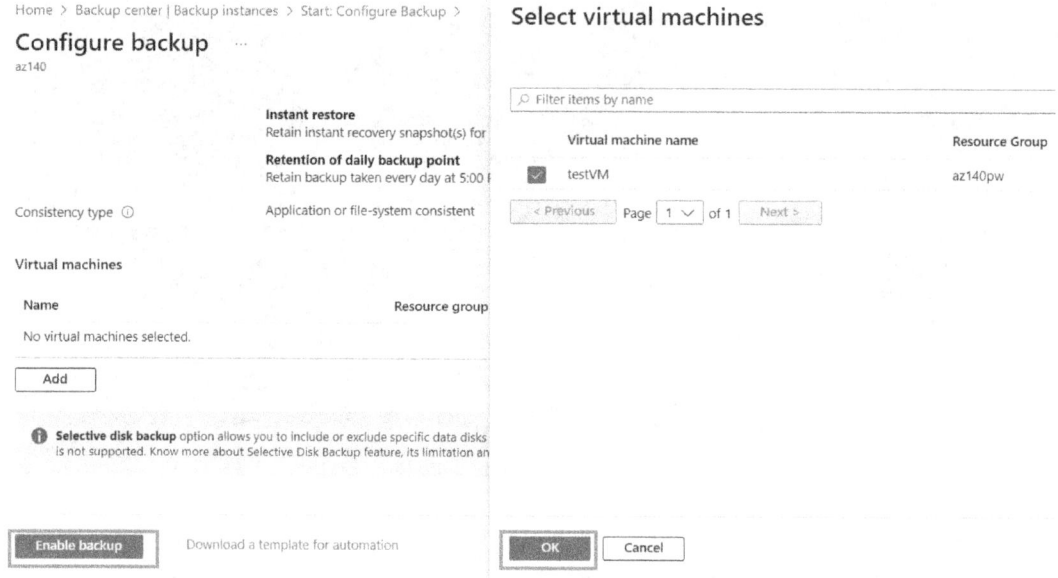

Figure 16.21 – Selecting the virtual machine to back up

20. Once you have selected the backup policy and added the virtual machines, click **Enable Backup**.
21. You should now see the VM listed within backup jobs with the completed **Configure backup** operation. When the scheduled backup runs, you will also see the following within the backup jobs:

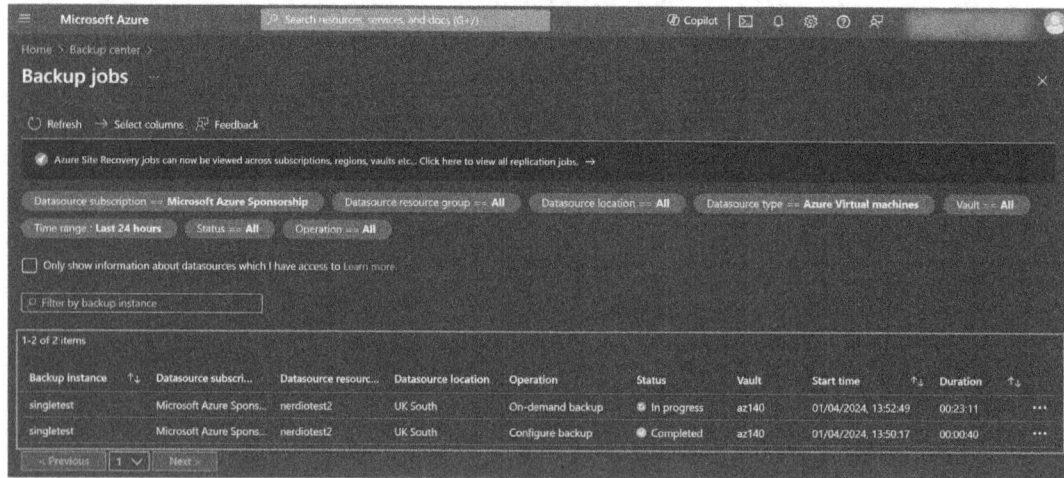

Figure 16.22 – Configuring a backup operation as completed

We now move on to take a look at restoring a virtual machine:

1. To restore a virtual machine, you will first need to navigate to the Azure Backup center.
2. On the main page of the Azure Backup center, click **Restore**, as shown in the following screenshot:

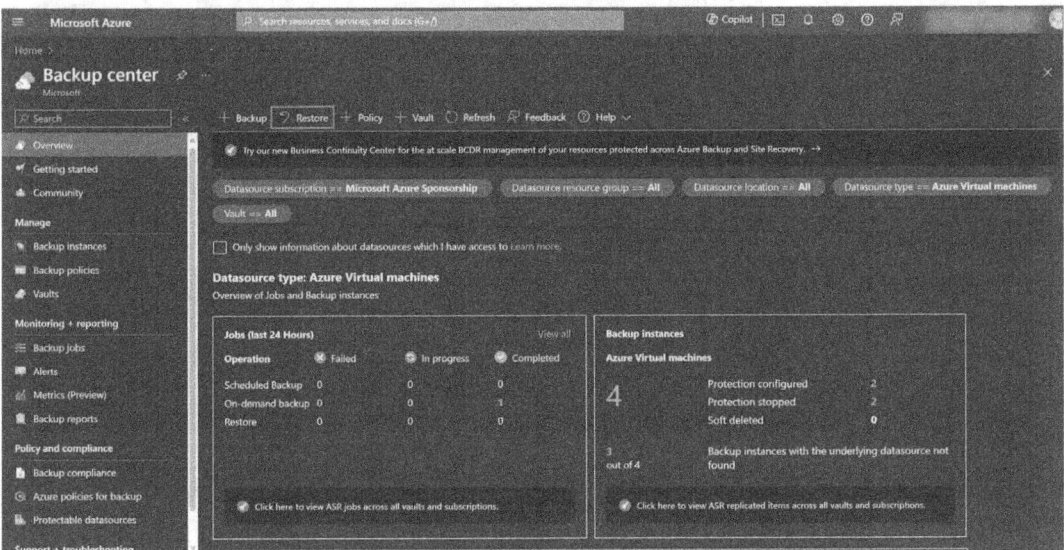

Figure 16.23 – Restore button highlighted in Backup center

3. You will now see the **Start: Restore** page appear. Select the virtual machine backup instance you require to restore.
4. In the **Restore Region** section, you have a choice of selecting a primary region or a secondary region. In this example, we will select **Primary Region**. Once you have completed the required fields on the page, click **Continue**.

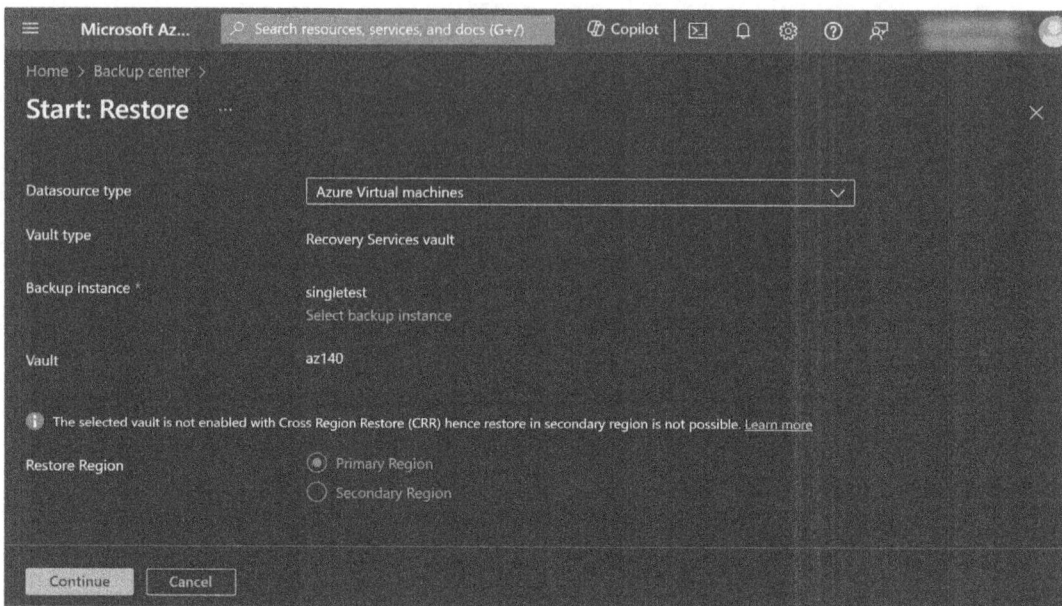

Figure 16.24 – Start: Restore page when restoring a virtual machine within Microsoft Azure

> **Important note**
>
> You can configure your **Recovery Vault for Cross Region Restore**, allowing the restoration of virtual machines to the secondary region. This is useful for those who are using a single host pool for both a primary and secondary region. In the event of a primary region outage, you can restore the virtual machine to the secondary region. You may want to consider configuring your disaster recovery within a paired region to take advantage of **Cross Region Restore** (**CRR**). You can find out more about region pairs here: https://docs.microsoft.com/azure/best-practices-availability-paired-regions#azure-regional-pairs.

5. You will now see the **Restore Virtual Machine** page appear. Click **Select**, which will launch the **Select restore point** blade.

Configuring backup and restore for FSLogix user profiles, personal virtual desktop infrastructures (VDIs), and golden images

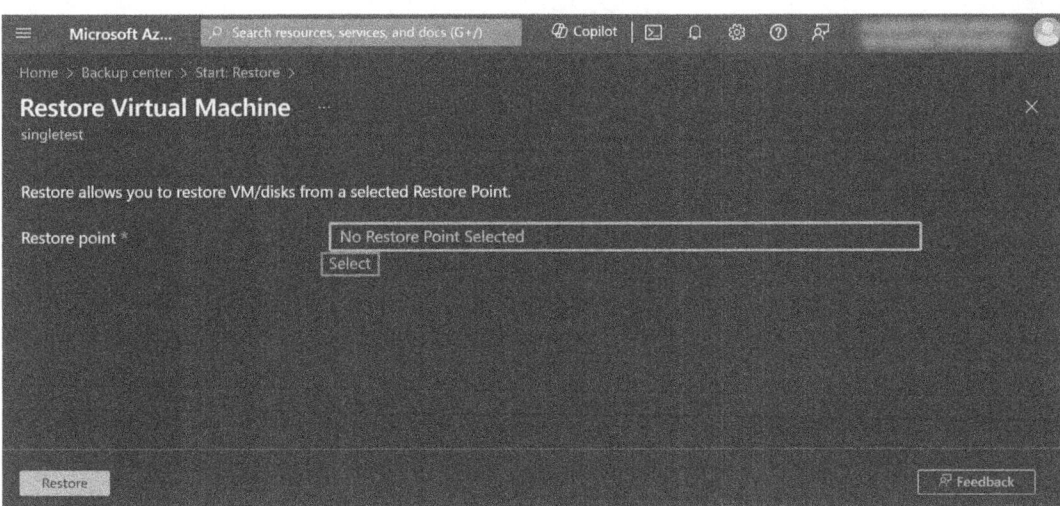

Figure 16.25 – Restore point page within the virtual machine restore wizard

6. Choose the required restore point and then click **Continue**.

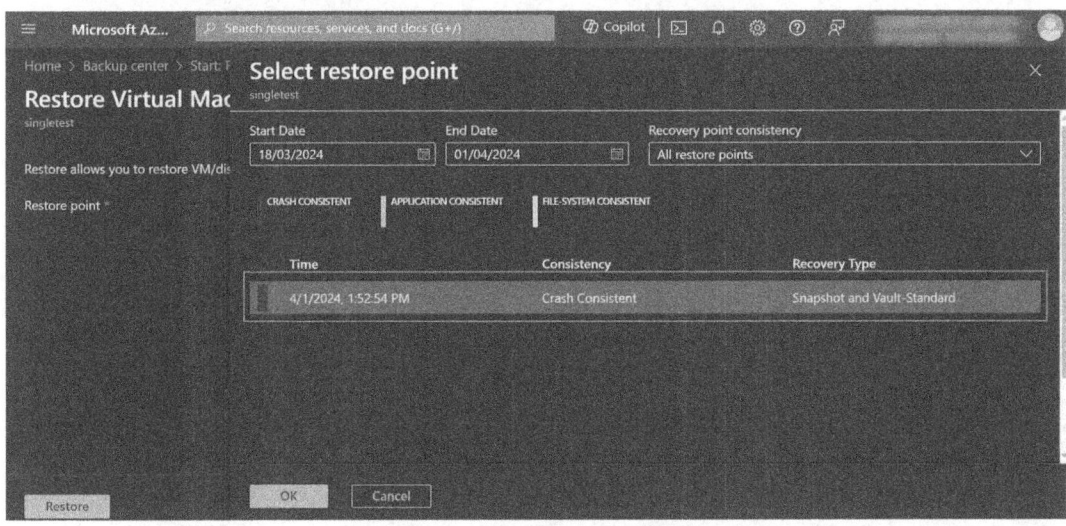

Figure 16.26 – Select restore point blade

7. The final part is to specify whether you want to create a new virtual machine or replace the existing one. In this example, we will select **Create new**.

8. Complete the **Virtual Machine** type fields, **Resource group**, **Virtual network**, and **Subnet**, and select a staging location.
9. Once all the fields have been completed, click **Restore**.

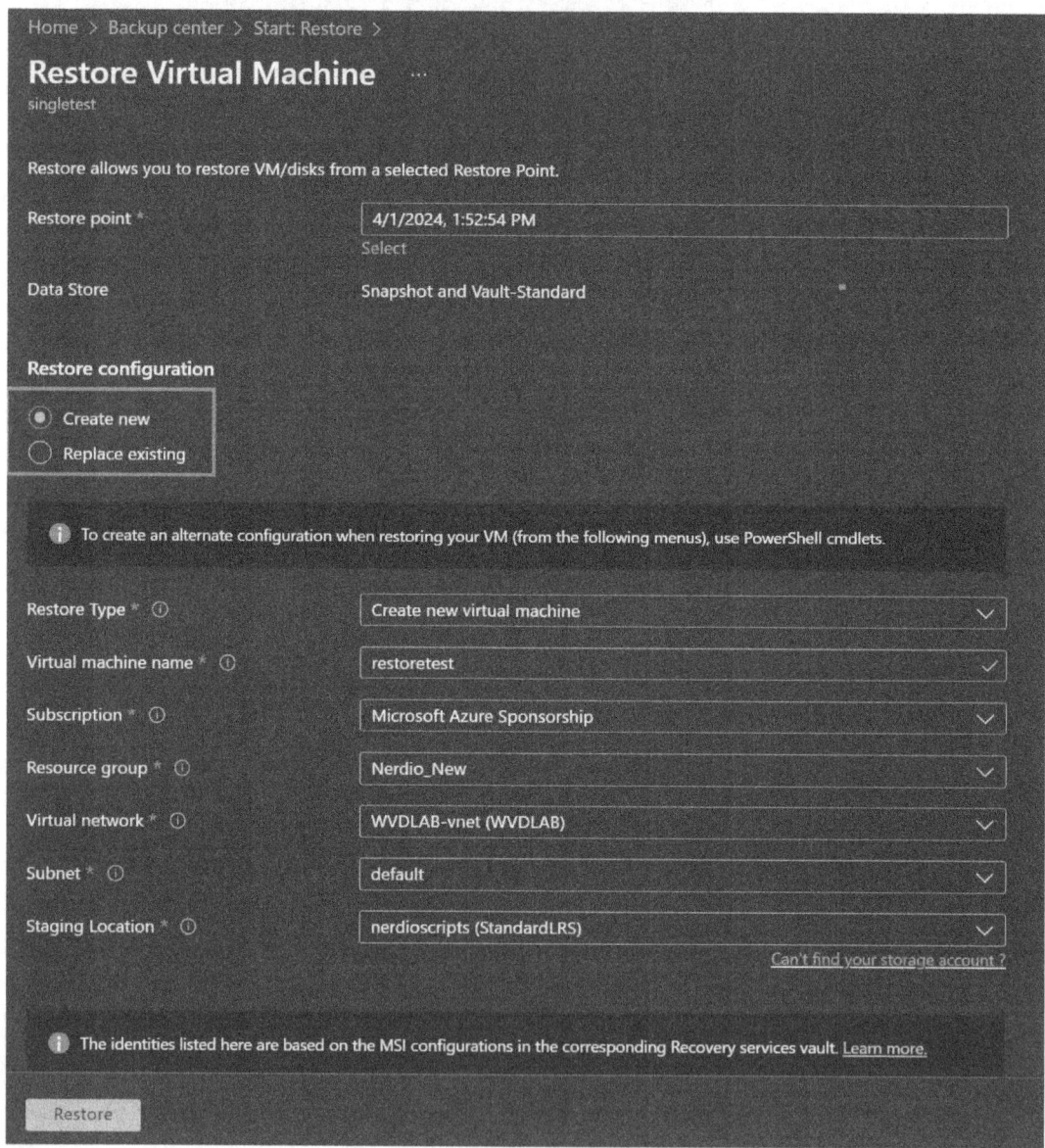

Figure 16.27 – Restore Virtual Machine page

10. Once the restore job has been started, you should see the restore operation within the **Backup jobs** section in Backup center:

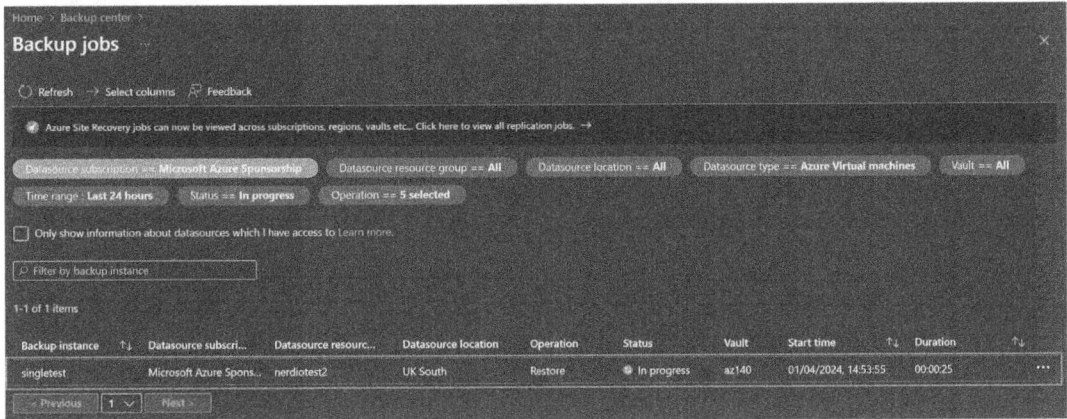

Figure 16.28 – Restore operation on the Backup jobs page within Backup center

11. Once the restore operation has been completed, you will also see that the operation is completed on the main page in Backup center. This information is pulled from the **Backup jobs** section of **Recovery Services vault**.

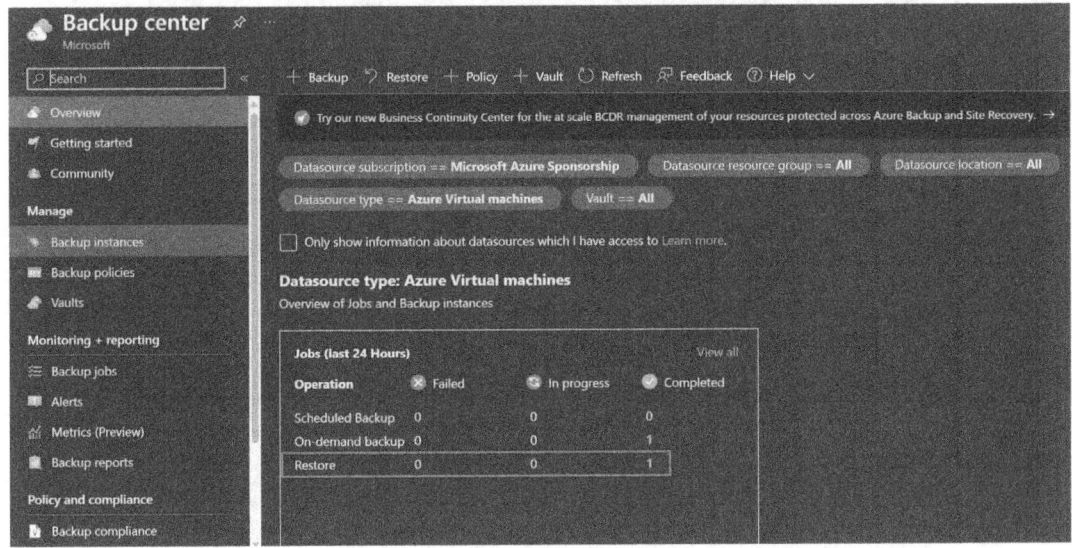

Figure 16.29 – The restore operation is complete under Jobs (last 24 Hours)

Now that we have looked at backing up virtual machines and restoring them, which is extremely useful for personal desktop host pools, we now move on to backing up Azure Files and restoring them.

Zone-redundant storage

The use of **zone-redundant storage** (**ZRS**) should be regarded as a redundancy option. ZRS essentially replicates your Azure-managed disk synchronously over three availability zones within your chosen Azure region.

Find out more about ZRS here: `https://learn.microsoft.com/azure/virtual-machines/disks-deploy-zrs?tabs=portal`.

Azure file backup and restore

It is recommended that you back up your FSLogix profile containers for Azure Virtual Desktop. You can do this in a similar way to how you would back up a virtual machine. We will now take a look at the steps for backing up FSLogix Profile container stores:

1. Navigate to Backup center.
2. Select **Backup** on the main page.

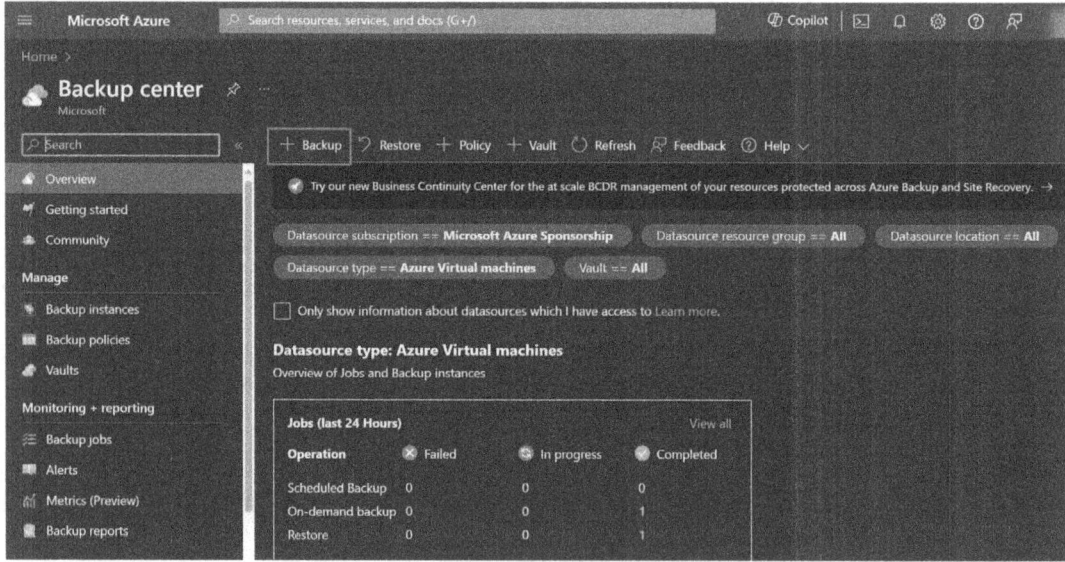

Figure 16.30 – The Backup button in Backup center

3. Within the **Start: Configure Backup** section, choose **Azure Files (Azure Storage)** and the required vault. In this example, we choose **az140** for **Vault**. Once all the fields have been populated, click **Continue**.

Configuring backup and restore for FSLogix user profiles, personal virtual desktop infrastructures (VDIs), and golden images | 545

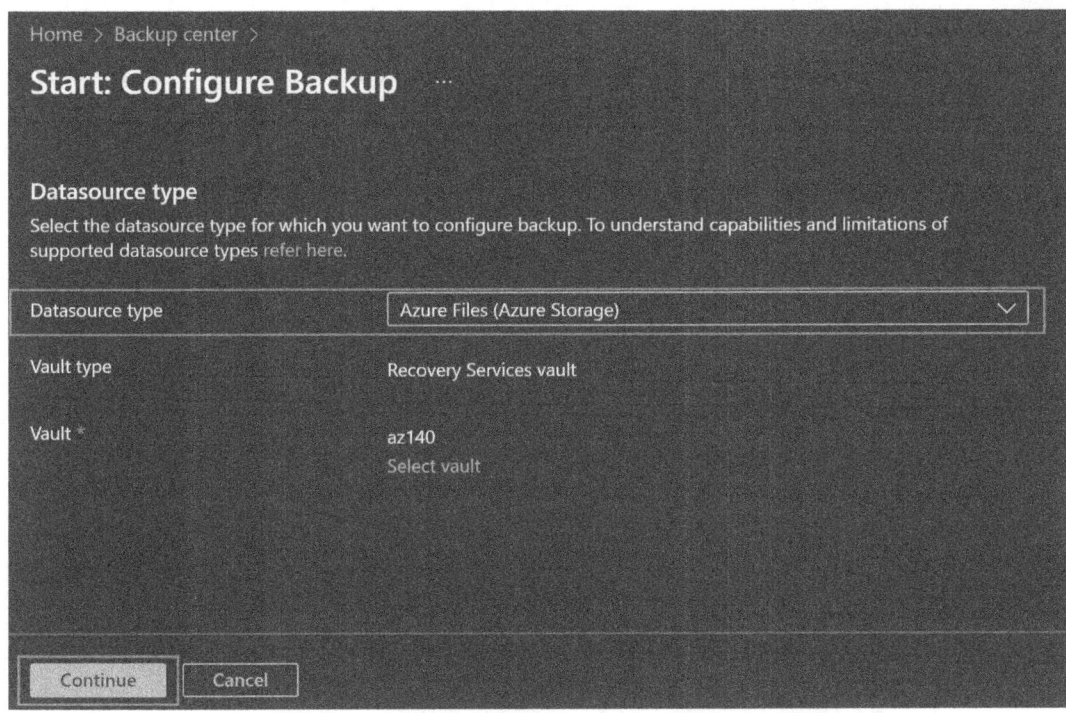

Figure 16.31 – Azure Files and the vault when configuring Azure Files for Backup

4. You will now see the **Configure Backup** page. Within this section, you need first to select **Storage Account**. Once this is selected, in the **Select storage account** blade, click **OK**.

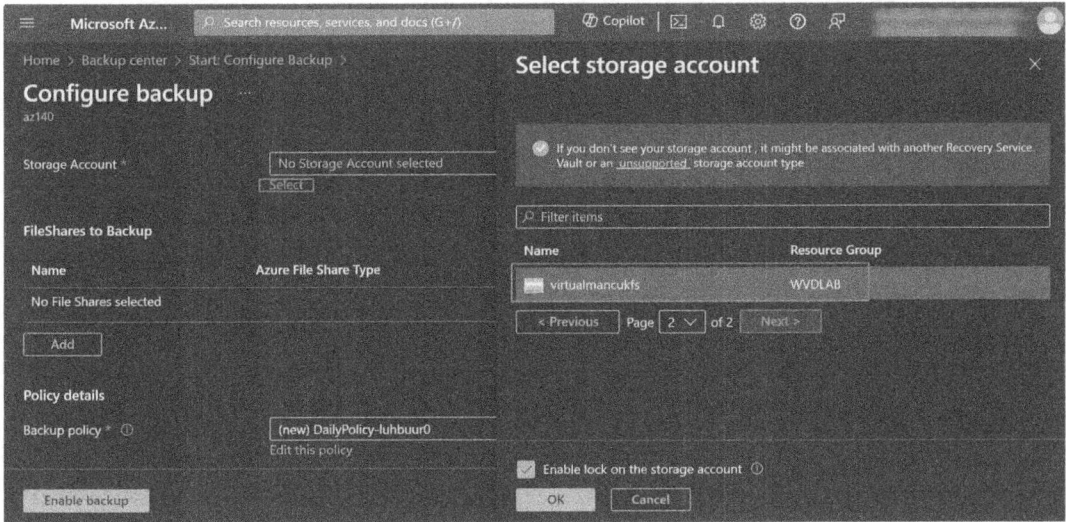

Figure 16.32 – Storage account selection for Backup

5. You will then need to add the required shares on the same page by clicking the **Add** button within the **FileShares to Backup** section. Once you have selected the required file shares, click **OK** within the **Select file shares** blade.

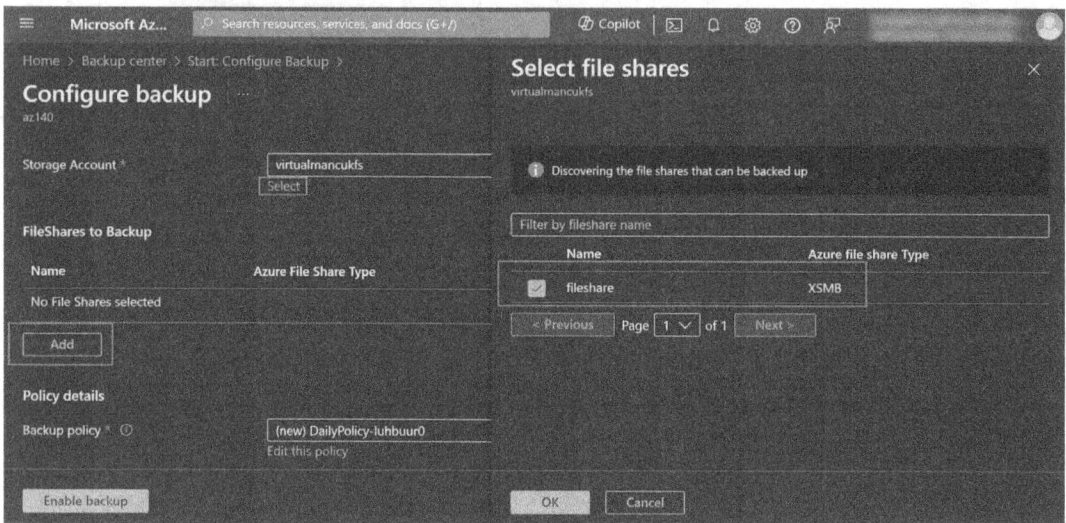

Figure 16.33 – The file share selected that you want to back up

6. Use the **default policy details** section or create a specific backup policy to suit your requirements within the **Policy details** section. Once you are finished, click **Enable** to perform a backup.

7. Once a backup has been taken, you should see the last backup status as being successful. You can find this information by navigating to the recovery vault and selecting **Backup Items** and **Azure Storage (Azure Files)**.

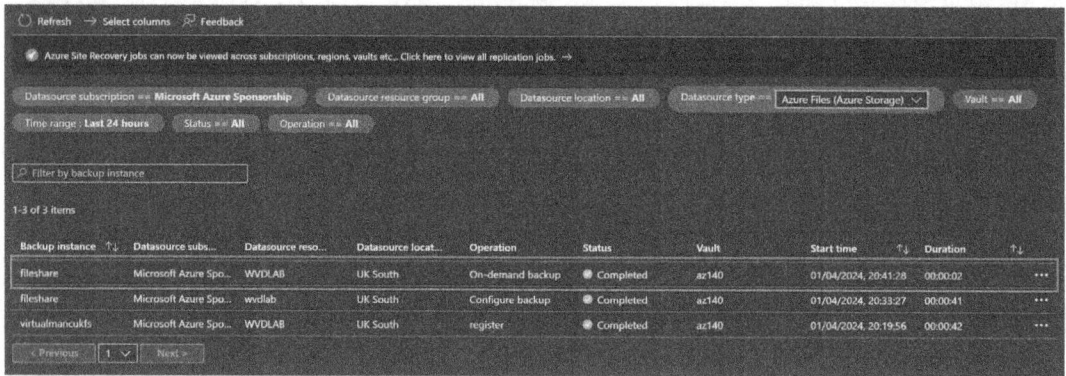

Figure 16.34 – The first backup job has been completed

Now that we have taken a look at backing up an Azure Files share, let's now recover an Azure Files share:

1. Navigate to **Backup center**.
2. Click **Restore** on the main page, as shown in the following screenshot:

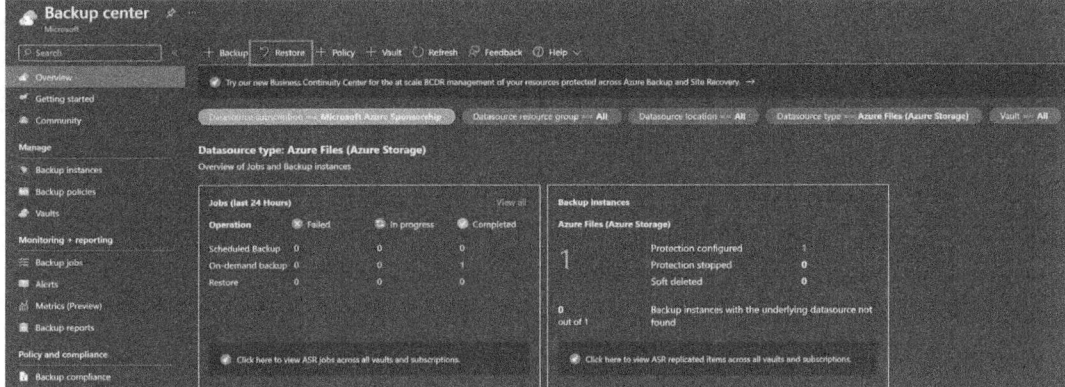

Figure 16.35 – Restore button highlighted within Backup center

3. Select **Azure Files (Azure Storage)** for **Datasource type**.
4. Select the file share you wish to restore.

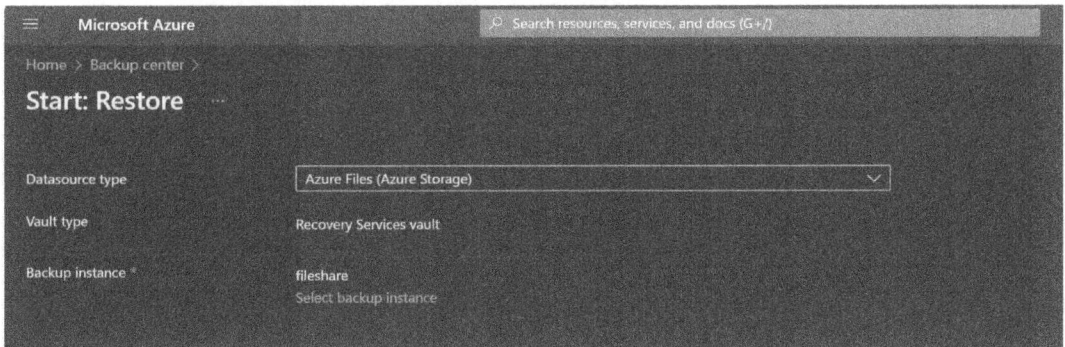

Figure 16.36 – Start: Restore populated for Azure Files

5. Within the **Restore** page, select a restore point for the backed-up file share.
6. Then you need to select a restore destination. In this example, we will choose **Original Location**; however, you can choose an alternative location if required.

7. In the **In case of Conflicts** section, you can choose to **Skip** or **Overwrite**. In this example, we will skip. Once you have finished, click **Restore**.

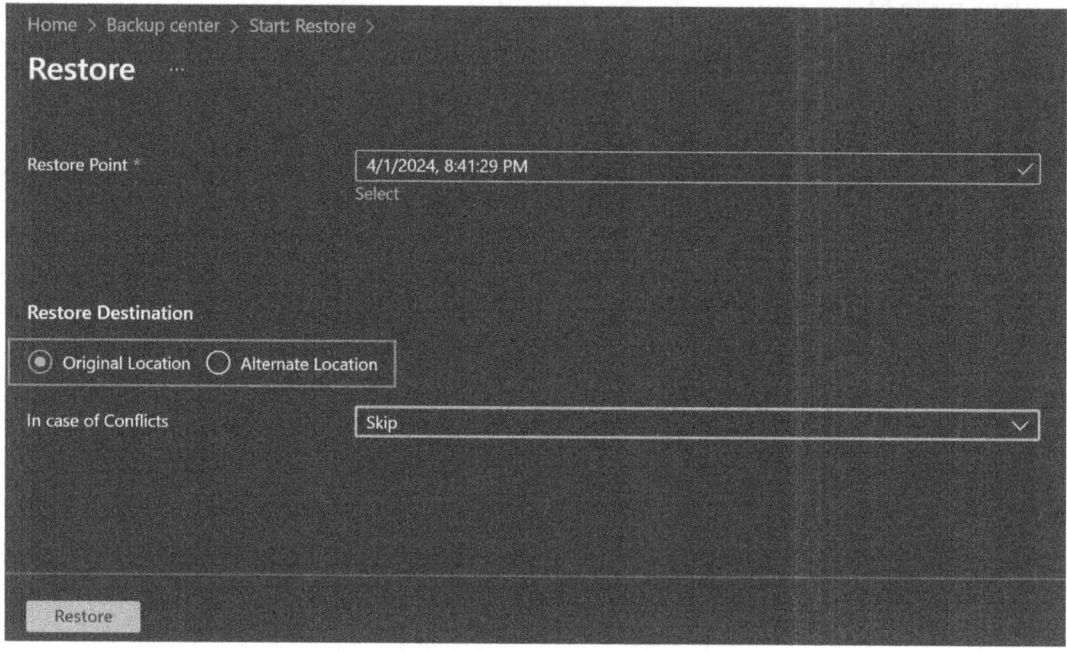

Figure 16.37 – Restore page

8. You can then view pending, in progress, and completed restore jobs within the **Recovery services** vault under **Backup Jobs**, as shown in the following screenshot:

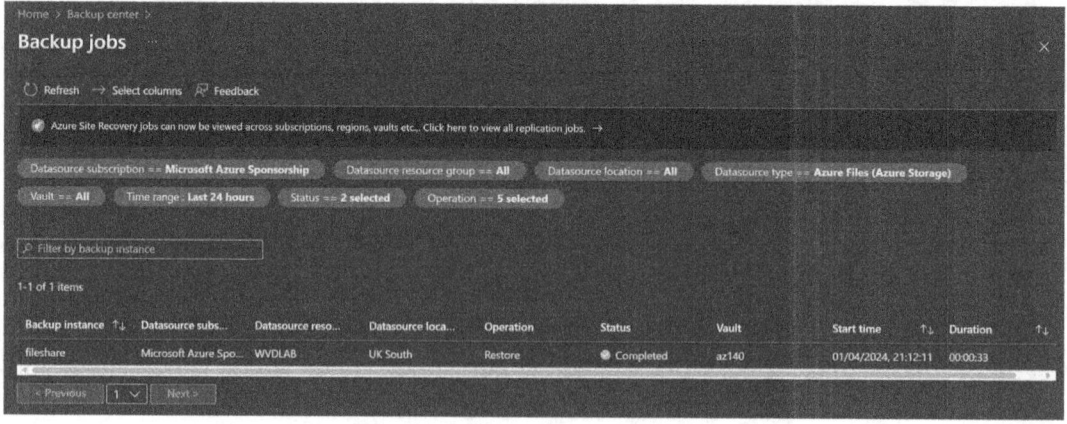

Figure 16.38 – The complete restore job for the Azure Files file share

That concludes backing up and restoring an Azure Files file share. We now move on to the final section of the chapter, where we briefly look at replicating images across multiple images.

Replicating virtual machine images between regions

In *Chapter 9*, *Creating and Managing Session Host Images*, we looked at Azure Compute Galleries. An Azure Compute Gallery enables you to replicate your images across multiple regions. When working with pooled desktop session host pools, you may want to consider using the replication feature within shared image galleries to distribute the image template across multiple regions. In the event of an outage, you can spin up new virtual machines at the failover.

As shown in the following screenshot, you can see updated image versions target replications. This enables you to push out a central operating system template image to multiple regions quickly. You should consider image replication when implementing a disaster recovery solution:

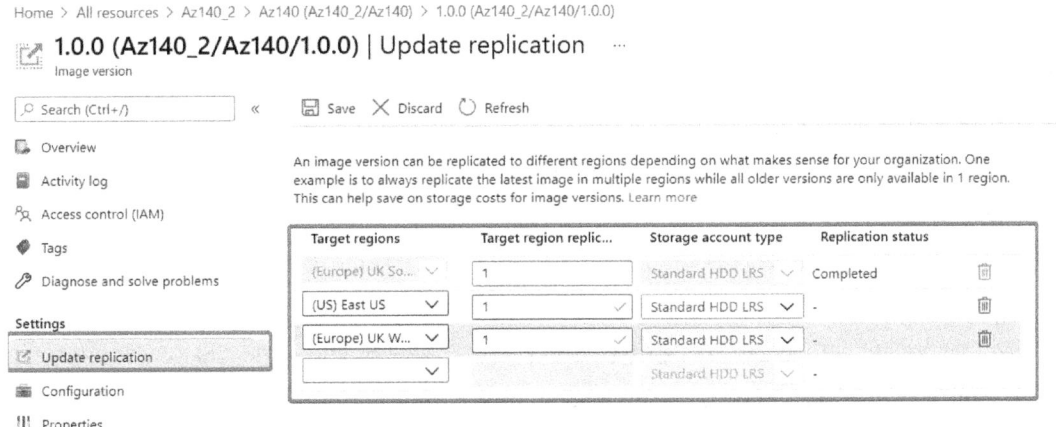

Figure 16.39 – Replication update for an image version within a Azure Compute Gallery

This concludes the short section on image replication. For more information on setting up a Azure Compute Gallery and using image versions, please refer to the *Azure Compute galleries* section in *Chapter 9*, *Creating and Managing Session Host Images*.

Summary

This chapter looked at planning and implementing business continuity and disaster recovery for Azure Virtual Desktop.

We started by looking at designing a backup strategy for Azure Virtual Desktop. We then moved on to look at the five critical components of an Azure Virtual Desktop environment, which are virtual networks, virtual machines, user identities, configuring user and app data, and application dependencies.

We then looked at how to back up and restore virtual machines and Azure Files. We then finished the chapter by reiterating the benefits of using shared image galleries to replicate a gold image/image template across multiple Azure regions.

In the next chapter, we will take a look at automating Azure Virtual Desktop management tasks.

Questions

Here are a few questions to test your understanding of this chapter:

1. When configuring a VNet for an Azure Virtual Desktop environment, what should you ensure you have checked before deploying virtual machines?

    DNS settings have been configured on the VNet.

2. Which Azure service would you use to ensure that your Azure Virtual Desktop template/master image is resilient?

    Use Azure Compute Galleries to replicate your virtual machine image across multiple regions.

3. Which feature within FSLogix can be used to provide resilience if there is a storage failure?

    FSLogix Cloud Cache.

17
Automating Azure Virtual Desktop Management Tasks

In this chapter, we will look at automating Azure Virtual Desktop by taking repeated processes and automating them on a schedule. By using an Automation account, we will then dive into the automation management of host pools, session hosts, and user sessions.

In this chapter, we will take a look at the following topics:

- Creating an Automation account for Azure Virtual Desktop
- Automating the management of host pools, session hosts, and user sessions using PowerShell
- Logging off all users automatically
- Scaling plans
- Scaling plans for personal host pools

Creating an Automation account for Azure Virtual Desktop

In this section, we will look at creating an Automation account for Azure Virtual Desktop. An Automation account is used to create **runbooks** that contain scripts that can automate processes to help you manage your Azure Virtual Desktop environment. Runbooks are extremely useful for automating repetitive tasks and batch processing, which can be quite time-consuming when carrying out such tasks manually.

Let's get started by first creating an Automation account:

1. Sign in to the Azure portal: https://portal.azure.com/.
2. From the top menu, enter automation in the search bar and click **Automation Accounts**:

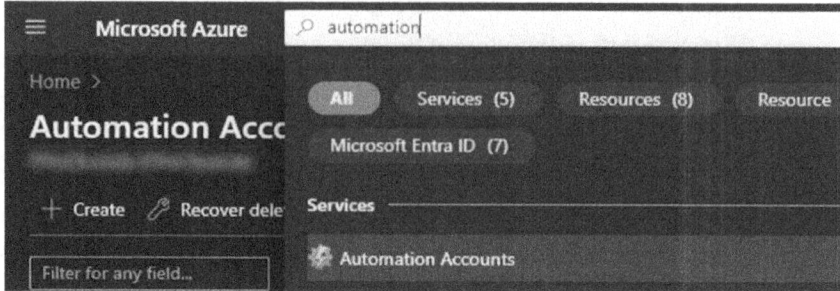

Figure 17.1 – Search result showing Automation Accounts

3. On the **Automation Accounts** page, click **Create**:

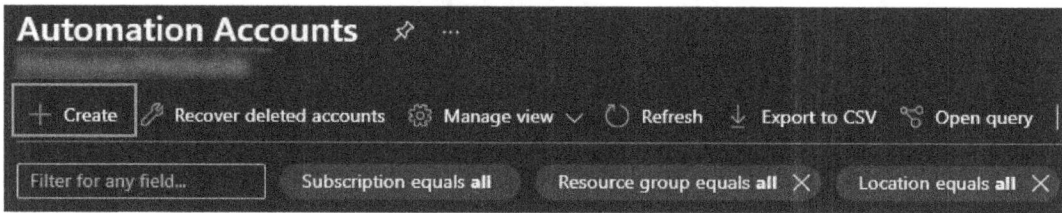

Figure 17.2 – The Create button highlighted within Automation Accounts

4. Within the **Create an Automation Account** blade, provide a name, select a **Subscription** option, create a resource group or select an existing one, and choose a **Region** option. Then, click **Next**:

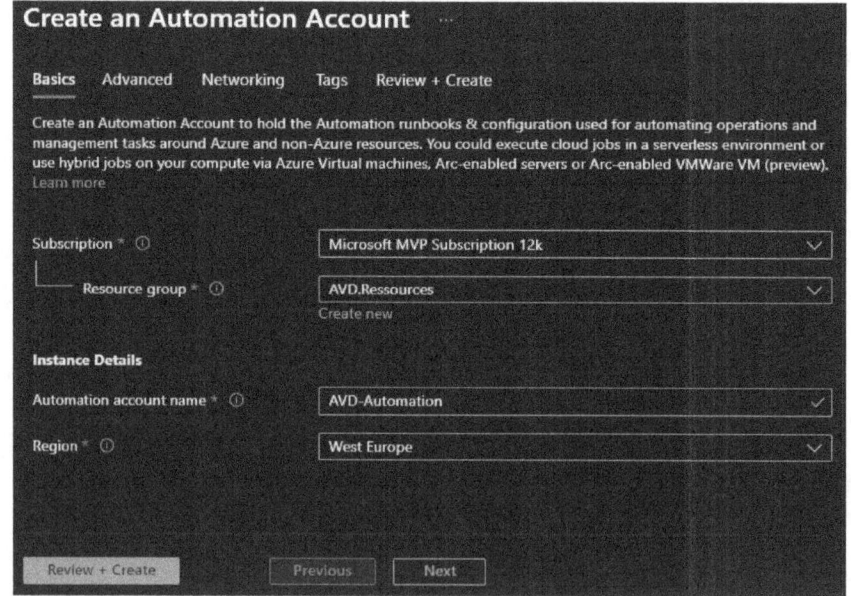

Figure 17.3 – Create an Automation Account blade

5. You can configure the identity for your automation on the **Advanced** tab. We will use a **System assigned** identity:

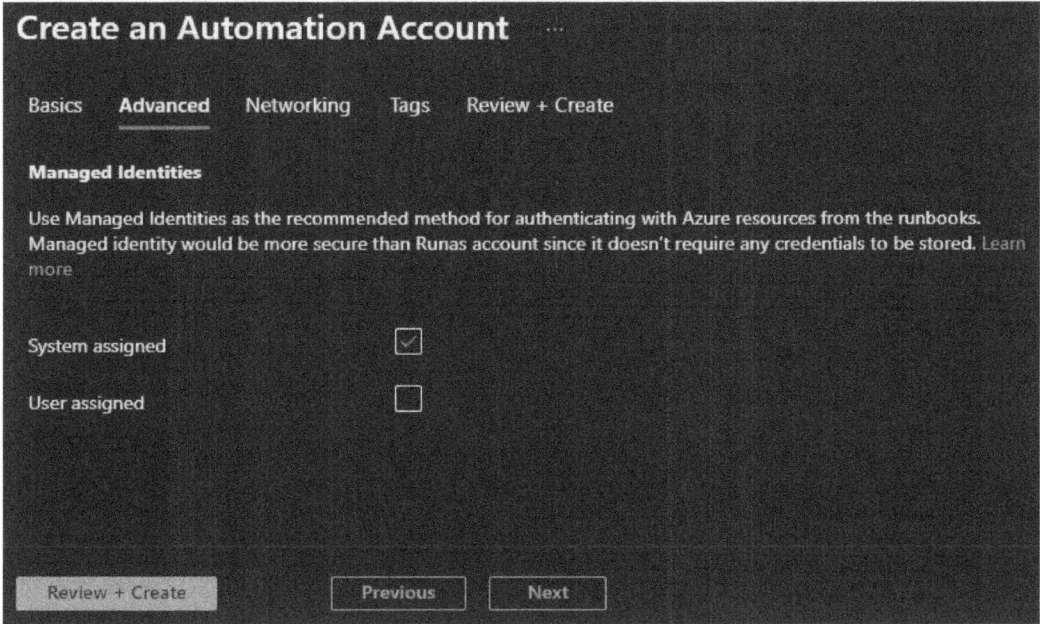

Figure 17.4 – Create an Automation Account – Advanced

> **Important note**
> While *RunAs* accounts are no longer supported, we're using a system-assigned identity. This means that our Automation account has its own identity, and we can give this identity permission to Azure resources such as a user or group.

6. While the other tabs are optional, we can directly click on **Review + Create** and then on **Create** to deploy our new Automation account.
7. Once created, you will see the new Automation account on the **Automation Account** page after clicking on **Go to resource**:

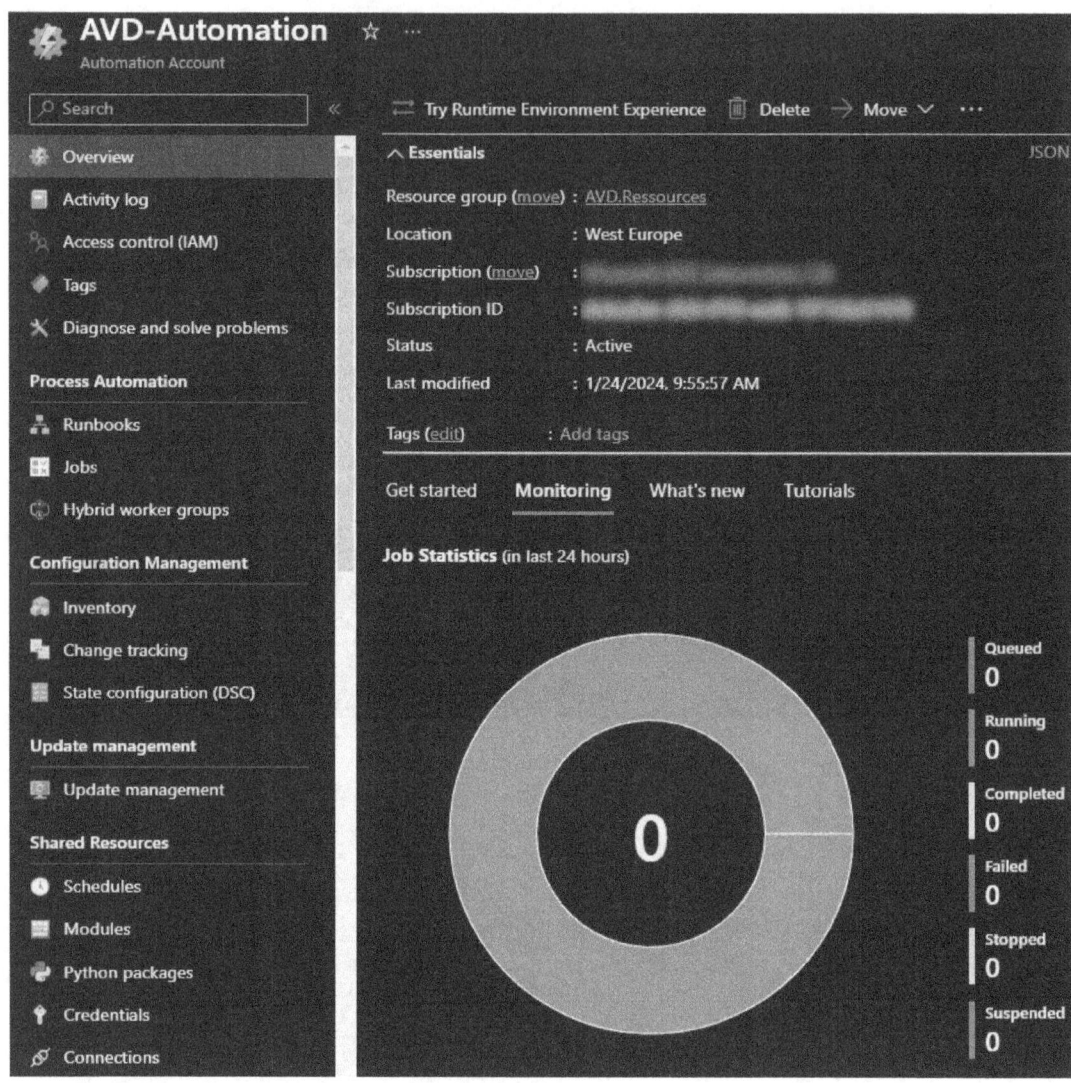

Figure 17.5 – Newly created Automation account

There you have it – we have created an Azure Automation account ready for use with Azure Virtual Desktop. The next section looks at some of the automation tasks you can use to simplify Azure Virtual Desktop management.

Giving the Automation account permissions

The Automation account's managed identity needs permission from the resources in the subscription to work with them. As we have learned in the *Managing Access* chapter, we can achieve that by giving the identity the correct permission to the resource groups containing our AVD-related resources.

Open the resource group containing your AVD resources (do this for all groups) in the Azure portal. Click on **Access control (IAM)** | **Add** | **Add role assignment**. Select a suitable role. I would go for **Desktop Virtualization Contributor** to give the Automation account access to all AVD and VM-related resources:

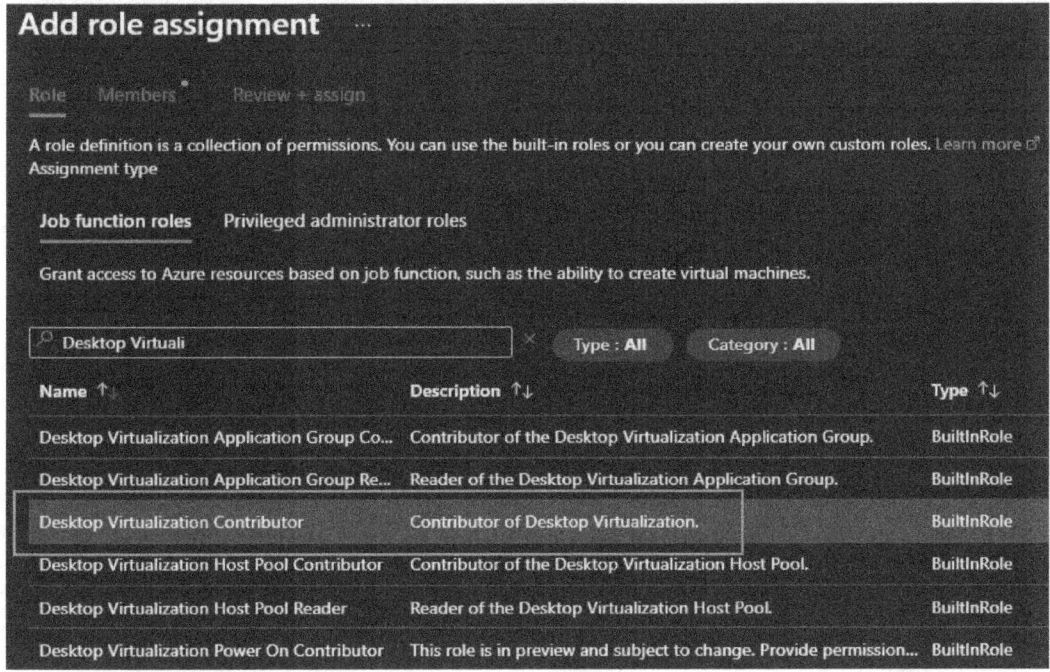

Figure 17.6 – Selecting a suitable role for the Automation account

Clicking on **Next** opens the **Members** page to select the identity of the new role assignment. To select our managed identity, we have to first select **Managed identity** and then click on **+ Select members**. Select **Automation Account** in the drop-down list and search for the name of your Automation account (in this case, **AVD-Automation**):

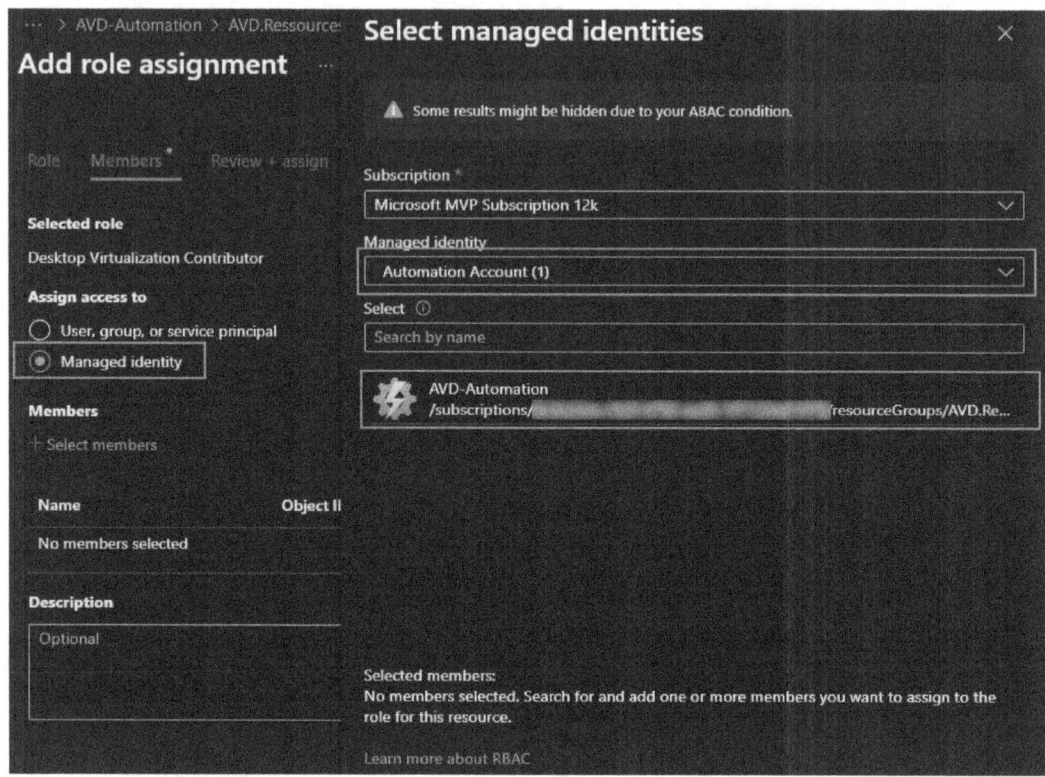

Figure 17.7 – Selecting the managed identity of the Automation account

Click on the managed identity and then on **Select**. Finalize the process by clicking **Review + assign**.

Repeat the step for the other resource groups containing your AVD-related resources.

> **Important note**
> Always follow the least privilege approach: give identities the fewest needed permissions on the scope containing the resources. Scopes are the resource itself, resource group, subscription, and management group.

Automating the management of host pools, session hosts, and user sessions using PowerShell

This section uses the Azure **Command-Line Interface** (**CLI**) and Azure Automation accounts to repeat tasks and simplify tasks such as logging off multiple users in a batch.

Configuring an Azure automation runbook

This subsection takes you through creating an Azure automation runbook, configuring the runbook, testing, and configuring a runbook schedule.

Importing PowerShell modules into the Automation account

First, we must prepare the Automation account with the correct PowerShell modules, which can be imported using the module gallery. If you have deployed a new Automation account, there is a good chance that the modules are already imported. If so, you can skip this step.

The following modules need to be imported:

- `Az.DesktopVirtualization`
- `Az.Accounts`
- `Az.Resources`

The following screenshot shows the Automation account's module gallery, where you would add PowerShell modules for use within runbooks. Click on **Modules** and **Browse gallery**:

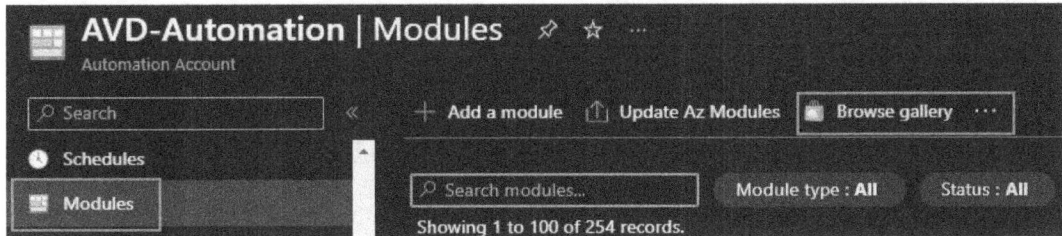

Figure 17.8 – PowerShell modules gallery where you can add the required PowerShell modules

You can search for missing modules on the next screen:

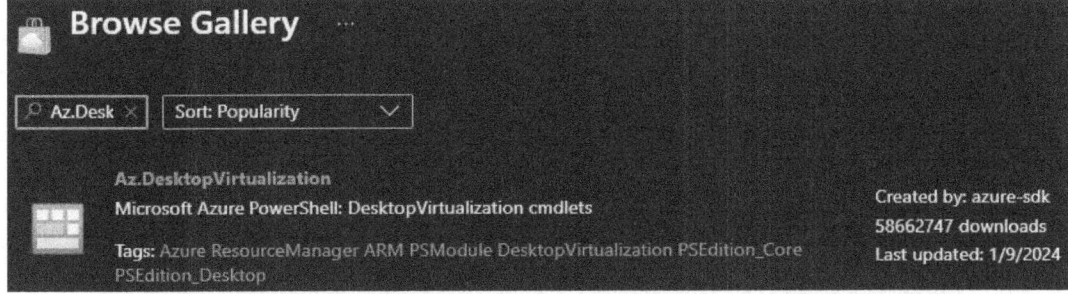

Figure 17.9 – PowerShell modules gallery: searching for modules

Select the module to add it to the Automation account. Finally, select the newest runtime and click on **Import**.

While we are mostly working with the Az modules, we can update all imported modules from the **Modules** page by clicking **Update Az Modules** and selecting the correct runtime (recommended) and the newest module versions:

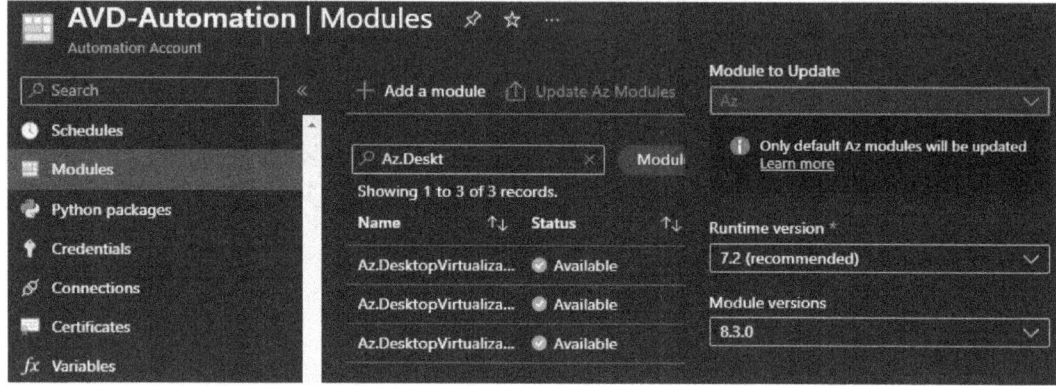

Figure 17.10 – PowerShell modules gallery: Update Az modules

> **Important note**
> Updating the PowerShell module can cause your automation to stop working if some breaking changes are in newer modules. Please always validate the function after an update.

Once you have begun importing the modules, you will see the modules in the overview:

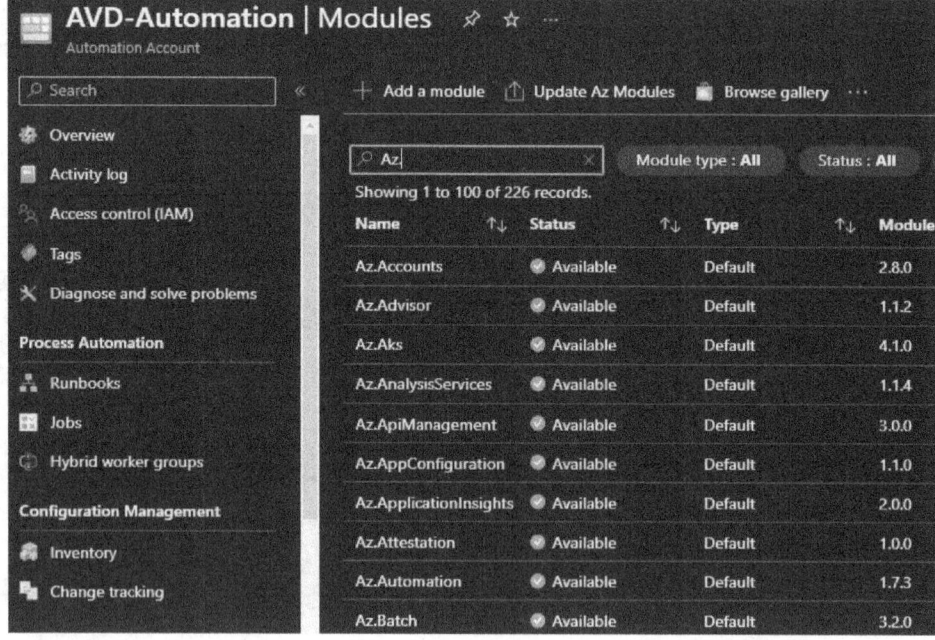

Figure 17.11 – Imported PowerShell modules in an Automation account

Automating the management of host pools, session hosts, and user sessions using PowerShell

The next part of this process is to create a PowerShell runbook within the Azure Automation account that we have just created.

Creating a PowerShell runbook

We'll now look at creating a PowerShell runbook within the Automation account.

To create a runbook, navigate to the Automation account. Under **Process Automation**, click **Runbooks**; once on the **Runbooks** page, click on **Create a runbook**:

Figure 17.12 – Creating a new runbook

You will then see the **Create a runbook** blade appear. Enter a name for the runbook and select the type – in this case, it's a **PowerShell** runbook. Provide a description as follows:

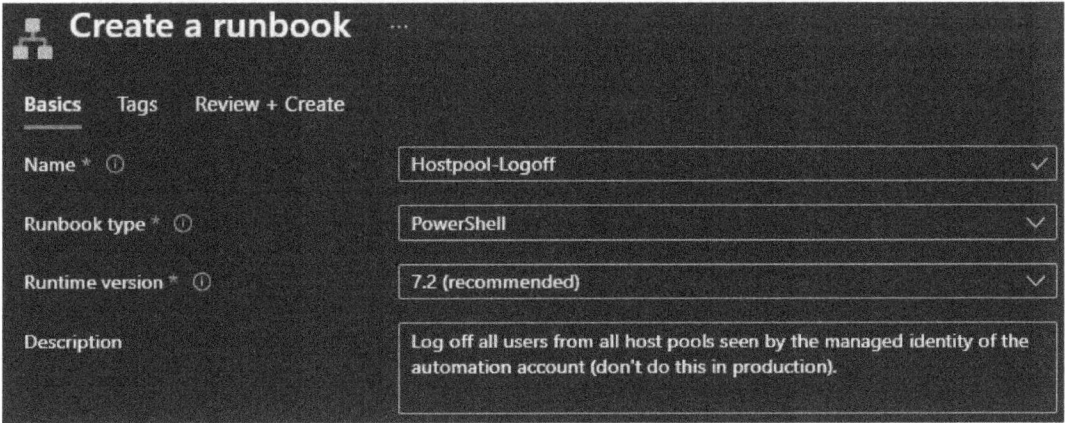

Figure 17.13 – The Create a runbook blade

Once the new runbook is created, it will open the edit page to enter the PowerShell code:

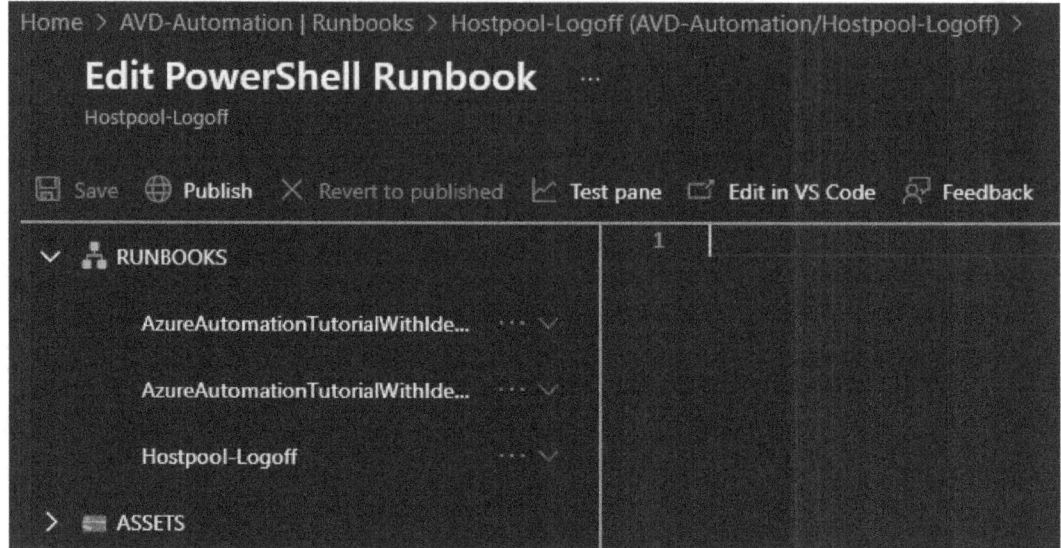

Figure 17.14 – Entering code into the runbook

Now that we have created a runbook, we can proceed with configuring it with the required script. In this example, we will be creating a host pool logoff runbook to simplify the logging off of multiple users spread across several session hosts within all host pools seen by the identity of the Automation account.

> **Important note**
> This is an example of how to create automations. Don't do this in a production environment unless you need the automation of logging off all user sessions.

Adding your PowerShell script to a runbook

Let's now take a look at adding the PowerShell script to the runbook. You should now see the runbook scripting interface; this is where you need to enter the PowerShell cmdlets.

Before entering the scripting and testing, I want to briefly run through the host pool logoff example. This is broken down into multiple sections so that I can discuss the key parts of the script.

The first part of the script is the header, detailing information about the script and any imports you may need for the script:

```
# Mastering Azure Virtual Desktop Log Off User example
Import-Module Az.DesktopVirtualization
```

The next part of the script is the authentication piece. This is needed to ensure that the runbook can communicate with the resources. While we are using the managed identity of the runbook, the process is straightforward:

```
# Login into Azure with the managed identity of the automation account
try
{
    "Login into Azure with the managed identity of the automation account..."
    Connect-AzAccount -Identity
}
catch {
    Write-Error "Login failed: $_"
    throw $_.Exception
}
```

Now that we have authenticated with Azure, the next part of the script will call the logoff function. The script then counts the user sessions within the host pools and runs `Remove-AzWvdUserSession` for each session on each host in each pool. This automates the logoff process for host pools, which, in turn, simplifies the IT admin's effort to carry out maintenance and so on:

```
# Start AVD Task

# Get all resource groups seen by the identity
$ResourceGroups = @(Get-AzResourceGroup)

# Enumerate through the resource groups
foreach ($ResourceGroup in $ResourceGroups)
{
    # Get all host pools in the resource group
    $HostPools = @(Get-AzWvdHostPool -ResourceGroupName $ResourceGroup.ResourceGroupName)

    # Enumerate through the host pools
    foreach ($HostPool in $HostPools)
    {
        Write-Output "Found host pool: $($HostPool.Name) in resource group $($ResourceGroup.ResourceGroupName)"
        $Sessions = @(Get-AzWvdUserSession -ResourceGroupName $ResourceGroup.ResourceGroupName -HostPoolName $HostPool.Name)
        Write-Output "The host pool $($HostPool.Name) has $($Sessions.Count) sessions"
        foreach ($Session in $Sessions)
        {
        $Session
```

```
            Write-Output "Logging off user $($Session.
UserPrincipalName) from host pool $($HostPool.Name)"
            $SessionHost = $Session.Id.Split("/")[10]
            $SessionId = $Session.Id.Split("/")[12]
            Remove-AzWvdUserSession -ResourceGroupName $ResourceGroup.
ResourceGroupName -HostPoolName $HostPool.Name -SessionHostName
$SessionHost -Id $SessionId -Force
        }
    }
}
```

You can download the full script here: https://github.com/PacktPublishing/Mastering-Azure-Virtual-Desktop-2nd-Edition/blob/main/Logoff%20all%20sessions%20from%20all%20pools.ps1.

Now that we have run through the host pool logoff script structure, we will proceed with configuring the runbook and publishing it:

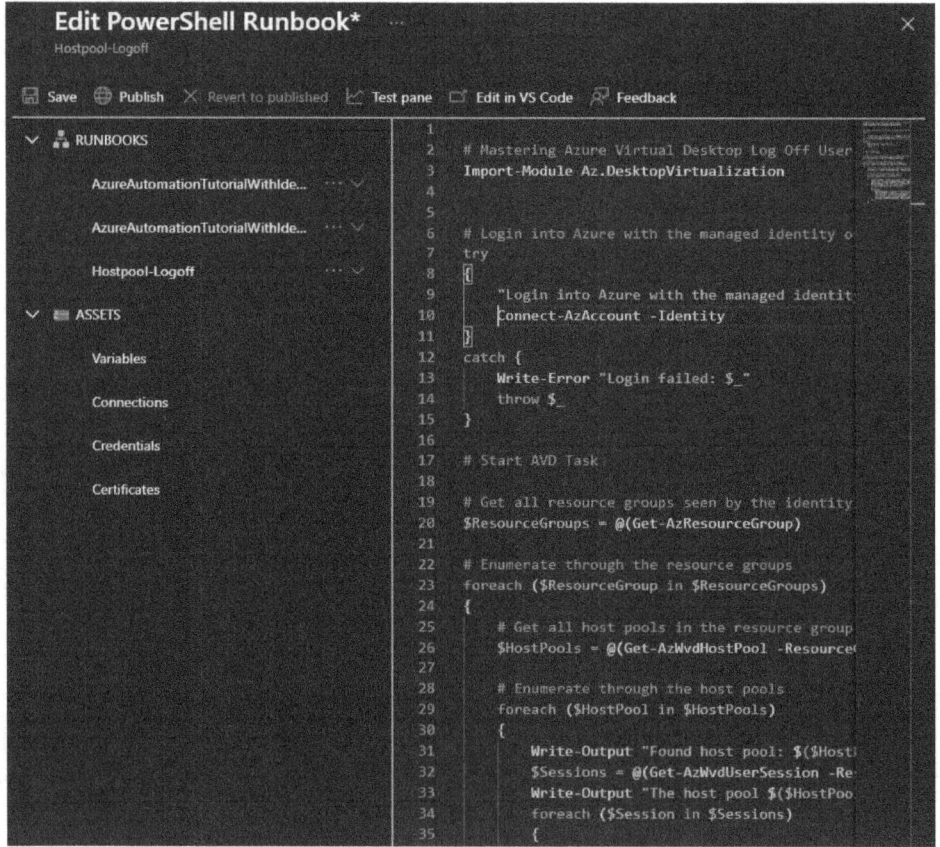

Figure 17.15 – PowerShell scripting added to the PowerShell runbook

Once you have copied the script into the PowerShell runbook, you can proceed with testing by clicking **Test pane**, as shown in the following screenshot:

Figure 17.16 – Location of the Test pane button for testing a PowerShell runbook

In the next section, we take a look at testing the script before publishing.

Testing a PowerShell runbook in Azure

This section runs through the basics of using the **Test** pane to test the script we created for logging off Azure Virtual Desktop users within a host pool.

To test the script, all you need to do is click the **Start** icon, as shown in *Figure 17.17*:

1. Click **Start** on the **Test** page:

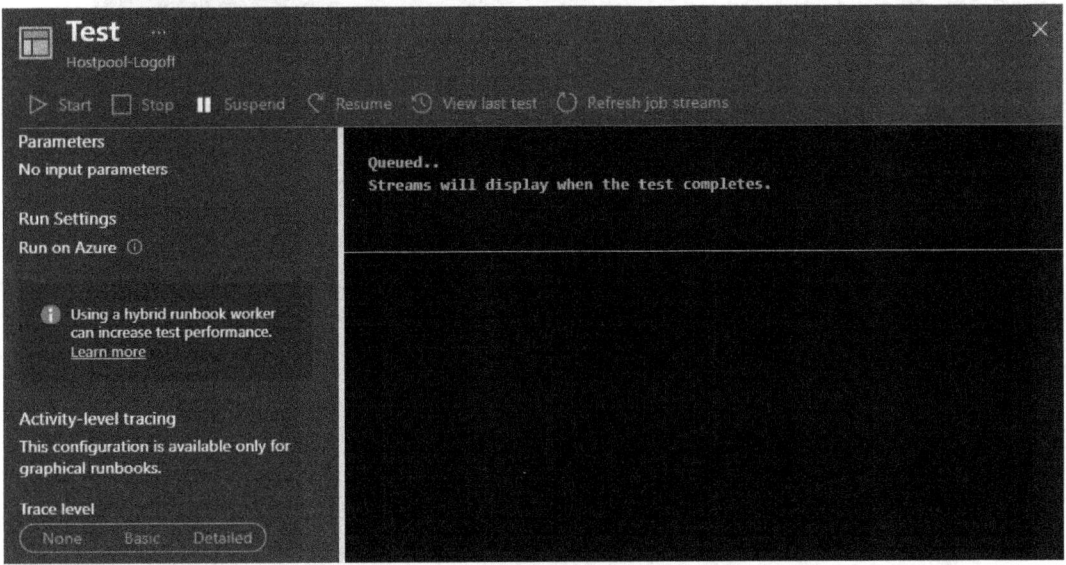

Figure 17.17 – The Test page of a PowerShell runbook

2. The black screen shown in the preceding screenshot shows the script's output and any errors that may occur, requiring you to make the appropriate changes if required.

3. As you can see from the test results in the following screenshot, the script worked as required and logged off the two user sessions:

Figure 17.18 – Test complete within a PowerShell runbook

4. Once you have finished the test and you have confirmed that the script is working and there are no errors, you can then go ahead and close the **Test** pane:

Figure 17.19 – Closing the test pane

5. After this, publish the runbook, as shown in the following screenshot:

Figure 17.20 – The Publish button and confirmation message bar

6. Once the runbook has been published, close **Edit PowerShell Runbook** by clicking the cross in the upper-right corner. You may have to refresh the view by clicking **Refresh**.
7. You will now see several options shown as accessible, including **Link to schedule**. You can find these within the menu bar, as shown in the following screenshot:

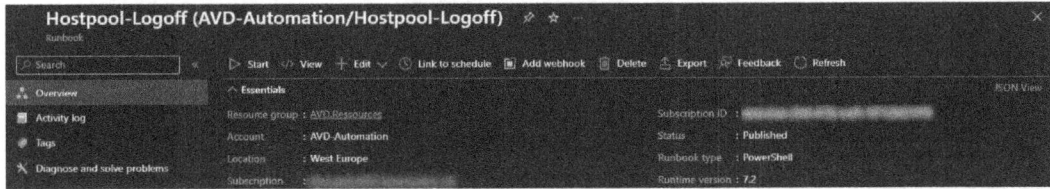

Figure 17.21 – The published runbook

In the next section, we look at creating a schedule and assigning the created runbook to it.

Creating a schedule

Follow these steps to create a schedule:

1. Within the runbook, you can create a schedule. First, you would need to navigate down to the section in the left-hand menu called **Resources**:

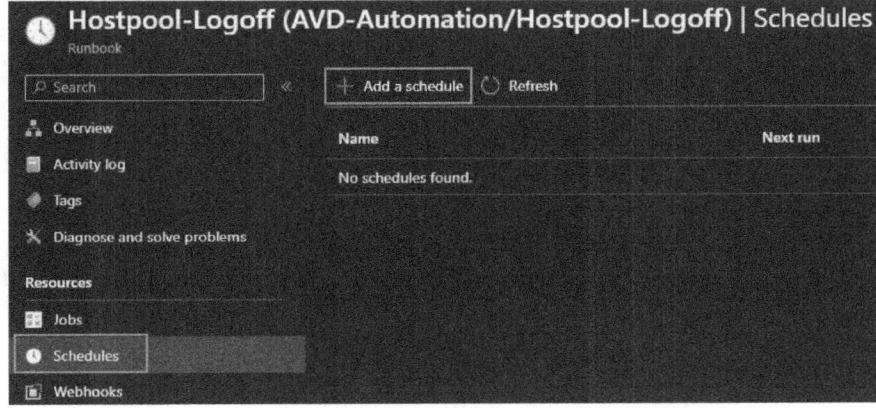

Figure 17.22 – The Schedules page within a runbook

2. Click on **Add a schedule** and then click **Link a schedule to your runbook**:

Figure 17.23 – The Schedule Runbook page

3. Then, click **Add a schedule**, and click on **+ Add a schedule** blade will appear:

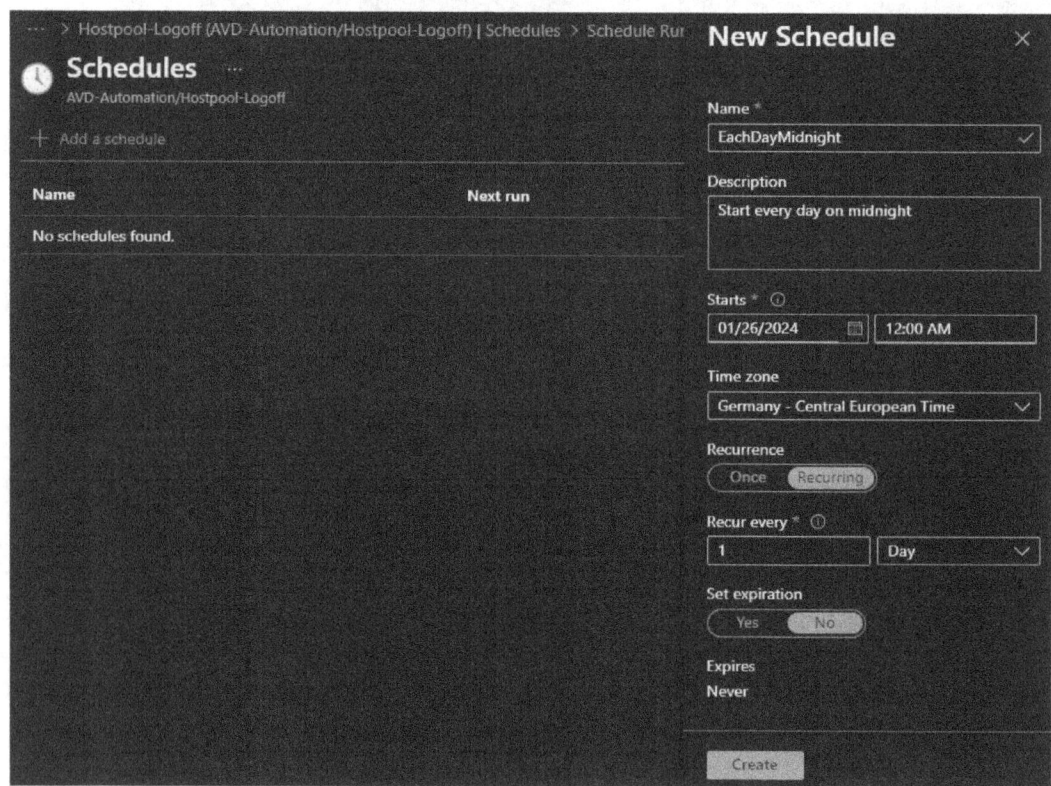

Figure 17.24 – The Schedules page and New Schedule blade

4. Provide a name, description, and a **Starts** time and date, ensuring you choose the correct recurrence: either **Once** or **Recurring**. Once you have entered the required information, click **Create**.

5. You will then see the newly created schedule on the **Schedule Runbook** page, as shown in the following screenshot:

Figure 17.25 – New schedule added to the runbook on the Schedule Runbook page

6. You will now see that the runbook is scheduled to run, as shown in the following screenshot:

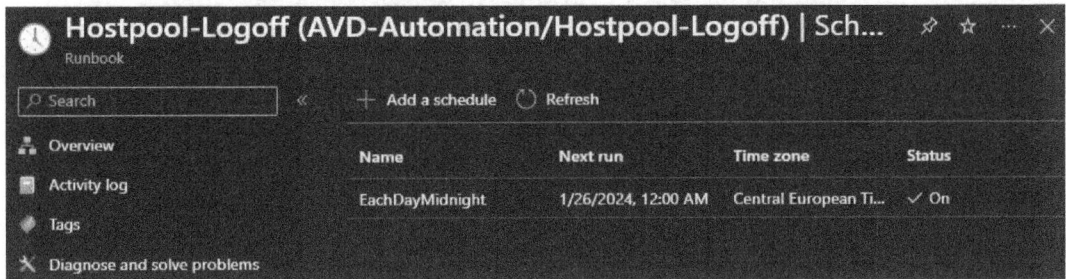

Figure 17.26 – Configured schedule within the runbook under Schedules

In this section, we looked at configuring an Azure automation runbook. First, we covered the setup of a system-assigned identity and the importing of required PowerShell modules for an Azure Virtual Desktop PowerShell script to run. We then created the runbook, added the PowerShell script to the runbook, and tested it. We then finished the section by creating an automated schedule for the runbook to run at a specific time.

In the next section, we look at implementing autoscaling in Azure Virtual Desktop using an Automation account.

Autoscale – scaling plans

Please note, at the time of writing, autoscale within the Azure portal is in public preview.

The built-in autoscale feature with Azure Virtual Desktop simplifies automating scale based on time and session limits per session host. It also allows the scaling of session hosts in a personal pool.

> **Important note**
> Ephemeral disks are not supported when using the scaling plans for Azure Virtual Desktop.

Giving Microsoft access to start and stop VMs

Before we can start configuring a scaling plan, we first need to give Microsoft's service principal permission to start and stop VMs. We can configure the right role in the IAM of each resource group containing the VMs.

Navigate to the resource group (and repeat this for additional resource groups if needed) and click **Access control (IAM)**, **+ Add**, and **Add role assignment**:

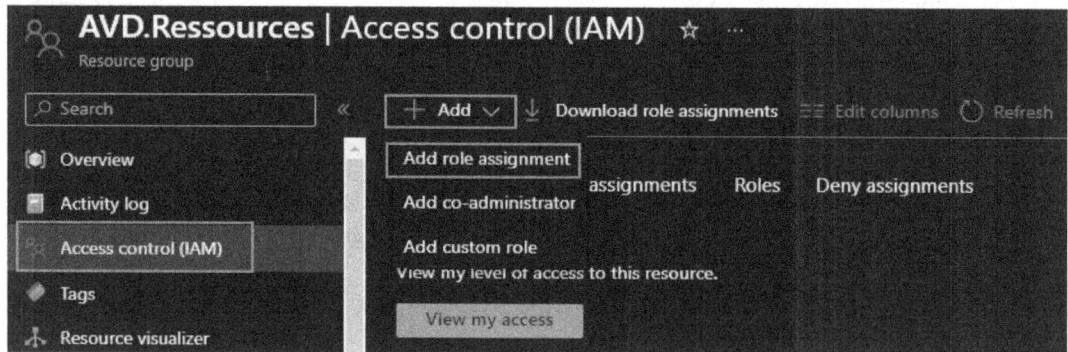

Figure 17.27 – Access control: Add role assignment

Select the **Desktop Virtualization Power On Off Contributor** role from the list:

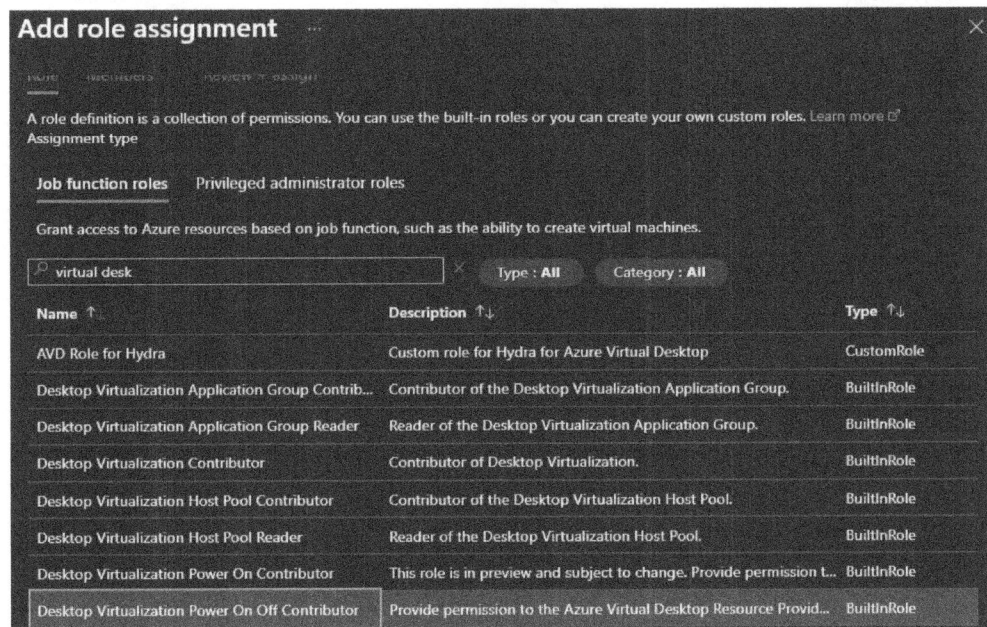

Figure 17.28 – Access control: choosing the Power On Off Contributor role

After clicking **Next**, you can select **User, group, or service principal** by clicking **+ Select members**. Search for and add the **Windows Virtual Desktop** or **Azure Virtual Desktop** service principal (the principal is written differently depending on the age of your tenant):

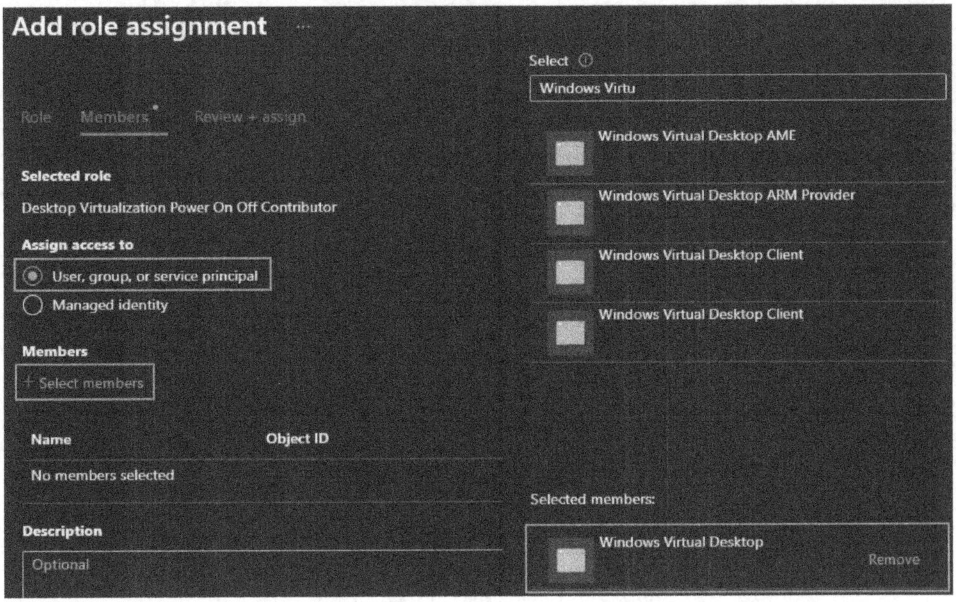

Figure 17.29 – Access control: adding Microsoft's service principal

In this section, we looked at giving Microsoft permission to start our VMs in specific resource groups for autoscaling. We can now proceed with looking at creating our first scaling plan.

Creating a pooled scaling plan (multiuser)

Before creating a scaling plan, you first need to ensure you have an existing pooled host pool. All your host pools must have a configured `maxSessionLimit` parameter.

> **Important note**
> You can update the `MaxSessionLimit` parameter using the following PowerShell cmdlet: `New-AZWvdHostPool` or `Update-AZWvdHostPool`.

Let's get started:

1. Log in to the Azure portal.
2. Navigate to the **Azure Virtual Desktop** page.
3. Select **Scaling plans**:

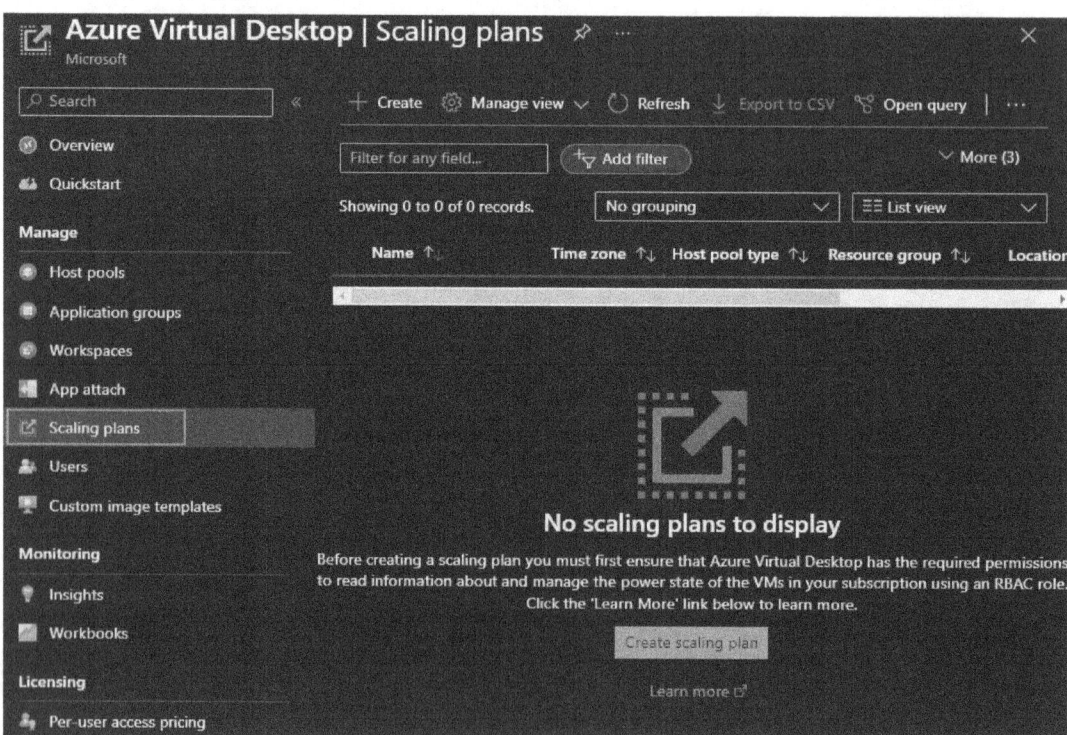

Figure 17.30 – The Scaling plans page in Azure Virtual Desktop

4. Within **Scaling plans**, click **Create**.
5. Enter the required resource group or select **Create new** if required.
6. Enter a name for the scaling plan.
7. Select the required Azure region for the scaling plan.
8. Select the required time zone for the scaling plan.
9. Select **Pooled** in **Host pool type** to scale hosts in a multiuser environment.
10. Enter a tag for the VMs that you do not want to be included in scaling operations within the **Exclusion tag** section:

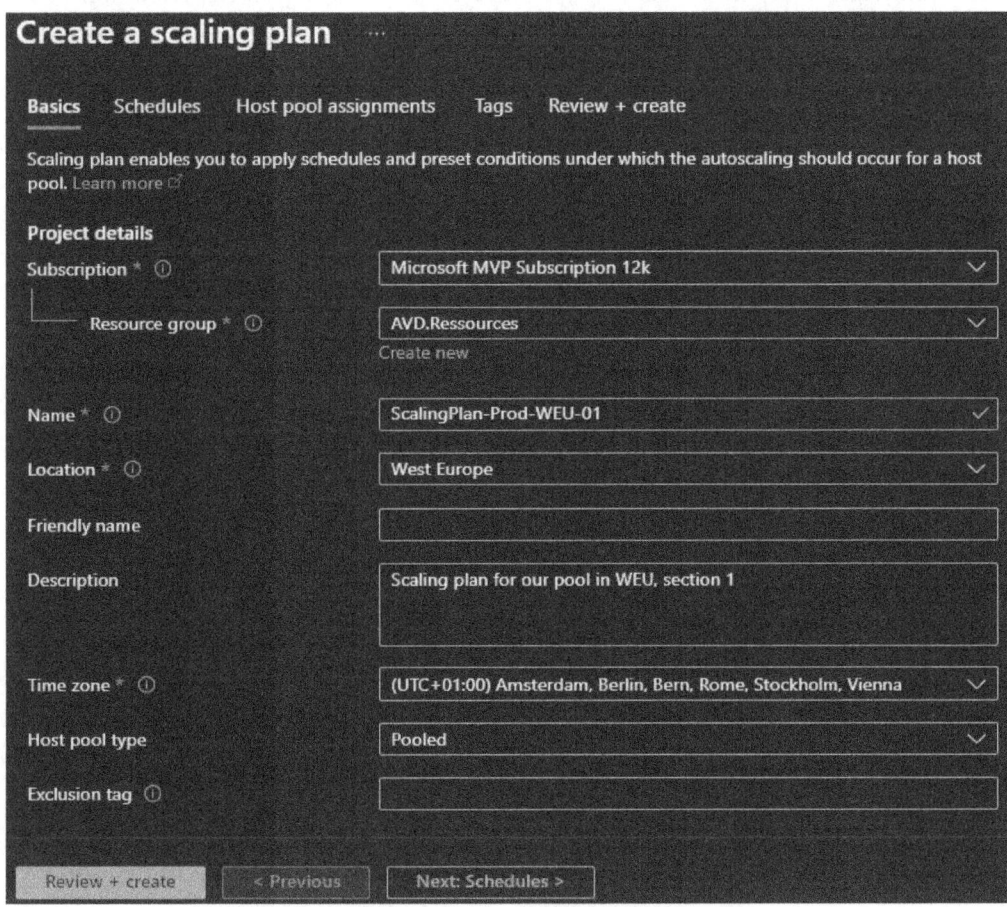

Figure 17.31 – Creating a pooled scaling plan page

11. Click **Next: Schedules**.

We'll now move on to look at the configuration of a schedule within a scaling plan.

Configuring a schedule

Schedules are essentially the configuration settings for autoscale. This includes the ramp-up and ramp-down modes, which will activate based on schedules throughout the day.

The following table details the different terms and meanings:

Term	Description
Start time	This is used to start preparing VMs for peak business hours or, as some refer to them, normal working hours.
Load balancing algorithm	The default setting is the breadth-first algorithm. This is recommended as it will distribute user sessions across existing VMs to ensure minimal delay to access times.
Peak hours	The start time for when the usage rate is expected to be the highest during the working day. This is also the same time that is used for the ramp-up phase.
Minimum percentage of session host VMs	You would set the required session host resources required for the ramp-up and peak hours.
Capacity threshold	The capacity threshold is the percentage of host pool usage that triggers the startup of the ramp-up and peak phases.
Load balancing (under peak hours)	This can be set as breadth-first or depth-first load balancing. Breadth-first is recommended for those who want to distribute new sessions across all available sessions within a host pool.

Table 17.1 – Properties for pooled schedules

To configure or change a schedule, follow these steps:

1. Within the **Schedules** tab, you can add a schedule. Go ahead and click **Add schedule**:

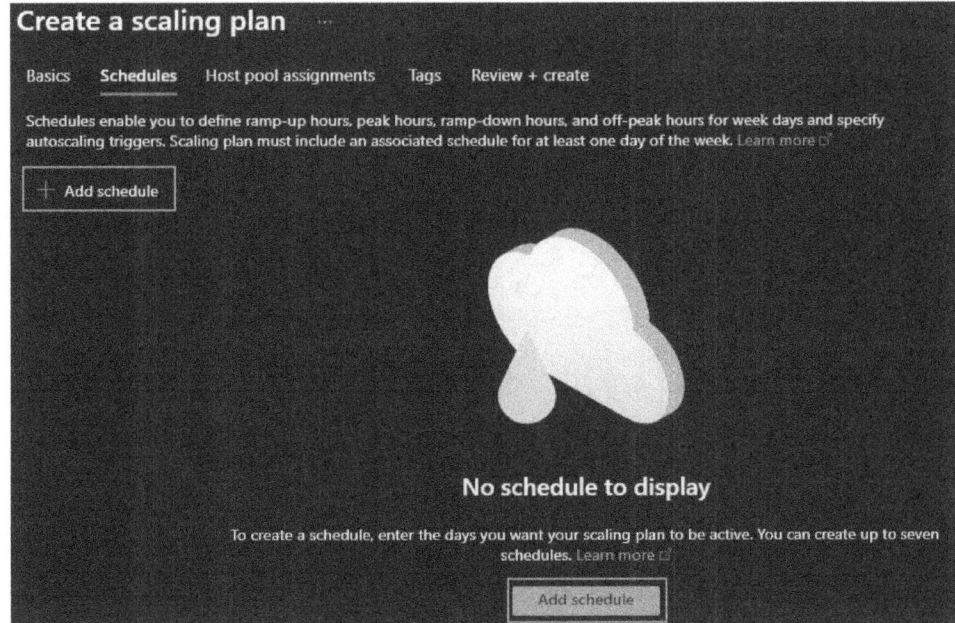

Figure 17.32 – The Schedules tab on the Create a scaling plan page

2. You will then see the **Add a schedule** blade appear with five tabs – **General**, **Ramp-up**, **Peak hours**, **Ramp-down**, and **Off-peak hours**:

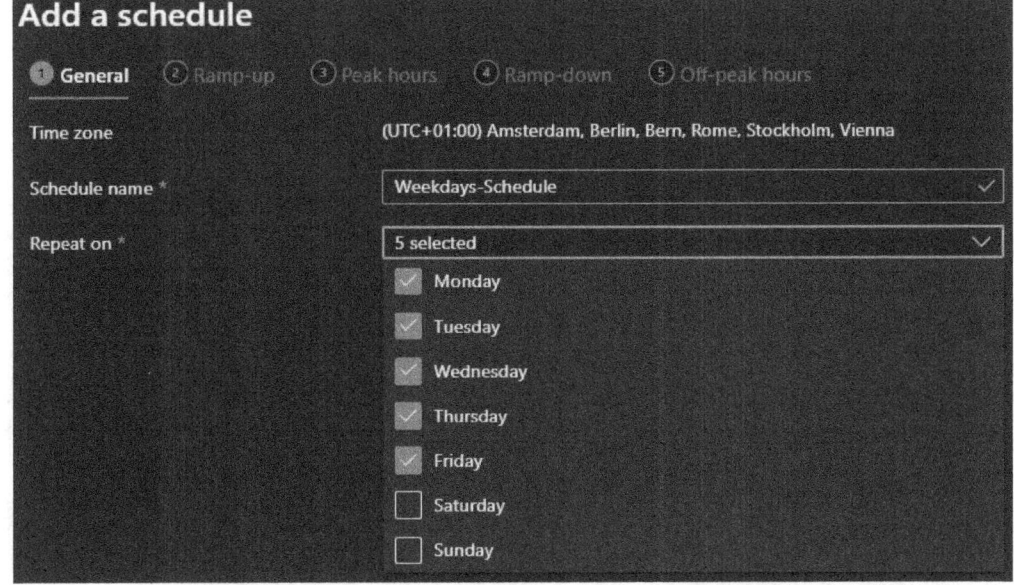

Figure 17.33 – The General tab on the Add a schedule blade

3. Enter the schedule name within the **General** tab and set the days you want the schedule to repeat. Then, click **Next**.

4. You can set the start time, load balancing algorithm, minimum percentage of hosts, and capacity threshold percentage within the **Ramp-up** tab. Once you have configured the settings you want, click **Next**:

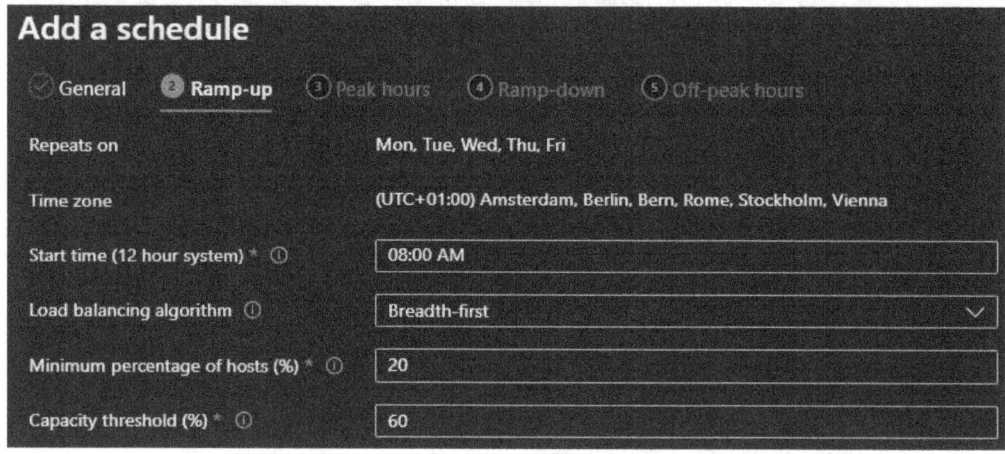

Figure 17.34 – The pooled Ramp-up tab on the Add a schedule blade

5. You can set the start time for peak hours and the load balancing algorithm within the **Peak hours** tab. Once set, click **Next**:

> **Important note**
> When applying load balancing preferences within a scale plan schedule, you will override the original host pool load balancing setting.

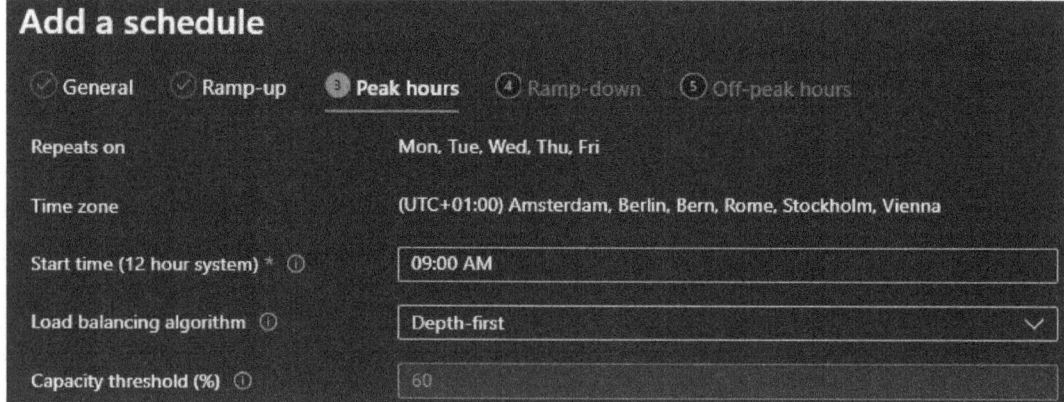

Figure 17.35 – The pooled Peak hours tab on the Add a schedule blade

6. Within the **Ramp-down** tab, you can specify the start time, load balancing algorithm, minimum number of hosts as a percentage, capacity threshold, and the ability to force logoff users and send a message. Once you have configured this tab, click **Next**.

   For **Ramp-down**, you will see similar values to the **Ramp-up** section. It is important to note that, in this instance, it will be for reduced host pool usage, also known as drop-offs. The **Ramp-down** section includes **Force logoff users**, **Delay time before logging out users and shutting down VMs (min)**, and the ability to send a message to the user before they are logged off:

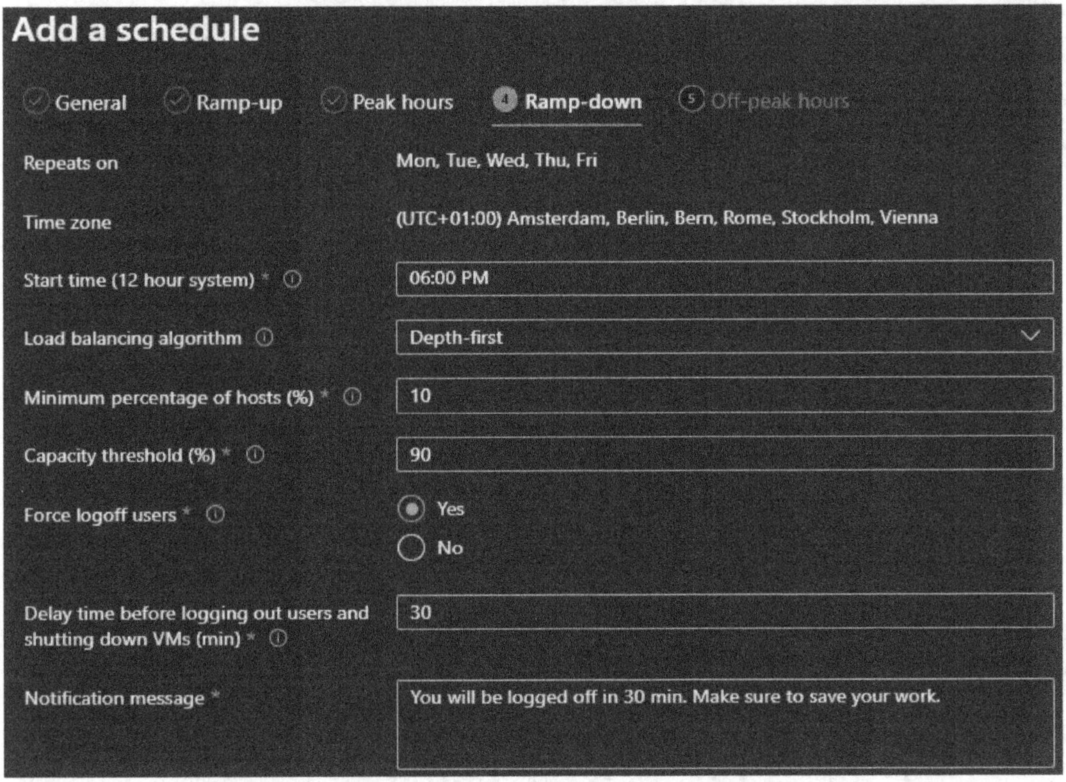

   Figure 17.36 – Pooled Ramp-down tab on the Add a schedule blade

7. The final tab on the **Add a schedule** page is **Off-peak hours**. You can set the time and the load balancing algorithm. Once configured, click **Add**.

   Off-peak hours work in a similar way to peak hours. The difference is that off-peak hours gradually reduce the number of session hosts based on the user sessions on each session host:

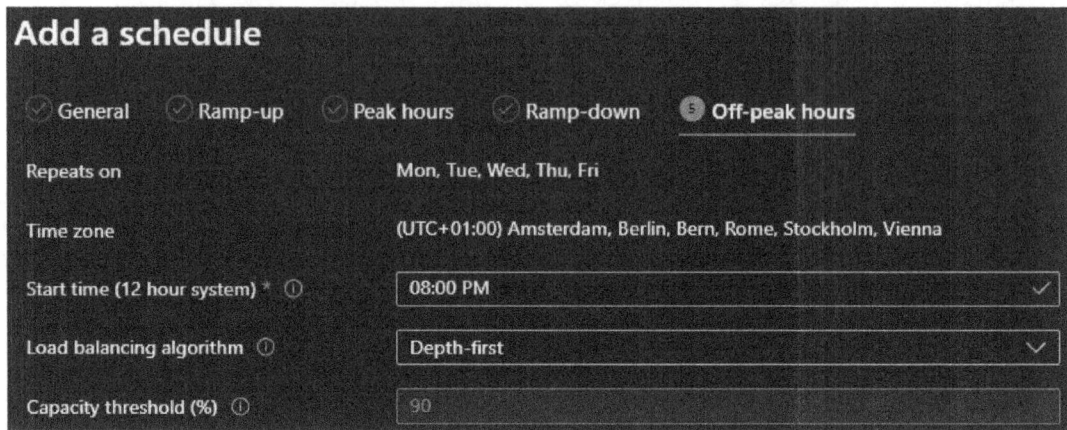

Figure 17.37 – Pooled Off-peak hours tab on the Add a schedule blade

8. Click **Add** to add the schedule to the plan. Now that we have configured the schedule, we can now assign host pools to the scaling plan.

Assigning host pools

The following steps look at how to assign the host pool and complete the creation of the scaling plan:

1. Now that we have created the schedule, we can move on to the host pool assignments. This is where you would select the session host pools you want to assign to the scaling plan.

2. Select the required host pools within the **Host pool assignments** tab and ensure that **Enable autoscale** is checked:

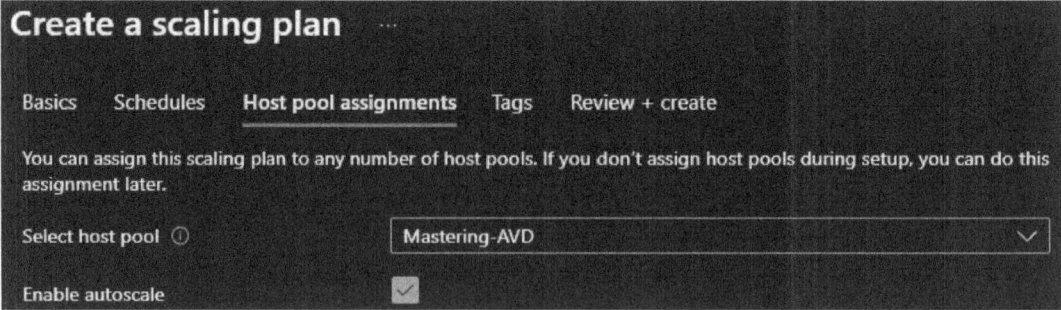

Figure 17.38 – The Host pool assignments tab on the Create a scaling plan page

3. Under the **Tags** tab, specify any tags you may want to set, then click **Next**.
4. Review your settings and click **Create**:

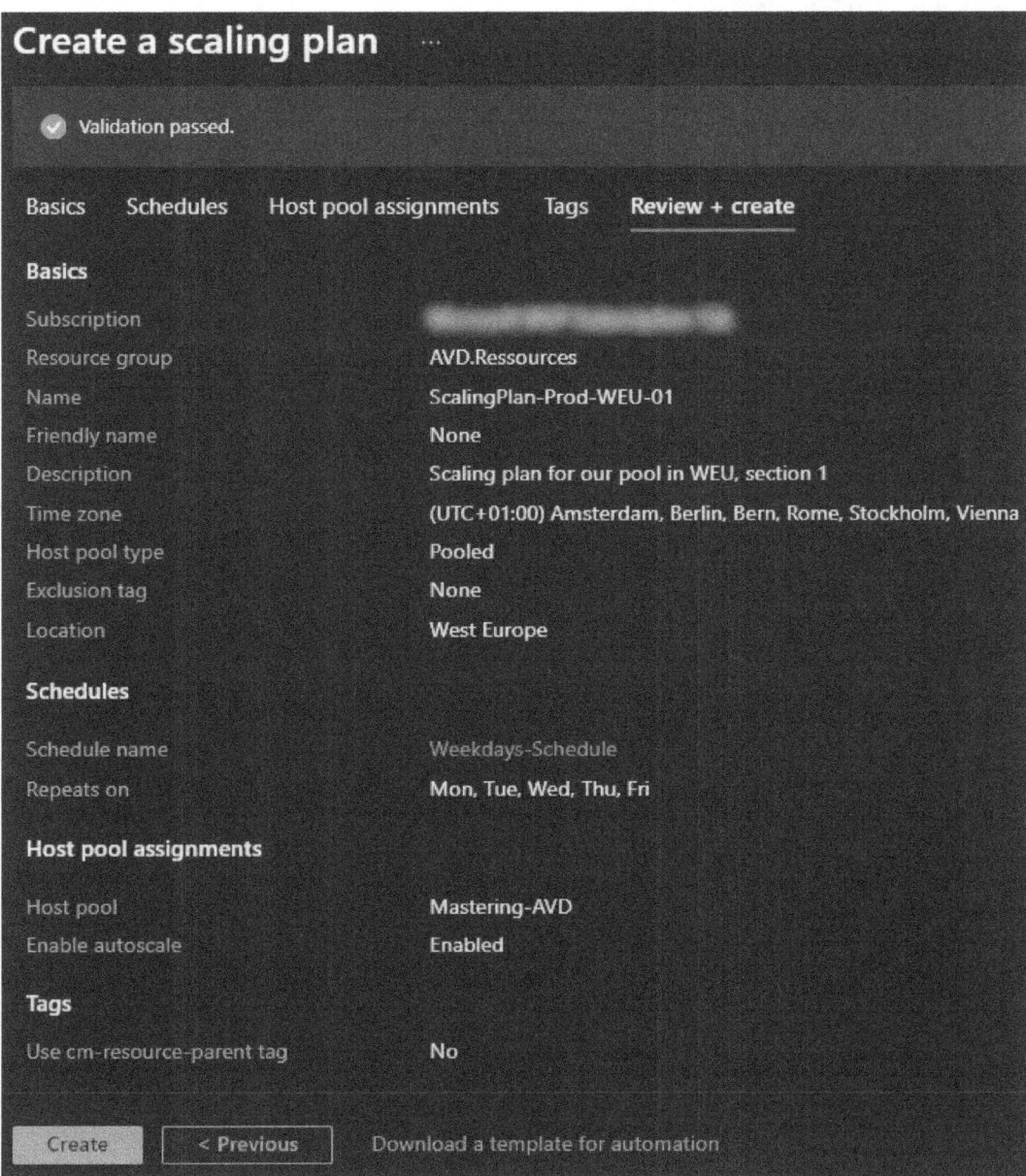

Figure 17.39 – The Review + create tab of the Create a pooled scaling plan page

There you have it – in this section, we looked at the built-in autoscale feature where you can configure scaling plans that meet your specific requirements for pooled host pools (multiuser).

Creating a personal scaling plan (assigned user)

Personal hosts can also be scaled by Azure. The scaling plans and schedules are different from the scaling of pooled hosts.

Let's get started:

1. Log in to the Azure portal.
2. Navigate to the **Azure Virtual Desktop** page.
3. Select **Scaling plans**:

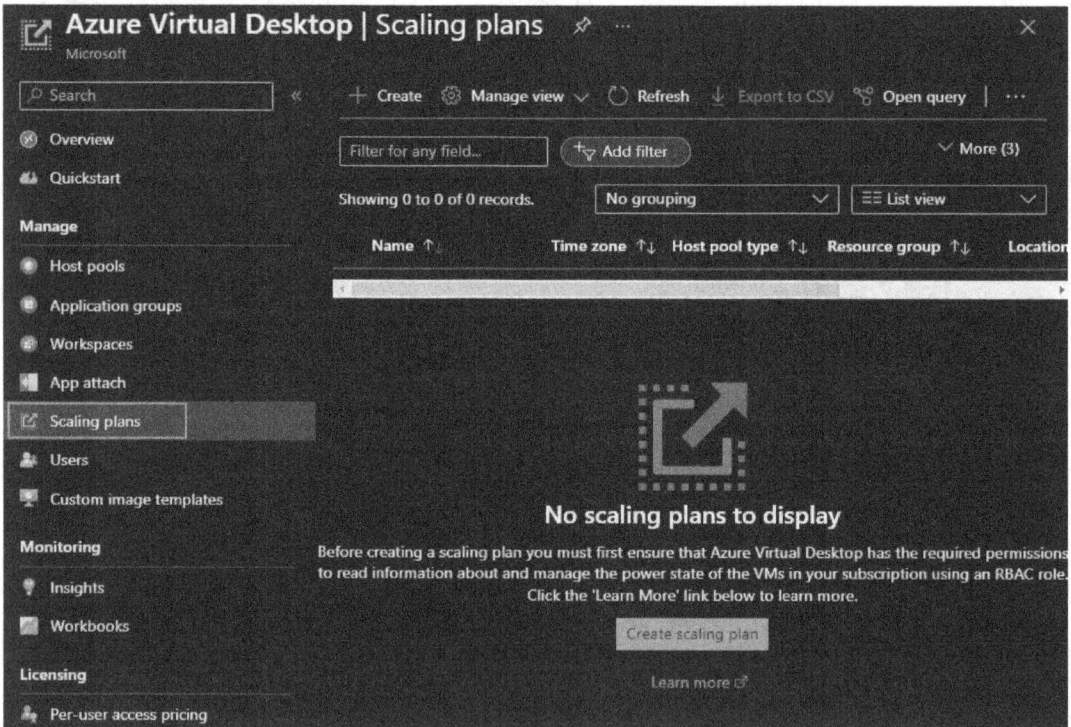

Figure 17.40 – The Scaling plans page in Azure Virtual Desktop

4. Within **Scaling plans**, click **Create**.
5. Enter the required resource group or select **Create new** if required.
6. Enter a name for the scaling plan.
7. Select the required Azure region for the scaling plan.
8. Select the required time zone for the scaling plan.
9. Select **Personal** in **Host pool type** to scale hosts in a multiuser environment.

10. Enter a tag for the VMs that you do not want to be included in scaling operations within the **Exclusion tag** section:

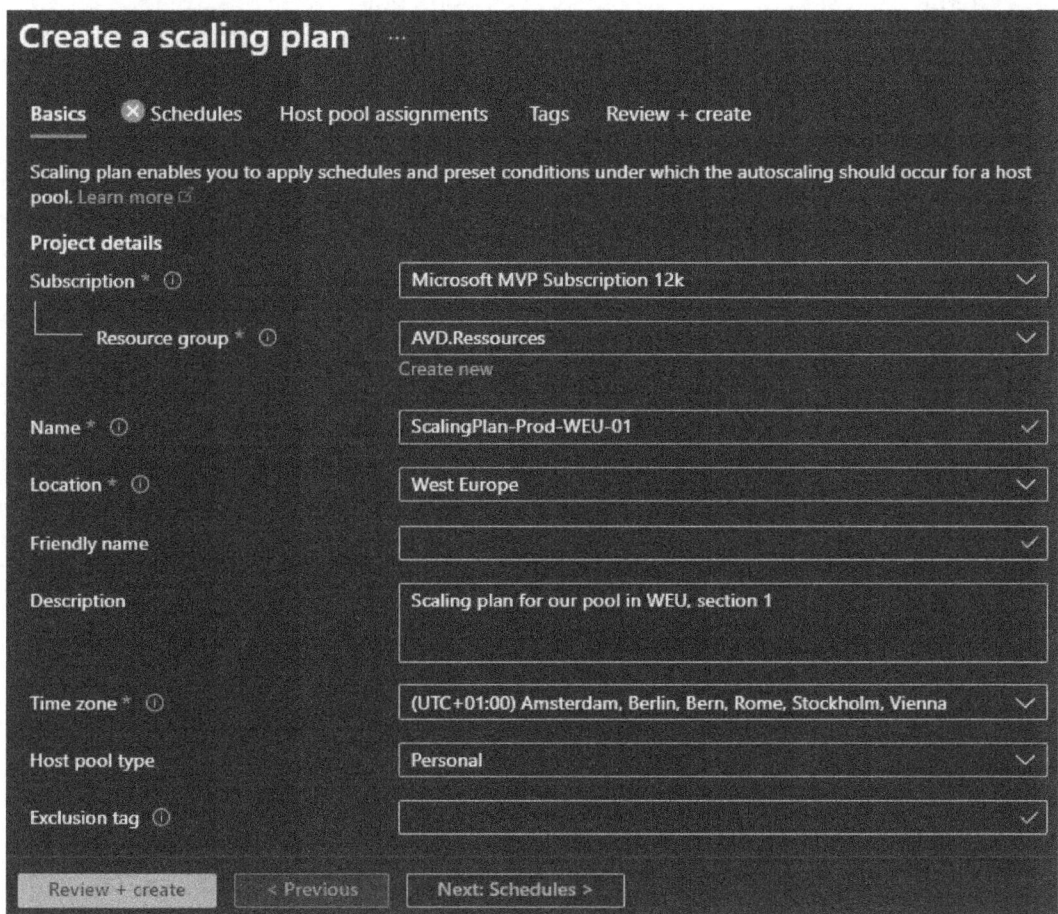

Figure 17.41 – Creating a personal scaling plan

11. Click **Next: Schedules**.

We'll now move on to look at the configuration of a schedule within a scaling plan.

Configuring a schedule

Schedules are essentially the configuration settings for autoscale. This includes the ramp-up and ramp-down modes, which will activate based on schedules throughout the day.

The following table details the different terms and meanings:

Term	Description
Start time	This is used to start preparing VMs for peak business hours or, as some refer to them, normal working hours.
Start VM on Connect	**Yes**: If the user starts a connection to its personal host, the host is started by Microsoft automatically.
VMs to start	You can configure, at the start of the ramp-up phase, whether Microsoft should start hosts in the pool automatically. There are three options: - **Assigned VMs only** - **Assigned and unassigned VMs** - **Don't turn VMs on at start time**
When disconnected for (min)	Performance of the selected action if the user session is disconnected for the defined number of minutes. Actions are selectable in the drop-down list: - **None** - **Shutdown the VM** - **Hibernate (preview) the VM if possible** Note: hibernation only works for specific images and VM sizes.
When logged off for (min)	Performance of the selected action if the user session is logged off for the defined number of minutes. Actions are selectable in the drop-down list: - **None** - **Shutdown the VM** - **Hibernate (preview) the VM if possible** Note: hibernation only works for specific images and VM sizes.
Peak hours	The start time for when the usage rate is expected to be the highest during the working day. This is also the same time that is used for the ramp-up phase.

Table 17.2 – Properties for personal schedules

To configure or change a schedule, follow these steps:

1. Within the **Schedules** tab, you can add a schedule. Go ahead and click **Add schedule**:

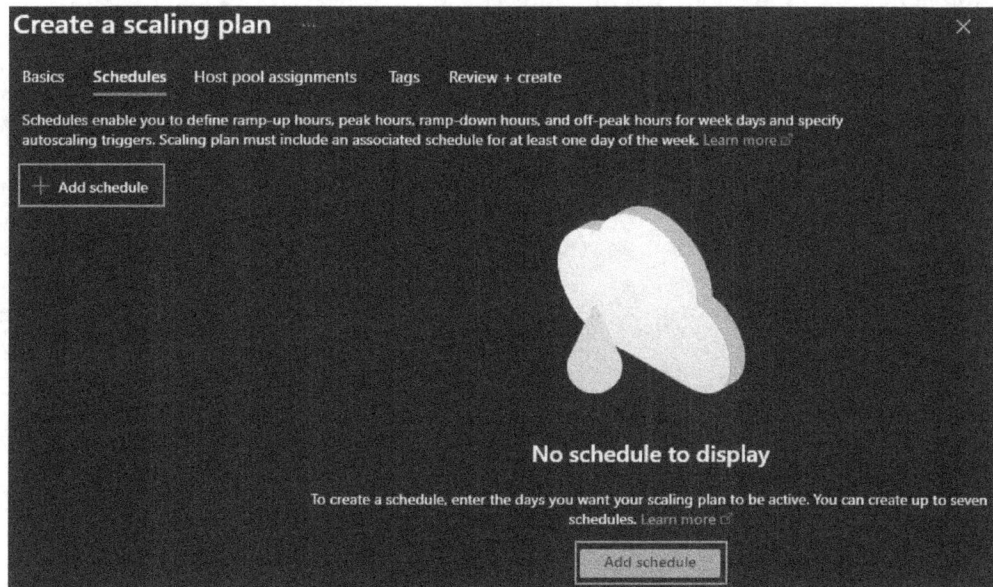

Figure 17.42 – The Schedules tab on the Create a scaling plan page

2. You will then see the **Add a schedule** blade appear with five tabs – **General**, **Ramp-up**, **Peak hours**, **Ramp-down**, and **Off-peak hours**:

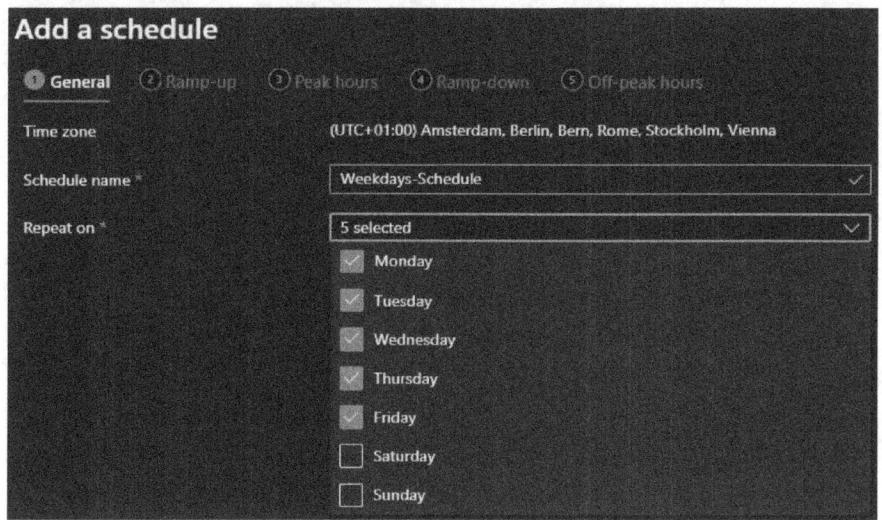

Figure 17.43 – The General tab on the Add a schedule blade

3. Enter the schedule name within the **General** tab and set the days you want the schedule to repeat. Then, click **Next**.

4. You can set **Start time**, **Start VM on Connect**, **VMs to start**, and **Disconnect** and **Log off settings** within the **Ramp-up** tab. Once you have configured the settings you want, click **Next**:

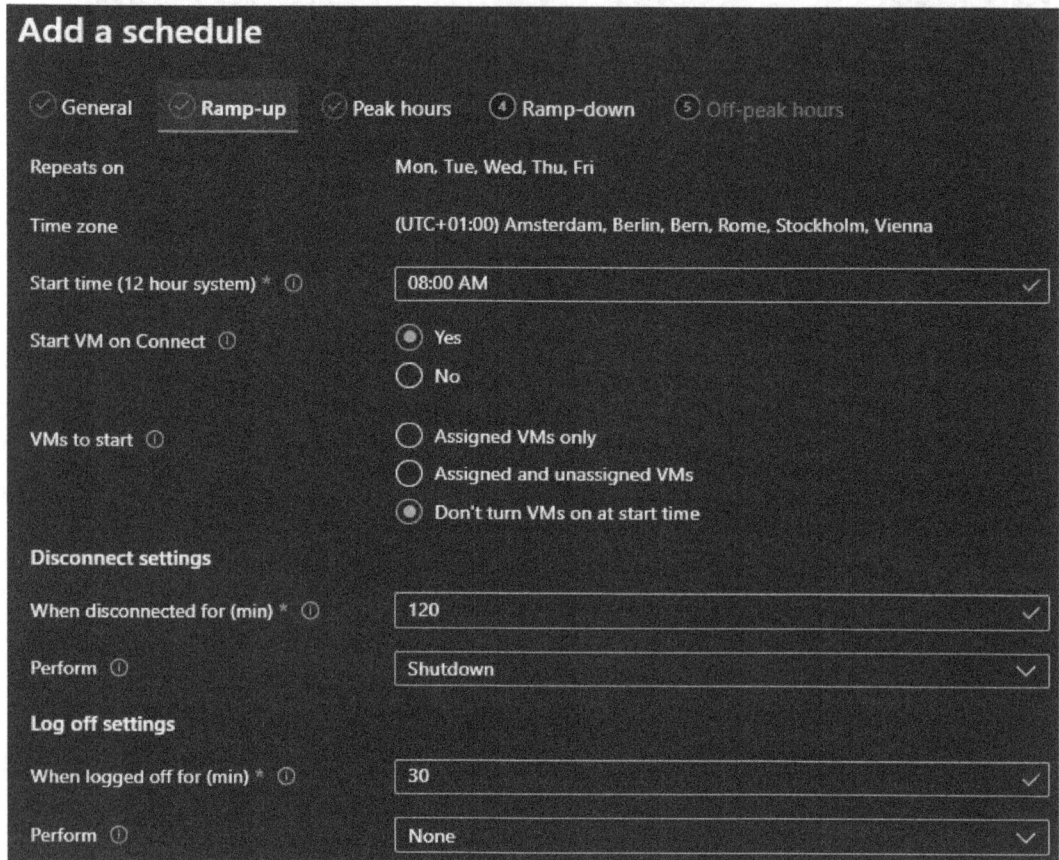

Figure 17.44 – The personal Ramp-up tab on the Add a schedule blade

5. You can then set **Start time**, **Start VM on Connect**, and **Disconnect** and **Log off settings** within the **Peak hours** tab. Once set, click **Next**:

> **Important note**
> When configuring **Start VM on Connect** within a scale plan schedule, you will override the original host pool configuration for **Start VM on Connect**.

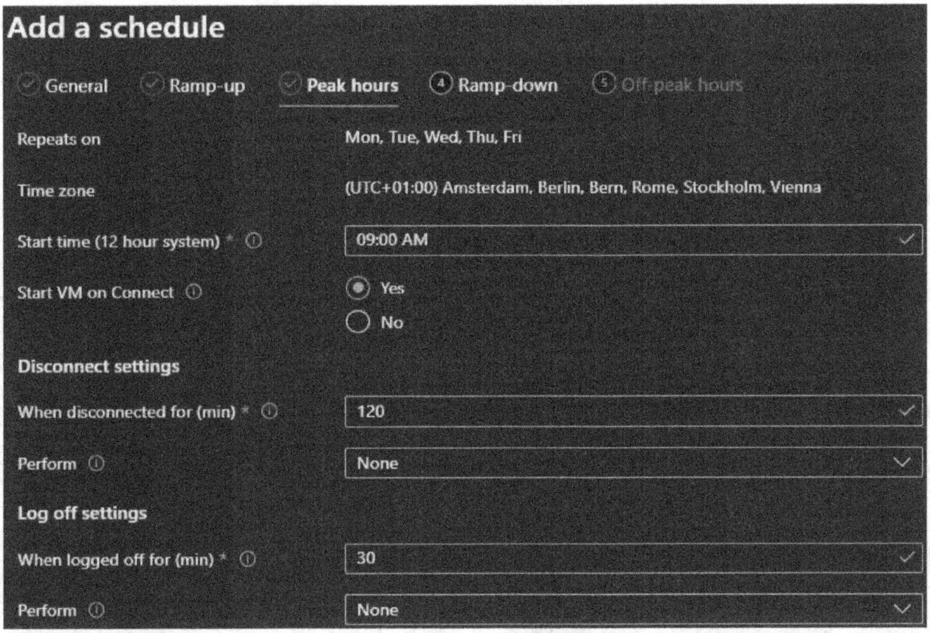

Figure 17.45 – The personal Peak hours tab on the Add a schedule blade

6. Within the **Ramp-down** tab, you can specify **Start time**, **Start VM on Connect**, and **Disconnect** and **Log off settings**. Once you have configured this tab, click **Next**:

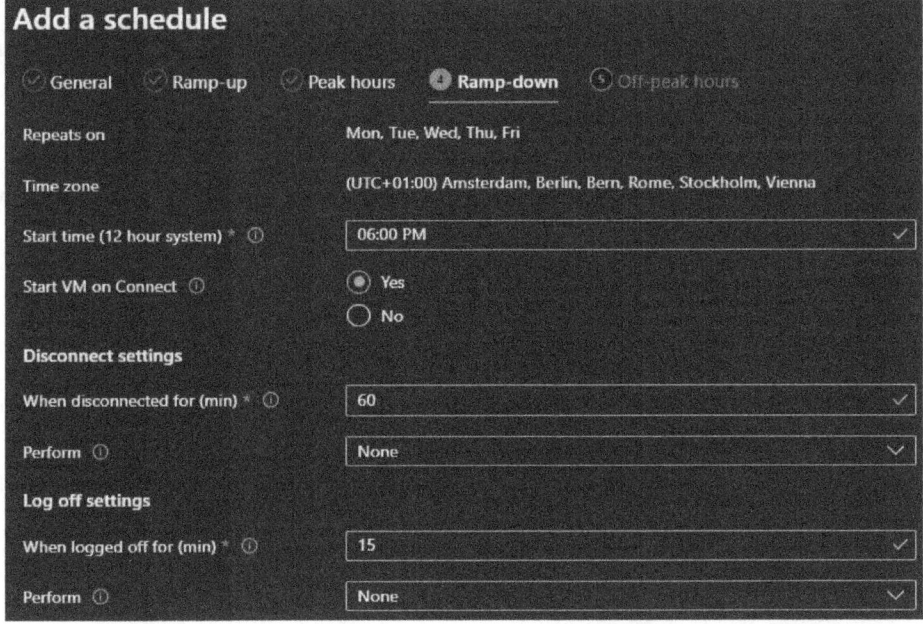

Figure 17.46 – The personal Ramp-down tab on the Add a schedule blade

7. The final tab on the **Add a schedule** page is **Off-peak** hours. You can set the time and the load balancing algorithm. Once configured, click **Add**.

Off-peak hours work in a similar way to peak hours. The difference is that off-peak hours start later:

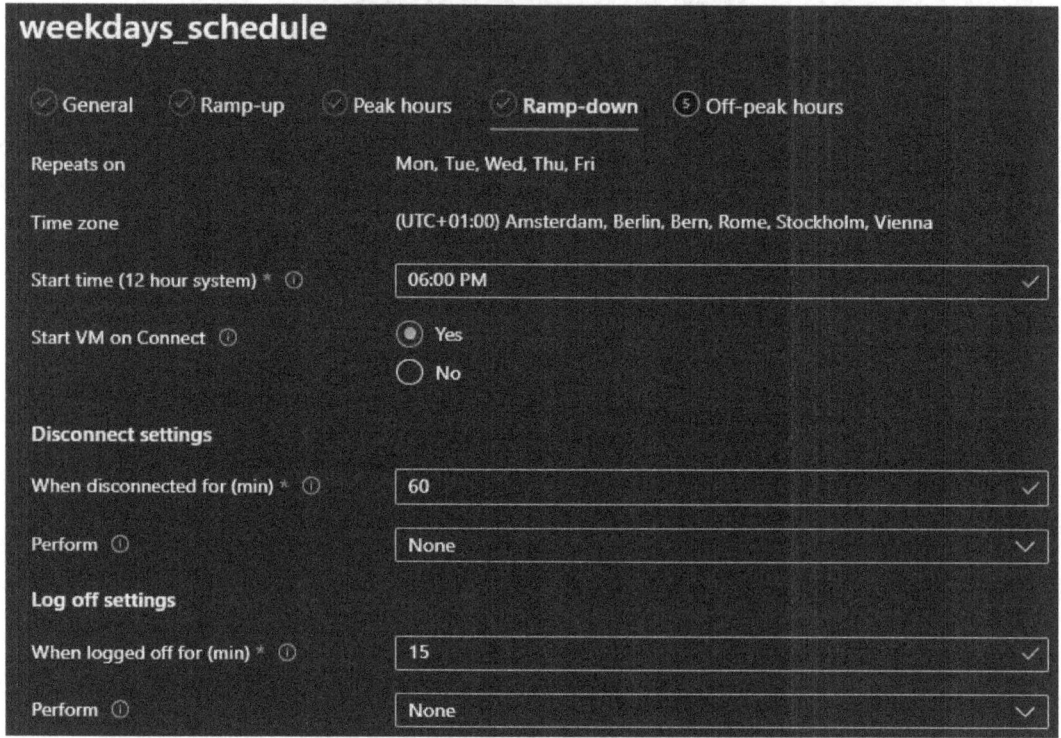

Figure 17.47 – Personal Off-peak hours tab on the Add a schedule blade

Click **Add** to add the schedule to the plan. Now that we have configured the schedule, we can now assign host pools to the scaling plan.

Assigning host pools

The following steps look at how to assign the host pool and complete the creation of the scaling plan:

1. Now that we have created the schedule, we can move on to the host pool assignments. This is where you would select the session host pools you want to assign to the scaling plan.

2. Select the required host pools within the **Host pool assignments** tab and ensure that **Enable autoscale** is checked:

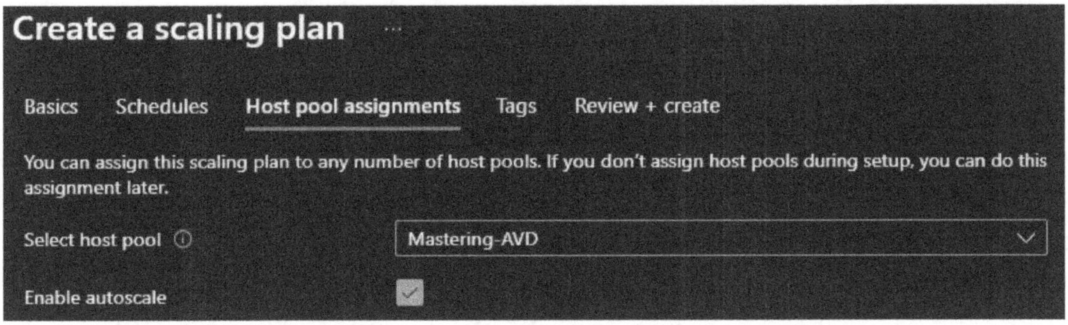

Figure 17.48 – The Host pool assignments tab on the Create a scaling plan page

3. Under the **Tags** tab, specify any tags you may want to set, then click **Next**.
4. Review your settings and click **Create**.

This concludes the setup process for creating a scaling plan in Azure Virtual Desktop.

There you have it – in this section, we looked at the built-in autoscale feature where you can configure scaling plans that meet your specific requirements for personal host pools (assigned users).

For the latest updates and information on the autoscale solution, visit the following Microsoft documentation link: `https://docs.microsoft.com/azure/virtual-desktop/autoscale-scaling-plan`.

Summary

In this chapter, we started by configuring an Azure Automation account and then looked at automating Azure Virtual Desktop tasks such as logging multiple sessions of session hosts within all host pools as an example. Finally, we ran through the setup of a scaling plan, configuring a schedule, and assigning host pools to a scaling plan.

In the next chapter, we will look at the topic of monitoring and managing the performance and health of an Azure Virtual Desktop environment.

Questions

Here are a few questions to test your understanding of this chapter:

1. You want to automate the logging off of multiple Azure Virtual Desktop users. Which Azure service would you use to achieve this objective?

    An Azure Automation runbook

2. Before you can create scaling plans, what must you configure?

   Giving Microsoft permission to start and stop VMs

3. Why can you no longer use a run-as account within Azure Automation?

   They are no longer supported and available.

18
Monitoring and Managing Performance and Health

In this chapter, we will look at the monitoring, performance, health, and management of **Azure Virtual Desktop** (**AVD**). Monitoring your AVD environment helps you spot issues and optimize the configurations for a good **user experience** (**UX**). In this chapter, we will look at the three key areas of monitoring, performance, and health. The key topics we'll cover in this chapter are outlined here:

- Configuring Azure Monitor for AVD
- Using AVD Insights
- Setting up alerts using alert rules
- Introduction to **Kusto Query Language** (**KQL**)
- Using Azure Advisor for AVD

Configuring Azure Monitor for AVD

Azure Monitor for AVD is essentially a built-in dashboard built using Azure Monitor workbooks. This helps an **information technology** (**IT**) administrator understand the current state of the environment and enables the troubleshooting of some of the issues that may occur within an AVD environment.

Before you can get started with Azure Monitor for AVD, you need to make sure of the following:

- One Log Analytics workspace is configured
- You have enabled the data collection of your AVD environment

Once you have met the criteria to proceed, we can move on to look at creating a Log Analytics workspace.

Creating a Log Analytics workspace

The first thing to do is deploy Log Analytics to configure it to collect data from AVD. To do this, you will first need to open the **Log Analytics workspaces** page using the Azure search bar, as shown in the following screenshot:

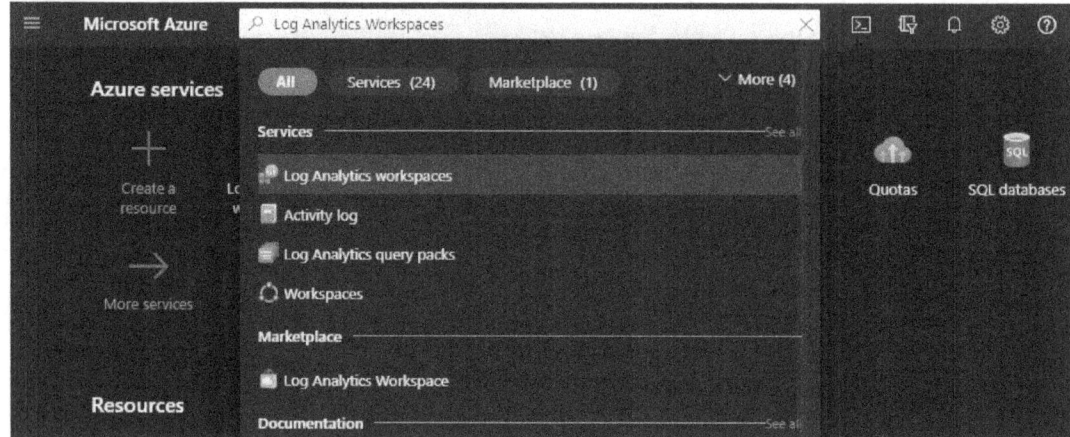

Figure 18.1 – Search bar and the Log Analytics workspaces service

Once we have opened the **Log Analytics workspaces** page, we can progress to creating a new Log Analytics workspace. Here are the steps to do this:

1. Within the **Log Analytics workspaces** page, click **Create**.
2. Select a **Resource group** type within the **Create Log Analytics workspace** page, enter a **Name** value for the instance, and choose an Azure **Region** type. Once complete, click **Review + Create**. The following screenshot provides an overview of this:

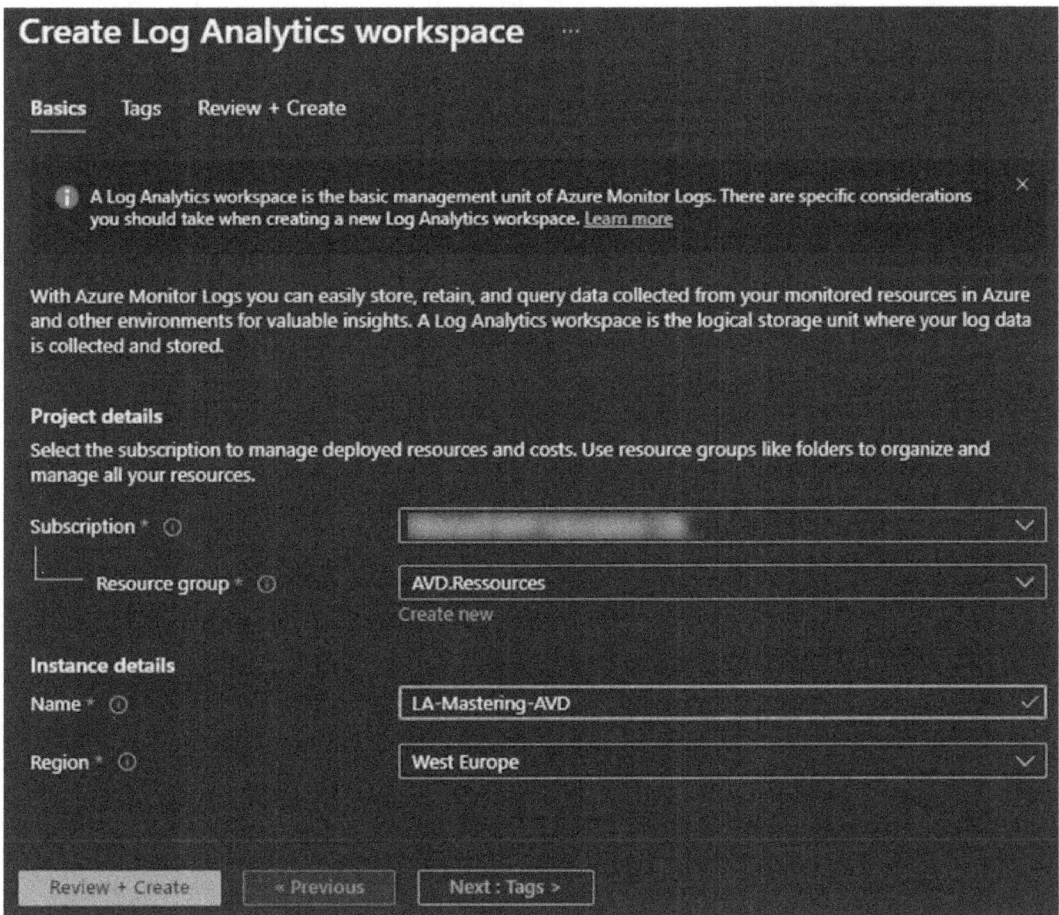

Figure 18.2 – Create Log Analytics workspace page

3. Review the configuration and click **Create**, as illustrated in the following screenshot:

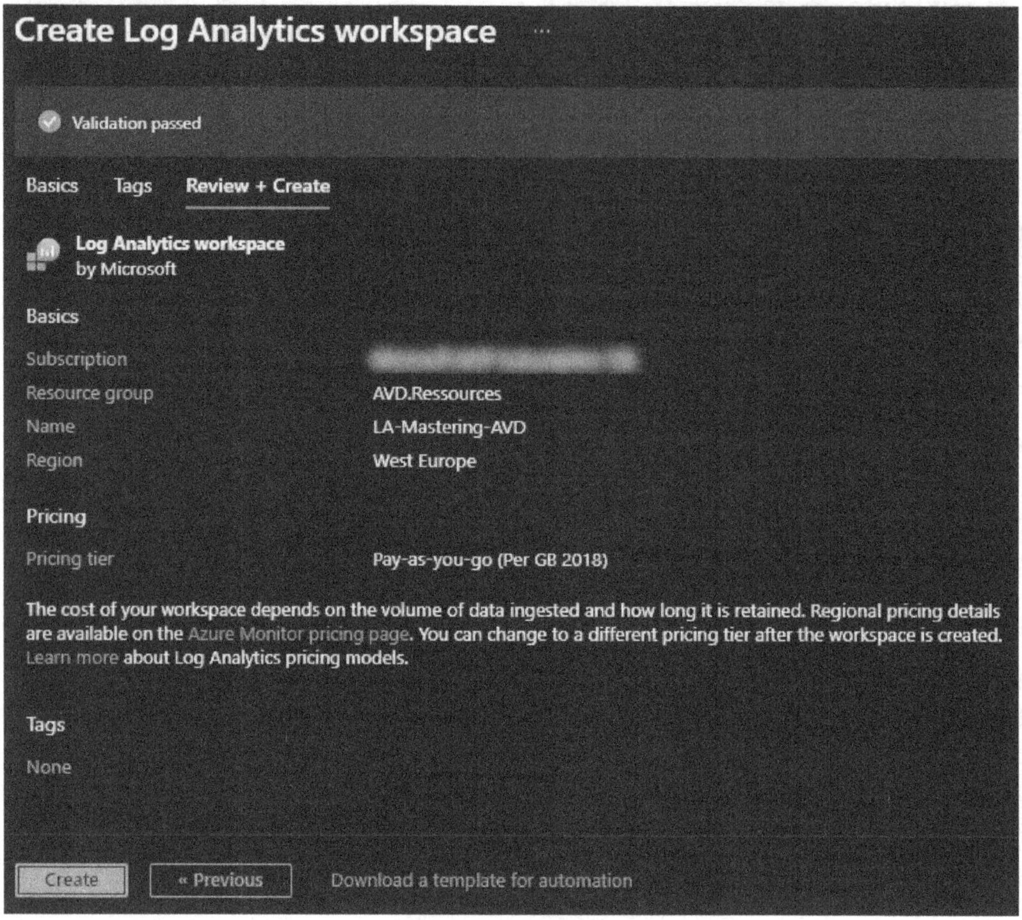

Monitoring and Managing Performance and Health

Figure 18.3 – The Review + Create tab on the Create Log Analytics workspace page

Once the deployment has finished, you should see the Log Analytics workspace you created on the **Log Analytics workspaces** page, as shown in the following screenshot:

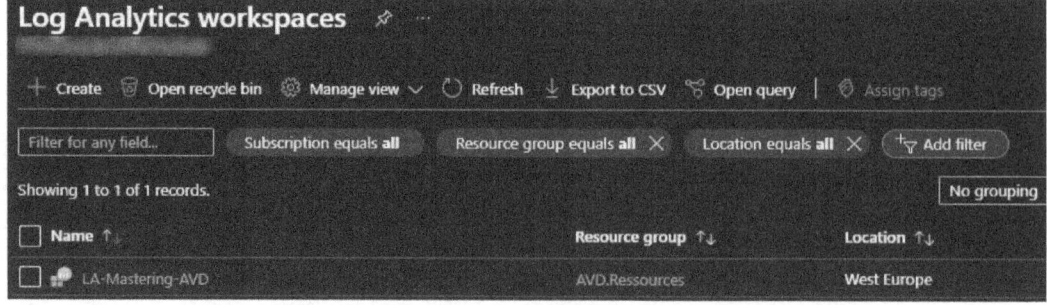

Figure 18.4 – Created Log Analytics workspace, LA-Mastering-AVD

> **Important note**
> The default pricing tier for Log Analytics will be configured, as shown in the preceding screenshot. You will not incur any charges until you collect sufficient amounts of data. You can cap data using the **Daily cap** feature located under **Usage and estimated costs** in the **General** menu section.

The following screenshot shows where the **Daily cap** button is located:

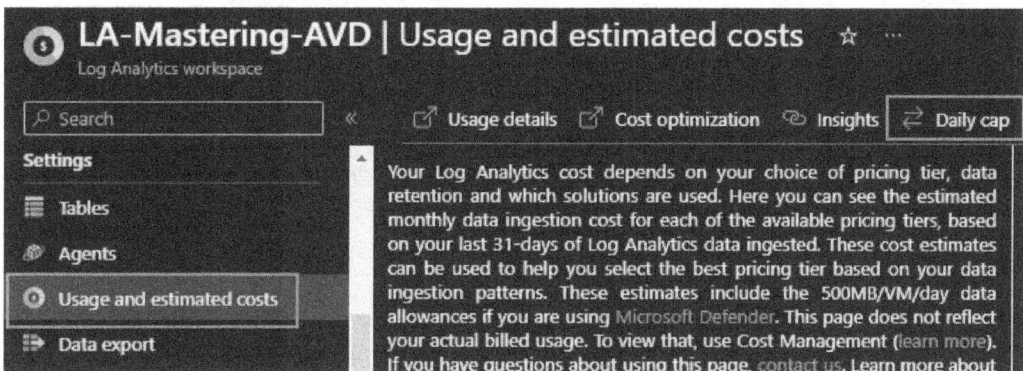

Figure 18.5 – The Usage and estimated costs menu within a Log Analytics workspace

The following screenshot shows the **Daily cap** blade, which can be used to control the daily ingestion of data:

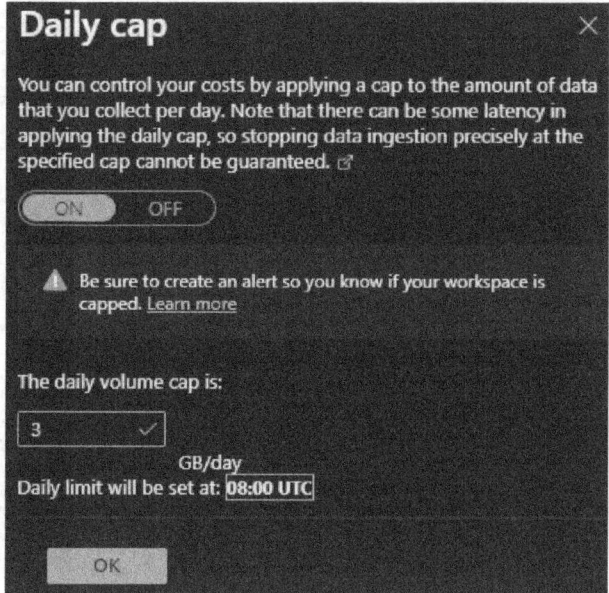

Figure 18.6 – The Daily cap blade

Now that we have created our Log Analytics workspace, we can proceed with configuring monitoring in AVD.

Configuring the monitoring of AVD

In this section, we will take a look at setting up the required monitoring components for AVD. To do this, we will navigate to the **Azure Virtual Desktop** page and then click **Insights** under the **Monitoring** section.

> **Important note**
> Microsoft will deprecate Log Analytics Agent on August 31st, 2024. Therefore, we are using the next-generation *Azure Monitoring Agent* for VM-related monitoring. The monitoring page on a host pool will show two different items: **Insights (Legacy)** for the old Log Analytics Agent and **Insights** for *Azure Monitoring Agent*-related data.

There are multiple different ways to configure Azure Monitor for AVD. In this example, we will use **Insights** on the **Host Pool** page. However, we first need to configure the workbook for AVD, as follows:

1. To get started, click **Open Configuration Workbook**, as highlighted in the following screenshot:

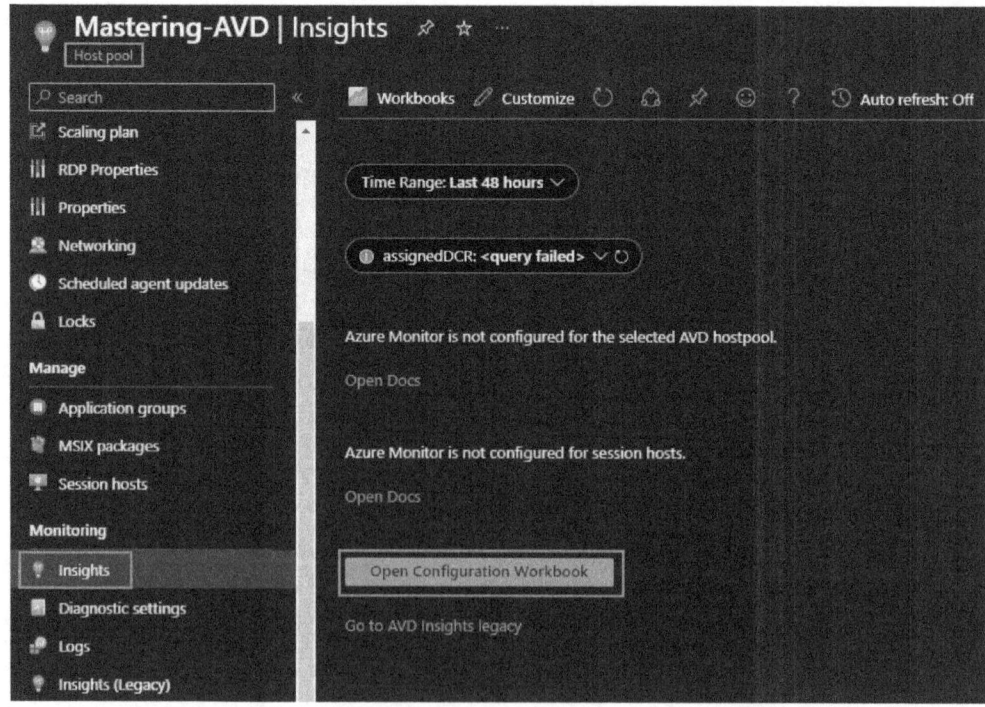

Figure 18.7 – AVD Insights workspace not configured

2. You will now see the **CheckAMAConfiguration** heading, which has three tabs: **Resources diagnostic settings**, **Session host data settings**, and **Data Generated**. You need to ensure that you have configured all the required components within **Resources diagnostic settings** and **Session host data settings**.

3. Select the required Log Analytics workspace and then click **Configure host pool**, as shown in the following screenshot:

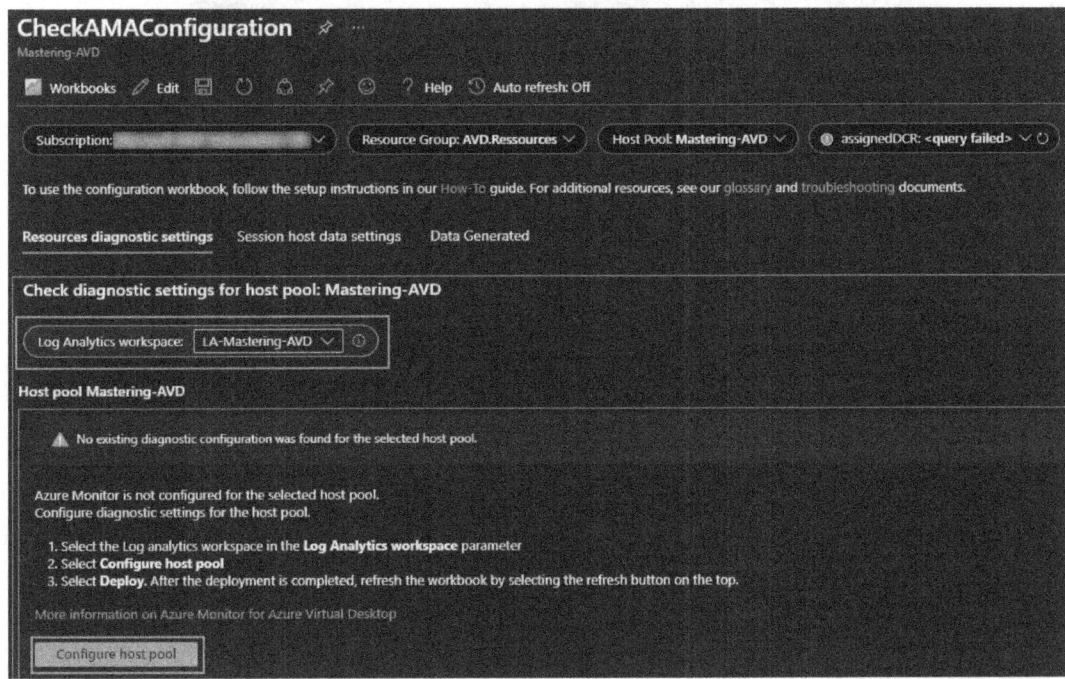

Figure 18.8 – The CheckAMAConfiguration workspace and the Configure host pool button highlighted

4. Once you have clicked the **Configure host pool** button, you will see the **Deploy Template** page appear, which is used to configure the host pool diagnostic settings.

5. Click **Deploy**, as shown in the following screenshot, and wait for the deployment to finish:

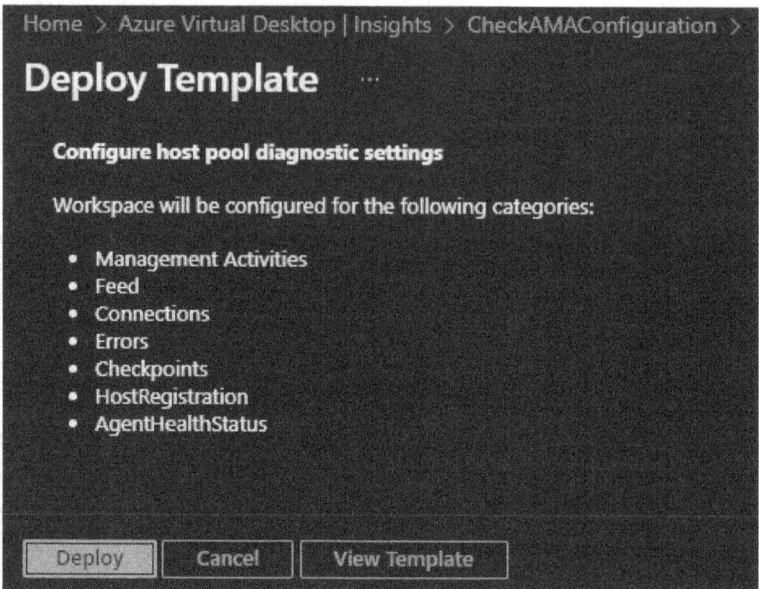

Figure 18.9 – The Deploy Template page for configuring host pool diagnostic settings

6. Once deployed, wait a while and use the **Refresh** icon to see that the host pool is configured. Scroll down and do the same for the workspace highlighted in the screenshot:

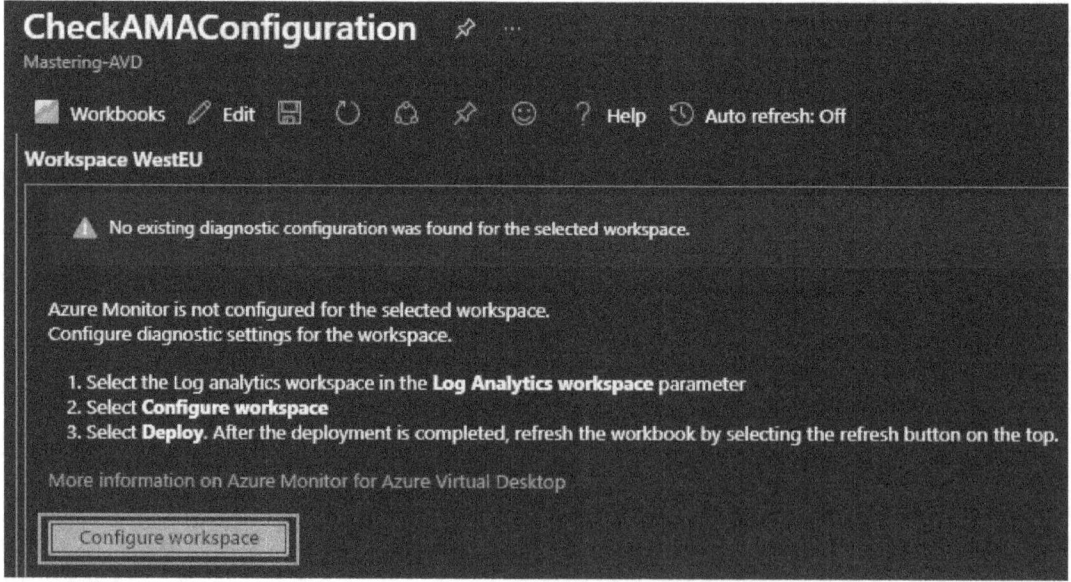

Figure 18.10 – The CheckAMAConfiguration workspace and the Configure workspace button highlighted

7. Click **Deploy** and wait for the deployment to finish.
8. Once you have finished configuring the diagnostics for both host pools and the workspace, we can proceed with **Session host data settings**.
9. Create a data collection rule. Azure Monitoring Agent uses the data collection rule to target a Log Analytics workspace and to get the configuration about what data should be collected and stored. That is different from the deprecated Log Analytics Agent. Data collection rules can be reused for other host pools.
10. Select the Log Analytics workspace in **Workspace destination** and a target resource group for the data collection rule, as shown in the screenshot:

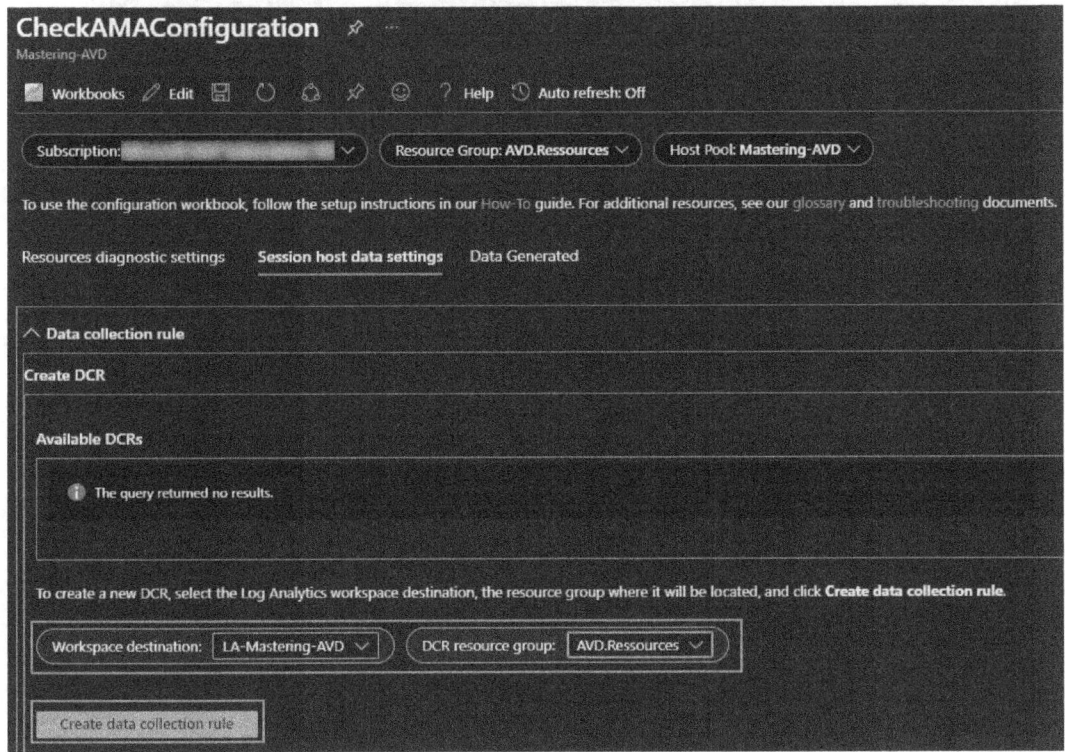

Figure 18.11 – Creating a data collection rule

11. Click **Deploy** and wait for the deployment to finish.
12. When the deployment has finished, we can select the data collection rule with the name **microsoft-avdi-westeurope** (the name can be different) and click on **Deploy Association**, as shown in the screenshot:

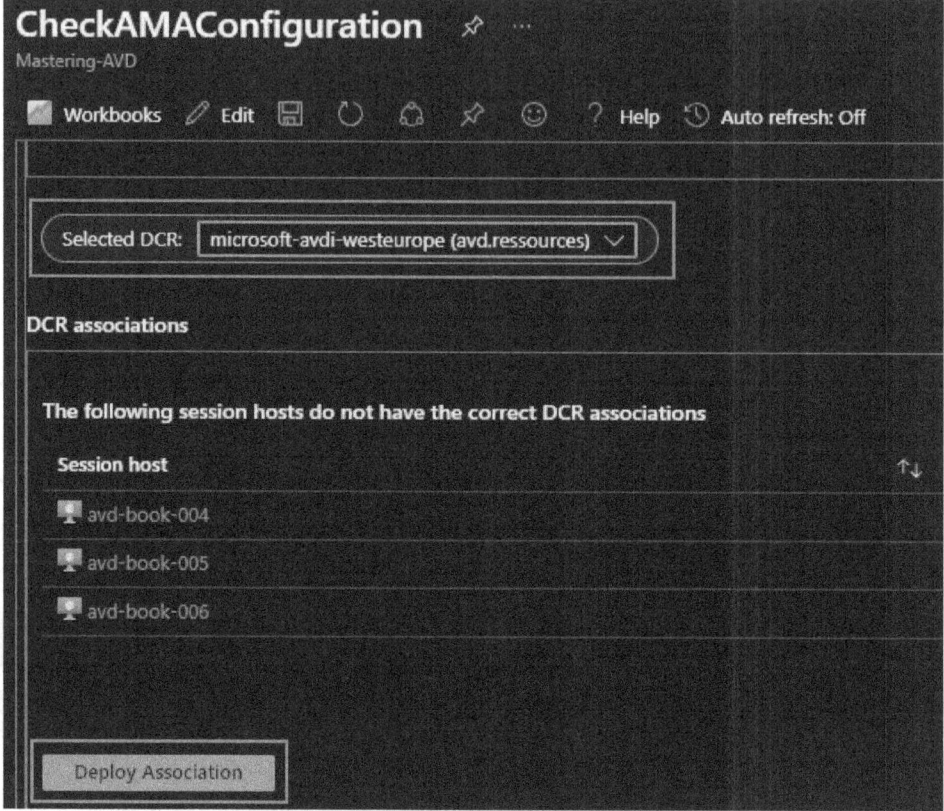

Figure 18.12 – Associating the existing hosts

13. Once you have clicked **Deploy Association**, you will see the **Deploy Template** page appear. You can see this page in the following screenshot:

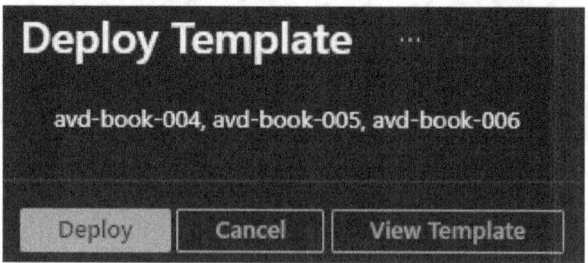

Figure 18.13 – Deploying the association

14. In the next step, we add the Azure Monitoring Agent extension in **Session hosts missing Azure Monitor extension** to our running hosts, as illustrated in the following screenshot:

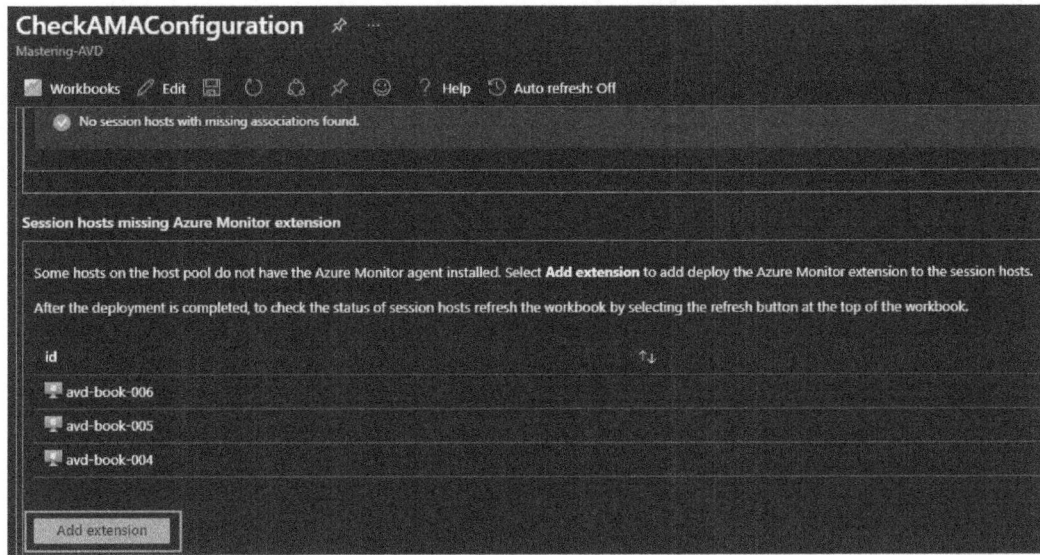

Figure 18.14 – Adding the extension to the existing hosts

15. Once you have clicked **Add extension**, you will see the **Deploy Template** page appear. You can see this page in the following screenshot:

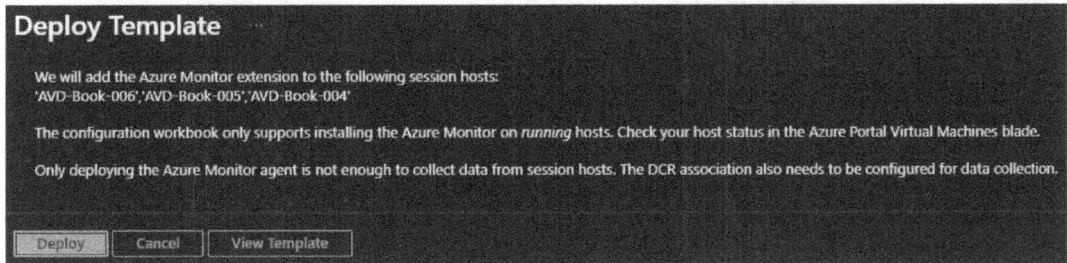

Figure 18.15 – The Deploy Template page for adding the extension to the running hosts

Once deployed, you can then progress with configuring performance counters.

16. We must add a managed identity to the hosts to finalize the Azure Monitoring Agent installation. In the portal, a system-managed identity will be created. Keep in mind that Microsoft recommends using a **User assigned** managed identity for large-scale deployments (>1000 hosts). We can add the identity directly in the configuration workbook shown in the following screenshot:

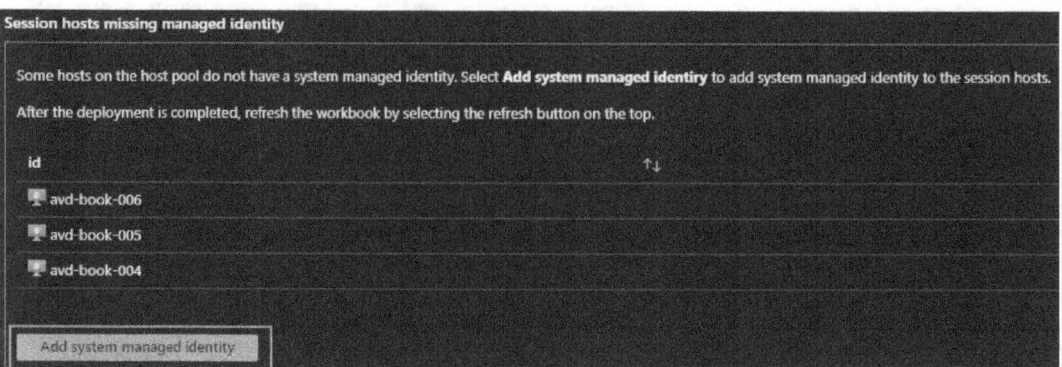

Figure 18.16 – Adding a system-managed identity

17. Once you have clicked **Add system managed identity**, you will see the **Deploy Template** page appear. You can see this page in the following screenshot:

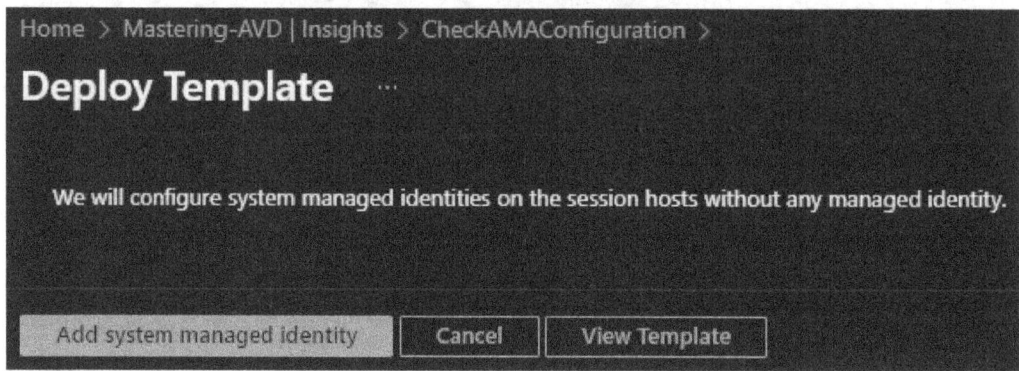

Figure 18.17 – The Deploy Template page for adding the system-managed identity

Once deployed, you can then progress with configuring performance counters.

Configuring performance counters and event logs

For the AVD dashboard to display the correct information, we need to ensure that all the correct counters are enabled for each session host. While we created the data collection rule from the insights of the host pool, performance counters and event logs are automatically configured. You can modify or validate the settings on the data collection rule. Navigate to **Data collection rules** in the Azure portal, as shown in the screenshot:

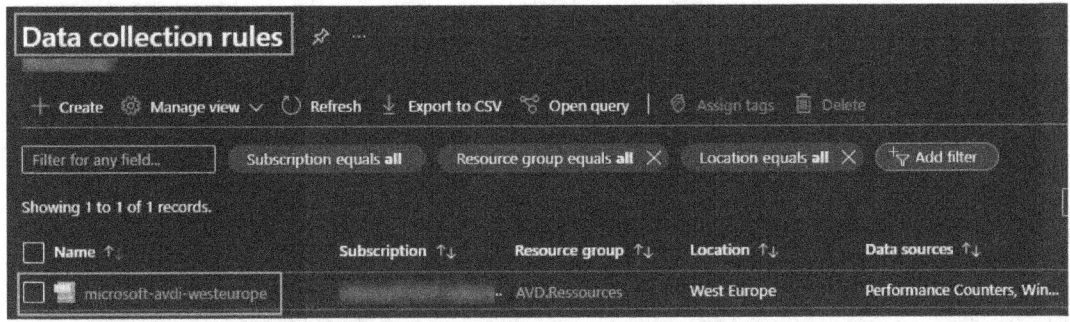

Figure 18.18 – Data collection rules

Click on the collection rule and navigate to **Data sources**. Performance counters and Windows event logs are configured. Clicking on **Performance Counters | Custom** shows the performance counters and intervals. Please note that only checked performance counters are used, as shown in the screenshot:

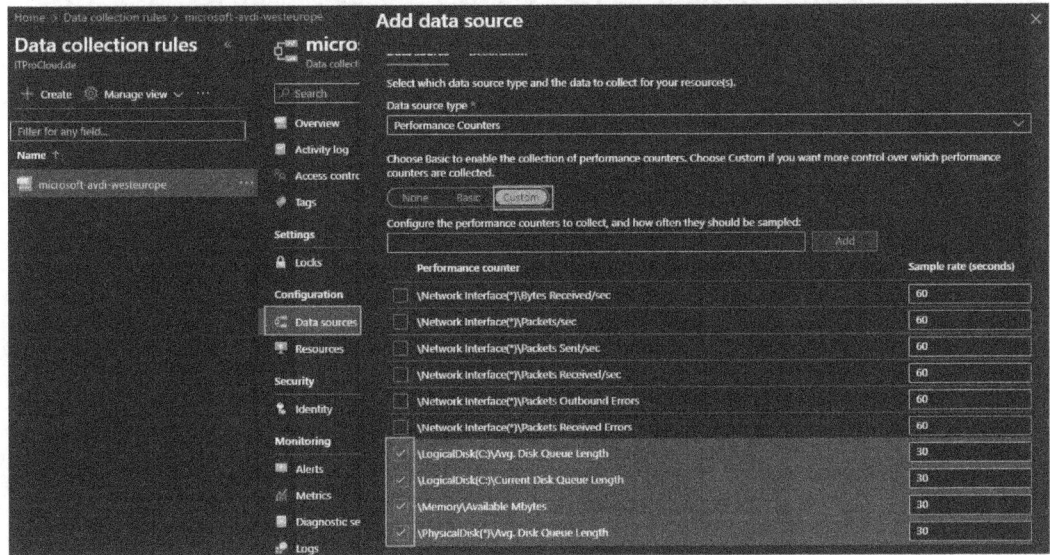

Figure 18.19 – Configuration of the created collection rule – Performance counters

The configuration of the event logs is similar. Switch to **Custom** to show the effective configuration of collecting event logs. In this case, AVD and FSLogix-related events are collected. See the following screenshot:

Monitoring and Managing Performance and Health

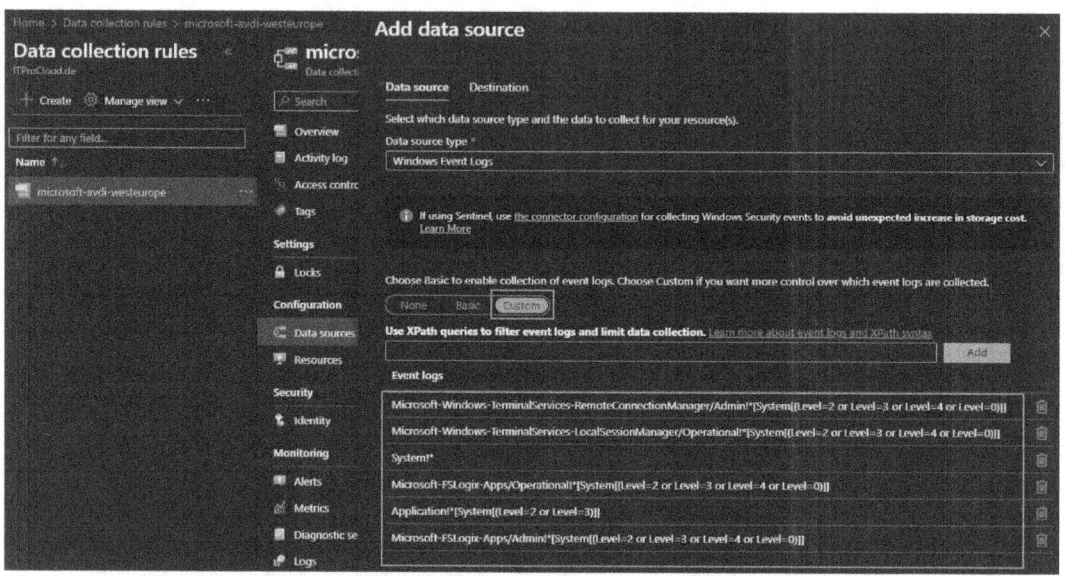

Figure 18.20 – Configuration of the created collection rule – Event logs

You can now click on the **Data Generated** tab and validate whether hosts are sending information, as illustrated in the following screenshot:

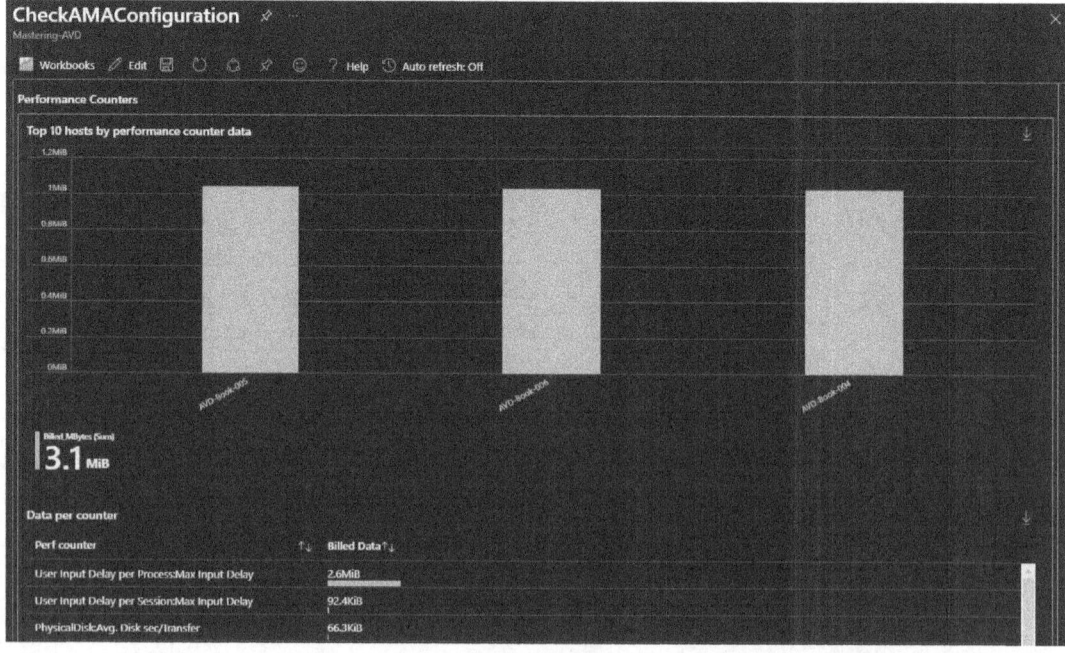

Figure 18.21 – Hosts are sending data to the workspace using the data collection rule

Now that we have finished configuring Azure Monitor for AVD, we can take a look at the different areas of the monitoring workspace for AVD.

Using Insights

In this section, we will take a look at how to use Azure Monitor to spot issues and view the current state of an AVD environment.

> **Important note**
> Log Analytics is essentially the data used when visualizing ingested AVD data. AVD Insights is a templated dashboard that uses the configured log analytics and counters to provide IT administrators with information about the organization's AVD environment.

Differences between AVD Insights and host pool insights

There are two types of insights at the time of writing this book: one is an item in the menu of AVD, and the other is a separate menu item at a specific host pool.

The AVD insights can show data for multiple host pools but are currently missing the host performance and host diagnostic view. This will probably be fixed and is a result of the deprecation of Azure Monitoring Agent.

The insights item on a host pool shows any kind of monitoring, including the host-related data captured by Azure Monitoring Agent. The view is limited to the host pool itself.

> **Important note**
> While we want to work with the complete monitoring data, including performance counters and event logs, we will start the insights on the host pool level.

Using the host pool insights

Within the host pool **Insights** page, there are nine tabs, as follows:

- **Overview**
- **Connection Diagnostics**
- **Connection Performance**
- **Host Diagnostics**
- **Host Performance**
- **Users**

- Utilization
- Clients
- Alerts

The **Overview** tab, as highlighted in the following screenshot, provides an overview of the host pool, which looks at **Host pool details**, **Connection Diagnostics**, **Host Performance**, **Utilization**, and **Alerts**. We can look at the specific workspace tabs to drill down for more detailed information:

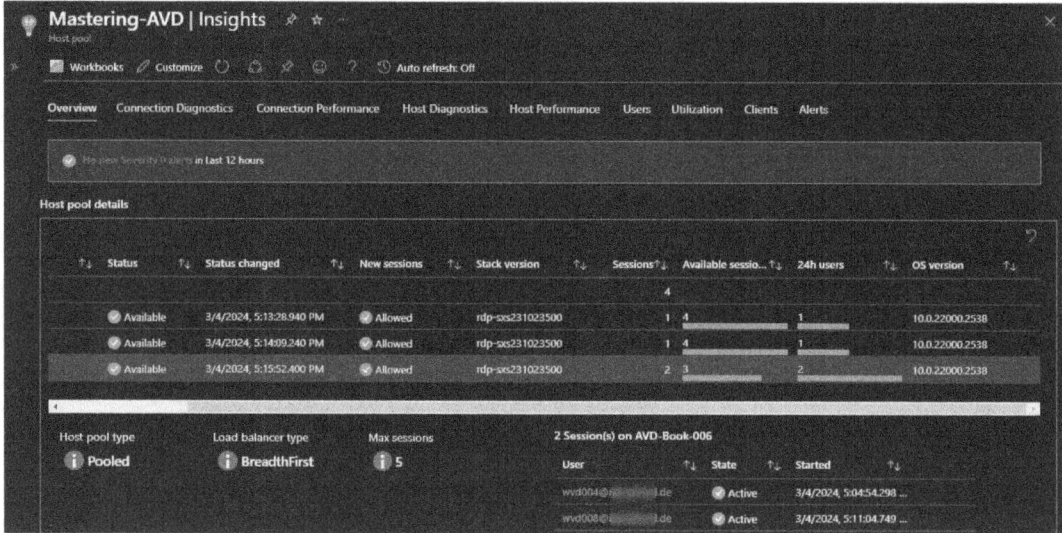

Figure 18.22 – The Overview tab of the Insights workbook

The **Connection Diagnostics** tab provides details on connections. This allows you, as the IT administrator, to review any alerts and investigate any problems. In this example, I noticed there was an **FSLogix** error relating to the storage path not being found, which shows up in the **Connection Diagnostics** tab:

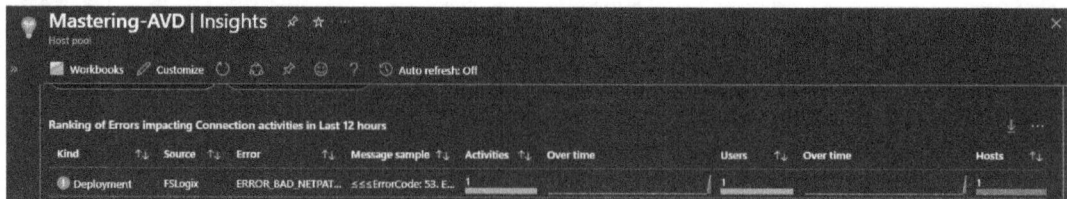

Figure 18.23 – List of errors impacting connections

The following screenshot shows a drill-down of errors that provides more granularity:

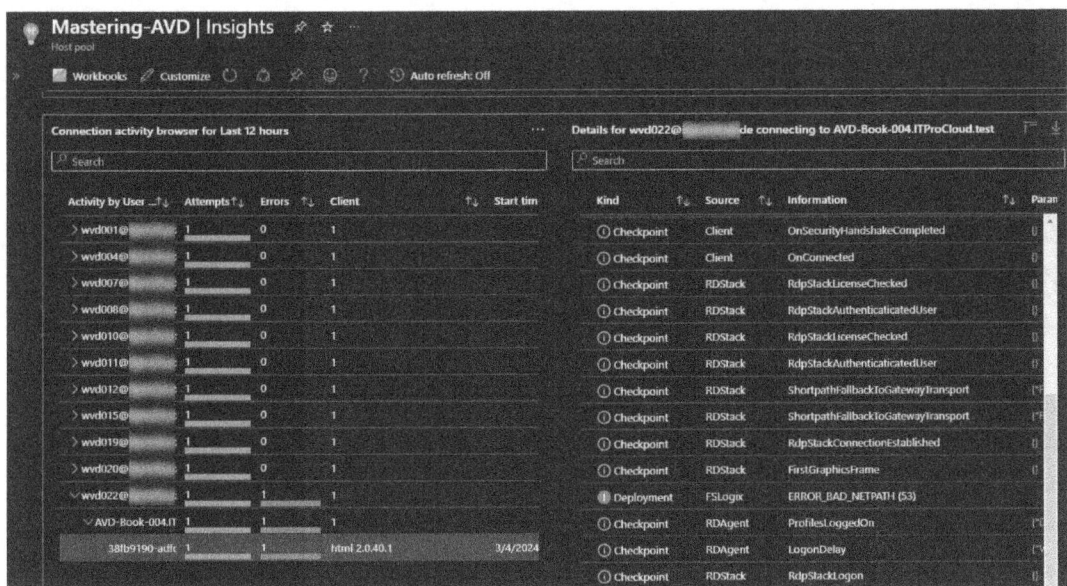

Figure 18.24 – A more detailed breakdown of connection events and associated errors

The **Connection Performance** tab, shown in the following screenshot, provides information on new and existing sessions, which enables IT administrators to review slow sign-in times and diagnose possible login issues:

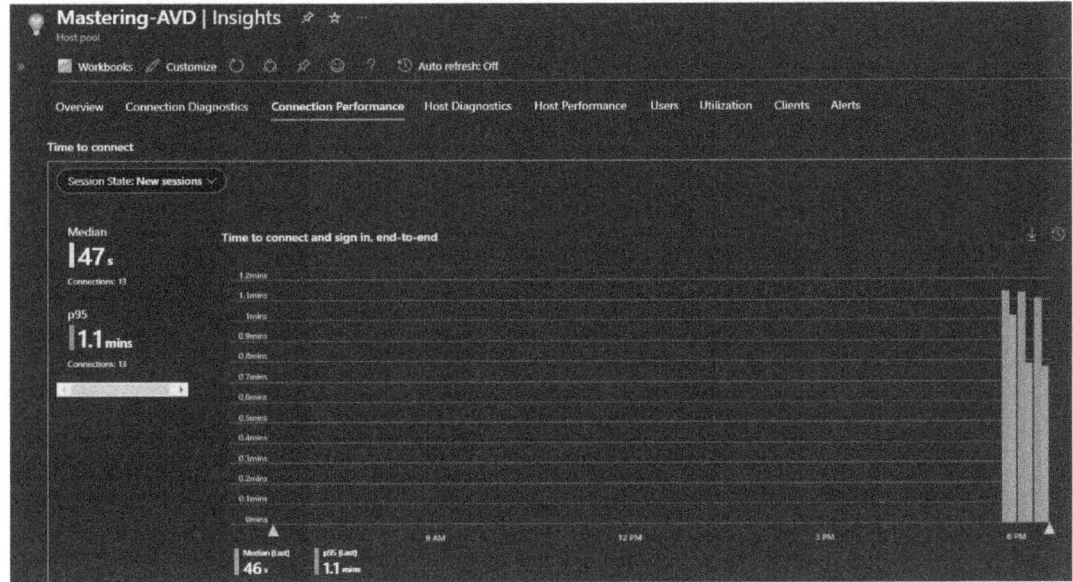

Figure 18.25 – The Connection Performance tab helps understand any performance/communication issues relating to the user or host

The **Host Diagnostics** tab, shown in the following screenshot, provides information on the host pool configuration, performance counters, events, and any errors. This can help you pinpoint any issues related to a session host and monitor the **central processing unit** (**CPU**) and memory usage:

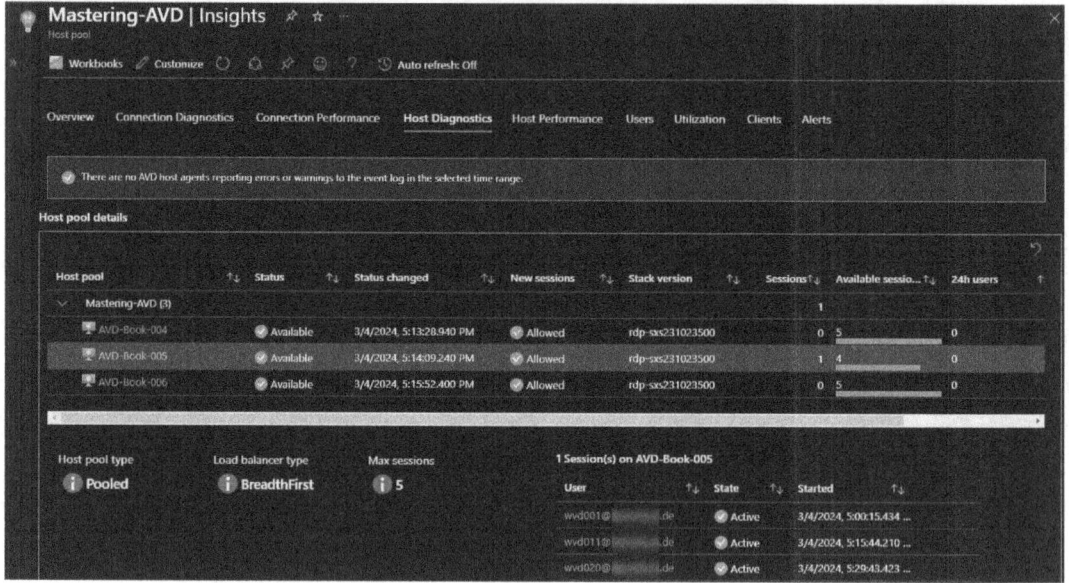

Figure 18.26 – The Host Diagnostics tab

The **Host Performance** tab, shown in the following screenshot, provides insights into the overall performance and enables IT administrators to drill down into possible issues with processes, CPU, memory, and disk queuing. This helps to identify the host saturation and any applications that may be consuming large resources:

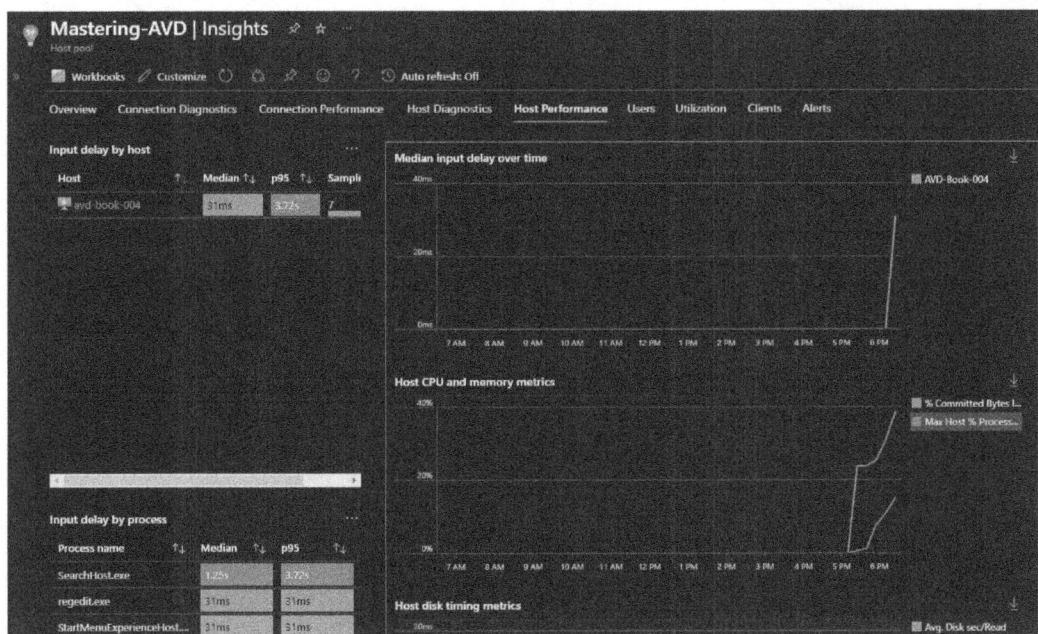

Figure 18.27 – The Host Performance tab

The **Users** tab, shown in the following screenshot, provides a detailed output on user performance and any errors relating to a specific user. This tab helps to identify specific user issues and allows IT administrators to quickly understand the client device in use, client version, and any errors during connectivity:

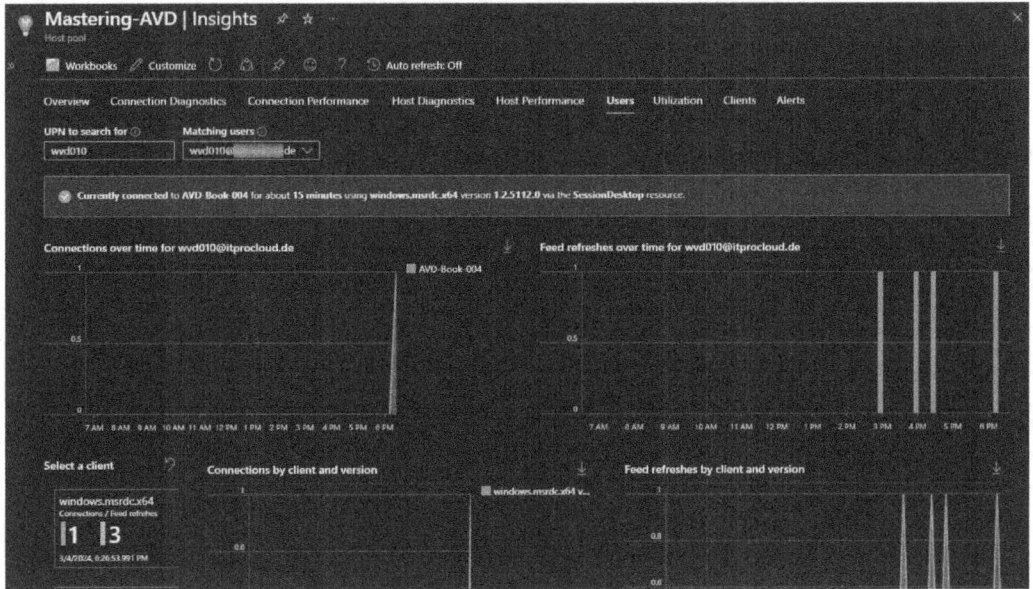

Figure 18.28 – The Users tab, which details user connection and client details

The **Utilization** tab shows the current utilization metrics of your AVD environment. This is particularly useful for capacity management and understanding any potential performance degradation of the environment.

The **Clients** tab, illustrated in the following screenshot, shows connections and feed refreshes and the version of the client in use. IT administrators can get a full picture of which clients are being used within the AVD environment from here:

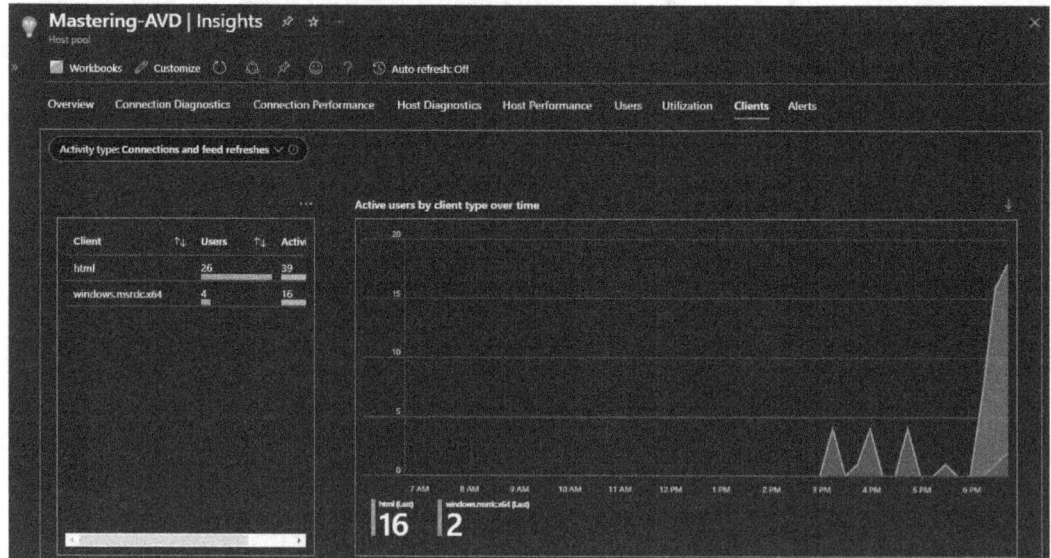

Figure 18.29 – The Clients tab, which provides details on client and version usage

The final tab, **Alerts**, is used to show the number of alerts raised over a period of time and the severity of those alerts. The tab can be seen in the following screenshot. We will cover the configuration of alerts under the *Setting up alerts using alert rules* section:

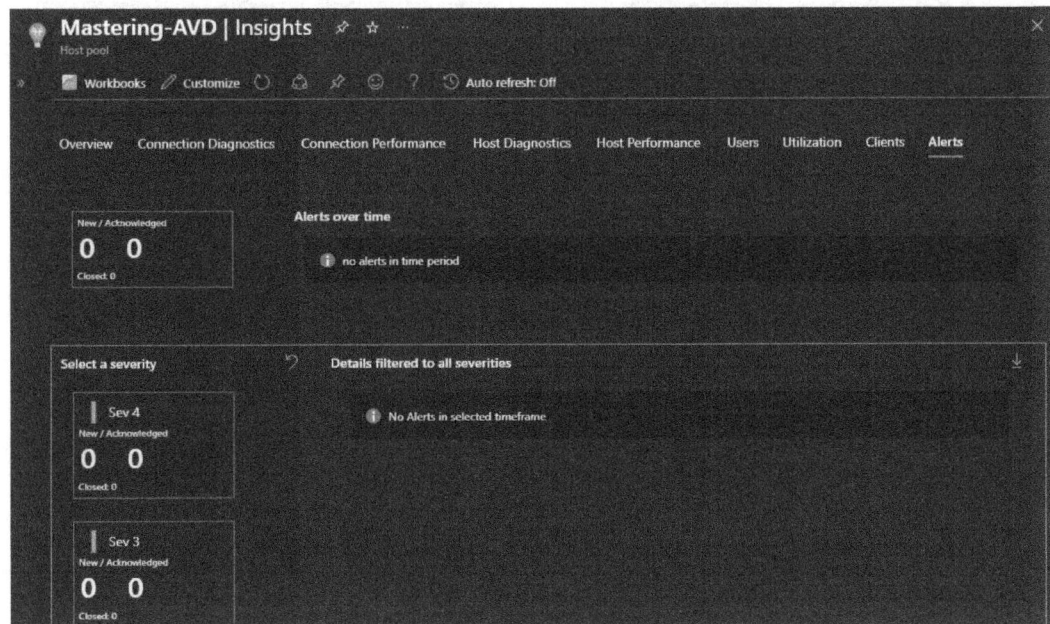

Figure 18.30 – The Alerts tab

In this section, we took a quick look at the **Insights** workspace and the different tabs and learned how IT administrators can use Azure Monitor to quickly diagnose and resolve AVD issues. In the next section, we take a look at setting up alerts based on outputs from a query.

Setting up alerts using alert rules

Within Log Analytics, you can query logs and set a frequency. You can also set an alert based on the output of the query. Rules can be triggered using one or more actions.

> **Tip**
> Using alerts to notify administrators or specified users about issues within your AVD environment can be helpful for those who are not continually monitoring the **Insights** page. This can also be useful for IT administrators who may be on call or need to receive a text message or other notification of a possible issue/failure.

To set up an alert, proceed as follows:

1. Start by typing `monitor` in the Azure search bar and click **Monitor** from the search results that appear, as illustrated in the following screenshot:

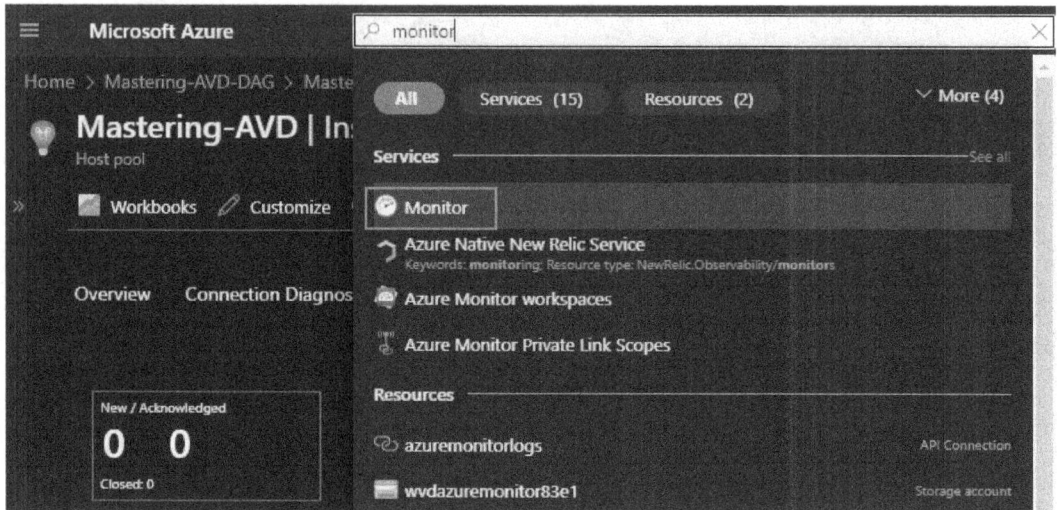

Figure 18.31 – Azure Monitor service within the search bar

2. Once within the Azure **Monitor** page, click on the **Alerts** icon located within the menu on the left, as illustrated in the following screenshot:

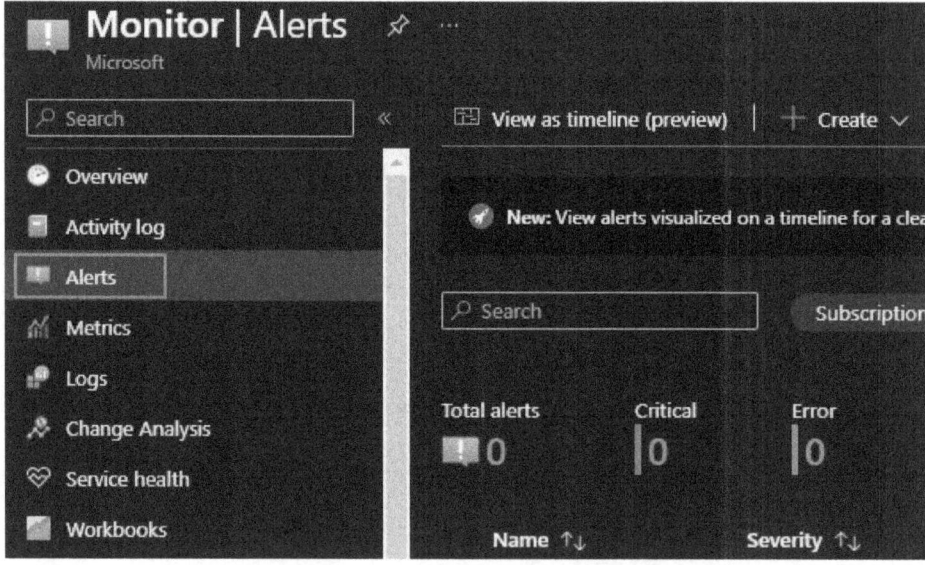

Figure 18.32 – The Alerts page icon in Azure Monitor

3. Click the **+ Create** and **Alert rule** buttons, as highlighted in the following screenshot, to load the **Create an alert rule** page:

Using Insights   609

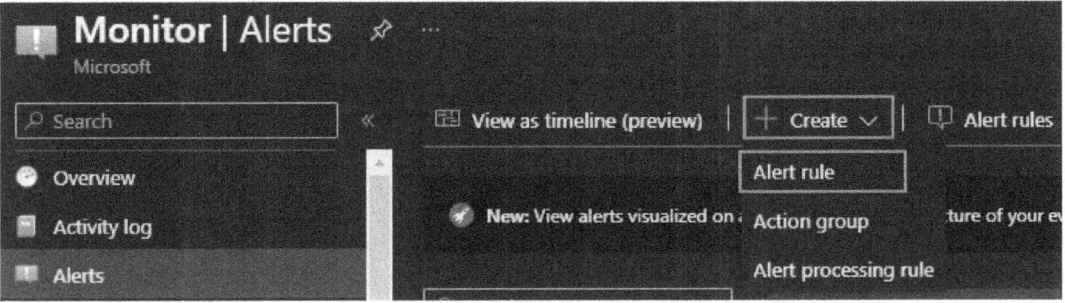

Figure 18.33 – The Alerts page and the buttons to create a new alert rule

4. In this example, we are going to create a simple alert for FSLogix disk-related issues. Within the **Select a resource** page, click the Log Analytics workspace and click **Apply**, as illustrated in the following screenshot:

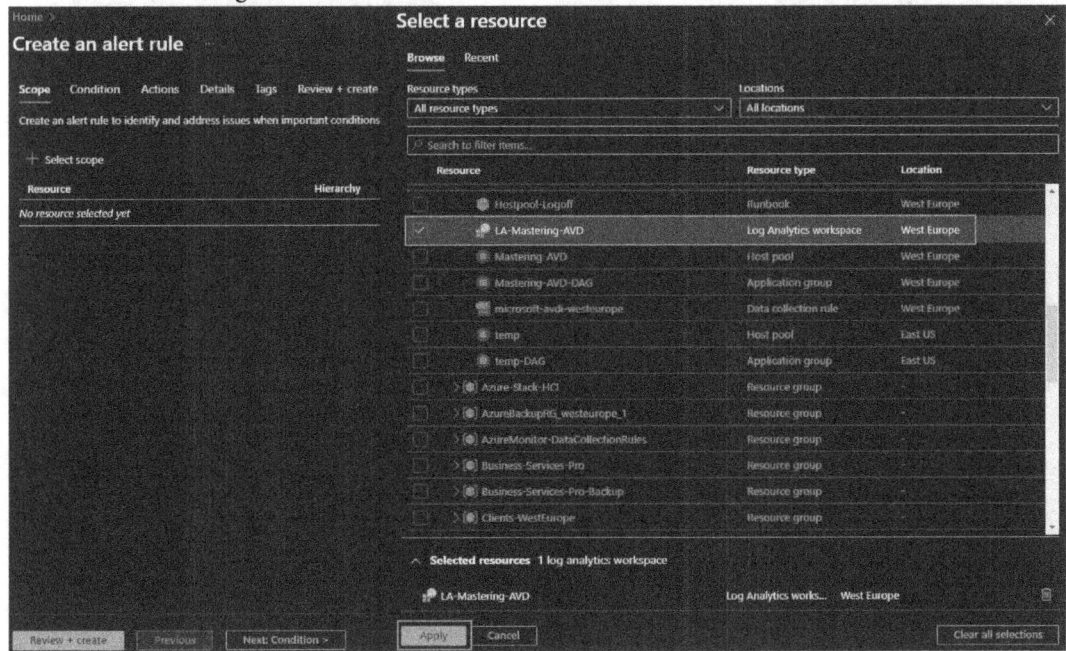

Figure 18.34 – Selecting a resource within the Select a resource section

5. Once complete, we need to add a condition. We need to click the **Select a signal** option to add a condition, as shown in the following screenshot:

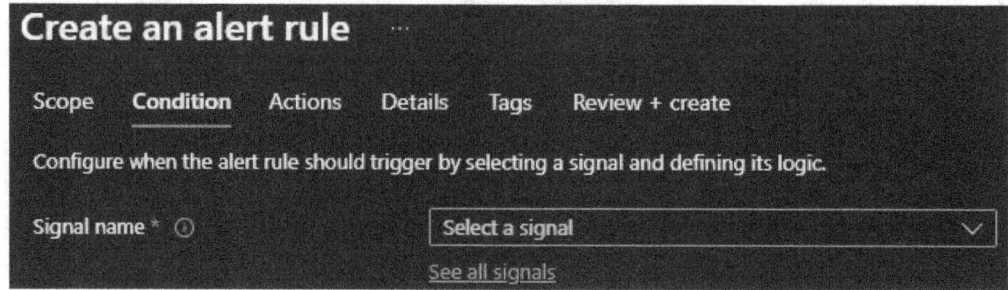

Figure 18.35 – Select a signal on the Create an alert rule page

6. Within the **Select a signal** blade, we click on **See all signals** and choose the required signal name. In this example, we will select the **Event** signal, as shown in the following screenshot:

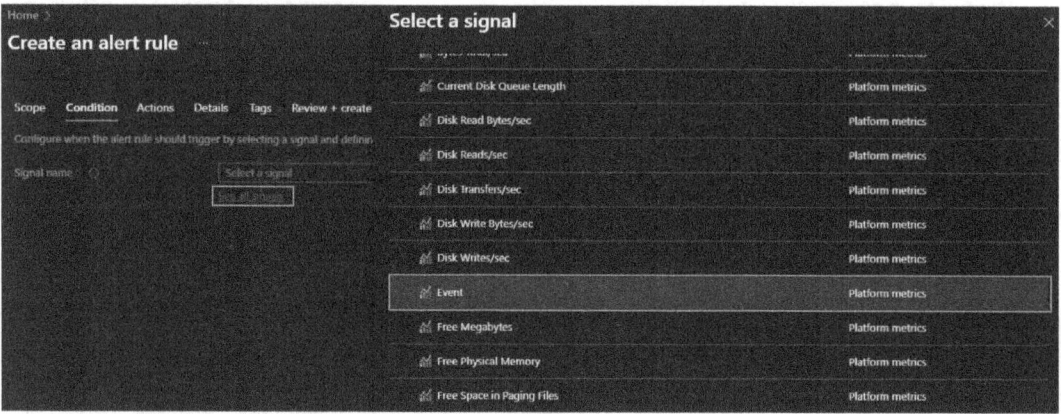

Figure 18.36 – The Select a signal page with Event selected

7. Once you have selected the **Event** signal and clicked **Apply**, you will see the **Configure signal logic** page appear. Within this page, you will set the `Microsoft-FSLogix-Apps/Operational` event log and the `EventID` value as `26`, which is a common event, as shown in *Figure 18.37*.

> **Important Note**
> The `EventID` value of `26` is related to the failure to load an `FSLogix` profile.

8. You then need to set the **Alert logic** field to a **Static** threshold, the **Operator** field to **Greater than or equal to**, the **Aggregation type** field to **Total**, the **Threshold value** field to 5, and the **Unit** value to **Count**.

Once complete, click the **Next: Actions** button, as illustrated in the following screenshot:

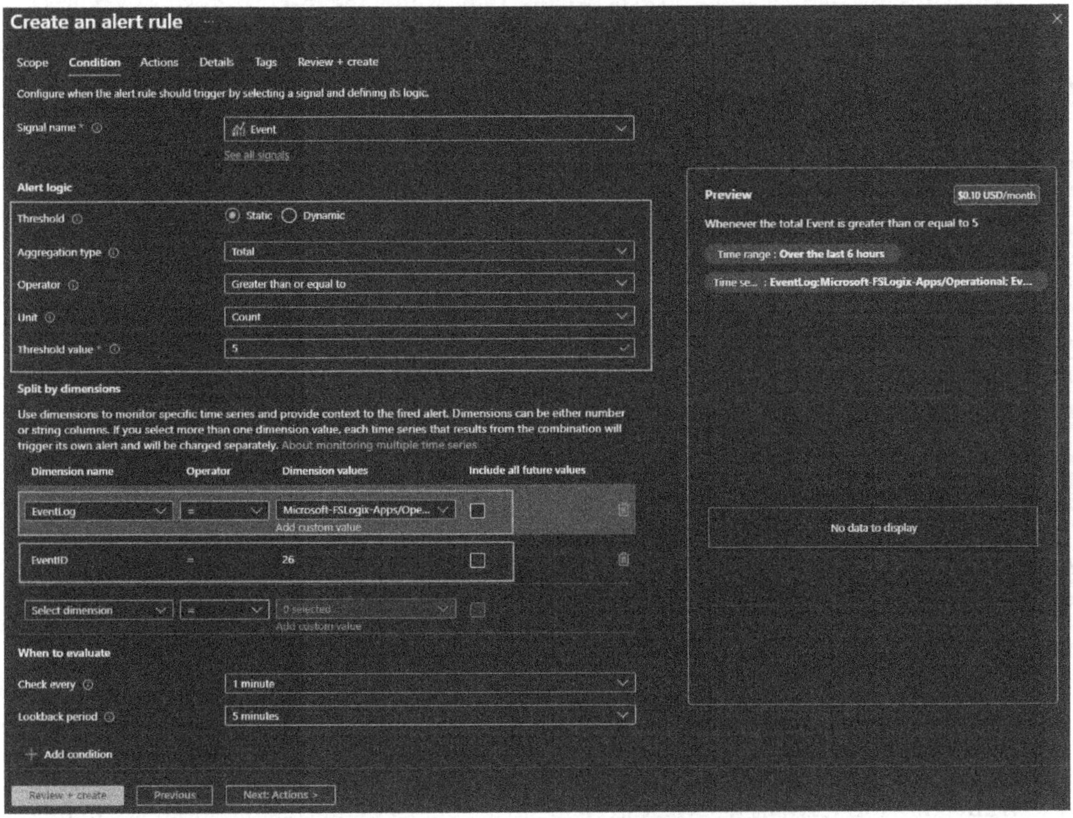

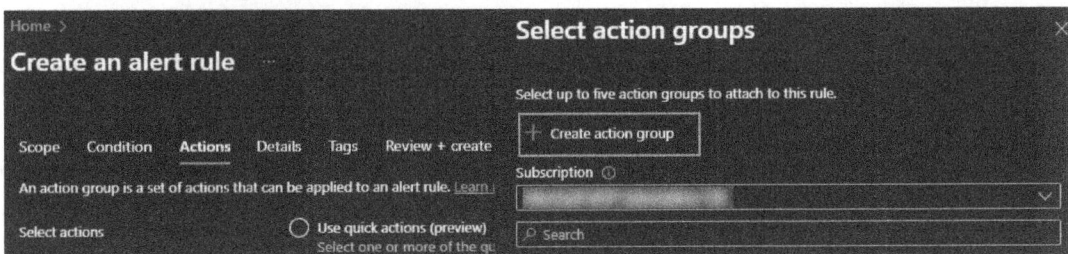

Figure 18.37 – The Configure signal logic page

The next step is to create an action group, and in this example, we only need a notification. You can configure actions in more advanced configurations. Proceed as follows:

1. Click on the **Create action group** button within the **Select action groups** page, as highlighted in the following screenshot:

Figure 18.38 – The Create action group button

2. Once the **Create action group** page appears, select a **Resource group** type, then choose an **Action group name** value and a **Display name** value within the **Instance details** section, as highlighted in the following screenshot:

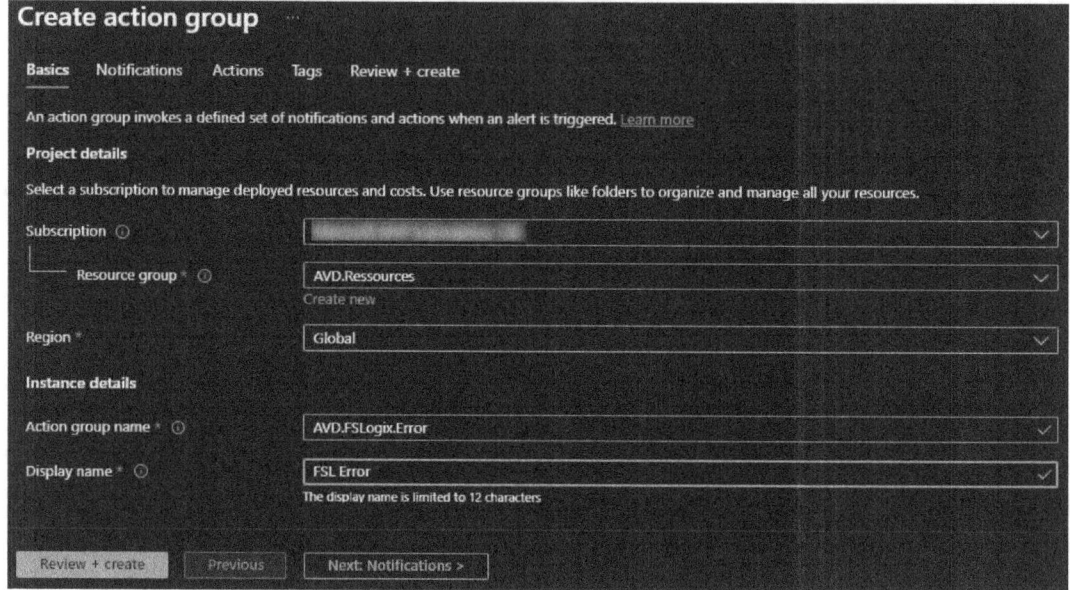

Figure 18.39 – The Create action group page

3. Once you have configured all the required fields on the **Basics** tab, click the **Notifications** tab.
4. Within the **Notifications** tab, you can configure a number of different notifications. In this example, we will configure email only. As shown in the following screenshot, we have selected the **Notification type** value as **Email/SMS message/Push/Voice** and set the **Name** value as `Admin email`:

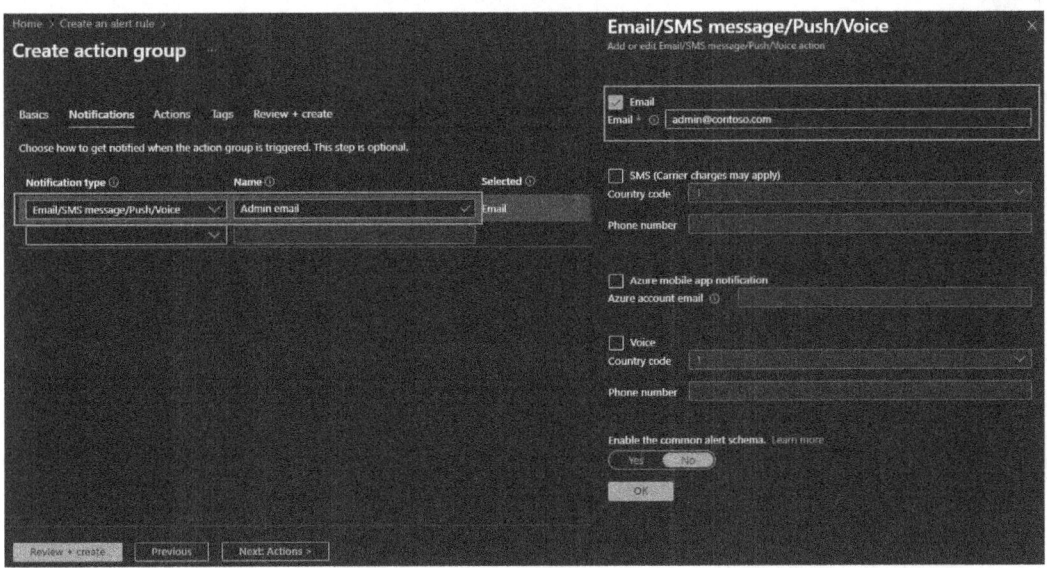

Figure 18.40 – Email configuration within the Notifications tab on the Create action group page

5. Once the notifications have been configured, proceed with the **Review + create** tab. Check the configuration and click **Create**, as illustrated in the following screenshot:

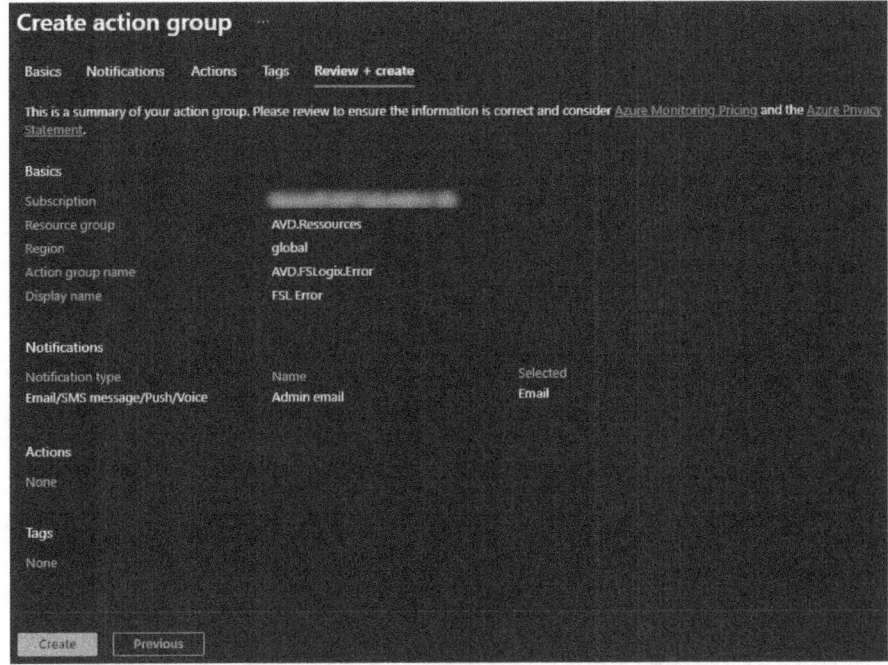

Figure 18.41 – The Review + create tab

6. You should now be able to see the new action group name we just created under the **Actions** section within the **Create an alert rule** page, as highlighted in the following screenshot:

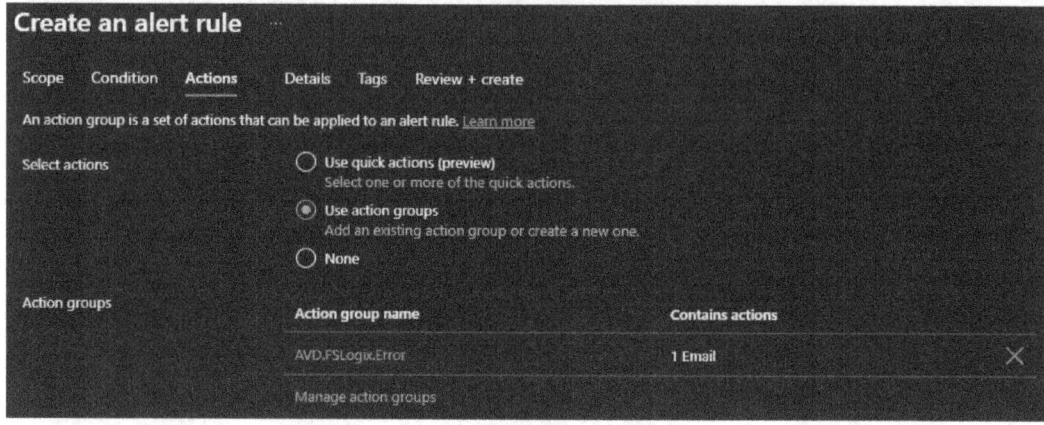

Figure 18.42 – Added action group name

7. The last action on the **Create an alert rule** page is to complete the alert rule details in the **Details** tab. Enter values in the **Alert rule name**, **Description**, and **Resource group** fields, and enter a **Severity** ranking, as shown in the following screenshot. Also, uncheck **Automatically resolve alerts**. Once finished, click **Review+create**:

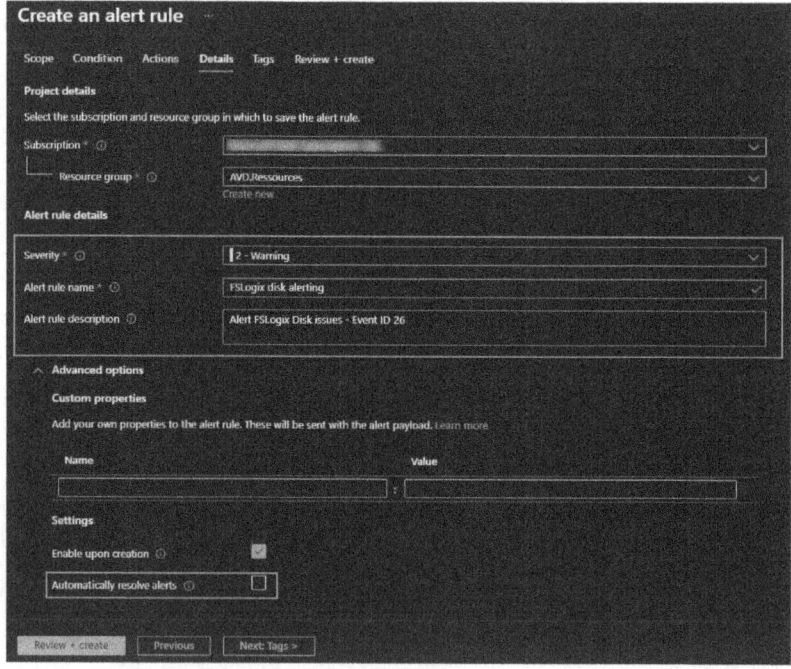

Figure 18.43 – Alert rule details within the Create an alert rule page

8. You will now be able to see the new alert within **Monitor | Alerts | Alert rules**, as shown in the following screenshot:

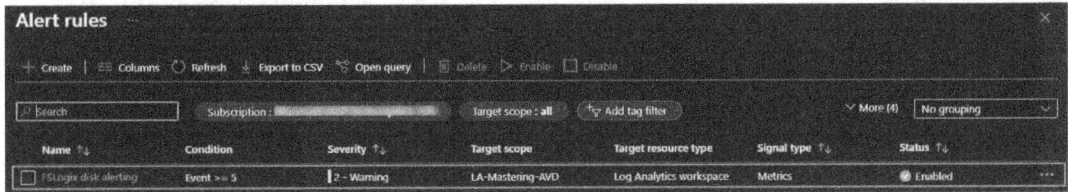

Figure 18.44 – Newly created FSLogix alert rule

This section looked at creating an alert rule to notify IT administrators of a specific issue or problem with the AVD environment. The example I showed was a simple `FSLogix` alert using event logs; however, you can create custom alerts specific to your environment.

In the next section, we take a look at Kusto and how you can use this query language to query AVD to diagnose issues and pull useful information specific to your environment.

Introduction to Kusto

Kusto Query Language (**KQL**) is a *read-only* language used to query datasets within Microsoft Azure. Similar to **Structured Query Language** (**SQL**), Kusto can be used to query data, but it can't update or delete as SQL can. Kusto can be used when querying AVD services and other related components, and you can create custom queries to output information that is important to you.

You can use Kusto with the following Azure services:

- Azure Application Insights
- Azure Resource Graph
- Azure Log Analytics
- Azure Monitor Logs
- Azure Data Explorer
- Microsoft Defender for Endpoint
- Microsoft Sentinel

> **Fun fact**
>
> It is understood that the internal code name *Kusto* was named after Jacques Cousteau, as a reference to "*exploring the ocean of data.*" You may notice that reference when launching the Kusto.Explorer tool, which we will look at shortly. The development of Kusto was focused on addressing the need for fast and scalable log analytics.

Connecting Log Analytics to Kusto.Explorer

Before we start writing a basic query for AVD, we first need to look at how to use Kusto.Explorer.

Kusto.Explorer is a free tool that you can download from the **Microsoft Docs** page here: `https://aka.ms/ke`.

Once you have downloaded and installed Kusto.Explorer, you'll need to connect to your Azure Log Analytics workspace using the following (cluster connection): `https://ade.loganalytics.io/subscriptions/<subscription-id>/resourcegroups/<resource-group-name>/providers/microsoft.operationalinsights/workspaces/<workspace-name>`

Remember to change the subscription **identifier** (**ID**), resource group name, and workspace name within the preceding string (cluster connection).

To connect your Log Analytics workspace to Kusto.Explorer, you will need to add a connection, as shown in the following screenshot:

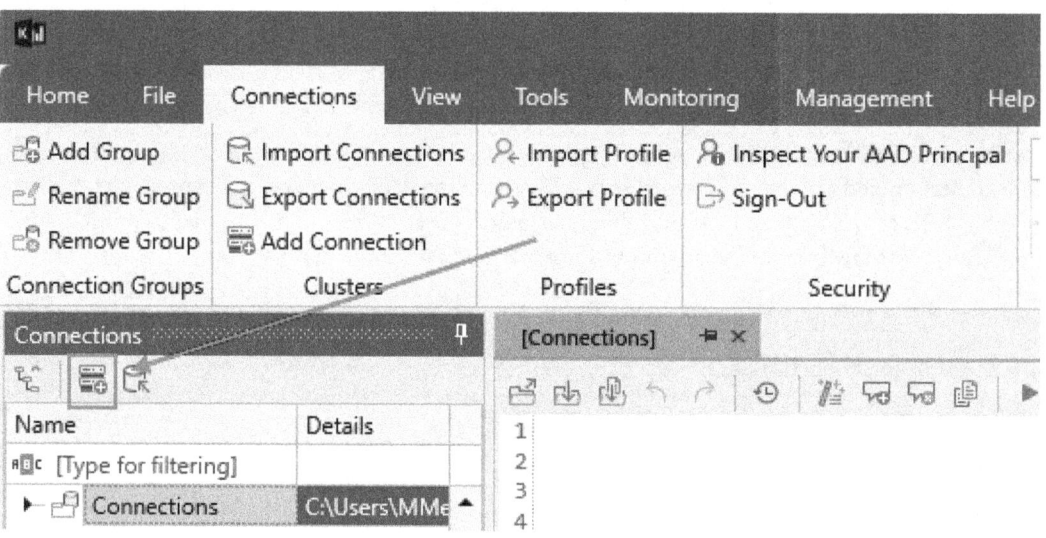

Figure 18.45 – The Add connection icon within Kusto.Explorer

Enter the cluster connection (the URL for Log Analytics), as illustrated in the following screenshot:

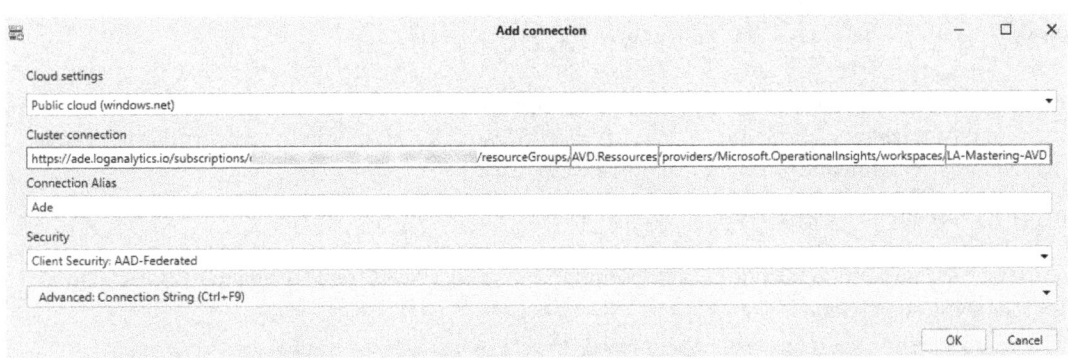

Figure 18.46 – Adding a cluster connection within Kusto.Explorer

Once connected, you should be able to see the **Connections** tab and a list of tables within the tree, as shown in the following screenshot:

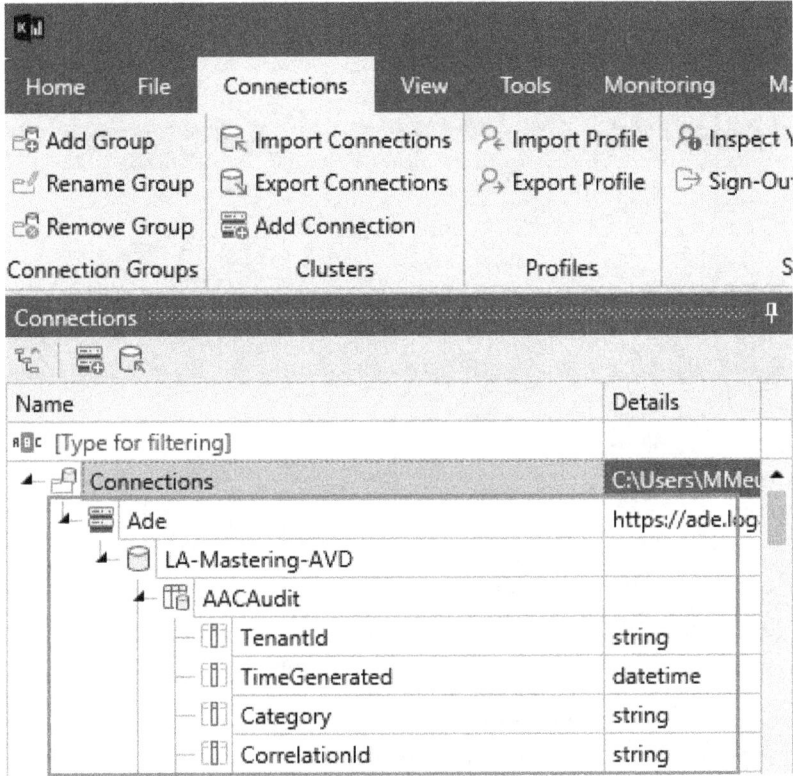

Figure 18.47 – Connection to the Log Analytics workspace

Now that we have configured Kusto.Explorer, we can proceed with creating queries for AVD.

Creating queries for AVD using Kusto.Explorer

In this section, we will take a look at a few quick queries you can build using Kusto.Explorer. You can create and customize queries within Azure; however, Kusto.Explorer allows you to work on them in a nice client application and produce graphs as well.

Let's get started with a basic AVD error query.

Within the **Connections** tab, right-click on **Connections** and select **Open In New tab**, as highlighted in the following screenshot:

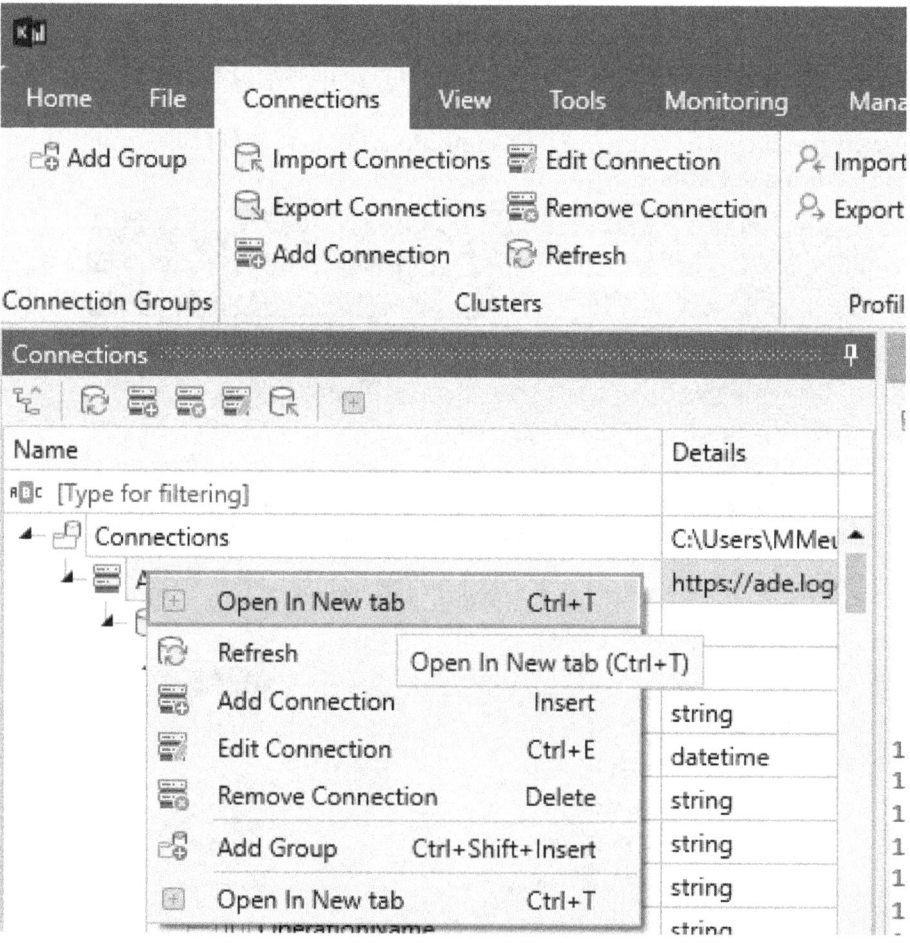

Figure 18.48 – New tab (Ctrl + T) used for creating a new query tab

Once you have clicked the new tab, you should be able to see a new tab created, as shown in the following screenshot:

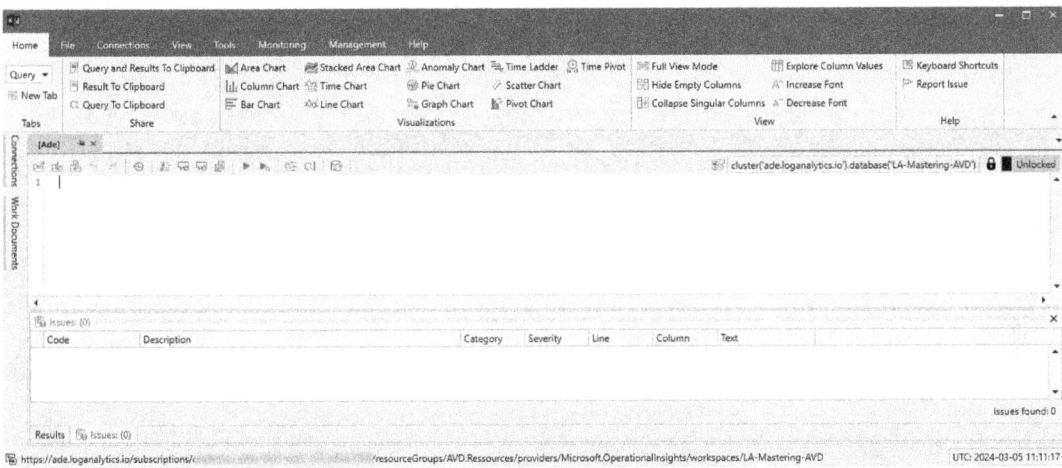

Figure 18.49 – New tab

A straightforward query with no filtering would be `WVDErrors` – this will collect any recorded errors within the `WVDErrors` table and display them within the output panel, as shown in the following screenshot:

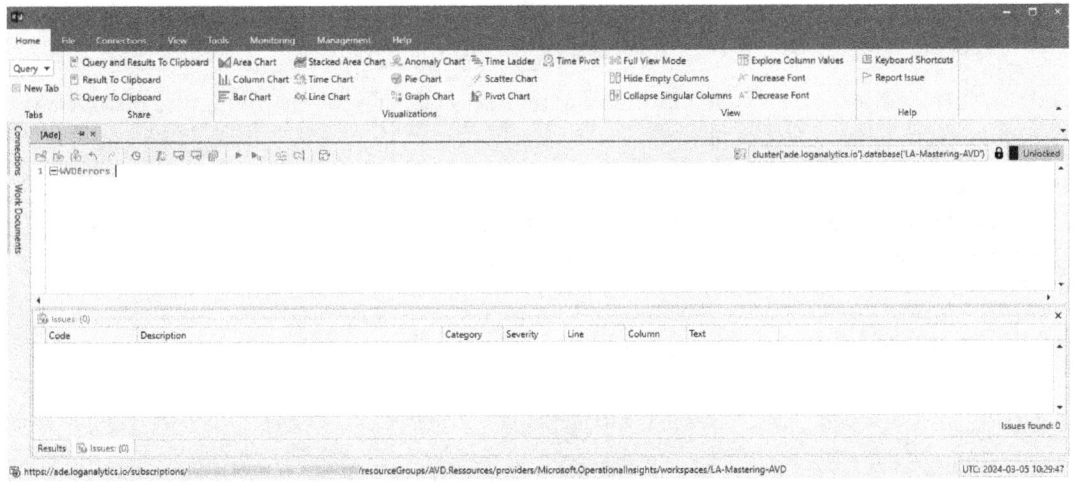

Figure 18.50 – Output of running the WVDErrors basic query within Kusto.Explorer

You can then add a filter by using the `where` Boolean expression to pull a specific time or time range, which may be helpful when reviewing the logs for a specific issue.

The following query shows the usage of where:

```
WVDErrors
| where TimeGenerated > (datetime(2024-03-05T10:00:00.0000000Z) - 24h)
```

Here's the output:

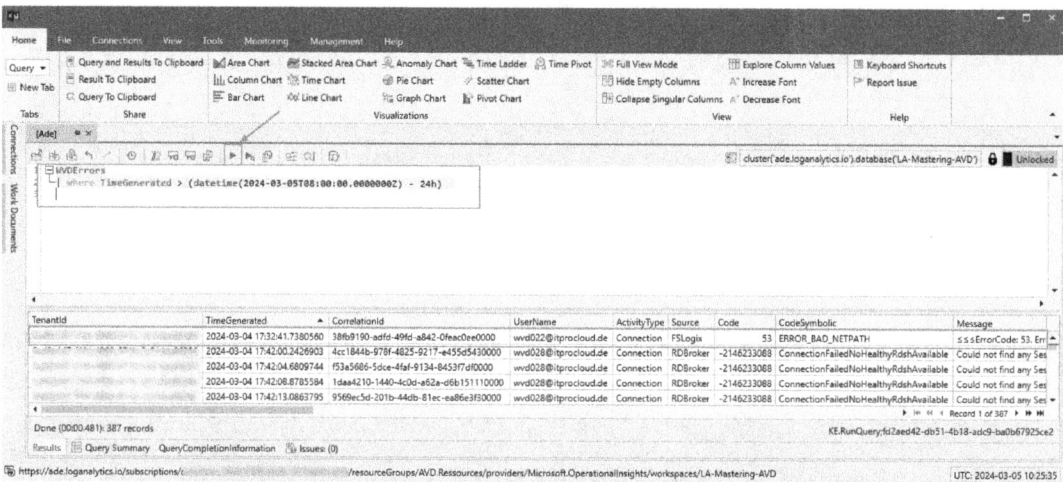

Figure 18.51 – The where Boolean expression for the WVDErrors table

In this example, I wanted to show you where to specify specific columns using the **project** operator. The project operator allows you to pick out specific columns you require.

I used the WVDConnections table this time and then selected a few of the columns to provide an easy-to-read output. Here's an example:

```
WVDConnections
| where TimeGenerated > (datetime(2024-03-05T10:00:00.0000000Z) - 24h)
| project UserName, State, SessionHostName, TimeGenerated, ConnectionType
```

The following screenshot shows the usage of Kusto.Explorer, using the preceding query, which uses the project operator:

Introduction to Kusto 621

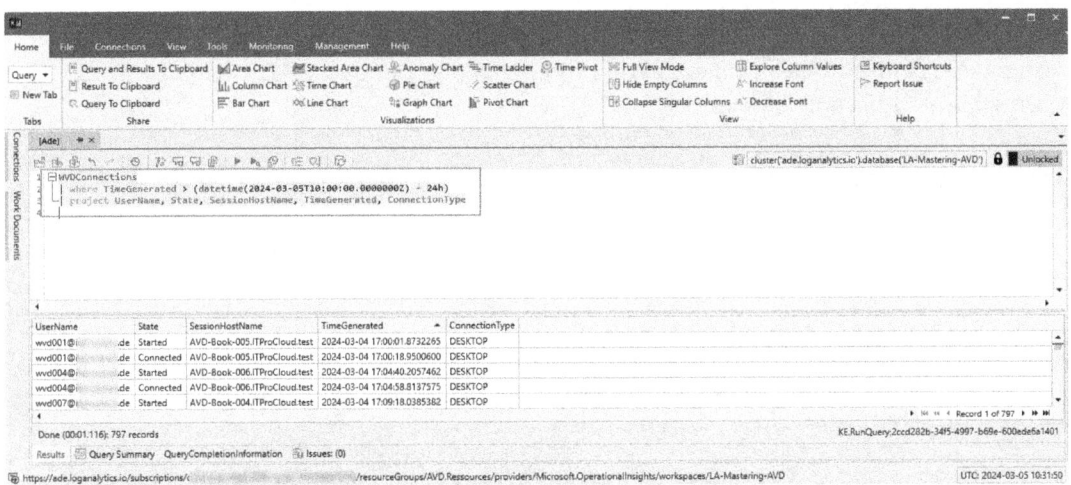

Figure 18.52 – The WVDConnections query using the project operator to select a few columns

In this final example, I wanted to show you how to filter using `CodeSymbolic`, which allows you to filter on a specific message. In this example, all session hosts are started and under full load. So, no additional user was able to log in anymore. I wanted to find out which users did not successfully log in because of the lack of resources. To do this, I used the following Kusto query:

```
WVDErrors
| where CodeSymbolic == " ConnectionFailedNoHealthyRdshAvailable"
```

The following screenshot shows the results of the preceding query:

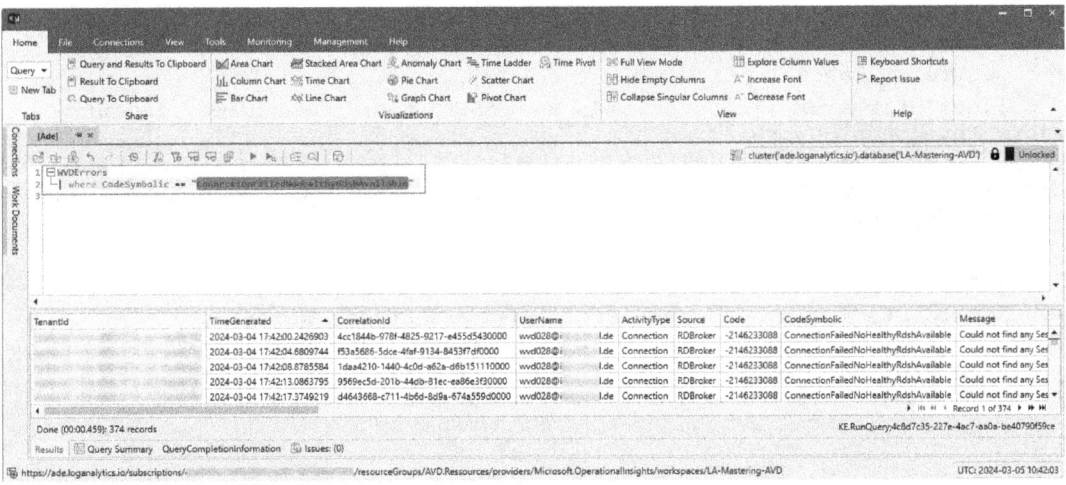

Figure 18.53 – Use of CodeSymbolic filtering

In this section, we looked at querying a Log Analytics workspace using Kusto.Explorer. We also looked at a couple of examples to help you get started with querying your own AVD environment. Let us now list some of my favorite queries.

Some additional Kusto queries

Kusto is a very powerful tool to get a lot out of your data. Let me list some other example queries for different approaches. Lines starting with // are comments:

```
// Start and end time of a session
WVDConnections
| summarize Start=min(TimeGenerated), End=max(TimeGenerated) by User-
Name, CorrelationId

// Shows temporary logins (user profile failed)
WVDCheckpoints
| where Name=="LogonDelay" and Parameters.["LogonType"]=="Temporary-
Session"

// Getting the min, max, avg login time per host
WVDCheckpoints
| where Name=="LogonDelay"
| extend LogonDurationInSeconds=Parameters.WinLogon_Total/1000.0
| extend HostPool=split(_ResourceId,"/")[8]
| join kind=leftouter (
    WVDConnections
    | where State == "Started"
    | project SessionHostName, CorrelationId
) on CorrelationId
| project UserName, TimeGenerated, HostPool, SessionHostName, LogonDu-
rationInSeconds, CorrelationId
| summarize MinLoginSeconds=min(LogonDurationInSeconds),MaxLoginSec-
onds=max(LogonDurationInSeconds),AvgLoginSeconds=avg(LogonDurationIn-
Seconds) by SessionHostName

// Getting profile data from the workspace by host pool and user
WVDCheckpoints | where (Name=="ProfileLoggedOff" or Name=="ODF-
CLoggedOff") and (Source=="RDAgent" or Source=="FSLogix")
| extend HostPool=tostring(split(_ResourceId,"/")[8]), Profile-
Type=iff(Name=="ProfileLoggedOff","Profile","ODFC")
```

```
| summarize arg_max(TimeGenerated, *) by UserName, _ResourceId, Profi-
leType
| extend ["VHD Size On Disk"]=todouble(replace_string(replace_
string(tostring(Parameters.VHDSizeOnDisk),",",""),".","")),["VHD 
Free Space"]=todouble(replace_string(tostring(Parameters.VHD-
FreeSpace),",","."))),["VHD Max Size"]=todouble(replace_string(to-
string(Parameters.MaxVHDSize),",","."))
| where ["VHD Size On Disk"]!=""
| extend ["VHD Free Space"]=iff(["VHD Free Space"]>["VHD Max 
Size"],["VHD Free Space"]/1024.0,["VHD Free Space"])
| project HostPool, UserName, TimeStamp=TimeGenerated,ProfileType,
["VHD Size On Disk"], ["VHD Free Space"], ["VHD Max Size"], Us-
age=100*(["VHD Max Size"]-["VHD Free Space"])/["VHD Max Size"]
| where ["VHD Size On Disk"]>0
| order by ["Usage"] desc
```

Log Analytics also allows the creation of custom workbooks. Workbooks visualize Kusto queries directly in a Log Analytics workbook. You can also import a pre-built community workbook into your environment using this blog post: https://blog.itprocloud.de/AVD-Azure-Virtual-Desktop-Error-Drill-Down-Workbook/.

After installing the workbook, you can open it in the Log Analytics workspace, as shown in the screenshot:

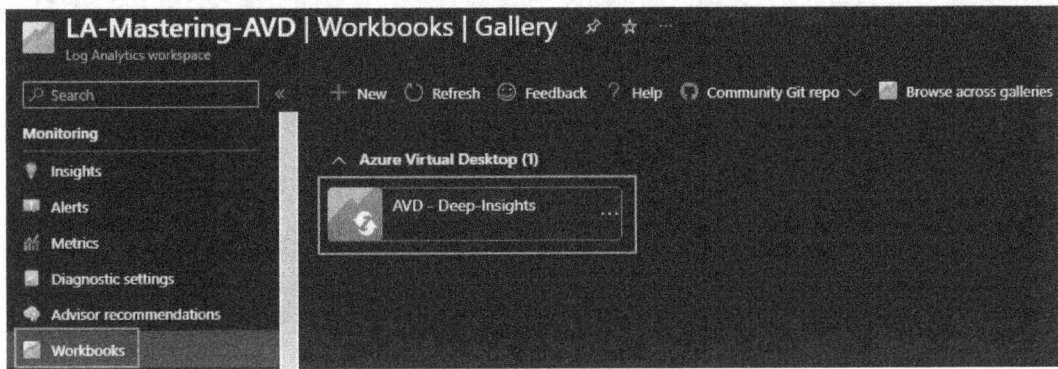

Figure 18.54 – Opening a custom workbook

In the **Deep-Insights** workbook, there are different aspects shown:

- **Connections**
- **Errors**
- **Session Bandwidths & Latencies**

- **Graphic Performance**
- **Logon Timing**
- **FSLogix**
- **Water Markings**
- **Resources**

For example, you can directly see some information such as bandwidth and latency, as shown in the following screenshot:

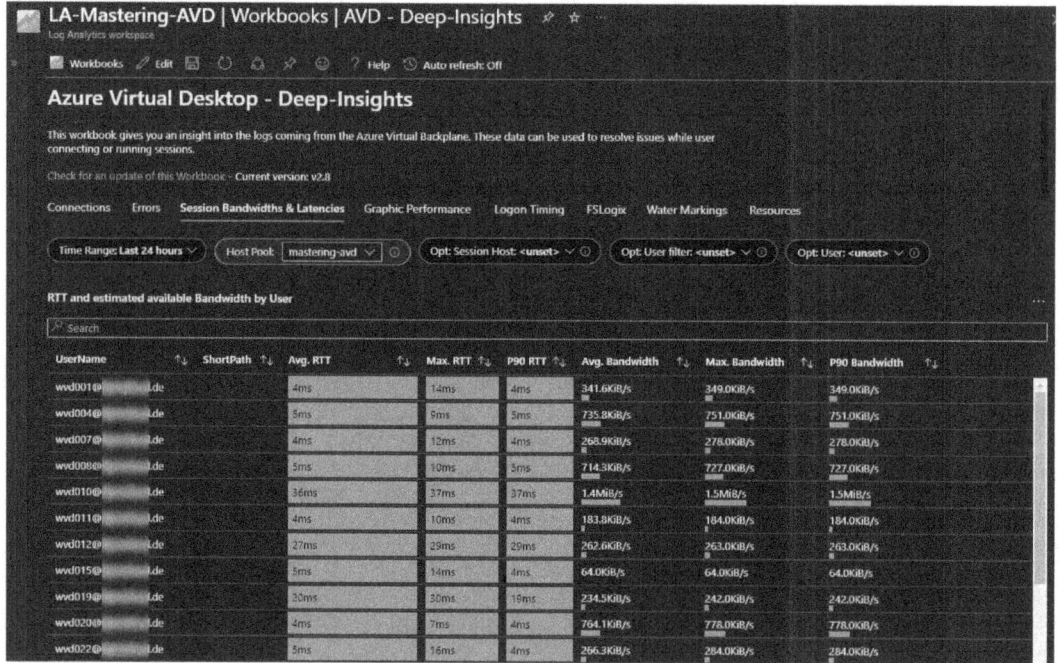

Figure 18.55 – Bandwidths and latencies

Another tab shows the usage of the FSLogix profiles of the users. That can be helpful to figure out some action tasks easily if profiles are growing unexpectedly. Here is a view of the **FSLogix** tab:

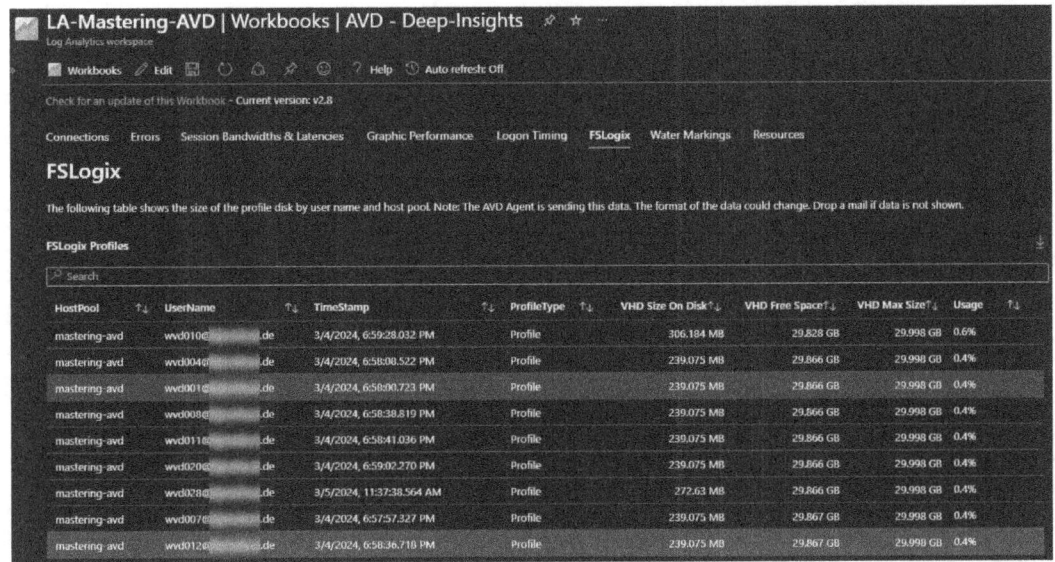

Figure 18.56 – FSLogix profile usage

As mentioned before, Kusto is very powerful if you connect different data.

In the next section, we will look at using Azure Advisor for AVD.

Using Azure Advisor for AVD

Azure Advisor can be used to help resolve common issues, and it also provides recommendations. Azure Advisor's recommendations include resource reliability, security, operational excellence, performance, and cost.

To get started with Azure Advisor, enter `advisor` into the Azure search bar and select the service that appears, as illustrated in the following screenshot:

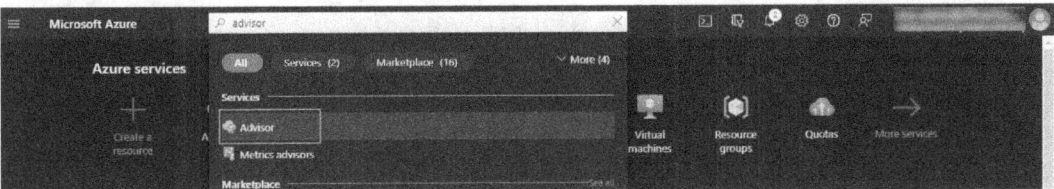

Figure 18.57 – The Advisor service within the Azure portal search bar

You will then be presented with several advisories within five categories, as follows:

- **Cost**
- **Security**
- **Reliability**
- **Operation excellence**
- **Performance**

The following screenshot shows the Azure Advisor **Overview** page:

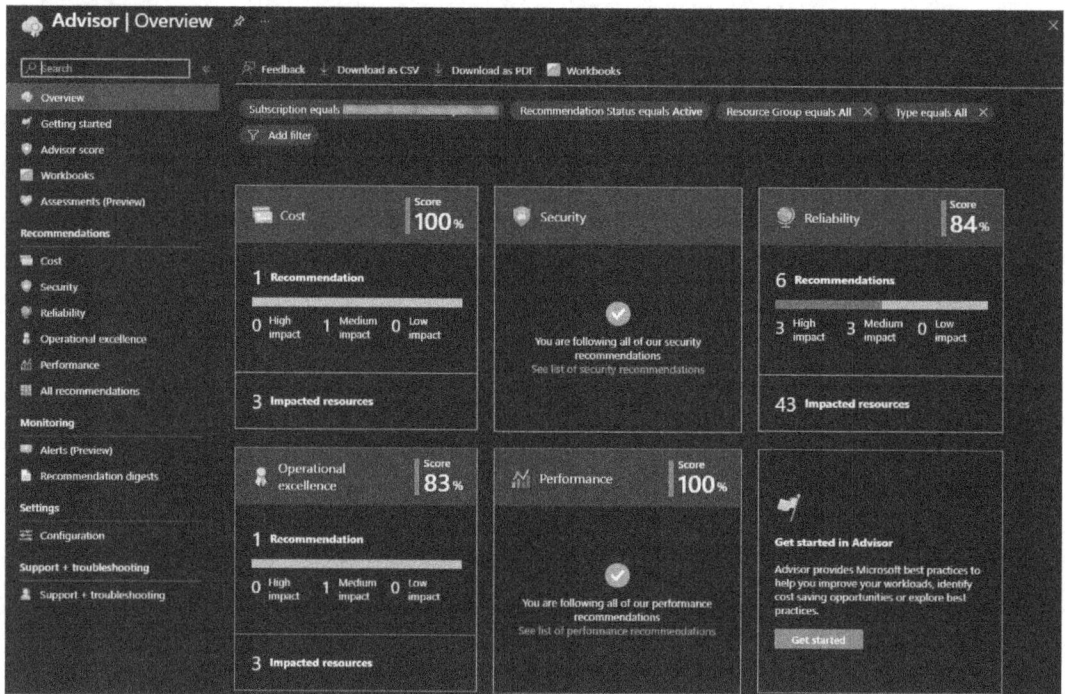

Figure 18.58 – Five recommendation categories Overview page

Use the recommendations to enhance your configuration and relatability and reduce the cost.

In this environment, Advisor detected three unattached disks. See the following screenshot:

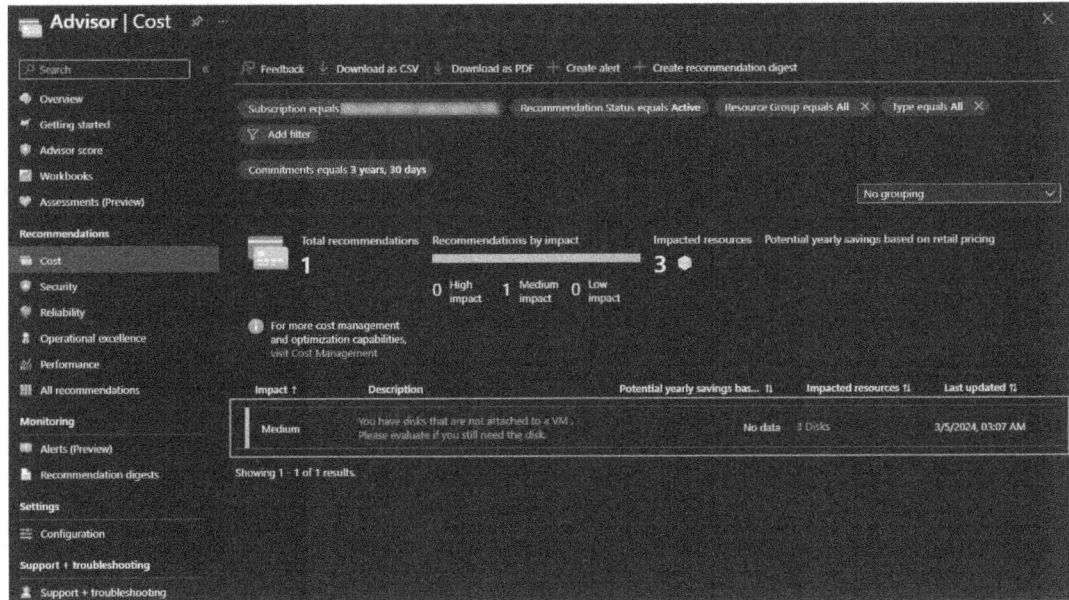

Figure 18.59 – Advisor found three unattached disks

In this case, I have to evaluate whether I still need the disks. If not, I would delete the disks to avoid spending money on orphan disks.

We looked at Azure Advisor in this section and briefly looked at the value it can bring to your AVD environment.

Summary

This chapter looked at setting up and configuring Azure Monitor for AVD and using the Insights workbook. We then moved on to setting up custom alerts using alert rules, which will notify IT administrators of a specific error or issue. We then looked at KQL and a few examples of querying a Log Analytics workspace. We then finished off the chapter by briefly looking at Azure Advisor for AVD.

The next chapter will show an easy deployment of a complete and preconfigured AVD environment with the getting started feature.

Questions

Here are a few questions to test your understanding of this chapter:

1. How can a user limit data ingestion to Log Analytics?

   Configure the **Daily cap** *setting found under the* **Usage and estimated costs** *page within the Log Analytics workspace.*

2. What is KQL?

   A read-only language used to query datasets

3. What does the `EventID` value of `26` commonly refer to when reviewing `FSLogix` profile logs?

   The failure to load an FSLogix profile

4. What is the name of the feature used for viewing metrics and troubleshooting issues with AVD?

   Insights

5. Why should you no longer use Log Analytics Agent to collect data from virtual machines?

   The agent is deprecated – use Azure Monitoring Agent instead.

19
Azure Virtual Desktop's Quickstart Feature

In this chapter, we'll look at using the **Quickstart** feature to deploy Azure Virtual Desktop. This feature provides an easy way to deploy and configure an Azure Virtual Desktop environment.

By the end of this chapter, you should be able to use the Azure Virtual Desktop Quickstart feature to deploy Azure Virtual Desktop environments.

In this chapter, we will cover the following topics:

- How the Quickstart feature works
- Using the Quickstart feature with Microsoft Entra ID DS
- Using the Quickstart feature without an identity provider
- Post-deployment cleanup
- Troubleshooting the Quickstart feature

How the Quickstart feature works

The purpose of the Quickstart feature is to address the challenges associated with deploying Azure Virtual Desktop environments. It removes the multi-step processes and makes deploying Azure Virtual Desktop simple.

Two key benefits of using the Quickstart feature are as follows:

- You can remove complex multi-step processes, including FSLogix profile container setup and configuration, which includes Azure Files and permissions.
- You can create session hosts and configure Azure Virtual Desktop core components, including host pools, workspaces, application groups, and validation user accounts.

There are two options you can choose from:

- **Existing Setup**: This is for organizations that already have an Azure tenant and subscription, including Active Directory and **Microsoft Entra ID Domain Services (Entra ID DS)**.
- **New Subscription (Empty)**: This is for subscriptions without Active Directory or Entra ID DS.

The Quickstart feature is essentially a wizard that enables you to deploy an Azure Virtual Desktop environment within a matter of hours. It lets you rapidly deploy small environments that can be used in production, testing, or lab settings.

The Quickstart feature uses nested templates to deploy the required Azure resources for validation and to automate the deployment of Azure Virtual Desktop. The Quickstart feature creates two or three resource groups that are dependent on the identity provider option that's selected within the wizard, which we will cover shortly.

Prerequisites

Before you get started, you need to ensure that you have an Entra ID tenant. Ensure the account you are using has global admin permissions on the subscription within Entra ID.

> **Important Note**
> At the time of writing, the Quickstart feature does not support accounts with **multi-factor authentication (MFA)**. Ensure the account being used has MFA turned off. Additionally, **managed service accounts (MSA)**, **business-to-business (B2B)**, and guest accounts are not supported.

For those using an environment with **Entra ID DS**, take note of the following:

- You need to have the Entra ID DS domain admin credentials to hand.
- You must configure Entra ID connect on your subscription and make sure the `USERS` container is syncing with Entra ID. You can check this by viewing the Entra ID page within the Azure portal.
- You need to ensure you have a domain controller deployed within the required region you plan to deploy Azure Virtual Desktop.
- The domain controller that you deploy to Azure must not have DSC extensions of the `Microsoft.Powershell.DSC` type.

For those who plan to deploy Azure Virtual Desktop without an identity provider, take note of the following:

- You need to ensure that the AD domain join's **user principal name** (**UPN**) does not include specific keywords such as admin, server, or support. The full list can be found here: `https://docs.microsoft.com/azure/virtual-machines/windows/faq#what-are-the-username-requirements-when- creating-a-vm-`.
- You must create a new host pool to add session hosts when using the Quickstart feature while using the **without an identity provider** option; if you're trying to deploy a session host in an existing host pool, it will fail.

Now, let's look at using the Quickstart feature wizard.

Using the Quickstart feature with Entra ID Domain Services (Entra ID DS)

This section details how to use the Quickstart feature to deploy an Azure Virtual Desktop environment in a subscription that already has **Entra ID Domain Services (Entra ID DS)**:

1. First, you need to open the Azure portal.
2. Once you have signed into Azure, open the **Azure Virtual Desktop** page using the search bar or navigate through the services.

3. Within the **Azure Virtual Desktop** page, select the **Quickstart** tab:

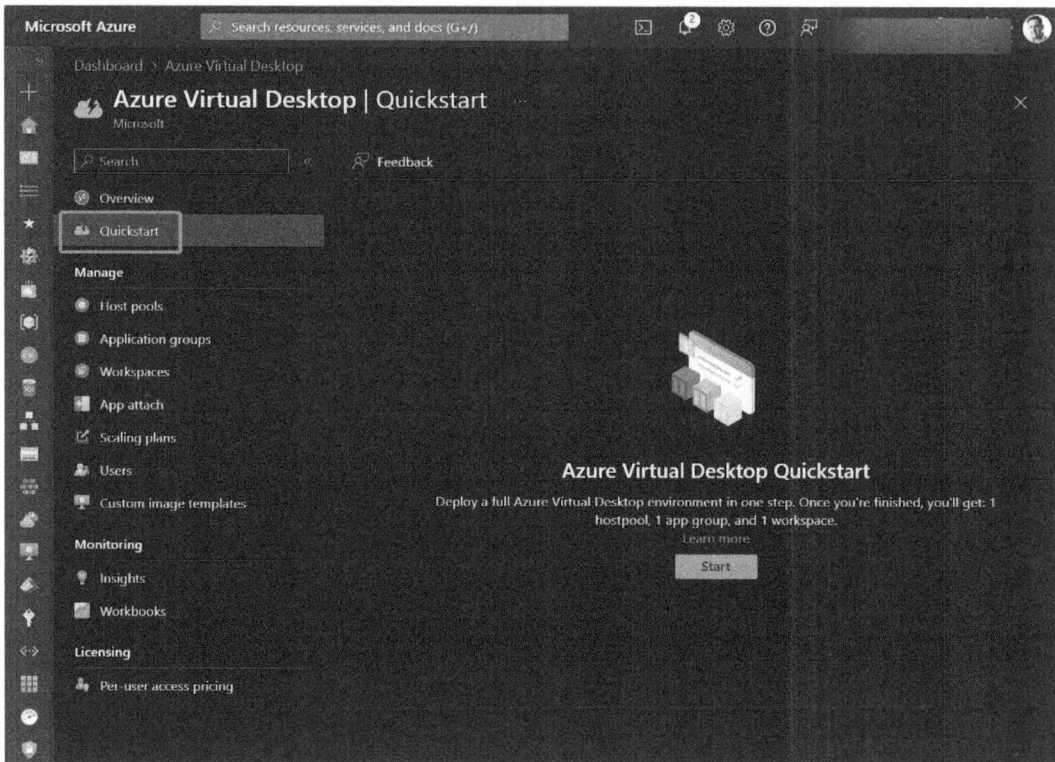

Figure 19.1 – The Quickstart tab within the Azure Virtual Desktop page of Azure

4. Click the **Start** button.
5. Select the required subscription and select **Existing active directory** in the **Identity provider** section.
6. For **Identity service type**, select **Active Directory** or **Entra ID Domain Services**. For this example, we will choose **Active Directory**.

7. Enter a **Resource group prefix** name:

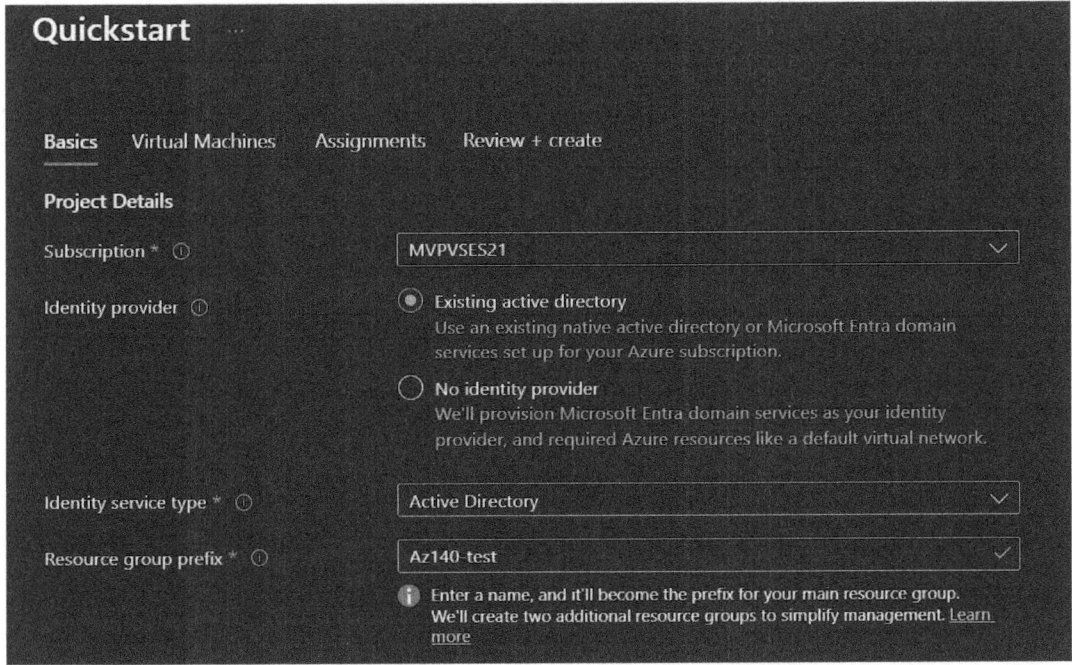

Figure 19.2 – The Quickstart wizard using an existing Active Directory identity provider

8. For **Location**, select the Azure region you wish to deploy your Azure Virtual Desktop resources to. This example uses **UK South**.
9. Select the required **Virtual network** and **Subnet** values:

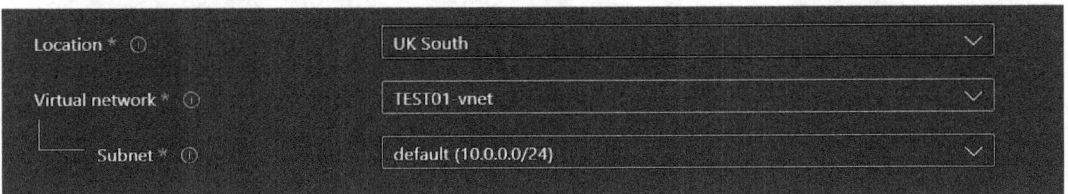

Figure 19.3 – The Location, Virtual network, and Subnet options within the Quickstart wizard page

10. Enter the required Azure user credentials. This is the full user principal name and ensures the account has the owner permissions on the Azure tenant.

11. Enter the required domain administrator credentials. You will need to ensure you enter the full user principal name:

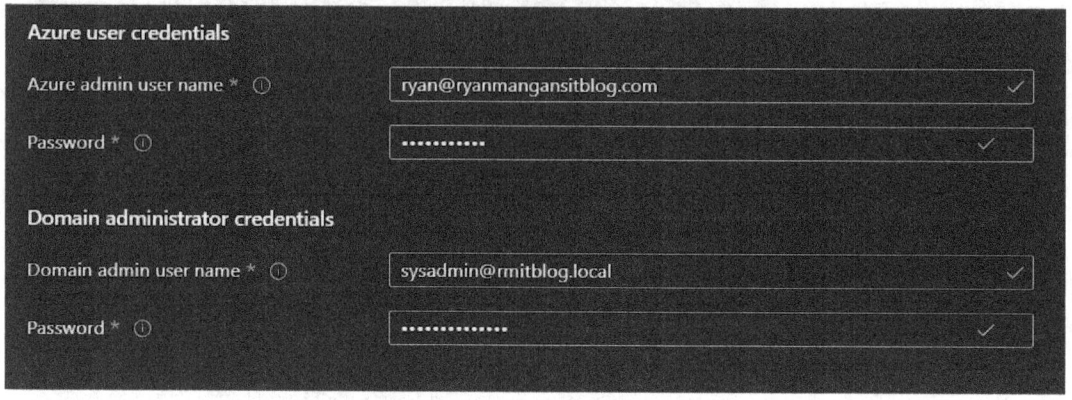

Figure 19.4 – Both the Azure user credentials and Domain administrator sections within the Quickstart wizard

12. Click **Next: Virtual Machines** to continue to the **Virtual Machines** tab.
13. Within the **Virtual Machines** tab, select the required option for users per virtual machine. For this example, we will choose multiple users.
14. Select an image type, and then an image. In this example, we will use the default of **Gallery** for **Image type** and **Windows 11 Enterprise multi-session + Microsoft 365 Apps** for **Image**.
15. Select the required virtual machine size and SKU you would like to deploy:

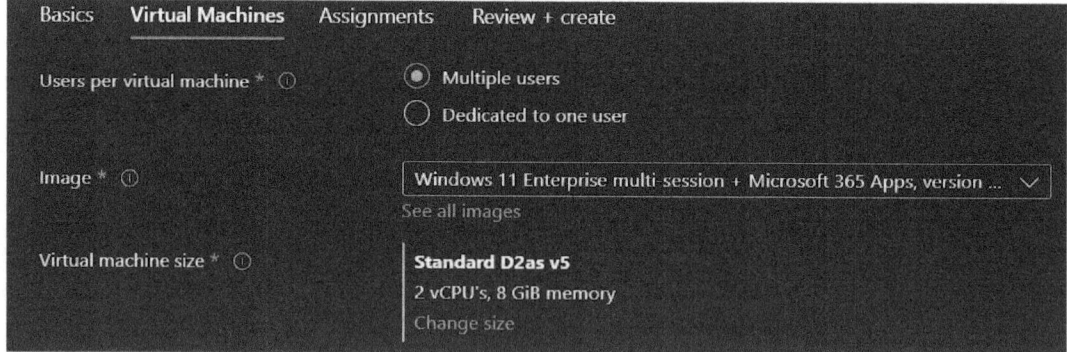

Figure 19.5 – The first four fields within the Virtual Machines tab within the Quickstart wizard

16. Enter a **Name prefix** to name the session hosts that will be deployed.

17. Under **Number of virtual machines**, specify the number of virtual machines you would like to deploy:

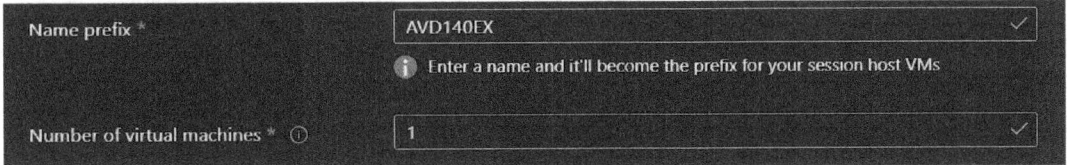

Figure 19.6 – The Name prefix and Number of virtual machines fields

18. **Specify domain or organizational unit**: This option enables you to specify a domain to join and the organizational unit path you wish to use. For this example, we will select **No**:

Figure 19.7 – The Specify domain or organizational unit section

19. Select a domain controller resource group. This is the resource group where the domain controller will reside.
20. Select a domain controller virtual machine:

Figure 19.8 – The Domain controller resource group and Domain controller virtual machine dropdown fields

21. You can also chain a custom **Azure Resource Manager** (**ARM**) template. This allows you to insert specific customizations into the deployment process. You can download an example template customization here: https://github.com/Azure/RDS-Templates/tree/master/wvd-sh/arm-template-customization. In this example, we will skip linking to an Azure template:

Figure 19.9 – The Link Azure template section within the Virtual Machines tab when using the Quickstart wizard

22. Click **Next: Assignments** to move on to the **Assignments** tab.

23. Within the **Assignments** tab, you can create a test user account. This will be used as a validation user account to test your deployment. In this example, we will skip the **Create test user account** section. Make sure to **uncheck** the box for **Create test user account**:

![Quickstart wizard Assignments tab showing Create test user account fields]

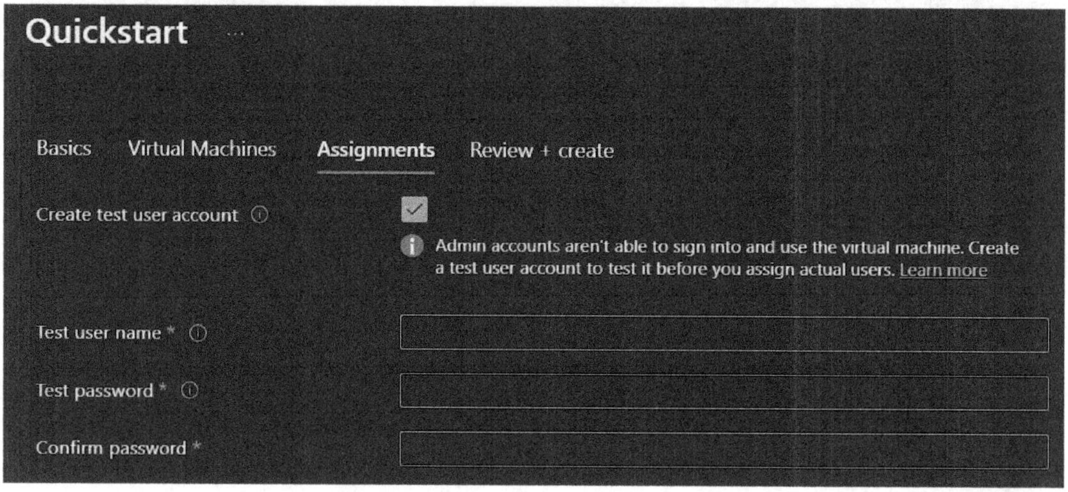

Figure 19.10 – The Create test user account fields within the Assignments tab in the Quickstart wizard

24. Within the **Assign existing users or groups** field, add the users or groups you wish to add to this Azure Virtual Desktop deployment:

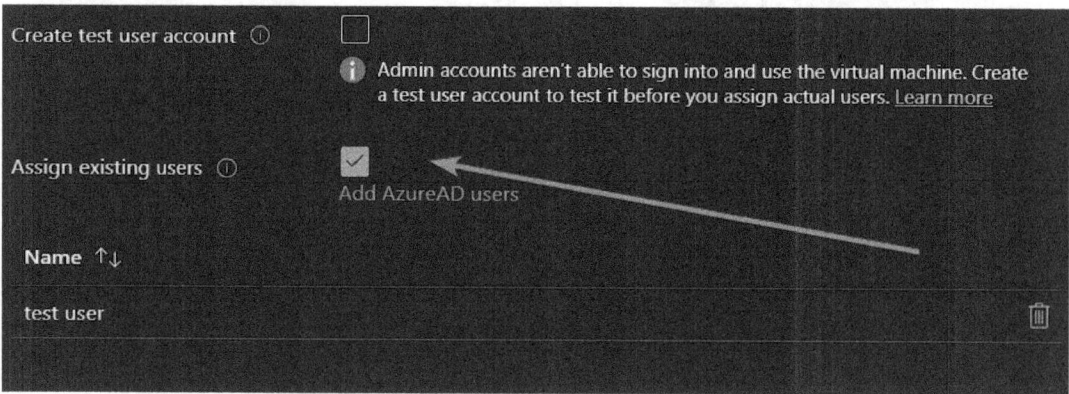

Figure 19.11 – Button to add users and groups to the Quickstart wizard deployment

25. Once you've finished adding users and groups, click **Review + create** to progress to the **Review + create** tab:

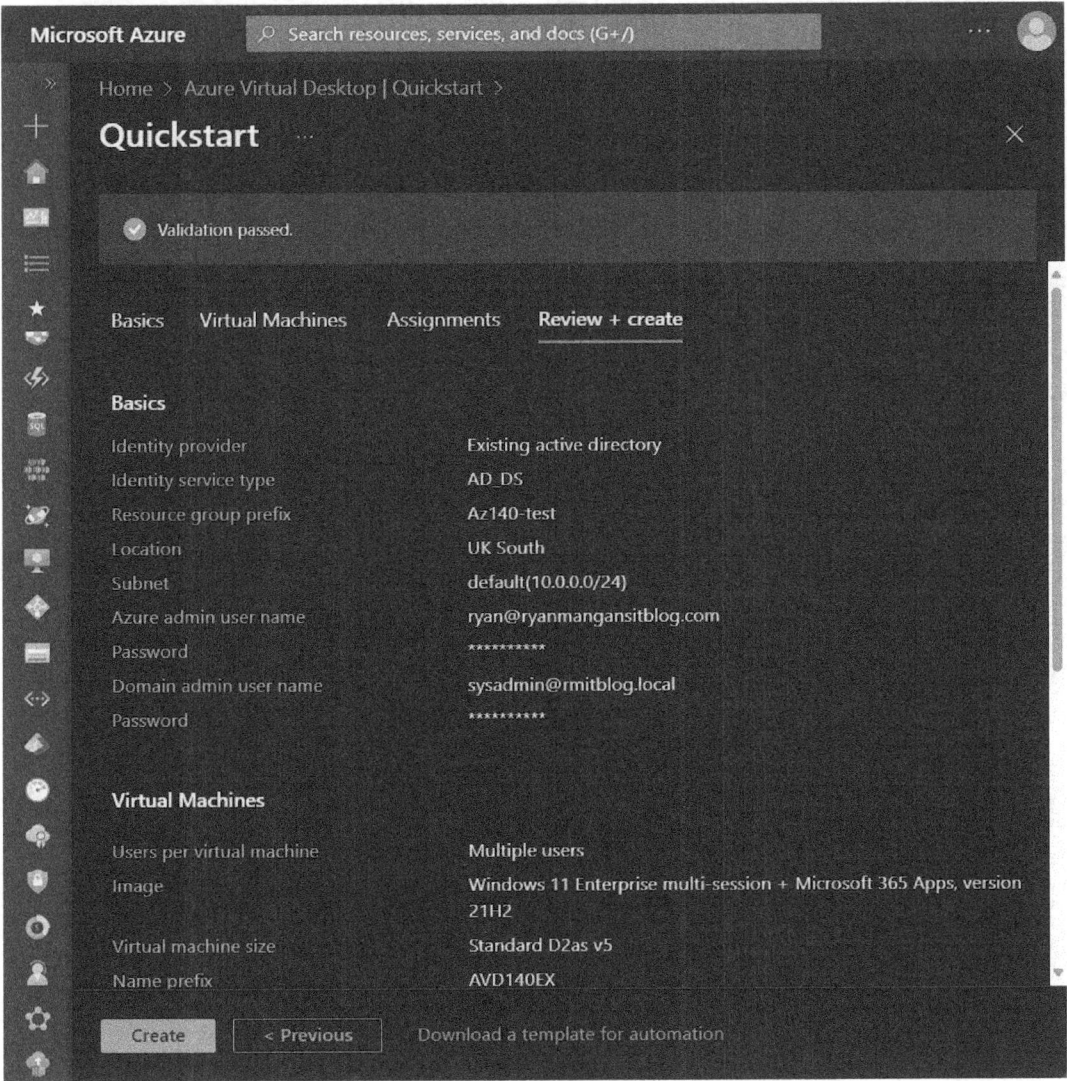

Figure 19.12 – The Quickstart wizard – validation has passed

26. Once validation has passed, click **Create**.

Once the deployment has finished, you should see the following:

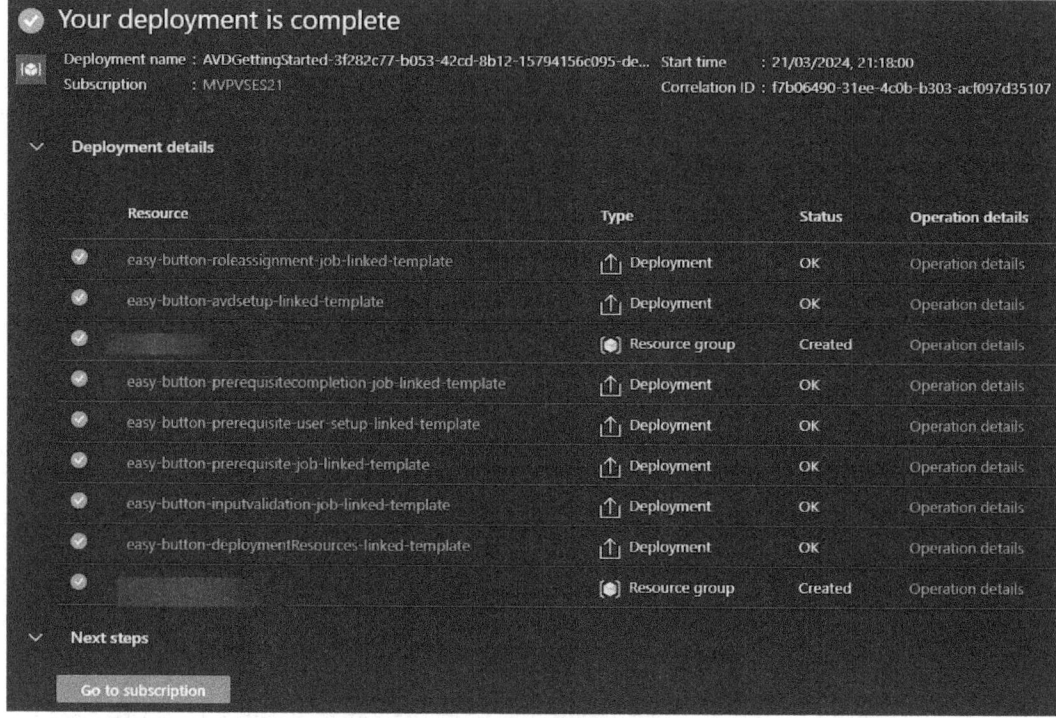

Figure 19.13 – Your deployment is complete

In the next section, we will look at using the Quickstart feature without an identity provider.

Using the Quickstart feature without an identity provider

Interestingly, when deploying Azure Virtual Desktop using the Quickstart feature without an identity provider, the wizard has fewer options. The reason for this is that the Quickstart wizard deploys the **Microsoft Entra Domain Services** infrastructure as part of the wizard and takes care of the majority of the configurations. Let's take a look at using this feature without the identity provider settings:

1. Within the Quickstart wizard, in the **Basics** tab, under **Identity provider**, you need to select **No identity provider**.

2. Under the **Identity service type** section, you will see that the only option you have is **Microsoft Entra Domain Services**; select it:

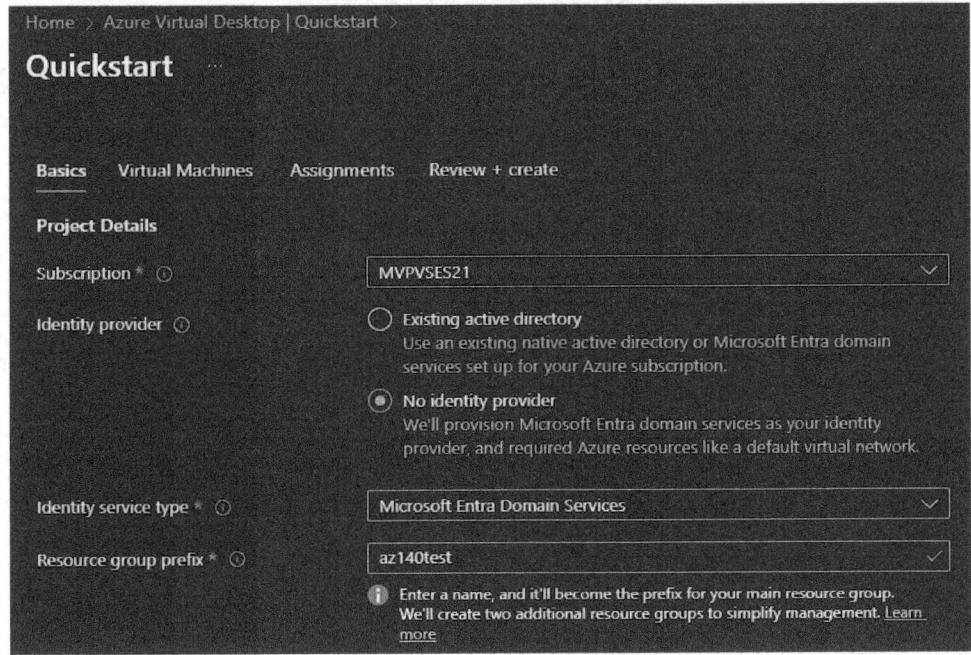

Figure 19.14 – Setting the No identity provider option

3. You need to make sure that you set **Domain admin user name** and **Password** values as these credentials will be used with the deployed Entra ID DS resource:

Figure 19.15 – The Domain administrator credentials section within the Quickstart wizard

4. You may also note that there is less to configure within the **Virtual Machines** tab:

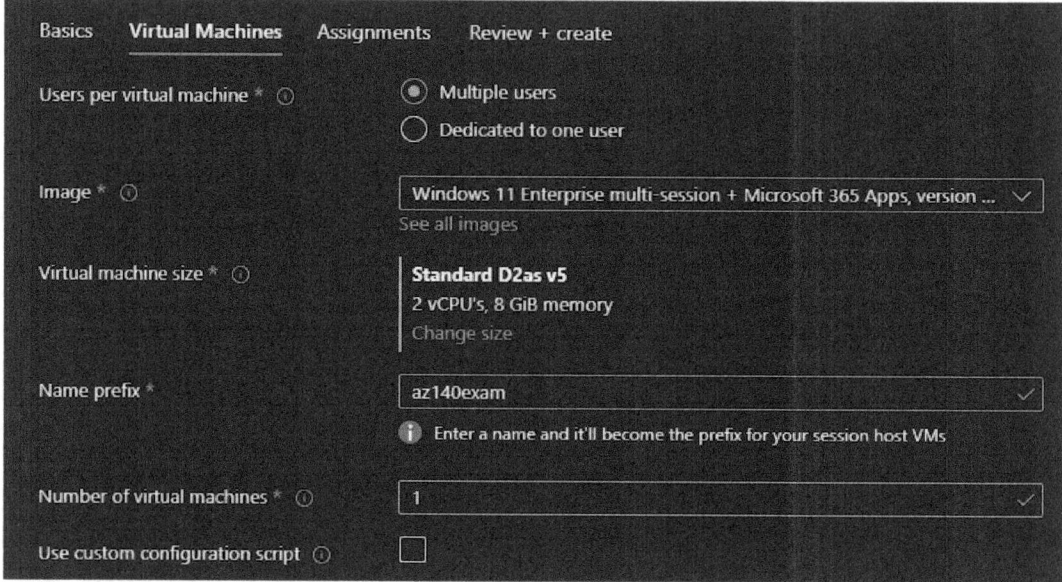

Figure 19.16 – The difference when using the No identity provider option within the Quickstart wizard

Now, let's look at the post-deployment cleanup.

Post-deployment cleanup

Once you have deployed Azure Virtual Desktop, you will see that two resource groups have been created – one marked with a prefix of -avd, which specifies all the resources required for Azure Virtual Desktop to function, and another marked with a prefix of -deployment, which specifies all the resources that will be used to automate the deployment of the Azure Virtual Desktop environment within your subscription:

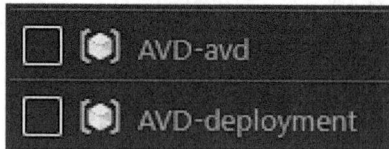

Figure 19.17 – Two resource groups created by the Quickstart feature

If you have used the Quickstart feature to deploy an Azure Virtual Desktop environment without an identity, such as when using a new Entra ID DS, then you will see two resource groups marked with the -deployment and -avd prefixes.

In the following screenshot, you can see the resources that have been used to deploy the Azure Virtual Desktop environment:

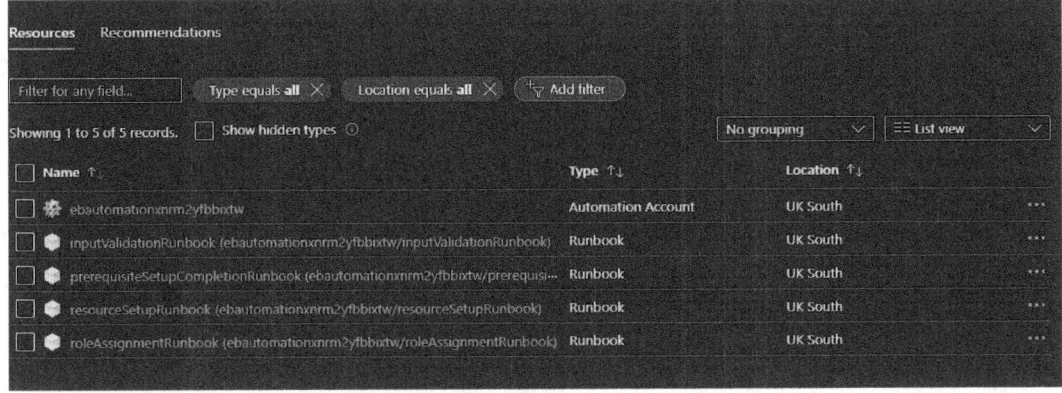

Figure 19.18 – The -deployment resource group's contents

The resource group marked with the prefix of prerequisite contains the virtual network, network security group, and the Entra ID DS resource. The following screenshot shows what you will find within this resource group:

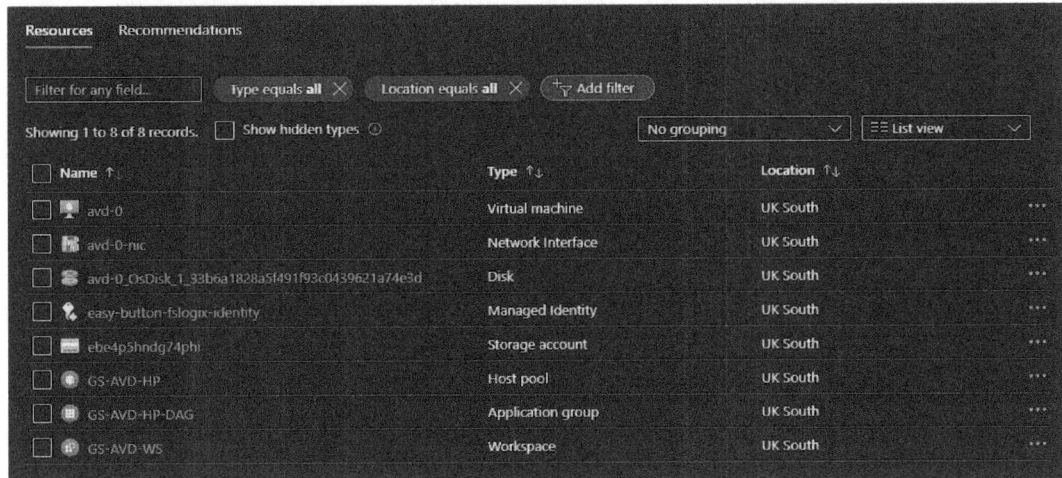

Figure 19.19 – The resource group deployed as part of the Quickstart
wizard when the No identity provider option is selected

You can go ahead and delete the resource group marked as `-deployment` as it's no longer required. If you have finished with the deployment, you can remove the two/three resource groups to safely delete the deployed Azure Virtual Desktop environment.

Now, let's look at some of the errors you may encounter and how to solve them while troubleshooting the Quickstart feature.

Troubleshooting the Quickstart feature

In this section, we'll look at some of the issues and errors you may come across when using the Quickstart wizard.

The following table details some of the most common issues:

Error	Description
No subscriptions	When opening the Quickstart feature wizard, you will see an error message stating *no subscriptions*. This is because you need to activate the Azure subscription. You should also ensure that the user account that's being used has the Owner **role-based access control** (**RBAC**) applied to the subscription.
The username must not include reserved words	Make sure that you are not using any of the words that have been reserved by Microsoft. The following link lists these reserved words: `https://learn.microsoft.com/azure/azure-resource-manager/troubleshooting/error-reserved-resource-name`.
The value must be between 12 and 72 characters long	This error message occurs when you have a password less than 12 or more than 72 characters long. You also need to ensure that the password will comply with Azure's password length and complexity requirements.
You don't have permission	This issue is related to the user account not having the owner permissions applied to the active Azure subscription.
The Microsoft.Powershell. DSC extension is installed on the Active Directory Domain controller **VM has reported a failure when processing extension Microsoft. Powershell.DSC.**	Uninstall the `Microsoft.powershell.DSC` extension.

Table 19.1 – Issues with the Quickstart feature

You can read more about troubleshooting the Azure Virtual Desktop Quickstart feature here: `https://docs.microsoft.com/azure/virtual-desktop/troubleshoot-quickstart`.

Summary

In this chapter, we looked at using the Quickstart feature to simplify the deployment of Azure Virtual Desktop. We started by looking at the benefits before looking at the prerequisites. We then ran through the full Quickstart wizard and looked at the subtle differences between using an existing identity provider and no identity provider. Finally, we learned how to troubleshoot deployment errors when using the Quickstart wizard feature.

Questions

Answer the following questions to test your knowledge of this chapter:

1. Which resource groups are created when deploying an Azure Virtual Desktop environment when creating an Entra ID DS environment as a part of the deployment?

    -prerequisites, -deployment, and -avd

2. When you're deploying an Azure Virtual Desktop environment using the Quickstart feature, what is the required RBAC role for the user account?

    Owner

3. Which resource group should you delete once you've deployed using the Quickstart wizard?

    -deployment

4. If the resource group with the `-prerequisites` prefix is present, which resources would you expect to find in there?

    Entra ID, virtual network, network security group

Final Assessment

Welcome to the study guide assessment test. The test questions are single-choice; however, they are not designed to resemble real-world **AZ-140** exam questions, as the exam includes different question types, such as multiple-choice, drag and drop, and case study questions. However, the topics covered in this assessment are in line with the same topics that are in the real-world exam. To make this as realistic as possible, it is recommended you attempt the questions *closed-book* and allocate time accurately. You can use notepaper or a sketchpad to jot down notes, although this will be different in a real test environment, as you will be using an approved Microsoft software-based testing solution. To take a look at the Microsoft Learn site for the AZ-140 exam, check out the following URL: `https://docs.microsoft.com/learn/certifications/exams/az-140`.

- Number of questions: 60
- Length of test: 90 minutes
- Passing score: 75% (estimated)

Questions

1. You need to use hybrid identities for a new Azure Virtual Desktop deployment to allow access to legacy apps in your environment. What should you use to integrate Microsoft Entra with **Entra ID Domain Services (Entra ID DS)**?

    A. A site-to-site VPN

    B. ExpressRoute

    C. Microsoft Entra Connect (correct answer)

2. Which diagnostic destination setting do you need to use when monitoring Azure Virtual Desktop with Azure Virtual Desktop Insights?

    A. Log Analytics (correct answer)

    B. Event Hubs

    C. Azure Queue Storage

3. Your company needs to limit user sessions to two hours to meet new security regulations on your on-premises, domain-joined Azure Virtual Desktop session hosts. What should you use to configure this requirement?

   A. Intune

   B. Group Policy (correct answer)

   C. Configuration Manager

4. You need to configure the location for FSLogix to mitigate the risk of the primary location being unavailable. How should this be configured?

   A. Configure general-purpose version 2 (GPv2) storage accounts.

   B. Configure a secondary file share location in the FSLogix configuration.

   C. Configure `FileStorage` storage accounts.

5. Your company wants to use MSIX images with MSIX app attach. How should you store the images for a file share?

   A. Azure NetApp Files

   B. Azure Compute Gallery

   C. MSIX app packages (correct answer)

6. Your company wants to stop users from copying and pasting information from `RemoteApp` in Azure Virtual Desktop to their local devices. What should you configure to meet this requirement?

   A. Configure the RDP properties of the host pool (correct answer).

   B. Implement an Intune compliance policy.

   C. Implement multi-factor authentication.

7. In a non-persistent Azure Virtual Desktop environment, which container solution should you use to allow customized taskbar settings to roam between session hosts?

   A. Kubernetes

   B. Profile containers (correct answer)

   C. Office containers

8. Your company has an Azure Virtual Desktop deployment with a site-to-site VPN connecting the on-premises network to the virtual network. You need to prevent users from connecting to the session hosts from outside the on-premises network. What should you configure?

    A. Conditional Access policy assignments (correct answer)

    B. An Intune Configuration policy

    C. A **Group Policy Object** (**GPO**)

9. What should you do to grant a group of users access to Azure Virtual Desktop VMs via the Azure portal?

    A. Add the group of users to the host pool assignment.

    B. Assign the group the relevant role (correct answer).

    C. Make the group of users local admin.

10. Your company has two offices, one located in London, using the UK South Azure region, and one located in Paris, using the France Central Azure region. What service can you use to facilitate Azure Virtual Desktop image sharing?

    A. Azure Compute Gallery (correct answer)

    B. Azure Marketplace

    C. Intune

11. You have created an Azure Virtual Desktop host pool as a **Proof of Concept** (**PoC**) and manually assigned users to allow them to access the host pool. What can you configure to automate this manual task?

    A. Direct assignment

    B. An Intune configuration policy

    C. Automatic assignment (correct answer)

12. What does Microsoft recommend using for an Azure Virtual Desktop profile solution?

    A. FSLogix (correct answer)

    B. Enterprise State Roaming

    C. Roaming profiles

13. Which Azure Virtual Desktop service that uses the TCP protocol is utilized to create a remote session and transfer RDP traffic?

    A. A session host communication channel

    B. Reverse connect transport (correct answer)

    C. A feed subscription

14. Which service can you utilize to estimate the connection round-trip time for a users location through the **Azure Virtual Desktop** (**AVD**) service to an Azure region?

    A. AVD Experience Estimator (correct answer)

    B. The Remote Desktop Services Diagnostic tool

    C. Network Watcher

15. Which Microsoft Cloud service does Azure Virtual Desktop use to connect remote users from any internet-connect device running an AVD client?

    A. Web Access Service

    B. RD Connection Broker

    C. The Remote Connection Gateway service (correct answer)

16. You have deployed AVD VMs with a Start/Stop during off-hours policy. What solution can you deploy to isolate the VMs that should not be stopped by this policy?

    A. Remote Desktop Diagnostics

    B. An Intune Configuration policy

    C. An Azure Automation account variable (correct answer)

17. What service should you use to replicate AVD session hosts to a different location?

    A. Azure Backup

    B. Azure Site Recovery (correct answer)

    C. ExpressRoute

18. You have been asked to grant access to a user that will allow them to manage all services of an AVD host pool and enable access to resources. What role should you assign?

    A. Host Pool Contributor (correct answer)

    B. Application Group Reader

    C. Workspace Contributor

19. Your manager has asked you to recommend a solution to manage language needs for users of a Windows 11 Enterprise multi-session AVD image. What solution should you recommend?

    A. Create a different image for each language.
    B. Deploy the Azure Marketplace images for your host pools.
    C. Customize the images so that they allow you to select whichever language you need (correct answer).

20. What is used to register VMs to the AVD host pool?

    A. The Log Analytics agent
    B. The AVD agent (correct answer)
    C. The Defender for Endpoint agent

21. You want to deploy AVD but are not sure which region to deploy it in. Which service can you use to help you decide the best region for your deployment?

    A. AVD Experience Estimator
    B. Azure Calculator
    C. An Azure datacenter map (correct answer)

22. You have an on-premises Active Directory Federations Services infrastructure that you need to integrate with Microsoft 365 services and decide to use Microsoft Entra Connect to configure hybrid IAM. Which Microsoft Entra feature should you use?

    A. Federation integration (correct answer)
    B. Password hash
    C. Pass-through authentication

23. What service can you use to securely connect to your AVD session hosts without exposing the RDP/SSH ports of the VMs?

    A. Azure Bastion (correct answer)
    B. **Network Security Groups (NSGs)**
    C. **Secure File Transfer Protocol (SFTP)**

24. You have been assigned the Desktop Virtualization Application role to allow you management control of a new application group. Which additional role do you need to allow you to assign users/groups to the new application group?

    A. Security Administrator
    B. User Access Administrator (correct answer)
    C. Privileged Access Administrator

25. What cloud service can you use to get recommendations on how to improve the security posture of your AVD deployment?

    A. Microsoft Defender for Cloud (correct answer)
    B. Microsoft Defender for Apps
    C. Microsoft Defender Vulnerability Management

26. What can you use to configure the control panel setting on Microsoft Entra joined Windows 10 session hosts?

    A. Group Policy
    B. Intune (correct answer)
    C. Configuration Manager

27. Which registry value should you use to set the location for Cloud Cache for **Server Message Block (SMB)**?

    A. VHDLocations
    B. CCDLocations (correct)
    C. Enabled

28. Which **operating system (OS)** does the OneDrive sync app on Virtual Desktop NOT support?

    A. Windows Server 2022 (correct answer)
    B. Windows 10
    C. Windows Server 2008 R2

29. What application should you use to help create a local image?

    A. Configuration Manager
    B. Intune
    C. Hyper-V Manager (correct answer)

30. Which Azure service should you use to facilitate communication between two AVD host pools that are deployed in different virtual networks?

    A. Networking peering (correct answer)

    B. A point-to-site VPN

    C. Bastion

31. Which storage solution should you use for your AVD deployment if you need ultra-performance, locally redundant, and 100 TiB per volume?

    A. Azure NetApp Files (correct answer)

    B. Storage Spaces Direct

    C. Azure Files

32. Which topic does Azure Advisor NOT give you recommendations about?

    A. Security

    B. Capacity (correct answer)

    C. Cost

33. What level of access to the AVD subscription, resource group, and Log Analytics workspace will you need to set up Azure Monitor monitoring for your AVD resources?

    A. Contributor

    B. Owner

    C. Reader (correct answer)

34. What subscription access rights do you require to deploy the Azure Automation scaling for your AVD deployment?

    A. Reader

    B. Contributor (correct answer)

    C. Owner

35. Which service do you need to install on your session host VM image before installing Microsoft Teams?

    A. The WebSockets service (correct answer)

    B. The Teams service

    C. The IIS service

36. Where can you suggest features for AVD with Azure Resource Manager integration?

    A. Microsoft Support Service

    B. Microsoft Fastrack Service

    C. The Tech Community (correct answer)

37. What can you use to hide applications from a specific group of users on an AVD session host?

    A. FSLogix application masking (correct answer)

    B. FSLogix profile containers

    C. MSIX app attach

38. Which AVD security need is the customer's responsibility?

    A. Physical hosts

    B. Network controls (correct answer)

    C. Virtualization Control Plane

39. Which Azure RBAC built-in role should you assign at the file share level to allow users to successfully utilize FSLogix profile containers when logging into AVD? (Your answer should adhere to the least privilege access best practice.)

    A. Storage File Data SMB Share Elevated Contributor

    B. Storage File Data SMB Share Reader

    C. Storage File Data SMB Share Contributor (correct answer)

40. Which resource provider do you NOT need to register when creating an AVD image using VM Image Builder?

    A. `Microsoft.Storage`

    B. `Microsoft.Compute`

    C. `Microsoft.ClassicStorage` (correct answer)

41. Which PowerShell cmdlet should you run to create a registration token to authorize a session host to join a host pool?

    A. `New-AzWvdRegistrationInfo` (correct answer)

    B. `New-AzWvdHostPool`

    C. `Get-AzWvdRegistrationInfo`

42. When joining your session hosts to Microsoft Entra, which **Mobile Device Management (MDM)** platform can you auto-enroll the VMs into?

    A. Configuration Manager

    B. Intune (correct answer)

    C. MaaS 360

43. Which identity-based authentication should you integrate with Azure File shares for FSLogix if your Microsoft Entra users need to access using Kerberos authentication?

    A. Microsoft Entra Domain services authentication

    B. Microsoft Entra Kerberos for hybrid identities (correct answer)

    C. AD Kerberos authentication for Linux clients

44. Which AVD Microsoft-hosted service queries a database for resources assigned to a user during the login process?

    A. Gateway services

    B. The Web Access service

    C. The broker service (correct answer)

45. Which AVD feature is based on the **Universal Rate Control Protocol (URCP)** and allows you to establish a direct UDP-based transport between the Remote Desktop client and session host?

    A. RDP Shortpath (correct answer)

    B. TLS 1.2

    C. Reverse connect transport

46. Which AVD resource can be described as a logical grouping of application groups?

    A. A published application

    B. A workspace (correct answer)

    C. A host pool

47. Which Network Watcher feature can you use to diagnose traffic-filtering problems to or from your AVD session host?

    A. Next hop

    B. VPN Diagnostics

    C. IP flow verify (correct answer)

48. Which OS does not officially support the AVD web client?

    A. Android (correct answer)

    B. macOS

    C. Linux

49. Which hybrid identity option should you use if you need to enable on-premises multi-factor authentication with AVD?

    A. Password hash synchronization

    B. Pass-through authentication

    C. Active Directory Federated Services (AD FS) (correct answer)

50. Which agent should you deploy on a session host to collect data about the processes running on the VM and its dependencies?

    A. The Log Analytics agent

    B. The Dependency agent (correct answer)

    C. The Telegraf agent

51. Which Microsoft-managed AVD component is an event-based aggregator that marks each user or administrator action on the AVD deployment as a success or failure?

    A. Remote Desktop diagnostics (correct answer)

    B. Extensibility components

    C. Connection Broker

52. Which AVD updating option should you choose if you have both Windows Server and Windows 10 host pools?

    A. Azure Update Manager

    B. Configuration Manager (correct answer)

    C. Windows Update for Business

53. What Microsoft VM deployment can you use together with Azure Hybrid Benefit to save the most on your AVD session host costs?

    A. Session host load balancing

    B. Azure Reserved Instances (correct)

    C. Spot Instances

54. What setting do you need to configure to collect information, including the error, connection, and agent health status, on your AVD host pool?

    A. Diagnostics (correct answer)
    B. Logs
    C. RDP properties

55. How many Azure Automation accounts do you need to configure autoscaling on five host pools?

    A. 5
    B. 10
    C. 1 (correct answer)

56. What role should you assign to the session host (VM object) to allow it to read and execute permissions on an Azure file share storing your MSIX application?

    A. Storage File Data SMB Share Elevated Contributor (correct answer)
    B. Storage File Data SMB Share Reader
    C. Storage File Data SMB Share Contributor

57. What property should you set to enable USB device redirection in an AVD host pool?

    A. `usbdevicestoredirect:s:`
    B. `usbdevicestoredirect:s:*` (correct answer)
    C. `usbdevicestoredirect:s:1`

58. Which storage option should you use if you have a capacity requirement of 10 TiB per subscription for your FSLogix profile containers?

    A. Storage Spaces Direct
    B. Azure Files
    C. Azure NetApp Files (correct answer)

59. Which Microsoft Identity solution does not support Intune management for your AVD deployment?

    A. A Microsoft Entra joined device
    B. A Microsoft Entra Domain Services joined devices (correct answer)
    C. A hybrid joined device

60. What tool should you use to install Office in AVD session hosts with the Windows 10 Enterprise multi-session OS?

    A. The Office Deployment Tool (correct answer)

    B. Configuration Manager

    C. Intune

Answers

1. **C.** You should use Microsoft Entra Connect to integrate Microsoft Entra with Entra ID DS so that you can use hybrid identities.

    Here is the external reference: https://learn.microsoft.com/entra/identity/hybrid/connect/whatis-azure-ad-connect

2. **A.** You need to set Log Analytics as the diagnostic destination setting when using AVD Insights to monitor AVD.

    External reference: https://learn.microsoft.com/azure/virtual-desktop/insights?tabs=monitor

3. **B.** You need to use Group Policy to limit user sessions to two hours when the AVD session hosts are on-premises domain-joined.

    Here is the external reference: https://learn.microsoft.com/azure/virtual-desktop/administrative-template?tabs=intune

4. **B.** You need to deploy the FSLogix agent with a secondary location path in the main region to mitigate the risk of the primary region being unavailable.

    Here is the external reference: https://learn.microsoft.com/fslogix/how-to-install-fslogix

5. **C.** You should store MSIX images as MSIX app packages for a file share.

    Here is the external reference: https://learn.microsoft.com/azure/virtual-desktop/app-attach-overview?pivots=msix-app-attach

6. **A.** You should configure the RDP properties of the host pool to stop users from copying and pasting information from a remote app in AVD to their local devices.

    Here is the external reference: https://learn.microsoft.com/azure/virtual-desktop/rdp-properties

7. **B.** You should use profile containers in a non-persistent AVD environment to allow user settings to roam between session hosts.

   Here is the external reference: https://learn.microsoft.com/azure/virtual-desktop/fslogix-containers-azure-files

8. **A.** You need to use Conditional Access policy assignments to prevent users from connecting to a session host from outside the on-premises network.

   Here is the external reference: https://learn.microsoft.com/entra/identity/conditional-access/concept-conditional-access-policies

9. **B.** You should assign a group a relevant role to grant a group of users access to AVD VMs.

   Here is the external reference: https://learn.microsoft.com/azure/virtual-desktop/rbac

10. **A.** You can use the Azure Compute Gallery to facilitate image sharing for AVD.

    Here is the external reference: https://learn.microsoft.com/azure/virtual-machines/azure-compute-gallery

11. **C.** You can use automatic assignment to automate the task of assigning users to AVD host pools.

    Here is the external reference: https://learn.microsoft.com/azure/virtual-desktop/configure-host-pool-personal-desktop-assignment-type?tabs=azure%2Cazure2

12. **A.** Microsoft recommends using FSLogix for AVD profile solution.

    Here is the external reference: https://learn.microsoft.com/azure/virtual-desktop/fslogix-containers-azure-files

13. **B.** The Reverse Connect Transport service is used to create a remote session and transfer RDP traffic.

    Here is the external references: https://learn.microsoft.com/azure/virtual-desktop/network-connectivity

14. **A.** You should use the AVD Experience Estimator tool to estimate the round-trip time for a location through the AVD service.

    Here is the external reference: https://azure.microsoft.com/products/virtual-desktop/assessment/

15. **C.** AVD uses the Remote Connection Gateway service to connect remote users from any internet-connected device running an AVD agent.

    Here is the external reference: https://learn.microsoft.com/azure/virtual-desktop/network-connectivity

16. **C.** You can deploy an Azure Automation variable to isolate VMs that should not be topped by a start/stop policy.

    Here is the external reference: `https://learn.microsoft.com/azure/automation/shared-resources/variables?tabs=azure-powershell`

17. **B.** You should use Azure Site Recovery to replicate AVD session hosts to a different location.

    Here is the external reference: `https://learn.microsoft.com/azure/site-recovery/site-recovery-overview`

18. **A.** You should assign the Host Pool Contributor role to a user who needs to manage all the AVD services.

    Here is the external reference: `https://learn.microsoft.com/azure/virtual-desktop/rbac`

19. **C.** You should customize the AVD images, allowing you to select whichever language you need if you need to manage language settings on Windows 11 Enterprise multi-session VMs.

    Here is the external reference: `https://learn.microsoft.com/azure/virtual-desktop/windows-11-language-packs`

20. **B.** The AVD agent is used to register VMs to an AVD host pool.

    Here is the external reference: `https://learn.microsoft.com/azure/virtual-desktop/agent-overview`

21. **C.** You should use the Azure datacenter map service to help you decide the best region for your AVD deployment.

    Here is the external reference: `https://datacenters.microsoft.com/globe/explore/`

22. **A.** You should use the federation integration feature in Microsoft Entra to integrate Microsoft 365 and AD FS.

    Here is the external reference: `https://learn.microsoft.com/entra/identity/hybrid/connect/whatis-fed`

23. **A.** You can use the Azure Bastion service to securely connect to your AVD session hosts without exposing the RDP/SSH ports of the VM.

    Here is the external reference: `https://learn.microsoft.com/azure/bastion/bastion-overview`

24. **B.** You need the User Access Administrator role to assign users/groups to a new application group.

    Here is the external reference: `https://learn.microsoft.com/azure/role-based-access-control/rbac-and-directory-admin-roles`

25. **A.** You should use Microsoft Defender for Cloud to get recommendations on how to improve the security posture of your AVD deployment.

    Here is the external reference: https://learn.microsoft.com/azure/defender-for-cloud/defender-for-cloud-introduction

26. **B.** You can use Intune to configure control panel settings on Microsoft Entra joined Windows 10 session hosts.

    Here is the external reference: https://learn.microsoft.com/mem/intune/fundamentals/azure-virtual-desktop

27. **B.** You should use the CCDLocations registry value to see the location of Cloud Cache for SMB.

    Here is the external reference: https://learn.microsoft.com/fslogix/concepts-fslogix-cloud-cache

28. **A.** The OneDrive sync app does not support the Windows Server 2022 OS.

    Here is the external reference: https://learn.microsoft.com/azure/virtual-desktop/onedrive-remoteapp

29. **C.** You should use Hyper-V Manager to create a local AVD image.

    Here is the external reference: https://learn.microsoft.com/azure/virtual-desktop/set-up-customize-master-image

30. **A.** You should use the networking peering service to facilitate communication between two AVD host pools that are deployed in different virtual networks.

    Here is the external reference: https://learn.microsoft.com/azure/virtual-network/virtual-network-peering-overview

31. **A.** You should an Azure NetApp Files solution if you need ultra-performance, locally redundant, and 100 TiB per volume AVD storage.

    Here is the external reference: https://learn.microsoft.com/azure/azure-netapp-files/azure-netapp-files-introduction

32. **B.** Azure Advisor does not give you recommendations about capacity.

    Here is the external reference: https://learn.microsoft.com/azure/advisor/advisor-overview

33. **C.** You need to have Reader access to the AVD subscription, resource group, and Log Analytics workspace to set up Azure Monitor monitoring for AVD resources.

    Here is the external reference: https://learn.microsoft.com/azure/virtual-desktop/insights?tabs=monitor

34. **B.** You require Contributor access rights to the subscription where AVD is hosted to deploy the Azure Automation scaling feature.

    Here is the external reference: https://learn.microsoft.com/azure/virtual-desktop/set-up-scaling-script

35. **A.** You need to install the WebSockets service on your session host VM before you install Microsoft Teams.

    Here is the external reference: https://learn.microsoft.com/azure/virtual-desktop/teams-on-avd

36. **C.** You can suggest AVD features with Azure Resource Manager integration on the Tech Community website.

    Here is the external reference: https://techcommunity.microsoft.com/

37. **A.** You can use FSLogix application masking to hide specific applications from a specific group of users on an AVD session host.

    Here is the external reference: https://learn.microsoft.com/fslogix/tutorial-application-rule-sets

38. **B.** Network control security is the customer's responsibility.

    Here is the external reference: https://learn.microsoft.com/azure/security/fundamentals/shared-responsibility

39. **C.** You should assign the Storage File Data SMB Share Contributor role at the file share level to allow users to successfully use FSLogix profile containers when logging into AVD.

    Here is the external reference: https://learn.microsoft.com/azure/virtual-desktop/fslogix-profile-container-configure-azure-files-active-directory?tabs=adds

40. **C.** You do not need to register the Microsoft.ClassicStorage provider when creating an AVD image using VM Image Builder.

    Here is the external reference: https://learn.microsoft.com/azure/virtual-machines/windows/image-builder-vnet

41. **A.** You should run the New-AzWvdRegistrationInfo PowerShell cmdlet to create a registration token to authorize a session host to join a host pool.

    Here is the external reference: https://learn.microsoft.com/powershell/module/az.desktopvirtualization/new-azwvdregistrationinfo?view=azps-12.0.0

42. **B.** Intune can auto-enroll VM session hosts when joining Microsoft Entra.

    Here is the external reference: `https://learn.microsoft.com/mem/intune/fundamentals/azure-virtual-desktop`

43. **B.** You should integrate Microsoft Entra Kerberos for hybrid identities with Azure File shares for FSLogix if Microsoft Entra users need to access AVD using Kerberos authentication.

    Here is the external reference: `https://learn.microsoft.com/azure/storage/files/storage-files-identity-auth-hybrid-identities-enable?tabs=azure-portal`

44. **C.** The broker service queries a database for resources assigned to a user during the login process.

    Here is the external reference: `https://learn.microsoft.com/azure/virtual-desktop/service-architecture-resilience`

45. **A.** RDP Shortpath allows you to establish a direct UDP-based transport between the Remote Desktop client and session host, and it is based on the Universal Rate Control Protocol.

    Here is the external reference: `https://learn.microsoft.com/azure/virtual-desktop/rdp-shortpath?tabs=managed-networks`

46. **B.** The workspace resource is described as a logical grouping of application groups.

    Here is the external reference: `https://learn.microsoft.com/azure/virtual-desktop/terminology`

47. **C.** You can use the IP flow verify feature of Network Watcher to diagnose traffic-filtering problems to or from your AVD session host.

    Here is the external reference: `https://learn.microsoft.com/azure/network-watcher/network-watcher-overview`

48. **A.** The Android operating system does not officially support the AVD web client.

    Here is the external reference: `https://learn.microsoft.com/azure/virtual-desktop/users/connect-web`

49. **C.** You should use AD FS to enable on-premises multi-factor authentication with AVD.

    Here is the external reference: `https://learn.microsoft.com/windows-server/identity/ad-fs/operations/configure-ad-fs-and-azure-mfa`

50. **B.** You should deploy the dependency agent on a session host to collect data about the processes running on the VM and its dependencies.

    Here is the external reference: `https://learn.microsoft.com/azure/azure-monitor/vm/vminsights-dependency-agent-maintenance`

51. **A.** The Remote Desktop Diagnostic component is an event-based aggregator that marks each user or administrator action on the AVD deployment as a success or failure.

    Here is the external reference: `https://techcommunity.microsoft.com/t5/security-compliance-and-identity/remote-desktop-services-diagnostic-tool/ba-p/247997`

52. **B.** Configuration Manager will allow you to update both Windows Server and Windows 10 host pools.

    Here is the external reference: `https://learn.microsoft.com/azure/virtual-desktop/configure-automatic-updates`

53. **B.** You can use Azure Reserved Instances together with Azure Hybrid Benefit to allow up to 80% savings on your AVD session host costs.

    Here is the external reference: `https://azure.microsoft.com/pricing/reserved-vm-instances`

54. **A.** You should configure the diagnostic setting to collect information, including the error, connection, and agent health status on your AVD host pool.

    Here is the external reference: `https://learn.microsoft.com/azure/virtual-desktop/diagnostics-log-analytics`

55. **C.** You need to configure one Azure Automation account to configure autoscaling on five host pools.

    Here is the external reference: `https://learn.microsoft.com/azure/automation/overview`

56. **A.** You should assign the Storage File Data SMB Share Elevated Contributor role to allow a session host to read and execute permissions on an Azure file share storing your MSIX Applications.

    Here is the external reference: `https://learn.microsoft.com/azure/virtual-desktop/app-attach-overview?pivots=msix-app-attach`

57. **B.** You should use the `usbdevicestoredirect:s:*` property to enable USB device redirection on an AVD host pool.

    Here is the external reference: `https://learn.microsoft.com/azure/virtual-desktop/configure-device-redirections`

58. **C.** You should use Azure NetApp Files if you have a capacity requirement of 10 TiB per subscription for your FSLogix profile containers.

    Here is the external reference: `https://learn.microsoft.com/azure/azure-netapp-files/azure-netapp-files-resource-limits`

59. **B.** A Microsoft Entra Domain Services joined device does not support Intune management on your AVD deployment.

    Here is the external reference: `https://learn.microsoft.com/mem/intune/fundamentals/azure-virtual-desktop`

60. **A.** You should use the Office Deployment Tool to install Office in AVD session hosts with a Windows 10 Enterprise multi-session OS.

    Here is the external reference: `https://learn.microsoft.com/deployoffice/overview-office-deployment-tool`

Appendix

Microsoft Resources and Microsoft Learn

The following table details a number of additional resources that you may find useful. The links provided include technical information, new releases, and information relating to fixes and implementing new features for Azure Virtual Desktop:

Name	Website
Virtual Desktops Community	https://avdcommunity.com/
MS Learn – Azure Virtual Desktop	https://docs.microsoft.com/learn/paths/m365-wvd/
Azure Virtual Desktop Forum	https://techcommunity.microsoft.com/t5/azure-virtual-desktop/bd-p/AzureVirtualDesktopForum
Microsoft Docs – Azure Virtual Desktop	https://docs.microsoft.com/azure/virtual-desktop/
Azure Virtual Desktop Feedback	https://techcommunity.microsoft.com/t5/azure-virtual-desktop-feedback/idb-p/AzureVirtualDesktop

Azure Virtual Desktop community shout-outs!

The IT industry is full of very smart people who spend their own time-solving problems and creating software, scripts, and other solutions for the community that is free for all. The following list features some of the community members you should check out for information on Azure Virtual Desktop:

Name	Social channel/website
Andrew Taylor	https://andrewstaylor.com
Simon Binder	www.kneedeepintech.com
Dominiek Verham	https://techlab.blog/
Freel Berson	https://microsoftplatform.blogspot.com/
Christiaan Brinkhoff	www.christiaanbrinkhoff.com
Marcel Meurer	https://blog.itprocloud.de/
Neil McLoughlin	www.virtualmanc.co.uk
Anoop Chandran	www.howtomanagedevices.com
Bas Van Kaam	www.bassvankaam.com
Bernhard Tritsch	https://eucscore.com
Patrick köhler	https://avdlogix.com/
Shabaz Darr	https://iamitgeek.com/
Stefan Dingemase	https://stefandingemanse.com/
Sander Rozemuller	https://rozemuller.com/
Stefan Beckmann	https://www.beckmann.ch/
Niels Kok	https://www.nielskok.tech/
Travis Roberts	https://www.ciraltos.com/
Ryan Mangan	https://www.democratising-clouds.com/
Johan Vanneuville	https://johanvanneuville.com/
James Kindon	https://jkindon.com/
Esther Barthel	https://www.virtues.it/
Tom Hickling	https://xenithit.blogspot.com/
Thomas Poppelgaard	https://www.poppelgaard.com
Jim Moyle	https://www.youtube.com/jimmoyle

Name	Social channel/website
Dean Cefola	`https://www.youtube.com/AzureAcademy`
Marco Moioli	`https://twitter.com/marco_moioli`
Micha Wets	`https://www.cloud-architect.be/`

Cool vendors

Here is a list of cool vendors that you can use with Azure Virtual Desktop:

Vendor name	Description	Link
AVD Hydra	Hydra is the solution to manage Azure Virtual Desktop for one or more tenants. Hydra's web platform allows administrators to deploy new session hosts, configure an auto-adapt scaling, maintain session hosts and pools automatically, and much more.	`www.itprocloud.com/hydra`
Login VSI	Understand the impact of change on your production environment	`www.loginvsi.com`
ControlUp	ControlUp equips IT teams with the tools to quickly resolve issues, proactively prevent tickets, and minimize costs, built on our **digital employee experience** (**DEX**) management platform.	`www.controlup.com`
EtherAssist	This AI technical assistant equips IT teams with the knowledge and tools to manage and support Azure Virtual Desktop environments.	`www.etherassist.co.uk`
IGEL	An edge operating system for cloud workspaces	`www.igel.com`
Rimo3	Analyze, modernize, and manage applications at scale and quickly in any Windows environment	`www.rimo3.com`
Nerdio	Quickly and easily deploy Azure Virtual Desktop and Windows 365, manage all environments from one simple platform, and optimize costs by saving up to 75% on Azure compute and storage.	`getnerdio.com`

Vendor name	Description	Link
Appventix	Deployment and application life cycle management for App-V and MSIX	`appventix.com`
Parallels RAS	Extend Azure Virtual Desktop capabilities by integrating, configuring, and unifying all virtual workloads and resources from a centralized console.	`www.parallels.com`
Omnissa	Elevate the digital workspace experience with the efficient and secure delivery of virtual desktops and apps from on-premises to the cloud. This was perviously VMware Horizon. It was sold in 2023 when Broadcom acquired VMware.	`www.omnissa.com`
TMEditX	Allows you to leverage the Microsoft MSIX Packaging Tool for what it is good at, and then analyze the results and fix up your package without having to repackage again.	`www.tmurgent.com`
Master Packager	Master Packager is an application packaging tool to create and edit **Microsoft Windows Installer** (**MSI**) files and repackage other installations to the MSI format.	`www.masterpackager.com/`
EUC Score	EUC Score measures and quantifies the perceived end-user experience in remote application and digital workspace environments, both on-premises and in the cloud	`https://eucscore.com`

Introducing EtherAssist – the premier AI technical assistant

EtherAssist sets the benchmark for AI-driven technical assistance, uniquely tailored for IT professionals and departments. This advanced tool not only optimizes IT operations but also substantially reduces overhead for both end-user customers and managed services partners.

Specialized support for Azure Virtual Desktop

EtherAssist equips Azure Virtual Desktop administrators with the expertise to efficiently design, build, maintain, and support deployments, ensuring top-notch service delivery.

The core features of EtherAssist include the following:

- **Error code troubleshooting**: Swiftly identify and resolve technical issues with intelligent diagnostics
- **Document automation**: Generate templates and essential documentation with precision and ease
- **Language translation**: Overcome language barriers in real time to ensure clear and effective communication
- **Technical validation**: Ensure your configurations and deployments meet the highest standards of accuracy and efficiency
- **PowerShell integration**: Leverage the power of AI with our specialized PowerShell module to automate and enhance scripting
- **API accessibility**: Integrate EtherAssist's capabilities into your code base and external products seamlessly

> Transform your IT operations today
>
> Embark on a journey of enhanced productivity and streamlined IT management with EtherAssist. Visit `www.etherassist.co.uk` to sign up and discover the power of AI-enhanced IT assistance.

Level up at AVD TechFest

Ready to level up your AVD skills? Join us at AVD TechFest, a premier international event that unites industry leaders, vendors, and community speakers to delve deep into Microsoft **Azure Virtual Desktop** (**AVD**). This two-day festival is the epicenter of AVD best practices and a top-tier opportunity to connect with others and exchange ideas.

Why attend AVD TechFest?

New skills. New connections. New opportunities. This is where it happens:

- **Networking**: Forge valuable connections within the European AVD community, engaging in discussions that span enterprise-level AVD deployment strategies, seamless integration techniques, and opportunities for organizational expansion.
- **For technologists**: Designed exclusively for AVD specialists, this platform is a conduit to share groundbreaking innovations, exchange battle-tested best practices, and establish a robust community knowledge base.
- **For decision makers**: Elevate your strategic comprehension of AVD's implementation dynamics, aligning them seamlessly with your existing infrastructure, tools, and organizational protocols to unlock tangible business benefits and drive transformative outcomes.

You can register now at `https://avdtechfest.com/`.

> **Note**
> AVD TechFest is presented by ControlUp.

Summary

This short chapter provided you with a number of Microsoft and community resources that you can turn to for further reading. I hope you enjoyed this book and found it useful, and I want to say a big thank you for buying a copy.

Index

A

access control conditions
  examples 291
Access Control Lists (ACLs) 60
active-active option 82
Active Directory Domain Services (AD DS) 32, 51, 99, 114, 274, 410
  Getting started feature, using with 631-638
  using, considerations 630
active-standby mode 82
alert rules
  used, for setting up alerts 607-615
antivirus exclusions
  configuring 358, 359
app attach 474
  versus MSIX app attach 475-477
  working 475
app attach package
  creating 477-480
application dependencies 521, 529, 530
application masking 484
  rule types 484
application security
  within session hosts 211
assistive technology 23

authentication 312
automatic assignment
  configuring 198-200
automatic updates
  disabling 247
Automation account
  creating, for Azure Virtual Desktop 551-553
  permissions 555, 556
  PowerShell modules, importing into 557, 558
autoscale 568
availability sets 154
availability zones 154
AVD client issues
  connectivity, testing 417, 418
  Remote Desktop client, displaying no resources 419, 420
  Remote Desktop client, resetting 419
  troubleshooting 416
AVD delegated access
  reference link 297
AVD environment
  application groups 18, 19
  considerations 36, 37
  deployments, assessing 14
  physical and virtual desktop environments, assessing 14

user personas 16-18
workloads, testing 42
AVD host pool
creating, with PowerShell 167-173
AVD implementation
OS, identifying 29
AVD Insights
versus host pool insights 601
AVD metadata location
configuring 38
AVD network connectivity 88
flow 89
host pool outbound access to internet 99
NSGs 94-96
NVAs 100
Private Link with AVD 93, 94
process steps 88
RDP-Shortpath 89-92
reverse connect 89
service tags 94-96
AVD queries
creating, for Kusto.Explorer 618-622
AVD session hosts
groups, managing 306, 307
local roles, managing 305
rights assignments, managing 307
AVD TechFest
benefits 670
skills, leveling up 670
AVD with Intune, to configure policies and controls
reference link 309
AzCopy 459
URL 459
Azure Active Directory
deployment options, comparing 61, 62

Azure Advisor
using, for AVD 625-627
Azure automation runbook
configuring 557
Azure Bastion 100
setting up 101-103
used, for connecting to VM 104, 105
Azure Compute Gallery (ACG) 233, 259
creating 260-262
image, capturing 263-268
image definition, creating 269, 270
Azure Data Lake Storage Gen2 (ADLS Gen2) 119
Azure Files 50, 51, 110-112
backup 544-546
best practices 113, 114
configuring, for MSIX app attach 446-454
integration, with ADDS 59
MSIX images, uploading to 458-460
NTFS permissions 59
restore 547, 548
role permissions 59
tiers 52, 53, 112, 113
Azure Firewall 80, 97-99
for application-level protection 99
using 87
Azure Firewall Basic 98
Azure Kubernetes Service (AKS) 78
Azure landing zone 18
Azure MFA
working 313
Azure Migrate 15
Azure Monitor for AVD
configuring 587
event logs, configuring 598-601
Log Analytics workspace, creating 588-592
monitoring components, setting up 592-598
performance counters, configuring 598-601

Index

Azure NAT gateway
 reference link 87
 using 87
Azure NetApp Files 50, 51, 110-112
Azure network security 88
Azure network traffic
 filtering 80
 routing 80
Azure portal 458
 custom role, creating 300-305
 used, for configuring Start VM
  on Connect 401-404
**Azure Private Link with Azure
  Virtual Desktop**
 reference link 94
Azure RBAC 291, 292
Azure region 17
**Azure Resource Manager (ARM)
  subscription 37**
**Azure Resource Manager (ARM)
  template 135, 161, 275, 635**
Azure services
 integrating, into VNet 81
Azure Site Recovery (ASR) 522
 using, to replicate domain controller 527
Azure subscription 37
Azure Virtual Desktop (AVD) 4, 177, 587
 backup strategy, designing 519-521
 best user experience 4
 community members, features 666
 components 8
 connecting to 71
 deploying, without identity provider 631
 enhanced security 5
 hibernation, integrating into 405, 406
 license requirements 6
 performance management 5
 per-user access license 355

 PowerShell, setting up for 164-166
 prerequisites for deployment 74
 screen capture protection, enabling 406
 simplifying management 5
 supported web browsers 71
 vendors 667, 668
 versus Windows 365 6
 working 7
Azure Virtual Desktop deployment
 licensing model, selecting 45, 46
**Azure Virtual Desktop for Azure
  Stack HCI (AVD HCI) 10**
 benefits 10
 limitations 10
Azure Virtual Desktop host
 OS and application updates,
  applying on 204
Azure Virtual Desktop licensing
 applying, to virtual machines 47-49
**Azure Virtual Desktop quickstart,
  troubleshooting**
 reference link 643
Azure Virtual Desktop sessions
 internet access, managing for 512, 513
Azure virtual network (VNet) 78
 Azure resources, communicating through 78
 communicating, with on-premises
  networks 79
 required URLs verifying 105
Azure VM Image Builder
 reference link 240

B

backup strategy
 designing, for Azure Virtual
  Desktop 519-521

best practices, FSLogix profile container 372, 373
bitsadmin util and setieproxy
 reference link 381
border gateway protocol (BGP) 80
breadth-first load balancing 152, 191, 192
breadth mode 5
broker 4
built-in roles
 reference link 293
Business Continuity and Disaster Recovery (BCDR) 56
business continuity (BC) 368
business-to-business (B2B) 630

C

central processing unit (CPU) 604
Client Access Licenses (CALs) 30
client logs
 accessing 70
Cloud Cache 368, 523
 configuring 369-371
 storage configuration types 368
Cloud Security Posture Management (CSPM) 328
Cloud Workload Protection Platform (CWPP) 328
code-signed certificate
 importing 455-458
components, Azure Virtual Desktop (AVD)
 broker 8
 diagnostics 8
 gateway 8
 load balancing 8
 management 8
 web client 8

Composite File System (CimFS) image 425, 442, 443
Composite Image File System (CimFS) image 425
Conditional Access 314
 components 315
 creating, for MFA 320-326
 decisions 316
 enforcements 316
 reference link 316
 signals 315
connection broker 4
connectivity
 managing, to internet 81
 managing, to on-premises networks 81
continuous network detection 27
CPU troubleshooting 278-281
 tips 280
Create a host pool tab
 VMs, creating within 154-160
Cross Region Restore (CRR) 540
customer-managed keys (CMK) 126
custom image template
 creating 240-246
custom role
 creating, with Azure portal 300-305

D

default outbound access for VMs
 reference link 87
delegated access model 296
depth-first load-balancing 153, 193
Desired State Configuration (DSC) 275
desktop virtualization 4
Desktop Virtualization Application Group Contributor 295

Desktop Virtualization Application Group Reader 295
Desktop Virtualization Contributor 294
Desktop Virtualization Host Pool Contributor 295
Desktop Virtualization Host Pool Reader 295
Desktop Virtualization Power On Contributor 296
Desktop Virtualization Power On Off Contributor 296
Desktop Virtualization Reader 294
Desktop Virtualization Session Host Operator 296
Desktop Virtualization User Session Operator 295
Desktop Virtualization Virtual Machine Contributor 296
Desktop Virtualization Workspace Contributor 295
Desktop Virtualization Workspace Reader 295
device redirections
   reference link 211
DevSecOps 328
digital estate assessment 15
direct assignment
   configuring, with PowerShell 201-204
disaster recovery considerations
   for MSIX app attach 529
disaster recovery (DR) 355
disk performance troubleshooting 283, 284
disks
   configuring 132, 133
distinguished name (DN) 160
domain controller
   deploying, in failover region 526
   replicating, with ASR 527

domain name server (DNS) 88, 274, 523
dynamic disks
   versus fixed disks 145
dynamic gateway 82
dynamic virtual channels (DVCs) 23

E

email discovery
   setting up, to subscribe to Azure Virtual Desktop feed 72, 73
endpoint detection and response (EDR) 207
enhanced security
   enabling, for AVD 335-339
enterprise content management (ECM) 135
Enterprise State Roaming (ESR) 53
Entra ID 527
Entra ID Domain Services (AADDS) 61
Entra ID join 216
Entra ID-joined host pool
   deploying 216-220
   user access, enabling 221-223
Entra ID-joined session hosts
   connecting, with Remote Desktop client 223-225
Entra ID join for Azure Virtual Desktop 215
   prerequisites 215
Entra ID MFA 312
ephemeral OS disks 134, 135
EtherAssist 669
   features 669
   support for AVD 669
Event Trace Log (ETL) files 71
ExpressRoute 79, 84-86, 522

F

failover region
  domain controller, deploying in 526
file shares
  configuring 128-131
firewall rules
  for Azure Virtual Machines 64
fixed disks
  versus dynamic disks 145
FSLogix 31, 49
  architecture 354
  benefits 54
  capabilities 54, 55
  filter driver architecture 55
  installing 356-358
  Microsoft Teams integration 372
  performance requirements 57, 58
  product portfolio 54
  profile containers, versus Office containers 56
  requirements 55
  storage best practices 58, 59
  Teams exclusions 372
FSLogixAppRuleEditorSetup.exe application
  download link 484
FSLogix Apps RuleEditor
  rule, creating 485-492
FSLogix Codes
  reference link 415
FSLogix profile containers 54
  and MSIX app attach 426
  best practices 372, 373
  capabilities 355
  configuring 360-363
  configuring, with Microsoft Intune 365, 366
  configuring, with registry 363, 364
  exclude list 366, 367
  include list 366
  license requirements 354, 355
  steps, for configuring 356
  storage options 58, 110-112
  user account permissions, for folder access 368
FSLogix profile issues
  troubleshooting 414-416
FSLogix profile logs
  reference link 415
fully qualified domain name (FQDN) 99

G

general-purpose version 2 (GPv2) storage account types 112
Getting started feature 629
  benefits 629
  Existing Setup 630
  issues 642
  New Subscription (Empty) 630
  post-deployment cleanup 640, 641
  prerequisites 630, 631
  using, with Microsoft Entra ID Domain Services (Entra ID DS) 631-638
  using, without identity provider 638-640
  working 629, 630
graphic processing unit (GPU) 16, 280
greenfield deployment 14
Group Policy 308
  enabling, via Intune 413, 414
  used, for enabling watermarking 411, 412
Group Policy Central Store 509
Group Policy Object (GPO) 362

H

Hard Disk Drive (HDD) 50, 111
headless UI mode
  enabling 348, 349
hibernation 404
  integrating, into AVD 405, 406
  prerequisites for enabling 405
  supported VM sizes 405
hibernation for Azure virtual machines
  reference link 405
high availability (HA) 154
host pool insights
  using 601-607
  versus AVD Insights 601
host pool outbound access
  to internet 99
host pools 5, 32, 149
  app groups 33
  creating 150-153
  DesktopVirtualization resource provider, registering 34, 35
  end users 34
  personal desktops 520
  personal host pool type 32
  pooled desktops 521
  pooled host pool type 32
  prerequisites 149
  provider, registering with PowerShell 35, 36
  users, assigning to 194
host pool settings
  automatic assignment, configuring 198-200
  configuring 182
  direct assignment, configuring with PowerShell 201-204
  OS and application updates, applying on Azure Virtual Desktop host 204
  personal desktop, re-assigning 200, 201

  PowerShell, using to configure load-balancing methods 191
  PowerShell, using to customize RDP properties 186
  RDP properties, customizing 182-186
  security and compliance settings, applying to session hosts 207
  users, assigning to host pools via PowerShell 198
  validation pool, configuring 205-207
Hybrid AD/Microsoft Entra ID environment, enabling on Universal Print
  reference link 387

I

identity and access management (IAM) 114
image 9
  capturing, in ACG 263-268
image definition 259
  creating, from ACG 269, 270
image gallery 259
image optimization 250
image source 259
image template
  capturing 256, 258
  creating 240
image version 259
  creating 271-273
information technology (IT) administrator 587
Infrastructure-as-a-Service (IaaS) file server 427
input/output operations per second (IOPS) 40, 283

Index

Insights
  AVD Insights, versus host pool insights 601
  host pool insights, using 601-607
  using 601
Integrated Software Vendor (ISV) 430
internet
  access, managing for Azure Virtual Desktop sessions 512, 513
  connectivity, managing to 81
internet access and outbound connections, options
  Azure Firewall, using 87
  Azure NAT gateway, using 87
  public IP addresses, for Azure VMs 87
Internet Protocol Security/Internet Key Exchange (IPsec/IKE)
  reference link 84
internet service provider (ISP) 84

K

Key Management Services (KMS) 30
Kusto 615
  additional queries 622-625
  with Azure services 615
Kusto.Explorer
  Log Analytics, connecting to 616, 617
  reference link 616
  used, for creating AVD queries 618-622
Kusto Query Language (KQL) 615

L

Lakeside 16
Language Interface Pack (LIP) 247
language ISO files
  reference link 248

language packs
  installing 247-249
  reference link 248
latency 22
latest security updates
  obtaining 344
licensing model
  selecting, for Azure Virtual Desktop deployment 45, 46
line-of-business (LOB) applications 31
load-balancing methods
  breath-first load balancing method 152, 191, 192
  configuring 190
  configuring, with PowerShell 191
  depth-first load balancing method 153, 193
load testing 42
local admin access
  configuring 229-231
local image
  creating 144
locally redundant storage (LRS) 133
Log Analytics
  connecting, to Kusto.Explorer 616, 617
Log Analytics workspace
  creating 588-592
logical unit number (LUN) 272

M

managed service accounts (MSA) 630
management and policies
  customer responsibilities 9
maximum inactive time and disconnection policies
  configuring 208-210
MaxSessionLimit 191

Index

Microsoft
  access, giving to start and stop VMs 568
  managed services 7
  resources 665
Microsoft Defender Antivirus
  configuring, for session hosts 339
  features, configuring 341-343
  notifications, suppressing 346-348
  versus Microsoft Defender for Endpoint 340, 341
Microsoft Defender for Cloud 326, 327
  and AVD 332-334
  AVD, securing 330, 331
  core security requirements 328
  reference link 328
  security areas 331
  security score 329
Microsoft Deployment Toolkit (MDT) 135
Microsoft Entra authentication
  enabling, for RDP 225-229
Microsoft Entra Connect 32
  benefits 62
  working 63
Microsoft Entra Domain Services (MEDS) 9
Microsoft Entra ID 5, 9
Microsoft Entra ID domain 32
Microsoft Entra ID Domain Services (Entra ID DS) 160, 630
  Getting started feature, using with 631-638
Microsoft Entra joined session hosts in Azure Virtual Desktop
  reference link 224
Microsoft Entra Kerberos
  for hybrid identities 60
Microsoft Entra Plans & Pricing
  reference link 314

Microsoft Intune
  FSLogix profile containers, configuring with 365, 366
  reference link 217
  screen capture protection, enabling via 410
  used, for configuring user settings on AVD 394-401
  watermarking, enabling via 413, 414
Microsoft Intune admin center 6
Microsoft-managed keys (MMK) 126
Microsoft Online Services
  outbound TCP ports 64, 65
Microsoft Security Policy Advisor
  reference link 211
Microsoft Teams AV redirection
  implementing 503-505
  loading of media optimizations, verifying 506, 507
  managing 503-505
Microsoft Teams integration, FSLogix 372
Microsoft Terminal Services Client (MSTSC) 240
Mobile Application Management (MAM) 395
Mobile Device Management (MDM) 339, 395
Movere 15
MSIX 422
  benefits 422, 423
MSIXAA Community Tooling application
  URL 445
MSIX app
  publishing, to RemoteApp application group 465-470
MSIX app attach 422, 424, 461, 493, 529
  and FSLogix profile containers 426
  Azure Files, configuring for 446-454
  benefits 425

configuring 461-464
disaster recovery considerations 529
prerequisites 428, 429
troubleshooting 472-474
versus app attach 475-477
working 427

MSIX app attach process stages
delayed or deferred registration 427
deregistration 427
destage 427
register 427
stage 427

MSIX container
simple application, packaging 431

MSIX image 442
creating 443, 444
uploading, to Azure Files 458-460

MSIX Log Explorer
download link 473

MSIXMGR tool
download link 443
URL 445

MSIX package
core contents 424
creating 430-440
structure 423

MSIX troubleshooting
reference link 473

MsMmrHostMsi installer 508

multi-factor authentication (MFA) 5, 160, 274, 311, 630
Conditional Access policy, creating for 320-326
implementing 317, 318
planning 317, 318

multi-factor user states configuration
reference link 319

multimedia redirection (MMR) 507, 508
implementing 507
managing 507
states 512
testing 511, 512
usage, controlling for websites 509, 510

multiple connection deployments
limitations 57

multiple profile connections 57

multi-session OS
reference link 309

multi-site VPN connection 82

N

name resolution
configuring, for AD and Microsoft Entra Domain Services 32
planning, for AD and Microsoft Entra Domain Services 32

network connectivity
monitoring 105
troubleshooting 105

Network File System (NFS) v3 120

networking limits
reference link 81

networking troubleshooting 285-287

network interface card (NIC) 140

network requirements assessment, AVD
applications 20, 21
bandwidth utilization, estimating 24
bandwidth utilization, limiting with throttle rate limiting 27
client connection sequence 28
connection security 29
display resolutions 21
dynamic bandwidth allocation 27

experience estimator 22
RDP bandwidth requirements 23
remote graphics bandwidth,
 estimating 24-27
reverse connect transport 27
session host communication channel 28
**Network Security Groups
 (NSGs) 80, 94, 96, 140, 512, 523**
network service provider 84
**Network Virtual Appliances
 (NVAs) 47, 80, 100, 328, 514**

O

OneDrive for Business
 implementing, for multi-session
  environment 500-502
 managing, for multi-session
  environment 500-502
on-premises domain controller
 using 526
on-premises networks
 Azure virtual network (VNet),
  communication with 79
 connectivity, managing to 81
organizational unit (OU) 159, 398
OS capabilities
 limiting 211, 212
OS, identifying for AVD implementation
 supported Azure OS images 29, 30
 Windows 11 multi-session 30
OS issues, related to AVD
 troubleshooting 274-277

P

Package editor feature 441

Partner Integrations
 reference link 379
password requirements, when creating VM
 reference link 236
personal desktop 520
 re-assigning 200, 201
personal host pool 32
personal scaling plan
 creating 578, 579
 host pools, assigning 584, 585
 schedule, configuring 579-584
Platform as a Service (PaaS) model 6
point of presence (POP) 122
point-to-site (P2S) VPN connections 79, 83
policy-based and route-based VPN gateways
 reference link 82
pool-based virtual desktop offerings 6
pooled desktops 521
pooled host pool 32
pooled scaling plan
 creating 570, 571
 host pools, assigning 576
 schedule, configuring 572-576
PowerShell
 AVD host pool, creating with 167-173
 custom RDP properties, resetting 189, 190
 roles, assigning via 299, 300
 setting up, for AVD 164-166
 used, for assigning users to host pools 198
 used, for configuring direct
  assignment 201-204
 used, for configuring exclusions 359
 using, to configure
  load-balancing methods 191
 using, to customize RDP properties 186
PowerShell modules
 importing, into Automation
  account 557, 558

PowerShell runbook
  creating 559, 560
  testing, in Azure 563, 564
PowerShell script
  adding, to runbook 560-563
  manual download and unpacking 344, 345
printers
  permissions, assigning 388-391
  registering, with Universal Print connector 385-388
  sharing 388-391
Printer Shares
  reference link 389
Private Link with AVD 93, 94
Process Explorer
  download link 278
profile container, creating with Azure Files and Microsoft Entra ID
  reference link 216
project operator 620
PsPing
  download link 275, 417
public IP addresses, for Azure VMs 87

Q

quality of service (QoS) policy 27
quick scans
  configuring 345, 346

R

RAM challenges 281-283
RBAC roles
  assigning, to IT admins 297-299
RD Agent 106

RDP properties
  customizing 182-186
  customizing, with PowerShell 186
  custom RDP properties, resetting with PowerShell 189, 190
  multiple custom RDP properties, adding 188
  multiple custom RDP properties, editing 188
  single RDP property, adding 186, 187
  single RDP property, editing 186, 187
RDP settings
  reference link 212
RDP-Shortpath 89-92
  reference link 92
RDS Client Access License (CAL) 6, 46, 354
  reference link 182
RDS Subscriber Access License (SAL) 354
recommendations for AVD 39
  general recommendations, for VMs 41
  multi-session recommendations 39, 40
  recommendations, on sizing VMs 40
redirection support
  reference link 183
redirections.xml file
  reference link 372
region pairs
  reference link 540
registry
  FSLogix profile containers, configuring with 363, 364
RemoteApp application 9
  application, deploying as 493-500
RemoteApp application group
  MSIX app, publishing to 465-470
RemoteApp group
  creating 493-500
Remote Desktop application 9

Remote Desktop client
  displaying, no resources 419, 420
  resetting 419
  used, for connecting to Entra ID-joined session hosts 223-225
Remote Desktop Protocol (RDP) 5, 19, 88, 143, 375
  Microsoft Entra authentication, enabling 225-229
Remote Desktop (RD) licensing role 178
Remote Desktop Services (RDS) 4, 46
required FQDNs and endpoints, for Azure Virtual Desktop
  references 96, 106, 417
resolve errors for reserved resource names
  reference link 642
resource group 37
reverse connect 89
Roaming User Profiles (RUPs) 53
role assignment 293
Role-Based Access Control (RBAC) 5, 221, 642
role definition 292
round-trip time (RTT) 22, 88, 286
route tables 80
rule
  creating, in FSLogix Apps RuleEditor 485-492
rule types, application masking
  hiding rule 484
  redirection rule 484
  specify value rule 484
runbooks 551
  PowerShell script, adding to 560-563
  schedule, creating 565-567

S

S2S and ExpressRoute
  coexisting, connections 86
scheduled task
  setting, to run PowerShell script 344
scope 292
screen capture protection
  configuring 407-409
  enabling, for Azure Virtual Desktop (AVD) 406
  enabling, via Intune 410
  prerequisites 407
screen capture protection prerequisites
  reference link 407
Secure Shell (SSH) 100
security and compliance settings
  application security, within session hosts 211
  applying, to session hosts 207
  endpoint detection and response (EDR) 207
  endpoint protection 207
  maximum inactive time and disconnection policies, configuring 208-210
  OS capabilities, limiting 211, 212
  session screen locks 208
  threat and vulnerability management 207
security defaults 313, 314
  reference link 314
security identifier (SID) 263
security posture 331
security posture manager (SPM) 332
security principle 292
Server Message Block (SMB) 59, 119, 215
Service Level Agreement (SLA) 41, 154
Service Level Agreements (SLA) for Online Services
  reference link 85

service tags 94, 96
session host image
    modifying 246
session hosts
    Microsoft Defender Antivirus, configuring for 339
    security and compliance settings, applying to 207
session host security tips
    reference link 207
session screen locks 208
Shared Image Galleries (SIGs) 259
Single Sign-On (SSO) 215, 376
Single Sign-On (SSO), configuring for Azure Virtual Desktop
    reference link 229
site-to-site (S2S) VPN 79, 82
Software as a Service (SaaS) model 6
Software Assurance (SA) 46
Solid-State Drive (SSDs) 50, 111, 281
Start VM on Connect 401
    configuring, with Azure portal 401-404
    enabling 404
    hibernation mode 404
storage accounts
    advanced settings, configuring 118-120
    basics, configuring 116, 117
    configuring 114
    creating 114, 115
    data protection, configuring 122-125
    encryption, configuring 125-128
    networking, configuring 120-122
Storage Explorer 460
    URL 460
storage feature
    comparing 51
storage platforms
    comparing 50, 51

Storage Spaces Direct 110-112
Storage Spaces Direct (S2D) 50, 51
Structured Query Language (SQL) 615
STUN
    direct connection via 90
Sysprep
    reference link 256

T

Teams exclusions, FSLogix 372
TMEditX 430
    reference link 422
Transport Layer Security (TLS) 28, 100, 119
TURN
    indirect connection via 90

U

Universal Naming Convention (UNC) 368
Universal Print 376
    component architecture 376, 377
    licensing requirements 377
    prerequisites 378
    print data, storing 378
    reference link 377
    service plan, verifying 377, 378
    setting up 380
Universal Print administrator roles 379
    Printer Administrator 379
    Printer Technician 379
    reference link 380
Universal Print connector
    functions 380
    installing 381-385
    prerequisites 380
    used, for registering printers 385-388

Index   685

Universal Print printer
  adding, to Windows device 391-394
URL check tool
  using 106, 107
user and app data 521
  configuring 528
user-defined route (UDR) 99
user experience (UX) 587
user identities 521
  managing 526
  planning for 61
user personas 16
  grouping criteria 16, 17
User Principal Name (UPN) 159, 631
User Profile Disks (UPDs) 53
user profiles 53
  previous user profile, challenges 53
user restrictions
  configuring, by using Entra ID Domain
    Services group policies 309
users
  assigning, to host pools 194
  assigning, to host pools via PowerShell 198
  assigning, to session host in
    Azure portal 194-197
user settings on AVD
  configuring, with Microsoft Intune 394-401

V

validation pool 205
  configuring 205-207
VHD image
  creating 135
virtual central processing unit (vCPU) 16
Virtual Desktop Infrastructure (VDI) 4
virtual desktop optimization tool 250-254

Virtual Hard Disk/Hyper-V Hard
  Disk (VHD/VHDX) 49
virtual hard disk (VHD) 110
Virtualize Windows 11 and 10 46
Virtualize Windows Server 46
virtual machine images
  replicating, between regions 549
Virtual Machine Scale Sets (VMSS) 78
virtual machines (VMs) 4, 14, 77, 111,
    149, 216, 291, 375, 521, 523
  active-active option 523
  active-passive option 523, 524
  availability set 524
  availability zone 525
  Azure Virtual Desktop licensing,
    applying to 47-49
  backup 530-538
  basic settings 137
  connecting to 237-239
  connecting, with Azure Bastion 104, 105
  creating 136-143, 233-237
  creating, within Create a host
    pool tab 154-160
  joining, to domain 63
  network settings 140
  restore 539-543
virtual network 521-523
virtual network, connecting to ExpressRoute
    circuits with Azure portal
  reference link 85
virtual network gateway 526
virtual Trusted Platform Modules
    (vTPMs) 155
VM applications 515
  reference link 515
VNet Peering 78
VNet service endpoint 78

VPN, types
  ExpressRoute 84-86
  multi-site VPN connection 82
  point-to-site (P2S) VPN connections 83
  S2S and ExpressRoute, coexisting connections 86
  S2S VPN gateway connection 82
  VNet peering 84
  VNet-to-VNet connections 84

W

watermarking
  enabling 410
  enabling, with Group Policy 411, 412
  prerequisites 411
  reference link 411
web content filtering
  reference link 208
Windows 11 multi-session 30
  image, customizing 31
  profile management solution 31
  supported versions 31
Windows 11, version 21H2 Language and Optional Features ISO
  reference link 247
Windows 11, version 22H2 and 23H2 Inbox Apps ISO
  reference link 248
Windows 365 6
  versus Azure Virtual Desktop 6
Windows App
  for Windows Store 68
  reference link 68
Windows device
  Universal Print printer, adding to 391-394

Windows NT 4.0 Terminal Server Edition 23
Windows Remote Desktop client 65
  installing 66-68
Windows Server session host licensing 178-182
Windows updates 255
Windows Virtual Machines, frequently asked question
  reference link 631
workspace 154
  information 161-163
  subscribing to 69, 70
WVDAgentURL tool 107

Z

zone-redundant storage (ZRS) 133, 544
  reference link 544

packtpub.com

Subscribe to our online digital library for full access to over 7,000 books and videos, as well as industry leading tools to help you plan your personal development and advance your career. For more information, please visit our website.

Why subscribe?

- Spend less time learning and more time coding with practical eBooks and Videos from over 4,000 industry professionals
- Improve your learning with Skill Plans built especially for you
- Get a free eBook or video every month
- Fully searchable for easy access to vital information
- Copy and paste, print, and bookmark content

Did you know that Packt offers eBook versions of every book published, with PDF and ePub files available? You can upgrade to the eBook version at packtpub.com and as a print book customer, you are entitled to a discount on the eBook copy. Get in touch with us at customercare@packtpub.com for more details.

At www.packtpub.com, you can also read a collection of free technical articles, sign up for a range of free newsletters, and receive exclusive discounts and offers on Packt books and eBooks.

Other Books You May Enjoy

If you enjoyed this book, you may be interested in these other books by Packt:

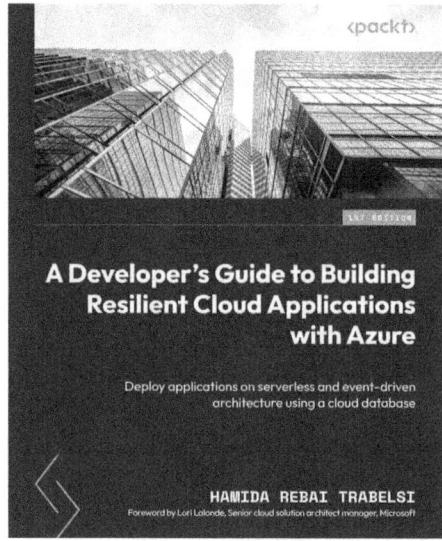

A Developer's Guide to Building Resilient Cloud Applications with Azure

Hamida Rebai Trabelsi

ISBN: 978-1-80461-171-5

- Understand the architecture of Azure Functions and Azure Service Fabric
- Explore Platform-as-a-Service options for deploying SQL Server in Azure
- Create and manage Azure Storage and Azure Cosmos DB resources
- Leverage big data storage in Azure services
- Select Azure services to deploy according to a specific scenario
- Set up CI/CD pipelines to deploy container applications on Azure DevOps
- Get to grips with API gateway patterns and Azure API Management

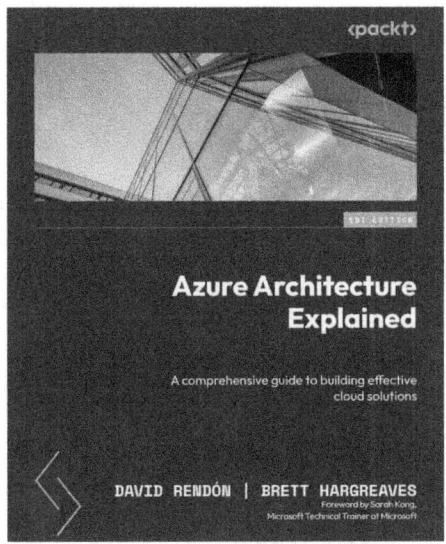

Azure Architecture Explained

David Rendón, Brett Hargreaves

ISBN: 978-1-83763-481-1

- Implement and monitor cloud ecosystem including, computing, storage, networking, and security
- Recommend optimal services for performance and scale
- Provide, monitor, and adjust capacity for optimal results
- Craft custom Azure solution architectures
- Design computation, networking, storage, and security aspects in Azure
- Implement and maintain Azure resources effectively

Packt is searching for authors like you

If you're interested in becoming an author for Packt, please visit `authors.packtpub.com` and apply today. We have worked with thousands of developers and tech professionals, just like you, to help them share their insight with the global tech community. You can make a general application, apply for a specific hot topic that we are recruiting an author for, or submit your own idea.

Share Your Thoughts

Now you've finished *Mastering Azure Virtual Desktop*, we'd love to hear your thoughts! Scan the QR code below to go straight to the Amazon review page for this book and share your feedback or leave a review on the site that you purchased it from.

`https://packt.link/r/1-835-88415-6`

Your review is important to us and the tech community and will help us make sure we're delivering excellent quality content.

Download a free PDF copy of this book

Thanks for purchasing this book!

Do you like to read on the go but are unable to carry your print books everywhere?

Is your eBook purchase not compatible with the device of your choice?

Don't worry, now with every Packt book you get a DRM-free PDF version of that book at no cost.

Read anywhere, any place, on any device. Search, copy, and paste code from your favorite technical books directly into your application.

The perks don't stop there, you can get exclusive access to discounts, newsletters, and great free content in your inbox daily

Follow these simple steps to get the benefits:

1. Scan the QR code or visit the link below

https://packt.link/free-ebook/978-1-83588-414-0

2. Submit your proof of purchase
3. That's it! We'll send your free PDF and other benefits to your email directly

www.ingramcontent.com/pod-product-compliance
Lightning Source LLC
Jackson TN
JSHW060726070725
87115JS00012B/50